Eric Lichtfouse
Jan Schwarzbauer
Didier Robert

Environmental Chemistry

Green Chemistry and Pollutants in Ecosystems

Part I	Analytical Chemistry	Chapters 1–11
Part II	Toxic Metals	Chapters 12–25
Part III	Organic Pollutants	Chapters 26–35
Part IV	Polycyclic Aromatic Compounds	Chapters 36–41
Part V	Pesticides	Chapters 42–48
Part VI	Green Chemistry	Chapters 49–57
Part VII	Ecotoxicology	Chapters 58–69

Eric Lichtfouse
Jan Schwarzbauer
Didier Robert
(Editors)

Environmental Chemistry

Green Chemistry and Pollutants in Ecosystems

With 289 Figures

Editors

Dr. Eric Lichtfouse

INRA
17, rue Sully
21000 Dijon, France

Dr. Jan Schwarzbauer

Institute of Geology and Geochemistry of Petroleum and Coal
RWTH Aachen University
Lochnerstrasse 4–20
52056 Aachen, Germany

Dr. Didier Robert

Head of the Chemical Department (IUT)
Laboratory of Chemical Applications
University of Metz
rue Victor Demange
57500 Saint-Avold, France

Library of Congress Control Number: 2004110949

ISBN 3-540-22860-8 Springer Berlin Heidelberg New York

This work is subject to copyright. All rights are reserved, whether the whole or part of the material is concerned, specifically the rights of translation, reprinting, reuse of illustrations, recitations, broadcasting, reproduction on microfilm or in any other way, and storage in data banks. Duplication of this publication or parts thereof is permitted only under the provisions of the German Copyright Law of September 9, 1965, in its current version, and permission for use must always be obtained from Springer. Violations are liable to prosecution under the German Copyright Law.

Springer is a part of Springer Science+Business Media
springeronline.com
© Springer-Verlag Berlin Heidelberg 2005
Printed in Germany

The use of general descriptive names, registered names, trademarks, etc. in this publication does not imply, even in the absence of a specific statement, that such names are exempt from the relevant protective laws and regulations and therefore free for general use.

Cover design: Erich Kirchner, Heidelberg
Typesetting: Büro Stasch (stasch@stasch.com), Bayreuth
Production: Luisa Tonarelli
Printing: Mercedes-Druck, Berlin
Binding: Stein & Lehmann, Berlin

Printed on acid-free paper   30/3111/LT – 5 4 3 2 1   SPIN 11593614

Preface

In 1889, the Nobel Prize winner Svante Arrhenius pointed out the existence of a "greenhouse effect" in which small changes in the concentration of carbon dioxide in the atmosphere could considerably alter the average temperature of a planet. About one century later, humans realise that most climate changes are correlated with the increase of the concentration of carbon dioxide in the atmosphere. A such prediction from Svante Arrhenius clearly highlights that more knowledge of environmental mechanisms is needed to cope with actual problems of pollution. Environmental Chemistry is a fast emerging discipline aiming at understanding the fate of pollutants in ecosystems and at designing novel processes that are safe for ecosystems. Past pollution should be cleaned. Future pollution should be predicted and avoided.

The 69 chapters of this book have been arranged into seven topics that form the core of Environmental Chemistry: Analytical Chemistry, Toxic Metals, Organic Pollutants, Polycyclic Aromatic Hydrocarbons, Pesticides, Green Chemistry, and Ecotoxicology. Most chapters have designed to include *(1)* a review on the actual knowledge and *(2)* cutting-edge research results. Thus this book will be useful to students and decision-makers who wish to learn rapidly the essential background of a specific topic, and to scientists who wish to locate the actual frontiers of science in a specific domain.

We wish here to thank all authors for providing high quality manuscripts. We are indebted to Armin Stasch, Luisa Tonarelli and Marion Schneider from Springer for technical assistance. We thank Dr. Christian Witschel, Executive Editor of Geosciences at Springer for having accepted our project to design this book. Last but not least, we thank very much Brigitte Elbisser who has been from 2000 to 2003 the key staff of the European Association of Chemistry and the Environment (ACE), producing Newsletters, taking care of budget and memberships, organising annual meetings, and helping at the book preparation.

Drs. Eric Lichtfouse, Jan Schwarzbauer and Didier Robert

Contents

Part I
Analytical Chemistry 1

1. In-situ Method for Analyzing the Long-Term Behavior of Particulate Metal Phases in Soils 3
2. Analysis of Toxic Metals by Micro Total Analytical Systems (μTAS) with Chemiluminescence 13
3. Diffuse Infrared Fourier Transform Spectroscopy in Environmental Chemistry 19
4. Detection of Biomarkers of Pathogenic Bacteria by Matrix-Assisted Laser Desorption/Ionization Time-of-Flight Mass Spectrometry 31
5. Multi-Isotopic Approach (^{15}N, ^{13}C, ^{34}S, ^{18}O and D) for Tracing Agriculture Contamination in Groundwater 43
6. ^2H and ^{18}O Isotopic Study of Ground Waters under a Semi-Arid Climate 57
7. $^{13}C/^{12}C$ Ratio in Peat Cores: Record of Past Climates 65
8. Isotopic Composition of Cd in Terrestrial Materials: New Insights from a High-Precision, Double Spike Analytical Method 75
9. Organic Petrology: A New Tool to Study Contaminants in Soils and Sediments .. 89
10. The Comminution of Large Quantities of Wet Sediment for Analysis and Testing with Application to Dioxin-Contaminated Sediments from Lake Ontario 99
11. Study on the Large Volume Stacking Using the EOF Pump (LVSEP) for Analysis of EDTA by Capillary Electrophoresis 107

Part II
Toxic Metals 119

12. A Framework for Interpretation and Prediction of the Effects of Natural Organic Matter Heterogeneity on Trace Metal Speciation in Aquatic Systems .. 121
13. Binding Toxic Metals to New Calmodulin Peptides 133
14. Leaching of Selected Elements from Coal Ash Dumping 145
15. Storm-Driven Variability of Particulate Metal Concentrations in Streams of a Subtropical Watershed 153
16. A Model for Predicting Heavy Metal Concentrations in Soils 177
17. Phytoremediation of Thallium Contaminated Soils by Brassicaceae 187

18	Mercury Recovery from Soils by Phytoremediation	197
19	Effect of Cadmium and Humic Acids on Metal Accumulation in Plants	205
20	Selection of Microorganisms for Bioremediation of Agricultural Soils Contaminated by Cadmium	215
21	Electrodialytic Remediation of Heavy Metal Polluted Soil	223
22	Electrodialytic Removal of Cu, Cr and As from Treated Wood	235
23	Treatment of Wastewater Contaminated by Mercury by Adsorption on the Crandallite Mineral	243
24	Low Cost Materials for Metal Uptake from Aqueous Solutions	251
25	Removal of Copper(II) and Cadmium(II) from Water Using Roasted Coffee Beans	259

Part III
Organic Pollutants ... 267

26	Bioremediation for the Decolorization of Textile Dyes – A Review	269
27	Degradation of the Indigo Carmine Dye by an Anaerobic Mixed Population	289
28	Biodegradation of Benzothiazoles by *Rhodococcus* Bacteria Monitored by ^{1}H Nuclear Magnetic Resonance (NMR)	295
29	Biotransformation of Nonylphenol Surfactants in Soils Amended with Contaminated Sewage Sludges	305
30	Quantification of in-situ Trichloroethene Dilution versus Biodegradation Using a Novel Chloride Concentration Technique	317
31	Anthropogenic Organic Contaminants Incorporated into the Non-Extractable Particulate Matter of Riverine Sediments from the Teltow Canal (Berlin)	329
32	Behaviour of Dioxin in Pig Adipocytes	353
33	Control of Halogenated By-Products During Surface Water Potabilisation	361
34	Organic Pollutants in Airborne Particulates of Algiers City Area	371
35	A Reactive Transport Model for Air Pollutants	383

Part IV
Polycyclic Aromatic Compounds ... 391

36	Analysis of High-Molecular-Weight Polycyclic Aromatic Hydrocarbons by Laser Desorption-Ionisation/Time-of-Flight Mass Spectrometry and Liquid Chromatography/Atmospheric Pressure Chemical Ionisation Mass Spectrometry	393
37	Atmospheric Polycyclic Aromatic Hydrocarbons (PAHs) in Two French Alpine Valleys	409
38	Evaluation of the Risk of PAHs and Dioxins Transfer to Humans via the Dairy Ruminant	419
39	Polycyclic Aromatic Hydrocarbons (PAHs) Removal during Anaerobic and Aerobic Sludge Treatments	431
40	Photodegradation of Pyrene on Solid Phase	441
41	Degradation of Polycyclic Aromatic Hydrocarbons in Sewage Sludges by Fenton's Reagent	449

Part V
Pesticides 461

42 Pesticide Mobility Studied by Nuclear Magnetic Resonance 463
43 Photo- and Biodegradation of Atrazine
    in the Presence of Soil Constituents 473
44 Behaviour of Imidacloprid in Fields. Toxicity for Honey Bees 483
45 Impact of a Sulfonylureic Herbicide on Growth
    of Photosynthetic and Non-Photosynthetic Protozoa 495
46 Abiotic Degradation of the Herbicide Rimsulfuron on Minerals and Soil 505
47 Binding of Endocrine Disrupters and Herbicide Metabolites
    to Soil Humic Substances 517
48 Potential Exposure to Pesticides during Amateur Applications
    of Home and Garden Products 529

Part VI
Green Chemistry 539

49 Carbon Dioxide, a Solvent and Synthon for Green Chemistry 541
50 Mechanochemistry:
    An Old Technology with New Applications to Environmental Issues.
    Decontamination of Polychlorobiphenyl-Contaminated Soil
    by High-Energy Milling in the Solid State with Ternary Hydrides 553
51 Development of a Bioreactor for Cometabolic Biodegradation
    of Gas-Phase Trichloroethylene 561
52 Enhanced Solubilization of Organic Pollutants
    through Complexation by Cyclodextrins 569
53 Chemical Samples Recycling:
    The MDPI Samples Preservation and Exchange Project 585
54 Photodecomposition of Organic Compounds in Aqueous Solution
    in the Presence of Titania Catalysts 591
55 Depollution of Waters Contaminated by Phenols and Chlorophenols
    Using Catalytic Hydrogenation 601
56 Treatment of Wastewater Containing Dimethyl Sulfoxide (DMSO) 615
57 Productive Use of Agricultural Residues:
    Cements Obtained from Rice Hull Ash 621

Part VII
Ecotoxicology 629

58 Environmental Metal Cation Stress and Oxidative Burst in Plants.
    A Review 631
59 The LUX-FLUORO Test as a Rapid Bioassay
    for Environmental Pollutants 645
60 Effects of Two Cyanotoxins, Microcystin-LR and Cylindrospermopsin,
    on *Euglena gracilis* 569

61	A New Bioassay for Toxic Chemicals Using Green Paramecia, *Paramecium bursaria*	673
62	Detection of Toxic Pollution in Waste Water by Short-Term Respirometry	681
63	Environmental Biosensors Using Bioluminescent Bacteria	691
64	Evaluation of Water-Borne Toxicity Using Bioluminescent Bacteria	699
65	Bacteria-Degraders Based Microbial Sensors for the Detection of Surfactants and Organic Pollutants	707
66	Study of Cr(VI) and Cd(II) Ions Toxicity Using the Microtox Bacterial Bioassay	725
67	Cultured Human Cells as Biological Detectors for Assessing Environmental Toxicity	735
68	Genotoxic Impact of Erika Petroleum Fuel on Liver of the Fish *Solea solea*	743
69	Heavy-Metal Resistant Actinomycetes	757
	Index	769

Contributors

Abate, C. M.
PROIMI (Planta Piloto de Procesos Industriales Microbiológicos)
Avenida Belgrano y Pasaje Caseros
4000 Tucumán, Argentina
e-mail: cabate@proimi.edu.ar

Aguer, J. P.
Laboratoire de Photochimie Moléculaire et Macromoléculaire, UMR 6505 CNRS
Université Blaise Pascal
63177 Aubière Cedex, France

Albarracín, V. H.
PROIMI (Planta Piloto de Procesos Industriales Microbiológicos)
Avenida Belgrano y Pasaje Caseros
4000 Tucumán, Argentina
e-mail: virginia@proimi.org.ar

Alekseeva, T.
Laboratoire des Matériaux Inorganiques
UMR 6002 CNRS, Université Blaise Pascal
63177 Aubière Cedex, France

Alinsafi, A.
Faculté des Sciences Semlalia
Avenue Prince Moulay Abdellah, B.P. 511
40000 Marrakech, Morocco

Al-Najar, H.
Institute of Plant Nutrition (330)
University of Hohenheim
70593 Stuttgart, Germany

Alonso, M. A.
Escuela Universitaria Politécnica de Almadén
Chemical Engineering Department
University of Castilla-La Mancha
Plaza Manuel Meca, 1
13400 Almadén (Ciudad Real), Spain

Amat, A.
Ecole Nationale Supérieure d'Agronomie de Toulouse
Laboratoire de Toxicologie et Sécurité Alimentaire
1, Av de l'Agrobiopole
31326 Auzeville Tolosane, France

Amoroso, M. J.
PROIMI (Planta Piloto de Procesos Industriales Microbiológicos)
Avenida Belgrano y Pasaje Caseros
4000 Tucumán, Argentina

Aravena, Ramón
Department of Earth Sciences
University of Waterloo
Ontario, Canada N2L 3G1

Aresta, Michele
METEA Research Centre
University of Bari
via Celso Ulpiani 27
70126 Bari, Italy
e-mail: aresta@metea.uniba.it

Arun, Marimuthu
Centre for Advanced Water Technology
Singapore Utilities International
Singapore 637723

Azambre, B.
Laboratoire de Chimie et Applications, EA3471
Université de Metz
Rue Victor Demange
57500 Saint Avold, France

Bagot, Didier
I.U.T. de Colmar, Université de Haute Alsace
29 Rue de Herrlisheim, B.P. 568
68008 Colmar Cedex, France
e-mail: didier.bagot@uha.fr

Contributors

Baldoni-Andrey, Patrick
Total, Groupement de Recherches de Lacq
RN 117, B.P. 34
64170 Lacq, France
e-mail: patrick.baldoni-andrey@total.com

Ballivet-Tkatchenko, D.
Laboratoire de Synthèse et Electrosynthèse
Organométalliques (LSEO), Université de
Bourgogne, UMR 5188 CNRS-UB, B.P. 47870
21078 Dijon Cedex, France
e-mail: ballivet@u-bourgogne.fr
web: http://www.u-bourgogne.fr/LSEO/

Baraud, F.
Equipe de Recherche en Physico-Chimie et
Biotechnologies, Batiment Sciences 2 Campus 2
Bd du Maréchal Juin
14032 Caen Cedex, France
e-mail: f.baraud@vire.iutcaen.unicaen.fr

Baumstark-Khan, C.
DLR, Institute of Aerospace Medicine
Radiation Biology
Linder Höhe
51147 Köln, Germany

Beltran, V. Laudato
The Getty Conservation Institute
1200 Getty Center Drive, Suite 700
Los Angeles, CA 90049-1684, USA

Bengsch, E. R.
Centre de Biophysique Moléculaire
CNRS & Université Orléans
45071 Orléans Cedex 02, France

Benhammou, A.
Faculté des Sciences Semlalia
Avenue Prince Moulay Abdellah, B.P. 511
40000 Marrakech, Morocco

Benito, J. M.
PROIMI (Planta Piloto de Procesos Industriales
Microbiológicos)
Avenida Belgrano y Pasaje Caseros
4000 Tucumán, Argentina

Bermond, Alain
Institut National Agronomique Paris-Grignon
Laboratoire de Chimie Analytique
UMR Environnement et Grandes Cultures
16 Rue Claude Bernard
75231 Paris Cedex 05, France

Bernard, C.
USM0505, Ecosystèmes et interactions toxiques
Museum National d'Histoire Naturelle
12 Rue Buffon
75005 Paris, France

Besombes, J.-L.
LCME, Ecole Supérieure d'Ingénieur de Chambéry
Université de Savoie
73376 Le Bourget du Lac, France

Besse, P.
Laboratoire de Synthèse Et Etude de Systèmes à
Intérêt Biologique
UMR 6504 du CNRS
Université Blaise Pascal
63177 Aubière Cedex, France
e-mail: besse@chimtp.univ-bpclermont.fr

Besse, P.
Laboratoire de Synthèse Et Etude de Systèmes à
Intérêt Biologique
UMR 6504 du CNRS
Université Blaise Pascal
63177 Aubière Cedex, France
e-mail: besse@chimie.univ-bpclermont.fr

Binet, Françoise
CNRS, UMR 6553 ECOBIO
"Ecosystèmes, Biodiversité et Evolution"
Université de Rennes I
Campus de Beaulieu, Bât 14 CAREN
35042 Rennes Cedex, France
e-mail: fbinet@univ-rennes1.fr

Birkefeld, Andreas
Institute of Terrestrial Ecology
Swiss Federal Institute of Technology
ETH Zürich
Grabenstrasse 3
8952 Schlieren, Switzerland
e-mail: birkefeld@env.ethz.ch
web: www.birkefeld.com

Bonjean, Nicolas
Center of Research on Prion Proteins
University of Liège
CHU – Tour de Pharmacie
4000 Liège, Belgium

Bonmatin, J. M.
Centre de Biophysique Moléculaire
CNRS & Université Orléans
45071 Orléans Cedex 02, France

Borisov, Vyacheslav A.
Tula State University
Av. Lenina, 92
Tula, 300600, Russian Federation

Boscher, A.
Laboratoire de Chimie Moléculaire et
Environnement, ESIGEC
Université de Savoie
73376 Le Bourget du Lac, France

Bouchaou, L.
Laboratory of Applied Geology and
Geoenvironment
Hydrogeology Team
Faculty of Science
Ibn Zohr University, B.P. 8106
80000 Agadir, Morocco
e-mail: lbouchaou@yahoo.fr

Brault, Agathe
Unité de Phytopharmacie et Médiateurs
Chimiques
Institut National de la Recherche Agronomique
Route de Saint-Cyr
78026 Versailles Cedex, France
e-mail: brault@versailles.inra.fr

Briand, J.
Laboratoire de Géochimie des Eaux
Université Paris 7
case courrier 7084, 2 place Jussieu
75251 Paris, France

Bufo, Sabino Aurelio
Dipartimento di Scienze dei Sistemi Colturali,
Forestali e dell'Ambiente
Università della Basilicata
Campus Macchia Romana
85100 Potenza, Italy
phone: +390971205232; fax +390971205378
e-mail: bufo@unibas.it

Buncel, E.
Department of Chemistry and School of
Environmental Studies
Queen's University
Kingston, Ontario, Canada, K7L 3N6

Burgeot, T.
Ifremer
Laboratoire Dept Polluants Chimiques
B.P. 21105
44311 Nantes Cedex 03, France

Camel, Valérie
Institut National Agronomique Paris-Grignon
Laboratoire de Chimie Analytique
UMR Environnement et Grandes Cultures
16 Rue Claude Bernard
75231 Paris Cedex 05, France
e-mail: camel@inapg.inra.fr

Camy, S.
Laboratoire de Génie Chimique
UMR 5503 CNRS-INPT-UPS
B.P. 1301
31106 Toulouse Cedex 1, France
web: http://www.ensiacet.fr

Canals, Àngels
Departament de Cristal.lografia, Mineralogia
i Dipòsits Minerals
Facultat de Geologia
Universitat de Barcelona
Martí i Franquès s/n
08028 Barcelona, Spain

Carnicer, Angel
Departamento de Ingenieria Quimica
Escuela Universitaria Politecnica
Universidad de Castilla-La Mancha
Plaza de Manuel Meca, 1
13400 Almaden, Ciudad Real, Spain

Castegnaro, M.
Ecole Nationale Supérieure d'Agronomie
de Toulouse
Laboratoire de Toxicologie et Sécurité
Alimentaire
1 Av de l'Agrobiopole
31326 Auzeville Tolosane, France
and
Consultant
Les Collanges
07240 Saint-Jean Chambre, France

Cavret, S.
Laboratoire de Sciences Animales
INRA-INPL-UHP
B.P. 172
54505 Vandoeuvre Cedex, France

Cecinato, Angelo
Istituto sull'Inquinamento Atmosferico-C.N.R.
Area della Ricerca di Roma
Via Salaria Km 29.3
C.P. 10
00016 Monterotondo Scalo Rome, Italy

Charvet, R.
Centre de Biophysique Moléculaire
CNRS & Université Orléans
45071 Orléans Cedex 02, France

Chauveheid, Eric
Water Quality Department
Brussels Water Company (C.I.B.E.)
Chaussée de Waterloo, 764
1180 Brussels, Belgium
e-mail: eric.chauveheid@cibe.be

Coffinet, S.
Laboratoire de Géochimie des Eaux
Université Paris 7
case courrier 7084
2 place Jussieu
75251 Paris, France

Colin, M. E.
Unité de Zoologie et d'Apidologie
INRA
84914 Avignon Cedex 09, France

Combourieu, B.
Laboratoire de Synthèse Et Etude de Systèmes à Intérêt Biologique
UMR 6504 du CNRS
Université Blaise Pascal
63177 Aubière Cedex, France
e-mail: Bruno.Combourieu@chimie.univ-bpclermont.fr

Commarieu, A.
Total, Groupement de Recherches de Lacq
RN 117, B.P. 34
64170 Lacq, France

Condoret, J. S.
Laboratoire de Génie Chimique
UMR 5503 CNRS-INPT-UPS
BP 1301
31106 Toulouse Cedex 1, France
e-mail: JeanStephane.Condoret@ensiacet.fr
web: http://www.ensiacet.fr

Cristensen, Iben V.
Department of Civil Engineering
Kemitorvet
Building 204
The Technical University of Denmark
2800 Lyngby, Denmark
e-mail: ic@byg.dtu.dk

David, Bernard
Laboratoire de Chimie Moléculaire
et Environnement, ESIGEC
Université de Savoie
73376 Le Bourget du Lac, France
e-mail: bernard.david@univ-savoie.fr

de Bellefon, Claude
Laboratoire de Génie
des Procédés Catalytiques
CNRS UMR 2214, CPE Lyon
43 bld du 11 novembre 1918, B.P. 2077
69616 Villeurbanne Cedex, France

De Carlo, E. Heinen
Department of Oceanography
SOEST
University of Hawai'i at Manoa
1000 Pope Road
Honolulu, HI, USA, 96822
e-mail: edecarlo@soest.hawaii.edu

De Pauw, Edwin
Mass Spectrometry Laboratory
University of Liège
3 allée de la chimie B6c
4000 Liège, Belgium

De Pauw-Gillet, Marie-Claire
Laboratory of Histology-Cytology
University of Liège
3 allée de la chimie B6c
4000 Liège, Belgium

De Wever, H.
VITO
Boeretang 200
2400 Mol, Belgium

Delgado Núñez, Lourdes
Laboratoire de Génie des Procédés Catalytiques
CNRS UMR 2214, CPE Lyon
43 bld du 11 novembre 1918, B.P. 2077
69616 Villeurbanne Cedex, France
e-mail: ldn@lgpc.cpe.fr

Delgenes, Jean-Philippe
Institut National de la Recherche Agronomique (INRA)
Laboratoire de Biotechnologie
de l'Environnement (LBE)
Avenue des Etangs
11100 Narbonne, France

Delort, A. M.
Laboratoire de Synthèse Et Etude de Systèmes
à Intérêt Biologique
UMR 6504 du CNRS
Université Blaise Pascal
63177 Aubière Cedex, France
e-mail: amdelort@chimie.univ-bpclermont.fr

Delteil, Corine
Institut National Agronomique Paris-Grignon
Laboratoire de Chimie Analytique
UMR Environnement et Grandes Cultures
16 Rue Claude Bernard
75231 Paris Cedex 05, France

Dibenedetto, Angela
METEA Research Centre
University of Bari
via Celso Ulpiani 27
70126 Bari, Italy

Djordjevic, Dragana
IChTM-Chemistry Center
Njegoseva 12
11000 Belgrade, Serbia and Montenegro

Dubroca, Jacqueline
Unité de Phytopharmacie et Médiateurs
Chimiques
Institut National de la Recherche Agronomique
Route de Saint-Cyr
78026 Versailles Cedex, France
e-mail: dubroca@versailles.inra.fr

Duval, E.
Laboratoire de Géochimie des Eaux
Université Paris 7
case courrier 7084
2 place Jussieu
75251 Paris, France

Čáslavský, Josef
Institute of Analytical Chemistry
Czech Academy of Sciences
Veveøí 97
61142 Brno, Czech Republic
e-mail: caslavsky@iach.cz

El Moualij, Benaissa
Center of Research on Prion Proteins
University of Liège
CHU – Tour de Pharmacie
4000 Liège, Belgium

Emmelin, Corinne
L.A.C.E., Université Claude Bernard
Boulevard du 11 novembre 1918
69622 Villeurbanne Cedex, France
phone: +33(0)4 72444214; fax: +33(0)4 72448114
e-mail: Corinne.Emmelin@univ-lyon1.fr

Emnéus, Jenny
Department of Analytical Chemistry
Center for Chemistry and Chemical Engineering
Lund University, P.O. Box 124
22100 Lund, Sweden

Fabre, Bernard
Gestion Risque Environnement
Laboratory in Colmar
Université de Haute Alsace
29 Rue de Herrlisheim, B.P. 568
68000 Colmar Cedex, France
e-mail: bernard.fabre@uha.fr

Fan, Teresa W.-M.
Department of Land, Air and Water Resources
University of California
One Shields Ave.
Davis, CA 95616-8627, USA
Present address:
Department of Chemistry, University of Louisville
2320 S. Brook St.
Louisville, KY 40208, USA
e-mail: teresa.fan@louisville.edu

Feidt, C.
Laboratoire de Sciences Animales
INRA-INPL-UHP
Ecole Nationale Supérieure d'Agronomie et des
Industries Alimentaires (ENSAIA)
2 avenue de la Forêt de Haye, B.P. 172
54500 Vandoeuvre-Lès-Nancy, France

Filella, Montserrat
Department of Inorganic, Analytical and
Applied Chemistry
University of Geneva
Quai Ernest-Ansermet 30
1211 Geneva 4, Switzerland

Fiol, Nuria
Dept of Chemical Engineering
Universitat de Girona
Avda. Lluís Santalo, s/n
17071 Girona, Spain
e-mail: nuria.fiol@udg.es

Fischer-Colbrie, G.
Institute of Environmental Biotechnology
Graz University of Technology
Petersgasse 12
8010 Graz, Austria

Fleche, C.
Laboratoire de Pathologie des Petits Ruminants
et des Abeilles, AFSSA
06902 Sophia Antipolis Cedex, France

Flotron, Vanina
Institut National Agronomique Paris-Grignon
Laboratoire de Chimie Analytique
UMR Environnement et Grandes Cultures
16 Rue Claude Bernard
75231 Paris Cedex 05, France

Forano, C.
Laboratoire des Matériaux Inorganiques
UMR 6002 CNRS, Université Blaise Pascal
63177 Aubière Cedex, France
e-mail: claude.forano@univ-bpclermont.fr

Frades, J. M.
Escuela Universitaria Politécnica de Almadén
Chemical Engineering Department
University of Castilla-La Mancha
Plaza Manuel Meca, 1
13400 Almadén, Ciudad Real, Spain

Frolova, L.
Kazan State University
Kremlevskaya St. 18
Kazan, 420008, Russia

Fulladosa, Elena
Dept of Chemical Engineering
University of Girona
Avda. Lluís Santalo, s/n
17071 Girona, Spain
e-mail: elena.fulladosa@udg.es

Gaubin, Yolande
Lab. of Cell Biology, Faculty of Medicine, UPS
37 Allées Jules Guesde
31073 Toulouse, France

Gieren, Birgit
Institute of Geology and Geochemistry of
Petroleum and Coal
RWTH Aachen University
Lochnerstraße 4–20
52056 Aachen, Germany

Glass, Richard
Central Science Laboratory
Pesticides and Veterinary Medicines Group
Sand Hutton
York YO41 1LZ, UK
e-mail: R.Glass@csl.gov.uk

Grathwohl, Peter
Center for Applied Geoscience, Applied Geology
University of Tübingen
Sigwarstraße 10
72076 Tübingen, Germany
e-mail: peter.grathwohl@uni-tuebingen.de

Greenway, Gillian M.
Department of Chemistry, University of Hull
Hull, HU6 7RX, UK
e-mail: G.M.Greenway@hull.ac.uk
web: www.analyticalsciencehull.org

Greenwood, Paul. A.
Department of Chemistry, University of Hull
Hull, HU6 7RX, UK
e-mail: P.A.Greenwood@hull.ac.uk
web: www.analyticalsciencehull.org

Grova, N.
Laboratoire de Sciences Animales
INRA-INPL-UHP, B.P. 172
54505 Vandoeuvre Cedex, France

Gu, Man Bock
Department of Environmental Science and
Engineering and Advanced Environmental
Monitoring Research Center (ADEMRC)
Europe Satellite Laboratory, Gwangju Institute of
Science and Technology (GIST)
1 Oryong-dong, Buk-gu
Gwangju 500-712, Korea
e-mail: mbgu@kjist.ac.kr
and:
National Research Laboratory on Environmental
Biotechnology, Gwangju Institute of Science and
Technology (GIST)
1 Oryong-dong, Buk-gu
Gwangju 500-712, Korea
e-mail: mbgu@kjist.ac.kr

Guebitz, Georg M.
Department of Environmental Biotechnology
Graz University of Technology
Petersgasse 12
8010 Graz, Austria
e-mail: guebitz@tugraz.at

Guittonneau, S.
Laboratoire de Chimie Moléculaire
et Environnement
ESIGEC
Université de Savoie
73376 Le Bourget du Lac, France

Hansen, Peter D.
Institute for Ecotoxicology
Technical University of Berlin
Sekretariat KEP2
Keplerstraâe 4–6
10589 Berlin, Germany

Haroune, N.
Laboratoire de Synthèse Et Etude de Systèmes
à Intérêt Biologique
UMR 6504 du CNRS
Université Blaise Pascal
63177 Aubière Cedex, France

Harrington, Paul
Environmental Biology Group (EBG)
Central Science Laboratory
Sand Hutton
York YO41 1LZ, UK
e-mail: P.Harrington@CSL.GOV.UK

Heinen, Ernst
Center of Research on Prion Proteins
University of Liège
CHU – Tour de Pharmacie
4000 Liège, Belgium

Heintz, O.
Laboratoire de Chimie et Applications EA3471
Université de Metz
Rue Victor Demange
57500 Saint Avold, France

Higashi, R. M.
Center for Health and the Environment
John Muir Institute of the Environment
University of California
One Shields Ave.
Davis, CA 95616-8627, USA
e-mail: rmhigashi@ucdavis.edu

Higueras, P.
Escuela Universitaria Politécnica de Almadén
Geological Engineering Department
University of Castilla-La Mancha
Plaza Manuel Meca, 1
13400 Almadén, Ciudad Real, Spain

Höllrigl-Rosta, Andreas
Institute for Environmental Biology and
Chemodynamics (former Department of
Biology V– Environmental Chemistry)
Aachen University
Worringerweg 1
52056 Aachen, Germany
Current address:
Federal Environmental Agency, Department IV 1.3
Seecktstraße 6–10
13581 Berlin, Germany
e-mail: Andreas.Hoellrigl-Rosta@uba.de

Horneck, G.
DLR, Institute of Aerospace Medicine
Radiation Biology
Linder Höhe
51147 Köln, Germany

Hosoya, Hiroshi
Department of Biological Science
Graduate School of Science, Hiroshima University
Higashi-Hiroshima 739-8526, Japan
e-mail: hhosoya@sci.hiroshima-u.ac.jp
and:
PRESTO
Japan Science and Technology Corporation (JST)
Higashi-Hiroshima 739-8526, Japan

Hosoya, Natsumi
School of Social Information Studies
Otsuma Women's University
Tokyo 260-8540, Japan

Hsissou, Y.
Laboratory of Applied Geology and
Geoenvironment, Hydrogeology Team
Faculty of Science, Ibn Zohr University, B.P. 8106
80000 Agadir, Morocco
e-mail: hsissouy@yahoo.fr

Iinuma, K.
Department of Quantum Science and Energy
Engineering, Graduate School of Engineering
Tohoku University
Sendai, Miyagi, 980-8579, Japan
e-mail: koichi.iinuma@qse.tohoku.ac.jp

Iliasov, Pavel V.
G. K. Skryabin Institute of Biochemistry and
Physiology of Microorganisms
Russian Academy of Sciences
Pr. Nauki, 5
Pushchino, 142290, Russian Federation

Inacio, J.
Laboratoire des Matériaux Inorganiques
UMR 6002 CNRS
Université Blaise Pascal
63177 Aubière Cedex, France

Innocent, C.
BRGM, ANA – ISO
3 Avenue Claude Guillemin, B.P. 6009
45060 Orléans Cedex 2, France
e-mail: c.innocent@brgm.fr

Irigaray, P.
Laboratoire de Sciences Animales
INRA-INPL-UHP
Ecole Nationale Supérieure d'Agronomie et des
Industries Alimentaires (ENSAIA)
2 avenue de la Forêt de Haye, B.P. 172
54 500 Vandoeuvre-Lès-Nancy, France
e-mail: Philippe.Irigaray@ensaia.inpl-nancy.fr

Ishizaka, Yukiko
Department of Chemistry and Bio Science
Faculty of Science
Kagoshima University
Kagoshima 890-0065, Japan

Jaffrezo, J.-L.
LGGE, CNRS
Université Joseph Fourier
38402 St Martin d'Hères, France

Jêdrysek, Mariusz-Orion
Laboratory of Isotope Geoecology
University of Wroclaw
Cybulskiego 30
50-205 Wrocław, Poland
e-mail: morin@ing.uni.wroc.pl

Jezequel, Karine
Gestion Risque Environnement
Laboratory in Colmar
Université de Haute Alsace
29 Rue de Herrlisheim, B.P. 568
68008 Colmar Cedex, France
e-mail: karine.jezeque@uha.fr

Jolivalt, Claude
Laboratoire de Synthèse Sélective Organique et
Produits Naturels, UMR CNRS 7573
Ecole Nationale Supérieure de Chimie de Paris
11, Rue Pierre et Marie Curie
75231 Paris Cedex 05, France
e-mail: Claude-Jolivalt@enscp.jussieu.fr

Josselin, Nathalie
CNRS, UMR 6553 ECOBIO
"Ecosystèmes, Biodiversité et Evolution"
Université de Rennes I
Campus de Beaulieu, Bât 14 CAREN
35042 Rennes Cedex, France
e-mail: nathalie.josselin@univ-rennes1.fr

Kadono, Takashi
Department Biological Science
Hiroshima University
Higashi-Hiroshima, Japan
Present address:
Graduate School of
Environmental Engineering
The University of Kitakyushu
Kitakyushu 808-0135, Japan

Kandelbauer, Andreas
Department of Environmental Biotechnology
Graz University of Technology
Petersgasse 12
8010 Graz, Austria

Karapanagioti, Hrissi K.
Department of Marine Sciences
University of the Aegean
81100 Mytilene, Greece
e-mail: hrissi@marine.aegean.gr

Kawano, Nakako
Unité Mixte de Recherche INRA-UHP
Interactions Arbres/Micro-organismes
Institut National de la Recherche Agronomique
54280 Champenoux, France

Kawano, Tomonori
Unité Mixte de Recherche INRA-UHP
Interactions Arbres/Micro-organismes
Institut National de la Recherche Agronomique
54280 Champenoux, France
Present address:
Graduate School of
Environmental Engineering
The University of Kitakyushu
Kitakyushu 808-0135, Japan
e-mail: kawanotom@env.kitakyu-u.ac.jp

Keller, Ralf
Institute of Geology and Geochemistry of
Petroleum and Coal
RWTH Aachen University
Lochnerstraße 4–20
52056 Aachen, Germany

Kerhoas, Lucien
Unité de Phytopharmacie et Médiateurs Chimiques
Institut National de la Recherche Agronomique
Route de Saint-Cyr
78026 Versailles Cedex, France
e-mail: kerhoas@versailles.inra.fr

Kersanté, A.
Laboratoire Fonctionnement des Ecosystèmes et Biologie de la Conservation
UMR 6553 CNRS
Université de Rennes I
Campus de Beaulieu, Bât 14
35042 Rennes Cedex, France

Kiem, Rita
Lehrstuhl für Bodenkunde
Department für Ökologie
Wissenschaftszentrum Weihenstephan für Ernährung, Landnutzung und Umwelt
Technische Universität München
85350 Freising-Weihenstephan, Germany

Kim, Byoung Chan
Department of Environmental Science and Engineering and Advanced Environmental Monitoring Research Center (ADEMRC)
Europe Satellite Laboratory, Gwangju Institute of Science and Technology (GIST)
1 Oryong-dong, Puk-gu
Gwangju 500-712, Korea

Kleineidam, Sybille
Center for Applied Geoscience
Applied Geology
University of Tübingen
Sigwarstraße 10
72076 Tübingen, Germany
e-mail: sybille.kleineidam@uni-tuebingen.de

Kollmann, Albert
Unité de Phytopharmacie et Médiateurs Chimiques
Institut National de la Recherche Agronomique
Route de Saint-Cyr
78026 Versailles Cedex, France
e-mail: kollmann@versailles.inra.fr

Koroleva, T.
Institute of Ecology of Natural Landscapes
Tatarstan Academy of Sciences
Daurskaya St. 28
Kazan, 420087, Russia

Korzhuk, Nikolay L.
Tula State University
Av. Lenina, 92
Tula, 300600, Russian Federation

Kosaka, Toshikazu
Department of Biological Science
Graduate School of Science
Hiroshima University
Higashi-Hiroshima 739-8526, Japan

Kotlaříková, Pavla
Technical University
Faculty of Chemical Technology
Purkyòova 118
612 00 Brno, Czech Republic
e-mail: kotlarikova@fch.vutbr.cz

Krimissa, M.
Isotope Hydrology and Geochemistry Laboratory (LHGI), Faculty of Sciences of Orsay
University of Paris-Sud
15 Rue Georges Clémenceau
91405 Orsay Cedex, France
Formerly member of the
Isotopic Hydrology Section, CNESTEN
5 Tensift Street
Rabat, Morocco
e-mail: krimissa@yahoo.fr

Krimissa, S.
Research Team 2642
"Geoscience: Deformation, Flow, Transfer"
Faculty of Science
University of Franche-Comte
16, Gray road
25030 Besançon, France

Kunimoto, Manabu
Environmental Health Science Division
The National Institute for Environmental Studies
Tsukuba 305-8506, Japan
Present address:
Department of Public Health and Molecular Toxicology, School of Pharmaceutical Science
Kitasato University
Tokyo 108-8641, Japan

Lapeyrie, Frédéric
Unité Mixte de Recherche INRA-UHP
Interactions Arbres/Micro-organismes
Institut National de la Recherche Agronomique
54280 Champenoux, France
e-mail: lapeyrie@nancy.inra.fr

Laurent, C.
Laboratoire de Sciences Animales
INRA-INPL-UHP
B.P. 172
54505 Vandoeuvre Cedex, France

Laurent, F.
Laboratoire de Sciences Animales
INRA-INPL-UHP
Ecole Nationale Supérieure d'Agronomie
et des Industries Alimentaires (ENSAIA)
2 avenue de la Forêt de Haye, B.P. 172
54505 Vandoeuvre-Lès-Nancy, France

Lavédrine, B.
Laboratoire de Photochimie Moléculaire
et Macromoléculaire
UMR 6505 CNRS
Université Blaise Pascal
63177 Aubière Cedex, France

Le Bonté, S.
Laboratoire des Sciences du Génie Chimique
CNRS-ENSIC-INPL
1 Rue Grandville, B.P. 451
54001 Nancy Cedex, France

Le Clainche, L.
Département d'Ingénierie et d'Etude
des Protéines
CEA Saclay
91191 Gif sur Yvette Cedex, France
e-mail: leclainche@dsvidf.cea.fr

Lebeau, Thierry
Gestion Risque Environnement
Environnement Laboratory in Colmar
Université de of Haute Alsace
29 Rue de Herrlisheim, B.P. 568
68008 Colmar Cedex, France
e-mail: thierry.lebeau@uha.fr

Ledent, Philippe
Réalco S.A.
Louvain-La-Neuve, Belgium

Lee, Eun Yeol
Department of Food Science and Technology
Kyungsung University
Busan 608-736, Korea
e-mail: eylee@star.ksu.ac.kr

Lewis, Rachel
Address unknown

Ligouis, Bertrand
Laboratories for Applied Organic Petrology
(LAOP)
Institute for Geosciences
Sigwartstraße 10
72076 Tübingen, Germany
e-mail: bertrand.ligouis@uni-tuebingen.de

Lin, Shu-Kun
Molecular Diversity Preservation International
(MDPI)
Matthaeusstrasse 11
4057 Basel, Switzerland
e-mail: lin@mdpi.org
web: http://www.mdpi.org/lin
and:
Molecules Editorial Office
Ocean University of China
Qingdao 266003, Shandong Province, China

Littke, Ralf
Institute of Geology and Geochemistry
of Petroleum and Coal
RWTH Aachen University
Lochnerstraße 4–20
52056 Aachen, Germany

Lodge, Keith B.
Department of Chemical Engineering
University of Minnesota Duluth
1303 Ordean Court
Duluth, Minnesota 55812-3025, USA
e-mail: klodge@d.umn.edu

Lopez-Bellido, Francisco Javier
Departamento de Produccion Vegetal y
Tecnologia Agraria, Escuela Universitaria de
Ingenieria Tecnica Agricola
Universidad de Castilla-La Mancha
Ronda de Calatrava 7
13003 Ciudad Real, Spain

Maier, J.
Institute of Environmental Biotechnology
Graz University of Technology
Petersgasse 12
8010 Graz, Austria

Malinowska, B.
Wroclaw University of Technology, Institute of
Chemistry and Technology Petroleum and Coal
ul. Gdańska 7/9
50-344 Wrocław, Poland
e-mail: Malinowska@nafta1.nw.pwr.wroc.pl

Marchand, N.
LCME, Ecole Supérieure d'Ingénieur
de Chambéry
Université de Savoie
73376 Le Bourget du Lac, France

Martínez, M.
Department of Chemical Engineering
Universitat Politècnica de Catalunya
Avda. Diagonal, 647
08028 Barcelona, Spain
e-mail: rosario.martinez@upc.es

Masclet, P.
LCME, Ecole Supérieure d'Ingénieur
de Chambéry
Université de Savoie
73376 Le Bourget du Lac, France

Mathers, James
Central Science Laboratory
Pesticides and Veterinary Medicines Group
Sand Hutton, York YO41 1LZ, UK
e-mail: J.Mathers@csl.gov.uk

Meallier, Pierre
L.A.C.E.
Université Claude Bernard
Boulevard du 11 novembre 1918
69622 Villeurbanne Cedex, France
phone: +33(0)4 72448000; fax: +33(0)4 72448114
e-mail: pierre.meallier@club-internet.fr
Retired 2000

Mejean, L.
Laboratoire de Sciences Animales
INRA-INPL-UHP
Ecole Nationale Supérieure d'Agronomie et des
Industries Alimentaires (ENSAIA)
2 avenue de la Forêt de Haye, B.P. 172
54500 Vandoeuvre-Lès-Nancy, France

Meklati, Brahim Youcef
Centre de Recherche Scientifique et Technique
en Analyses Physico-Chimiques (C.R.A.P.C)
B.P. 248 Alger RP
16004 Algiers, Algeria

Minamisawa, Hiroaki
Department of Basic Science & High Technology
Center, College of Industrial Technology
Nihon University
Izumi-cho, Narashino, Chiba, 275-8576, Japan
e-mail: minami@mmm.cit.nihon-u.ac.jp

Minamisawa, Mayumi
Tokyo College of Medico-Pharmaco
Technology
6-5-12, Higashikasai, Edogawa-ku
Tokyo, 153-8530, Japan

Miralles, N.
Chemical Engineering Department
Universitat Politècnica de Catalunya
Avda. Diagonal, 647
08028 Barcelona, Spain

Moineau, I.
Centre de Biophysique Moléculaire
CNRS & Université Orléans
45071 Orléans Cedex 02, France

Monteagudo, Jose Maria
Chemical Engineering Department
Escuela Universitaria Politécnica de Almadén
University of Castilla-La Mancha
Plaza Manuel Meca, 1
13400 Almadén, Ciudad Real, Spain
e-mail: josemaria.monteagudo@uclm.es

Mougin, Christian
Unité de Phytopharmacie et Médiateurs
Chimiques
Institut National de la Recherche Agronomique
Route de Saint-Cyr
78026 Versailles Cedex, France
e-mail: Christian.Mougin@versailles.inra.fr

Mudry, J.
Research Team 2642
"Geoscience: Deformation, Flow, Transfer"
Faculty of Science
University of Franche-Comte
16, Gray road
25030 Besançon, France
e-mail: jacques.mudry@univ-fcomte.fr

Murat, Jean-Claude
Lab. of Cell Biology
Faculty of Medicine, UPS
37 Allées Jules Guesde
31073 Toulouse, France
e-mail: murat@cict.fr

Nakajima, Sugiko
Tokyo College of Medico-Pharmaco
Technology
6-5-12, Higashikasai
Edogawa-ku, Tokyo, 153-8530, Japan

Niemz, Claudia
Laboratories for Applied Organic Petrology (LAOP)
Straße der Freundschaft 92
02991 Lauta, Germany
e-mail: info@laop-consult.de

Nishihara, Naohisa
Department of Biological Science
Graduate School of Science
Hiroshima University
Higashi-Hiroshima 739-8526, Japan

Nowack, Bernd
Institute of Terrestrial Ecology
Swiss Federal Institute of Technology
ETH Zürich
Grabenstrasse 3
8952 Schlieren, Switzerland
e-mail: nowack@env.ethz.ch

Ottosen, Lisbeth M.
Department of Civil Engineering, Building 204
The Technical University of Denmark
2800 Lyngby, Denmark
e-mail: lo@byg.dtu.dk

Paiva, L. B.
Universidade de Mogi das Cruzes, Centro
Interdisciplinar de Investigação Bioquímica (CIIB)
Av. Dr. Cândido Xavier de Almeida Souza, 200
Centro Cívico
Mogi das Cruzes, SP, Brazil, CEP: 08780-911
e-mail: flaviorodrigues@yahoo.com

Pastore, Tiziano
INCA, Unit of Bari
METEA Research Centre
via Celso Ulpiani 27
70126 Bari, Italy

Patureau, Dominique
Institut National de la Recherche Agronomique
(INRA), Laboratoire de Biotechnologie
de l'Environnement (LBE)
Avenue des Etangs
11100 Narbonne, France
e-mail: patureau@ensam.inra.fr

Pedersen, Anne J.
Department of Civil Engineering
Kemitorvet, Building 204
Technical University of Denmark
2800 Lyngby, Denmark
e-mail: ajp@byg.dtu.dk

Perez Duran, Sonia
Central Science Laboratory
Pesticides and Veterinary Medicines Group
Sand Hutton, York YO41 1LZ, UK
e-mail: S.Perez-Duran@csl.gov.uk

Pfohl-Leszkowicz, A.
Ecole Nationale Supérieure d'Agronomie
de Toulouse, Laboratoire de Toxicologie et
Sécurité Alimentaire
1 Av de l'Agrobiopole
31326 Auzeville Tolosane, France

Pierard, Olivier
Center of Research on Prion Proteins
University of Liège
CHU – Tour de Pharmacie
4000 Liège, Belgium

Plançon, C.
Laboratoire des Sciences du Génie Chimique
CNRS-ENSIC-INPL
1 Rue Grandville, B.P. 451
54001 Nancy Cedex, France

Plisson-Saune, S.
Total, Groupement de Recherches de Lacq
RN 117, B.P. 34
64170 Lacq, France

Poch, J.
Applied Mathematics Department
Universitat de Girona
Avda. Lluís Santaló, s/n
17071 Girona, Spain
e-mail: poch@ima.udg.es

Polic, Predrag
Department of Chemistry, University of Belgrade
POB 158
11001 Belgrade, Serbia and Montenegro

Pons, M. N.
Laboratoire des Sciences du Génie Chimique
CNRS-ENSIC-INPL
1 Rue Grandville, B.P. 451
54001 Nancy Cedex, France
e-mail: Marie-Noelle.Pons@ensic.inpl-nancy.fr

Popovic, Aleksandar
Department of Chemistry, University of Belgrade
POB 158
11001 Belgrade, Serbia and Montenegro
e-mail: apopovic@chem.bg.ac.yu

Potier, O.
Laboratoire des Sciences du Génie Chimique
CNRS-ENSIC-INPL
1 Rue Grandville, B.P. 451
54001 Nancy Cedex, France

Rabbow, E.
RWTH Aachen
Lehrstuhl für Flugmedizin
Pawelstraße 30
52057 Aachen, Germany

Radmanovic, Dubravka
Department of Chemistry, University of Belgrade
POB 158
11001 Belgrade, Serbia and Montenegro

Recreo, Fernando
Departamento de Impacto Ambiental de la Energia
Comportamiento Ambiental de Contaminantes
en Sistemas Geologicos
Centro de Investigaciones Energeticas,
Medioambientales y Tecnologicas
Avda. Complutense, 22
28040 Madrid, Spain

Reshetilov, Anatoly N.
G. K. Skryabin Institute of Biochemistry and
Physiology of Microorganisms
Russian Academy of Sciences
Pr. Nauki, 5
Pushchino, 142290, Russian Federation

Rettberg, P.
DLR, Institute of Aerospace Medicine
Radiation Biology
Linder Höhe
51147 Köln, Germany
e-mail: petra.rettberg@dlr.de

Ribeiro, Alexandra B.
Departamento de Ciências e Engenharia do
Ambiente, Faculdade de Ciências e Tecnologia
Universidade Nova de Lisboa
Quinta da Torre
2825-114 Caparica, Portugal
e-mail: abr@fct.unl.pt

Richard, C.
Laboratoire de Photochimie Moléculaire
et Macromoléculaire
UMR 6505 CNRS
Université Blaise Pascal
63177 Aubière Cedex, France

Richard, Dominique
Laboratoire de Génie des Procédés Catalytiques
CNRS UMR 2214, CPE Lyon
43 bld du 11 novembre 1918, B.P. 2077
69616 Villeurbanne Cedex, France
e-mail: dri@lobivia.cpe.fr

Ricking, Mathias
Department of Earth Sciences
Free University of Berlin
Malteserstraße 74–100, House B
12249 Berlin, Germany

Robert, Didier
Laboratoire de Chimie et Applications (EA, 3471)
Université de Metz
Rue Victor Demange
57500 Saint-Avold, France
e-mail: Didier.Robert@iut.univ-metz.fr

Robra, K. H.
Institute of Environmental Biotechnology
Graz University of Technology
Petersgasse 12
8010 Graz, Austria

Rodrigues, F. A.
Universidade de Mogi das Cruzes, Centro
Interdisciplinar de Investigação Bioquímica (CIIB)
Av. Dr. Cândido Xavier de Almeida Souza, 200
Centro Cívico
Mogi das Cruzes, SP, Brazil, CEP: 08780-911
e-mail: flaviorodrigues@yahoo.com

Rodriguez, Luis
Chemical Engineering Department
Faculty of Environmental Sciences
University of Castilla-La Mancha
Avenida Carlos III, s/n
45071 Toledo, Spain
e-mail: Luis.RRomero@uclm.es

Römheld, V.
Institute of Plant Nutrition (330)
University of Hohenheim
70593 Stuttgart, Germany
e-mail: roemheld@uni-hohenheim.de

Ruelle, Virginie
Mass Spectrometry Laboratory
University of Liège
3 allée de la chimie B6c
4000 Liège, Belgium
e-mail: v.ruelle@skynet.be

Rychen, G.
Laboratoire de Sciences Animales
INRA-INPL-UHP
Ecole Nationale Supérieure d'Agronomie et des
Industries Alimentaires (ENSAIA)
2 avenue de la Forêt de Haye, B.P. 172
54500 Vandoeuvre-Lès-Nancy, France

Sancelme, M.
Laboratoire de Synthèse Et Etude de Systèmes
a Intérêt Biologique
UMR 6504 du CNRS
Université Blaise Pascal
63177 Aubière Cedex, France

Satoh, Y.
Department of Quantum Science and Energy
Engineering, Graduate School of Engineering
Tohoku University
Sendai, Miyagi, 980-8579, Japan

Schäffer, Andreas
Institute for Environmental Biology and
Chemodynamics (former Department of
Biology V – Environmental Chemistry)
Aachen University
Worringerweg 1
52056 Aachen, Germany

Schulin, Rainer
Institute of Terrestrial Ecology
Swiss Federal Institute of Technology
ETH Zürich
Grabenstrasse 3
8952 Schlieren, Switzerland
e-mail: schulin@env.ethz.ch

Schulz, R.
Institute of Plant Nutrition (330)
University of Hohenheim
70593 Stuttgart, Germany

Schwab, R.
Institut für Geologie und Mineralogie
Schloßgarten 5A
91054 Erlangen, Germany

Schwarzbauer, Jan
Institute of Geology and Geochemistry
of Petroleum and Coal
RWTH Aachen University
Lochnerstraße 4–20
52056 Aachen, Germany
e-mail: schwarzbauer@lek.rwth-aachen.de

Schweich, Daniel
Laboratoire de Génie des Procédés Catalytiques
CNRS UMR 2214, CPE Lyon
43 bld du 11 novembre 1918, B.P. 2077
69616 Villeurbanne Cedex, France

Scrano, Laura
Dipartimento di Scienze dei Sistemi Colturali,
Forestali e dell'Ambiente; Università della
Basilicata, Campus Macchia Romana
85100 Potenza, Italy
phone: +390971205231; fax +390971205378
e-mail: scrano@unibas.it

Semenchuk, Irina N.
Institute of Biocolloidal Chemistry
Vernadsky bv.
42 Kiev, Ukraine

Serarols, J.
Applied Mathematics Department
Universitat de Girona
Avda. Lluís Santaló, s/n
17071 Girona, Spain

Shirin, S.
Department of Chemistry and School
of Environmental Studies, Queen's University
Kingston, Ontario, Canada, K7L 3N6

Siñeriz Louis, M.
PROIMI (Planta Piloto de Procesos Industriales
Microbiológicos)
Avenida Belgrano y Pasaje Caseros
4000 Tucumán, Argentina

Skandrani, Dalila
Lab. of Cell Biology, Faculty of Medicine, UPS
37 Allées Jules Guesde
31073 Toulouse, France

Skrzypek, Grzegorz
Laboratory of Isotope Geoecology
University of Wroclaw
Cybulskiego 30
50-205 Wroclaw, Poland
e-mail: buki@ing.uni.wroc.pl

Soler, Albert
Departament de Cristal.lografia, Mineralogia
i Dipòsits Minerals, Facultat de Geologia
Universitat de Barcelona
Martí i Franquès, s/n
08028 Barcelona, Spain

Stolarski, M.
Wroclaw University of Technology
Institute of Chemistry and Technology
Petroleum and Coal
ul. Gdańska 7/9
50-344 Wrocław, Poland

Takai, Nobuharu
Department of Biotechnology
College of Science and Engineering
Tokyo Denki University
Hatoyama, Hikigun, Saitama, 350-0394, Japan

Tallos, Alberto
Departamento de Impacto Ambiental de la
Energia, Comportamiento Ambiental de
Contaminantes en Sistemas Geologicos
Centro de Investigaciones Energeticas,
Medioambientales y Tecnologicas
Avda. Complutense, 22
28040 Madrid, Spain

Tanaka, Miho
Department of Biological Science
Graduate School of Science
Hiroshima University
Higashi-Hiroshima 739-8526, Japan
Present address:
INRA UMR BiO3P, B.P. 29
35653 Le Rheu Cedex, France

Taranova, Ljudmila A.
Institute of Biocolloidal Chemistry
Vernadsky bv.
42 Kiev, Ukraine

Taviot-Guého, C.
Laboratoire des Matériaux Inorganiques
UMR 6002 CNRS
Université Blaise Pascal
63177 Aubière Cedex, France

Tosuji, Hiroaki
Department of Chemistry and Bio Science
Faculty of Science, Kagoshima University
Kagoshima 890-0065, Japan

Touton, Isabelle
Unité de Phytopharmacie et Médiateurs
Chimiques
Institut National de la Recherche Agronomique
Route de Saint-Cyr
78026 Versailles Cedex, France
e-mail: itouton@versailles.inra.fr

Town, Raewyn M.
School of Chemistry
The Queen's University of Belfast
Belfast BT9 5AG, Northern Ireland

Trably, Eric
Institut National de la Recherche Agronomique
(INRA), Laboratoire de Biotechnologie
de l'Environnement (LBE)
Avenue des Etangs
11100 Narbonne, France

Uchida, S.
Department of Quantum Science and Energy
Engineering
Graduate School of Engineering
Tohoku University
Sendai, Miyagi, 980-8579, Japan

vanLoon, G. W.
Department of Chemistry and School of
Environmental Studies
Queen's University
Kingston, Ontario, Canada, K7L 3N6
e-mail: vanloon@chem.queensu.ca

Villaescusa, Isabel
Dept of Chemical Engineering
University of Girona
Avda. Lluís Santalo, s/n
17071 Girona, Spain
e-mail: isabel.villaescusa@udg.es

Villumsen, Arne
Department of Civil Engineering
Kemitorvet, Building 204
The Technical University of Denmark
2800 Lyngby, Denmark
e-mail: av@byg.dtu.dk

Vinken, Ralph
Institute for Environmental Biology and
Chemodynamics (former Department of
Biology V – Environmental Chemistry)
Aachen University
Worringerweg 1
52056 Aachen, Germany

Vita, Claudio
Département d'Ingénierie et d'Etude
des Protéines
CEA Saclay
91191 Gif sur Yvette, France
e-mail: claudio.vita@cea.fr

Vitòria, Laura
Departament de Cristal.lografia, Mineralogia
i Dipòsits Minerals
Facultat de Geologia
Universitat de Barcelona
Martí i Franquès, s/n
08028 Barcelona, Spain
e-mail: lvitoria@ub.edu

Walecka-Hutchison, Claudia
University of Arizona
Department of Soil, Water, and Environmental
Science
429 Shantz Building, #38
1200 E. South Campus Dr.
Tucson, Arizona 85721-0038, U.S.A.
e-mail: Claudiah@ag.arizona.edu

Walendziewski, J.
Wroclaw University of Technology
Institute of Chemistry and Technology
Petroleum and Coal
ul. Gdañska 7/9
50-344 Wrocław, Poland
e-mail: Walendziewski@nafta1.nw.pwr.wroc.pl

Walworth, James L.
University of Arizona
Department of Soil, Water, and Environmental
Science
429 Shantz Building, #38
1200 E. South Campus Dr.
Tucson, Arizona 85721-0038, U.S.A.
e-mail: Walworth@ag.arizona.edu

Weber, J. V.
Laboratoire de Chimie et Applications
(EA, 3471)
Université de Metz
Rue Victor Demange
57500 Saint-Avold, France

Yang, Zhaoguang
Centre for Advanced Water Technology
Singapore Utilities International
Singapore 637723

Yassaa, Noureddine
Faculté de Chimie
Université des Sciences et de la Technologie
Houari Boumediene
B.P. 32 El-Alia, Bab-Ezzouar
16111 Algiers, Algeria
e-mail: nyassaa@mailcity.com

Yoshida Shoichiro
Institute of Industrial Science
University of Tokyo
4-6-1, Komaba, Meguro-ku
Tokyo, 153-8505, Japan

Zakirov, A.
Kazan State University
Kremlevskaya St. 18
Kazan, 420008, Russia

Zawadzki, J.
Department of Chemistry, University of Torun
Gagarina 7
87-100 Torun, Poland

Zhang, Lifeng
Centre for Advanced Water Technology
Singapore Utilities International
Singapore 637723
e-mail: lfzhang@cawt.sui.com.sg

Zhu, Zhiwei
Centre for Advanced Water Technology
Singapore Utilities International
Singapore 637723

Zorzi, Daniele
Center of Research on Prion Proteins
University of Liège
CHU – Tour de Pharmacie
4000 Liège, Belgium

Zorzi, Willy
Center of Research on Prion Proteins
University of Liège
CHU – Tour de Pharmacie
4000 Liège, Belgium

Part I
Analytical Chemistry

Part I

Chapter 1

In-situ Method for Analyzing the Long-Term Behavior of Particulate Metal Phases in Soils

A. Birkefeld · R. Schulin · B. Nowack

Abstract

Soils can act as a sink for anthropogenic and naturally released heavy metals. Among these are heavy metal oxides and sulfides, which are emitted e.g. by mining industry and metal smelting. The dissolution and transformation behavior of these heavy metal phases specifies their fate in the soil and determines whether the metals become bioavailable or could contaminate the groundwater. To gain more information about these dissolution reactions in soils, in-situ methods are needed. We present here a method to fix particulate metal phases on an inert support. This method allows us to expose and recover metal phases in the environment under controlled conditions.

Acrylic glass was chosen as inert polymer substrate for the heavy metal phases as it is stable to weathering. Epoxy resin was used as adhesive film between the acrylic glass support and the heavy metal coating. The fine-grained heavy metal phases are applied onto the epoxy resin using a dust spray gun. The heavy metal coated polymer platelets can be inserted in a controlled way into selected soil profiles and be recovered after definite time intervals. Qualifying and quantifying analysis can be carried out on every single polymer support.

Key words: in-situ method; metal phase transformation; soil pollution; heavy metals

1.1 Introduction

Soils represent an important sink for anthropogenic and natural heavy metals (Alloway and Ayres 1997; Schachtschabel et al. 1998). The fate of heavy metals in the environment is of elemental concern due to potential contamination risks of water, soils and sediments and due to toxicity of heavy metals to plants, animals and humans via the food chain. Soils near urban or industrial settlements have often been polluted by particulate heavy metal phases, mainly through atmospheric transport (Ge et al. 2000). Major anthropogenic heavy metal phases, emitted e.g. by mining and smelting operations, are sulfides (Ketterer et al. 2001) and oxides (Dudka and Adriano 1997). The dissolution and phase transformations of these particulate phases over time influences the bioavailability of the metals in the soil and the transport into the groundwater (Dudka et al. 1995). Knowledge about the physico-chemical phase transformations of heavy metal phases in the long-term aspect is therefore crucial (Ge et al. 2000; Fengxiang and Banin 1997; Li and Thornton 2001). Specific heavy metal phases like oxides can be present in soils even after long time periods (Li and Thornton 2001).

Methods have been established to classify heavy metal phases based on their dissolution behavior from solid phases in soils (Tessier et al. 1979; Brümmer et al. 1986; Ure and Davidson 1995). These sequential extraction methods classify the heavy metals into different groups, e.g. water soluble, exchangeable, organically bound, bound to iron and manganese oxides and bound in silicate structures. These procedures were applied in several studies investigating soils contaminated by mining and smelting activities (Li and Thornton 2001) and on unpolluted soils where heavy metals were added (Ma and Uren 1998). Conversely the exact detection of these phases is not straight forward (Manceau et al. 1996) and generated different approaches for sequential extractions (Ure and Davidson 1995). However, identifying the heavy metal species especially the one from anthropogenic sources is of major concern to assess their toxic behavior in soils (Ford et al. 1999; Henderson et al. 1998; Brümmer 1986).

In the last years extended X-ray absorption fine structure (EXAFS) analysis has also been increasingly used to identify heavy metal phases in soils (Welter et al. 1999; Roberts et al. 2002; Fendorf et al. 1994) but this method is still limited to high metal concentrations. Distinct heavy metal phases have been detected in soils, e.g. lead and zinc in the vicinity of industrial and mining facilities (Welter et al. 1999; Manceau et al. 1996).

To investigate the dissolution and transformation behavior of metal phases directly in the soil an in-situ method is needed. The in-situ dissolution of (soil)-minerals inserted into soil or groundwater horizons has been investigated (Righi et al. 1990; Ranger et al. 1991; Hatton et al. 1987; Bennett et al. 2001). The minerals e.g. vermiculite, were placed in porous bags in the soils and recovered after different time intervals. But these experiments were only designed to investigate soil forming processes. Tsaplina (1996) has reported the behavior of heavy metal oxides mixed into the topsoil of selected locations (PbO, ZnO 500 mg kg^{-1} and CdO 50 mg kg^{-1}). The samples for this study were collected after four and eight years. Fractions of the applied phases could be recovered by separation from the soil by density fractionation and were then identified by X-ray diffraction (Tsaplina 1996). The final sample analysis showed that about 40 wt% to 50 wt% of the initial heavy metal oxides were still present in the soil.

The sparse amount of literature about the in-situ behavior of heavy metal phases clearly shows that there are methodical and analytical difficulties in this area. In this article we would like to introduce a new in-situ method to investigate the behavior of heavy metal phases in soils (Birkefeld et al. 2004). The concept is to fix the heavy metal phases on a support which can be placed in the soil and which can easily be recovered. This approach allows direct analysis of the heavy metal particles after recovery and determination of their physico-chemical changes. A reinsertion into the soil for further investigation is possible. The method has been developed using heavy metal oxides and sulfides as model substances.

1.2
Experimental

To insert heavy metal phases of interest into the soil with the ability to recover them after certain time segments and to possibly re-apply them, we need the aid of a carrier material. The material should have good mechanical properties for sawing and mill cutting. We choose polymethacrylate (Plexiglas®) because this transparent polymer is resistant to weathering and ageing (Schwarz 2000).

A thin (0.025 g cm^{-2}) layer of epoxy resin (Bisphenol-Epichlorhydrin resin, Suter AG Switzerland) (Suter 1999) was used to mount the metal phases onto the polymer support. Epoxy resin was used due to its resistance against decomposition (Suter 2001). Prior to the application of the resin and the heavy metals, the supports were cut into sections of 2.2 × 2.2 cm. This size was chosen due to the limited space in the sample holder of the X-ray fluorescence (XRF) analytical instrument. 20 polymer supports were clamped into an acrylic glass frame to hold them together (11 × 11 cm). With a commercially available paint roll with a width of 5 cm and a diameter of 3 cm the epoxy resin was applied onto the surface of the polymer supports (0.12 g resin/support). The frame holder with the polymer supports was placed in a dust sealed chamber with fume extraction hoses. Before the application of the metal phases the epoxy resin covered supports were allowed to dry for 5 min in order to enhance the viscosity of the resin and to avoid complete capillary covering of the applied heavy metal particles. With a commercially available compressed air dust spray gun (SAV 030 025, Schneider Druckluft, Germany) the heavy metal phases were applied from a distance of approx. 1 m (air pressure: 1.5 kg cm^{-2}) onto the epoxy resin covered surface of the supports. The application time (spraying) was 10 seconds. After an overnight drying period at 50 °C in a laboratory drying cabinet the heavy metal coated polymer supports were cleaned with a soft brush in deionized water (Nanopure, >18 MΩ; 5 ppb TOC) to remove loose excess heavy metal particles from the support surface. After an additional 24 h-drying period at 40 °C the polymer supports were ready for use.

The model substance for our method development was a lead oxide from a commercial metal oxide manufacturer (PbO LOX 150 Pennarroya Oxide, Germany). This oxide was produced by direct oxidation of metallic lead and has a minimum content of PbO of 99.8%. Its density is 9.5 g cm^{-3} and the average particle size is about 0.1 mm diameter.

To determine the uncovered, reactive surface area of the particles, adsorption experiments were carried out with phenylphosphonic acid (PPA) (analytical grade; Fluka, Switzerland) in a 0.1 mmolar acetate buffer (pH 4.6). PPA was chosen because of its well-known adsorption characteristics onto a variety of different surfaces e.g. aluminum oxide (Laiti and Öhman 1996). The concentration of the PPA was 10–400 µM. Blank measurements with Plexiglas® and with epoxy resin covered supports showed that they did not adsorb PPA. The initial concentration and the final concentration of PPA after reaction times of 5, 10, 15, 20 and 30 min were measured by ion chromatography analysis (DX 100 ion chromatograph, Dionex USA). The maximum surface capacity of the heavy metal phases was also determined in suspension under the same conditions (10–400 µM PPA in 0.1 molar acetate buffer, pH 4.6). The amount of PPA adsorbed in suspension in mol m^{-2} was then used to calculate the exposed surface area of the resin-immobilized metals. A dissolution experiment with ethylene diamine tetraacetic acid (EDTA) was carried out to get an overview about the dissolution behavior of the mineral substances. The analytical conditions were EDTA 1 mM at pH 4.7 with a reaction time of 1 h.

To establish the amount of heavy metals on each polymer support X-ray fluorescence analysis (XRF) was made on every support before further use (X-Lab 2000, Spectro Germany). Because the XRF analysis requires fine grained samples, the method was externally calibration checked with data from atomic absorption spectrometry (AAS) measurements (SpectrAA 220, Varian Australia) after complete dissolution with EDTA of the heavy metals on the support. For the EDTA dissolution the same samples were used which were analyzed previously with XRF. The coated polymer supports were placed in Teflon® vessels and 10 ml 0.1 M EDTA (analytical grade, Merck Germany) was added. The vessels were placed in a microwave digestion system (ETHOS, Milestone Italy) and underwent a temperature program with constant pressure control (1. step 400 W for 7 min; 2.step 300 W for 3 min; 3. step 200 W for 40 min). The temperature was held at 160 °C over the whole dissolution process. The polymer supports and the resin were not altered by the EDTA and only the oxide minerals were dissolved. Scanning electron microscopic (SEM) investigations were carried out with a Cam Scan CS44, CamScan – United Kingdom.

1.3
Results and Discussion

1.3.1
Particle Appearance on Support Surface

To show the applicability of the approach several preleminary examinations were carried out. Optical and scanning electron microscopic examinations of the heavy metal covered supports showed the allocation density of the phases (PbO) on the epoxy resin. The metal oxides form almost a single layer on the resin (Fig. 1.1) which gives the conclusion that the cleaning step after hardening of the resin is complete. All of the particles are in contact with the epoxy resin and are therefore attached to the surface of the support. There is no sign of loose particles which could fall off during handling the supports and while resting in the soil profile. This is important to avoid material loss before the proper experiment or during placing or retrieving the supports. The microscopic examination showed that the heavy metal particles were not immersed in the epoxy resin film (Fig. 1.2). Only a certain percentage of their surface is in contact with the epoxy resin. This gives a high amount of uncovered mineral surface which can interact with the soil environment. The epoxy resin used had a high viscosity which avoids a complete coverage of the particles due to the capillary

Fig. 1.1.
SEM surface picture of a polymer support coated with lead oxide. SEM: scanning electron microscope

Fig. 1.2.
Lead oxide particles on a polymer support, picture taken at an angel of 70°

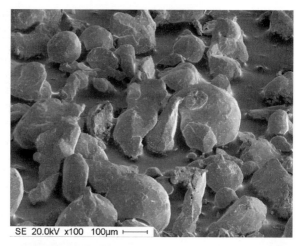

Fig. 1.3.
Substance application chamber: In front the dust spray gun (*1*) and in the chamber the Plexiglas support (*2*). The tube is the dust extraction unit (*3*) (for working safety reasons)

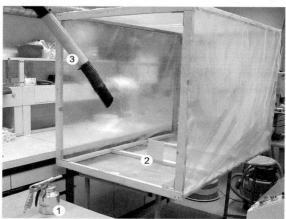

effect of the resin. This effect can be additionally controlled if after preparing the epoxy resin (mixing with the hardener substance), it is allowed to pre-react (hardening) for a few minutes which increases the viscosity. It could be shown that the used technique to attach particulate matter onto a support is suitable.

1.3.2
Coating Process

The coating process itself is not difficult to handle. A constant distance between the dust spray gun and the epoxy covered supports should be used. Also a constant air flow/pressure and a constant spray time should be kept. These are the most favorable conditions to produce polymer supports that are uniformly covered with heavy metal phases. It is self-evident that the application step can only be carried out in rooms with appropriate dust and air extraction units. A fine-dust face mask and an overall should be used to comply with the working safety regulations (Fig. 1.3).

To estimate the amount of applied minerals, every polymer support is weighed before and after the mineral application with a laboratory balance. The mass of the epoxy resin onto the supports was calculated by simple subtraction of the support mass before and after pure resin application on a support test set ($N = 30$).

1.3.3
Support Testing prior Insertion

The mechanical stability and the amount of resin-free particle surface was estimated with special experiments. Simple dissolution with ethylenediaminetetraacetic acid (EDTA) showed that the heavy metal phases can be dissolved without falling off from the surface. This dissolution experiment was carried out to simulate a dissolution process which could take place in a soil. Therefore it was important to see how the heavy metal covered support reacts mechanically if it is going to be recovered from

Fig. 1.4.
Polymer support with lead oxide particles after a dissolution experiment with 1 mM EDTA over 1 h at pH 4.7

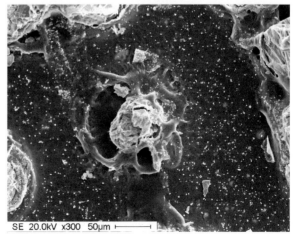

Fig. 1.5.
Polymer support with lead oxide particles after a dissolution experiment with 1 mM EDTA over 1 h at pH 4.7

the soil. It can be seen from Fig. 1.4 that the grains show signs of dissolution but are still attached to the support. Figure 1.5 shows a grain that has been dissolved almost completely. The former contact area of the particle with the support is still visible. The small remaining particle is still attached to the support. This experiment showed the good adhesion stability of the attached particles and their resistance to the cleaning steps. The estimation of the uncovered mineral surface which can directly interact with the soil environment can be determined with adsorption experiments. Preliminary experiments with phenylphosphonic acid (PPA) in an acetate buffered environment (pH 4.7) showed promising results, indicating that PPA is adsorbing onto the mineral surfaces but not onto the carrier material. Comparing the adsorption capacity of the bulk mineral material to the results with the mineral covered support will yield the amount of free reactive surface of the mineral supports. Together with the surface area (analyzed with the Brunauer-Emmet-Teller (BET) nitrogen sorption method) of the bulk mineral substance, we can then calculate the reactive surface area of the supports. This method to estimate the resin-free particle surface is useful to get an overview about the amount of "reactable" particle area.

1.3.4
Analytical Methods

The nondestructive X-ray fluorescence (XRF) analysis is the method of choice for quantifying the metal concentration of the covered polymer supports. With this analysis technique we have a useful tool to monitor e.g. the dissolution behavior of the heavy metals in the soil over different time scales. The supports can be reinserted into the soil after XRF analysis to undergo further reactions. A comparison of the XRF analysis and atomic absorption spectroscopic (AAS) analysis of the same supports has to be performed for each material that is used. The intention of this comparison is to find out whether the XRF results that are obtained using a calibration for powdered metal phases can be used to determine the metal concentration on the supports.

The electron optical analysis with the Scanning electron microscope (SEM) keeps track of the surface changes of the minerals on the supports. Traces of dissolution processes or precipitations on the mineral surfaces can be detected with this method (Bennett et al. 2001). Figures 1.1, 1.2, 1.4, and 1.5 show the application of this method to our samples. As already discussed above this method allows the direct visualization of the dissolution process. Additional coupling with the electron dispersive X-ray probe (EDX) can show the spatial elemental distribution on the supports or even of a mineral grain and its surrounding vicinity. With the use of an "environmental scanning electron microscope, ESEM" there is no need to coat the supports with a conducting surface e.g. gold or carbon. It will provide the same data as a conventional SEM (with attached EDX probe) but will have the same nondestructive benefits like the XRF method with a possible further use of the samples.

A more mineral specific analytical method is the infrared Raman spectroscopy in conjunction with a infrared microscope. This technique can detect certain mineral phases (Sobanska et al. 1999) on the support. Raman spectroscopy could give an indication of occurred phase transformations of the minerals. It might not have the power of the EXAFS technique but it is compact and can be performed in a normal laboratory environment. Because it is also a non destructive method the supports can be examined before and after insertion into soils. The use of nondestructive analytical methods should be favorized to keep the samples unchanged and open for other examinations.

1.4 Conclusions

The establishment of a method for the in-situ analysis of the reactions of heavy metal phases in soils was successfully carried out. With lead oxide as a model substance it could be shown that the particles are strongly fixed on the polymer supports and survive even extended dissolution without falling off. The method might have the ability to be used with different substances e.g. heavy metal oxides and sulfides, iron oxides or clays. The method can also be used to expose particles to other environments, e.g. natural waters, sediments or technical systems such as water treatment plants. The designated analytical methods are all nondestructive and avoid alterations of the samples that can therefore be further used.

Acknowledgements

We are most grateful to the ETH Zürich which funded this project. We would also thank Dr. Andreas Wahl from PENNAROYA Oxide who donated to us the lead oxide. Thanks to Dr. Karsten Kunze Geological Institute ETH Zürich and Peter Wägli Laboratory for Solid State Physics ETH Zürich for making the SEM pictures possible. Susan Tandy and Paula Amunátegui we would like to thank for cross-reading the manuscript.

References

Alloway BJ, Ayres DC (1997) The behaviour of pollutants in the soil. In: Ayres DC (ed) Chemical principles of environmental pollution. Blackie, London, pp 395
Bennett PC, Rogers JR, Choi WJ (2001) Silicates, silicate weathering, and microbial ecology. Geomicrobiol J 18:3–19
Birkefeld A, Nowack B, Schulin R (2004) A new in-situ method to monitor mineral particle reactions under field conditions. Environ Sci Technol (submitted)
Brümmer GW (1986) Heavy metal species, mobility and availability in soils. In: Sadler PJ (ed) The importance of chemical "speciation" in environmental processes. Springer-Verlag, Berlin
Brümmer GW, Gerth J, Herms U (1986) Heavy metal species, mobility and availlability in soils. Z Pflanz Bodenkunde 149:382–398
Dudka S, Adriano DC (1997) Environmental impacts of metal ore mining and processing: a review. J Environ Qual 26:590–602
Dudka S, Ponce-Hernandez R, Tate G, Hutchinsons TC (1995) Forms of Cu, Ni and Zn in soils of Sudbury, Ontario and the metal concentrations in plants. Water Air Soil Poll 90:531–542
Fendorf SE, Sparks DL, Lamble GM, Kelley MJ (1994) Applications of X-ray absorption fine structure spectroscopy to soils. Soil Sci Soc Am J 58:1583–1595
Fengxiang H, Banin A (1997) Long-term transformations and redistribution of potentially toxic heavy metals in arid-zone soils incubated: I. under saturated condtions. Water Air Soil Poll 95:399–423
Ford RG, Scheinost AC, Scheckel KG, Sparks DL (1999) The link between clay mineral weathering and the stabilization of Ni surface precipitates. Environ Sci Technol 33:3140–3144

Ge Y, Murray P, Hendershot WH (2000) Trace metal speciation and bioavailability in urban soils. Environ Pollut 107:137–144

Hatton A, Ranger J, Robert M, Nys C, Bonnaud P (1987) Weathering of a mica introduced into four acidic forest soils. J Soil Sci 38:179–190

Henderson PJ, McMartin I, Hall GE, Percival JB, Walker DA (1998) The chemical and physical characteristics of heavy metals in humus and till in the vicinity of the base metal smelter at Flin Flon, Manitoba, Canada. Environ Geol 34:39–58

Ketterer ME, Lowry JH, Simon J, Humphries K, Novotnak MP (2001) Lead isotope and chalcophile element compositions in the environment near a zinc smelting-secondary zinc recovery facilitie, Palmerton, Pennsylvania, USA. Appl Geochem 16:207–229

Laiti E, Öhman LO (1996) Acid/base properties and phenylphosphonic acid complexation at the boehmite/water interface. J Colloid Interf Sci 183:441–452

Li X, Thornton I (2001) Chemical partitioning of trace and major elements in soils contaminated by mining and smelting acticities. Appl Geochem 16:1693–1706

Ma YB, Uren NC (1998) Transformations of heavy metals added to soil – application of a new sequential extraction procedure. Geoderma 84:157–168

Manceau A, Boisset M-C, Sarret G, Hazemann J-L, Mench M, Cambier P, Prost R (1996) Direct determination of lead speciation in contaminated soils by EXAFS spectroscopy. Environ Sci Technol 30:1540–1552

Ranger J, Dambrine E, Robert M, Righi D, Felix C (1991) Study of current soil-forming processes using bags of vermiculite and resins placed within soil horizons. Geoderma 48:335–350

Righi D, Bravard S, Chauvel A, Ranger J, Robert M (1990) In-situ study of soil processes in oxisol-spodsol sequence of Amazonia (Brazil). Soil Sci 150:438–445

Roberts DR, Scheinost AC, Sparks DL (2002) Zinc speciation in a smelter-contaminated soil profile using bulk and microspectroscopic techniques. Environ Sci Technol 36:1742–1750

Schachtschabel P, Blume HP, Brümmer GW, Hartge KH, Schwertmann U, Auerswald K, Beyer L, Fischer WR, Kögel-Knabner I, Renger M, Strebel O (1998) Anthropogene Veränderungen und Belastungen. In: Scheffer F (ed) Lehrbuch der Bodenkunde. Enke, Stuttgart, pp 494

Schwarz O (2000) Polymethylmethacrylat [PMMA], Acrylglas. In: Schwarz O (ed) Kunststoffkunde. Vogel-Verlag, Würzburg

Sobanska S, Ricq N, Laboudigue A, Guillermo R, Bremard C, Laureynes J, Merlin JC, Wignacourt JP (1999) Microchemical investigations of dust emitted by a lead smelter. Environ Sci Technol 33:1334–1339

Suter (1999) Epoxydharz L – Sicherheitsdatenblatt

Suter (2001) Epoxydharz C – Technical Data Sheet

Tessier A, Campbell PGC, Bisson M (1979) Sequential extraction procedure for the speciation of particulate trace metals. Anal Chem 51:844–851

Tsaplina MA (1996) Transformation and transport of lead, cadmium and zinc oxides in a sod-podzolic soil. Eurasian Soil Sci 28:32–40

Ure AM, Davidson CM (1995) Chemical speciation in the environment. Blackie, London

Welter E, Calmano W, Mangold S, Tröger L (1999) Chemical speciation of heavy metals in soils by use of XAFS spectroscopy and electron microscopical techniques. Fresenius J Anal Chem 364:238–244

Part 1

Chapter 2

Analysis of Toxic Metals by Micro Total Analytical Systems (µTAS) with Chemiluminescence

P. A. Greenwood · G. M. Greenway

Abstract

A portable low cost micro total analytical system (µTAS) is described for the environmental analysis of metal cations in natural waters using luminol chemiluminescence. Every element of the analysis can be performed on these devices therefore producing a portable laboratory that can be used for *in-field* analysis. Miniaturising these devices brings several advantages including low reagent consumption, low volumes of waste produced and an increased in performance. The ability to scale out these devices and produce many low cost µTAS creates a network of in-situ sensors that can provide greater spatial and temporal information for an environmental system. Chemiluminescence is a promising method of detection for µTAS due to its high sensitivity and the simplicity of the measurement technique.

Key words: microTAS, metal cations, cobalt, chemiluminescence, 'in-field' analysis

2.1 Introduction

There are many methods available for the analysis of metals in environmental analysis. Atomic spectroscopic methods such as atomic absorption spectrometry and inductively coupled plasma mass spectrometry are most widely used and generally meet the requirements of sensitivity and specificity (Fifield 1995). These methods however are not suited to measurements in the field and require the sample to be brought to the machine. This can cause time delays, increased probability of errors and sample degradation. Low cost portable but sensitive detection methods for metals are urgently required for environmental analysis. Here we report a portable device for sensitive measurements of several metal cations.

2.1.1 Micro Total Analytical Systems

The idea of a total chemical analysis system was devised to create one device that included sample pre-treatment, removal of interferences, subjected to analysis, quantification and the data produced is converted to an electrical signal during a data acquisition step. With the incorporation of all these necessary elements to perform the chemical analysis, automated stand-alone devices can be envisaged for use at the point of analysis. This step was taken further by miniaturising devices creating micro total analytical systems (µTAS) (Manz et al. 1990). The µTAS generally takes place using flowing systems of a microlitre or less volume to achieve analysis in the order of seconds rather than many minutes. Other advantages of miniaturising these devices are an increased performance include more rapid heat transfer, more mass transfer across the width of the channels (diffusion), reduced reagent consumption/waste, and the improved separation efficiency leading to faster separations of analytes. This reduces the band broadening of the peaks and increased pressure caused by turbulent flow. These devices allow for portable, low cost, and low reagent consumption analysis that is ideal for the environmental analysis of many species. Although only small volumes of sample are being analysed within the system which generally leads to unrepresentative samples, these devices could be developed into a network of many in-situ sensors producing more spatial and temporal information about an environmental system than is currently possible. This is achieved by scaling out the devices (e.g. producing many smaller devices rather than one large sampling device) rather than scaling up the size of the device.

2.2
Experimental

The micro reactor (Fig. 2.1) was fabricated in house (McCreedy) 2001 using photolithography and wet etching techniques. Superwhite Crown B70 borosilicate glass was obtained precoated with a chromium film and photo resist layers (Alignrite, Bridgend, Wales). A mask of the required channels is placed over the glass and photolithography, using UV light was carried out. This process transfers the required pattern onto the chip where it is developed and the remaining photoresist is removed. Wet etching of the silica microstructure using 1% $HF-NH_3$ for 1 h at 70 °C produced the required channels identical to the prepared mask, with geometries of 200 µm wide and 100 µm deep. Thermal bonding was used to fuse the thick top plate of Superwhite Crown B70 borosilicate glass (Instrument Glasses, Enfield, UK) incorporating reservoirs of 17 mm depth and 3 mm i.d.

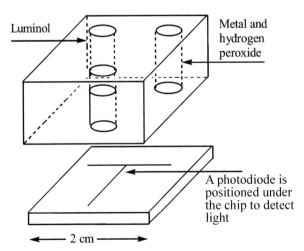

Fig. 2.1.
A schematic diagram of the micro reactor

Fig. 2.2.
The apparatus of the prototype µTAS. The micro reactor is placed on top of the detector. The laptop computer using Labview (National Instruments) is used to collect the data

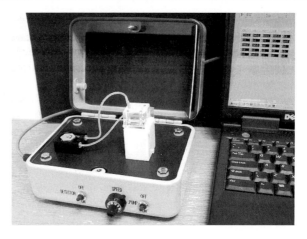

The µTAS was contained in a custom built, light tight, insulation box. For the initial assessment of this work a sensitive miniaturised photomultiplier tube (H5784, Hamamatsu Photonics, Enfield, UK) with a 12 V microcomputer power supply (RS Components, Northants, UK) was used to detect the chemiluminescence. Light was captured using a fibre optic attachment connected to the photomultiplier tube in which the end of the fibre optic was placed under the micro reactor. The analogue output from the photomultiplier tube detector was connected to a laptop computer using a labview controller (National Instruments) where the emission intensity was recorded. The pressure pumping was achieved with a peristaltic pump and flexible calibrated polyvinyl chloride tubing (Gilson Minipuls 3, Anachem, Luton).

A simple 3 reservoir T design chip was utilised. After optimisation of the system, one reagent reservoir contained luminol in carbonate buffer (pH 10.4) and cobalt nitrate standard solutions (in the range 10^{-11}–10^{-3} mol l^{-1}). The second reservoir contained hydrogen peroxide (0.1 mol l^{-1}). Reagents were loaded into the reservoirs by a micropipette (Sealpette, Jencons, Leighton Buzzard, UK). The third reservoir was used as a waste reservoir and was also the reservoir to which a negative pressure was applied to move the liquids through the micro reactor.

Figure 2.2 shows a prototype portable systems currently being evaluated. In that system a simple low cost photodiode is used to detect the light produced and flow is achieved with a miniaturised battery powered (+9 V) peristaltic pump.

2.3
Results and Discussion

2.3.1
Chemiluminescence as a Detection Technique

Chemiluminescence is a promising method of detection for µTAS due to its high sensitivity, minimal apparatus and the simplicity of the measurement technique. The low reagent consumption in µTAS is advantageous in chemiluminescence reactions

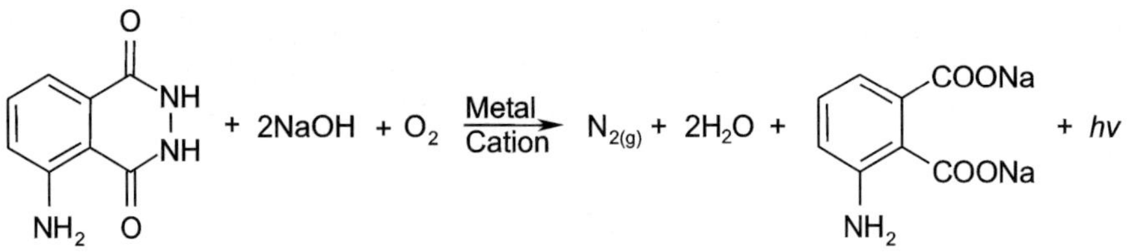

$$\text{luminol} + 2\text{NaOH} + O_2 \xrightarrow{\text{Metal Cation}} N_{2(g)} + 2H_2O + \text{3-aminophthalate} + h\nu$$

Fig. 2.3. The luminol reaction

because many of the reactions are irreversible and in conventional flow injection systems there is high reagent consumption, with large amounts of waste produced.

Chemiluminescence (CL) reactions are chemical reactions that result in the emission of electromagnetic radiation. These reactions have been widely used for sensitive and selective analysis especially over the last decade. Although there are only a limited number of reactions that produce chemiluminescence, the number of applications for analysis is ever increasing.

The luminol reaction was first reported in 1928 (Albrecht 1928) and has since become the most widely studied chemiluminescent reaction. Luminol exhibits direct chemiluminescence when oxidised in the presence of certain metal cations e.g. Co^{2+}, Fe^{2+}, and Cu^{2+}, and the reaction can therefore be used to sensitively detect these metals. Luminol (5-aminophthalhydrazide) is oxidised in aqueous alkaline conditions, usually by hydrogen peroxide, although permanganate and hypochlorite can also be used, to produce 3-aminophthalate, nitrogen and an intense blue-violet emission (λ_{max} = 425 nm) (see Fig. 2.3).

The luminol reaction could therefore quantify hydrogen peroxide in rainwater samples however this study concentrates on the analysis of metals. Cobalt(II) was chosen as the analyte for this study as it is a very efficient metal which enhances the luminol reaction producing an intense chemiluminescence emission (Burdo et al. 1975).

Gas phase reactions have been used in µTAS (Wilson et al. 2000) therefore this approach can also be extended to gas analysis where NO_2 reacts with luminol in alkaline conditions producing chemiluminescence providing an analysis with detection limits below 30 ppt (Laird 1995).

2.3.2
The Effects of pH for Mobilising Reagents

Micro-total analytical systems are particularly effective when electrokinetic flow is used to transport reagents in the channels at lower electric field strengths (below 1 000 kV). This provides a flexible and robust method of moving fluids through microchannels of typically less than 200 µm internal width, 100 µm depth. Electrokinetic flow however is based on bulk electro-osmotic flow and requires the pH of the solution being moved to be between pH 4–9 and this present several disadvantages when detecting species using the luminol chemiluminescence reaction (Nelstrop et al. 2001). The optimum pH for the luminol reaction is 10.4 and the reaction produces nitrogen gas which can disrupt the continuity of electroosmotic flow. Pressure pumping was therefore used in this work because the pH of the solution was not relevant and the nitrogen bubbles did not affect the movement of the solution. Negative pressure pumping was applied in this system for mobilisation of the reagents because this produces negligible backpressure on the microreactor channels and non discriminatory movement of reagents under varying reaction conditions. The negative pressure was applied to the waste reservoir and sucked the solutions through the channels. The three reagents met at the intersection of the T design and were mixed together along the length of the channel.

2.3.3
Mixing of Reagents within the Microreactor

Generally, the pattern of flow of fluids can be described as either laminar or turbulent. In order to determine the flow, it is common practice to evaluate the Reynolds number (Reynolds 1883) and compare it to the transitional number 2000. This number can be calculated using the equation below.

$$Re = vl\rho / \Im \qquad (2.1)$$

where, v is the velocity of the liquid flow, l is the diameter of the capillary, ρ is the density of the liquid, and $\Im$ is the viscosity of the liquid.

Due to the small channel dimensions in microfluidic devices mixing generally relies on laminar flow with a Reynolds number of approximately 10. Molecular diffusion is the only mechanism for transport of particles across the boundaries of adjacent fluid domains. This mixing time is proportional to d^2/D, d being the width of the channel and D the diffusion coefficient. Therefore the quality of the mixing is strongly influenced by the reduction of the channel width. The channel widths of the reported microreactors are typically 50 μm which allow complete mixing in a time scale of approximately 300 ms. The fibre optic position under the channel was optimised for where the mixing was just fully completed and the reaction was producing the most intense light emission.

The flow rate of solutions through the channel was optimised to be 5 μl s^{-1} resulting in a total volume of reagents required for the reaction to be 15 μl.

2.3.4
Analytical Measurements in Water

A calibration curve of cobalt concentration was obtained in the range 3×10^{-11} to 2×10^{-3} mol l^{-1}. This was plotted as log(chemiluminescence intensity in mV) versus log(cobalt conc. in mol l^{-1}) and contained nine points ($n = 3$) ($y = 0.55x - 5.43$, $R^2 = 0.99$). The sensitivity of the detection system was good and comparable with conventional flow injection analysis systems (Chang 1980) with the relative standard deviation being 4.5% ($n = 3$). The limit of detection was calculated using the linear portion of the graph in the range 10^{-10}–10^{-8} mol l^{-1} where the equation of the line was $y = 1.7x + 18.3$ with y being the chemiluminescence intensity (mV) and x the cobalt conc. (mol l^{-1}). The limit of detection was evaluated using the blank + 3 times $s_{y/x}$ (the error calculated in the y-direction) as defined by Miller and Miller (1984) and was found to be 3×10^{-11} mol l^{-1} Co(NO$_3$)$_2$ at the 95% confidence level. The sample run time was approximately 3 seconds, which resulted in a quicker analysis and much lower reagent consumption compared with conventional flow injection analysis (FIA). After each run, the products of the reaction were removed by syringe and the chip was flushed with buffer. This resulted in an average overall throughput time of 1 min.

2.4
Conclusions

A micro total analytical system has been developed that would be suitable for the analysis of metal ions in aqueous environmental samples. A calibration over seven orders of magnitude with a detection limit for Co(NO$_3$)$_2$ of 3×10^{-11} mol l^{-1} was successfully achieved using negative pressure pumping and chemiluminescence detection. The micro total analytical system can be seen to have several advantages over conventional techniques for field analysis. Currently work is being carried out with the prototype shown in Fig. 2.2, the high sensitivity of the method means that an inexpensive photodetector can be used to achieve acceptable sensitivity (usually the photodetectors are a factor of 10 less sensitive than a photomultiplier tube). The plan is to immobilised or seal reagents into the chip devices and to automate measurements. Filtering systems will have to be incorporated into the design for the analysis of natural waters. This system can also easily be adapted to monitor hydrogen peroxide or nitrogen dioxide.

References

Albrecht HO (1928) Über die Chemiluminescenz des Aminophthalsäurehydrazides. Z Phys Chem 136:321
Burdo TG, Seitz WR (1975) Mechanism of cobalt catalysis of luminol chemiluminescence. Anal Chem 47:1639
Chang CA, Patterson HH (1980) Halide ion enhancement of chromium(III), Fe(II), and cobalt(II) catalysis of luminol chemiluminescence. Anal Chem 52:653
Fifield FW (1995) The analysis of water. In: Fifield FW, Haines (eds) Environmental analytical chemistry. Blackie Academic & Professional, Glasgow, UK, pp 381
Laird CK (1995) The analysis of atmospheric sampling. In: Fifield FW, Haines (eds) Environmental analytical chemistry. Blackie Academic & Professional, Glasgow, UK, pp 307
Manz A, Graber N, Widmer HM (1990) Miniaturised total chemical analysis systems: a novel concept for chemical sensing. Sens Actuators B1:244–248
McCreedy T (2001) Rapid prototyping of glass and PDMS microstructures for micro total analytical systems and micro chemical reactors by microfabrication in the general laboratory. Anal Chim Acta 427:39
Miller JC, Miller JN (1994) Statistics for analytical chemists, 3rd edn. Ellis Harwood, London
Nelstrop LJ, Greenwood PA, Greenway GM (2001) An investigation into electroosmotic flow and pressure pumped luminol chemiluminescence detection for cobalt analysis by a miniaturised total analytical system. Lab on a Chip 1:138
Reynolds O (1883) An experimental investigation of the circumstances which determine whether the motion of water shall be direct or sinuous, and of the law of resistance in parallel channels. Philos Trans 1883, 174:935; ibid. 1895, 186:123
Wilson NG, McCreedy T (2000) On chip catalysis using a lithographically fabricated glass microreactor – the dehydration of alcohols using sulphated zirconia. Chem Commun 9:733

Chapter 3

Diffuse Infrared Fourier Transform Spectroscopy in Environmental Chemistry

B. Azambre · O. Heintz · D. Robert · J. Zawadzki · J. V. Weber

Abstract

The experimental implementation of Diffuse Infrared Fourier Transform Spectroscopy (DRIFTS) to the study of carbonaceous materials and catalysts is discussed in detail. Analytical methods used to sample and optimize the quality of infrared spectra processing are outlined. Several examples demonstrate the feasibility of this technique in the field of adsorption and catalysis: *(1)* An in-situ spectroscopy is applied to the elucidation of the oxidation mechanisms occurring during the treatment of a cellulose char under air; *(2)* the type of interactions occurring between adsorbed methanol or impregnated copper salts with carbonaceous materials was considered with respect to char surface chemistry; *(3)* the mode of adsorption of several aqueous organic adsorbates on titanium oxide was investigated.

Key words: DRIFTS; in-situ treatments; adsorption; carbon; catalysts

3.1 Introduction

Among the various Fourier transformed infrared (FTIR) techniques, infrared spectroscopy used in diffuse reflectance mode (DRIFTS) seems to be the most versatile one, since it can provide both chemical and structural informations for almost all types of solids. Thus, materials such as common powders, crystals, gemstones, polymers, fibers, solids with rough surfaces, minerals and biomass products can be analyzed (Chamers et al. 1985; Coleman 1993). Recently DRIFTS has been applied in environmental chemistry, as shown by the increasing number of papers published this last decade. The diffuse reflectance technique has been successfully applied to biological systems in order to investigate decomposition dynamics in forest soils (Haberhauer et al. 1999) and to monitor wood biodegradation (Ferraz et al. 2000). An attempt was also made to determine quantitatively cations in water by spot tests reactions (Ghauch et al. 2000).

Nevertheless, most literature is devoted to the study of supported catalysts and adsorbents (Azambre et al. 2000; Dandekar et al. 1999; Lee et al. 1998; Koch et al. 1998). Since these materials are heterogeneous in nature and are usually characterized by a strong absorber behavior towards all kinds of radiations, an arsenal of complementary analytical methods is often required to investigate them. These techniques must be considered as complementary, and no attempt will be made in this work to compare the potentialities of diffuse reflectance spectroscopy in this domain to those of X-ray Photoelectron Spectroscopy (XPS), Nuclear Magnetic Resonance (NMR), Electron Spectroscopy for chemical Analysis (ESCA), and X-Ray Diffraction (XRD).

Nonetheless, infrared spectroscopy has proven to be an effective and sometimes unique tool for the elucidation of the chemical mechanisms occurring at the interface between two phases, which is crucial, if one would like to understand the performances of an adsorbent or a catalyst. Moreover, the characterization of carbonaceous adsorbents which are commonly used to remove hazardous compounds from environment (water or air) or as supports for precious metal particles is not usually considered as an easy task, because of the ill-defined character of carbon and the co-existance of many kinds of oxygenated groups on the surface. Direct information on the structure of the bulk of carbon materials, the presence of various surface functional groups, and on mineral absorption and electronic absorption may be obtained by the way of infrared spectroscopic studies.

An extended review of IR spectra of oxidized and non-oxidized carbon films, including the effects of surface chemistry on the adsorption of a large variety of molecules probes, has been presented by Zawadzki in 1998. However, due to the behavior of carbon as a black body absorber, several experimental difficulties arise when trying to characterize powdered carbonaceous materials by conventional transmission methods. The use of diffuse reflectance spectroscopy instead of transmission can solve, at least partly, some of these problems.

Applications to heterogeneous catalysts studies become still more numerous (Dandekar et al. 1999; Lee et al. 1998; Bollinger et al. 1996; Weckhuysen et al. 1999). Despite the crucial importance of these materials in industry, there is little knowledge about factors governing their final properties. For instance, for fifty years their preparation has been mainly subjected to empirical recipes, because of the lack of efficient analytical tools to investigate it.

The field of application of diffuse reflectance spectroscopy can be greatly extended by the incorporation of an environmental cell into the analytical device. Thus, it is possible to follow in situ some catalytic reactions in realistic conditions of temperature and pressure (Dandekar et al. 1999; Griffiths et al. 1986). In many cases, DRIFTS seems to be an ideal method for the characterization of intermediate adsorbates. Organic materials and adsorbed compounds are usually studied in the mid-infrared range, whereas both d-d and charge transfer transitions of metal particles can be analyzed in the UV-visible range (Weckhuysen et al. 1999). DRIFTS has been successfully used to investigate the pathways of nitrogen oxides (NO_x) decomposition under aerobic or non-aerobic conditions for a large variety of catalysts (Klingenberg et al. 1999; Meunier et al. 2000). Furthermore, the use of carbon monoxide (CO) as a probe molecule, traditionally used in IR to gain knowledge about the oxidation state and dispersion of the supported metals can be combined with the great sensitivity of DRIFTS. In this way, supported catalysts characterized by metal loadings as low as 0.05% can be investigated, which is far below the range of metal loadings used in real commercial catalysts (Stakheev et al. 1999).

However, despite the great advances made recently in the design of analytical devices allowing work under in-situ conditions, diffuse reflectance studies connected with the preparation of heterogeneous catalysts and adsorption of organic molecules on them are still scarce. Experimental methodologies dealing with the characterization of such systems by DRIFTS are beyond the scope of this paper. However, no attempt will be made here to give an exhausting list of references, and we will focus only to the mid-infrared region. In the first part of this study, we will recall some of the fundamental aspects of this technique, and we will give an overview of the preparation procedures required to record good-quality spectra in the case of highly absorbing samples.

Emphasis will be given to the applicability of quantitative analysis on carbon-supported metals and opaque samples in general. In the experimental section, some examples of the research carried out in our laboratory will be presented to outline the feasibility of this technique when applied to the characterization of carbonaceous supports and Cu/C catalysts. Results obtained after the chemisorption of some carboxylic acids and phenols on TiO_2 will be also analyzed, likewise those related to the in-situ adsorption of methanol on cellulose chars. In this last section, we will focus mainly on the methodologies that can be used to optimize the exploitation of spectroscopic data.

3.2
Theory

3.2.1
Basic Principles of DRIFTS

The potentialities of Diffuse Reflectance Infrared Spectroscopy (DRIFTS) are closely linked to diffusion phenomena of light in solid media. When an infrared radiation reaches a sample surface, one or several processes can occur: it can be absorbed, it can be reflected from the surface (the so-called specular reflectance, see Fig. 3.1), or it can penetrate the sample before being scattered. If the powdered particles are randomly oriented, this penetration is isotropic and constitutes the diffuse reflectance phenomenon. Generally, the light intensity scattered at a given wavelength from an "infinitely thick" solid layer is compared with that scattered from an infinitely thick layer of a non-absorbing reference. The need to use a reference originates from the fact that diffuse reflectance spectra of pure samples tend to have very intense bands, with little difference in intensity between spectral regions usually characterized by strong or weak absorptions (Schoonheydt 1984).

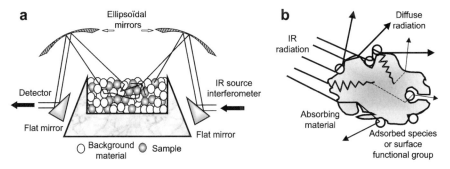

Fig. 3.1. Principle of the diffuse reflection phenomenon. **a** Scheme of the diffuse reflectance accessory. **b** Course of the infrared radiation through an heterogeneous absorbing material

The most applicable procedure for the rapid identification and quantification of common powders is to dilute samples in an alkali halide matrix, like KBr or KCl. Nevertheless, owing to the ionic character of these species, undesirable catalytic or solid-state reactions can occur with the studied sample, and other reference materials such as CaF_2 and diamond powder will be preferred in that case (Ventura et al. 1999). Interactions of the infrared radiation with an heterogeneous powdered matrix are schematized on Fig. 3.1.

3.2.2
DRIFTS versus Transmittance Spectroscopy

Since the optical phenomena connected with DRIFTS are different from those involved in transmittance spectrometry, spectra obtained by either infrared method can not be considered as equivalent. Moreover, infrared spectrometry used in diffuse reflectance mode has many advantages compared with the conventional transmission method. On first approximation, DRIFTS can be considered as a fast and non-destructive technique because samples can be directly analyzed in their powdered form. On the contrary, for transmission measurements, preparation procedures are generally more time-consuming and morphological changes can arise as a result of the grinding or hot compression molding procedures adopted in the design of pellets.

The second point is that the DRIFT technique is better suited to the analysis of strongly absorbing materials, whose main characteristics are a very low signal and sloping baselines when analyzed in transmission. Moreover, since the effective path length of the infrared radiation is increased many fold by scattering, the sensitivity of DRIFTS is comparatively enhanced, and detection limits of the order of the nanogram can be achieved for some absorbing materials (Fuller et al. 1978). This is obviously important for some industrial applications in which the signal provided by the common pellet technique used to be insufficient, as can be the case for the identification of low-quantity fractions separated during chromatographic runs.

Another relevant aspect is connected with the information depth of the diffuse reflectance technique. Although the most commonly encountered theories depict the solid medium as a single continuum rather than individual particles, and thus require an infinitely thick layer, the greatest contribution to DRIFT spectra originates in the first few layers of samples (Fuller et al. 1978). This implies the fact that adsorbed species can be effectively probed using this technique, which seems to be a necessary condition for the study of heterogeneous reactions occurring at catalysts surfaces.

3.3
Practical Considerations

3.3.1
Influence of Practical Sampling on the Reproducibility of DRIFTS Spectra

Interactions of light with matter are more complex to handle in diffuse reflection spectroscopy than in transmission. Particle size, size distribution, sample cup orientation and position, surface state and packing density are the main factors that affect both the reproducibility and quality of solid samples spectra. The influence of alkali halide particle size on diffuse reflectance spectra was first considered by Fuller et al. 1978. It was shown that particle size has a considerable influence on the reflectance of KCl at high wave numbers (between 2 000 and 4 000 cm^{-1}). The scattering properties of the analyzed material are also strongly influenced by adsorbed species on the external surfaces (Schoonheydt 1984). The removal of adsorbed water and steam from both the sample and the enclosure of the spectrometer is crucial, because infrared bands of gaseous and adsorbed species can overlap some analytical regions of interest and somewhat hide or distort the spectral information. Thus, sample and reference, e.g. potassium bromide (hygroscopic) have to be very dry and neat before recording the spectrum.

3.3.2
Influence of the Data Processing Parameters

The optical and physical properties of samples are not the only parameters that affect the quality of infrared spectra. Even if the preparation procedures are carefully controlled, it is also relevant to optimize the instrumental parameters involved in infrared spectra processing. Details about the theoretical significance of these parameters and the procedures used to investigate their effects on spectral characteristics can be found elsewhere (Griffiths et al. 1986; Reklat et al. 1997). Although common to the different FTIR techniques, factors such as resolution, apodization and zero filling have to be optimized in respect with the physical state of the material to be analyzed.

For instance, bandwidths are intrinsically broad in solid state, and there is no need to increase inconsiderately the resolution for DRIFT spectroscopy. Otherwise, the signal to noise ratio of the spectrum will be lowered (according to the trading rules), without gaining further spectral informations. During a previous work (Azambre et al. 1999), an experimental design allowed us to determine the values required for some instrumental parameters in the case of DRIFTS studies achieved on conventional powders: triangular apodization, $R = 2$ cm^{-1}, ZFF = 2 or 4. However, it should be remembered that the values selected by the user must fit to the type of experiment. An

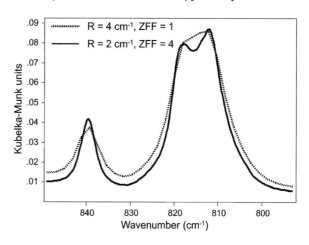

Fig. 3.2.
Influence of resolution and zero filling on the DRIFTS spectrum of anthron (an organic pollutant) in the region 850–790 cm^{-1}

example of the effect induced by a change in the resolution and zero filling values is given on Fig. 3.2. Note the high influence of resolution for the separation of the two bands in the region 800–830 cm^{-1}.

For instance, when one needs to precise the initial estimates before curve-fitting a spectrum, which is the common procedure used to acquire semi-quantitative data on mixtures or overlapping regions, the most important criterion is to optimize the separation between bands before running the curve-fitting program (Maddams 1980). Thus, it seems reasonable to work with a sufficiently high resolution (2 cm^{-1}).

3.4
Quantitative Analysis

It is well-established that DRIFT spectroscopy is intrinsically quantitative. However, due to the difficult experimental implementation of the sampling procedures, it's not an easy task to obtain accurate calibration curves. Nevertheless, quantitative studies have been successfully applied to the analysis of drugs, dyes, fibers and most scarcely to heterogeneous catalysts. Generally, the quantitative studies that can be found in literature are achieved on a large set of measurements (multiple calibrations) for each sample concentration, in order to average the differences between spectra due to the uncontrolled variation of some physical properties (Liang et al. 1999). The use of chemiometric methods can also be of great help to statistically correlate the infrared data to some macroscopic properties of the studied materials (Alciaturi et al. 1996; Kokot et al. 1996).

3.5
Experimental Studies on Adsorbents and Heterogeneous Catalysts

3.5.1
Oxidation of a Carbonaceous Material

Since the removal of gaseous or liquid organic pollutants and heavy metals from environment has become a priority these last years, there is a constant need to find cheap and new materials able to trap or degrade them. Due to their very high specific surface and unique properties, several kinds of carbon and carbon-supported catalysts can be used to circumvent at least, partially, these problems. In addition to the pore size distribution, the surface chemistry of carbon has often to be tailored, because oxygenated and nitrogeneous surface groups can interact directly with some of these hazardous compounds by the intermediate of their polar groups (see an example on Fig. 3.3).

In this section, we discuss the methods that can be used to monitor by DRIFTS the functional evolution of a carbonaceous material during its oxidation under air. For this purpose, we used the following experimental procedure: microcristalline cellulose was heated from room temperature to 450 °C inside a sealed crucible and a semi-coke characterized by few functional groups but easily oxidizable structures were obtained. In-situ investigations on the chemical processes involved in the oxidation of the char were made possible by the adjunction of an environmental chamber to the DRIFTS accessory. The experimental device allows to treat thermally a sample at

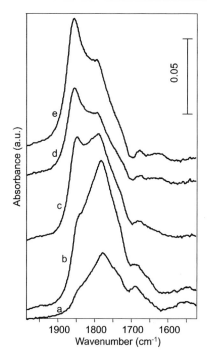

Fig. 3.3.
Evolution of infrared spectrum of absorbing species in the C=O stretch region during the oxidative treatment of the char in function of the temperature range. During the heating to 280 °C: *a* between room temperature and 210 °C; *b* between 210 and 280 °C. During the isotherm at 280 °C: *c* between 0 and 50 min; *d* between 50 min and 3 h; *e* between 3 and 6 h. We can see the evolution of the band at 1950 cm^{-1} during the thermal treatment

controlled rate and pressure up to 500 °C under flowing or static atmospheres. The in-situ spectra of the char were recorded in presence of potassium bromide and pure KBr as the background reference. Typical experiments consist in recording an interferogram every 3 min (which is equivalent to every 15 °C when heating rate of 5 °C min^{-1} is used), so accurate monitoring of the chemical events can be achieved. Due to the incomplete aromatization of the char, the IR absorbing species are clearly visible. Bands assignments and detailed characterization of the mechanisms related to the oxidation of the cellulose char are provided elsewhere (Azambre et al. 2000).

In our case, the functional evolution of the char in the region corresponding to the C=O stretching vibrations was examined step by step, by directly subtracting the spectrum obtained at a given temperature T_1 to the one obtained at a lower temperature T_0. Using this procedure, only the IR absorbing species formed or degraded in a neighboring temperatures range are investigated, without interference of the non-specific absorption of the carbon or by an uneven change in the bands intensities ratio, as it is the case for ex-situ studies. Therefore, changes in the mechanisms occurring during oxidation likewise change the appearance of new functional groups which can be readily detected by using this data manipulation. Spectra obtained in this way are shown in Fig. 3.3.

The chemisorption of oxygen atoms on the surface of the char is displayed prominently by the appearance of a positive band in the spectrum corresponding to the beginning of the thermal treatment under air. In this temperature range, oxidation proceeds via the formation of acidic groups, as it can be seen by an increase in the absorption of the region corresponding to the C=O stretching vibrations of carboxyles (1 730–1 680 cm^{-1}), quinones (1 670 cm^{-1}) and cyclic anhydrides (2 bands at 1 845 and

1780 cm^{-1}). These trends show that the 5 or 6 members ring anhydride groups are progressively converted into open chains anhydrides. The same kind of spectra recorded in the region corresponding to the C-H bending and stretching vibrations (not shown here) allowed us to establish that oxidation proceeds on both aromatic and aliphatic sites of the carbonaceous material.

3.5.2
Adsorption of a Volatile Organic Compound

In order to examine the possible existence of interactions between the oxygenated groups of carbon and the polar groups of methanol, an in-situ adsorption of methanol vapor in static mode was carried out into the environmental cell in presence of the oxidized (after 30 min at 280 °C) and the non-oxidized char (Fig. 3.4). In a second step, the sample was contacted with methanol vapor so as to obtain a pressure of 2 KPa in the cell and the single-beam spectrum was once again recorded. The absorbance spectra representative of the interactions between the adsorbed species and the surface were obtained after rationing the two single-beam spectra and subtracting the gaseous phase spectrum of methanol. Spectra corresponding to adsorbent-adsorbate interactions and gas phase methanol (for comparison) are displayed in Fig. 3.4.

The physical adsorption on the porosity of the chars is evidenced by the disappearance of the fine structure corresponding to the gas phase rotational spectrum of methanol and the appearance of new bands at 2 975, 2 940, 2 909 and 2 829 cm^{-1} (aliphatic C-H stretching vibrations) and 1 025 cm^{-1} (C-O stretching vibration of molecular methanol). The increase of the absorption in the 3 500–2 600 cm^{-1} region shows the formation of hydrogen bonds. The negative band at 3 650 cm^{-1} is attributed to the decrease of the O-H stretch corresponding to free hydroxyl groups. For the oxidized char, the occurrence of negative bands at 1 845, 1 780 and 1 760 cm^{-1} and the coccurrence of a new band at 1 710 cm^{-1} suggest that the C=O functions of the anhydride groups interact with methanol, or react with the alcohol via an esterification mechanism.

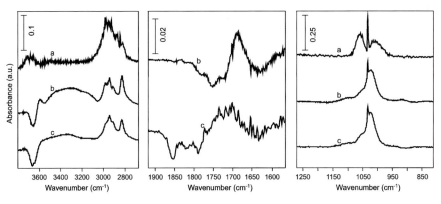

Fig. 3.4. Spectra of methanol: *a* in gas phase; *b* adsorbed on the non-oxidized char after subtraction of the spectrum of the bare support; *c* adsorbed on the oxidized char after subtraction of the bare support

3.5.3
Interactions between Chars and Copper Salts

The example provided here is connected with the preparation of carbon-supported catalysts (Fig. 3.5). Although the impregnation method using an aqueous metallic precursor is a classical technique for the preparation of catalysts, little knowledge is usually acquired on the form of the deposited salts. In our case, we used the following procedure. The chars were impregnated with a copper nitrate solution so as to obtain a copper loading of 8.2 wt%, and then were dried in an oven overnight at 120 °C. Because of solid state reactions occurring between KBr and the copper salt, no diluting material was used to study the changes occurring in the supports chemistry after impregnation. Spectra corresponding to copper salts deposited on the chars are compared with the spectrum of the pure salt treated in a similar fashion (Fig. 3.5).

The basic nitrate structure of the thermally modified salt is evidenced by the presence of bands in the 1410–1340 and 860–800 cm^{-1}, 1600–1500 and 1350–1200 cm^{-1} (-NO_2), 3700–2700 cm^{-1} regions assigned respectively to the asymmetric and symmetric stretching vibrations of -NO_3, -NO_2 and O-H bonds. In comparison with Cu/C-Ox, the spectrum of Cu/C is closer to the one of the non-supported salt, indicating that a greater fraction of the metallic precursor has been deposited on the non-oxidized support on its bulk or non-dispersed form. This interpretation seems to be logical in respect with the weaker wettability and higher hydrophibicity of the non-oxidized char. For the Cu/C sample, positive bands at 1670 and 1735 cm^{-1} have to be attributed to the oxidation of the highly reactive char by the impregnation solution. On the contrary, the main evolution for the oxidized char is related to the occurrence of both negative bands at 1856 and 1801 cm^{-1} and a positive band at 1542 cm^{-1}. These spectral features may be tentatively assigned to the formation of metallic carboxylates, because of the possibility of specific interactions to occur between functional groups and copper ions during impregnation.

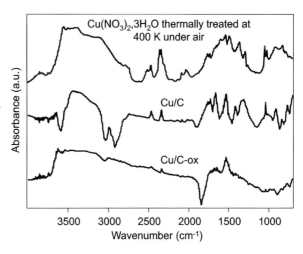

Fig. 3.5.
Spectra of copper nitrate salts deposited on the non-oxidized (C) and oxidized (C-Ox) chars after subtraction of the corresponding unloaded support. Comparison with the spectrum of the bulk salt thermally treated at 400 K under air

3.5.4
Chemisorption of Acids and Phenols on TiO_2

Due to its ability to be either an efficient carrier for precious metals or a photocatalyst for the degradation of pollutants in wastewater, titanium dioxide has attracted much attention these last years. It is well known that the type of interactions between the organic molecules in solution and the hydroxyl surface groups of the solid photocatalyst has an important influence on both the mechanisms of degradation and catalytic performances. Thereby, the ex-situ characterization by DRIFTS of aqueous adsorbates on TiO_2 can provide useful informations not only on the way by which a molecule is bonded to the surface, but also on the strength of the bonding.

In this study, we have chosen to consider the adsorption of a carboxylic diacid and a substituted phenol on a commercial TiO_2 P-25 photocatalyst. The experimental procedure was that of Robert et al. (2000). Comparison with the spectra of the pure organic compounds diluted in KBr allows us to identify the bonds that were affected during the chemisorption experiments (Fig. 3.6).

In the case of p-chlorophenol adsorption, major changes are related to the disappearance of O-H bending bands (1440–1430 and 1370–1350 cm^{-1}), the decrease of the intensities and shift of the C-O stretching bands (1240 and 1090 cm^{-1}) and the strong decrease of the band attributed to the skeletal ring breathing mode between 1600 and 1500 cm^{-1}. Since the intensity of the latter band is affected to some extent by the electronegativity of the substituting group, the observed evolution seem to

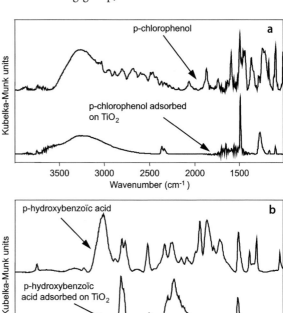

Fig. 3.6.
Spectra of organic compounds adsorbed or non-adsorbed on TiO_2. **a** p-chlorophenol; **b** p-hydroxybenzoic acid

point out the formation of phenolate and/or flat adsorption on TiO_2. On the other hand, it can be seen on the spectrum of adsobed p-hydroxybenzoïc acid, the disappearance of the intense C=O stretching band (1678 cm^{-1}) and the strong increase of a band at 1610–1580 cm^{-1} corresponding to carboxylate bonds. These results show that the bifunctional acid is completely dissociated on the surface and seem to confirm the hypothesis of formation of cyclic chelate by exchange of two hydroxyl ligands for one titanium atom, as it was outlined by other authors (Tunesi et al. 1991).

3.6 Conclusions

In this study, the applicability of diffuse reflectance spectroscopy pertaining to the characterization of adsorbents, namely carbon, and heterogeneous catalysts was considered. The common preparation and data handling procedures required to record high quality spectra were discussed in respect with the properties of the studied materials. We have shown that DRIFTS can be a useful and sometimes unique tool for the characterization and study of carbon-supported catalysts. With the help of an environmental cell designed for in-situ studies, it is possible to monitor the functional evolution of a carbonaceous support all along its manufacture in realistic conditions of atmosphere, temperature and pressure. Moreover, we have demonstrated that this technique can be successfully used both to gain knowledge of the type of interactions occurring between a support and an aqueous metallic precursor and to characterize the mode of adsorption of various organic molecules. However, quantitative aspects pertaining to the study of absorbing and heterogeneous materials have been little investigated to date, and more studies are required in the future to explore the potentialities of Diffuse Reflectance Infrared Fourier Transform Spectroscopy (DRIFTS) in this domain.

References

Alciaturi CE, Escobar ME, Vallejo R (1996) Prediction of coal properties by derivatives drift spectroscopy. Fuel 75:491–499
Azambre B, Heintz O, Schneider M, Krzton A, Weber JV (1999) Optimization of some instrumental factors in diffuse reflectance infrared fourier transform spectroscopy. Talanta 50:359–365
Azambre B, Heintz O, Krzton A, Zawadzki J, Weber JV (2000) Cellulose as a precursor of catalyst support: new aspects of thermolysis and oxydation-IR, XPS and TGA studies. Anal Appl Pyrolysis 55:105–112
Bollinger MA, Vannice MA (1996) A kinetic and drifts study of low temperature carbon monoxide oxidation over Au-TiO$_2$ catalyst. Appl Cat B Environ 8:417–443
Chalmers JM, Mackenzie MW (1985) Some industrial applications of FT-IR diffuse reflectance spectroscopy. Appl Spectroscopy 39:634–641
Coleman P (1993) Practical sampling for infrared analysis. CRC Press, London, pp 54–83
Dandekar A, Baker RTK, Vannice MA (1999) Carbon-supported copper catalysts. II Crotonaldehyde hydrogenation. J Catal 184:421–439
Ferraz A, Baeza J, Rodriguez J, Freer J (2000) Estimating the chemical composition of biodegraded pine and eucalyptus wood by Drift spectroscopy and multivariable analysis. Biores Technol 74: 201–212
Fuller MP, Griffihs PR (1978) Diffuse reflectance measurements by infrared fourier transform spectrometry. Anal Chem 50:1906–1910

Ghauch A, Turnar C, Fachinger C, Rima J, Charef A, Suptil J, Martin-Bouyer M (2000) Use of Drift in spot test reaction for quantitative determination of cation in water. Chemosphere 40:1327–1336

Griffiths PR, de Haseth JA (1986) Fourier transform infrared spectrometry. John Wiley & Sons Ltd., New York

Haberhauer G, Gerzabek MH (1999) Drift and transmission FT-IR spectroscopy of forest soils and appoach to determine decomposition processes of forest littier. Vib Spectroscopy 19:413–417

Klingenberg B, Vannice MA (1999) NO adsorption and decomposition on La_2O_3 studied by Drift. Appl Cat B Environ 21:19–33

Koch A, Krzton A, Finqueneisel G, Heintz O, Weber JV, Zimny T (1998) A study of carbonaceous char oxidation in air by semi-quantitative FTIR spectroscopy. Fuel 77:563–569

Kokot S, Yang P, Gilbert C (1996) Prediction of physical properties of fabrics using Drift spectroscopy and thermal analysis. Anal Chim Acta 332:105–110

Lee SJ, Kim K (1998) DRIFT of stearic acid self assembled on fine silver particles. Vib Spectroscopy 18:187–201

Liang Y, Christy AA, Nyhus AK, Hagen S, Schanche J, Kvalheim OM (1999) Pre-treatment of Drift spectra for quantitative analysis of macro porous polymer particles. Vib Spectroscopy 20:47–57

Maddams WF (1980) The scope and limitations of curve fitting. Appl Spectroscopy 34:245–257

Meunier FC, Zuzaniuk V, Breen JP, Olsson M, Ross JRH (2000) Mechanistic differences in the selective reduction of NO by propene over cobalt and silver promoted alumina catalyst: kinetic and in-situ Drifts studies. Catal Today 59:287–304

Reklat A, Bessau W, Kohl A (1997) Systematic errors of FTIR transmission spectra. Mikrochim Acta 14:307–314

Robert D, Parra S, Pulgarin C, Krzton A, Weber JV (2000) Chemisorption of phenols and acids on TiO_2 surface. Applied Surf Scien 167:51–58

Schoonheydt RA (1984) Diffuse reflectance spectroscopy. In: Delannay F (ed) Characterization of heterogeneous catalysts. Marcel Dekker, Inc., New York Basel pp 125–160

Stakheev AY, Kustov LM (1999) Effect of the support on the morphology and electronic properties of supported metal clusters: modern concepts and progress in 1990s. Appl Cat A General 188:3–35

Tunesi S, Anderson M (1991) Influence of chemisorption on the photodecomposition of salicylic acid and related compounds using suspended TiO_2 ceramic membranes. J Phys Chem 95:3399

Ventura C, Papini M (1999) Drift study of granular materials and their mixtures in the mid-infrared spectral range. Vib Spectroscopy 21:17–31

Weckhuysen BM, Shoonheydt RA (1999) Recent progress in diffuse reflectance spectroscopy of supported metal oxide catalyst. Catal Today 49:441–451

Zawadzki J (1988) Chemistry and physics of carbon. 21. P.A. Thrower Ed., Marcel Dekker, New York

Chapter 4

Detection of Biomarkers of Pathogenic Bacteria by Matrix-Assisted Laser Desorption/Ionization Time-of-Flight Mass Spectrometry

V. Ruelle · B. El Moualij · W. Zorzi · D. Zorzi · P. Ledent · O. Pierard · N. Bonjean
M. C. De Pauw-Gillet · E. Heinen · E. De Pauw

Abstract

In recent years, various mass-spectrometry procedures have been developed for identifying bacteria. The accuracy and speed with which data can be obtained by Matrix-assisted Laser Desorption/Ionization Time-of-flight Mass Spectrometry (MALDI-TOF MS) make this an advantageous technique for environmental monitoring. However, minor variations in the sample preparation can influence the mass spectra significantly. In the present study, we have introduced a procedure to prepare bacteria by microextraction and we have optimized experimental parameters for rapid identification by MALDI-TOF MS of whole bacterial cells isolated from environmental samples such as wastewater and soil.

Key words: MALDI-TOF MS; bacteria; biomarker; activated sludge; wastewater treatment plant; optimization; salt; solvent

4.1 Introduction

In areas such as environmental monitoring, food processing, assessment of public health hazards and disease diagnosis, it is important to screen rapidly for the presence of pathogenic microorganisms, and especially to identify them accurately. For this purpose, several methods have been investigated (Hamels et al. 2001) but they lack reproducibility and are time consuming. As an alternative, we propose in this work an original procedure based on mass spectrometry. This method is quick, simple and reproducible. It could be used to detect minor variations between pathogenic and non-pathogenic bacterial strains.

Recently, various studies have used mass spectrometry in bacterial chemotaxonomy. The idea is to use one or several mass peaks that appear to be specific of certain type of bacteria as genus-, species- or strain-specific biomarkers. E.g. bacteria were characterized by analysis of glycerophospholipids and ubiquinones (Anhalt and Fenselau 1975). Because mass spectrometry is swift and precise, other studies were quick to follow using techniques such as pyrolysis mass spectrometry (Snyder et al. 1996), fast-atom bombardment (Nichols and McMeckin 2001) or electrospray ionization (Krishnamurthy et al. 1999). In 1987, Karas et al. developed a new mass spectrometric method entitled matrix-assisted laser desorption/ionization time-of-flight mass spectrometry (MALDI-TOF MS). MALDI-TOF MS is a soft ionization method allowing desorption of intact peptides (Dai et al. 1999) and proteins from whole bacterial cells (Holland et al. 1996; Krishnamurthy et al. 1996; Easterling et al. 1998; Wang et al. 1998; Arnold et al. 1999; Holland et al. 1999; Lynn et al. 1999; Evason et al. 2001; Ryzhov and Fenselau 2001) or cell fractions without extensive separation (Chong et al. 1997; Nilsson 1999). The advantages of MALDI-TOF MS include high sensitivity, speed, reproducibility (Saenz et al. 1999), and desorption of intact high-mass proteins (Chong et al. 1997; Madonna et al. 2000). Different investigations have shown applications for characterizing intact bacteria cells (Saenz et al. 1999), bacterial spores (Ryzhov et al. 2000; Birmingham et al. 1999; Demirev et al. 2001; Elhanany et al. 2001) and other microorganisms such as viruses (Thomas et al. 1998; Birmingham et al. 1999) and fungi (Welham et al. 2000).

The quantification and the identification of viable microorganisms, including pathogenic forms, are important to evaluate the biological quality of wastewater. To extract live bacteria from wastewater, we have developed and patented an original procedure (D. Zorzi et al., in preparation) based on repeated sonications of activated sludge followed by an enzymatic treatment. Activated sludge consists of stable aggregated flocs connected by biopolymers present in the environment or produced by the lysis of microorganisms. Indeed, it has been shown that the treatment of sludge by non-aggressive physical means improves the yield of bacterial extraction (Banks et al. 1977). Sonication separates components of the flocs, thus promoting the release of living microorganisms and small entities about 2.5 microns in diameter (Jorand et al. 1995). We further improved this microextraction procedure and consequently the yield of bacterial extraction by combining sonication with enzymatic digestion of the biopolymers. The enzyme cocktail contains cellulases, proteases, amylases and lipases.

In the present study, we used Matrix-assisted Laser Desorption/Ionization Time-of-flight Mass Spectrometry to analyze bacteria isolated from activated sludge. We focused on a drawback of this technique: minor variations in sample preparation can influence the mass spectra significantly (Domin et al. 1999; Wang et al. 1998). Here we focused on optimizing the experimental parameters and tested this optimized protocol on bacterial strains isolated from the environment. Strains isolated by microextraction from a wastewater processing plant in Embourg (Belgium) were compared with reference strains of *Salmonella* sp., *Escherichia coli*, and wild-type *Acinetobacter* sp., species viewed as indicators of the biological quality of wastewater and sewage sludge.

The work consisted in analyzing mass-peak patterns obtained from whole bacterial cells in the mass range from 2 to 20 kDa. The mass peaks emerging as reproducibly species-specific were selected as biomarkers for the reference and extracted strains.

4.2
Experimental

4.2.1
Chemicals

Materials were purchased from commercial sources and used without further purification. The culture medium and Api 20E gallery were obtained from available sources. Formic acid; trifluoroacetic acid; isopropanol; ethanol; methanol; acetonitrile; α-cyano-4-hydroxycinnamic acid (α-cyano); 3,5-dimethoxy-4-hydroxycinnamic acid (sinapinic acid); ferulic acid; 2-(4-hydroxyphenylazo) benzoic acid; 5-methoxysalicilic acid; 2,5-dihydroxybenzoic acid and cytochrome c were obtained from SIGMA (Sigma-Aldrich Inc., St. Louis, MO, USA). The water was obtained from a Milli-Q water filtration system (Millipore Co., Bedford, MA, USA).

4.2.2
Bacterial Sample Preparation

All the wild-type reference strains used in this study were obtained from the American Type Culture Collection: ATCC11775 and ATCC25922 for *Escherichia coli*, ATCC13076 and ATCC13311 for *Salmonella* and ATCC17924 for *Acinetobacter*. Environmental bacteria derived by microextraction from the activated sludge of Embourg-Belgium, were isolated by the ISO 9308-1:2000 method for *Escherichia coli* (8 strains, 1B1 to 1B8); the Me1/028/V2 – ISO 6340:1995 method for *Salmonella* sp. (8 strains, 2B1 to 2B8). Six *Acinetobacter* sp. strains (14B1 to 14B6) were isolated on the selective medium Herella and the identification was confirmed in an Api 20E gallery.

The culture media tested, as described in Difco Manual (11th edition – 1998), were Luria-Bertani agar; Mueller-Hinton; Plate Count agar; Violet Red Bile agar; Endo agar; M9 minimal medium; Terrific broth; 2XYT; Rappaport Vassiliadis; Kliger and Simmons. The incubation times tested were 24, 48, and 72 h. The bacteria were grown aerobically at 37 °C and cells were collected.

The bacterial sample was suspended in an extraction solvent to obtain a preparation containing about 5×10^8 whole bacterial cells per ml. The extraction solvents tested for dissolving proteins from bacteria were H_2O; 0.1% trifluoroacetic acid; 40% ethanol and various solvent mixtures such as, in v/v: acetonitrile-0.1% trifluoroacetic acid (66/33); formic acid-methanol-H_2O (10/45/45); formic acid-isopropanol-H_2O (17/33/50) and 0.1% trifluoroacetic acid-ethanol (50/50).

The cell suspension was vortexed for about 1 min and the bacterial concentration was determined by measuring the optical density at 600 nm with a SmartSpec 3000 spectrometer (Bio-Rad Laboratories Inc., Hercules, CA, USA). For desalting, the bacterial sample was washed three times with sterile water prior to extraction.

To adjust most of the experimental parameters we used *Escherichia coli* strain ATCC11775 grown for 24 h on LB agar. Effects of culture time and medium were studied using the reference strains ATCC11775 and ATCC25922 (*Escherichia coli*) and ATCC13076 and ATCC13311 (*Salmonella*). Each experiment was triplicated to test the reproducibility of this technique.

4.2.3
MALDI Sample Preparation

The matrices tested were sinapinic acid (10 mg ml^{-1}); α-cyano (10 mg ml^{-1}); ferulic acid (12.5 mg ml^{-1}); 2-(4-hydroxyphenylazo) benzoic acid (3.5 mg ml^{-1}) and a mixture of 1 mg of 5-methoxysalicilic acid and 9 mg 2,5-dihydroxybenzoic acid per milliliter. Each matrix was dissolved in methanol and different solvent mixtures such as, in v/v: 0.1% trifluoroacetic acid-acetonitrile (90/10); 0.1% trifluoroacetic acid-acetonitrile (60/40); ethanol-acetonitrile (50/50) and formic acid-acetonitrile-H_2O (17/33/50).

Three sample deposits were used. In the one-layer method, 1 µl of bacterial preparation was mixed with 3 µl of a solution containing 2 volumes saturated matrix solution and 1 volume cytochrome c solution. 1 µl of this mixture was applied to the MALDI-TOF sample plate and crystallized by air-drying. In the two-layer method,

1 µl bacterial preparation was applied to the MALDI-TOF sample plate and dried. Then 3 µl of the 2:1 matrix-cytochrome c mixture was applied to the MALDI-TOF sample plate and crystallized by air-drying. In the three-layer method, 1 µl of bacterial preparation was applied to the MALDI-TOF sample plate and dried. This first layer was overlaid with 1 µl of ethanol and dried. For the last layer, 3 µl of the 2:1 matrix-cytochrome c mixture was placed on top of the second layer and dried at room temperature before analysis.

The two- and three-layer methods were used only to study the effects of the ethanol treatment and of the sample deposit. In all these experiments, cytochrome c (MW = 12 360 Da) was used as an internal standard for mass calibration at a concentration of 10 pmol µl^{-1} in 0.1% trifluoroacetic acid. Calibration was done using the peaks of matrix and cytochrome c. Each sample was deposed in triplicate.

4.2.4
Microextraction Procedures

The sludge was diluted in extraction buffer containing 2 mM Na_3PO_4, 4 mM NaH_2PO_4, 9 mM NaCl, 1 mM KCl, pH 7 (Frolund et al. 1995). The optimal dilution was calculated to obtain an SV30 (Sludge Volume after 30 min of sedimentation, expressed in terms of percentage of the total volume) of about 1.6 in a final volume of 360 ml. Forty milliliters of diluted sludge was dispersed for 15 s at 37 W with an ultrasound generator (Labsonic 1510, standard 19 mm dia probe) in 50-ml falcon tubes (Greiner; 227261). The sample tube was kept on ice for 15 min before physical treatment. The probe was immersed in the liquid for 15 min.

Following the sonicated samples were digested for 20 min at room temperature and shaken gently with 0.04% (v/v) of an enzyme cocktail containing amylases (BAN 800MG; Aquazym 120L; Fungamyl 800L and AMG 300L); proteases (Neutrase 0,5L; Flavourzyme 1000L; savinase 16L; Novocor ABL; Flavourzym 500MG and alcalase 2,5L); cellulases (Celluzym 0,7L; Novozym 188L; Viscozyme L and Denimax 991 L) and lipases (Lipolase 100L; Novocor ADL and Novozym 868). This enzyme cocktail was kindly provided by Realco SA Belgium.

4.2.5
Cultures of Microorganisms

Escherichia coli: Samples were filtered through a sterile membrane (0.45 µm) and seeded into TBX medium. Bacteria were counted after a 24-h incubation at 37 °C. Confirmation of the strain was done by indole production according to the ISO 9308-1:2 000 methods. *Salmonella: Salmonella* bacteria were counted in successive steps according to ME1/028/V2-ISO6340:1995. *Acinetobacter:* Samples were seeded into Herella medium. After a 24-h incubation at 37 °C and 42 °C, putative *Acinetobacter* colonies were isolated and seeded into Herella and PCA media in order to confirm the identification on the basis of oxidase and catalase activity. The genus was identified by means of an Api 20E gallery.

4.2.6
Instrumentation

All mass spectra were generated on a TofSpec 2E MALDI-TOF mass spectrometer (Micromass UK Ltd., Manchester, UK) equipped with a nitrogen laser (337 nm) and operating in positive linear mode with an accelerating voltage of 20 kV. Each spectrum was the sum of the ions from 100 laser shots coming from different regions of the same well and was analyzed in the mass range from 2 to 20 kDa. Only mass peaks having a relative intensity exceeding 5% were considered. The criteria for estimating spectrum quality were: signal intensity, resolution, the signal/baseline ratio, reproducibility, and the number of mass peaks. Resolution was evaluated at 50% peak height and the signal/baseline ratio was calculated as peak intensity to mean of baseline.

4.3
Results and Discussion

4.3.1
Introduction

In this work we used MALDI-TOF MS as an original protocol for typing bacterial strains on the basis of mass spectra. The identification was based on screening for characteristic peaks seen as biomarkers for the identification of a genus, species, or bacterial strain.

Minor variations in experimental factors can affect the general appearance of the observed mass spectra. This makes it necessary to identify experimental parameters giving rise to a spectrum of reproducibly good quality. We therefore compared the effects of several parameters known to influence spectrum quality: matrix, extraction solvent, culture time and medium, salt content, and the type of deposit on the MALDI plate.

4.3.2
Optimisation of Experimental Parameters

The choice of the matrix is a critical step in mass spectrometry because it can influence the nature of the compounds detected, e.g. phospholipids, peptides and proteins. Different matrices and matrix solvents were screened for compatibility with analysis of whole bacteria (see Fig. 4.1).

Among the matrix solutions used, only sinapinic acid diluted in 0.1% trifluoroacetic acid-acetonitrile (60/40) gave mass peaks for low and high molecular weight. With this matrix, it was possible to observe the ion peak at m/z 9 060 with a resolution of 600 and a signal/baseline ratio near 100, as reported in Fig. 4.1. The α-cyano matrix also yielded a quality spectrum but we did not observe peaks of high mass with this matrix. The signal/baseline ratio of peak 9 060 was five times lower with α-cyano than with sinapinic acid. Ferulic acid gave rise to high-mass peaks, but the signal was weaker than with sinapinic acid (data not shown). The other matrices did not give signal higher than the baseline in the mass range observed (data not shown).

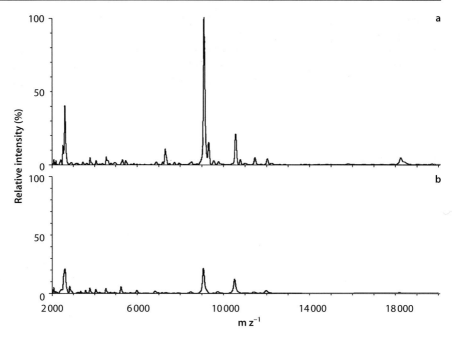

Fig. 4.1. Mass spectra of *Escherichia coli* (ATCC11775), illustrating the effect of the matrix on the general appearance of mass spectra. The matrices used were: **a** sinapinic acid in 0.1% trifluoroacetic acid–acetonitrile (60/40) and **b** α-cyano in 0.1% trifluoroacetic acid-acetonitrile (60/40)

The ion peak at m/z 9 060 was used to compare signal intensity, resolution and signal/baseline ratio. This ion is a biomarker of *Escherichia coli* strain ATCC11775 which corresponds to the acid-resistant precursor protein HdeB (Holland et al. 1999). This comparison showed that the sinapinic acid improves the number and the intensity of mass peaks in the studied mass range (2–20 kDa). This matrix is mostly used to analyze proteins (Haag et al. 1998) whereas α-cyano promotes the ionization of low-mass compounds such as peptides (Dai et al. 1999).

The conditions of protein extraction can also affect the general appearance of the spectrum. Among the studied extraction solvents, only formic acid-isopropanol-H_2O (17/33/50; v/v/v) gave the most relevant signal/baseline ratio. As seen in Fig. 4.2, this solvent allows the detection of a high number of mass peaks, particularly in the mass range 2–12 kDa and the signal/baseline was approx. 140 for the ion peak at m/z 6 600. Trifluoroacetic acid gave similar results but the signal intensity was approx. 20% lower (data not shown). With water, no high-mass peaks appeared.

It was shown that the extraction solvents have various effects on the mass spectra. It can be explained by their action on the bacteria cell surface or by their proteins solubility properties. Some of these solvents induce small holes in the bacterial cell surface allowing the accessibility of bacterial compounds to mass spectral analysis. Moreover, the solubility of proteins and peptides in a given solvent varies according to their degree of hydrophobicity and isoelectric point (Ryzhov and Fenselau 2001).

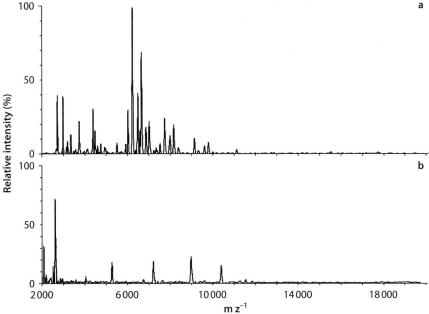

Fig. 4.2. Mass spectra of *Escherichia coli* (ATCC11775), illustrating the effect of the extraction solvent on the general appearance of the observed mass spectra. **a** Formic acid-isopropanol-H_2O (17/33/50) and **b** water was used as extraction solvent

In order to improve the analysis of high-mass proteins (>12 kDa), extraction methods using ethanol were tested (data not shown). The one- and two-layer methods (see Sect. 4.2.2 and 4.2.3) did not give reproducible mass peaks, probably because it interferes with the matrix (one-layer). Moreover, the inability of bacteria to dissolve in ethanol makes the solution heterogeneous (two-layer).

In the three-layer method, the bacterial preparation (in formic acid-isopropanol-H_2O as extraction solvent) was homogenous, and because the ethanol was completely evaporated before the matrix was applied, it could not interfere with matrix. This treatment improved mass spectrum quality in terms of both signal intensity and the number of mass peaks.

The ethanol treatment produces nonspecific pores through the lipid layer, allowing translocation of proteins exceeding 15 kDa while maintaining the bacterial cell whole. Furthermore, ethanol treatment promotes uniform dispersion of the bacteria in the matrix/sample crystal by decreasing their tendency to clump together or adhere to the probe surface (Madonna et al. 2000).

The salt content can also affect spectrum quality. We therefore compared the results of experiments performed using the three-layer method, with and without prior desalting of the bacterial sample (data not shown). More mass peaks were obtained with the desalted sample than with the salt-containing sample. All peaks were in both spectra, but many of them were masked by the baseline in the spectrum obtained without desalting. Salt influences the general appearance of the observed mass spectra because it modifies the desorption/ionization proprieties of proteins so that some are detected preferentially to others (Wang et al. 1998).

The type of sample deposit can also influence the appearance of the spectrum. Indeed, we have shown that the quality of spectra (intensity and number of reproducible mass peaks) was better when the three-layer method was applied (data not shown). This technique combines good dissolution (in the extraction solvent) with good extraction (by the extraction solvent and ethanol).

The choice of the culture medium can also affect the general appearance of mass spectra. Indeed, after comparison of mass spectra obtained from the most used culture media, we have proved that the fingerprints observed for the same bacterial strain could show some differences (data not shown). This probably reflects the fact that nutrients present in the medium can induce or repress the synthesis of certain proteins. Mueller-Hinton medium yielded quality spectra for all strains tested, in terms of both signal intensity and number of peaks. Other culture media such as Plate Count agar also yielded quality spectra, but not for all bacterial strains. The goal being to choose a culture medium suitable for all the tested strains, we retained Mueller-Hinton medium.

In order to test the effect of culture time, we compared mass spectra obtained from culture of the same strains after 24, 48 and 72 h of incubation (data not shown). We observed an influence of the incubation time on the spectra obtained. Of the three incubation times tested, 48 h gave the higher number of peaks. A shorter incubation time (24 h) resulted in fewer peaks, these being in a lower mass range. A longer incubation time (72 h) led to the disappearance of peaks in the mass range observed. Considering these results another imperative should be taken into account: the identification procedure should not take too much time.

This first study allowed us to establish parameters leading to good-quality mass spectra and good reproducibility with reference strains. Accordingly we recommend a 48 h incubation of the bacteria on Mueller-Hinton medium, desalting of the bacterial sample, and the three-layer method including ethanol treatment, with formic acid-isopropanol-H_2O as the extraction solvent and sinapinic acid in 0.1% trifluoroacetic acid-acetonitrile (60/40) as the matrix.

4.3.3
Bacterial Identification with MALDI-TOF MS

The optimized protocol described above (see Sect. 4.3.2) was applied to the identification of specific bacterial biomarkers.

In the first approach, we compared mass spectra obtained from reference strains of *Escherichia coli* (ATCC11775 and ATCC25922), *Salmonella* (ATCC13076 and ATCC13311) and *Acinetobacter* (ATCC17924) to find their respective biomarkers. The presence of these different species/genus in wastewater and sewage sludge, are considered as indicators of the biological quality. In the field of this study, various mass peaks appeared to be species-specific biomarkers (data not shown). Two peaks emerged as specific to *Escherichia coli* (ion peaks at m/z 9 240 and 9 540), two to *Salmonella* (m/z 9 240 and 9 740) and one to *Acinetobacter* (m/z 5 800). Other mass peaks appeared to be strain-specific biomarkers. Two peaks emerged as specific to *Escherichia coli*

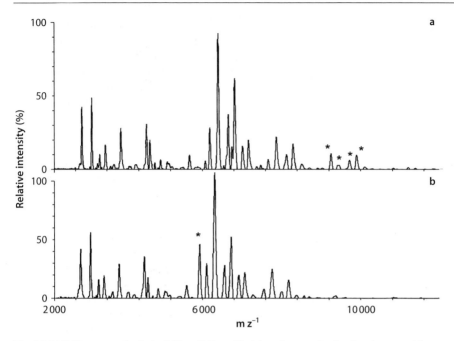

Fig. 4.3. MALDI spectra of **a** *Escherichia coli* 1B1 and **b** *Acinetobacter* 14B5 showing the potential species- or strain-specific biomarkers

strain ATCC11775 (ion peaks at m/z 9060 and 9740) and one to *Salmonella enteritidis* strain ATCC13076 (m/z 5000). It should thus be possible to use these mass peaks as species- or strains-specific biomarkers in the bacterial taxonomy. It is worth mentioning that the same mass/charge value appearing in two different biomarker sets does not necessarily correspond with the same protein (Krishnamurthy et al. 1999).

In the second approach, we extracted live strains of *Escherichia coli* (8), *Salmonella* (8) and *Acinetobacter* (6) from sewage sludge by a procedure combining sonication and enzymatic digestion. The spectra obtained were compared with each of other environmental strains and with reference bacteria. This comparison showed that the four peaks emerging as specific to *Escherichia coli* strain ATCC11775 (ion peaks at m/z 9060, 9240, 9540 and 9740) were also found in the mass spectra of *Escherichia coli* 1B1, as seen in Fig. 4.3. The others environmental strains of *Escherichia coli* (1B2 to 1B8) exhibited only the species-specific biomarkers at m/z 9240 and 9740 (data not shown). The specific biomarker of *Acinetobacter* strain ATCC17924 (at m/z 5800) was also observed in the mass spectra of others *Acinetobacter* strains, e.g. 14B5 (as seen in Fig. 4.3). The species-specific biomarkers of *Salmonella* were also observed in mass spectra of other *Salmonella* strains isolated from the environment (data not shown).

Therefore, the different biomarkers of reference strains were also identified in the mass fingerprint of environmental strains.

In the absence of more detailed structural information, we assume that the observed mass peaks are peptides or proteins, but this should be confirmed by MS/MS (Fenselau and Demirev 2001). The biomarkers 9060, 9240, 9540 and 9740 of *Escheri-*

chia coli reference strain ATCC11775 have already been identified. The 9 740 and 9 060 ions correspond respectively to acid-resistant precursor proteins HdeA and HdeB after proteolytic cleavage by peptidase Lep (Holland et al. 1999). The 9 540 and 9 240 ions correspond to the α- and β-subunit of a DNA-binding protein HU, DBHA and DBHB respectively (Ryzhov et al. 2000).

4.4
Conclusion

In this study, we have shown that the general appearance of a MALDI spectrum obtained from whole bacteria cells can be influenced by a number of experimental factors. We have tested several of these factors and have determined conditions that reproducibly yield good-quality spectra for the bacterial species tested: a matrix solution composed of sinapinic acid, use of an extraction solvent enabling extraction of more and higher-mass cell compounds, and ethanol treatment, which further increased protein extraction and favors a more homogenous deposit.

This optimized protocol was applied to identify reference and environmental strains of *Escherichia coli*, *Salmonella* and *Acinetobacter*. The identification has been realized by the observation of their respective species- and strains-specific biomarkers. This taxonomy of bacteria could be completed within few minutes, with minimal sample preparation, provided that sufficient bacteria have been collected prior the MALDI-TOF analysis. This study showed that the use the MALDI-TOF MS technique could be applied to characterize whole bacteria cells by identification their biomarkers. This approach is then proposed as a complement to classical bacterial identification methods.

Acknowledgements

The work has been funded by the Region Wallonne (Grants RW 981/3799 and RW 114713). Enzyme cocktail was kindly provided by Realco SA Belgium. We thank Nadine Burlion from ISSeP "Institut Scientifique de Services Publics, Belgium", for many fruitful discussions on this work and for technical assistance in the laboratory. The pathogenic bacteria enumerations were performed at ISSeP. Virginie Ruelle is supported by a fellowship from A.R.C 99/04-245.

References

Anhalt JP, Fenselau C (1975) Identification of bacteria using mass spectrometry. Anal Chem 47:219–225
Arnold RJ, Karty JA, Ellington AD, Reilly JP (1999) Monitoring the growth of bacteria culture by MALDI-MS of whole cells. Anal Chem 71:1990–1996
Banks CJ, Walker I (1977) Sonication of activated sludge flocs and the recovery of their bacteria on solid media. J General Microbiol 98:363–368
Birmingham J, Demirev P, Ho YP, Thomas J, Bryden W, Fenselau C (1999) Corona plasma discharge for rapid analysis of microorganisms by mass spectrometry. Rapid Commun Mass Spectrom 13:604–606
Chong BE, Wall DB, Lubman DM, Flynn SJ (1997) Rapid profiling of *E. coli* proteins up to 500 kDa from whole cell lysates using MALDI-TOF MS. Rapid Commun Mass Spectrom 11:1900–1908

Dai Y, Li L, Roser DC, Long SR (1999) Detection and identification of low-mass peptides and proteins from solvent suspensions of *Escherichia coli* by HPLC fractionation and MALDI-TOF MS. Rapid Commun Mass Spectrom 13:73–78

Demirev PA, Ramirez J, Fenselau C (2001) Tandem mass spectrometry of intact proteins for characterization of biomarkers from *Bacillus cereus* T spores. Anal Chem 73:5725–5731

Difco Laboratories (1998) Difco manual, 11th edn. Difco Laboratories, Division of Becton Dickinson and company, Sparks, Maryland, USA

Domin MA, Welham KJ, Ashton DS (1999) The effect of solvent and matrix combinations on the analysis of bacteria by MALDI-TOF MS. Rapid Commun Mass Spectrom 13:222–226

Easterling ML, Colangelo CM, Scott RA, Amster IJ (1998) Monitoring protein expression in whole bacterial cells with MALDI-TOF MS. Anal Chem 70:2704–2709

Elhanany E, Barak R, Fisher M, Kobiler D, Altboum Z (2001) Detection of specific *Bacillus anthracis* spore biomarkers by MALDI-TOF MS. Rapid Commun Mass Spectrom 15:2110–2116

Evason DJ, Claydon MA, Gordon DB (2001) Exploring the limits of bacterial identification by intact cell-mass spectrometry. J Am Soc Mass Spectrom 12:49–54

Fenselau C, Demirev PA (2001) Characterization of intact microorganisms by MALDI mass spectrometry. Mass Spectrometry Reviews 20:157–171

Frolund B, Palmgren R, Keiding K, Nielsen P (1995) Extraction of extracellular polymers from activated sludge using a cation exchange resin. Water Res 30(8):1749–1758

Haag AM, Taylor SN, Johnston KH, Cole RB (1998) Rapid identification and speciation of *Haemophilus* bacteria by MALDI-TOF MS. J Mass Spectrom 33:750–756

Hamels S, Gala JL, Dufour S, Vannuffel P, Zammatteo N, Remacle J (2001) Consensus PCR and microarray for diagnosis of the genus *Staphylococcus*, species, and methicillin resistance. Bio Techn 31:1364–1372

Holland RD, Wilkes JG, Rafii F, Sutherland JB, Persons CC, Voorhees KJ, Lay JO (1996) Rapid identification of intact whole bacteria based on spectral patterns using MALDI-TOF MS. Rapid Commun Mass Spectrom 10:1227–1232

Holland RD, Duffy CR, Rafii F, Sutherland JB, Heinze TM, Holder CL, Voorhees KJ, Lay JO (1999) Identification of bacterial proteins observed in MALDI TOF mass spectra from whole cells. Anal Chem 71:3226–3230

Jorand F, Zartarian F, Thomas F, Block JC, Bottero JY, Villemin G, Urbain V, Manen J (1995) Chemicals and structural (2D) linkage between bacteria within activated sludge flocs. Water Res 29(7):1639–1647

Karas M, Bachmann D, Bahr U, Hillenkamp F (1987) Matrix-assisted ultraviolet laser desorption of non volatile compounds. Int J Mass Spectrom Ion Processes 78:53–68

Krishnamurthy T, Ross PL, Rajamani U (1996) Detection of pathogenic and non-pathogenic bacteria by MALDI-TOF MS. Rapid Commun Mass Spectrom 10:883–888

Krishnamurthy T, Davis MT, Stahl DC, Lee TD (1999) Liquid chromatography/microspray mass spectrometry for bacterial investigations. Rapid Commun Mass Spectrom 13:39–49

Lynn EC, Chung MC, Tsai WC, Han CC (1999) Identification of Enterobacteriaceae bacteria by direct MALDI mass spectrometric analysis of whole cells. Rapid Commun Mass Spectrom 13:2022–2027

Madonna AJ, Basile F, Ferrer I, Meetani MA, Rees JC, Voorhees KJ (2000) On-probe sample pretreatment for the detection of proteins above 15 kDa from whole cell bacteria by MALDI-TOF MS. Rapid Commun Mass Spectrom 14:2220–2229

Nichols DS, McMeckin TA (2002) Biomarker techniques to screen for bacteria that proceduce polyunsaturated fatty acids. J Microbial Methods 48(2–3):161–170

Nilsson CL (1999) Fingerprinting of *Helicobacter pylori* strains by MALDI mass spectrometric analysis. Rapid Commun Mass Spectrom 13:1067–1071

Ryzhov V, Fenselau C (2001) Characterization of the protein subset desorbed by MALDI from whole bacterial cells. Anal Chem 73:746–750

Ryzhov V, Hathout Y, Fenselau C (2000) Rapid characterization of spores of *Bacillus cereus* group bacteria by MALDI-TOF MS. Appl Environ Microbiol 66(9):3828–3834

Saenz AJ, Petersen CE, Valentine NB, Gantt SL, Jarman KH, Kingsley MT, Wahl KL (1999) Reproducibility of MALDI-TOF MS for replicate bacterial culture analysis. Rapid Commun Mass Spectrom 13:1580–1585

Snyder AP, Thornton SN, Dworzanski JP, Meuzelaar HLC (1996) Detection of the picolinic acid biomarker in *Bacillus* spores using a potentially field-portable pyrolysis-gas chromatography-ion mobility spectrometry system. Field Anal Chem & Techn 1:49–59

Thomas JJ, Falk B, Fenselau C (1998) Viral characterization by direct analysis of capsid proteins. Anal Chem 70:3863–3867

Wang Z, Russon L, Li L, Roser DC, Long SR (1998) Investigation of spectral reproducibility in direct analysis of bacteria proteins by MALDI-TOF MS. Rapid Commun Mass Spectrom 12:456–464

Welham KJ, Domin MA, Johnson K, Jones L, Ashton DS (2000) Characterization of fungal spores by laser desorption/ionization time-of-flight mass spectrometry. Rapid Commun Mass Spectrom 14:307–310

Chapter 5

Multi-Isotopic Approach (^{15}N, ^{13}C, ^{34}S, ^{18}O and D) for Tracing Agriculture Contamination in Groundwater

L. Vitòria · A. Soler · R. Aravena · À. Canals

Abstract

Groundwater in the Maresme area (NE Spain) is characterised by high concentrations of nitrate, sulphate and chloride. Chemical and isotope data presented in this paper were collected in order to characterise the main sources of nitrate contamination. This study has also provided information about sources of sulphate and chloride in the area. The nitrate groundwater contamination is related to agriculture activities such as the intense use of synthetic fertilisers. The high rates of fertilisation and re-circulation of the shallow groundwater used for irrigation explain the high concentration of nitrates in the groundwater with values as high as 482 mg l^{-1}. The isotope composition of the groundwater nitrate ranges between +6.8‰ and +9.4‰ for $\delta^{15}N$ and from +5.1‰ to +10.2‰ for $\delta^{18}O$. Fertilisers used in the area show values close to 0‰ and +23‰ for $\delta^{15}N$ and $\delta^{18}O$, respectively. The more enriched $\delta^{15}N$ values in the groundwater compared to the fertilisers is associated to volatilisation of the ammonia component of the fertilisers. The ^{18}O pattern in the groundwater implies that nitrate from nitrogenous fertilisers is recycled in the soil where it becomes ^{18}O depleted. Based on the $\delta^{15}N_{NO3}$ and $\delta^{18}O_{NO3}$ data, a significant denitrification was discarded in the study site. Groundwater dissolved sulphate has $\delta^{34}S_{SO4}$ values between +5.8 to +7.0‰ suggesting that the main source of sulphate might not be related to seawater intrusion, which in turn questioned the origin of chloride previously related to a seawater intrusion.

Key words: isotopes, nitrate contamination, groundwater, fertilisers

5.1 Introduction

Groundwater contamination by nitrates derived from agricultural activities is a serious problem in many countries. Maximum nitrate concentrations permitted by the European Directive 98/83/CE in waters for human consumption is 50 mg l^{-1}. However, in some areas of Spain, groundwaters with concentrations even ten times higher than the permitted level have been reported (IGME 1985; Navarro et al. 1989). Ingestion of high nitrate concentrations cause methahemoglobine disease in children as well as babies, and, additionally, there is evidence that nitrogenous compounds formed inside the stomach act as human cancer promoters (Magee and Barnes 1956).

The main sources of nitrate in groundwater are fertilisers (either organic or inorganic), manure piles, septic systems, cesspool, sewage lagoons and soil nitrogen. Conventional chemical analyses allow us to know the degree of nitrate contamination, but they do not provide information about its origin. Numerous research programs have been carried out over the last 15 years in order to characterise isotopically the different sources of nitrate contamination, and also to assess the usefulness of ^{15}N

and ^{18}O as tracers to evaluate the fate of nitrate in groundwater (Böttcher et al. 1990; Wassenaar 1995; Aravena and Robertson 1998; Pauwels et al. 2000).

The aim of this study is to characterise chemically and isotopically nitrate contaminated groundwater in an area polluted by the use of synthetic fertilisers due to a very intensive agriculture activity (mainly flowers and horticulture). This study is part of a research project aiming to apply the isotope approach to study the origin of nitrate in contaminated groundwater throughout Spain. The approach includes the isotope characterisation of nitrate in contaminated sites with well-known nitrate sources in order to elucidate the nitrate contribution in other polluted areas where nitrate comes from different sources. This study was carried out in the Maresme region situated close to the Mediterranean Sea shoreline. For more than 50 years, groundwaters in this region have been intensively exploited for agricultural purposes, causing an alteration of the aquifer hydrodynamics and therefore a progressive seawater intrusion (Custodio et al. 1976; Bayó 1987). Now, farmers must treat groundwaters before using them in order to reduce their salinity, which may cause diseases to their crops (chlorosis, poor vegetative growth) and a decrease in the agricultural production. The treatment they employ involves groundwater dilution with rainwater, or the use of small reverse osmosis plants. In both cases, they dilute or remove the nutrients from waters and force them to use more fertilisers. This practice results in a high concentration of nitrates in waters.

5.2
Study Area

The Maresme area is one of the sites classified by the Catalan Water Agency (ACA) of the Catalan Government as a "Vulnerable zone by nitrate contamination of agricultural sources" (Fig. 5.1). It is located between the Catalan Coastal Range and the Mediterranean Sea, between the towns of Vilassar de Mar and Premià de Mar (25 km

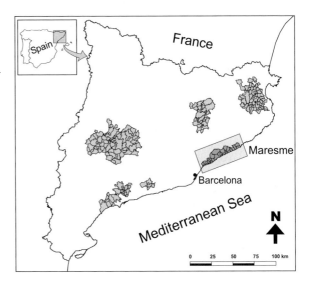

Fig. 5.1.
"Vulnerable areas of contamination by nitrates from agricultural sources" (*grey areas*) classifieds by the Catalan Government (inside the *box*: the studied area)

Chapter 5 · Multi-Isotopic Approach for Tracing Agriculture Contamination in Groundwater

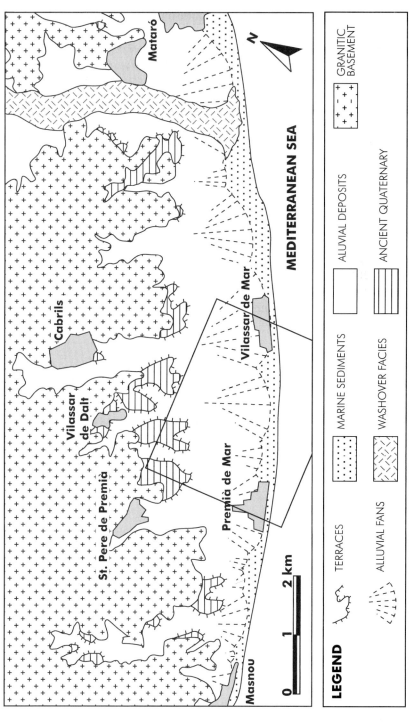

Fig. 5.2. Geological map of the study area (modified after Villarroya 1986). Sampling site is situated inside the *box* corresponding to Fig. 5.4

to the NE of Barcelona). The sampled area is 3 km², and consists geologically of Holocene alluvial alluvial deposits of coarse sands derived from the weathered granodiorite that forms the Catalan Coastal Range (Fig. 5.2). The main hydrogeological units are an unconfined sandy aquifer underlayed by an aquitard composed of silts and clays and a confined sandy aquifer. The thickness of these units is 5–40 m, 5–15 m and 15–20 m, respectively. The unconfined aquifer is the only one affected by groundwater extractions and its water table varies between 4 and 30 m in depth (Guimerà et al. 1995).

Flowers, fruit and vegetable crops are the main agricultural products of the area, where about half of them grow under greenhouse conditions. Fertilisation is carried out with inorganic fertilisers usually injected through trickle irrigation systems that use groundwater extracted from partially-penetrating wells (5–40 m deep).

The soil type in this area is usually coarse sand with a low organic matter content (<3% of dry soil) and a low C:N ratio of approximately 1:2 (Guimerà et al. 1995). This low natural fertility, together with the low water-holding capacity of the soil, requires high fertiliser and irrigation applications, resulting in high nutrient leaching in the upper zone of the aquifer affected by groundwater withdrawals. Re-circulation of the shallow groundwater by the irrigation system causes high concentrations of nitrates in the groundwater (higher than 450 mg l^{-1}).

5.3
Methodology

Eight groundwater samples were collected during October 2000. Sampling was done from old private, partially-penetrating wells, 1 to 2 m in diameter and 10 to 40 m in depth, equipped with pumps. Eleven of the most frequently used fertilisers in the region were also analysed which correspond to a major scale chemical and isotopic fertiliser characterisation published in Otero et al. 2004 and Vitòria et al. 2004. Sampling locations are shown in Fig. 5.4. Conductivity, pH and temperature were measured in situ.

Before chemical analysis, water samples were filtered with 0.45 µm Millipore® filter. Major anions contents (NO_3^-, NO_2^-, PO_4^{3-}, SO_4^{2-} and Cl^-) were determined by Liquid Chromatography, ammonium (NH_4^+) by colorimetry, total organic carbon (TOC) by the matter organic combustion and alkalinity (HCO_3^-) was measured by titration.

For nitrogen and oxygen isotopic analysis, the dissolved nitrates were concentrated using anion-exchange columns filled with Bio Rad® AG 1-X8(Cl^-) 100–200 mesh resin. Nitrates were then eluted with HCl and converted to $AgNO_3$ by the addition of silver oxide following a modified method from Silva et al. (2000). The purified silver nitrate solution was then freeze-dried. For sulphur isotope analysis ($\delta^{34}S_{SO4}$), dissolved sulphates were precipitated as $BaSO_4$ by adding $BaCl_2 \cdot 2H_2O$ after acidifying with HCl. The samples were boiled simultaneously in order to prevent $BaCO_3$ precipitation. Unfiltered splits were treated with $NaOH-BaCl_2$ solution to precipitate carbonates and analyse the $\delta^{13}C$ of the total inorganic dissolved carbon (TIDC).

Nitrogen isotopic compositions were analysed with a Carlo Erba elemental analyser (EA) coupled with a Isochrom Continuous Flow isotope ratio mass spectrometer

(IRMS). Sulphur and carbon isotopic compositions with a Carlo Erba EA coupled with a Finnigan Mat Delta C IRMS. The oxygen isotopic compositions of nitrates were determined by on–line pyrolysis with a Eurovector TC/EA (high temperature conversion-elemental analyser) unit coupled with a Micromass Isoprime Continuous Flow IRMS. Oxygen and hydrogen isotopes of water with a Finigan Mat Delta S IRMS after equilibration with H_2 and CO_2. All $\delta^{15}N$, $\delta^{34}S$, $\delta^{13}C$, $\delta^{18}O$ and δD values presented are normalised to AIR (atmospheric N_2), VCTD (Vienna Canyon Diablo Troilite), PDB (Peedee Belemnite) and VSMOW (Vienna Standard Mean Ocean Water), respectively. The isotope ratios were calculated using purchased international (IAEA-N1, IAEA-N2, NBS-127, IAEA-S1, IAEA-S3, IAEA-CH7, IAEA-CH6, USGS-24) and internal laboratory standards. Reproducibility of the samples calculated from standards systematically interspersed in the analytical batches were ±0.4‰, ±0.3‰, ±0.7‰, ±0.3‰, ±0.3‰, ±0.4‰ ±1.5‰ and ±0.2‰ for $\delta^{34}S$, $\delta^{15}N_{NO3}$, $\delta^{18}O_{NO3}$, $\delta^{15}N_{Ntotal}$, $\delta^{13}C_{HCO3}$, $\delta^{13}C_{Ctotal}$, δD and $\delta^{18}O_{H2O}$, respectively.

Each fertilisers was homogenised, dissolved in water (stirred and filtered) and then treated as the groundwater samples for chemical analyses and nitrate isotopic (^{15}N and ^{18}O) determinations. Because fertilisers may have ammoniacal compounds, $\delta^{15}N_{Ntotal}$ and also $\delta^{13}C_{Ctotal}$ were analysed from the powdered fertilisers.

Chemical and isotopic analyses were carried out at the Serveis Cientificotècnics of the University of Barcelona (Spain), except for the $\delta^{15}N_{Ntotal}$, $\delta^{15}N_{NO3}$ and $\delta^{18}O_{NO3}$ analyses, which were performed at the Environmental Isotope Laboratory of the University of Waterloo (Canada).

5.4
Results and Discussion

5.4.1
Chemical Data

Chemical data for water samples and fertilisers are shown in Table 5.1. In the Maresme region, non-contaminated or low-contaminated groundwaters ($NO_3^- < 50$ mg l^{-1} and $SO_4^{2-} < 200$ mg l^{-1}) are bicarbonate calcium type waters (Villarroya 1986) (Fig. 5.3) with total dissolved solids (TDS) between 600 and 900 mg l^{-1}. In the study area, due to the agricultural activity, groundwaters are chloride sulphate calcium type waters with high concentrations of nitrates and sulphates and the TDS around 2 000 mg l^{-1} (Table 5.1).

The chloride contour lines show two zones with higher concentrations (Fig. 5.4a). These high chloride concentrations derive either from a marine intrusion or from the influence of fertilisers. In addition, these areas have the maximum concentration of nitrates (Fig. 5.4b). This correlation might be related to the highly intensive agricultural activity in the study area. This activity involves high rates of groundwater extraction promoting seawater intrusion and the use of a high load of fertilisers by the irrigation systems that may cause accumulation of chloride of fertiliser origin in soil and groundwater. Moreover, high nitrate leaching, which seems to be correlated to high rates of irrigation (Guimerà et al. 1995), could explain the high nitrate concentration observed in the groundwater.

Table 5.1. Ammonium, major anions and total organic carbon (TOC) concentrations, and total dissolved solids (TDS) of the Maresme groundwaters and fertilisers used in the area. Fertiliser data are from Otero et al. 2004

Sample	Name	NH_4^+	Cl^-	NO_2^-	NO_3^-	PO_4^{3-}	SO_4^{2-}	HCO_3^-	TOC	TDS
Maresme groundwaters		in mg l^{-1}								
PM-031	Ribas-Mateu	0.2	266.1	<0.5	271.2	<29	421.7	337.6	<1	1799.3
PM-038	Ca n'Agustí	0.1	172.0	<0.5	343.2	<25	340.8	319.0	<1	1599.9
PM-046	Josep Lloveres	<0.1	588.8	<0.7	333.0	<36	435.0	208.5	<1	2228.8
PM-056	Genís Lloberes	<0.1	262.1	<0.5	235.4	<25	346.9	340.3	<1	1679.0
PM-061	Can Pitxot	<0.1	320.3	<0.6	482.0	<29	400.2	277.9	<1	2049.5
PM-123	Roses Floriach	0.1	368.8	<0.7	435.8	<37	407.9	358.7	1.9	2156.1
PM-145	Can Llanas-1	<0.1	392.9	<0.6	294.2	<30	487.3	260.6	<1	2010.9
PM-159	Joan Ytchart-2	0.2	264.3	<0.6	269.0	<29	449.9	385.5	<1	1906.9
Fertilisers		in mg kg^{-1}								
F2	NH_4NO_3	218000	9000	<50	780000	<2500	20000	–	300	–
F8	K_2SO_4	<100	11000	<100	<100	<5000	542000	–	<100	–
F7	$Ca(NO_3)_2$	<100	6000	<100	698000	<5000	<5000	–	<100	–
F9	NPK 13-0-46	<100	8000	<50	609000	<2500	5000	–	1600	–
F10	NPK 15-5-30	49000	3000	<100	443000	65000	127000	–	7200	–
F11	NPK 17-6-18	149000	<2000	<100	243000	84000	345000	–	3400	–
F12	NPK 15-10-15	117000	4000	<100	243000	128000	297000	–	2400	–
F13	NPK 0-52-34	<100	<2000	<100	<100	666000	<5000	–	<100	–
F14	NPK 12-61-0	155000	<2000	<100	200	712000	5000	–	<100	–
F15	NPK 10-10-18	129000	111000	<100	<100	17000	239000	–	30900	–
F17	NPK 16-8-12	52000	4000	<100	1000	70000	257000	–	25000	–
F20	NPK 12-12-17	79000	6000	<100	242000	84000	245000	–	400	–
F24	$Mg(NO_3)_2$	300	<2000	<100	362000	<5000	<5000	–	<100	–
F25	$MgSO_4$	<100	2000	<50	<100	<2500	401000	–	<100	–

Fig. 5.3.
Chemical trilinear diagrams for contaminated groundwaters in the study area compared with non-contaminated waters (from Villarroya 1986) in the Maresme region. Fertiliser data are from Vitòria et al. 2004

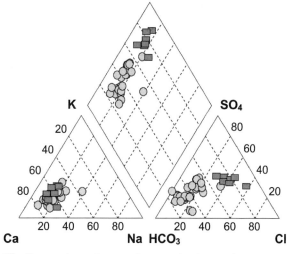

■ Contaminated groundwaters in the study area
○ Non-contaminated groundwaters in the Maresme region (NO_3 < 50 mg/L, SO_4 < 200 mg/L)

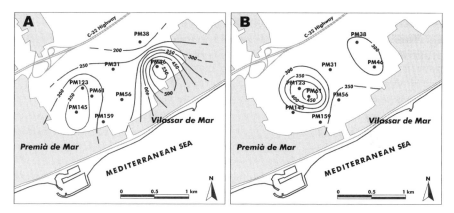

Fig. 5.4. *Black dots* in **a** and **b** represent the sampled wells of the study area with their code-number. *Contour lines* of chloride (**a**) and nitrate (**b**) concentrations (in mg l^{-1}) in groundwaters are drawn showing two areas of anomalous higher concentrations of both elements in the area

5.4.2
Isotope Data

^{18}O and ^{2}H stable isotope analyses from groundwater were also performed to determine the origin of the groundwater recharge and to evaluate possible evaporation during irrigation (Table 5.2). The isotope composition of groundwater is very similar in all samples (Fig. 5.5) and plots very close to the Local Meteoric Water Line (LMWL). These values are also similar to the weighted mean isotope composition of the local

Table 5.2. Isotopic compositions of the Maresme groundwaters and fertilisers used in the area

Sample	Name	δD_{H_2O}	$\delta^{18}O_{H_2O}$	$\delta^{13}C_{HCO_3}$	$\delta^{13}C_{Ctotal}$	$\delta^{15}N_{Ntotal}$	$\delta^{15}N_{NO_3}$	$\delta^{34}S_{SO_4}$	$\delta^{18}O_{NO_3}$
Maresme groundwaters		in ‰							
PM-031	Ribas-Mateu	−32.3	−5.78	−14.2	–	–	7.7	5.4	7.2
PM-038	Ca n'Agustí	−34.1	−5.83	−14.4	–	–	6.8	4.9	7.0
PM-046	Josep Lloveres	−33.3	−5.29	−13.8	–	–	6.9	6.8	5.8
PM-056	Genís Lloberes	−33.5	−5.66	−14.7	–	–	9.4	6.1	6.0
PM-061	Can Pitxot	−33.7	−5.49	−14.3	–	–	6.8	5.6	11.0
PM-123	Roses Floriach	−32.3	−5.59	−15.3	–	–	7.6	5.8	9.3
PM-145	Can Llanas-1	−27.7	−5.44	−13.4	–	–	7.1	6.1	6.5
PM-159	Joan Ytchart-2	−31.4	−5.23	−14.7	–	–	8.1	5.6	7.2
Fertilisers		in ‰							
F2	NH_4NO_3	–	–	–	–	2.5	5.6	0.7	25.1
F7	$Ca(NO_3)_2$	–	–	–	–	3.9	4.5	–	22.1
F8	K_2SO_4	–	–	–	–	–	–	7.1	–
F9	NPK 13-0-46	–	–	–	−23.6	−0.5	−1.0	3.8	48.5
F10	NPK 15-5-30	–	–	–	−28.0	0.1	0.3	1.4	22.1
F11	NPK 17-6-18	–	–	–	−28.1	0.3	2.0	−6.5	23.1
F12	NPK 15-10-15	–	–	–	−30.0	0.3	1.7	1.1	24.2
F13	NPK 0-52-34	–	–	–	–	–	–	–	–
F14	NPK 12-61-0	–	–	–	−34.3	−1.1	–	−0.2	–
F15	NPK 10-10-18	–	–	–	−29.9	0.8	–	6.8	–
F17	NPK 16-8-12	–	–	–	–	−1.0	–	9.0	–
F20	NPK 12-12-17	–	–	–	−27.4	0.4	0.8	5.8	21.0
F24	$Mg(NO_3)_2$	–	–	–	–	0.0	0.8	–	22.4
F25	$MgSO_4$	–	–	–	–	–	–	21.4	–

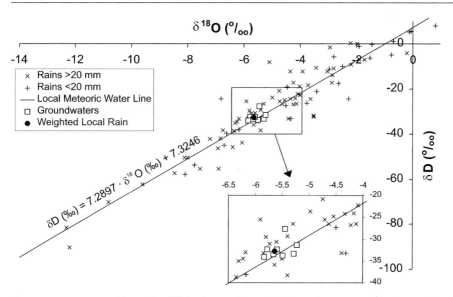

Fig. 5.5. Isotope compositions (δD, $\delta^{18}O$) of groundwaters in the study area are very close to the weighted local rainwater (calculated from the Barcelona Station GNEIP Database since 1985) showing that waters are not affected by the evaporation process during irrigation

rain calculated using the precipitation data of the station in Barcelona (IAEA/WMO 2001), indicating that rainwater is the main recharge of the aquifer. On the other hand, the high nitrate concentrations in the aquifer suggest that the irrigation water is another source of recharge in the area. These waters are not affected by evaporation because trickle irrigation in greenhouses prevents this process. Moreover, although a marine intrusion has been described in the region (Custodio et al. 1976; Bayó 1987), the δD and $\delta^{18}O$ data of groundwaters do not show any significant variations as would be expected from a mixing trend between seawater and groundwaters.

In order to characterise the nitrate contamination by the use of synthetic fertilisers, a multi-isotopic analysis of fertilisers and groundwater nitrate was performed. Results are shown in Table 5.2. Nitrate isotopic compositions of fertilisers ($\delta^{15}N_{Ntotal}$ and $\delta^{18}O_{NO3}$) and groundwaters ($\delta^{15}N_{NO3}$ and $\delta^{18}O_{NO3}$) are plotted in Fig. 5.6. The isotope values of the fertilisers range between −1.1‰ and +3.9‰ for $\delta^{15}N_{Ntotal}$ and between +21.0‰ and +24.8‰ for $\delta^{18}O_{NO3}$. They are close to the atmosphere isotope composition ($\delta^{15}N = 0$‰ and $\delta^{18}O = +23.5$‰) as they are manufactured using N_2 and O_2 from the air. An exception is the F18 fertiliser that has the typical $\delta^{18}O$ value (+48.5‰) reported for Chilean nitrate deposits (+37.8 to +50.4‰; Böhlke et al. 1997).

Compared to fertilisers, dissolved nitrate in the Maresme groundwaters has a very different isotopic composition. Their $\delta^{18}O_{NO3}$ (+5.1‰ and +10.2‰) and $\delta^{15}N_{NO3}$ (+6.8‰ and +9.4‰) values are lower and higher than the fertilisers values, respectively. This pattern must be related to the reactions that affect fertiliser nitrogen compounds when they are applied over the fields. Depending on the ambient conditions (pH, dissolved oxygen, microbial activity) these compounds are affected by processes and reactions that can change their isotopic compositions. The higher $\delta^{15}N$ values of

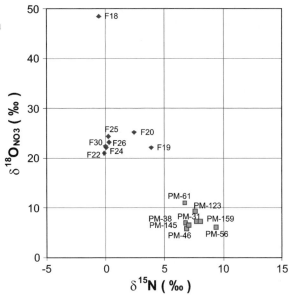

Fig. 5.6.
$\delta^{18}O_{NO3}$ vs. $\delta^{15}N_{NO3}$ and $\delta^{15}N_{Ntotal}$ from the groundwater samples and fertilisers, respectively.
PM: groundwater, F: fertilizer

the groundwater can be related to those fertilisers containing ammonium in their chemical composition. During irrigation, ammonium-bearing fertilisers are affected by the hydrolysis of the urea and consequently ammonia is lost through volatilisation in the unsaturated zone. During this reaction:

$$CO(NH_2)_2 + H_2O \longrightarrow CO_2 + 2NH_3 \begin{smallmatrix}\nearrow NH_3\uparrow \\ \longleftrightarrow NH_4^+\end{smallmatrix}$$

the residual ammonium in waters is enriched in ^{15}N compared to the fertilisers. Then, in the unsaturated zone of the aquifer, this isotope signal is transmitted to the nitrate during nitrification:

$$2NH_4^+ + 3O_2 \longrightarrow 2NO_2^- + 2H_2O + 4H^+$$

$$2NO_2^- + O_2 \longrightarrow 2NO_3^-$$

About half of the fertilisers used in this region are ammonium containing fertilisers. Nitrification is thought to be a complete reaction and to produce nitrates with a $\delta^{15}N_{NO3}$ value equal to the former ammonium molecules affected by volatilisation, justifying the nitrogen isotopic values in groundwaters. As for the $\delta^{18}O_{NO3}$ value, the $\delta^{18}O$ in groundwater is controlled by the $\delta^{18}O$ of the water and the $\delta^{18}O$ of the dissolved O_2 (i.e. Kumar et al. 1983), according to the following reaction:

$$\delta^{18}O_{NO3} = 2/3\delta^{18}O_{H2O} + 1/3\delta^{18}O_{O2}$$

The $\delta^{18}O_{H2O}$ mean value for the Maresme groundwater is −5.5‰ (Table 5.2). Assuming that the dissolved oxygen is in equilibrium with the atmospheric O_2 ($\delta^{18}O_{atm}$ = +23.5‰, Horibe et al. 1973), nitrates formed by the previous reaction would have a $\delta^{18}O_{NO3}$ value of +4.2‰. However, most of the samples have slightly higher (+5.1‰ and +6.9‰) values and two of them (PM-123 and PM-61), have values of +8.8‰ and +10.2‰, significantly higher than the theoretically expected value. These samples also show the higher nitrate concentration (Table 5.1). This pattern might be explained by a small contribution of the nitrate from the nitrogenous fertilisers, which are characterised by $\delta^{18}O_{NO3}$ values close to +23‰ (Table 5.2). Therefore, $\delta^{18}O_{NO3}$ data, showing no significant input in groundwaters of the fertilizer isotopically enriched nitrate, suggests that most of the fertiliser nitrate is recycled in the soil.

Denitrification may also affect the isotope composition of the nitrate (Delwiche 1970). This reaction, which occurs under anaerobic conditions, can be mediated by organic matter (o.m.) (Reaction 5.1) or sulphur (e.g. pyrites) oxidation (Reaction 5.2):

$$4NO_3^- + 5C + 2H_2O \longrightarrow 2N_2 + 4HCO_3^- + CO_2 \qquad (5.1)$$

$$14NO_3^- + 5FeS_2 + 4H^+ \longrightarrow 7N_2 + 10SO_4^{2-} + 5Fe^{2+} + 2H_2O \qquad (5.2)$$

During denitrification, an increase in $\delta^{15}N$ and $\delta^{18}O$ of the residual NO_3 by a relation 2:1 can be observed (Amberger and Schmidt 1987; Böttcher et al. 1990, among others).

In the Maresme area, denitrification does not seem to affect groundwaters. Firstly, as re-circulation of waters through the unsaturated zone is very important, they remain constantly in an aerobic medium. Secondly, soils in this area are poor in organic matter and granodioritic sands do not contain large enough amounts of sulphides to allow the Reaction 5.2 to take place. Thirdly, the expected isotopic trend for denitrification is not observed in the groundwater data (Fig. 5.7a,b). A slight enrichment trend in ^{15}N is observed as nitrate concentrations decrease in Fig. 5.7a. However, the positive correlation between $\delta^{18}O_{NO3}$ and NO_3^- concentration (Fig. 5.7b) discards the possibility of denitrification and represents the mixing of nitrates that are being nitrified from ammoniacal fertilisers (low $\delta^{18}O_{NO3}$ values) and nitrates from nitrogenous fertilisers enriched in ^{18}O. Samples PM-61 and PM-123 that are the most ^{18}O enriched samples and have the higher concentrations of nitrates seem to have a higher influence from nitrogenous fertilisers, which are not affected by the volatilisation and nitrification reactions.

The carbon isotope data of dissolved inorganic carbon may provide information about carbon cycling that could be related to nitrogen cycling. Data obtained in the study area show a negative correlation between $\delta^{13}C_{HCO3}$ and HCO_3^- concentration (Fig. 5.7c). This pattern can be explained by denitrification as shown by Reaction 5.1 (which has already been discarded) or by the re-circulation of water in the unsaturated zone of the aquifer. In agricultural soils, CO_2 concentration is usually high, with a $\delta^{13}C$ between −22‰ and −24‰ (Aravena and Suzuki 1990). Then, during recharge

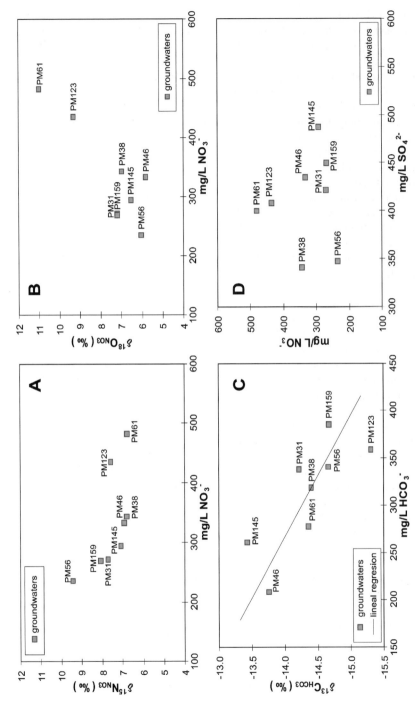

Fig. 5.7. Groundwaters in the Maresme area. **a** $\delta^{15}N_{NO3}$ vs. nitrate concentrations in mg l^{-1}; **b** $\delta^{18}O_{NO3}$ vs. nitrate concentrations in mg l^{-1}; **c** $\delta^{13}C_{HCO3}$ vs. bicarbonate concentrations in mg l^{-1}; **d** Nitrate vs. sulphate concentrations, both in mg l^{-1}

events, water dissolves the soil CO_2 which is involved in carbonate dissolution and becomes part of the DIC pool. If this process is repeated over successive irrigation, the HCO_3^- concentration increases and its isotopic composition decreases by the influence of the soil CO_2.

The groundwater in the study area is also characterised by high concentration of sulphates. The origin of this sulphate can be related to the seawater intrusion or to irrigation waters with high loads of fertilisers recharging the aquifer. The $\delta^{34}S_{SO4}$ values of groundwater, ranging from +5.8 to +7.0‰, discards a significant marine influence. Sulphate of marine origin has a $\delta^{34}S_{SO4}$ value close to +21‰ (Rees et al. 1978). Therefore, although there is no correlation between concentration of nitrates and sulphates in groundwaters (Fig. 5.7d), fertilisers could be the source of sulphates in this area. Dissolved sulphates in groundwaters have a homogenous isotopic composition while fertilisers have a wide range of $\delta^{34}S_{SO4}$ values ranging from –6.5 to +21‰ and a mean value of +6.3‰. Although this mean value is in agreement with the $\delta^{34}S_{SO4}$ of groundwaters, the concentration in sulphate of each fertiliser and the relative consumption of each fertiliser in the region should be taken into account to calculate the weighted mean isotopic value of fertilisers. Furthermore, the $\delta^{34}S$ data question the origin of the chloride previously associated to seawater intrusion.

5.5
Conclusions

The Maresme area is characterised by an intensive agricultural activity that uses groundwater for irrigation with high loads of synthetic fertilisers and results in an overexploitation of the aquifer. The high rates of fertigation, nitrate leaching and recirculation of the shallow groundwaters produces a high accumulation of nitrates and sulphates in groundwaters, up to 482 mg l^{-1} and 487 mg l^{-1}, respectively. Chemical and isotope data of these waters are used to identify and evaluate the nitrate contamination, as well as the sources of sulphate and chloride in the area.

The enriched $\delta^{15}N_{NO3}$ values (+6.8‰ and +9.4‰) obtained in the groundwater compared to the $\delta^{15}N_{Ntotal}$ values of fertilisers (–1.1‰ and +3.9‰) indicate the occurrence of ammonia volatilisation and nitrification reactions in the study site. The $\delta^{18}O_{NO3}$ values in groundwater, ranging between +5.1‰ and +10.2‰, do not show a significant contribution of the fertilisers enriched $\delta^{18}O_{NO3}$ signatures (+21.8‰ and +24.8‰, with one value of +48.5‰), except for two samples that seem to have a larger influence of nitrogenous fertilisers. The ^{18}O pattern suggests that the fertilisers nitrate component is recycled in the unsaturated zone. Denitrification reactions have not been observed due to the re-circulation of waters in oxidising environments that do not permit this process to take place.

The $\delta^{34}S_{SO4}$ of groundwaters (from +5.8 to +7.0‰) discards a significant contribution of seawater (+23.0‰) as a source of sulphate. Also, δD and $\delta^{18}O$ in groundwaters do not show an influence of the marine isotopic signature. Therefore, the high chloride concentrations also found in groundwaters could be associated to salt accumulation from the heavy use of fertilisers, thus questioning the role of seawater intrusion as a source of chloride in the area.

Acknowledgements

This project has been financed by the CICYT project HID99-0498 of Spanish Government, and partially by the 2001SGR-00073 project from the Catalonian Government. We would like to thank J. Fraile and J. A. Ginebreda from the Catalan Water Agency (Environmental Department of Catalonian Government) for their collaboration, the Serveis Cientificotècnics (University of Barcelona) and the Department of Earth Sciences of the University of Waterloo (Canada).

References

Amberger A, Schmidt HL (1987) Natürliche Isotopengehalte von Nitrat als Indikatoren für dessen Herkunft. Geochim Cosmochim Acta 51:2699–2705
Aravena R, Robertson WD (1998) Use of multiple isotope tracers to evaluate denitrification in groundwater: Study of nitrate from a large-flux septic system plume. Ground Water 36:975–981
Aravena R, Suzuki O (1990) Isotopic evolution of rivers in Northern Chile. Water Resour Res 26:2887–2895
Bayó A (1987) Acuíferos en deltas y costas lineales en Catalunya. Tecnología del Agua 32:102–112
Böttcher J, Strebel O, Voerkelius S, Schmidt HL (1990) Using isotope fractionation of nitrate-nitrogen and nitrate-oxygen for evaluation of microbial denitrification in sandy aquifer. J Hydrol 114:413–424
Custodio E, Batista E, Bayó A (1976) Intrusión marina en los acuíferos del litoral catalán. II Asamblea General de Geodesia y Geofísica: programa y resúmenes 13–17 diciembre Barcelona. Instituto Geográfico y Catastral, Barcelona, 174 pp
Delwiche CC (1970) The nitrogen cycle. Sci Am 233:136–146
Guimerà J, Marfà O, Candela L, Serrano L (1995) Nitrate leaching and strawberry production under drip irrigation management. Agric Ecosys Environ 56:121–135
Horibe Y, Shigehara K, Takakuwa Y (1973) Isotope separation factors of carbon dioxide-water system and isotopic composition of atmospheric oxygen. J Geophys Res 78:2625–2629
IAEA/WMO (2001) Global network for isotopes in precipitation. http://isohis.iaea.org
IGME (1985) Calidad y contaminación de las aguas subterráneas de España. Informe de síntesis. Instituto Geológico y Minero de España, Madrid
Kumar S, Nicholas DJD, Williams EH (1983) Definitive ^{15}N NMR evidence that waters serves as a source of O during nitrite oxidation by *Nitrobacter agilis*. FEBS Lett 152:71–74
Magee PN, Barnes JM (1956) The production of malignant hepatic tumor in the rat by feeding dimethyl-nitrosamine. Br J Cancer 10:114–122
Navarro A, Fernández A, Doblas, JG (1989) Las aguas subterráneas de España: estudio de síntesis. Instituto Tecnológico Geominero de España, Madrid
Otero N, Vitòria L, Soler A, Canals A (2004) Fertilizer characterisation: major, trace and rare earth elements. Agric Ecosys Environ, submitted
Pauwels H, Foucher JC, Kloppmann W (2000) Denitrification and mixing in a schist aquifer: influence on water chemistry isotopes. Chem Geol 168:307–324
Rees CE, Jenkins WJ, Monster J (1978) The sulphur isotopic composition of ocean water sulphate. Geochim Cosmochim Acta 42:377–381
Silva SR, Kendall C, Wilkison DH, Ziegler AC, Chang CCY, Avanzino RJ (2000) A new method for collection of nitrate from fresh water and analysis of nitrogen and oxygen isotope ratios. J Hydrol 228:22–36
Villarroya M (1986) Estudio hidrogeológico del acuífero costero del Maresme Sur. PhD. Dissertation, Facultat de Geologia, Universitat de Barcelona, 226 pp
Vitòria L, Otero N, Soler A, Canals A (2004) Fertilizer characterisation: isotopic data (N, S, O, C and Sr). Environ Sci Technol 38:3254–3262
Wassenaar LI (1995) Evaluation of the origin and fate of nitrate in Abbotsford aquifer using the isotopes of ^{15}N and ^{18}O in NO_3^-. Appl Geochem 10:391–405

Chapter 6

^{2}H and ^{18}O Isotopic Study of Ground Waters under a Semi-Arid Climate

L. Bouchaou · Y. Hsissou · M. Krimissa · S. Krimissa · J. Mudry

Abstract

This paper summarizes the application of isotope hydrological tools to infer water sources in different parts of the Souss-Massa region of Morocco. The oxygen-18 (^{18}O) isotopic data show a variation between −7‰ upstream and −4‰ downstream, with intermediate values in the medium part of the plain. The upstream watershed, which is the place of condensation and the beginning of the Atlas Mountain, shows more characteristic ^{2}H and ^{18}O-depleted waters. This finding can be explained by the altitude and the continental effects. On the other hand, ^{2}H and ^{18}O-enriched waters values towards the ocean show an evaporation effect near the condensation source or the irrigation returns, notably in the irrigated perimeters. The rain isotope values indicate a main recharge from the Atlasic Mountain, whereas the contribution of the local rains is negligible in downstream. The ^{2}H-^{18}O relationship displays straight lines with variable slopes on an upstream-downstream movement. The slopes, which are below 8 in certain areas, represent the evaporation during the infiltration either by runoff or by irrigation returns. Besides, the different values of slopes correspond to the variables isotopic values observed at a regional scale within the basin.

Key words: environmental isotope, groundwater, semi-arid climate, recharge, Morocco

6.1
Introduction

Isotope hydrological methods are increasingly used in the investigations of groundwater resources and are particularly useful tools for groundwater studies in semi-arid and arid regions (Wolfgang and Mebus 1999). Environmental stable isotopes are commonly used in regional groundwater studies to identify flow regimes and recharge sources (Fritz et al. 1979; Yurtsever and Payne 1979). Generally, groundwater preserves its stable isotopic signature (^{18}O/^{16}O, ^{2}H/^{1}H) unless diluted or mixed with waters of a different isotopic composition. Therefore, the determination of water stable isotopic composition can enable the identification of groundwater recharge from isotopically distinct and well characterized water sources, such as precipitation and surface water bodies (Darling and Bath 1988). The seasonal variations of the isotopic composition of precipitations are related to three main factors: temperature, evaporation, and change of preferential directions of the air masses (Craig 1961; Rozanski et al. 1993). In the Souss-Massa region of Morocco, several hydrogeological studies have been carried out in order to assess the behavior of the aquifer (Hsissou et al. 1997, 1999; Boutaleb et al. 2000). Actually the management and quality conservation of groundwater resources are of major concern in all Moroccan regions.

The aim of this paper is to report the use of ^{2}H and ^{18}O isotope analysis to infer water sources in the Souss-Massa basin. Indeed, the application of isotopic techniques should allow for a better understanding of hydrogeological processes taking place in the recharge area; to monitor the evolution of the signal through the aquifer system; and to test such tracers under an arid climate, with a variable topography and where classic hydrogeology is limited by the lack of systematic data. In this study, oxygen (^{18}O) and hydrogen (^{2}H) were used to understand the various contributions to groundwater within the Souss-Massa basin, which is characterized by a semi-arid climate.

6.2
Experimental

6.2.1
The Souss-Massa Basin

The Souss-Massa basin is a major watershed of Morocco (Fig. 6.1). Its resources are mainly agriculture, tourism and sea fishing. These various activities require important water resources. The area is characterized by a semi-arid climate and by a marked seasonal contrast. The main water resource is provided by the Souss-Massa Plio-Quaternary plain aquifer and by the reservoirs. The aridity of the climate, the over exploitation and the deterioration of the water quality, induce serious problems for a

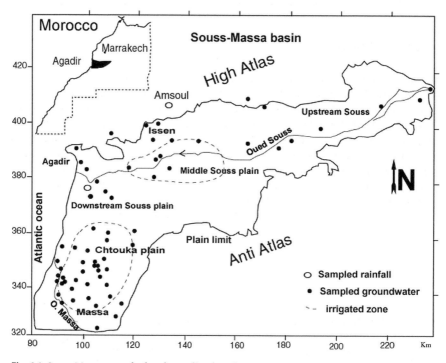

Fig. 6.1. Souss-Massa watershed and sampling location

sustainable water management. The sand and gravel aquifer, which was previously mainly exploited for irrigation, is becoming a source of increasing importance for the domestic supply of the Souss-Massa region. The studied aquifer is localized between two mountains, the High-Atlas to the north and the Anti-Atlas to the south, and it is limited by the Atlantic ocean to the west (Fig. 6.1). The rainfall amount is variable and irregular from year to year. The average amounts to 250 mm yr^{-1} in the plain area and 500 mm yr^{-1} in the mountain. Every river of the region, called "oued", has a temporary flow regime, because the drought period is very long (6 to 8 months) every year. The aquifer flows from the east to the west, towards the sea. The water quality is very variable and in some areas, it presents a high salinity exceeding 4 g l^{-1} (Hsissou et al. 1999; Boutaleb et al. 2000).

6.2.2
Analysis

Samples were collected from wells, reservoirs and springs (Fig. 6.1). Some rain samples were collected during the precipitation time. All samples were analysed at the Isotope Hydrology and Geochemistry Laboratory at the University of Paris-Sud (France). Oxygen and deuterium isotope data are expressed in standard delta (δ) notation in per mil, relative to the Standard Mean Ocean Water (SMOW). The $\delta^{18}O$ of the water samples was determined by the CO_2-H_2O equilibration method. The deuterium values of the water samples were obtained by reduction to H_2 over hot metallic zinc.

6.3
Results and Discussion

We studied the water bodies of the Souss-Massa watershed, using stable isotopes. To obtain an idea about the origin of water and the behavior of the isotopic signal in the basin, we used the relationship between delta ^{18}O and delta ^2H, and compared it with the line defined by Craig (1961) as follows: delta $^2H = 8\delta^{18}O + d$, where d is the so-called 'deuterium excess', usually around 10 per mil. This straight line, with a slope 8 for continental precipitation, is called Meteoric World Line (MWL). In spite of the scarcity of the measurements of the rain delta in the Souss-Massa area, it could be noted, that the few more or less dispersed values, show enriched values in the plain, and relatively impoverished values towards the mountain (–7.5‰ in the Amsoul weather center, Fig. 6.1).

The $\delta^{18}O$ and δ^2H values of water from wells, surface water and springs sampled in this study are plotted and compared with this Meteoric World Line in the following sections.

6.3.1
Whole Watershed

All the isotopic data in the Souss-Massa watershed are regrouped and plotted in Fig. 6.2. The values of the oxygen-18 contents vary respectively between –2.3 and –7.64‰, and deuterium between –13.6 to –51.6‰. This displays well a high variability of the isotopic input in the area. Figure 6.2 shows a close tendency between the data of Souss-

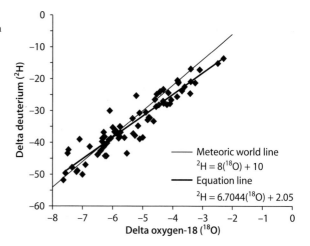

Fig. 6.2.
Water isotopic compositions in the Souss-Massa watershed

Massa and the World Meteoric Line, but a different correlation from the World Meteoric Line can be observed. The slope of the experimental line (6.7) is lower than the one of World Meteoric Line (8), which seems to demonstrate water evaporated from the aquifer system. In order to understand this difference and determine excess vs. the general correlation, we will examine separately the different areas of the Souss-Massa watershed.

The altitude gradient values which are given in literature vary between 0.15 with 0.5/100 m (Yurtsever and Gat 1981). Some computed values on the Atlas range from 0.14 to 0.26/100 m. Without a monitoring of stable isotopes in the precipitations of the region, an average altitude gradient of $0.26\delta^{18}O/100$ m was used. It was obtained in the Beni Mellal Atlas (Bouchaou et al. 1995). In groundwater, the oxygen-18 and deuterium values are related to the precipitation values in the Atlas Mountain (>700 m above sea level (a.s.l.)). The apparently more atlantic isotope tracing in many groundwater samples from wells is attributed to the contrasted change of the local climate between plain and mountain, but may also be attributed to an increasing evaporative isotope enrichment from irrigation return water. We notice a very clear differentiation between the impoverished rains of the rain gages −4‰ in the Agadir station (50 m a.s.l) and higher than −7‰ in the mountain station (>700 m a.s.l.).

6.3.2
Upstream Natural Behavior

The values of the upstream Souss are the most negative (Fig. 6.3). They are gathered around an average $\delta^{18}O$ of −7‰ which characterizes the isotopic signal of High Atlas Mountain with a rainfall value of −7‰ in Amsoul rain gage (Fig. 6.1). This impoverished signal reaches the aquifer quickly in this part of the area, according to the fractures in crystalline rocks and the high hydraulic conductivity in the alluvial materials. Thus, this upstream signal, near the recharge area, is well characterized by the most negative values which constitutes the inflow of the downstream aquifer.

Fig. 6.3.
Water isotopic compositions in the Souss upstream basin

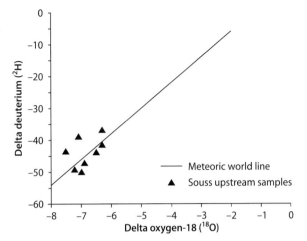

Fig. 6.4.
Water isotopic compositions in the Issen watershed

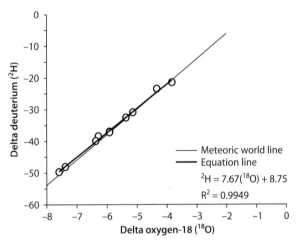

The same isotopic signal is observed in the Issen basin (Fig. 6.4) which belongs to the Atlas Mountain. The values are a perfect fitting on the Meteoric World Line with a correlation of 0.99.

The upstream Souss and the Issen watershed which are characterized by an altitude above 700 m a.s.l., constitute the condensation area and the beginning of the recharge chain. They display very clustered and impoverished values which can be explained by the altitude effect and the continental environment.

During previous studies, it could be noticed that the maximum infiltration occurred, close to the rivers coming from the High Atlas and in the upstream part. Differential gauging of the Souss River between Aoulouz (upstream) and Taroudant (middle stream), give a 80% infiltration rate, after evaporation. This is explained by a very permeable conglomeratic formation in this part of the river (Dijon 1969).

6.3.3
Souss Middle and Downstream Part

The delta oxygen-18 values in this part are distributed all along the Meteoric Water Line between -4 and -8‰ (Fig. 6.5). These reflect the measured isotopic signal in the rainfall at the Agadir station in the plain (-4‰) and the signal of the Atlas Mountains (>-7‰). The majority of samples present the values more impoverished than the rainfall in the plain. This indicates that the major contribution recharge by the local rains is negligible toward the Atlantic Ocean. The impoverished values in several wells of the downstream aquifer can be explained by their exploitation of very deep aquifers, whose recharge area is located towards the Atlas or their situation on privileged flow axis (Zaghloule et al. 1998; Hsissou et al. 1999). They are specially located within the limestone which occupy a considerable part of the Souss-Massa region. Groundwater from deep boreholes has prevailingly ^{18}O values of -7‰.

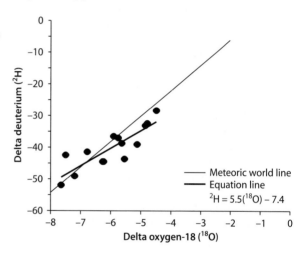

Fig. 6.5.
Water isotopic compositions in the middle and downstream Souss watershed

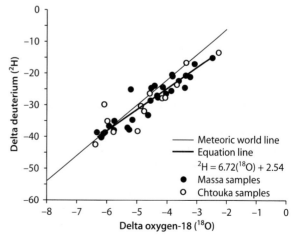

Fig. 6.6.
Water isotopic compositions in the Chtouka-Massa watershed

Tentatively, the deep groundwater is recharged by a relatively fast subsurface inflow from the high Atlas Mountain. The enriched values towards the ocean (outlet of the aquifer) result from an evaporation effect near the condensation source and irrigation returns, specially in the irrigated perimeter of Ouled Teima.

6.3.4
Chtouka and Massa Region

This area is an important irrigated perimeter. The values of the Chtouka-Massa are distributed between −6 and −2‰ (Fig. 6.6). Many samples are plotting below the Meteoric World Line which indicates an evaporation trend. In this sector the less impoverished values of ^{18}O is around −6‰ comparing to the Souss sector where the lowest value is −8‰. This enrichment toward the south is due to the decrease of rainfall on the Anti-Atlas Mountain (250 mm yr^{-1}) which is less watered than the High Atlas (500 mm yr^{-1}). It can be explained by the depletion of the altitude in the Anti Atlas compared to the High Atlas. This shows a climate aridity toward the south of the Souss-Massa watershed. The evaporated water is due to the presence of surface water (Massa River), the arid climate and the irrigation returns because in this part, we have an wide perimeter irrigated both by ground and surface waters.

The two sectors give the same variation. Generally, the slope which is slightly < 8 in certain cases, represents an evaporation during local infiltration, the streaming or the irrigation returns. Besides, the different values of slopes correspond to the variables observed at a regional scale.

It is also noted that in the different areas, the lines show variable slopes, which are the effect of streaming on the isotopic content contribution to the aquifer; they can depend on the flow velocity and the infiltration rate. The High Atlas range is relatively well-watered, compared to the remainder of the area. Therefore, the significance of the slopes quickly enables precipitation with an impoverished content of isotope. This impoverished input can reach the aquifer, without a notable evaporation phenomenon. This phenomenon is noticed particularly in the upstream part and in the vicinity of the rivers.

6.4
Conclusion

The isotope technique enabled us to determine the recharge mechanisms of the Souss-Massa aquifer. It shows how the Souss-Massa shallow aquifer is highly influenced by the contribution of the Atlas Mountain which has a high rainfall, particularly in its upstream part. The stable isotope composition of groundwater reflects a recharge from the Atlantic precipitation on the Atlas Mountain, with a deuterium excess higher than 10‰. The increase of isotope values from the east to the west of the region is attributed to the altitude and the continental effect. The regional distribution of the used stable isotopes may therefore be controlled by an isotopic enrichment due to the partial evaporation of water and the arid climate, indicated by a relatively low deuterium excess value. The effect of an admixture of irrigation return flow is noticed in the irrigated perimeter (middle Souss plain, Chtouka and Massa part) and during flood irrigation, water is submitted to evaporation and becomes isotopically enriched.

Acknowledgements

This work is supported by of the Moroccan Research Department (PARS n° SDU 54 project). We are grateful for the help by Isotope Hydrology and Geochemistry Laboratory (LHGI), Faculty of Sciences of Orsay (France) and CNESTEN of Rabat (Morocco) for the isotope analyses. Furthermore, we owe a special word of thanks to Mrs Sophie Maraux for the English version. We also would also like to thank Dr. J. L. Michelot for his help at various stages. Finally, we thank the reviewers for their constructive comments.

References

Bouchaou L, Michelot JL, Chauve P, Mania J, Mudry J (1995) Apports des isotopes stables a l'étude des modalités d'alimentation des aquifères du Tadla (Maroc) sous climat semi-aride. CR Acad Sci Paris, tome 320, No. 2a, pp 95–101

Boutaleb S, Bouchaou L, Mudry J, Hsissou Y, Mania J, Chauve P (2000) Hydrogeologic effects on the quality of water in the oued Issen watershed (Western Upper Atlas Mountains, Morocco). Hydrogeol J 8:230–238

Craig H (1961) Standard for reporting concentrations of deuterium and oxygen-18. Naturel Water Science 133:1833–1834

Darling WG, Bath AH (1988) Stable isotope study of recharge processes in the English chalk. J Hydrol 101:31–46

Dijon R (1969) Etude hydrogéologique et inventaire des ressources en eau de la vallée du Souss. Notes et Mémoires du Service Géologique. Maroc, No. 214

Fritz P, Hennings CS, Suzuki O, Salati E (1979) Isotope hydrology in Northern Chile. Isotope Hydrology 1978, Proceedings series. IAEA, Vienna, pp 525–544

Hsissou Y, Mudry J, Mania J, Bouchaou L, Chauve P (1997) Dynamique et salinité de la nappe côtière d'Agadir (Maroc), influence du biseau salé et des faciès évaporitiques. IAHS Publication 244:73–82

Hsissou Y, Mudry J, Mania J, Bouchaou, L, Chauve P (1999) Utilisation du rapport Br/Cl pour déterminer l'origine de la salinité des eaux souterraines: exemple de la plaine du Souss (Maroc). CR Acad Sci Paris 328:381–386

Rozanski K, Araguas-Araguas L, Gonfiantini R (1993) Isotopic patterns in modern global precipitation. In: Swart et al. (eds) Edition climate change in continental isotopic records. Am Geophys Univ monograph 78:1–36

Wolfgang, W, Mebus A (1999) Application of environmental isotope methods for groundwater studies in the ESCWA region. Geologisches Jahrbuch, Hanover 1999, 129 p

Yurtsever Y, Gat JR (1981) Atmospheric waters: in stable isotope hydrology. Technical reports series No. 210, AIEA, Vienna, pp 103–139

Yurtsever, Y, Payne BR (1979) Application of environmental isotopes to groundwater investigations in Qatar. In: Isotope Hydrology 1978, proceedings series, IAEA, Vienna, pp 465–490

Zaghloule Y, Lahrach A, Ben Aabidat L, Bouri S, Ben Dhia H, Khattach D (1998) Anomalies géothermiques de surface et hydrodynamisme dans le bassin d'Agadir (Maroc). JAES 27(1):71–85

Chapter 7

$^{13}C/^{12}C$ Ratio in Peat Cores: Record of Past Climates

G. Skrzypek · M.-O. Jedrysek

Abstract

Three carbon isotope profiles, from the raised Polish peat bogs Zieleniec, Szrenica, and Suche Bagno, representing the last millennium, have been analysed. $\delta^{13}C$ in the peat profiles varies from −31 to −22‰. $\delta^{13}C$ changes were found similar in the horizons of various cores. We suggest that variation in $\delta^{13}C$ of peat is dominantly governed by variations in temperature of vegetation period of *Sphagnum* composing given strata. It is also shown, that an increase of 1 °C of the vegetation temperature results in the of about −0.6‰ of $\delta^{13}C$. Based on $\delta^{13}C$ isotope calibration, the following sequence in the climate variations between AD 600 and 1950 in Poland is proposed: a cold period from AD 600 to 1050, a very cold period from AD 1050 to 1200, a very warm period corresponding to the "Little Climatic Optimum" (Matthes 1939) from AD 1200 to 1550, a very cold period corresponding to the "Little Ice Age" from AD 1550 to 1820 and a warm moderate period corresponding to the "Global Climatic Warming" from AD 1830 to 1960.

Key words: peat, carbon isotopes, climate, temperature

7.1 Introduction

The time period AD 1550–1850, generally known as the Little Ice Age, is the most marked cold period of the last 1 000 years. Isotope analysis is a valuable tool for paleoenvironmental reconstruction, but until now the Little Ice Age has been poorly documented by $\delta^{13}C$ analysis in peat profiles. Modern geochemical investigations of peat profiles may bring some light on this topic, which is crucial in paleoclimatological research. For example, they may answer the timely questions: were the climatic variations the result of natural processes only, and when and where did the cold and warm climates occur during the last millennium? In this paper we attempt to calibrate carbon isotopic variations in peat and speculate how far the $\delta^{13}C$ isotope values reflect paleoclimatic conditions (O'Leary 1981).

Previous environmental studies of peat based on carbon isotopic composition have been done by several authors. White et al. (1994) suggest that the $^{13}C/^{12}C$ ratios of total peat carbon ($\delta^{13}C(CO_2)_{atm}$) respond to variations in atmospheric CO_2 concentration. They compared their results, from Southern America peat profiles, to concentration of CO_2 in air inclusion in ice cores from Antarctica. The result of theses studies is a complex mathematical model based on a comparison of two very distant regions with extremely different conditions. In contrast to polar regions, where global factors control atmospheric CO_2 concentration and $\delta^{13}C(CO_2)_{atm}$, in peat bogs local factors (e.g. diurnal and seasonal cycling in assimilation/respiration (Szaran 1990) or methanogenesis/oxidation of methane seems to be crucial (Jedrysek 1995, 1999; Jedrysek et al. 1995). Other peat profiles show a good dependence of the δ^{13}C of the cellulose and total organic matter on metabolic pathways (C3, C4) of several plant species (Aucour et al. 1994). These authors obtained similar δ^{13}C values (-25.5 ±2.3‰) as compared to ours δ^{13}C (-26.12 ±3.54‰, exclusively C3 peat), however Aucour et al. (1994) studies were preformed on 3 relatively old profiles of 40 000 years (75 samples), 30 000 years (40 samples) 40 000 years (80 samples). Thus, those low resolution results cannot be compared our one millennium profile. Likewise, Sukumar et al. (1993) showed a good correlation between the δ^{13}C value of peat and the humidity of air. However, this was done on 22 samples only covering 10 thousand years old peat core (i.e. ca. 450 years per sample) – this is incomparable to our relatively high-resolution studies (ca. 50 years per sample). Likewise, Brenninkmeijer et al. (1982) and Dupont and Brenninkmeijer (1984), studied variation of δ^{13}C, δ^{18}O and δD in peat profiles. They suggest that the most important factor which determines the carbon isotopic composition of peat is the water level. This is in agreement with our sulphur and hydrogen isotope studies but contradicts our carbon isotope evaluations (Jedrysek et al. 1995; Skrzypek 1999, and results presented in this paper).

The δ^{13}C composition of organic matter (vegetation) is governed by a few factors: the isotopic composition of CO_2 assimilated, the photosynthetic pathway (C3, C4, CAM), temperature, species, salinity, light intensity, humidity (O'Leary 1981; Hemming et al. 1998). Many of these factors can be neglected here as each peat bog studied shows negligible biodiversity, (our calculation of isotopic temperature effect is -0.6‰ °C^{-1}) similar light intensity, hydrological regime, etc. (Skrzypek and Jedrysek 2001). Also, differences in temperature of vegetation and differences in atmospheric pressure in the two regions studied were probably similar. Numerous calibrations of δ^{13}C in plants and temperature, resulted in very contrasting values, i.e. the change in isotope value to change in temperature of vegetation varied from -1.2‰ °C^{-1} to +0.33‰ °C^{-1} (Smith et al. 1973; Whelan 1973; Lipp et al. 1991; Troughton and Card 1975). Therefore, we calculate in this paper, the isotope temperature effect, on the basis of two selected peat profiles.

7.2
Experimental

Carbon stable isotope analyses have been carried out in three Polish peat bogs (Fig. 7.1.). These are "Zieleniec" and "Szrenica" in the Sudety Mountains (Southwest Poland) and Suche Bagno in the Wigry Lake District (Northeast Poland). All were the raised type of peat bogs, with dominance of *Sphagnum* species, the *Ombro-Sphagnioni* type (Tolpa et al. 1967). This has been documented due to plant detritus analysis and hydrogeological assessment (Skrzypek 1999; Tolpa et al. 1967; Bajkiewicz-Grabowska 1992). The peat bogs exist under different climatic conditions. Zieleniec peat bog is on a wooded mountain ridge at about 800 m above sea level (a.s.l.), the mean summer temperature is 14.7 °C and rainfall is about 1 300 mm yr^{-1}. Szrenica peat bog is located on an exposed mountain pass above the timberline at 1 249 m a.s.l., the mean summer temperature is 12.8 °C and rainfall averages about 1 300 mm yr^{-1}. Suche Bagno peat bog is located in wooded lowland at about 140 m a.s.l. Summer average temperature is 15.4 °C

Chapter 7 · $^{13}C/^{12}C$ Ratio in Peat Cores: Record of Past Climates

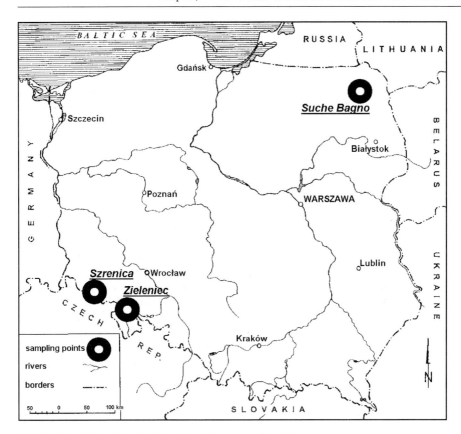

Fig. 7.1. Map of Poland: localisation of sampling areas of peat bog

and rainfall 600 mm yr^{-1} (Wiszniewski 1973; Bajkiewicz-Grabowska 1992). The distance between the mountainous and lake land peat bogs is about 600 km. Three cores from the Suche Bagno peat bog, one from Zieleniec and one from Szrenica were collected. The uninterrupted vertical peat cores, from 60–153 cm long, cover approximately the last two millennia. However, special attention is paid to the last millennium only in this paper. Samples have been collected digging with the spade (the layer up to 0.5 m) or by drilling with *Eijkelkamp Agrisearch Equipment* (Holland). These were divided into 3 to 5 cm thick intervals and ^{14}C dating was used to correlate corresponding profiles. The gas counting method has been used and $^{13}C/^{12}C$ correction have been done. Following dates have been obtained: *(i)* Suche Bagno S1: 680 BP (86 cm depth), 2 050 BP (138 cm) and *(ii)* Suche Bagno S2: 190 BP (44 cm depth), 2 100 BP (95 cm depth) and *(iii)* Suche Bagno S3: 1 170 BP (112 cm) and *(iv)* Zieleniec: 260 BP (21.5 cm depth), 430 BP (27.5 cm depth) and *(v)* Szrenica 1 110 BP (135.5 cm depth) (Skrzypek 1999). The approximate age of the other undated samples have been interpolated from the growing rate (distance in centimetres between two dated samples in the profile divided by the time span between the respective dated samples). The results of our interpolation are in agreement with palynological studies (Skrzypek 1999).

The samples were stored, immediately after collection, at −20 °C. Isotope preparation started with defrosting, then external organic material (for example roots of macrophytes) were removed. The residual material was washed and homogenised in 4% HCl (5 h), dried under vacuum and milled in a ceramic mortar. Afterwards, about 10 mg of material from each sample was combusted with CuO wire, in a quartz tube sealed under vacuum at 900 °C. The CO_2 gas was then introduced to a mass spectrometer and the carbon stable isotope ratio ($\delta^{13}C$) was measured. Values are quoted relative to the Pee Dee Belemnite, with a precision of ±0.05‰.

7.3
Results and Discussion

Each $\delta^{13}C$ value, in this paper, represents 3 to 5 cm thick peat horizon in uninterrupted peat profile. Three peat profiles analysed have been correlated using ^{14}C dating. All $\delta^{13}C$ data obtained, are presented in Fig. 7.2. The different distances between points on the x-axis (time scale) on the same profile are the result of different growth rates of peat.

All the $\delta^{13}C$ profiles show generally similar trends (Fig. 7.2). However, the mountainous cores from Szrenica and Zieleniec show ^{13}C enrichment, and lake land cores from Suche Bagno (Suche B.1, Suche B.2, Suche B.3) show general ^{13}C depletion. We believe that this is dominantly due to lower temperatures on mountainous Szrenica Mt. (higher altitudes, 1 249 m a.s.l.) and Zieleniec (lower altitudes, 800 m a.s.l.) peat bogs, as compared to lowland Suche Bagno peat bogs (140 m a.s.l.). Despite these differences, all the profiles show the Little Ice Age as elevated $\delta^{13}C$ values (Fig. 7.2). Despite this each sample represents about 3–4 cm interval in the profile, they cover different age spans. This is due to the different growing rate at the respective peat bogs.

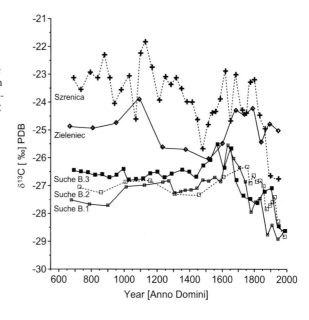

Fig. 7.2.
$\delta^{13}C$ values (raw data) of the organic matter in Polish peat bogs. Three profiles one peat-bog (Suche B1, B2 and B3), are more similar to each other than the Zieleniec and Szrenica profiles (see Fig. 7.1). Higher $\delta^{13}C$ values correspond to lower temperatures of the vegetation period

7.3.1
Calculations of the Isotopic Temperature Effect (‰ / 1 °C)

In this paper we have introduced new parameters (Fq) representing the ratio of isotope differences between two samples, to differences in temperature of vegetation period. Based on 10 years measurement of the temperature of vegetation period, one may calculate that the difference in temperature of the growing period between the Szrenica and Zieleniec is −1.9 °C. These results are based on average temperatures for June, July and August, as vegetation is remarkable during these months only. We assume also that, despite the variation in temperature, differences in temperature during growing period between the peat-bogs studied were rather similar throughout the last millennium. Thus, we have calculated the differences between average $\delta^{13}C$ values of the analysed profiles and the differences in modern temperatures of vegetation period in these regions. In the constructed equations following symbols have been used:

- T_{Ziel} – the current average temperature of summer months VI, VII, VIII calculated for 10 years (1976–1985) for Zieleniec (Bystrzyckie Mountains)
- T_{Szren} – the current average temperature of summer months VI, VII, VIII calculated for 10 years (1976–1985) for the ridge of the Karkonosze Mountains (area of Szrenica peat bog)
- $\delta^{13}C_{Ziel}$ – the average $\delta^{13}C$ value calculated for AD 1550–1850 period for cores from the Zieleniec peat bog
- $\delta^{13}C_{Szren}$ – the average $\delta^{13}C$ value calculated for the AD 1550–1850 period for core from Szrenica peat bog

The equation to calculate the current difference in temperature between Wigry and Karkonosze areas ($\Delta T_{SB-Szren}$) is as follows (see Eq. 7.1.):

$$T_{SB} - T_{Szren} = \Delta T_{SB-Szren} \quad (°C) \tag{7.1}$$

The ($\Delta T_{SB-Szren}$) corresponds to $\Delta^{13}C_{SB-Szren}$ which can be calculated due to following equation (see Eq. 7.2.):

$$\delta^{13}C_{SB} - \delta^{13}C_{Szren} = \Delta^{13}C_{SB-Szren} \quad (‰) \tag{7.2}$$

Thus, we can define the Fq value. The Fq represents the difference in $\delta^{13}C$ values corresponding to the difference in temperature of the growing seasons (‰ °C^{-1}). The difference in $\delta^{13}C$ between Suche Bagno and Szrenica cores may be shown as (see Eq. 7.3.):

$$\frac{\Delta^{13}C_{SB-Szren}}{\Delta T_{SB-Szren}} = Fq \quad (‰/°C) \tag{7.3}$$

The example of Fq value calculated has been shown in the Table 7.1. It represents the Little Ice period ca. AD 1550 to 1850. The Fq value is −0.57 for the Szrenia-Zieleniec pair. It means that a change of 1 °C results in the change of about −0.6‰ in carbon isotope composition of *Sphagnum*.

Table 7.1. Calculation of the Fq value (‰ °C^{-1})

Profile	Unit	Szrenica	Zieleniec	Difference		$\Delta^{13}C/\Delta T$ [‰/°C]
				$\Delta^{13}C$	ΔT	
Carbon isotope composition $\delta^{13}C$	[‰]	−23.88	−24.96	−1.08		−0.57
Temperature of vegetation T	[°C]	12.8	14.7		1.90	

7.3.2
The Last 1 400 year Climatic Variations Evidence for Poland

Each sample at each profile should represent the same time period, thus similar sampling resolution would be required to make the profiles comparable. However, The Szrenica profile shows much faster accumulation of the peat, as compared to Zieleniec (Fig. 7.2). This means that the average time span represented by three samples in the Szrenica profile, corresponds to one sample from Zieleniec. To make these profiles comparable, 3 point running average filter was drawn through the Szrenica profile. Likewise the mean $\delta^{13}C$ curve of three Suche Bagno $\delta^{13}C$ profiles (Fig. 7.2) has been calculated and is shown on Fig. 7.3. The calculated Szrenica profile and Zieleniec profile show remarkable similarity. Also, the Suche Bagno calculated profile shows a degree of similarity to these previous profiles. Based on these three $\delta^{13}C$ profiles and our Fq calibration, the following model of climatic history between AD 600 and 1950 in Poland is proposed:

1. ca. AD 600–1050, cold period.
   Little variations and high negative $\delta^{13}C$ values in the Zieleniec and Suche Bagno profiles suggest that the temperature was relatively stable and low. In the Szrenica profile, significant variations in $\delta^{13}C$ values suggest remarkable temperature variations during this period on the ridge of the Karkonosze Mountains. At about AD 700 the temperature first increased, followed by a decrease and then decreased again at about AD 900. The different shape of the curves in Fig. 7.3 might result from different atmospheric circulation, especially between Southwest Poland and Northeast Poland. However, we do not have any evidence for that.
2. AD 1050–1200, very cold period, Pessima XI–XII Centuries (Ps XI–XII).
   Between AD 1050 and 1200 the temperature in the Sudety Mountains (cores Szrenica and Zieleniec) was probably the lowest of the last 1 000 years as the $\delta^{13}C$ value shows its maximum, higher than during the "Little Ice Age". The maximum $\delta^{13}C$ value in the Szrenica profile corresponds to the maximum in the Zieleniec profile and a slight pessimum is also recorded in the Suche Bagno profiles. The difference between the period described and the next one "Little Climatic Optimum", the $\delta^{13}C$ values in the Szrenica and Zieleniec profiles is about 1‰. It can be considered to a decrease of about 2 °C in the average annual vegetation period temperature.
   In Northeast Poland this cold period was not so clearly visible, the change of $\delta^{13}C$ was below 0.5‰. This period can be compared to the "Little Climatic Optimum" in the English lowlands (Lamb 1977, 1985; Barber 1981). This suggests different atmospheric circulation patterns in the Northwest and Southwest part of Poland.

Fig. 7.3.
Calculated plots: Szrenica: three point running average filter; Zieleniec: raw data, Suche B: mean curve calculated from three profiles (Suche B.1, Suche B.2, Suche B.3)

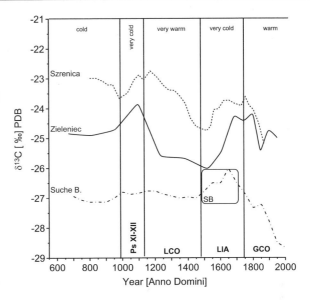

3. AD 1200–1550, very warm period, Little Climatic Optimum (LCO).
   The $\delta^{13}C$ decrease in the Szrenica and Zieleniec profiles is about 2‰. This may correspond to an increase in the temperature of the growing season by about 3.5 °C. This range and timing climatic optimum is generally in agreement with Lamb (1985) and Stachlewski (1978). However, the time between AD 1200–1550 was a relatively warm period in Poland. This means that the cold period described by Barber (1981) between AD 1320–1500 in England did not occur in Poland.

4. AD 1550–1830 very cold period, Little Ice Age (LIA).
   The cold period of "Little Ice Age" is very clear in all the peat cores analysed. The 1‰ (Szrenica and Suche Bagno) and 1.8‰ (Zieleniec) increase in the $\delta^{13}C$ value is rather rapid, and it can be regarded as a decrease in the growing season temperature of about 2–3 °C in the Sudety and Wigry areas. However, the shape of the $\delta^{13}C$ curves are different for both areas. In the case of the Szrenica profile especially, several short colder and warmer periods have been observed (seen clearly on Fig. 7.2). Moreover, it seems that took place at different time i.e.: AD 1550–1750 in Suche Bagno (rectangle SB Fig. 7.3.) and AD 1550–1830 in Sudety (Fig. 7.3). Very clear expression of it in both areas suggests that the atmospheric circulation during that time, was uniform in the larger scale. Lamb (1985) suggested that it started in England about AD 1400. This contradicts the results obtained for Poland. However, it correlates well to the theory (1550–1830) presented earlier by Lamb (1977) and Barber (1981), but it is not in agreement with Lamb (1985). There is good correlation in the timing of variation in temperature in Poland and England which suggests that the atmospheric circulation in the Western and Central European can be considered as the same system during the "Little Ice Age".

5. AD 1830–1960, warm moderate period, Global Climatic Warming
   Between 1830–1960, the climate became warmer. The temperature first increased (1830–1900), then slightly decreased (1900–1930) and increased again (1930–present).

This is visible in cores from Suche Bagno (Fig. 7.2). These observations are generally in agreement with the records of the meteorological station (started at AD 1826) from Jagiellonian University (Kraków) (Trepińska 1971; Jedrysek et al. 2003).

7.4 Conclusions

The results presented show that homogeneity of isotopic composition changes within one peat bog is good. Variations in $\delta^{13}C$ value of total organic matter in peat core is governed by temperature. The increase of air temperature during the growing season of about 1 °C results in a change of $\delta^{13}C$ value of organic matter for about -0.6‰.

$\delta^{13}C$ profiles in peat bogs could be a valuable tool when reconstructing past climates. Better quantitative understanding of the potential relations between carbon isotope variations of plants and the temperature of vegetation, may result in a new means of measuring paleoclimate reconstructions.

Acknowledgements

The authors are grateful to Dr. Debs Hemming for her critical reading of the manuscript and valuable remarks. G. Skrzypek is the fellow supported by The Foundation for Polish Science (FNP) ed. VII. This study was supported from the 6 PO4D 03011, 2022/W/ING/02 1017/S/ING/02-IX grants.

References

Aucour AM, Hillaire-Marcel C, Bonnefille R (1994) Late Quaternary biomass changes from ^{13}C Measurements in a highland peatbog from Equatorial Africa (Burundi). Ternary Research 41:225

Bajkiewicz-Grabowska E (1992) Physiography and climate of the Park. In: Zdanowski B (ed) Lakes in the Wigierski National Park. Voumes of the Polish Academy of Science 3:7, (in Polish)

Barber KE (1981) Peat stratigraphy and climatic change. Balkema, Rotterdam

Brenninkmeijer CAM, Geel B, Mook WG (1982) Variations in the D/H and $^{18}O/^{16}O$ ratios in cellulose extracted from a peat bog core. Earth Planet Sci Let 61:283–290

Dupont LM, Brenninkmeijer CAM (1984) Paleobotanic and isotopic analysis of the late Sub-Boreal and early Sub-atlantic peat from Engbertsdijksveen VII, The Netherlands. Rev Paleobot Palynol 41:241

Hemming DL, Switsur VR, Waterhouse JS, Heaton THE, Carter AHC (1998) Climate variation and the stable carbon isotope composition of tree ring cellulose: an intercomparison of Quercus robur, Fagus sylvatica and Pinus silvestris. Tellus 50B:25

Jedrysek MO (1995) Carbon isotope evidence for diurnal variations in methanogenesis in freshwater sediments. Geochim Cosmochim Acta 59:557

Jedrysek MO (1999) Spatial and temporal patterns in diurnal variations of carbon isotope ratio of early-diagenetic methane from freshwater sediments. Chem Geol 159:241

Jedrysek MO, Skrzypek G, Wada E, Doroszko B, Kral T, Pazdur A, Vijarnsorn P, Takai Y (1995) $\delta^{13}C$ and $\delta^{34}S$ analysis in peat profiles and global change. Przegląd Geologiczny 43:1004, (in Polish, English abstract and figures)

Jedrysek MO, Krapiec M, Skrzypek G, Kaluzny A, Halas S (2003) Air-pollution effect and Paleotemperature Scale versus $\delta^{13}C$ Records in Tree Rings and in a Peat Core (Southern Poland). Water Air Soil Poll 145(1):359

Lamb HH (1977) Climate: present, past and future, vol 2. Menthuen, London

Lamb HH (1985) Climate, history and the modern world. Menthuen, London

Lipp J, Trimborn P, Fritz H, Moser H, Becker B, Frenzel B (1991) Stable isotopes in tree ring cellulose and climatic change. Tellus 43B:322
Matthes FW (1939) Report of Commitee on Glaciers. Trans Amer Geophys Union
O'Leary MH (1981) Carbon isotope fractionation in plants. Phytochemistry 20(4):553
Skrzypek G (1999) Isotope record of environmental changes in selected Upper Holocene peat cores from Poland. PhD thesis University of Wroclaw, (in Polish, English abstract)
Skrzypek G, Jedrysek MO (2001) Conservation of organic matter in peat: $\delta^{13}C$ and δD in peat profiles from "Suche Bagno". Pol Tow Mineral Prace Spec 18:195
Smith BN, Herath HM, Chase JB (1973) Effect of growth temperature on carbon isotopic ratios in barley, pea and rape. Plant Cell Physiol 14:177
Stachlewski W (1978) The Climate. The past, present, future. PWN, Warszawa, (in Polish)
Sukumar R, Ramesh R, Pant RK, Rajagopalant G (1993) A $\delta^{13}C$ record of late Quaternary climate change from tropical peat in southern India. Nature 364:703
Szaran J (1990) $\delta^{13}C$ and CO_2 concentration in the air. In: Jedrysek MO (ed) Course-book of isotope geology. University of Wroclaw, Comm Mineral Sci Poland, p 161
Tolpa S, Janowski M, Palczynski A (1967) Genetic classification in the central European peat sediments. Volumes on Problems of Advance in Agricultural Sicences 76:9
Trepińska J (1971) The secular course of air temperature in Cracow on the basis of 140-year series of meteorogical observations (1826-1965) made at the Observatory of the Jagiellonian University. Acta Geophysica Polonica 19:277
Troughton JH, Card KA (1975) Temperature Effects on the Carbon-isotope Ratio of C3, C4 and Crasssulacean-acid-metabolism (CAM) Plants. Planta 123:185
White JWC, Ciais P, Figge RA, Kenny R, Markgraf A (1994) A high-resolution record of atmospheric CO_2 content from carbon isotopes in peat. Nature 367:153
Wiszniewski W (ed) (1973) Climatic atlas. PPWK, Warszawa

Part 1

Chapter 8

Isotopic Composition of Cd in Terrestrial Materials: New Insights from a High-Precision, Double Spike Analytical Method

C. Innocent

Abstract

A high-precision method for determining the isotopic composition of cadmium (^{110}Cd, ^{112}Cd, ^{114}Cd, ^{116}Cd) has been developed and applied to study natural and anthropogenic materials such as sedimentary rocks, soils, corals and mining waste. Preliminary results obtained on sedimentary rocks, e.g. jurassic limestone, shale and greywacke, suggest the existence of natural variations in Cd isotopic ratios. The origin of this variation is likely to result mainly from a mass-dependent isotopic fractionation process. It is suggested that variation in the isotopic composition of Cd may also occur in materials of anthropogenic origin. Very low Cd concentrations (about 10 ppbw) are reported for the clastic rocks, whereas the limestone displays a much higher concentration (75 ppbw), suggesting that Cd may be trapped during diagenesis.

Key words: cadmium; isotopic composition; heavy metals; stable isotope geochemistry; new isotopic systems

8.1 Introduction

Cadmium is a trace metal that has been extensively used as a geochemical tracer of both natural processes and anthropogenic pollution. It is one of the most extensively studied trace metals occurring in seawater, and its oceanic distribution resembles that of phosphorus (e.g. De Baar et al. 1994). As a consequence, it has been utilized since the early 1980s as a tracer of biological productivity and paleoproductivity (e.g. Boyle 1988). Cd is also known to be a toxic heavy metal that has become a major environmental and health concern. Sources for Cd release into the environment are coal, oil and wood combustion, mining activities, iron and steel manufacturing, refuse incineration, agricultural fertilizing (phosphates), and cement production (Ngriagu and Pacyna 1988). An outstanding issue is to determine whether human activity may induce isotopic fractionation of Cd occurring in natural and industrial compounds. This could indeed potentially allow to discriminate precisely the different sources of Cd pollution. Indeed, the isotopic composition of an element may constitute an excellent tracer of its source, because isotopes of a given element have very close physical and chemical properties. Isotopic ratios are thus much less variable than concentration ratios of different elements.

During the past few years, a number of metallic elements have been found to exhibit measurable isotopic variations, e.g. Fe, Cu, Zn, Mo, and Tl, owing to the increasing accuracy and precision of isotopic mass spectrometers. Cd may also display variable isotopic composition, due to its relatively low boiling point (767 °C), and also to the large mass

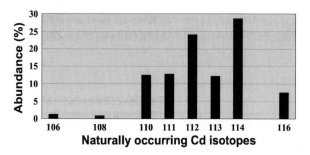

Fig. 8.1.
Cadmium isotopes. Average abundances are derived from IUPAC (1998)

range covered by its isotopes: 10 mass units, from 106 to 116 (average atomic mass 112.41 g mol^{-1}) (Fig. 8.1). Thus, isotopic fractionation processes related to the volatility of this element, such as a Rayleigh distillation mechanism, might occur. Indeed, such isotopic fractionation has been suggested to occur in some isotopically fractionated meteorites (Rosman et al. 1980; Rosman and De Laeter 1988) and in lunar soils (Sands et al. 2001). Furthermore, large Cd isotopic fractionation has been obtained experimentally by evaporating molten Cd under vacuum (Wombacher et al., unpublished data). Consequently, natural isotopic fractionation of Cd could occur, for example, during the outpouring of acidic volcanic magmas and/or the emplacement of granitoids, since the temperature of granitic magmas is close to the boiling point of Cd. On another hand, isotopic fractionation could also occur during human activities that involve a comparable range of temperatures: incineration, industrial manufacturing, for instance. However, it is worth noticing that, in contrast to its volatility, the very stable oxydation state of Cd (+2) does not ease a priori isotopic fractionation processes resulting from biological activity, although they cannot be completely ruled out. Whatever, even if man-induced variations of Cd isotopic compositions are explored, it is necessary, as a first step, to check whether natural materials display any significant Cd isotopic variations. If it is so, the systematics of Cd isotopes in natural terrestrial materials must be investigated in detail.

High variations of Cd isotopic composition have never been documented so far in terrestrial materials (Rosman and De Laeter 1975). The extent of Cd isotopic variation in natural materials would be likely rather limited compared to meteorites and lunar rocks. Clearly, difficulties to document any existing variation in the isotopic composition of terrestrial materials arise from the fact that measurements of Cd isotopes are associated to large analytical uncertainties. In this contribution, a high-precision, double spike, thermal ionisation mass spectrometry procedure for Cd isotopic analysis is described. This method constitutes a significant step forward in terms of analytical precision.

8.2
Experimental

8.2.1
Double Spike Analysis

Cd is an ultra-trace element that is typically present in terrestrial materials in amounts as low as some tens of ppbw (ng g^{-1}). Thus it is necessary to concentrate Cd prior to analysis (see below). Also, for precise ratios to be derived, a double spike procedure is required.

Thermal ionisation mass spectrometry allows very precise isotopic ratio measurements. One drawback, however, is that isotopic ratios that are measured by the mass spectrometer do not represent true ratios. Measured ratios are shifted as a result of the so-called mass fractionation bias (or instrumental bias). Light isotopes are preferentially emitted at the beginning of the analysis and/or at lower temperatures, whereas heavier isotopes are conversely emitted at the end of the run and/or at higher temperatures. Theoretical aspects of instrumental mass fractionation have been discussed in detail by Wasserburg et al. (1981). This effect that occurs during thermal ionisation may be calculated in three ways, using a linear, power or exponential law. The exponential law gives the most precise correction, but is much more complex than the two others. Thus, usually, the power or the linear law is utilized to correct for mass fractionation. The power law is:

$$A/B_n = A/B_m (1+d)^{a-b}$$

where A and B are two isotopes of a given element, of respective masses a and b. Subscripts n and m refer to the normalized and to the measured ratio. d (which can be positive or negative) is called the mass fractionation factor. In most cases, the absolute value of d is greatly inferior to 1, and the equation can be reasonably simplified to the linear law:

$$A/B_n = A/B_m [1+(a-b)d]$$

Mass fractionation effect is easily corrected for the elements in which only one radiogenic isotope may display a variable relative abundance (Sr, for example). In this case, all ratios are normalized to a stable ratio which may not (or at least is not considered to) be variable. For the elements in which there is no normalizing ratio, two alternatives are possible. The simplest way to correct for mass fractionation is to perform repeated analyses of a standard of known isotopic composition, from which an average, systematic mass fractionation effect is statistically derived. Then sample data are corrected for this derived value. However, the validity of applying such a correction procedure requires that physical and chemical parameters are strictly identical for standard and sample runs.

A far more precise correction method, though more complex to handle, relies on the use of a double spike, enriched in two isotopes A and B of the element of interest (with a typical A/B ratio close to 1). This double spike is added to a first aliquot of the sample, whereas the second one is processed without any spike addition. Both aliquots are analysed, which allows to correct accurately for mass fractionation, and also to measure precisely the concentration of the element of interest in the sample (usually the concentration is however measured first, as it is crucial to add a sufficient amount of double spike for the correction to be possible). The double-spiking theory has been described in a number of papers (e.g. Gale 1970; Hofmann 1971; Russell 1971; Hamelin et al. 1985; Galer 1999) and will not be developed here further.

In addition to the necessity of ensuring that a sufficient amount of the double spike is added to the sample for the analysis to be possible, it is also critical that the errors on the measured isotopic ratios result only from mass fractionation effect. In addition, this effect must be corrected using the linear law, as matrix calculations hold on the equation of the linear law (Hamelin et al. 1985). Fortunately, in most cases, the linear law is a good enough approximation to correct for mass fractionation (Galer 1999). Finally, it is important that the mass fractionation factor d does not undergo a large evolution during the runs (Galer 1999). In this work, a double spike enriched in ^{111}Cd and ^{116}Cd (48.53% and 47.09% of the double spike, respectively) was used.

8.2.2
Cd Standard Analyses

The Cd laboratory standard has been prepared from a Sn and In-free Cd SO$_4$ solution as Sn and In may cause mass interferences (see below), purchased from Fisher laboratory. Repeated analyses of this standard have been carried out in order to optimize Cd isotopic analysis. The standard method for Cd loading (e.g. Rosman et al. 1980; Sands et al. 2001) is similar to that of Pb isotopic analysis (Akishin et al. 1957; Cameron et al. 1969). About 150 ng of Cd are loaded on a Re filament, with 5 µg of silicagel and 5 µl of 0.25N phosphoric acid. Analyses are done on a Finnigan MAT 262, in static multicollection mode. Typical Cd ionisation temperatures range from 1 100° to 1 200 °C. Total Cd ion beams are in the order of 10^{-11} A (best runs are obtained with ^{116}Cd ion beams higher than 10^{-12} A).

Standard raw data indicate that mass fractionation is the only measurable effect that affects peaks 110, 112, 114 and 116. In contrast, anomalously decreasing ion beams are sometimes measured on peaks 111 and 113 at high temperatures when the total beam decreases itself. Comparable observations have been reported for Pb (Thirlwall 2000; Doucelance and Manhès 2001). No satisfactory explanation has been proposed yet for this phenomenon. Thirlwall hypothetized that it results from an abnormal mass fractionation behaviour. Alternatively, it could also be due to the optics of mass spectrometers, at least for Cd. Indeed, the magnitude of this effect is apparently lower on the Finnigan MAT 262 at BRGM (Orléans) than on the VG Sector 54 mass spectrometer at Geotop (Montréal). Whatever that it may be, as ^{111}Cd is one of the two isotopes of the double spike, it is critical that the ion beam does not decrease too rapidly with time during the run, in order to avoid non linear effects on the isotope 111. Thus, for Cd isotope analyses, only 5 of the 8 peaks (110, 111, 112, 114 and 116) will be taken into account. Hydrocarbons could be another potential source of interferences (at least at the beginning of a run, at low temperatures), but hydrocarbons were never dectected at typical run temperatures.

8.2.3
Sample Processing

Cd has to be isolated from other elements before analysis. Cd loads have to be very pure to avoid ion beam inhibition. In particular, Zn is found to inhibit Cd emission and should be removed before Cd runs. In addition, mass interferences on Cd isotopes may come from three elements: Pd (106, 108, 110), In (113) and Sn (112, 114, 116). Pd is not critical, since it cannot be ionized at run temperatures. In contrast, In is very efficiently ionized during Cd emission. However, because ^{113}Cd is not measured, the presence of In does not constitute a major hurdle. The main problem comes from Sn, which causes interferences on peaks of interest, and may be ionized at run temperatures.

Ideally, a total amount of about 500 ng of Cd is required for the two runs (spiked and unspiked aliquots). For silicates, this implies that gram quantities of material have to be processed. Samples are first weighed, then placed in 10 ml of water overnight, in an ultrasonic vibrator. This allows to dissolve $Cd(NO_3)_2 \cdot 4H_2O$ that would be volatilized when heating the sample. This step is skipped over if this salt is not present and/or wanted.

Next, samples are heated at 600 °C in a furnace during one hour, in order to ease the dissolution of clay minerals. At this temperature, no Cd is lost. Then, samples are placed in Savillex® closed beakers filled with a 6N HCl–7N HNO_3 mixture (10 ml each), on a hot plate, during at least 4 days. The supernatant is removed, and the residue (if existing) is dissolved in a 22.6 HF–14N HNO_3 mixture (10 ml each), with 1 ml of 12.29N $HClO_4$. The resulting solution is evaporated and the evaporated residue is redissolved in a 12N HCl–14N HNO_3 mixture (5 ml each) to which the supernatant previously collected is added. The total sample is then evaporated to dryness. Finally, the sample is placed in 6N HCl (20 ml) and separated in two aliquots (50% each approximately). Both aliquots are weighed, and one of them is spiked with the ^{111}Cd-^{116}Cd double spike. After complete evaporation, the two aliquots are redissolved in 6N HCl (10 ml) and are ready for ion-exchange separation.

Several chemical procedures for Cd extraction have been published previously (Rosman and De Laeter 1974, 1975; Rosman et al. 1980; Sands and Rosman 1997), but they are dedicated to meteorites, lunar soils, and Cd-bearing minerals. In this paper, an alternative method for Cd extraction is presented: it is more time-consuming, but allows to obtain high-purity Cd loads from terrestrial materials (silicates, soils, limestones, plants, etc…). Three successive steps are necessary.

The first step involves 12 ml-columns filled with 1X8 100-200 mesh anion-exchange resin. The resin is first washed with 7N HNO_3 and water, successively, then conditionned in 6N HCl before loading the samples. Major and most trace elements are eluted with 30 ml of 6N HCl, followed by 30 ml of 1N HCl. Cd is eluted with some trace elements, including Zn and Sn, using 30 ml of 7N HNO_3. Most of In is removed at this stage.

Cd-enriched solutions are evaporated. The evaporated residues are redissolved in 2 ml of 1N HCl and loaded on 2 ml columns filled with 50WX12, 200-400 mesh, cation-exchange resin (previously washed with 6N HCl and water, successively, then conditionned in 1N HCl). Cd is not retained, and is thus immediately recovered with 5 ml of 1N HCl. Sn is eluted with Cd, whereas Zn remains adsorbed on the column. The eluate is redissolved in 1 ml of 1N HBr.

The final step involves a microcolumn (300–400 ml) filled with 1X8 200-400 mesh resin. After washing with 7N HNO_3 and water, then conditionning in 1N HBr, the sample is loaded on the column. Cd is very efficiently retained, and remaining impurities are removed with 4 ml of 1N HBr, then with 4 ml of 1N HCl. Finally, Cd is eluted with 4 ml of 5N HF, Sn remaining adsorbed on the column.

Chemical yields average 100%. This is critical in order to avoid any hypothetic isotopic fractionation process within the column. Total blanks are typically below the nanogram level, and can be neglected.

8.3
Results and Discussion

8.3.1
Cd Concentrations

Apart from the correction of mass fractionation, the double spike method also allows a precise determination of Cd concentrations by isotope dilution (see above). It is classically considered that Cd concentration in upper crust rocks averages ca. 100 ppb (Taylor and Mc Lennan 1995; Wedepohl 1995). Recently, comparable Cd concentrations have been measured in geostandard basalts and diabases (Sands and Rosman 1997). A peridotite has been found by these authors to be strongly Cd-depleted (16 ppb), whereas they measured a higher Cd concentration in a marine mud (205 ppb; ppb refers to ppbw or ng g^{-1}).

Cd concentrations and measured, raw isotopic compositions are reported in Table 8.1. Clastic sedimentary rocks are very depleted in Cd (10 ppb). Similarly, the quaternary coral displays a low Cd concentration (7 ppb). The two carbonate sedimentary rocks (jurassic limestone and cretaceous chalk) display Cd contents higher than those of clastic rocks by one order of magnitude. The marble sample shows the highest Cd content (278 ppb) of all consolidated rocks investigated in this study, suggesting that diagenesis and metamorphism may favor Cd enrichment.

8.3.2
Raw Cd Isotopic Data

A striking feature exhibited by raw Cd isotopic data (Fig. 8.2) is that they plot on the mass fractionation straight line, as do the raw data obtained on the Cd standard (not shown). This suggests that the main process at the origin of Cd isotopic fractionation (if it occurs) in these samples should be also mass-dependent, like mass fractionation process in the spectrometer. Indeed, the relatively low boiling points of Cd (767 °C) and of some Cd coumpounds do not rule out isotopic fractionation processes resulting from a simple Rayleigh distillation law, for example during magmatogenesis. Certainly, anthropogenic processes could also fractionate Cd isotopically by such a mechanism (refuse incinerators), but it should be noticed that fractionated Cd measured in meteorites and in the moon is unlikely to originate from a Rayleigh distillation process only (Rosman and De Laeter 1988; Sands and Rosman 1997).

In addition, the raw isotopic data indicate that the use of the double spike technique provides the best mean of deciphering true isotopic variations, since they plot on or very close to the analytical mass fractionation straight line. The differences in the measured isotopic ratios may result either from the mass fractionation effect or/ and from true isotopic variations due to a natural, mass-dependent isotopic fractionation process. Moreover, it is worth noticing that recalculated isotopic ratios using standard double spike calculations should also plot on or close to this curve.

Data obtained on moon samples (Sands et al. 2001) have been reported in Fig. 8.2 for comparison. The much larger error bars suggest that terrestrial Cd isotopic hetero-

Table 8.1. Raw Cd isotopic data obtained on natural and anthropogenic terrestrial materials. Cd concentrations have been measured by isotope dilution. Geostandards TB and GSR 5 are referenced in Geostandard Newsletter XVIII Special Issue (1994)

Sample	Cd (ppb)	$^{112}Cd/^{110}Cd$	Error (2σ)	$^{114}Cd/^{110}Cd$	Error (2σ)	$^{116}Cd/^{110}Cd$	Error (2σ)
Upper Precambrian greywacke	8.08	1.922216	0.000173	2.278022	0.000251	0.591401	0.000111
Upper Precambrian shale	10.51	1.921560	0.000134	2.277084	0.000271	0.591441	0.000097
TB (Shale, former East Germany)	8.49	1.923975	0.000353	2.281646	0.000419	0.597880	0.000231
GSR 5 (Shale, P.R. China)	5.36	1.921892	0.000340	2.277221	0.000433	0.593871	0.000208
Modern coral	6.95	1.924832	0.000195	2.284785	0.000245	0.594252	0.000115
Dupl.		1.917802	0.000581	2.267645	0.000582	0.587753	0.000300
Dupl.		1.920973	0.000428	2.273665	0.000591	0.590303	0.000295
Dupl.		1.924491	0.000784	2.282676	0.001262	0.593170	0.000605
Jurassic reef limestone	74.78	1.924009	0.000843	2.281686	0.001000	0.592867	0.000490
Dupl.		1.921399	0.000110	2.276301	0.000171	0.590736	0.000060
Dupl.		1.921332	0.000250	2.276547	0.000295	0.590746	0.000129
Dupl.		1.920317	0.000164	2.273988	0.000190	0.589798	0.000064
Dupl.		1.921744	0.000075	2.277609	0.000105	0.591262	0.000040
Chalk	106.26	1.922214	0.000193	2.278170	0.000257	0.591314	0.000112
Overlying soil (openfield area)	372.53	1.916239	0.000124	2.264405	0.000214	0.586243	0.000073
Dupl.		1.915445	0.000291	2.262708	0.000653	0.585580	0.000245
Marble	278.41	1.923295	0.000105	2.280840	0.000124	0.592234	0.000063
Mining waste "PZ 10"		1.919210	0.000060	2.271720	0.000077	0.589032	0.000027
Mining waste "PZ 15"		1.920203	0.000057	2.273949	0.000099	0.589767	0.000040
Mining waste "PZ 17"		1.919756	0.000065	2.272843	0.000130	0.589367	0.000049

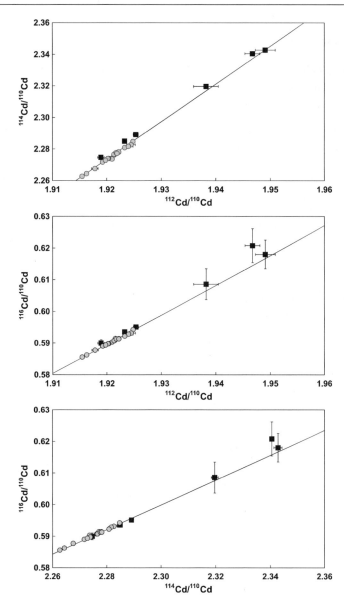

Fig. 8.2. $^{114}Cd/^{110}Cd$ vs. $^{112}Cd/^{110}Cd$, $^{116}Cd/^{110}Cd$ vs. $^{112}Cd/^{110}Cd$ and $^{116}Cd/^{110}Cd$ vs. $^{114}Cd/^{110}Cd$ diagrams, featuring raw Cd isotopic data (Table 8.1, reported as *grey circles*), and data from lunar soils (Sands et al. 2001). Raw isotopic data obtained on terrestrial materials plot on the theoretical mass fractionation straight line. See text for detailed explanations

geneities would be typically too small to be unambiguously illustrated using another technique than double spike analysis. This would explain why it has not been evidenced for earth materials in any prior study.

8.3.3
Double Spike Data

Here, 3 samples have been analysed by this technique. One of them, the jurassic limestone has been processed twice, in order to assess reproducibility of the *whole* process. The data indicate that reproducibility is good (Table 8.2, Fig. 8.3). It is in the same range than the analytical uncertainty, and lower than 100 ppm on the isotopic ratio (except for $^{116}Cd/^{110}Cd$ ratio, for which it probably results from the fact that 116 is the smallest of the 4 peaks of interest). An important fact that is worth noticing is that no evidence has been found for non linear effects on the isotope 111 for the 4 analyses.

The present dataset suggests that the recommended values for isotopic abundances of terrestrial Cd (IUPAC 1998) may be inaccurate. In addition, the true isotopic composition of the laboratory Cd standard remains unknown, as it has not been analysed yet using the double spike technique. To circumvent this problem, the means of the three double spike data points were used as reference isotopic ratios in order to express the results in delta notation. The three δCd values per a.m.u. are more or less constant for the jurassic limestone and, at a lesser degree, for the greywacke, as it can also be seen in Fig. 8.3. This is not observed for double spike data obtained on the shale sample. In addition, in contrast to the limestone and the greywacke, the shale sample tends to plot away from the mass fractionation straight line in Fig. 8.3.

Apart from radioactive disintegration, isotopic heterogeneities may result either from thermodynamic, kinetic and/or biological processes. For the limestone and the greywacke, the isotopic compositions may be explained by a mass-dependent process, possibly a simple Rayleigh distillation law. In other words, isotopic fractionation may be thermodynamically-induced. For the shale, ancient biological activity and/or kinetic effects during diagenesis could have also played a role in the isotopic fractionation. However, interference of ^{116}Sn may not be completely ruled out as an explana-

Table 8.2. Double spike isotope analyses for three natural terrestrial samples. δCd and δCd per a.m.u., calculated for each of the three isotopic ratios of interest, are also reported. See text for further explanations

Sample	$^{112}Cd/^{110}Cd$	Error (2σ)	$^{114}Cd/^{110}Cd$	Error (2σ)	$^{116}Cd/^{110}Cd$	Error (2σ)
Upper Precambrian greywacke	1.923659	0.000193	2.281443	0.000236	0.592728	0.000028
Upper Precambrian shale	1.923031	0.000157	2.280570	0.000180	0.592799	0.000074
Jurassic reef limestone	1.924275	0.000109	2.283604	0.000122	0.593598	0.000043
Dupl.	1.924363	0.000316	2.283611	0.000394	0.593526	0.000199

Sample	$\delta^{112}Cd$	(a.m.u.$^{-1}$)	$\delta^{114}Cd$	(a.m.u.$^{-1}$)	$\delta^{116}Cd$	(a.m.u.$^{-1}$)
Upper Precambrian greywacke	−0.01	0.00	−0.19	−0.05	−0.51	−0.08
Upper Precambrian shale	−0.33	−0.17	−0.57	−0.14	−0.39	−0.06
Jurassic reef limestone	0.31	0.16	0.76	0.19	0.96	0.16
Dupl.	0.36	0.18	0.76	0.19	0.84	0.14

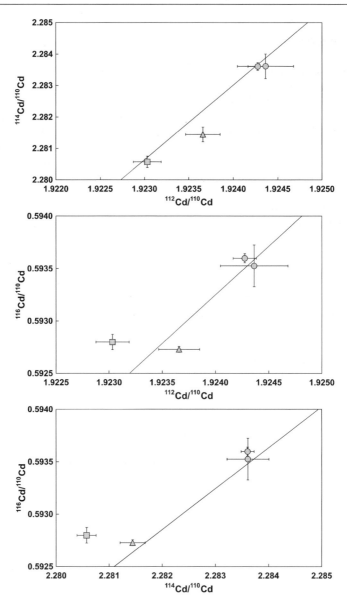

Fig. 8.3. Same Cd isotopic diagrams than in Fig. 8.2 for double spike Cd isotopic measurements (Table 8.2) on the limestone (*circles*), the shale (*square*) and the greywacke (*triangle*) investigated in this study. Data do not plot exactly on the mass fractionation straight line (Fig. 8.2), especially for the shale sample. See text for further explanations

tion for the different behaviour of the shale sample (as indicated by its ^{116}Cd/^{110}Cd ratio, Fig. 8.3), although no evidence for Sn emission was found. This potential problem will have to be investigated further in the future work.

8.3.4
Future Developments

The isotopic composition measured on the shale implies not only to check whether Sn is present before the beginning of the Cd run, at low temperature, but also to correct directly for Sn *during* the run, even if not detectable during the preliminary operations. This can easily be done using ^{118}Sn, since Sn is the only element present at this mass. However, one should keep in mind that an accurate Sn correction implies that only minute amounts of Sn (if any) are present: it is critical for the correction bias, due to mass fractionation of Sn in the mass spectrometer, to be negligible compared to the overall analytical uncertainty. Also, a ^{110}Cd-^{116}Cd double spike, displaying significant advantages over the ^{111}Cd-^{116}Cd double spike has been purchased from Oak Ridge National Laboratory. It constitutes an ideal double spike for Cd isotope analysis when peaks at masses 110, 112, 114, and 116 are measured. Indeed, in order to minimize the analytical uncertainty, the two isotopes of a double spike should be preferably the lowest peak of interest and the normalizing isotope: this is the reason why Cd ratios are normalized to ^{110}Cd, instead of ^{112}Cd or ^{114}Cd in the literature. This point is discussed in detail by Galer (1999), on the basis of Pb isotopes. Another advantage of using this new double spike is, of course, that measurement of the ^{111}Cd peak, which has sometimes been found to behave anomalously at high temperatures (see above), is avoided. Finally, efforts should aim at reducing the quantity of Cd required for a reliable measurement to the lowest possible level, in order to be able to analyse accurately samples which have very low Cd contents (such as waters), or which are available in minute quantities (forams, for example). At present, it is possible to measure precisely Cd amounts on the order of 10 ng. The challenge we are facing is to be able to perform measurements at a 1-ng level.

Once the analytical developments are achieved, the next step of this work will be to propose a preliminary survey of natural Cd isotope systematics in terrestrial materials, from the lithosphere, the biosphere and the atmosphere. The lithospheric issue will have to include analyses of mantle material (ultrabasic rocks, basalts) and of crust-derived magmatic rocks. The behaviour of Cd isotopes in the surficial and sedimentary cycle should be studied, from continental weathering to the deep ocean. Finally, diagenesis and metamorphism (s.l.) will be considered, in particular in the light of results obtained in this study, suggesting that diagenesis/metamorphism may lead to Cd enrichment.

Some authors refer to the word "anthroposphere" to describe all the interactions between the natural cycles and human activities. It is well established that the geochemical cycle of some heavy metals, such as Pb, is at present largely controlled by anthropogenic release at the earth's surface. Apart from industrial sources, agricultural fertilizers are known to cause severe enrichment of Cd in agricultural soils. It is also known that this "excess" Cd is trapped by plants (Hagemeyer and Lohrie 1995; Gawel et al. 1996), and, more generally, by the vegetable kingdom. Bacterias are now being used as Cd traps in order to remove this metal from contaminated soils (Bagot et al., unpublished data). It is likely that anthropogenic Cd is present in continental waters as well as in the ocean. Finally, Cd input due to fertilizing has been proven to be released in the atmosphere, as indicated by epiphytic lichens (Innocent et al., unpublished data). Obviously, the precise measurement of Cd isotopes in all these materials will help to quantify Cd budgets (including human-derived Cd) at the earth's surface.

8.4 Conclusion

This preliminary study suggests the existence of measurable Cd isotopic heterogeneities in terrestrial materials, as indicated by high precision data obtained by thermal ionisation mass spectrometry using the double spike techniques. Three sedimentary rocks display distinct Cd isotopic signatures considering the analytical uncertainty and the overall repeatability of the mesurements. These isotopic variations likely originate from a mass-dependent process, which would occur at temperatures close to the Cd boiling point (767 °C). This may allow Cd isotopic fractionation during natural (outpouring of acidic magmas) and human activities (incineration, industrial manufacturing). Nevertheless, further studies, as well as analytical improvements are required in order to better assess these assumptions, and also to investigate whether other processes (especially biological activity) may also trigger Cd isotopic fractionation.

Acknowledgements

Dr. C. Gariépy initiated this work during my Post-Doctoral Fellowship at GEOTOP, Montréal. Clem convinced me that Cd isotopic fractionation is possible in the earth, in spite of the fact that Cd isotopic heterogeneities had never been found yet in terrestrial materials. The first emission tests and the double spike (^{111}Cd-^{116}Cd) calibration were done at Montréal. This work owes a lot to him. Thanks are due to D. Landes, F. Lenain, and A. M. Gallas for their assistance in the mass spectrometry laboratory. Early drafts of this article benefitted from thorough readings by J. P. Girard and C. Guerrot (BRGM, ANA/ISO).

References

Akishin PA, Nikitin OT, Panchenkov GM (1957) A new effective emitter for the isotopic analysis of lead. Geokhimiya 5:425
Boyle EA (1988) Cadmium: chemical tracer of deepwater paleoceanography. Paleoceanography 3:471–489
Cameron AE, Smith DH, Waler RL (1969) Mass spectrometry of nanogram-size samples of lead. Anal Chem 41:525–526
De Baar HJW, Saager PM, Nolting RF, Van der Meer J (1994) Cadmium versus phosphate in the world ocean. Mar Chem 46:261–281
Doucelance R, Manhès G (2001) Reevaluation of precise lead isotope measurements by thermal ionization mass spectrometry: comparison with determinations by plasma source mass spectrometry. Chem Geol 176:361–377
Gale NH (1970) A solution in closed form for lead isotopic analysis using a double spike. Chem Geol 6:305–310
Galer SJG (1999) Optimal double and triple spiking for high precision lead isotopic measurement. Chem Geol 157:255–274
Gawel JE, Ahner BA, Friedland AJ, Morel FMM (1996) Role of heavy metals in forest decline indicated by phytochelatin measurements. Nature 381:64–65
Govindaraju (ed) (1994) Geostandards Newsletter XVIII Special Issue
Hagemeyer J, Lohrie K (1995) Distribution of Cd and Zn in annual xylem rings of young spruce trees [*Picea abies* (L.) Karst.] grown in contaminated soil. Trees 9:195–199
Hamelin B, Manhès G, Albarède F, Allègre CJ (1985) Precise lead isotope measurements by the double spike technique: a reconsideration. Geochim Cosmochim Acta 49:173–182

Hofmann A (1971) Fractionation corrections for mixed-isotope spikes of Sr, K and Pb. Earth Planet Sci Lett 10:397–402

IUPAC (1998) Isotopic composition of the elements 1997. Pure Appl Chem 70:217–235

Ngriagu JO, Pacyna JM (1988) Quantitative assessment of worldwide contamination of air, water and soils by trace metals. Nature 333:134–139

Rosman KJR, De Laeter JR (1974) The abundance of cadmium and zinc in meteorites. Geochim Cosmochim Acta 38:1665–1677

Rosman KJR, De Laeter JR (1975) The isotopic composition of cadmium in terrestrial minerals. Int J Mass Spectrom Ion Phys 16:385–394

Rosman KJR, De Laeter JR (1988) Cadmium mass fractionation in unequilibrated ordinary chondrites. Earth Planet Sci Lett 89:163–169

Rosman KJR, Barnes IL, Moore LJ, Gramlich JW (1980) Isotopic composition of Cd, Ca and Mg in the Brownfield chondrite. Geochem J 14:269–277

Russell RD (1971) The systematics of double spiking. J Geophys Res 76:4949–4955

Sands DG, Rosman KJR (1997) Cd, Gd and Sm concentrations in BCR-1, BHVO-1, BIR-1, DNC-1, MAG-1, PCC-1 and W-2 by isotope dilution thermal ionisation mass spectrometry. Geostand Newslett J Geostand Geoanal 21:77–83

Sands DG, Rosman KJR, De Laeter JR (2001) A preliminary study of cadmium mass fractionation in lunar soils. Earth Planet Sci Lett 186:103–111

Taylor SR, Mc Lennan SM (1995) The geochemical evolution of the continental crust. Rev Geophys 33:241–265

Thirlwall MF (2000) Inter-laboratory and other errors in Pb isotope analyses investigated using a ^{207}Pb-^{204}Pb double spike. Chem Geol 163:299–322

Wasserburg GJ, Jacobsen SB, De Paolo DJ, Mc Culloch MT, Wen T (1981) Precise determination of Sm/Nd ratios, Sm and Nd isotopic abundances in standard solutions. Geochim Cosmochim Acta 45:2311–2323

Wedepohl KH (1995) The composition of the continental crust. Geochim Cosmochim Acta 59:1217–1232

Part I

Chapter 9

Organic Petrology:
A New Tool to Study Contaminants in Soils and Sediments

B. Ligouis · S. Kleineidam · H. K. Karapanagioti · R. Kiem · P. Grathwohl · C. Niemz

Abstract

The contamination of soils and sediments by carbonaceous particles has been investigated by organic-petrological methods. Results from the study of soils from industrialised areas show that airborne contaminants like brown coal, hard coal, charcoal, and char have accumulated. The observed soil contamination is due to dust emission by open-cast brown coal mines, to the burning of brown coal and hard coal, and coking plants. It was found that soil contamination can occur over a distance of several dozen km from the contamination sources. The investigation of soils and sediments demonstrates the heterogeneous character of organic matter and proves the presence of coal and charcoal particles of fossil origin in the majority of the samples. The heterogeneity of the organic matter is found responsible for variations in the sorption behaviour of organic pollutants like polycyclic aromatic hydrocarbons in soils and sediments.

Key words: organic petrology, organic matter, carbonaceous materials, airborne contaminants, refractory C pool, coal, coke, char, sorption capacity, organic pollutants, soils, sediments

9.1
Introduction

The characterisation of soil and sediment organic matter (OM) has always been an important facet of soil and earth sciences to determine, for example, the biochemical reactivity of OM as well as the amount and type of hydrocarbons possibly generated during its burial and thermal maturation. The organic matter derives from a variety of organic precursors and debris such as planctonic biomasses, land plants, soil leaching, products of natural burning, and reworked material (Tyson 1995; Cope 1981; Jones and Chaloner 1991). Thus, the heterogeneity of OM is expected.

Routine tools for characterising OM are based upon chemical methods. These methods generally average the values derived from a complex mixture of different components and, therefore, are unable to assess the heterogeneity of the organic matter. In contrast, organic petrology, which uses optical microscopy involving incident light and fluorescence, enables the observation, identification, and quantification of the individual components, as well as the determination of the maturity of the organic matter and the reconstruction of its depositional environment (Taylor et al. 1998). This applies not only to the natural occurring heterogeneity in OM but also to airborne carbonaceous particles that accumulate in the soil compartment and altering the OM composition.

It is already known that these airborne contaminants produced by industrial activity significantly contribute to the amount of organic matter in soils (Emels 1987; Wik and Renberg 1987; Rumpel et al. 1998; Schmidt and Noack 2000; Schmidt et al. 2000;

Ghose and Majee 2000). Soils of industrialised areas are characterised by unusually high fractions of aromatic carbon due to the presence of airborne carbonaceous particles, such as coal and combustion residues (Kiem et al. 2003).

The contribution of charcoal deposited either by natural wildfires and/or atmospheric deposition to the soil OM has been clearly established by various authors (e.g. Schmidt et al. 2001; Glaser et al. 2002). All these particles are considered resistant to degradation (Schmidt and Noack 2000; Derenne and Largeau 2001) and, consequently, tend to accumulate in the refractory soil organic carbon compartment and to alter the amount and composition of the refractory C pool (Kiem et al. 2000, 2002).

In addition to a pure characterisation and quantification of the OM constituents, their interaction with organic pollutants are of broad interest for environmental questions such as transport processes (Estrella et al. 1993), remediation strategies, and clean-up efficiencies. Several studies in the past have maintained the heterogeneous nature of organic matter as being responsible for unexpected experimental results, such as high sorption capacity (K_{OC}) and strongly non-linear sorption isotherms ($1/n < 1$) in different soils and sediments (e.g. Grathwohl 1990; Weber et al. 1992; Xing and Pignatello 1997; Chiou et al. 1998; Xia and Ball 1999).

High surface area carbonaceous materials were suspected to show enhanced sorption properties (Kile et al. 1999). Recent studies have proven high sorption capacities and non-linear sorption isotherms for individual quasi-homogeneous constituents like different coals, coke and soot (Bucheli and Gustafsson 2000; Jonker and Koelmanns 2001; Kleineidam et al. 2002; Accardi-Dey and Gschwend 2002). A comprehensive review of sorption theories and sorption modelling in different geosorbents can be found in Allen-King et al. (2002).

In recent years, new areas for the application of organic petrology were developed to provide solutions to various complex problems in the field of contaminated soils and sediments (Spurny 1986; Depers and Bailey 1994; Kleineidam et al. 1999b; Karapanagioti et al. 2000). A careful microscopical examination of the composition and the fraction of organic matter had a major impact on the data interpretation obtained by chemical methods. This study summarizes earlier research results produced by the authors to show how organic petrology can be used to deal with organic contaminants in the environment. The following focuses on two basic applications:

1. to characterise airborne organic matter (contaminants) in arable soils from industrialised areas (Kiem et al. 2003)
2. to explain variations in sorption behaviour of polycyclic aromatic hydrocarbons (PAH) due to heterogenous organic matter in soils (Karapanagioti et al. 2000; Karapanagioti et al. 2001) and sediments (Kleineidam et al. 1999b; Kleineidam et al. 2002).

9.2
Materials and Methods

Different types of materials were investigated: two soils from the industrialised areas of Thyrow and Bad Lauchstädt in eastern Germany, a Canadian River alluvium with its different subsample and, finally, several lithocomponents separated from fluvial gravel deposits in SW Germany.

Thyrow lies approximately 20 km south of Berlin in the Federal state Brandenburg, while Bad Lauchstädt is located approximately 15 km southwest of Halle in Saxony-Anhalt. The soils investigated at Thyrow and Bad Lauchstädt derive from long-term agroecosystem experiments. Both areas are highly industrialised and were subject to brown-coal mining over many centuries. The soils at Thyrow have a sandy texture and are classified as luvisols, while the soil at the Bad Lauchstädt experimental station represents a chernozem with a high silt content. Organic carbon content is higher in the loamy soil of Bad Lauchstädt (15–16 g kg^{-1}) in comparison with the sandy soil of Thyrow (3.2 g kg^{-1}).

Each of the soil samples was investigated in the sand (2 000–63 μm) and the coarse silt fractions (63–20 μm) of the carbon depleted plots. In order to remove the minerals, the combined sand and coarse silt fractions were subjected to density fractionation using a sodium polytungstate solution at a density 1.9 g cm^{-3}. The organic matter isolated in the light subfraction represented on average 90% of the organic carbon in the size fractions. The light fraction material was embedded in epoxy resin and the mount was polished parallel to the base. Further details and a description of the sampling procedure are outlined in Kiem et al. (2003).

The Canadian River Alluvium sample was investigated in different grain size subsamples (0.5–0.25 mm, 0.25–0.125 mm, 0.125–0.063 mm, ≤0.063 mm) which were further divided into a magnetic and nonmagnetic fraction. Organic carbon fraction varied from 5.9% in the largest magnetic grain-size fraction to 0.1% in the nonmagnetic 0.125–0.063 mm fraction. The reasonably high fractions of organic carbon allowed the petrographic investigations to be carried out on the whole untreated subsamples. Subsamples were coated with epoxy resin on a glass slide, and the mount was polished. A more detailed sample characterisation is given in Karapanagioti et al. (2000).

The lithocomponents investigated were separated from the gravel fractions based on macroscopic investigations involving, for example, dark or light-coloured limestones and sandstones, and quartz or felspath minerals, jurassic and triassic limestone, triassic sandstone, and jurassic bituminous shale and buntersandstone. The organic carbon fraction varied from 40% for the bituminous shale to as low as 0.004% for the quartz minerals. Due to the rather low organic carbon fraction in some lithocomponents, the organic matter was isolated and concentrated together from the lithocomponents by a standard demineralisation procedure using hydrochloric and hydrofluoric acid. This extracted sample of organic matter was dispersed on a glass slide in a permanent mounting medium. A more in depth characterisation of the sample is provided by Kleineidam et al. (1999a,b).

Sorption experiments for the Canadian River samples and the lithocomponents were carried out with a batch technique using Phenanthrene as a chemical probe. Phenanthrene is a frequent organic contaminant in soils and groundwater. For more information about the conducted sorption experiments, readers are referred to the above mentioned publications (Kleineidam et al. 1999b; Karapanagioti et al. 2000).

All of the samples described in the paper were subject to microscopic analysis using a Leitz DMRX microscope photometer. Organic matter was identified and characterised using white light and UV illumination (blue-light irradiation) in transmitted and incident light mode. Maceral classification developed for coal petrography was used to describe and to classify the different organic particles (Taylor et al. 1998).

Most of the groups of organic matter are subdivided according to genetic, botanical, and microscopic criteria. Different terminologies are used according to the characteristics of each group: palynological kerogen terminology for recent organic matter, algae, and amorphous organic matter (AOM) (Tyson 1995); lithotype terminology for raw brown coal, hard coal terminology based on coal rank; terminology of coke textural components for hard-coal coke, terminology for combustion char type morphologies (Taylor et al. 1998).

The petrographic composition of the samples is either qualitative as in the case of the Canadian River sample and the lithocomponents or quantitative in nature as with the soils from Thyrow and Bad Lauchstädt. Quantification was carried out by a point counting method similar to that used in coal petrography (Taylor et al. 1998). The contribution of various constituents is expressed relative to the total volume of organic matter (vol.%).

9.3
Results and Discussion

The various groups of organic materials identified by organic petrographic analysis are presented in Table 9.1. Three major groups can be distinguished. First, the recent organic matter group consists of translucent phytoclasts, fungal phytoclasts, pollen, spores, and recent charcoal. The translucent and fluorescent phytoclasts are epidermal tissues, cortex tissues, and secondary xylem. Second, the fossil organic matter consists primarily of algae, spores, pollen, amorphous organic matter (AOM), xylite, coal, vitrite, and charcoal. Coal in this group corresponds to eroded and resedimented coal particles which occur principally in clastic sediments. Third, the group of airborne contaminants is composed of particles of raw brown coal, hard coal, charcoal, brown-coal coke, hard-coal coke, char, and asphalt.

In the soil samples from eastern Germany, the dominant group is defined by recent organic matter which consists firstly of resistant plant remains to be followed by airborne contaminants. The proportion of airborne contaminants are higher in the fractions of Bad Lauchstädt (Table 9.1: see under chernozem) than those from Thyrow (Table 9.1: see luvisol). At Thyrow, most contaminants are hard coal, hard-coal coke, and charcoal (Fig. 9.1a). There are minor amounts of raw brown coal, char, and traces of asphalt; the brown-coal cokes are absent. In contrast, at Bad Lauchstädt the raw brown coal (Fig. 9.1b) dominates the contaminants, which is composed of a wide range of hard coals, brown-coal cokes (Fig. 9.1c), and hard-coal cokes (Fig. 9.1d). A higher contribution of char and asphalt is also recorded in the Bad Lauchstädt sample. However, the contribution of charcoal to organic matter is smaller in Bad Lauchstädt than at Thyrow.

Thyrow is not surrounded by nearby industrial plants or by coal mines and, thus, is further away from potential sources of organic contaminants. This is corroborated by the lower proportions of contaminants found at Thyrow as compared to Bad Lauchstädt and indicates that part of the airborne contaminants found at Thyrow were transported over a distance of at least several dozen km. In contrast, Bad Lauchstädt is located in a region with a long tradition of brown coal mining and high industrial activity. The observed soil contamination reflects air pollution containing

Chapter 9 · Organic Petrology: A New Tool to Study Contaminants in Soils and Sediments

Table 9.1. Composition of the organic matter related to particle-size fraction and to $\log K_{OC}$ in arable soils from industrialised areas and in sediments, as revealed by organic petrographic analysis (volume%). *OM:* organic matter; *AOM:* amorphous organic matter; *m:* magnetic fraction; *nm:* nonmagnetic fraction; *BS:* Buntersandstone; *SS:* Triassic sandstone; *Qz:* quartz monominerals; *Jk:* Jurassic limestone; *LL:* light-coloured limestone; *MsK:* Triassic limestone; *DL:* dark-coloured limestone; *Le:* Jurassic bituminous shale; *DS:* dark-coloured sandstone. For detailed $\log K_{OC}$ values of the lithocomponents see Kleineidam et al. (1999b) (adapted from Kiem et al. 2003; Kleineidam et al. 1999b and Karapanagioti et al. 2000)

SAMPLE	PARTICLE-SIZE FRACTION (μm)	$\log K_{OC}$	RECENT ORGANIC MATTER			RAW BROWN COAL		HARD COAL			CHARCOAL: RECENT & FOSSIL	ALGAE, POLLEN & SPORES (FOSSIL OM)	AOM OF DIFFERENT MATURITY (FOSSIL OM)	BROWN-COAL COKE	HARD - COAL COKE				CHAR		ASPHALT
			Phytoclast ± translucent and fluorescent	Fungal phytoclast	Pollen and spores	Matrix coal	Xylite	Sub-bituminous coal	High volatile bituminous coal	Medium volatile bituminous coal	Vitrite				Isotropic matrix	Circular anisotropic matrix	Lenticular / elongated anisotropic matrix	Mixed : coke (anisotropic) and coal	Pyrolytic carbon	Sphere and network (anisotropic)	Solid (isotropic)	
			Quantitative petrographic analysis : composition determined by automatic point count method (vol.%)																			
(1) Luvisol Unmanured plot	2000 - 63		48.8	2.6	0.2	2.8	3.2	0.2		6.8	13.1				19.7					1.6	1.0	
	63 - 20		18.4	9.0	28.7	3.3	6.6	0.4		11.1	3.7				12.9			0.2	3.0	2.5	0.2	
(2) Chernozem Unmanured plot	2000 - 63		51.0	7.7	2.3	10.3	1.0	2.7		3.0	1.3			2.3	2.3	9.2	2.5			0.8	3.6	
	63 - 20		24.4	6.8	4.6	26.4	6.6	6.8	0.4	4.2	0.6			4.0	3.6	2.6	3.2			1.2	4.2	0.4
Bare fallow	2000 - 63		35.0	2.9	0.2	16.7	2.2	1.0	2.7	1.0	6.9				18.1	1.6	2.7	2.5	1.8		4.3	0.4
	63 - 20		26.7	5.9	3.1	32.4	2.0	3.3	0.8	2.2	1.2			4.9	1.4	2.0	2.2	3.3		1.4	6.8	0.4
			Qualitative petrographic analysis based on microscopical examination																			
Canadian River Alluvium	500 (m)	5.7	x				x					x										
	500 (nm)	6.3	x				x					x										
	250 (m)	5.3	x				x				x											
	250 (nm)	5.2					x				x											
	125 (m)	5.1	x				x				x	x										
	125 (nm)	4.7										x										
	63 (m)	5.6	x								x	x										
	63 (nm)	5.4					x					x										
Lithocomponents	BS, SS, Qz	< 4.7										x										
	Jk, LL	4.9- 5.6				x					x											
	Msk, DL, Le	5.6- 6.1										x	x									
	DS, DL	6.3- 6.7				x		x			x											

carbonaceous particles emitted undoubtedly in the surrounding region from opencast coal mines and briquette factories. This is underscored by the high amount of brown coal and coke dust. The noted presence of asphalt in the soils suggests mechanical erosion of this material probably from paving located in the vicinity of the investigated soils and also from wind transportation.

Char is present at both sites but it reflects a minor amount as compared to coal and coke. Airborne particles of coke and char might be emitted by different sources in industrial society. Coke pollution may be due to coking plants, steel smelters, or gas manufactured plants, whereas charred particles are by-products of all combustion processes of fossil fuels in industries, power plants, and vehicles. The burning of brown coal has been the major energy source over the last century in the eastern part of Germany, the location of the soils investigated in this study. This may have strongly contributed to the contamination of the soils by dust particles from combustion products.

The main consumers of energy coals are thermal power plants. Pulverized coal combustion produces emissions of flyash and gas. Flyash, which contributes greatly to airborne contaminants, is made of very fine solid particles often less than 10 μm in diameter (Taylor et al. 1998). These flyash particles are composed of inorganic matter which are composed of thermal transformation products found in minerals contained in feed coal. In cases of incomplete combustion, flyash particles also exhibit unburnt carbon in the form of char and unburnt coal particles.

In contrast to the above described samples, the soil and sediments samples from the Canadian River alluvium and the lithocomponents separated from the aquifer sediments show no impact from airborne organic compounds (cokes, char, asphalt). The organic matter in those samples is characterized by natural occurring OM including detritic coal particles and fossil charcoal. A qualitative analysis of those samples is summarized in Table 9.1. All samples exhibit a heterogeneous contribution of OM constituents indicating the complexity of OM formation in soils and sediments.

The Canadian River sample contains recent organic matter (phytoclasts) coal and charcoal particles. The latter which dominate the OM are believed to come initially from coal seams in New Mexico, where the South Canadian River originates. Microscopic examination has shown that coal particles can also be found in clay matrices of quartz aggregates. Immature AOM is also present within the clay matrices. Pictures of the OM and further details on OM characterisation are provided by Karapanagioti et al. (2000).

Not surprisingly, based on the different sedimentary history, the organic matter in the lithocomponents is highly variable in terms of nature, alteration, and maturation. It should be pointed out that most samples – except bituminous shale – were heterogeneous in terms of the various types of organic matter occurring within a specific lithocomponent (Table 9.1). In contrast to the recent Canadian River sample, the organic matter of the lithocomponents is tertiary, jurassic, and triassic age and consequently reveals different fluorescence colours and intensity. Both light and dark-coloured sandstones are dominated by opaque particulate organic matter such as charcoal, coal, and wood (xylite), as well as by fluorescent (UV/blue-light irradiation) organic matter particles in the form of algae, spores, pollen or cuticles. The limestones always contain considerable amounts of AOM that show an orange to brown fluorescence. For individual quartz grains, AOM was detected as a coating and exhibited a green fluorescence.

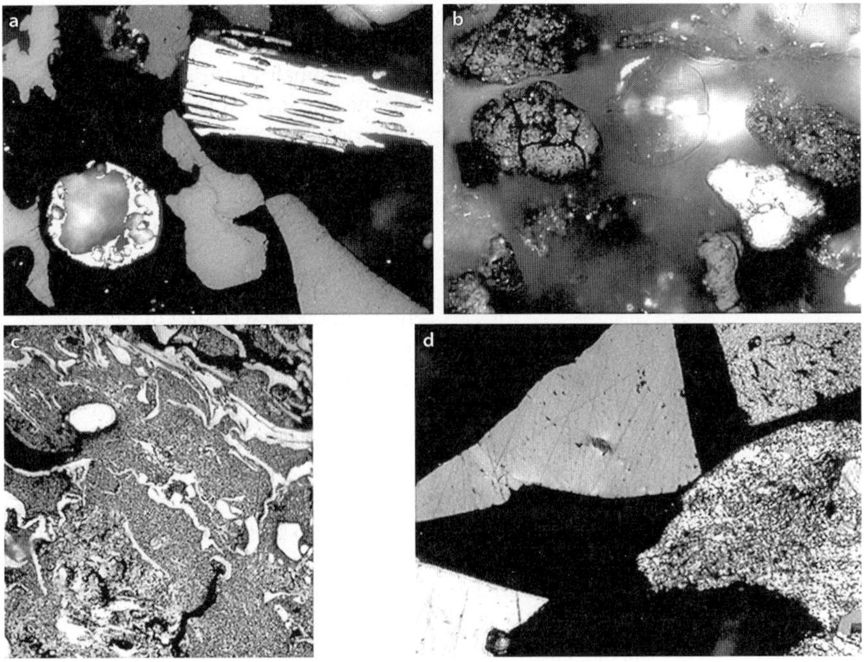

Fig. 9.1. Photomicrographs of different airborne contaminants in soils. All photomicrographs taken of the polished sections are prepared from densimetric concentrates using oil immersion. **a** Thyrow: luvisol, unmanured plot; coarse silt fraction. Airborne contaminants isolated from the soil organic matter as observed by incident light microscopy. *Grey particles* are hard-coal coke: see isotropic matrix in Table 9.1; *white particles* are charcoal (*upper right*) and char (*bottom left*). Field 165 µm wide. **b** Bad Lauchstädt: chernozem, unmanured plot, coarsed silt fraction. Recent organic matter and airborne contaminants isolated from the soil organic matter as observed by incident light microscopy. Recent organic matter is represented by a translucent spore (*centre*) and a phytoclast (*upper right*). The *fine heterogeneous grey particles* are strongly weathered raw brown coal with characteristic desiccation cracks. The *white particle* is hard-coal coke. Field 165 µm wide. **c** Bad Lauchstädt: chernozem, bare fallow, sand fraction. Brown-coal coke as observed by incident light microscopy. Note the very high reflectivity of the fragments of tissues in the coke matrix. Field 115 µm wide. **d** Bad Lauchstädt: chernozem, unmanured plot, sand fraction. Airborne contaminants isolated from the soil organic matter as observed by incident light microscopy and crossed polarizers. Examples of isotropic (*upper* and *lower left side*) and circular anisotropic (*upper* and *lower right side*) hard-coal coke. Field 170 µm wide

The organic matter characterisation of the Canadian River sample and the lithocomponents was correlated to the sorption properties of the samples (Fig. 9.2: $\log K_{OC}$ values and sorption non-linearity). The K_{OC} represents here the different sorptivities for phenanthrene – a chemical probe for hydrophobic organic compounds – at the trace concentration level of $1\ \mu g\ l^{-1}$. Based on the experimental results for sorption experiments, subsamples containing predominately detritic coal particles, and charcoal revealed the highest K_{OC}, and the highest non-linearity of sorption isotherms (e.g. dark-coloured lithocomponents as DL and DS, and the magnetic grain size 500 µm). Both types of particles are considered responsible for the highest observed $\log K_{OC}$ values and the extreme non-linearity of the sorption isotherms (upper end member). The

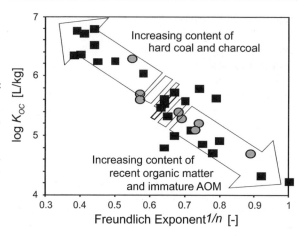

Fig. 9.2.
Log K_{OC} vs. $1/n$ for all samples showing the increase of K_{OC} and non-linearity for samples with OM dominated by coal and charcoal particles. Low K_{OC} and more linear sorption isotherms are found in samples where AOM dominates the OM. *Squares:* lithocomponents; *circles:* grain size fractions (after Kleineidam et al. 1999b and Karapanagioti et al. 2000). *OM:* organic matter, *AOM:* amorphous organic matter

lower end member is characterised by samples with organic matter in the form of organic coatings around the quartz grains. They demonstrate the lowest K_{OC} and the most linear sorption isotherms (QZ, nonmagnetic grain size fraction 125–63 µm).

The remaining subsamples are mixtures of the two groups and can include AOM coatings and detritic coal particles or charcoal and, thus, show intermediate sorption capacities and intermediate non-linearity. Due to the high sorption capacity of the coaly particles, even the presence of a small fraction of the composite organic content (<3%) causes K_{OC} higher than expected values for soil organic matter exhibited in empirical relationships (e.g. K_{OC}-K_{OW} relationships).

The light coloured lithocomponents with $\log K_{OC}$ values in the range of 4.9–5.6 contain organic matter consisting mainly of spores, pollen, cuticles, algae, and xylite particles. The organic matter of triassic limestone, dark limestone, and bituminous shale with $\log K_{OC}$ values around 6.0, consists mainly of marine algae and large grumose amorphous organic matter derived from phytoplankton and bacteria. This organic matter shows orange to brown fluorescence.

9.4
Conclusion

The research results presented in this study prove that organic-petrological techniques can be combined with chemical analyses as an effective tool for use in the environmental sciences.

In conclusion, organic petrology is a precise tool for

- the characterisation of heterogeneous organic matter present in soils and sediments.
- the identification and classification of airborne contaminants in topsoils which makes it possible both to characterise the emission sites and to gain insight into transport distance and transport processes.
- the determination of airborne particle size.
- a possible correlation between organic matter and sorption properties with reference to the hydrophobic organic contaminants.

Acknowledgements

Funding was provided by the Collaborative Research Center 275, the U.S. National Science Foundation through the EPSCorR Program, by the National Research Foundation (DFG) through grants (Ko 1035/10-1, Gr 971/15-1,2), and by Dr. D. A. Sabatini for H. K. Karapanagioti. Finally, the authors would like to acknowledge the collaboration of Prof. M. Körschens (Bad Lauchstädt, Germany) and the assistance of M. Baumecker (Thyrow, Germany) in obtaining the soil samples.

References

Accardi-Dey A, Gschwend PM (2002) Assessing the combined roles of natural organic matter and black carbon as sorbents in sediments. Environ Sci Technol 36:21–29
Allen-King R, Grathwohl P, Ball WP (2002) New modeling paradigms for the sorption of hydrophobic organic chemicals to heterogeneous carbonaceous matter in soils, sediments, and rocks. Adv Water Res 25:985–1016
Bucheli TD, Gustafsson Ö (2000) Quantification of the soot water distribution coefficient of PAHs provides mechanistic basis for enhanced sorption observations. Environ Sci Technol 34:5144–5151
Chiou CT, Kile DE (1998) Deviations from sorption linearity on soils of polar and nonpolar organic compounds at low relative concentrations. Environ Sci Technol 32:264–269
Cope MJ (1981) Products of natural burning as a component of the dispersed organic matter of sedimentary rocks. In: Brooks (ed) Organic maturation studies and fossil fuel exploration. London, New York, pp 89–109
Depers AM, Bailey JG (eds) (1994) ICCP Commissions I/II/III. Working group on environmental applications of coal petrology. White paper and abstracts. Department of Geology, University of Wollongong, Wollongong, N.S.W. and Department of Geology, University of Newcastle, Callaghan, N.S.W
Derenne S, Largeau C (2001) A review of some important families of refractory macromolecules: composition, origin, and fate in soils and sediments. Soil Sci 166:833–847
Emels KC (1987) A note on fly ash from coal combustion characteristics and possible usefulness as a tracer substance. Mitt Geol Paläont Inst Univ Hamburg, SCOPE/UNEP 62:99–108
Estrella MR, Brusseau ML, Maier RS, Pepper IL, Wierenga PJ, Miller RM (1993) Biodegradation, sorption, and transport of 2,4-dichlorophenoxyacetic acid in saturated and unsaturated soils. Appl Environ Microbiol 59:4266–4273
Ghose MK, Majee SR (2000) Assessment of dust generation due to opencast coal mining – an Indian case study. Environ Monit Assess 61:255–263
Glaser B, Lehmann J, Zech W (2002) Ameliorating physical and chemical properties of highly weathered soils in the tropics with charcoal – a review. Biol Fert Soils 35:219–230
Grathwohl P (1990) Influence of organic matter from soils and sediments from various origins on the sorption of some chlorinated aliphatic hydrocarbons: implications on K_{OC} correlations. Environ Sci Technol 24:1687–1693
Jones TP, Chaloner WG (1991) Fossil charcoal, its recognition and paleoatmospheric significance. Paleogeography, Paleoclimatology, Paleoecology (Global and Planetary Change Section) 97:39–50
Jonker MTO, Koelmanns AA (2001) Polyoxymethylene solid phase extraction as a partitioning method for hydrophobic organic chemicals in sediment and soot. Environ Sci Technol 35:3742–3748
Karapanagioti HK, Kleineidam S, Sabatini DA, Grathwohl P, Ligouis B (2000) Impacts of heterogeneous organic matter on phenanthrene sorption: equilibrium and kinetic studies with aquifer material. Environ Sci Technol 34:406–414
Karapanagioti HK, Childs J, Sabatini DA (2001) Impacts of heterogeneous organic matter on phenanthrene sorption: different soil and sediment samples. Environ Sci Technol 35:4684–4690
Kiem R, Knicker H, Körschens M, Kögel-Knabner I (2000) Refractory organic carbon in C-depleted arable soils, as studied by ^{13}C NMR spectroscopy and carbohydrate analysis. Org Geochem 31:655–668

Kiem R, Knicker H, Kögel-Knaber I (2002) Refractory organic carbon in particle-size fractions of arable soils. I: Distribution of refractory carbon between the size fractions. Org Geochem, 33:1683–1697

Kiem R, Knicker H, Ligouis B, Kögel-Knabner I (2003) Airborne contaminants in the refractory organic carbon fraction of arable soils in highly industrialized areas. Geoderma 114:109–137

Kile DE, Wershaw RL, Chiou CT (1999) Correlation of soil and sediment organic matter polarity to aqueous sorption of nonionic compounds. Environ Sci Technol 33:2053–2056

Kleineidam S, Rügner H, Grathwohl P (1999a) Impact of grain scale heterogeneity on slow sorption kinetics. Environ Toxicol Chem 18:1673–1678

Kleineidam S, Rügner H, Ligouis B, Grathwohl P (1999b) Organic matter facies and equilibrium sorption of phenanthrene. Environ Sci Technol 33:1637–1644

Kleineidam S, Schüth C, Grathwohl (2002) Solubility-normalized combined adsorption-partitioning sorption isotherms for organic pollutants. Environ Sci Technol 36:4689–4697

Rumpel C, Knicker H, Kögel-Knabner I, Skjemstad JO, Hüttl RF (1998) Airborne contamination of immature soil (Lusatian mining district) by lignite-derived material: its detection and contribution to the soil organic matter budget. Wat Air Soil Poll 105:481–492

Schmidt MWI, Noack AG (2000) Black carbon in soils and sediments: Analysis, distribution, implications, and current challenges. Glob Biogeochem Cycles 14:777–793

Schmidt MWI, Knicker H, Hatcher PG, Kögel-Knabner I (2000) Airborne contamination of forest soils by carbonaceous particles from industrial coal processing. J Environ Qual 29:768–777

Schmidt MWI, Skjemstad JO, Czimczik CI, Glaser B, Prentice KM, Gelinas Y, Kuhlbusch TAJ (2001) Comparative analysis of black carbon in soils. Glob Biogeochem Cycles 15:163–167

Spurny KR (1986) Review of applications. In: Spurny (ed) Physical and chemical characterization of individual airborne Particles. Ellis Horwood Limited, Chichester, UK, pp 400–412

Taylor GH, Teichmüller M, Davis A, Diessel CFK, Littke R, Robert P (1998) Organic petrology. Gebrüder Borntraeger, Berlin

Tyson RV (1995) Sedimentary organic matter: organic facies and palynofacies. Chapman & Hall, London

Weber Jr. WJ, McGinley PM, Katz LE (1992) A distributed reactivity model for sorption by soils and sediments. 1. Conceptual basis and equilibrium assessments. Environ Sci Technol 26:1955–1962

Wik M, Renberg I (1987) Distribution in forest soils of carbonaceous particles from fossil fuel combustion. Wat Air Soil Poll 33:125–129

Xia G, Ball WP (1999) Adsorption-partitioning uptake of nine low-polarity organic chemicals on a natural sorbent. Environ Sci Technol 33:262–269

Xing B, Pignatello JJ (1997) Dual-mode sorption of low polarity compounds in glassy polyvinylchloride and soil organic matter. Environ Sci Technol 31:792–799

Chapter 10

The Comminution of Large Quantities of Wet Sediment for Analysis and Testing with Application to Dioxin-Contaminated Sediments from Lake Ontario

K. B. Lodge

Abstract

When preparing field samples for analysis and testing, it is invariably necessary to split the sample into representative subsamples. If the mass of the original sample is of the order of 1 000 kg (dry-mass basis), then subsamples that are about 100 000–1 000 000 000 times smaller are required for chemical analysis. With such large reductions in size, the representability of the subsample is a concern. Here, a systematic method for preparing representative subsamples from large quantities of wet sediment is described. The sediment is passed through a commercially available slurry sampler that provides 1/20th splits. Further splits are obtained with a custom-built churn splitter. The procedure was applied to sediments collected from Lake Ontario, USA, and from the lower Fox River and Green Bay, Wisconsin, USA. Its efficacy was evaluated by dioxin and total-organic-carbon analysis for the sediments from Lake Ontario and by analysis of the ammonia contained in the pore-water of the sediments from the Fox River and Green Bay.

Key words: trace analysis, dioxins, organic pollutants, sediments, representability, subsampling, ammonia, pore water

10.1 Introduction

There are various stages in the evaluation and testing of sediments collected from the field and brought into the laboratory. First, there are considerations relating to the collection of samples from the field (Mudroch and MacKnight 1991; Keith 1987). Second, there are matters relating to the treatment and storage of samples in the laboratory before testing (Othoudt et al. 1991). Certain studies require large quantities because the sediments are subjected to a battery of tests including chemical analysis (O'Keefe et al. 1990), bioassays (Batterman et al. 1989) and the evaluation of technologies for remediation (Timberlake and Garbaciak 1995). The principle concern here is the preparation of representative subsamples from a large bulk sample.

If testing requirements can be met with dry samples, then the whole sample could be dried, ground and riffled to prepare a range of representative subsamples. There is a range of riffling or splitting equipment available for handling bulk dry materials. Methods using such equipment are routinely applied in the mining and mineral industry. An example is provided by the preparation of reference materials for particle-size analysis (Wilson et al. 1980).

However many circumstances arise under which it is necessary to use wet samples in order to preserve, as much as possible, the sample's physical, chemical and biological integrity. Sediment testing with bioassays and sediment characterisation with

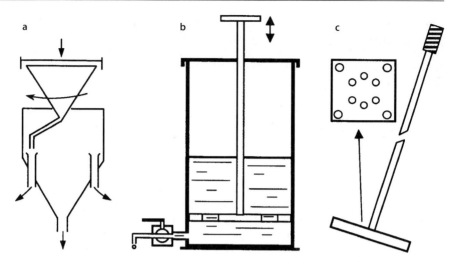

Fig. 10.1. The slurry sampler (**a**) (the body of the sampler is 36 in tall and 18 in wide), the churn splitter (**b**) (the body of the splitter is 12 in tall and 6 in wide) and the mud masher (**c**) (it is about 4 ft tall; the bottom plate is 6 in square)

particle-size analysis are examples for which wet samples are required. For splitting large bulk wet samples, the literature does not appear to contain descriptions of methods for which evaluation data have been presented. Here a method is described and data, by which its efficacy may be judged, are presented.

The method incorporates basic principles of sampling. Two golden rules are stated for the sampling of powders (Allen 1981), and there is no reason to believe that these rules would not also apply to sediment slurries. Reformulated for slurries, they are: *(1)* the slurry should be sampled when in motion, and *(2)* the whole stream of slurry should be sampled for many short increments of time in preference to part of the stream being taken for the whole time. The principle item of equipment that incorporates these two principles in its design is a slurry sampler, shown in Fig. 10.1a. A secondary item, the churn splitter shown in Fig. 10.1b, also exploits these principles.

10.2
Experimental

10.2.1
Equipment

Slurry Sampler. The Carpco Rotary Slurry Sampler (Model WS-220, Outokumpu Technology) is the principle device used in these procedures; a similar model, if not identical, is available from Gilson (Model SP-220). Figure 10.1a shows a schematic of the slurry sampler; it comprises a cone rotated by an electric motor. The sediment slurry is fed through a screen (with openings of 3.5 mm) atop of the cone; the cone's rotation causes the slurry to be expelled through the spigot at the bottom of the cone in a continuous stream. This flows over two ports that are designed to collect $1/20^{th}$ of the feed; the remainder of the slurry passes to the main sump. The slurry sampler is mounted on top of a stand that aided in the manipulation of the collection vessels. A Guzzler hand pump (Cole Parmer 07090-12, 2-in ports, 3 strokes/gal) is used to deliver the slurry to the top of the cone.

Churn Splitter. The slurry sampler was used to prepare representative subsamples that were about 1 gal or less (Fig. 10.1a). To obtain samples from these, a custom-built churn splitter (capacity 1 gal), fabricated of stainless steel, was used (Fig. 10.1b). The mixing is provided by the disc, containing holes in a specific pattern, on a rod that extends through the cover. A stainless steel ball valve (Whitey SS 65TF12, orifice 0.875 in) is used to dispense samples while the disc is kept in motion by hand. This splitter is similar to those supplied by Bel-Art; these have been used for sampling dilute sediment suspensions collected from river water (Horowitz 1986).

Containers. The availability and choice of containers very much dictates the procedures used. Samples arrived from the field in 30-gal drums or 5-gal plastic buckets. Thirty-gallon polyethylene drums, mounted on dollies, were used to hold and collect bulk samples while processing them with the slurry sampler. Five-gallon plastic buckets and 1-gal jars were used to collect the $1/20^{th}$ splits. Smaller glass jars, 500 and 250 ml, were used for samples dispensed from the churn splitter.

Mixing Tools. Upon extended storage, sediments will consolidate. To prepare a thixotropic slurry, consolidated sediments were broken up with a custom-built mud "masher"; this was made of steel and is shown in Fig. 10.1c. To mix thixotropic sediments, an electric drill fitted with a high-efficiency paddle assembly (Cole Parmer 04541-30, 5 in) proved adequate.

10.2.2
General Procedures

Bulk sediments and wet subsamples were stored in a cold room held at 4 °C. When feasible, sediments were processed as soon as possible after arrival in the laboratory. Large material (twigs, pebbles, etc.) was removed by passing the sediments through a piece of expanded metal placed over a 30-gal drum. Samples were then further homogenized with the mud masher; this served also to prepare a thixotropic slurry. In a few cases, water from Lake Superior was added to achieve this; the volume of water added was no more than 2% of the original volume. Sediment slurries had a solids' content in the range of 35–45%.

The whole sediment sample was then passed through the slurry sampler; upon this, the first pass, additional debris >3.5 mm was collected from the screen atop of the cone. The exact procedure followed from this point depended upon the volume of original sample and upon the volume of subsamples required. It is of paramount importance to keep track of what constitutes a representative sample. This point is best illustrated by considering the typical cases shown in Fig. 10.2. In stage A before

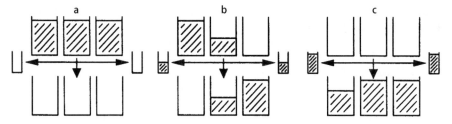

Fig. 10.2. Stages in splitting a sediment slurry

splitting, 3 drums of sediment slurry are stationed by the pump inlet; these 3 drums constitute the original sample. By stage B, about half the original sample has been passed through the sampler. Material has been collected at each of the 1/20th ports and from the sampler's sump; this material does not represent the original sample. The entire original sample has been passed through the sampler in stage C. Now, the material collected at each of the 1/20th ports and from the sampler's sump, constituting 18/20th of the original sample, does represent the original sample. The material collected from the sump is necessarily distributed over 3 drums; the sample in the drums taken together constitutes a representative sample, but the material in any one drum, taken alone, does not.

The three drums, containing the material collected from the sump during the first pass, are moved to the pumping station. This sample is then passed through the slurry sampler; this constitutes the second pass of the original sample through the sampler. As more passes of the original sample are made, the volume decreases, as do the volumes of the 1/20th splits. To obtain samples for chemical analysis, a 1/20th split of a gallon or less is desired. In reaching this stage, storage of splits becomes a problem. However, the combination of two representative samples is itself representative of the original sample, so splits are often combined. Small samples for chemical analysis are dispensed from the churn splitter; samples are collected in aluminium trays and air-dried.

10.2.3
Treatment of Samples from the Fox-River and Green Bay

Bulk sediments were collected using an Eckman dredge from 13 sites and brought to the laboratory in 5-gal plastic pails; the total sample volume for each site was ≤100 gal. The splitting procedure was used to obtain representative wet samples in 5-gal pails, 1-gal glass jars and 1-quart glass jars. Samples that were dispensed from the churn splitter for chemical analysis were air-dried, ground and riffled with a spinning riffler (Quantachrome Model SRR-2). Samples were distributed to 6 research groups involved in an integrated assessment using bioassays and chemical analysis (Ankley et al. 1992).

In the early stages of splitting, the masses of 1/20th splits were measured; this enabled an evaluation of the physical performance of the slurry splitter to be made. Two research groups, one at the U.S. EPA's Mid-Continent Ecology Division in Duluth and the other at the Department of Fisheries and Wildlife at Michigan State University, East Lansing, measured pore-water ammonia in separate 1-gal subsamples for each site. From these analyses, some assessment of the representability of subsamples can be made.

10.2.4
Treatment of Samples from Lake Ontario

About 200 gal of sediment was collected with a 0.25-sq meter box core from two locations, Stations 208 and 210 (Onuska et al. 1983), near the mouth of the Niagara River. The samples were brought to the laboratory in 30-gal steel drums. The splitting was done in two stages.

The purpose of the first stage was to produce a representative sample whose volume is about 20 gal. The whole sample was passed through the splitter repeatedly and the $1/20^{th}$ splits are combined until a representative sample of the requisite volume was obtained. In the second stage, the representative sample, whose volume was about 20 gal, was passed repeatedly through the slurry sampler. The $1/20^{th}$ splits were transferred to the churn splitter and samples for bioassays (Batterman et al. 1989), chemical analysis and K_{OC} experiments (Lodge and Cook 1989; Lodge 2002) were dispensed. With each pass the volumes of the $1/20^{th}$ splits decrease; when they became $\leq\frac{1}{2}$ gal, they were combined in the churn splitter.

From the first $1/20^{th}$ split that went to the churn splitter a splitting-evaluation sample of type A, 500 or 250 ml, was obtained. From this point on, the volume of the $1/20^{th}$ splits decreased until a stage was reached when further splitting was impractical. This was reached when the volume of the combined $1/20^{th}$ splits became $\leq\frac{1}{2}$ gal. At this stage a splitting-evaluation sample of type B was collected. A selection of the splitting-evaluation samples of both type A and B was analysed for dioxin and total organic carbon.

10.2.5
Chemical Analysis

The analytes, with sources of methods, are ammonia (Ankley et al. 1990), dioxin and total organic carbon (Lodge and Cook 1989; Lodge 2002). A few milligrams of dried sediment were used for total organic carbon analysis; 10 g of dried sediment were used for dioxin analysis.

10.3
Results and Discussion

10.3.1
Splitting of Sediments from the Fox River and Green Bay

The mass of a $1/20^{th}$ split after the nth pass is given by

$$M_n = p^{-1}(q/p)^{n-1} M_0$$

where M_0 is the mass of the original sample, $1/p$ is the fraction of the feed collected at the smaller ports ($p \sim 20$) and q/p is the fraction of the feed collected from the splitter's sump ($q \sim 18$). It is inconvenient to measure M_0 directly. Using $M_0 = pM_1$, then

$$\log(M_n/M_1) = (n-1)\log(q/p)$$

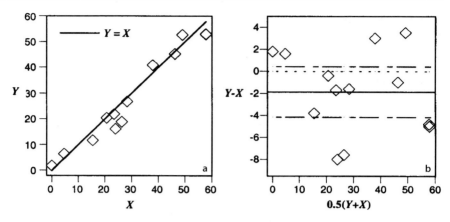

Fig. 10.3. Evaluation of the splitting procedures for sediments from the Fox River and Green Bay: measurements of ammonia in pore water. **a** y-axis: pore-water ammonia measured at the U.S. EPA's Mid-Continent Ecology Division in Duluth, Y (mg l^{-1}); x-axis: pore-water ammonia measured at Michigan State University, X (mg l^{-1}); **b** y-axis: difference in measurements $Y - X$ (mg l^{-1}); x-axis: mean of measurements $0.5(Y + X)$ (mg l^{-1}); *dotted line:* $Y - X = 0$; *semi dotted line:* 95% confidence interval, *line:* mean value

where M_1 is the mass of the first 1/20th split. Results for the splitting of the sediments from the Fox River and Green Bay were fitted to this equation using linear regression. The value of q/p is 0.8766 ±0.0005 (± the standard deviation, $N = 94$); this should be compared with the design value of $q/p = 18/20 = 0.9$. The measured value of q/p corresponds to $1/p \sim 1/16$ in contrast to the design value of $1/20$; the effect on the splitting is not significant.

The results of the pore-water ammonia analysis for the samples from the Fox River and Green Bay are shown in Fig. 10.3; the line in graph A represents the ideal condition in which both laboratories obtain exactly the same results. Inter-laboratory variability is a part of this picture. For each site, the difference in the two measurements is plotted against the corresponding mean in graph B. This shows the mean of the differences (−1.8 mg l^{-1}) and the 95% confidence interval. The confidence interval contains the difference of zero, so it is concluded there is no significant difference in the measurements.

10.3.2
Splitting of Sediments from Lake Ontario

For the splitting evaluation samples taken during the sampling of the Lake Ontario sediments, the results of the dioxin and total-organic-carbon analyses are shown in Table 10.1. There are insufficient data for a rigorous statistical examination. However, the F-test and Dioxin's Q-test, applied to those data with duplicate analyses, show no significance difference in the measurements. Our dioxin analyses of a similar sediment sample (O'Keefe et al. 1990), repeated ten times, gave a level of 48 ±12 pg g^{-1}, a relative standard deviation of 25%. The relative standard deviations of the dioxin data for the splitting evaluation samples presented here range from 6–11%. The relative standard deviations for the total-organic-carbon results range from 4–8%.

Table 10.1. Levels of dioxin and total organic carbon (TOC) in splitting-evaluation samples for Lake Ontario sediments

2nd stage sample	Type	Dioxin (pg g^{-1})	TOC (%)	Comments
Sediment from station 210				
1	A	31.4	1.398	
2	A	39.1	1.481	
3	A	36.2	1.390	
3	B	33.4	1.346	
4	A	37.8	1.400	
4	B	32.8	1.232	
5	B	34.1	1.437	
5	B	32.2	1.385	Duplicate analysis
Average		35	1.38	
Standard deviation		3	0.07	
Sediment from station 208				
1	A	452	2.823	
3	A	434	2.611	
3	B	441	2.566	
5	B	546	2.441	
Average		470	2.6	
Standard deviation		50	0.2	
Sediment blend of stations 210 and 208 (nominally 10:1)				
1	A	40.1	1.446	
2	A	36.0	1.338	
2	A	35.1		Duplicate analysis
2	B	35.7	1.411	
3	B	33.5	1.444	
Average		36	1.41	
Standard deviation		2	0.05	

10.4 Conclusion

Analyses of samples taken at various stages of splitting give results in agreement with each other. From this observation, it is concluded that the procedures adopted here are successful in producing subsamples that are representative of the original sample. It should be noted that the procedure requires a thixotropic sediment slurry, and so the sample must contain sufficient clay. Splitting of exceptionally sandy sediments may not be possible by this method; these will probably settle out and consolidate too quickly. As a part of the U.S. EPA's Assessment and Remediation of Contaminated Sediments Program, these methods were applied to bulk sediments collected from the Grand Calumet River, Indiana; the Buffalo River, New York; the Saginaw River, Michigan; and the Ashtabula River, Ohio (Timberlake and Garbaciak 1995; Garbaciak and Miller 1995). However, splitting evaluations were not done with these samples.

Acknowledgements

The U.S. EPA provided financial support under cooperative agreements CR-813504 and CR-817486 with the University of Minnesota Duluth. Assistance is gratefully acknowledged from the following; at the U.S. EPA's Mid-Continent Ecology Division in Duluth, P. M. Cook, A. R. Batterman and G. T. Ankley; at the Natural Resources Research Institute of the University of Minnesota Duluth, J. R. Ludwig, G. Betts, J. Gordon, J. Cox, D. R. Marklund, C. Harper and S. W. Grosshuesch.

References

Allen T (1981) Particle size measurement, 3rd edn. Chapmann and Hall, London, pp 5
Ankley GT, Katko A, Arthur JW (1990) Identification of ammonia as an important sediment-associated toxicant in the lower Fox River and Green Bay, Wisconsin. Environ Toxicol Chem 9:313–322
Ankley GT, Lodge K, Call DJ, Balcer MD, Brooke LT, Cook PM, Kreis Jr. RG, Carlson AR, Johnson RD, Niemi GJ, Hoke RA, West CW, Giesy JP, Jones PD, Fuying ZC (1992) Integrated assessment of contaminated sediments in the lower Fox River and Green Bay, Wisconsin. Ecotox and Env Saf 23:46–63
Batterman AR, Cook PM, Lodge KB, Lothenbach DB (1989) Methodology used for a laboratory determination of relative contributions of water, sediment and food chain routes of uptake for 2,3,7,8-TCDD bioaccumulation by lake trout in Lake Ontario. Chemosphere 19(1–6):451–458
Garbaciak Jr. S, Miller JA (1995) Field demonstrations of sediment treatment technologies by the U.S. EPA's Assessement and Remediation of Contaminated Sediments (ARCS) program. In: Demars KR, Richardson GN, Young RN, Chaney RC (eds) Dredging, remediation, and containment of contaminated sediments. ASTM STP 1293, American Society for Testing and Materials, Philadelphia, pp 145–154
Horowitz AJ (1986) Comparison of methods for the concentration of suspended sediment in river water for subsequent chemical analysis. Environ Sci Technol 20:155–160
Keith LH (ed) (1987) Principles of environmental sampling. American Chemical Society, Washington DC
Lodge KB (2002) Desorption from contaminated sediment and the organic-carbon normalized partition coefficient, K_{OC}, for dioxin. Adv Environ Res 7:147–156
Lodge KB, Cook PM (1989) Partitioning studies of dioxin between sediment and water: the measurement of K_{OC} for Lake Ontario sediment. Chemosphere 19(1–6):439–444
Mudroch A, MacKnight SD (1991) CRC handbook of techniques for aquatic sediments sampling. CRC Press Inc, Boca Raton, Florida
O'Keefe P, Wilson L, Buckingham C, Rafferty L (1990) Quality assurance/quality control assessment for a collaborative study of 2,3,7,8-TCDD bioaccumulation in Lake Ontario. Chemosphere 20(10–12):1277–1284
Onuska FI, Mudroch A, Terry KA (1983) Identification and determination of trace organic substances in sediment cores from the western basin of Lake Ontario. J Great Lakes Res 9(2):169–182
Othoudt RA, Geisy JP, Grzyb KR, Verbrugge DA, Hoke RA, Drake JB, Anderson D (1991) Evaluation of the effects of storage time on the toxicity of sediments. Chemosphere 22(9–10):801–7
Timberlake DL, Garbaciak SG (1995) Bench-scale testing of selected remediation alternatives for contaminated sediments. J Air Waste Manag Assoc 45:52–56
Wilson R, Leschonski K, Alex W, Allen T, Koglin B, Scarlett B (1980) Certification report on reference materials of defined particle size, quartz (BCR No. 66, 67, 68, 69, 70). EUR 6825, Commission of The European Communities BCR Information

Chapter 11

Study on the Large Volume Stacking Using the EOF Pump (LVSEP) for Analysis of EDTA by Capillary Electrophoresis

L. Zhang · Z. Zhu · M. Arun · Z. Yang

Abstract

The widespread use of ethylenediaminetetraacetic acid (EDTA) has requested an urgent monitoring program regarding surface and drinking water. Analyzing EDTA at low-level concentrations such as µg l^{-1} in the environmental samples is quite complex using conventional methods. In this study, a simple, quick and sensitive capillary electrophoretic technique – large volume stacking using the EOF pump (LVSEP) – has been developed for determining EDTA in drinking water for the first time (EOF: electroosmotic flow). It is based on a precapillary complexation of EDTA with Fe(III) ions, followed by LVSEP and direct UV detection at 258 nm. The curve of peak response vs. concentration was linear between 5.0 and 600.0 µg l^{-1}, as well as between 0.7 and 30.0 mg l^{-1}. The regression coefficients were 0.9988 and 0.9990, respectively. The detection limit of current technique for EDTA analysis was 0.2 µg l^{-1} with additional 10-fold preconcentration procedure, based on the signal-to-noise ratio of 3. As opposed to the classical capillary electrophoresis (CE) method, a 1 000-fold concentration factor could be smoothly achieved on this LVSEP method. To the best of our knowledge, it represents the highest sensitivity for EDTA analysis via CE. Several drinking water samples were tested by this novel method with satisfactory results.

Key words: water, ethylenediaminetetraacetic acid, large volume sample stacking, capillary electrophoresis

11.1 Introduction

Since its commercialization in the early 1950s, the synthetic ethylenediaminetetraacetic acid (EDTA) has been widely used as chelating reagent for metal ions in detergent industrial, pharmaceutical and agricultural applications (Bersworth 1954; Randt et al. 1993). EDTA possesses strong capability to form stable complexes with 62 different metal cations as well as its low biodegradability (Sheppard and Henion 1997), it is generally believed that, EDTA may affect the distribution of metals within the aquatic ecosystem and may remobilize heavy metals from sediments decomposition by electrolytic oxidation. Particularly, it is used to deliver trace minerals in animal feeds and some human foods (Fishbein et al. 1970; Seegers 1967). Presently, EDTA has entered into various foods and water resources (Xue et al. 1995; Kari and Giger 1995). Therefore, the World Health Organization (WHO) committed to a preliminary standard upper limit of 200 µg l^{-1} because it contributes significantly to the loss in quality of drinking water (Schön et al. 1997). Consequently, the analytical determination of EDTA is very important for understanding its environmental fate.

Though determinations of EDTA can be done by either electrochemical (Schön et al. 1997) or chromatographic methods, traditional ones are a variety of chromatographic methods including gas chromatography (GC), high performance liquid chromatography (HPLC), and ion chromatography (IC). For instance, EDTA can be measured by gas chromatography in concentrations down to 1 μg l^{-1}. Despite the low detection limit, this method has some drawbacks such as a relatively large influence of the matrix on the accuracy and a tedious derivatization step of the carboxylic group by esterification (Randt et al. 1993). Alternatively, EDTA may also be determined as their negatively charged complexes by ion chromatography-UV spectrophotometric detection with the detection limit of 100 μg l^{-1} (Venezky and Rudzinski 1984). High performance liquid chromatography (HPLC) is more frequently used for the EDTA determination, mainly in reversed-phase ion-pair mode. Based on Loyaux-Lawniczak's method, the limit of detection can reach 5 μg l^{-1} (Kari et al. 1995; Loyaux-Lawniczak et al. 1999)

However, all these methods are complicated and time-consuming. Capillary electrophoresis (CE) is a new developed technique for the separation and determination of both organic and inorganic cations and anions. Compared with HPLC, IC and GC, CE has the advantages of higher efficiency, simpler chemistry, faster separation time, ease of automation, and smaller sample and reagent requirements. However, there are few reports being available regarding the determination of EDTA with capillary electrophoresis (Bürgisser and Stone 1997; Owens et al. 2000; Pozdniakova et al. 1999). This is because of the poor concentration limit of detection primarily as a consequence of the short optical path length within the detection cell and the extremely small sample volume that can be introduced into the CE capillary. For example, according to Pozdniakova's method, the detection limit achieved is only 600 μg l^{-1} (Pozdniakova et al. 1999). CE-MS (MS: mass spectrometry) was once used for the quantitative determination of EDTA in human plasma and urine via large volume injection on a special amine bonded capillary (Sheppard and Henion 1997). As the sample matrix had not been removed before the separation began, the stacking efficiency was limited despite the sensitive ion spray tandem mass spectrometry was used at the selected reaction monitoring (SRM) mode, the detection limit was 7 μg l^{-1} only. Additionally, CE-MS is beyond the scope of many analytical laboratories and has to be accepted as a routine method of analysis. Consequently, it is very important to develop an alternative CE technique that is compatible with the analysis of low sample concentrations.

Sample stacking is an inherent and exclusive feature of CE. It may take place when the sample compounds encounter isotachophoretic concentration at the interface between sample zone and buffer (isotachophoretic sample stacking, ITPSS) (Gebauer et al. 1992) or when the conductivity of the sample is smaller than that of the buffer (field-amplified sample stacking, FASS) (Chien and Burgi 1992). In this study, a novel sample injection technique – large volume stacking using the EOF pump (LVSEP) – was developed and applied for the first time in the detection of EDTA in drinking water by capillary electrophoresis. Hydroxyethylcellulose (HEC) was used to suppress the EOF under acidic conditions. The sample stacking and sample separation could be performed under the same negative voltage without causing loses of the stacked samples as no polarity switch was necessary. The whole procedure could be fully controlled by software. Compared with conventional capillary zone electrophoresis (CZE) method, this novel method provided 1 000-fold enhancement without loss of the high-resolution feature afforded by capillary electrophoresis. The detection limit of present method was as low as 0.2 µg l^{-1} or 2.0 µg l^{-1}, with or without additional 10-fold preconcentration procedure, respectively.

11.2 Experimental

11.2.1 Instrumental

Separations were performed on a P/ACE MDQ Capillary Electrophoresis System (Beckman Instruments Inc., Fullerton, CA) equipped with a PDA detector. Fused silica capillary (Polymicro Technology, Phoenix, AZ) of 75 µm i.d. and 39 cm length to the detector (49 cm total length) was used. Deionized water was obtained by passing tap water through an USF-ELGA option 15 system and an USF MAXIMA system (Vivendi Water, UK) with the resistance greater than 18.2 MΩ cm^{-1} and on-line TOC less than 2 ppb. Between all electrophoretic separations the capillary was rinsed with 0.1 M sodium hydroxide and deionized water for 3 min, and with running buffer for 5 min. All measurements were the average of at least three repeatable measurements.

11.2.2 Chemicals

All chemicals used were of analytical-reagent grade. Ethylenediaminetetraacetic acid disodium salt dihydrate (Na$_2$H$_2$EDTA · 2H$_2$O) and ethylenediaminetetraacetic acid iron(III)-sodium salt hydrate (Na[Fe^{III}EDTA] · H$_2$O) were obtained from Fluka. Phosphoric acid, boric acid, disodium tetraborate decahydrate, and hydroxyethylcellulose (HEC) were supplied by Merck. Ammonium-iron(III)-sulfate (NH$_4$Fe(SO$_4$)$_2$) and N,N'-dimethylformide (DMF) were purchased from Aldrich. A stock standard solution was prepared by dissolving 0.100 g of Na[Fe^{III}EDTA] · H$_2$O in 100 ml of deionized water. Standard solutions, ranging from 0 to 1 200 mg l^{-1} of EDTA, were daily prepared from this solution and stored in the dark. Electrophoretic buffer solutions were prepared from 50 mM sodium borate by adding 10% H$_3$PO$_4$ solution to adjust to the desired pH. All electrolyte solutions were filtered through a 0.45 µm polytetrafluoroethylene membrane filter (Whatman, UK).

11.2.3 Large Volumes Stacking Using the EOF Pump

0.05% HEC was added in the running buffer (pH 4.1). The sample solutions were injected in the hydrodynamic mode by overpressure (3.0 psi). After sample injection, a negative voltage (−20 kV) was applied to effect water removal and subsequent separation. The detection wavelength was 258 nm. The coolant temperature was controlled at 25 °C.

11.3
Results and Discussion

11.3.1
Theory

In capillary electrophoresis, electrophoretic mobility (μ_{ele}) is one of the most important parameters, which is proportional to the charge of the analyte divided by its frictional coefficient. This is approximately equal to the charge to mass ratio of the analyte. In a CE separation, as the analyte mixture migrates through the capillary due to the applied electric field, different electrophoretic mobilities drive each of the components into discrete bands. Meanwhile, the electroosmotic flow (EOF) mobility (μ_{eof}) is inherent in any CE setup that uses bare-fused silica capillaries or capillaries coated with a charged group, which is proportional to the dielectric constants of the buffer and zeta potential, and inversely proportional to the viscosity of buffer. Consequently, the size and direction of apparent migrational mobility (μ_{app}) of a component depends on the sum of its electrophoretic mobility and the EOF mobility.

$$\mu_{app} = \mu_{ele} + \mu_{eof} \tag{11.1}$$

In large volume stacking using the EOF pump (LVSEP), the electrophoretic mobility of the target ions must be in the direction opposite to that of the EOF in order that cations or anions could be focused at the rear or front of the sample zone. During the separation procedure, the electrophoretic velocities of interested ions (e.g. anions) must be greater than the EOF velocities so that the two procedures can proceed consecutively under the same voltage.

To inject [Fe^{III}EDTA]$^-$ with LVSEP into the capillary, the electrophoretic velocities of [Fe^{III}EDTA]$^-$ ($-V_{[Fe^{III}EDTA]^-}$) must be greater than the EOF velocity (V_{eof}), that is, $-V_{[Fe^{III}EDTA]^-} > V_{eof}$. Here, the negative sign indicates that the electrophoretic velocity must be in the direction opposite to that of the electroosmotic flow. This is required for both removing the sample matrix from the capillary using EOF pump and the subsequent separation.

At pH 3–6, [Fe^{III}EDTA]$^-$ can exist stably. At the same time, the EOF can be suppressed by addition of HEC to the running buffer. So, in this region, $\mu_{eof} \approx 0$, thus,

$$\mu_{app} \approx \mu_{ele} \tag{11.2}$$

and

$$-V_{[Fe^{III}EDTA]^-} > V_{eof} \tag{11.3}$$

When negative voltage is applied from the cathodic inlet toward the anodic outlet, the anions will move to the outlet. Thus, LVSEP and water removal can be carried out under the negative voltage. On the basis of this, a scheme was designed for the stacking and separation of [Fe^{III}EDTA]$^-$, as illustrated in Fig. 11.1.

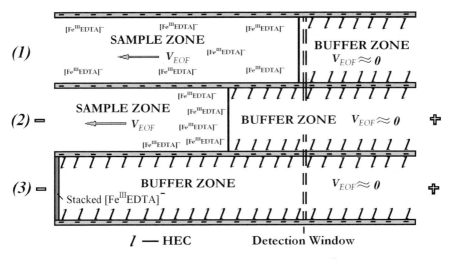

Fig. 11.1. Schematic illustration of the matrix removal and stacking of [Fe^{III}EDTA]$^-$. **a** Sample injection; **b** sample stacking; **c** ready to separate

When large sample volumes were introduced into the capillary by pressure injection, the whole capillary was separated into two zones. One was sample zone without HEC, and the other was running buffer zone where HEC was adsorbed onto the inner wall of capillary. After sample injection, a negative voltage was applied and the sample stacking began. Due to its poor conductivity, the electric field strength was mainly applied in the sample zone. The negatively charged complexes [Fe^{III}EDTA]$^-$ moved forwards rapidly in the sample zone until they encountered the electrolyte boundary where they experienced a lower applied field and their migration rate slowed down. During this process, these complexes were stacked at the interface between the low-conductivity sample zone and the running buffer zone. In the meantime, the EOF pump pushed out the water in the sample zone. The final result was that the stacked complexes were focused at the start point of capillary. Once the water in the sample zone was totally pushed out, the subsequent separation began. As the EOF in the running buffer had been completely suppressed by the presence of HEC, these concentrated [Fe^{III}EDTA]$^-$ moved towards the anode until they were detected when passing through the capillary window. Therefore, this stacking resulted in an improved sensitivity and high peak efficiency.

11.3.2
Choice of Metal Chelate

EDTA in the aqueous sample would be preliminarily chelated to nickel, calcium, magnesium and iron(III). To obtain the highest possible CE-UV signal intensity, it is advantageous controlling all available EDTA to a chromophoric metal complex that is highly stable. There are many competing factors that determine which EDTA chelate is most readily formed, while copper and iron have both been commonly used for the analytical determination of EDTA in water. The iron(III)-EDTA complexes has the highest formation constant of all these cations. After heating the water sample at 90 °C with Fe(III) for three hours, at least 95% $[Ni^{II}EDTA]^{2-}$ and most of other formation of EDTA would be converted to $[Fe^{III}EDTA]^-$ (Nowack et al. 1996).

$$MeEDTA + Fe(III)_{aq} = Fe^{III}EDTA + Me_{aq} \tag{11.4}$$

It was reported that $[Ni^{II}EDTA]^{2-}$ could be used for the determination of EDTA by CE-MS (Sheppard and Henion 1997). However, as its optimal UV wavelength is at 214 nm, strong interference will cause a high background and poor sensitivity. The complex ion $[Fe^{III}EDTA]^-$ could be detected at 258 nm, which is a common used UV wavelength with little background absorbency.

Consequently, iron(III) was chosen as the complex cation. To avoid photolysis of the $[Fe^{III}EDTA]^-$, the samples were stored in the dark between steps.

11.3.3
Effect of Hydroxyethylcellulose

In large volume sample injection, in order to achieve the sample stacking without sacrificing the high resolution of CE, the sample matrix should be removed without causing loss of the stacked sample prior to separation. For this purpose, Burgi et al. (1993) used a polarity switching to remove the large water plug (Chien and Burgi 1992). Pálmarsdóttir et al. applied a backpressure to compensate the EOF while stacking (Pálmarsdóttir and Edholm 1995). Another convenient technique is adding a modifier in the background electrolyte to reduce or reverse EOF (Zhu and Lee 2001). Lee even suppressed the EOF by adjusting the pH in the acidic range (He and Lee 1999). In our work, a nonionic hydroxylic polymer – hydroxyethylcellulose (HEC) – was added so that both sample stacking and subsequent separation could proceed consecutively under the same voltage.

HEC has been used to modify the inner wall of the capillary and to suppress the EOF at low pH condition (Zhu and Lee 2001). At pH 4.1, with addition of 0.05% HEC, the neutral marker N,N'-dimethylformide (DMF) was injected and separated at 30 kV, but it was not detected after 60 min. The DMF peak was then pushed through the capillary window by pressure and was detected as it passed by the detection window. This showed that the DMF was moving toward the negative electrode under electroosmotic force, but the EOF was negligible.

Consequently, 0.05% HEC was added into the running buffer to suppress the EOF.

11.3.4
Effect of pH

During the pH range of 3.5–5.0, 100% of iron(III)-EDTA complexes could be deprotonated. Considering that iron(III) hydroxide could form readily at pH > 5, thereby reducing the iron available for the EDTA complexation, buffer containing 50 mM disodium tetraborate, adjusted to pH 4.1 with 10% phosphoric acid, was used to protect precipitation of iron, where at least 95% iron(III)-EDTA complexes existing in the deprotonated forms.

Additionally, pH 4.1 was also a suitable condition for HEC to suppress the EOF.

11.3.5
Effect of Sample Volume

Aqueous samples containing [Fe^{III}EDTA]$^-$ (equivalent to 1.0 mg l^{-1} EDTA) were injected into the capillary at various injection time (0, 10, 20, 30, 40, 50, 60, 70 s) at 3.0 psi inlet pressure. After sample injection, a negative voltage (−20 kV) was applied for both sample stacking and subsequent separation. Figure 11.2 compares the impacts of injection time to UV-absorbent response peak and CE current. In Fig. 11.2a, the peak areas were proportional to the injection time between 10–50 s. If the injection time was longer than 50 s, the sample zone would surpass the capillary window and even be pumped out of the outlet of the capillary during the injection step. Small peaks 6' and 7' at the beginning of the separation indicated that part of [Fe^{III}EDTA]$^-$ had surpassed the detection window during the injection. When they moved backwards during the stacking process, they were detected. Generally, as the instrument will do an auto zero before starting running, the surpassing of the analytes into the detection cell before running will result in a high background, which can diminish the stacked analyte signals. That is the reason why the response areas of peaks 6 and 7 were smaller than that of peak 5. Thus, in present LVSEP method, the maximum injection volume was the capillary volume between the inlet and the detection window. In our following experiment, injection time of 50 s at 3.0 psi was used.

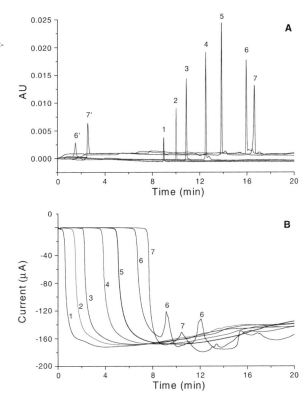

Fig. 11.2.
Effect of injection time on the peak reponse and current. Injection at 3.0 psi was performed. Injection time: (1)–(7) corresponding to 10–70 s with a step of 10 s. All other conditions are described in the Experimental Section

As far as the electric current was concerned, it was initially low due to the high resistance of aqueous sample solution. Taking example of 50 s as the injection time, with the water being pluged out, the current gradually increased from 0.5 to 1 µA during the first 4 min after power application. Then the current ascended sharply. A flat was shown in current-time plot (Fig. 11.2b), which was named as "flat time" by us, with a point of tangency followed it directly. The flat time was just the time that it took for the water to be removed out of the capillary. It also could be considered as the start point of the separation of the stacked analytes. Interestingly, despite different inject volume of analytes being applied, the final currents after stacking were almost the same, which means that the final running current had no relationship with the sample volume.

11.3.6
Calibration

To test the working range of present method, calibration curve was carried out as follows. A series of calibration solutions were prepared with EDTA concentrations ranging from 2.0 µg l^{-1} to 200.0 mg l^{-1}. The peak response vs. analyte concentration was linear between 5.0 and 600.0 µg l^{-1} as well as between 0.7 and 30 mg l^{-1} (Fig. 11.3). Regression coefficients were 0.9988 and 0.9990, respectively. The regression equations of calibration curves were $y = 6.8 \times 10^4 x + 5.6 \times 10^2$ and $y = 5.6 \times 10^4 x + 4.8 \times 10^4$, respectively, where y was the peak area (Au min) and x was the concentration of EDTA (mg l^{-1}). Based on a signal-to-noise ratio of 3, the detection limit was as low as 2.0 µg l^{-1}. For sample concentrations lower than 100 µg l^{-1}, random standard deviations (RSDs) were determined to be 3–6% during a working day, showing this method possessing excellent precision.

11.3.7
Comparison with Classical CZE Method

The negatively charged complex [Fe^{III}EDTA]$^-$ could be determined by using classical CZE method. The optimal condition was as follows. The pressure injection (0.5 psi, 5.0 s) was used to introduce the sample into the capillary. A positive voltage (+30 kV)

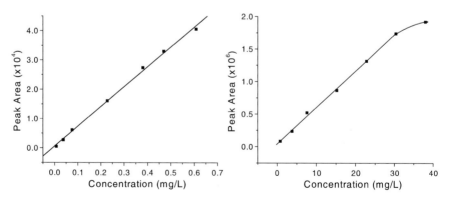

Fig. 11.3. Calibration curves

Fig. 11.4.
Electrophoregrams obtained under a classical CZE method of 200 mg l^{-1} EDTA in the presence of 200 mg l^{-1} DMF and **b** LVSEP method of 0.5 mg l^{-1} EDTA in the presence of 200 mg l^{-1} DMF. Peak identities: (1) DMF; (2) EDTA

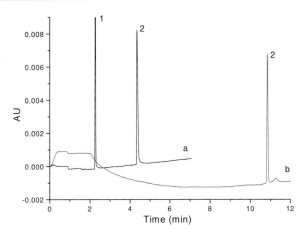

was applied for separation. Running buffer was 50 mM sodium borate solution (pH 8.5). The peak area was proportional to the concentration of EDTA ranging from 5.0 to 200.0 mg l^{-1} with a detection limit of 2.0 mg l^{-1}. In comparison, the LVSEP method has much lower detection limit of 2.0 μg l^{-1} and wider linear range, which indicated that over 1 000-fold enhancement had been achieved on LVSEP method. Figure 11.4 compared the peak response of [Fe^{III}EDTA]$^{-}$ under classical CZE method and LVSEP method. The DMF peak could not be detected by using the LVSEP technique, which resulted from the EOF being suppressed by HEC in this case.

11.3.8
Analytical Application

Because of the salt and other substances in river water and wastewater, the conductivity of the matrix is much higher than that of ultra pure water. This would present some difficulties in stacking the anions in these high-contaminated water samples. However, to the drinking water that has relatively clean matrix, present novel LVSEP technique could be applied successfully. This could be proved by using LVSEP method to stack the [Fe^{III}EDTA]$^{-}$-spiking tap water sample. When both of the tap water sample and deionized water sample were spiked with 100 μg l^{-1} of [Fe^{III}EDTA]$^{-}$ and analyzed by this LVSEP technique, the electropherograms showed that their response peak areas were almost the same.

To analyze the EDTA in real samples, all free EDTA and its metal complexes should be converted to [Fe^{III}EDTA]$^{-}$ firstly. 50 ml of tap water sample and EDTA-spiking tap water samples were put in opaque polyethylene (PE) bottles to avoid photolysis of [Fe^{III}EDTA]$^{-}$. 1.5 ml of 1 mM NH$_4$Fe(SO$_4$)$_2$ was added to each of these samples. After the obtained sample solutions were heated in water bath at 90 °C for 4 h, they were ready for LVSEP CE analyses. Additional preconcentration step could also be performed by evaporation of the sample solution to 5 ml with 10-fold concentration factor.

Table 11.1 listed the analysis results of the tap water sample and EDTA-spiking tap water samples. These results indicated that present LVSEP method could be applied in the determination of EDTA in drinking water with excellent accuracy (Recov-

Table 11.1. Determination of EDTA drinking water

Spiked (µg l⁻¹)	Found (µg l⁻¹)	Recoveries (%)	RSD ($n = 3$) (%)
0	0.4	–	5.0
1.0	1.2	80	6.3
2.0	1.9	75	7.2

ery > 70%) and precision (RSD < 8%). It should be mentioned that fresh samples should always be used for each injection when using LVSEP method. Depletion occurs after several prolonged stacking procedures, and the concentration of the sample would decrease if a single sample vial was used.

11.4
Conclusion

A novel and simple LVSEP technique for determination of EDTA in drinking water was developed. This method offered high sensitivity and precision, and could be performed on commercial instrumentation fully automatically. Compared with conventional CZE method, a 1 000-fold concentration factor could be smoothly achieved using this LVSEP method. To the best of our knowledge, it represents the highest sensitivity for EDTA analysis via CE. This LVSEP method could be used for the determination of EDTA in real drinking water samples.

References

Bersworth Chemical Co. (1954) The versenes. Powerful organic chelating agents for the exacting chemical control of cations. Framingham, MA
Burgi DS et al. (1993) Anal Chem 65:3726–3729
Bürgisser CS, Stone AT (1997) Determination of EDTA, NTA, and other amino carboxylic acids and their Co(II) and Co(III) complexes by capillary electrophoresis. Environ Sci Technol 31:2656–2664
Chien RL, Burgi DS (1992) On-column sample concentration using field amplification in CZE. Anal Chem 64:489A–496A
Fishbein L, Flamm WG, Falk HL (1970) Chemical mutagens, environmental effects on biological systems. Academic Press, New York, pp 249–251
Gebauer P, Thormann W, Boèek P (1992) Sample self-stacking in zone electrophoresis – theoretical description of the zone electrophoretic separation of minor compounds in the presence of bulk amounts of a sample component with high mobility and like charge. J Chromatogr 608:47–57
He Y, Lee HK (1999) Large-volume sample stacking in acidic buffer for analysis of small organic and inorganic anions by capillary electrophoresis. Anal Chem 71:995–1001
Kari FG, Giger W (1995) Modeling the photochemical degradation of ethylenediaminetetraacetate in the River Glatt. Environ Sci Technol 29:2814–2827
Kari FG, Hilger S, Canonica S (1995) Determination of the reaction quantum yield for the photochemical degradation of Fe(III)-EDTA – implications for the environmental fate of EDTA in surface waters. Environ Sci Technol 29:1008–1017
Loyaux-Lawniczak S, Douch J, Behra P (1999) Optimisation of the analytical detection of EDTA by HPLC in natural waters. Fresenius J Anal Chem 364:727–731
Nowack B, Kari FG, Hilger SU, Sigg L (1996) Determination of dissolved and adsorbed EDTA species in water and sediments by HPLC. Anal Chem 68:561–566

Owens G, Ferguson VK, Mclaughlin MJ, Singleton I, Reid RJ, Smith FA (2000) Determination of NTA and EDTA and speciation of their metal complexes in aqueous solution by capillary electrophoresis. Environ Sci Technol 34:885–891

Pálmarsdóttir S, Edholm LE (1995) Enhancement of selectivity and concentration sensitivity in capillary zone electrophoresis by on-line coupling with column liquid chromatography and utilizing a double stacking procedure allowing for microliter injections. J Chromatogr 693:131–143

Pozdniakova S, Ragauskas R, Dikcius A, Padarauskas A (1999) Determination of EDTA in used fixing solutions by capillary electrophoresis. Fresenius J Anal Chem 363:124–125

Randt C, Wittlinger R, Merz W (1993) Analysis of nitrilotriacetic acid (NTA), ethylenediaminetetraacetic acid (EDTA) and diethylenetriaminepentaacetic acid (DTPA) in water, particularly wastewater. Fresenius J Anal Chem 346:728–731

Schön P, Bauer KH, Wiskamp V (1997) Indirect determination of the sum of chelons in drinking and ground-waters by anodic stripping voltammetry. Fresenius J Anal Chem 358:699–702

Seegers WH (1967) Blood clotting enzymology. Academic Press, New York, pp 364–366

Sheppard RL, Henion J (1997) Quantitative capillary electrophoresis-ion spray tandem mass spectrometry determination of EDTA in human plasma and urine. Anal Chem 69:2901–2907

Venezky DL, Rudzinski WE (1984) Determination of ethylenediaminetetraacetic acid in boiler water by liquid chromatography. Anal Chem 56:315–317

Xue H, Sigg L, Kari FG (1995) Speciation of EDTA in natural-waters – exchange kinetics of Fe-EDTA in river water. Environ Sci Technol 29:59–68

Zhu L, Lee HK (2001) Field-amplified sample injection combined with water removal by electroosmotic flow pump in acidic buffer for analysis of phenoxy acid herbicides by capillary electrophoresis. Anal Chem 73:3065–3072

Part II
Toxic Metals

Part II

Chapter 12

A Framework for Interpretation and Prediction of the Effects of Natural Organic Matter Heterogeneity on Trace Metal Speciation in Aquatic Systems

M. Filella · R. M. Town

Abstract

A comprehensive collection, critical analysis and interpretation of the data published for complexation of trace metals by natural organic matter (refractory humic-type substances and non degraded biota material) and filtered whole natural waters have been undertaken. Our interpretation framework considers the role of metal loading conditions and analytical detection windows on the complexation data in a systematic way. Its application has shown that the same patterns are observed for complexation of a range of trace metals with different types of natural organic matter: apparently stronger binding sites are utilised at lower metal ion loadings, progressively weaker sites contribute to metal complexation at higher loadings. Taking into account the detection window of the technique employed greatly improves the internal consistency of the overall data. Despite the widely recognised role played by natural organic matter in trace metal fate in aquatic systems, and the enormous number of publications in this field, there is in fact very few data available for metals other than copper, and that were determined at environmentally relevant pH values and metal loading conditions. This situation hinders reliable determination of complexation parameters with predictive value.

Key words: binding heterogeneity, analytical detection window, humic substances, natural organic matter, natural waters, trace metal speciation

12.1 Introduction

The physicochemical form in which a metal ion occurs (i.e. its speciation) determines its mobility, bioavailability and toxicity in the environment. The significance of natural organic matter (NOM) in determining trace metal speciation in aquatic systems has been frequently reported (e.g. Buffle 1988; Bidoglio and Stumm 1994; Town and Filella 2002b), but few details are known about aspects such as the relative contribution of different types of NOM, e.g. humic substances vs. biota exudates, to overall metal binding, or the relative degree to which different trace metals are bound by a given NOM type.

Natural organic matter, such as humic substances and biota exudates and cell walls, contains a range of binding site types that can interact with trace metals to varying degrees. These site types can be divided into so-called *(i)* major sites that represent up to ca. 90% of total sites, e.g. carboxylate and phenolate functional groups in humic substances, and *(ii)* minor sites that correspond to a small fraction of the total sites but comprise a relatively large number of site types, e.g. N- and S-containing functional groups in humic substances. It is the minor sites, with their associated wide range of complexation constants, that are proposed to be responsible for buffering trace metal concentrations in natural systems (Buffle et al. 1990b).

Progress in determination of reliable physicochemical parameters to describe metal complexation by NOM is hindered to some extent by the inherent physicochemical complexity described above. More than 20 years ago, the difficulties encountered when trying to compare complexation constants reported in the literature for trace metals – humic-type substances binding highlighted the key effect of the experimental metal:ligand ratio used on the value of the constant determined. Complexation parameters measured under conditions of low metal ion loading correspond to more stable complexes than do the parameters determined in the presence of lower ligand excess. This led to the postulation of the existence of an effective continuum of ligand sites with varying metal affinity in the humic molecules. That is, if such a ligand is titrated with metal ions, then the most strongly binding sites will initially be involved in binding, with the weaker sites making an increasing contribution as the metal ion concentration is increased. Two important consequences for the understanding of trace metal behaviour in aquatic systems arise from these observations on humic substances: *(i)* at the low levels of metal concentrations present in most natural systems and at the corresponding typical excess of NOM binding sites, the strongest (and least labile) sites will be those responsible for metal binding, and *(ii)* NOM is expected to play an important role in the buffering of free metal ion concentrations in natural systems.

Some features of, and consequent limitations to, current understanding of trace metal-NOM interactions are: *(i)* Existing data interpretation methods that take into account NOM binding heterogeneity as it pertains to metal ion loading have been infrequently applied to real systems. *(ii)* Most of the trace metal-NOM complexation studies have been performed under conditions far from those existing in real aquatic systems (typically, unrealistically high metal ion concentrations are used), and this will often lead to an underestimation of the real extent to which metals are bound by NOM in natural waters (only the most weakly binding sites are "seen"). *(iii)* Very few reliable studies have been performed on NOM other than isolated humic-type substances (which may not necessarily be representative or typical of all NOM components in natural waters). The nature and properties of NOM will vary across systems, and amongst size fractions of a given system.

We have undertaken a critical analysis and interpretation of all data published for complexation of trace metals by NOM in natural water systems with the aim of developing a robust framework that is applicable across systems and that has predictive value. This study is based on the compilation of data published over the past 35 years and on the reinterpretation of these data in terms of the binding heterogeneity of NOM. It includes collection of data on trace metal binding by humics, biota-derived substances (algae and bacteria) and whole water samples. This paper summarises the approach used in the data reinterpretation procedure and illustrates the insights attainable by application of this framework to trace metal binding by various types of NOM.

12.2 Methodology

12.2.1 Data Compilation

The metal titration data for a given individual system is usually modelled assuming complexation by discrete ligands. Ligand concentrations ($\tilde{C}c$) and corresponding conditional stability constants ($\tilde{K}$) are calculated by transformation of the metal titration curves using different linearisation methods. The most widely used are the graphical solutions derived by Scatchard (Scatchard plot) and the method derived independently by van den Berg and later by Ružiæ (van den Berg-Ružiæ plot). Both are easily derived from the expression of the conditional stability constant. The relative merits of each approach have been discussed (Apte et al. 1988).

For the Scatchard transformation (Scatchard 1949; Scatchard et al. 1957), the ratio c_{ML}/c_M is plotted against c_{ML}

$$\frac{c_{ML}}{c_M} = \tilde{K}\tilde{C}c - \tilde{K}c_{ML} \tag{12.1}$$

where c_M = free metal ion; $c_{ML} = c^*_{ML} - c_M$, c^*_{ML} = total metal concentration; $\tilde{C}c$ = total ligand concentration having a conditional stability constant $\tilde{K}$. The slope equals $\tilde{K}$, the y-intercept is the product of $\tilde{K}$ and $\tilde{C}c$ and the x-intercept is $\tilde{C}c$.

For the Ružiæ-van den Berg linearisation, sometimes called the Ružiæ method or van den Berg method (van den Berg and Kramer 1979; van den Berg 1982; Ružiæ 1982), the c_M/c_{ML} ratio is plotted against c_M,

$$\frac{c_M}{c_{ML}} = \frac{1}{\tilde{K}\tilde{C}c} + \frac{1}{\tilde{C}c}c_M \tag{12.2}$$

The reciprocal of the slope gives the total ligand concentration and the $\tilde{K}$ value can be calculated from the reciprocal of the y-intercept ($\tilde{K}\tilde{C}c$).

If the results are linear over the range of data, the titration data is modelled with one class of metal binding ligand. If several classes of ligands are present, the transformed plots of the titration data as defined above will not be linear. When two linear segments are obtained, each corresponds to a domain where one of the two types of site is dominant. Equations equivalent to Eq. 12.1 and Eq. 12.2 have been developed for such a case (Buffle 1988, page 231). For a multiligand system (more than 3 ligand classes), a curve is produced and the system of equations needs to be solved numerically by successive approximations.

We have compiled literature log $\tilde{C}c$ and log $\tilde{K}$ values reported for trace metal binding by different types of NOM. In some cases, the values included in our compilation were those published directly by the authors concerned by following the above-described data treatment methods, in others (and where the data was available) we scanned figures (Silk Scientific 1998) containing appropriate metal binding information (e.g. free metal as a function of bound metal concentration) and calculated the parameters ourselves from the corresponding Scatchard plots. Data collected will be freely available soon in the Internet database DOOM-MB (http://www.unige.ch/DOOM-MB/).

12.2.2
Data Reinterpretation: From Discrete Ligands to Continuous Functions

One of the few approaches used to describe the stability of metal complexes formed by humic substances that accounts for the metal loading dependence of NOM binding site affinity is the Differential Equilibrium Function (DEF). It is based on the definition of a differential intensity parameter, K^*, that is a weighted average mean of the equilibrium constants K_i, for all sites present in the humic substance (Gamble 1970; Altmann and Buffle 1988; Buffle et al. 1990a; Buffle and Filella 1995). The weighting factor depends on the mole fraction and degree of occupation (θ_i) of each site i. The DEF is not equivalent to a molecular complexation model that describes complexation at each individual site. However, it is most useful in making predictions of the complexation behaviour of the system as a whole. In theory, the DEF function can take any shape. In practice, for very heterogeneous ligands such as humic substances, $\log \theta = f(\log K^*)$ is approximately linear and can thus be described by two parameters: a slope and an intercept. The slope (Γ) is a measure of the degree of heterogeneity and buffering intensity of the system ($\Gamma = 1$ corresponds to the homogeneous case). For natural systems, Γ values are typically in the range 0.3–0.7 (Buffle et al. 1990b).

Although relatively few single NOM-metal titration curves have been interpreted by using continuous equilibrium functions, there is a plethora of such curves described in terms of the one or two site models described above. It is now well established that the so-called metal complexation capacity ($\widetilde{C}c$) of NOM and the corresponding average equilibrium quotient ($\widetilde{K}$) given by these methods depend on the analytical technique and the conditions used and that the $\log \widetilde{C}c$ and $\log \widetilde{K}$ values only have meaning when considered as a data couple (Buffle 1988; Town and Filella 2000a; Filella et al. 2002). It has been shown however that the ($\log \widetilde{C}c$, $\log \widetilde{K}$) couples are analogues of ($\log \theta$, $\log K^*$) couples (DEF curves) (Buffle et al. 1984; Buffle 1988) and thus plots of $\log \widetilde{C}c$ vs. $\log \widetilde{K}$ for similar types of NOM, and determined under the same conditions, will mimic the underlying DEF dependence.

We have included all published data in our compilation and evaluation of the $\log \widetilde{C}c$ and $\log \widetilde{K}$ values. Our analysis is based solely on evaluating trends in compiled published data, with no a priori assumptions or imposed models. There is a certain degree of dispersion in many cases, due to both natural variations and experimental limitations. Nevertheless, we consider this to be the most appropriate approach to elucidating the important unifying factors that determine the global behaviour of trace metals: any observable trends under these conditions are likely to be significant determinants.

12.3
Results and Discussion

12.3.1
Pb(II) Complexation by Humics

Figure 12.1a shows all compiled $\log \widetilde{K}$ values for Pb(II) complexation by humic-type substances as a function of pH. Different symbols correspond to different studies (see Filella and Town (2001) for details). As expected, a wide variation in $\log \widetilde{K}$ values as a function

Fig. 12.1.
a Collection of Pb(II) – humic substances complexation equilibrium quotients, $\log \tilde{K}$ (1:1 stoichiometry) as a function of pH. (**b**) Changes in the complexation capacity ($\log \tilde{C}c$) with the corresponding equilibrium quotient ($\log \tilde{K}$) for the complexation of Pb(II) with humic substances at two different pH values 4.5–5.0 and 6.8–8.0. Humic substance origin: △, seawaters; ▲, freshwaters; △, soils and sediments; ○, sewage sludge; ●, peat and coal. The linear regression through the pH 4.5–5.0 and pH 6.8–8.0 data is shown as a *solid* and a *dashed line*, respectively. See Town and Filella (2002a) for references of the individual data points

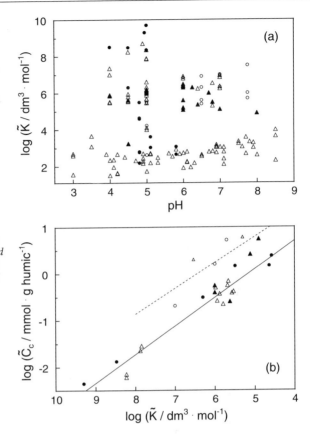

of pH is observed. A huge variation in $\log \tilde{K}$ values still remains when this source of variation is taken into account, i.e. by considering $\log \tilde{K}$ at only one pH value. However, most of this variability is easily explained when metal loading is considered (Fig. 12.1b) by plotting $\log \tilde{C}c$ vs. $\log \tilde{K}$ for a given pH range: linear correlations are obtained.

This type of behaviour has been observed so far for interactions of humic substances with Cu(II) (Buffle et al. 1984; Filella et al. 2002), Pb(II) (Filella and Town 2001; Town and Filella 2002a), and Ni(II) and Co(II) (Town and Filella 2002c) and can be anticipated for similar trace metals. As noted above, meaningful comparison of relative binding strengths of given metal-ligand systems must be made at the same metal loading, i.e. $\tilde{C}c$ value. Construction of $\log \tilde{C}c$ vs. $\log \tilde{K}$ plots for each metal-NOM system allows the predictive K^* and Γ values for the given conditions (e.g. pH) to be established. This approach can be readily incorporated into existing speciation codes (Huber et al. 2002).

Two important consequences that emerge from our data analysis approach are that: (*i*) as a consequence of the dependence of metal binding intensity for the trace metal-humic systems on the metal loading conditions, the effective binding at typical natural water concentrations (metals present at very low concentration; humic binding sites in excess) could be grossly underestimated (by several orders of magnitude) if binding is predicted by complexation parameters that have been determined at higher concentration levels. Unfortunately, this is the usual approach adopted by many speciation models (Huber et al. 2002); *(ii)* the wealth of information currently hidden amongst the vast amount of published individual $\tilde{C}c$ and $\tilde{K}$ data can be revealed when appropriate methods of data mining and treatment are used, i.e. as illustrated herein.

12.3.2
Not All Natural Organic Matter is Humics

In natural waters, the NOM present will be a combination of refractory NOM either from terrestrial (properties comparable to soil-derived humic substances) or aquatic sources and non-degraded biota material (bacteria, phytoplankton, and their exudates). The relative contribution of each source will depend on the particular system involved and on the spatial scale considered, i.e. the direct influence of biota may be more local in scale than the effects of the more refractory NOM components. However, the understanding of metal complexation in systems where the nature and flux of NOM is dominated by biological activity is far less developed than that for humic substances, yet the consequences can be of great ecological significance. For example, algae may dominate complexation under certain conditions, e.g. in freshwaters with low humic con-

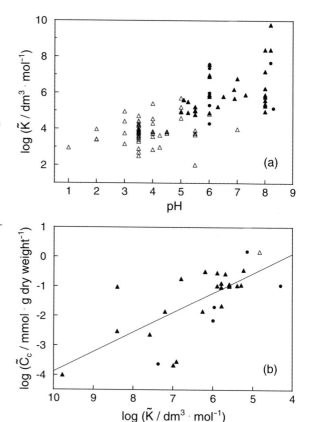

Fig. 12.2.
a Collection of Pb(II) – biota-derived substances complexation equilibrium quotients, log $\tilde{K}$ (1:1 stoichiometry) as a function of pH. △, cells; ▲, cells plus exudates; ●, exudates. **b** Changes in the complexation capacity (log $\tilde{C}c$) with the corresponding equilibrium quotient (log $\tilde{K}$) for the complexation of Pb(II) with biota-derived substances. pH 6.0–8.3. △, cells; ▲, cells plus exudates; ●, exudates. See Town and Filella (2002a) for references of the original publications. Reproduced with permission from Town and Filella (2002a)

tent and in seawater (Wangersky 1986). Microorganisms can provide a high buffer capacity for metal ions in solution (Kiefer et al. 1997). Different types of complexing substances need to be considered: alga and bacteria exudates and cell walls. Analogous to humic substances discussed above, extracellular ligands and cell walls are heterogeneous with respect to nature of functional groups and molecular weight (Gonçalves et al. 1987). Metal complexation by these substances can thus be analysed by an approach equivalent to that applied for the more refractory humic substances. When this approach is used (Town and Filella 2000a; Town and Filella 2002a), biota exhibit the same trends with respect to degree of complexation as a function of pH (Fig. 12.2a) and metal ion loading (Fig. 12.2b) as observed for humic substances. This behaviour has been observed so far for interactions of biota with Cu(II) (Town and Filella 2000a) and Pb(II) (Town and Filella 2002a), and it is anticipated to be applicable to other trace metals, although, in many cases there is a paucity of useful data available (Town and Filella 2002c).

The relative binding affinity of biota as compared to the more refractory humic substances can be assessed by comparison of the respective $\log \tilde{C}c$ vs. $\log \tilde{K}$ plots. On this basis, humic substances are shown to be more strongly binding than biota towards Cu(II) (Town and Filella 2000a) and Pb(II) (Town and Filella 2002a); see following section. It must be noted, however, that the particular NOM component that dominates complexation will always be system dependent (as noted above).

12.3.3
What Whole Water Complexation Data Can Tell Us

In addition to the data reported for isolated NOM fractions discussed above, there is also a large body of literature complexation data that corresponds to metal binding studies (titrations) on untreated non-fractionated water samples (usually samples are only filtered; ca. 0.45 μm). The NOM in this context is quantified as dissolved organic carbon (DOC). These studies are particularly popular among oceanographers. Analogous to the approach widely used for interpretation of titration data for isolated NOM, measurements on whole water samples are often interpreted in terms of the existence of one or two classes of binding sites (Town and Filella 2000b). We have shown (Town and Filella 2000a), via a critical analysis of the existing data, that it is not necessary to invoke the existence of any particular classes of ligands in order to explain the degree of metal complexation observed. Compiled published $\log \tilde{C}c, \log \tilde{K}$ data for whole waters exhibits the same linear relationship as observed above for isolated NOM components, as illustrated for Cu(II) and Pb(II) in Fig. 12.3a,b. The same behaviour is observed for Cd(II), Zn(II) (Town and Filella 2000a), Co and Ni (Town and Filella 2002c), and is anticipated to be generally applicable to complexation of many trace metals.

Comparison of the $\log \tilde{C}c$ vs. $\log \tilde{K}$ plots for whole waters with those for isolated humic substances and biota may offer some insights into the relative significance of various NOM fractions for trace metal binding in aquatic systems. This comparison is shown for Cu(II) and Pb(II) in Fig. 12.3a,b. All data shown in these figures were obtained at a similar pH and detection window (see below). The $\log \tilde{C}c$ values for the whole waters were converted to a common scale (mmol (g C)$^{-1}$) via the reported DOC concentrations for each freshwater system or by assuming a DOC of 850 μg dm^{-3} as typical for surface seawater (Guo and Santschi 1997)(DOC: dissolved organic carbon).

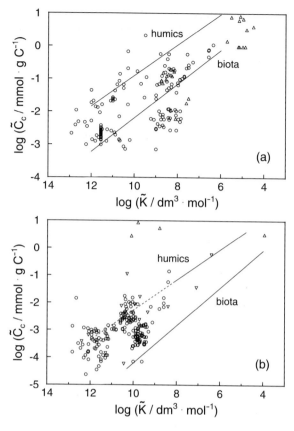

Fig. 12.3.
Comparison of **a** Cu(II) and **b** Pb(II) binding curves ($\log \tilde{C}c - \log \tilde{K}$) for isolated humic substances (Cu: pH 8.0; Pb: pH 6.8–8.0) and biota (Cu: pH 7.0–8.0; Pb: pH 6.0–8.3) with that for whole water systems. ○, seawaters; △, freshwaters; ▽, estuaries. The *solid lines* are best fits for the range of experimental data points for the isolated components. Analytical window of whole water samples, $\log \alpha = 1–3$. Data converted to mmol (g C)$^{-1}$ scale as explained in the text. Part **a** adapted from Town and Filella (2000a); part **b** reproduced with permission from Town and Filella (2002a)

The binding curves for both Cu(II) and Pb(II) in natural waters lie within those for isolated humic and biota complexants. The relative importance of both types of ligands for complexation in a given system will depend on the amounts of each present. Extensive parameters must always be considered in making any generalisations about trace metal behaviour in aquatic systems. For example, the nature and amount of NOM will vary across systems (e.g. freshwater and marine NOM have different compositions), and large seasonal variations can occur within a given system.

12.3.5
Sources of Dispersion in Reported Data across All Systems

As observed in Fig. 12.1–12.3, there is a certain amount of dispersion in most of the $\log \tilde{C}c$ vs. $\log \tilde{K}$ plots. The data shown in these figures all correspond to a narrow pH range, the presentation format takes the metal loading into account, and all values were measured by techniques with similar detection windows. Some dispersion is inevitable when different analytical techniques and authors are considered simultaneously. In the case of biota data, the higher variability observed as compared to humic compounds can be ascribed to *(i)* the large number of different organisms considered, *(ii)* the physico-chemical conditions being poorly controlled or not defined, *(iii)* the conditions under which each study was conducted not being directly comparable, and *(iv)* the dynamic nature of the organisms themselves, e.g. exudate production.

The ionic strength (I) at which measurements were made does not appear to have a significant influence on the complexation parameters. This is illustrated by comparing the data for whole seawaters (usually obtained at the marine ionic strength) with that for freshwaters (typically adjusted to $I =$ ca. 0.01 mol dm^{-3} prior to measurement), as shown for Cu(II) in Fig. 12.3a. No significance difference between the data from each source is observed. The effect of this factor is thus expected to be minor relative to the effect related to the dependence of binding strength on metal loading.

12.3.6
Significance of the Analytical Detection Window

In terms of experimental factors, by far the greatest influence on the measured complexation parameters is that of the analytical detection window of the technique employed. The analytical detection window determines the concentration of ligands and the relative stability of their metal complexes which can be detected in a sample by a given technique. One limit of this window (corresponding to the weakest detectable complexation sites) is determined by the ability of a particular technique to detect a decrease in the metal signal due to the complexation of the metal by a natural ligand, and the other limit (strongest detectable sites) is determined by the method's detection limit (see van den Berg et al. (1990) for a detailed explanation of this concept and its calculation). The position of this window (and hence the accessible binding parameters) may be varied by using the same method under different experimental conditions, or by using different analytical methods having different detection windows. A wide range of different techniques, with a corresponding range of detection windows, have been used in studies on complexation in whole water samples; there is much less variety amongst techniques used in studies on isolated humic substances or biota (detection window, $\log \alpha$, typically 1–3).

Log $\widetilde{C}c$-log $\widetilde{K}$ data grouped according to detection window should thus show a series of parallel lines. This is indeed the case, as shown for Cu(II) in Fig. 12.4. The data on the left most side of the plots as presented represents the more stable complexes which are detected at higher α values. Direct comparisons of complexation data are only valid if the values are obtained within the same detection window. This applies to e.g. assessing which components may be more strongly binding than others (e.g. biota vs. humics; Fig. 12.3), and also to elucidating which trace metals may be more strongly bound by a given NOM component, e.g. Cu(II) vs. Pb(II).

12.3.7
Kinetic Considerations

Interpretation and comparison of complexation parameters ($\log \widetilde{C}c$, $\log \widetilde{K}$) within the analytical detection window framework explained above implicitly assumes that there is no contribution from kinetic factors. For a given detection window, it is assumed that all of the complexes that fall within the measurable range will in fact be detected.

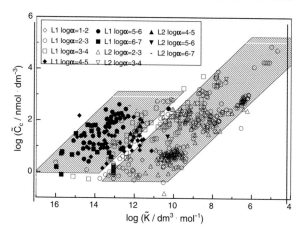

Fig. 12.4.
Effect of the analytical detection window, log α, on the binding parameters determined for Cu(II) in whole natural waters. The *shaded region on the left* encompasses most data in the log α range greater than 4; *shading on the right* encompasses log α < 4 data. L1 and L2 denote "strong" and "weak" binding sites as defined in the original publications; see Town and Filella (2000a) for details. Reproduced with permission from Town and Filella (2000a)

This implies that all such complexes are able to completely dissociate and be detected within the timescale of the analytical technique employed, i.e. are fully labile under the experimental conditions. This assumption may not always be valid. Since $\tilde{K} = k_a / \tilde{k}_d$, particular attention should be paid to the possible influence of kinetic factors when *(i)* the metal ion of interest has a low rate constant for water substitution in the inner coordination sphere, k_{-w}, e.g. Ni, $k_a = 3 \times 10^4$ s^{-1} vs. Cu, $k_a = 1 \times 10^9$ s^{-1} (Eigen 1963), and *(ii)* low metal-to-ligand ratios are used, at which the strongest and least labile binding sites are involved in complexation. For heterogeneous NOM ligands, there will be a distribution of complex dissociation rate constants that is dependent on the metal ion loading and parallels that of the thermodynamic stability constants (Fig. 12.5). Thus, complexes formed under typical environmental conditions (low metal loading) will be less labile than those formed under higher loading conditions (as generally used in laboratory determinations). This dependence of complex lability on metal loading has important implications both for the understanding of metal behaviour in natural systems and for the interpretation of experimental results.

12.4
Conclusions

Several key points are highlighted by our framework for interpretation of metal-NOM complexation parameters.

- Metal complexation data for isolated NOM components at environmentally relevant pH conditions are scarce. Trace metal-humic interactions have been the object of many studies, particularly for copper, but most of the data pertain to soil-derived humics and have been collected at pH values that are not relevant for natural systems. The same problem applies to metal-biota-derived ligand equilibria; studies on isolated cells in particular are typically conducted under relatively acidic conditions.
- Complexation data for isolated NOM components at metal loading conditions relevant to aquatic systems are scarce. Binding studies are typically performed at much higher concentrations than those found in the environment, and thus only the weaker

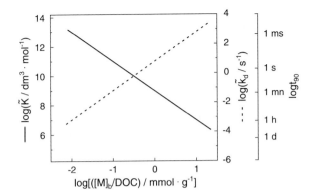

Fig. 12.5.
Schematic representation of the theoretical variation of log $\tilde{K}$ and log $\tilde{k}_d$ with metal loading. Based on Fig. 6.29 in Buffle (1988)

binding sites are seen. Direct application of such complexation parameters to prediction of metal binding under environmentally-relevant conditions is not valid. Speciation codes that incorporate metal-NOM complexation parameters determined at high metal loadings thus risk to seriously underestimate the extent of metal-NOM binding under environmentally relevant conditions (Huber et al. 2002).

- The significance of the analytical detection window must be recognised. The range of detection windows used in studies of whole waters is far greater, and often very high values are used, in contrast to the conditions employed for measurements on isolated NOM components. Comparison of complexation parameters across metals and across systems is only valid if the data were all collected within the same detection window.
- Kinetic factors, i.e. metal complex lability, must also be considered. Under environmentally relevant conditions of low metal concentrations and large ligand excess, the most stable and least labile binding sites will be involved in complexation (Filella and Town 2000; Town and Filella 2000c). Reliable data interpretation of complexation parameters determined under these conditions must consider the contribution of kinetic factors (i.e. the relative timescales of the analytical technique and the rate of dissociation of the complexes under study).

References

Altmann RS, Buffle J (1988) The use of differential equilibrium functions for interpretation of metal binding in complex ligand systems: its relation to site occupation and site affinity distributions. Geochim Cosmochim Acta 52:1505–1519

Apte SC, Gardner MJ, Ravenscroft JE (1988) An evaluation of voltammetric titration procedures for the determination of trace metal complexation in natural waters by use of computer simulation. Anal Chim Acta 212:1–21

Bidoglio G, Stumm W (1994) Chemistry of aquatic systems: local and global perspectives. Kluwer Academic, Dordrecht

Buffle J (1988) Complexation reactions in aquatic systems. An analytical approach. Ellis Horwood, Chichester

Buffle J, Filella M (1995) Physico-chemical heterogeneity of natural complexants: clarification. Anal Chim Acta 313:144–150

Buffle J, Tessier A, Haerdi W (1984) Interpretation of trace metal complexation by aquatic organic matter. In: Kramer CJM, Duinker JC (eds) Complexation of trace metals in natural waters. Martinus Nijhoff/Dr W. Junk. The Hague, pp 301–316

Buffle J, Altmann RS, Filella M, Tessier A (1990a) Complexation by natural heterogeneous compounds: site occupation distribution functions, a normalized description of metal complexation. Geochim Cosmochim Acta 54:1535–1553

Buffle J, Altmann RS, Filella M (1990b) Effect of physico-chemical heterogeneity of natural complexants. Part II. Buffering action and role of their background sites. Anal Chim Acta 232:225–237

Eigen M (1963) Fast elementary steps in chemical reaction mechanisms. Pure Appl Chem 6:97–115

Filella M, Town RM (2000) Determination of metal ion binding parameters for humic substances. Part 1. Application of a simple calculation method for extraction of meaningful parameters from reverse pulse polarograms. J Electroanal Chem 485:21–33

Filella M, Town RM (2001) Heterogeneity and lability of Pb(II) complexation by humic substances: practical interpretation tools. Fresenius J Anal Chem 370:413–418

Filella M, Town RM, Buffle J (2002) Speciation in fresh waters. In: Ure AM, Davidson CM (eds) Chemical speciation in the environment, 2nd edn. Blackwell Science, London, pp 188–236

Gamble DS (1970) Titration curves of fulvic acid: the analytical chemistry of a weak acid polyelectrolyte. Can J Chem 48:2662–2669

Gonçalves MLS, Sigg L, Reutlinger M, Stumm W (1987) Metal ion binding by biological surfaces: voltammetric assessment in the presence of bacteria. Sci Total Environ 60:105–119

Guo L, Santschi PH (1997) Composition and cycling of colloids in marine environments. Rev Geophys 35:17–40

Huber C, Filella M, Town RM (2002) Computer modelling of trace metal ion speciation: practical implementation of a linear continuous function for complexation by natural organic matter. Computers and Geosciences 28(5):587–596

Kiefer E, Sigg L, Schosseler P (1997) Chemical and spectroscopic characterization of algae surfaces. Environ Sci Technol 31:759–764

Ružiæ L (1982) Theoretical aspects of the direct titration of natural waters and its information yield for trace metal speciation. Anal Chim Acta 140:99–113

Scatchard G (1949) The attractions of proteins for small molecules and ions. Ann N Y Acad Sci 51:660

Scatchard G, Coleman JS, Shen AL (1957) Physical chemistry of protein solutions. VII. The binding of some small anions to serum albumin. J Am Chem Soc 79:12–20

Silk Scientific (1998) UN-SCAN-IT, automated digitizing system, Version 5.0

Town RM, Filella, M (2000a) Dispelling the myths: is the existence of L1 and L2 ligands necessary to explain metal ion speciation in natural waters? Limnol Oceanogr 45:1341–1357

Town RM, Filella M (2000b) A compilation of complexation parameters reported for trace metals in natural waters. Aquatic Science 62:252–295

Town RM, Filella M (2000c) Determination of metal ion binding parameters for humic substances. Part 2. Utility of ASV pseudo-polarography. J Electroanal Chem 488:1–16

Town RM, Filella M (2002a) Implications of natural organic matter binding heterogeneity on understanding lead(II) complexation in aquatic systems. Sci Total Environ 300:143–154

Town RM, Filella M (2002b) Size fractionation of trace metals in freshwaters: implications for understanding their speciation and fate. Rev Environ Sci Biotechnol 1(4):277–297

Town RM, Filella M (2002c) The crucial role of the detection window in metal ion speciation analysis in aquatic systems: the interplay of thermodynamic and kinetic factors as exemplified by nickel and cobalt. Anal Chim Acta 466:285–293

van den Berg CMG (1982) Determination of copper complexation with natural organic ligands in sea water by equilibration with MnO_2. I. Theory. Mar Chem 11:307–322

van den Berg CMG, Kramer JR (1979) Determination of complexing capacities of ligands in natural waters and conditional stability constants of the copper complexes by means of manganese dioxide. Anal Chim Acta 106:113–120

van den Berg CMG, Nimmo M, Daly P, Turner DR (1990) Effects of the detection window on the determination of organic copper speciation in estuarine waters. Anal Chim Acta 232:149–159

Wangersky PJ (1986) Biological control of trace metal residence time and speciation: a review and synthesis. Mar Chem 18:269–297

Chapter 13

Binding Toxic Metals to New Calmodulin Peptides

L. Le Clainche · C. Vita

Abstract

A 33-amino acid peptide corresponding to the helix-turn-helix motif of the calcium binding site I of *Paramecium tetraurelia* calmodulin has been synthesized and its binding loop stabilized by a specific disulphide bond. Analysed by electrospray mass spectrometry (ES-MS), circular dichroism (CD) and fluorescence, such a cyclic peptide is found to bind calcium, cadmium, terbium and europium ions with native-like affinity, and with 30 ±1 µM and 8 ±4 µM dissociation constants for calcium and cadmium ions, respectively. Metal binding induces an ordered conformation in the peptide, resembling that of the calmodulin site I. Interestingly, uranium, in the uranyl form, binds this peptide, as revealed by ES-MS, CD and fluorescence spectroscopy. Sequence mutation aiming to increase the binding cavity suppresses binding of calcium, cadmium and uranium ions, but preserves binding of lanthanide ions, showing that metal selectivity can be modulated by specific mutation in the binding loop. Such disulphide-stabilized peptide may represent a useful model to engineer new metal specificity in calmodulin variants. These novel proteins may be useful in the development of new biosensors to monitor metal pollution and to augment metal binding capability of bacterial and plant cells that can be used in biosorption techniques to (bio)remediate soils and waters contaminated by heavy metals.

Key words: calmodulin, metal chelation, heavy metals, uranium, fluorescence, biosensor, bioremediation

13.1 Introduction

Heavy metals represent a major source of important environment pollution in the industrialised society. In particular, nuclear power plants and other activities associated with nuclear research may become a source of environment pollution by particular metals, which may be intrinsically toxic to cells and living organisms, such as those coming from the nuclear fuel (uranium, plutonium, etc.), the fission products (strontium, selenium, technetium, cesium, iodine, etc.) or the activation products and liquid or gas wastes (cobalt, tritium, cesium, etc.) in nuclear reactors. To protect the environment it would be useful to develop specific analytical methods to monitor the presence of such toxic metals, but also effective options for metal remediation of soils and waters that might be contaminated in case of nuclear accidents.

Chemosensors or sensors based on fluorescence measurements, specific for several metals such as sodium, potassium, calcium and magnesium) are well known (Tsien 1993), particularly in aqueous media and within biological samples. New fluorescent protein biosensors specific for some metals have also been developed (Giuliano et al. 1995) and bacteria have been selected that contain appropriate promoters that can translate the presence of some pollutants into a bioluminescent signal (Belkin 1998; Gu and Chang 2001).

Chemical approaches are used for remediation of environments contaminated with toxic metals. However, they are often expensive and lack the appropriate specificity for the metals to be treated. The absorption of toxic metals of contaminated soils or waters by microbes and plants has been suggested as a viable strategy for environmental (bio)remediation (Raskin et al. 1997; Stephen and Macnaughton 1999). Biological approaches offer the potential for selective removal of toxic metals in a cost-effective remediation of large surfaces. Engineering proteins with increased metal binding capability in bacterial and plant cells should increase metal accumulation capability of such cells (Sousa et al. 1998) and will enhance the applicability of biosorption techniques over alternative metal removing techniques (Raskin et al. 1997; Stephen and Macnaughton 1999).

In order to develop new proteins binding strontium, uranium and cesium ions, which are potential contaminants derived from nuclear plants, we planned to change the calcium specificity of protein calmodulin to chelate such toxic metal ions.

Calmodulin (CaM) is a ubiquitous, calcium-binding regulatory protein. In response to an increase in the intracellular calcium concentration, it undergoes an important structural change that allows the protein to bind and activate various cellular protein targets. Calmodulin includes four Ca^{2+} binding sites, named EF-hand, in two distinct domains (Wilson et al. 2000; Linse et al. 1991). These sites consist of a highly conserved 12 residues loop flanked by two orthogonal helices, forming a characteristic helix-turn-helix motif (Fig. 13.1). The calcium ion is mainly co-ordinated by the oxygen atoms of the side-chains of the loop amino acids (Asp, Glu) and sits at the centre of a bi-pyramid with pentagonal base. Herein, we report the synthesis of CaM-derived peptides corre-

Fig. 13.1a. Amino acid sequence (in one letter code) of calmodulin peptides synthesized and studied in this work. Mutations of the native sequence are in bold and underlined

Fig. 13.1b. Drawing of calcium binding site I of calmodulin from *Paramecium tetraurelia* (pdb code, 1EXR)

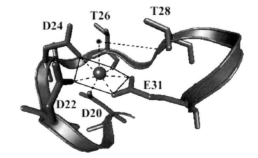

sponding to the isolated helix-turn-helix motif, their specific affinities for different metal ions, including uranium, and how a mutation can affect metal specificity, emphasising the potential use of calmodulin structures in toxic metal binding.

13.2 Experimental

13.2.1 Chemicals

All metals were used in their nitrate salt form (>99.9% purity, Aldrich, France). The stock solutions were acidified at pH 2.0 using nitric acid to avoid the formation of hydroxides.

13.2.2 Peptide Synthesis

Designed peptides were obtained by solid phase synthesis on an automatic peptide synthesizer (model 433A, Applied Biosystems, France) by using fluorenylmethyloxycarbonyl (Fmoc)-amino acid derivatives (Nova Biochem, France) and 1-N-hydroxy-7-azabenzotriazole (HOAt)/N,N'-dicyclohexylcarbodiimide (DCCI, Applied Biosystems) mediated coupling. The disulphide bonds were formed directly on the purified peptides by using 5,5'-dithio-bis(2-nitrobenzoic acid) (DTNB, Ellman's reagent, Sigma, France) in 100 mM Tris buffer solution, pH 8.0, as oxidant. The resulting oxidised products were purified by high performance liquid chromatography (HPLC). The concentration of the aqueous stock solutions of the peptides were determined by recording absorption spectra, by using the extinction coefficients of 1 280 $M^{-1} cm^{-1}$ for tyrosine, 5 690 $M^{-1} cm^{-1}$ for tryptophan and 120 $M^{-1} cm^{-1}$ for disulphide bond.

13.2.3 Mass Spectrometry Analysis

Electrospray mass spectrometric (ES-MS) detection of positive ions was performed by using a Q-TOF II (Micromass, Manchester, UK) (Q: quadrupole, TOF: time of flight). The sample was introduced into the source with a syringe pump (Harvard Apparatus, Cambridge, MA, USA). Nitrogen was employed as both the drying and spraying gas with a source temperature of 80 °C. The cone voltage was set to 30 V, the voltage applied on the capillary to 3 500 kV, and the sample solution flow rate was 5 µl min^{-1}. Spectra were recorded by averaging 40 scans from 400 to 3 000 m/z at a scan rate of 6 s $scan^{-1}$.

13.2.4 Fluorescence Spectroscopy

Fluorescence spectra were recorded on a Cary Eclipse (Varian, France) spectrofluorimeter equipped with a thermostated cell holder. A 285 nm excitation wavelength was used with a 250–395 nm excitation filter and a 430–1100 nm emission filter. A 20 nm excitation slit and a 10 nm emission slit were used. The spectrum was recorded from 450 nm to 650 nm in a 1.0 cm quartz cell. In a typical experiment, to a 25 µM solution of peptide CaM-D1 in 2-[N-Morpholino]ethanesulfonic acid 10 mM, pH 6.5, 0.1 equivalent aliquots of Tb(NO$_3$)$_3$ are added and a spectrum is recorded after each addition. A plot of the intensity at 545 nm vs. the concentration of added lanthanide ion was fitted with the binding isotherm to calculate the dissociation constant. The dissociation constant for Eu^{3+} was calculated in a competition experiment in the presence of 2 equivalents of Tb^{3+}. The constant was calculated at half-intensity of the 545 nm emission peak using the formula $K_d(Tb^{3+})/K_d(Eu^{3+}) = 3C_0/(2C_i - C_0)$, where C_0 is the initial peptide concentration and C_i is the europium concentration introduced.

13.2.5
Time Resolved Laser Induced Fluorescence Analysis (TRLIF)

A Nd-YAG laser (model minilite, Continuum, Santa Clara, California, USA) operating at 266 nm (quadrupled) and delivering about 1 mJ of energy in a 4 ns pulse with a repetition rate of 20 Hz was used as the excitation source. The laser output energy was monitored by a laser powermeter (Scientech, Boulder, Colorado, USA). The beam was directed into a 4 ml quartz cell. The laser beam was directed into the cell of the spectrofluorimeter "FLUO 2001" (Dilor, France) by a quartz lens. The radiation coming from the cell was focused at the entrance slit of a polychromator. The detection was performed by an intensified photodiode 1024 array cooled by the Peltier effect (–25 °C) and positioned at the polychromator exit. Recording of spectra was performed by integration of the pulsed light signal given by the photodiode intensifier, using an integration time of 0.5 s. Logic circuits, synchronized with the laser shot, allow the intensifier to be active with determined time delay (from 0.1 to 999 µs) and during determined aperture time (from 0.5 to 999 µs). A computer controls the whole system. In a typical experiment, aliquots of a solution of Tb(NO$_3$)$_3$ are added to a 20 µM solution of a peptide in a 10 mM MES buffer, pH 6.5. A spectrum is recorded after each addition and the lifetime of the fluorescent species is determined. The plot of the intensity of the maximum emitted fluorescence vs. the added concentration of lanthanide is fitted with the binding isotherm to calculate the dissociation constant.

13.2.6
Circular Dichroism Analysis

CD spectra were obtained with a CD6 dichrograph (Jobin Yvon, Longjumeau, France), equipped with a thermostatic cell holder and an IBM PC operating with a CD Max data acquisition and manipulation program. Peptides were dissolved at 50 µM in 10 mM MES buffer, pH 6.5. Spectra were run at room temperature from 180 nm to 250 nm using a 0.1 cm quartz cell. Each spectrum represents the average of four spectra, obtained with an integration time of 0.5 s every 0.5 nm. Spectra were smoothed by using the instrument software by taking a sliding average over nine data points.

13.3
Results and Discussion

13.3.1
Design of Metal-Binding Calmodulin Peptides

To investigate metal binding to an isolated EF-hand motif of calmodulin, we synthesized a 33-residue peptide CaM-M1, spanning region 7–39 of calmodulin from *Paramecium tetraurelia* and corresponding to the entire EF-hand motif of site I (Schaefer et al. 1987) (Fig. 13.1). However, this peptide, when tested in circular dichroism measurements and in the presence of excess calcium ions, did not present any ordered secondary structure (not shown). Formation of a precipitate at high peptide concentration (10^{-4} M) suggested that aggregation phenomena might be responsible for the observed lack of ordered structure. To overcome such phenomena and favour a native-like ordered structure a disulphide bond was engineered to bridge position 13 and 29. Molecular simulation in calmodulin showed that the mutation Phe13Cys, Val29Cys and the formation of 13–29 disulphide bridge did not modify the overall protein structure, and furthermore could stabilize the isolated metal binding site I. The disulphide-stabilized cyclic peptide, CaM-M1c, also contained the Thr20Tyr mutation in the metal binding loop, as a spectroscopic sensitive probe of metal binding (Fig. 13.1, *vide infra*). The affinity of CaM-M1c for various metal ions was studied by circular dichroism (CD), electrospray mass spectrometry (ES-MS) and time resolved laser induced fluorescence (TRLIF) spectroscopy.

13.3.2
Metal Binding

First, the circular dichroism spectrum of CaM-M1c (10 mM MES buffer, pH 6.5) was recorded in the absence or presence of four equivalents of metals (Fig. 13.2).

In the case of Mg^{2+}, Ba^{2+}, Sr^{2+} or Cs^+, no significant change in the spectrum was observed. With Ca^{2+}, Cd^{2+}, Tb^{3+}, Eu^{3+}, the α-helical content of the peptide increases

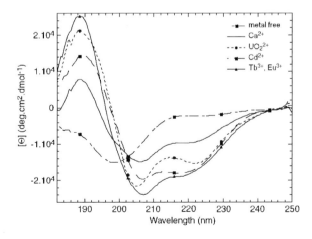

Fig. 13.2.
Circular dichroism (CD) spectrum of a 50 µM solution of CaM-M1c peptide in 10 mM MES buffer, pH 6.5, in the absence and in the presence of various divalent and trivalent metal ions. CD values are expressed in mean residue ellipticity (Θ)

in the presence of the metal, as it is attested by the increase of the ellipticity at 222 nm and by the shift of the minimum from 190 nm to 205 nm. The formation of ordered structure in the peptide by calcium and cadmium addition was monitored by circular dichroism (CD) and the values of ellipticity, at 222 nm, as a function of metal concentration, were fitted with the binding isotherm calculated for a 1:1 peptide:metal adduct, leading to 30 ±1 µM and 8 ±4 µM dissociation constants for calcium and cadmium, respectively. The 1:1 stoechiometry of the complexes was confirmed by mass spectrometry.

The calcium affinity presented by this peptide is close to that presented by the native protein and is higher than that reported by other synthetic peptides in aqueous media (Marsden et al. 1990). The crystallographic structure of calmodulin shows that the helices of site I are in strong hydrophobic interactions with the helices of site II. In a linear isolated EF-hand peptide, these intramolecular interactions are lost, but the newly exposed hydrophobic surfaces may associate to form unordered aggregates. However, our results indicate that the incorporation of the disulphide bond 13–29 in the CaM-M1c peptide may bridge the two helices in a native-like orientation and stabilize intramolecular interactions, thus preventing non specific interactions (aggregation) and resulting in the formation of a high affinity metal binding site.

13.3.3
Lanthanide Binding

Lanthanide ions (Tb^{3+}, Eu^{3+}) present typical emission fluorescence in the visible range and have often been used as calcium surrogates in the calcium-binding proteins. Therefore, their affinity for CaM-M1c was studied by time resolved laser-induced fluorescence (TRLIF) (Fig. 13.3). This spectroscopy is based on a pulse excitation of the fluorescent element followed by temporal resolution of the emission signal. Using a 260 nm wavelength excitation, the emission of the bound metal is observed through an energy transfer mechanism from the excited tyrosine to the metal. The plot of the fluorescence intensity vs. the concentration of added metal showed a 1:1 peptide:metal adduct, and the fit with the binding isotherm led to 1.5 ±0.6 µM and 0.6 ±0.2 µM dissociation constants for terbium and europium, respectively (Fig. 13.3).

13.3.4
Metal Binding to a Calmodulin Domain

As a comparison, we then studied a synthetic peptide CaM-D1, corresponding to the entire first domain of *Arabidopsis thaliana* calmodulin, where a tyrosine has been introduced in the loop of the first site, similarly to the CaM-M1c peptide (Fig. 13.1). TRLIF experiments with this peptide led to the dissociation constant 1.4 ±0.5 µM and 4.9 ±0.5 µM for the lanthanide terbium and europium, respectively. In this case, the peptide in the absence of the metal already presents a CD spectrum representative of an ordered structure, thus CD spectroscopy could not be used to determine binding affinity. Competition of terbium binding by calcium ions was attempted to determine calcium affinity. However, the fluorescence of terbium, added as two-fold excess to

Fig. 13.3.
a Time resolved laser-induced fluorescence (TRLIF) titration of a 20 µM solution of CaM-M1c peptide with terbium ions (0, 0.1, 0.3, 0.5, 1, 3 equivalents).
b Plot of the fluorescence intensity at 545 nm vs. the concentration of added terbium ions and the corresponding fit with the binding isotherm.
arb: arbitrary units

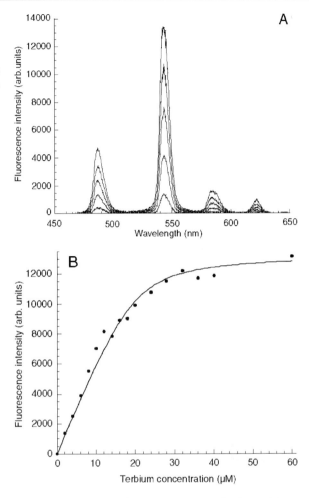

the peptide, did not change up to 60 equivalents of calcium. Since at higher calcium concentrations the peptide precipitates, we could only estimate that calcium binding dissociation constant could be superior to 28 µM.

The affinity constants for the lanthanide ions observed with CaM-M1c are about a hundred fold better than those reported in the literature for similar EF-hand isolated peptides (Marsden et al. 1990). This is in good agreement with the structure constraints introduced by the disulphide bridge. When metal is added, the cyclic CaM-M1c undergoes a more limited reorganisation of its structure between its free state and its metal loaded state, as compared to the linear CaM-M1 peptide; this leads to a smaller entropy barrier, which favours metal binding. In the peptide CaM-D1, the same phenomenon is observed: the peptide, in the absence of the metal ion, is already pre-organised and metal co-ordination does not lead to a significant change in the structure.

13.3.5
Design of Lanthanide Specificity

Lanthanide ions present co-ordination chemistry close to that of calcium, but a greater charge/size ratio. Therefore, the electrostatic interactions are stronger in the case of lanthanide(III) than in the case of calcium(II) ions. A new peptide, CaM-M2c, was thus synthesized, containing the Glu25Asp mutation (Fig. 13.1). This mutation provides a greater cavity in the binding loop, but a similar charge. CD, ES-MS and TRLIF experiments demonstrated that CaM-M2c has lost its affinity for calcium, in the studied concentration range 0.1 to 1 000 µM, but has retained the affinity for the two lanthanide ions, with 3.5 ±1 and 3.2 ±0.8 µM dissociation constants for terbium and europium, respectively. Thus, mutation Glu25Asp results in the loss of the affinity for calcium. This is most probably due to the larger binding cavity induced by the introduced mutation. The electrostatic interaction between the negative charges of the carboxylate groups and the positive charge of the metal cation is, in this case, the driving force for the co-ordination of the lanthanide ions.

13.3.6
Uranium Binding

In nuclear power plants, uranium is a source of highly toxic waste in its UO_2^{2+} uranyl form. The main difficulty in studying uranyl compounds stems from the fact that, in water and at pH 6.5, several uranium species coexist (Grenthe et al. 1992). In order to test the ability of calmodulin derived peptides to bind this metal, four equivalents of uranyl nitrate were added to a 5 µM solution of CaM-M1c in a H_2O/NH_3 solution, pH 6.5, and an ES-MS spectrum was then recorded before and after the addition of the metal. This analysis clearly shows the presence of new species in solution with 769, 988 and 1 317 mass-charge ratio (m/z) (Fig. 13.4). This is in good agreement with a compound of formula (CaM-M1c)(UO_2^{2+}) in its different charge states. In solution, the different uranium(VI) forms can be characterised by both the fluorescence spectrum and the corresponding fluorescence lifetime (Table 13.1).

The TRLIF experiment conducted in a 10 mM 2-[N-Morpholino]ethanesulfonic acid aqueous buffer, pH 6.5, yields an emission spectrum with maximum wavelength at 466, 481, 495, 517, 541 and 566 nm. The lifetime of the fluorescent species was 6 µs. The TRLIF spectrum is closed to that already reported for aqueous uranyl ion but with a different lifetime (Moulin et al. 1998). The corresponding CD spectrum of a solution of CaM-M1c in the presence of uranyl exhibits minima at 205 nm and 222 nm, suggesting the presence of helical ordered structure (Fig. 13.2). The same experiments carried out with CaM-D1 showed the formation of a 1:1, peptide: metal complex (not shown).

At pH 6.5, the major (~67%) species in solution is the hydroxo polynuclear complex $[(UO_2^{2+})_3(OH)_5]^+$. However, the introduction of peptide in solution leads to a new complex bearing a UO_2^{2+} core. The identification of this complex is established by the shape of the fluorescence spectrum, the short lifetime of the fluorescent species, and the ES-MS experiment. We suppose that the hydroxo polynuclear cluster observed in aqueous medium is the source of uranyl ion for the peptide chelatant, and that it decomposes when the peptide is added. A dissociation constant can hardly be determined because

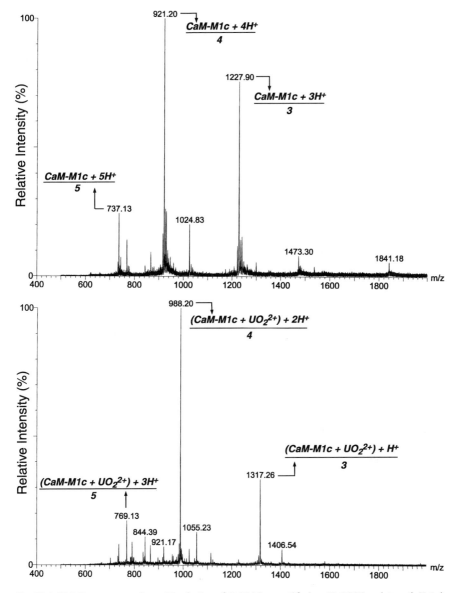

Fig. 13.4. ES-MS spectrum of a 5 μM solution of CaM-M1c peptide in a H_2O/NH_3 mixture (pH 6.5) before (*above*) and after (*below*) the addition of an excess of uranyl nitrate

of the coexistence in solution of various U(VI) species. With peptide CaM-M2c, however, no interaction of the compound with uranyl was detected, even when a large excess of metal was added. Thus, while CaM-M1c and CaM-D1 co-ordinate an uranyl ion, no such complex was detected with CaM-M2c. This is in good agreement with the larger cavity in the binding loop induced by the Glu25Asp mutation.

Table 13.1. Maximum wavelength of fluorescence emission and lifetime values of common aqueous species of uranium (Moulin et al. 1998)

Species	Maximum wavelength (nm)	Lifetimes (µs)
UO_2^{2+}	470-488-509-533-559-588	2
UO_2OH^+	480-497-519-544-570-598	80
$UO_2(OH)_2$	488-508-534-588	10–20
$UO_2(OH)_3^-$	482-499-519-543-567-594	0.8
$(UO_2)_2(OH)_2^{2+}$	480-497-519-542-570-598	9
$(UO_2)_3(OH)_5^+$	479-496-514-535-556-584	23
$(UO_2)_3(OH)_7^-$	503-523-547-574	230

13.4 Conclusion

We have described the synthesis and determined the metal affinity of a new class of cyclic peptides, corresponding to the isolated EF-hand of calmodulin site I. As revealed by circular dichroism experiments, one of these peptides, CaM-M1c, in the presence of calcium, cadmium, lanthanide ions, acquires in solution an ordered native-like conformation. The binding constants for these metal ions were determined by spectroscopic methods, including circular dichroism (CD) and time resolved laser induced fluorescence (TRLIF), and appeared to be similar to those described for the native protein. This was confirmed by metal binding experiments on the entire first domain of calmodulin. Interestingly, uranyl ions were found to tightly bind both the disulphide stabilized peptide and the entire first calmodulin domain. This is the first report of uranium binding to a calmodulin structure. Furthermore, we have shown that affinity for some ions (e.g. lanthanide ions) can be modulated by a single mutation in the peptide sequence. The study of new mutations to induce selectivity for uranium and other metals, potential pollutants from nuclear plants, is currently under study in our laboratory. The present data suggest that the calmodulin peptides here described are promising models for the binding of toxic metals and for the development of new specific biosensors or bioremediation approaches.

Acknowledgements

We wish to thank G. Plancque and C. Moulin (DEN/SPCA/LASO-CEA Saclay, France) for the TRLIF experiments, B. Amekraz for the ES-MS studies, C. Pradines-Lecomte and G. Peltier (DSV/DEVM/LEP – CEA Cadarache, France) for the synthesis of the calmodulin first domain.

References

Belkin S (1998) A panel of stress-responsive luminous bacteria for monitoring wastewater toxicity. Methods Mol Biol 102:247–258

Giuliano KA, Post PL, Hahn KM, Taylor DL (1995) Fluorescent protein biosensors: measurement of molecular dynamics in living cells. Annu Rev Biophys Biomol Struct 24:405–434

Grenthe J, Fuger JM, Konings RJ, Lemire AB, Muller C, Nguyen Trung, Wanner H (1992) Chemical Thermodynamics of Uranium. NEA-TDB, OECD, Nuclear Energy Agency Data Bank (North Holland, Amsterdam)

Gu MB, Chang ST (2001) Soil biosensor for the detection of PAH toxicity using an immobilized recombinant bacterium and a biosurfactant. Biosen Bioelectron 16:667–674

Linse S, Helmersson A, Forsen S (1991) Calcium binding to calmodulin and its globular domains. J Biol Chem 266:8050–8054

Marsden BJ, Shaw GS, Sykes BD (1990) Calcium binding proteins. Elucidating the contributions to calcium affinity from an analysis of species variants and peptide fragments. Biochem Cell Biol 68:587–601

Moulin V, Tits J, Moulin C, Decambox P, Mauchien P, de Ruty O (1992) Complexation behaviour of humic substances towards actinides and lanthanides studied by time-resolved laser-induced spectrofluorometry. Radiochim Acta 58:121–128

Moulin C, Laszak I, Moulin V, Tondre C (1998) Time-resolved laser-induced fluorescence as a unique tool for low-level uranium speciation. Appl Spectroscopy 52:528–535

Raskin I, Smith RD, Salt DE (1997) Phytoremediation of metals: using plants to remove pollutants from the environment. Curr Opin Biotechnol 8:221–226

Schaefer WH, Lukas TJ, Blair IA, Schultz JE, Watterson DM (1987) Amino acid sequence of a novel calmodulin from *Paramecium tetraurelia* that contains dimethyllysine in the first domain. J Biol Chem 262:1025–1029

Sousa C, Kotrba P, Ruml T, Cebolla A, De Lorenzo V (1998) Metalloadsorption by *Escherichia coli* cells displaying yeast and mammalian metallothioneins anchored to the outer membrane protein LamB. J Bacteriol 180:2280–2284

Stephen JR, Macnaughton SJ (1999) Developments in terrestrial bacterial remediation of metals. Curr Opin Biotechnol 10:230–233

Tsien RY (1993) Fluorescent chemosensors for ion and molecule recognition. Czarnik AW (ed), American Chemical Society, Washington, DC, pp 130–146

Wilson MA, Brunger AT (2000) The 1.0 A crystal structure of Ca(2+)-bound calmodulin: an analysis of disorder and implications for functionally relevant plasticity. J Mol Biol 301:1237–1256

Part II

Chapter 14

Leaching of Selected Elements from Coal Ash Dumping

A. Popovic · D. Radmanovic · D. Djordjevic · P. Polic

Abstract

Coal ash obtained by coal combustion in the "Nikola Tesla A" power plant in Obrenovac near Belgrade (Serbia) is suspended in river water then carried by a pipeline to a dump. In order to predict the leachability and possible environmental impact of selected elements due to ionic strength of river water, we extracted coal ash with distilled water and 0.002 M–2 M solutions of KNO_3. The results show that changes in river water ionic strength could significantly influence pollution by calcium, chromium and manganese ions, but not by zinc, nickel and copper ions. In the case of lead, magnesium, arsenic and iron ions it is difficult to predict the effects of ionic strength on pollution processes in the vicinity of the dump. Further, pollution by cadmium ions is unlikely because extractable cadmium is not detectable within the applied ionic strength range.

Key words: coal ash, leaching, microelements, major elements

14.1 Introduction

On the basis of investigations of coal genesis, its composition, as well as general characteristics of coal deposits, coal can be defined as a combustible sedimentary rock, originating mainly (some coals are algal) from residues of terrestrial and aquatic plants, and of minerals (<50%) (Wood et al. 1983). Chemical and physical characteristics of coal are predetermined by the nature of precursor plants, the amount of inorganic material, and by the nature, intensity and duration of the biochemical and geochemical processes that are responsible for coal formation. All elements in coal can be found in a variety of forms, which are responsible for the coal's technological, economical, but also ecological impact. A series of physico-chemical transformations take place during coal combustion in power plants, often changing solubility and association patterns of various elemental species. The environmental impact of coal ash production has at least two aspects: *(a)* emission and deposition of enormous amounts of coal ash, polluting air, water and soil with ash particles (including the problem of huge ash dumps); *(b)* leaching of microelements (including toxic heavy metals), but also major cations and anions from ash by atmospheric and surface waters. Here we report results obtained by extraction of coal ash with distilled water and 0.002 M–2 M solutions of KNO_3, in order to predict the leachability and possible environmental impact of selected elements due to ionic strength of river water used for transport of coal ash to the dump.

14.1.1
Minerals in Coal

The inorganic coal component consists of discrete mineral fragments and variously associated trace elements. Almost all natural elements have been found in coal (Finkelman 1993), except some very rare elements like polonium, astatine, francium and protactinium. In US coals, 79 elements have been identified, including about 63 weight% of carbon. The most abundant heavy metal is lead of 11 ppmw average concentration, while ruthenium and osmium have been detected at trace levels of about 1 ppbw (Finkelman 1993). Elements can be variously associated. For example, antimony, lead, cadmium and selenium can be associated with sulfides and other mineral and organic phases. Arsenic, cobalt and mercury, however, are commonly associated with one sulfide mineral, pyrite, as shown in eastern Kentucky coal. Further, association types may vary within the same deposit. Local irregularities occur in the distribution of arsenic, lead, fluorine, mercury and cadmium in deposit in East Siberia (Vyazova and Kryukova 1996), as well as for titanium (Kuehn and Kurzbach 1992). In coals containing less than 5% of ash, microelements mostly originate from organic complexes, e.g. biotic V and Ni, whereas in coals enriched in ash, microelements also originate from inorganic, mineral components.

14.1.2
Coal Combustion

During coal combustion, various transformations take place (Thompson and Argent 2002). The characteristics of ash depend on the nature of coal, on the temperature of combustion, and on the amount of air, or oxygen, used for combustion (Tomasek et al. 1995). For example, concentrations of vanadium, chromium, nickel and cobalt do not change, while concentrations of copper, zinc, lead and selenium decrease with the increase of combustion temperature (Tripathy and Sahu 1995). Most organic components are oxidized, while inorganic elements behave differently, depending on their physico-chemical characteristics and on types of their association in the coal. Some of the elements are predominantly evaporating, while others are mainly being concentrated in the ash like arsenic, copper, cadmium, lead and zinc (Alvarez Rodrigez et al. 1995).

14.1.3
Environmental Impact of Coal Treatment

Coal ash production raises two major concerns: first, the emission and deposition of huge amounts of coal ash, thus polluting air, water and soil with ash particle, notably in the case of ash dumps; and, second, the leaching of microelements (such as toxic heavy metal ions) and other elements from ash by atmospheric and surface waters. Coal ash can, thus, pollute soil in the vicinity of plant due to leaching of radionuclides and other inorganic and organic pollutants. It also represents a threat for water sources due to radioactive (Vukovic et al. 1996) and heavy metal contamination (Glazer et al.

1995). This pollution can have tragic consequences for the ecosystem, like in the accident that occurred in Slovenia (Tamse 1995). Leaching of trace and major elements from different substrates, including coal ash, can be influenced by different factors such as pH (Chen and Lin 2001) or complexing capability of the reagent (Sun et al. 2001). One of the mechanisms of heavy metal mobilization is due to changes of ionic strength, whose variations in the aquatic environment will necessarily cause ionic exchange reactions on the ash particles (Wang et al. 1996). It should be expected that the increase of ionic strength of the extractant will enhance ion-exchange processes.

14.1.4
Coal Ash Environmental Issues in Serbia

The Power Plants "Nikola Tesla" ("TENT-A", "TENT-B") in Obrenovac, Serbia, use lignite from the Kolubara Basin. The plants are located on the Sava River, 30–50 km upstream from Belgrade. The plants have a total power of 2.9×10^9 W. The lignite has lower caloric power between 6 and 8×10^6 J kg^{-1}, 45–53% average moisture and 10–23% ash content. The maximum daily consumption of coal in Obrenovac is about 9×10^7 kg, and under these conditions, the amount of obtained ash is estimated to be approx. 1.7×10^7 kg d^{-1}. The ash is being suspended in water (1/10, w/w) taken from River Sava, and transported through pipelines to the dump. The ionic strength of River Sava mostly varies between 0.005 and 0.02, depending mainly on discharge. Excess water leaves the dump after the transport and becomes infiltrated into the surrounding alluvial formation. The dump of TENT-A is divided into three cassettes, one active, currently being filled of 1.1 km^2, and two passive cassettes, encompassing a total area of 2.9 km^2. The cassettes of TENT-B encompass 2 km^2 each. We have studied the ash obtained by coal combustion in the "Nikola Tesla A" power plant from the ecochemical standpoint (Polic et al. 1998; Polic et al. 1999; Popovic et al. 1998; Popovic et al. 2000; Popovic et al. 2001). The major pollutants originating from the ash transport and dumping process are chromium, arsenic and nickel (Polic et al. 1998). Concerning trace element associations, we have found that nickel and chromium are primarily associated with magnesium alumosilicates, whereas copper, arsenic and lead are associated with both calcium and magnesium alumosilicates (Popovic et al. 1998). Oxides of iron and manganese seem to be the dominant substrates of chromium and nickel (Polic et al. 1999).

14.2
Experimental

Ash samples were extracted 18 h at 20 °C by distilled water and 0.002–2 M solutions of KNO$_3$ at pH 7. The solid:liquid ratio was 1/10, w/w. The residue was washed with water, the combined extracts and washings were filtered and analyzed by atomic absorption spectroscopy ("SpectrAA-20+, Varian") at the following wavelengths: 285.2 nm (Mg), 422.7 nm (Ca), 357.9 nm (Cr), 279.5 nm (Mn), 248.3 nm (Fe), 232.0 nm (Ni), 324.7 nm (Cu), 213.9 nm (Zn), 193.7 nm (As), 228.8 nm (Cd), 217.0 nm (Pb).

14.3
Results and Discussion

As shown in Table 14.1, all elements, with the exception of cadmium, were extracted in detectable concentrations, most of them already at ionic strength values which are common for river water. Also, the leachability of elements, in general, increases with ionic strength. For example, extractable calcium, manganese (Fig. 14.1) and chromium are elements showing logarithmic dependence on ionic strength over the entire examined range. There are also significant positive correlation between concentrations of these elements and applied KNO_3 concentrations at low ionic strengths, which clearly indicates that low ionic strengths (like in river water) are already responsible for extraction of these elements.

Leachable lead, nickel and zinc show an overall logarithmic dependence on the ionic strength (Fig. 14.2). This is always a relatively reliable indication of surface adsorption of trace elements on substrates. However, it should be noted, that in the range of ionic strength values common for Sava water, extracted lead is close to the detection limit, which aggravates drawing of conclusions concerning its behavior, while in cases of nickel and zinc, a logarithmic dependence is evident only for higher values.

The remaining four examined elements do not show regularities like lead, zinc and nickel, and especially not like calcium, chromium and manganese, but for most of them relatively low ionic strengths are sufficient for their extraction (except for cadmium). Magnesium concentrations approach asymptotic values at KNO_3 concentrations of 0.5 and higher. Concentrations of iron, copper and arsenic are linearly increasing starting with ionic strengths of 0.2 (iron and arsenic), or 0.5 (copper). At

Table 14.1. Concentrations of elements (ppmw) found in extracts of different ionic strength. The bold numbers represent ionic strengths occurring in river water; *n.d.*: not detectable concentrations below detection limit

Ionic strength	Element										
	Mg	Ca	Cr	Mn	Fe	Ni	Cu	Zn	As	Cd	Pb
0	n.d.	2190	0.6	0.60	n.d.	0.28	0.04	n.d.	2.61	n.d.	n.d.
0.002	9.2	2410	1.26	1.48	8.88	n.d.	0.20	0.40	3.49	n.d.	n.d.
0.005	9.2	2750	1.54	1.80	8.88	0.24	0.20	0.29	2.57	n.d.	0.60
0.010	5.2	2740	2.11	2.10	4.48	n.d.	0.20	0.36	6.45	n.d.	0.72
0.020	29.9	3320	2.62	2.96	7.68	0.24	0.13	0.34	1.83	n.d.	0.90
0.050	6.53	3270	2.41	2.60	7.68	0.48	0.13	0.36	2.33	n.d.	1.20
0.100	9.87	3340	3.20	4.31	10.01	0.90	0.47	0.52	3.27	n.d.	1.34
0.200	n.d.	3350	3.51	5.64	3.68	1.41	0.47	1.00	1.27	n.d.	3.57
0.500	49.9	3590	3.84	5.97	6.68	2.08	0.47	1.35	2.03	n.d.	5.91
1.000	53.2	3830	4.51	6.97	10.01	4.75	0.80	1.68	3.20	n.d.	8.24
2.000	56.6	3850	4.70	6.65	15.51	4.25	1.24	1.82	4.08	n.d.	9.12

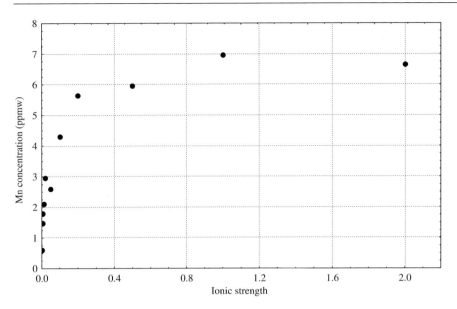

Fig. 14.1. Mn concentration extracted by aqueous solutions of KNO$_3$ at increasing ionic strength

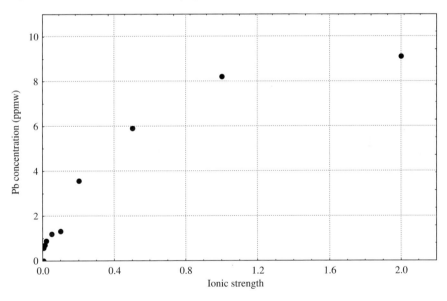

Fig. 14.2. Pb concentration extracted by aqueous solutions of KNO$_3$ at increasing ionic strength

lower concentrations of the extractant, it is hardly possible to give predictions on the extractability for iron and arsenic, while the concentration of copper becomes almost constant before the concentration of KNO$_3$ reaches 0.1.

Table 14.2. Estimation of pollution by various elements caused by river water ionic strength only

Element	Estimated amount of element leached (dry ash basis – ppm)	Estimated annual amount of element leached (t)
Ca	2750 – 3320	6050 – 7300
Cr	1.54 – 2.62	3.39 – 5.76
Mn	1.80 – 2.96	3.96 – 6.51
Ni	0 – 0.24	0 – 0.53
Cu	0.13 – 0.20	0.29 – 0.44
Zn	0.29 – 0.34	0.64 – 0.75
Pb	0.60 – 0.90	1.32 – 1.98

Knowing the annual amount of coal ash produced in the "Nikola Tesla A" power plant (2.2×10^6 t), the range of ionic strength values of river Sava (0.005–0.02), as well as the dependencies of element extractabilities on ionic strength, it might be possible to roughly estimate the pollution effects caused by ionic strength only (Table 14.2), but the overall pollution should be regarded as quantitatively very significant (except for cadmium).

14.4
Conclusions

On the basis of leaching experiments of coal ash from "Nikola Tesla A" power plant it can be stated that:

- calcium, chromium, manganese and lead leachabilities depend logarithmically on the ionic strength of the applied extractant. The amounts of extracted zinc, nickel and copper are close to constant for low ionic strength, while, in general, they reveal logarithmic (zinc, nickel), or linear dependence (copper). Concentrations of extracted magnesium, arsenic and iron do not show regularities at low ionic strength, while in general, there is an increase of element extractability;
- extractable cadmium is not detectable within the applied ionic strength range.
- correlations reveal different mechanisms of element sorption/desorption mechanisms from the ash matrix;
- in most cases, ionic strength similar to those existing in river water are sufficient for a significant mobilization of trace elements, hence, natural desorption processes should be responsible for a relevant pollution of surrounding surface and ground waters.

References

Alvarez Rodrigez R, Clemente C, Serrano C, de Marcos J, Velasco V (1995) Evaluation of the distribution of trace elements in the coal combustion products. Coal Sci Technol 24:1971–1974

Chen S-Y, Lin J-G (2001) Bioleaching of heavy metals from sediment: significance of pH. Chemosphere 44:1093–1102

Finkelman RB (1993) Trace and minor elements in coal. In: Macko, Engel (eds) Organic geochemistry. Plenum Press, New York, pp 593–607

Glazer Z, Soltyk W, Kuszneruk J (1995) Effect of the Bagno-Luben ash dumps on chemical characteristics of subsurface waters in the region of the Belchatow brown coal mine. Tech Poszukiwan Geol Geosynoptyka Geoterm 34:53–61

Kuehn W, Kurzbach H (1992) The chemical composition of the coal ash in the White Elster basin with special emphasis on the titanium content. Braunkohle (Düsseldorf) 44:27–32

Polic PS, Grzetic IA, Popovic AR, Djordjevic DS, Kisic DM, Zbogar ZM (1998) Microelements from coal ash: exchangeable fractions. Hem Ind 52:12–18

Polic PS, Grzetic I, Djordjevic D, Popovic A, Markovic D (1999) Association forms of heavy metals in fly ash from power plants. In: Borrell PM, Borrell P (eds) EUROTRAC-2, WIT Press, Southampton, volume 2, pp 296–300

Popovic A, Djordjevic D, Grzetic I, Polic P (1998) Calcium and magnesium leaching and associated microelement pollution from coal ash. In: Proceedings of 16th River and Estuarine Pollution, Fossil Fuel and Environmental Quality Conference, Derby, United Kingdom

Popovic A, Djordjevic D, Polic P (2000) Leaching of trace and major elements in coal ash dumps. Toxic Environ Chem 75:141–150

Popovic A, Djordjevic D, Polic P (2001) Trace and major element pollution originating from coal ash suspension and transport processes. Environ Int 26:251–255

Sun B, Zhao FJ, Lombi E, McGrath SP (2001) Leaching of heavy metals from contaminated soils using EDTA. Environ Pollut 113:111–120

Tamse M (1995) Deposition of ash, slag and products of additive desulfurization from the power plant Sostanj into the barrier between the lakes Velenje and Druzmirje. Rud-Metal Zb 42:59–68

Thompson D, Argent BB (2002) Thermodynamic equilibrium study of trace element mobilisation under pulverised fuel combustion conditions. Fuel 81:345–361

Tomasek V, Weiss Z, Klika Z, Konova J (1995) Contribution to the evaluation of leaching form ash. Acta Mont Ser B 5:119–137

Tripathy S, Sahu KC (1995) Morphology and mineral chemistry of coal ash from ash from Talcher thermal power station. Indian J Earth Sci 22:137–148

Vukovic Z, Mandic M, Vukovic D (1996) Natural radioactivity of ground waters and soil in the vicinity of the ash repository of the coal-fired power plant "Nikola Tesla A"- Obrenovac (Yugoslavia). J Environ Radioact 33:41–48

Vyazova NG, Kryukova VN (1996) Regularities of distribution of some toxic elements of the coals of East Siberia. Khim Tverd Topl 3:101–106

Wang Y, Ren D, Yin J, Li Y, Wang X, Xie H (1996) Study on the leaching experiments of minor and trace elements in coal and its burnt products. Huanjing Kexue 17:16–19

Wood GH, Kehn TM, Carter MD, Colbertson WC (1983) Coal resource classification system of the U.S. Geological Survey. US Geol Surv Circ, p 89

Part II

Chapter 15

Storm-Driven Variability of Particulate Metal Concentrations in Streams of a Subtropical Watershed

V. L. Beltran · E. H. De Carlo

Abstract

Extensive urbanization in Hawai'i, and Honolulu in particular, during the 20th century presents an opportunity to examine the effects thereof upon the storm-driven transfer of terrestrial material from the land to the ocean. This contribution focuses on the variability of Pb, Zn, Cu, Ba, Co, As, Ni, V, and Cr concentrations in streams during storm events in the small subtropical Ala Wai Canal Watershed in Honolulu, O'ahu. As expected, a comparison of metal loads for particulate and dissolved phases revealed the dominance of suspended particulate matter as a means of metal transport through the watershed. Particulate Pb, Zn, Cu, Ba, and Co displayed enhanced concentrations and elevated particulate loads during storm flow in the urbanized lower watershed. Enrichment of these metals likely derives from automotive or industrial-related sources. Agricultural fertilizer use in conservation areas, particularly the association of As with phosphate, appears to be responsible for an upper watershed enrichment of particulate As concentrations and loads. Storm-derived concentrations and loads of particulate Ni, V, and Cr exhibited a relative spatial invariance throughout the watershed, suggesting primarily mineralogical controls on their distributions. Principal components analysis (PCA) was applied to the particulate metal concentrations from samples collected in both the upper and lower portions of the watershed. PCA established eigenvalues explaining 77% of the total variance and separated particulate metals into two distinct factors. Factor 1 elements, including particulate Pb, Zn, Cu, Ba, and Co, were interpreted to represent metals exhibiting anthropogenic enrichment in the urban watershed. The association of particulate As, Ni, V, and Cr within Factor 2 likely denotes metals whose concentrations do not display enhancements in urban segments of the watershed. Examination of solid phase metal concentrations during a "Kona" storm (offshore low-pressure system) revealed that the downstream transport of relatively unimpacted upper watershed material during tradewind-derived rains results in an approximately 3-fold dilution in the urban concentrations of Pb, Zn, and Cu.

Key words: Hawai'i, subtropical, watershed, storm runoff, suspended particulate matter, trace metals, anthropogenic

15.1 Introduction

15.1.1 Background

Storm-driven freshwater pulses, or freshets, dominate the transfer of terrestrial matter from the land to the coastal ocean in subtropical islands. Associated with this highly episodic delivery of sediment may be anthropogenic material especially that which is derived from non-point source pollution. Potentially enriched in metals such as lead, copper, and zinc (e.g. Sansalone and Buchberger 1997; Sutherland and Tolosa 2000), urban particulate material can be mobilized by freshets and flushed to the near-shore environment, impacting biota through biomagnification and bioaccumulation (e.g. Tinsley 1979). Ingestion of solids may also pose greater dangers, as metals are more highly concentrated in this form (Luoma 1989).

Whereas the impact of anthropogenic inputs across the continental land-ocean interface in temperate climates has been well studied (e.g. Rivera-Duarte and Flegal 1994; Hollibaugh 1996), similar inputs to subtropical coastal environments have not been as clearly examined. A fundamental distinction between temperate environments and subtropical Hawai'i is the origin and nature of soil and the resultant sediment. Owing to the high weathering rates and erosion potential of basalt, the volcanic landscape in Hawai'i is characterized by heavy loads of suspended particulate matter (SPM) during storm events (Tomlinson and De Carlo 2001; Hoover 2002). Further, weathering under persistent warm and moist conditions produces lateritic soils rich in iron and aluminum oxides, which act as efficient scavengers of metals (e.g. Sposito 1984).

The pairing of high-relief topography and intense episodic rainfall in Hawai'i enables an efficient examination of the temporal and spatial evolution of material as it traverses the land-ocean divide. On the island of O'ahu, slopes greater than 20% occur on 45% of the land (Juvik and Juvik 1998) and rainfall ranges upwards of 700 cm per annum (Giambelluca et al. 1986) with intensities up to 76 mm in 1 h. Rapid changes in hydrographic conditions, due to high rainfall and the natural focusing effect created by steeply sloped watersheds, establish an ideal scenario for the study of event-based terrestrial mass transfer.

The intersection of watersheds with increasingly urbanized areas also provides an opportunity to investigate how land use affects the composition of SPM. Extensive urbanization of O'ahu, and Honolulu in particular, during the late 20th century magnifies the importance of freshets in transferring anthropogenic material to the coastal ocean. De Carlo and Spencer (1995, 1997) and De Carlo and Anthony (2002) have unraveled records of anthropogenic input to estuarine sediments in the Ala Wai Canal in Honolulu. Representing a potential flux into watershed streams, road-deposited sediments and soils in Honolulu have been found to contain elevated metal concentrations (Sutherland and Tolosa 2001; De Carlo and Anthony 2002). Finally, isotopic studies of Pb in suspended sediment and soils revealed shifts in isotopic composition consistent with changing contributions from natural and anthropogenic sources (Spencer et al. 1995; De Carlo and Spencer 1997).

Because of an economic reliance on the quality of the tropical coastline, local (City and County of Honolulu Department of Public Works), State (Department of Health-Clean Water Branch, Department of Land and Natural Resources), and Federal agencies (EPA, USGS, NOAA) are highly interested in quantifying and understanding the fate and potential impacts of terrestrial mass transfer on the coastal ocean. Although the transport and impact of nutrients on coastal Hawaiian waters has been thoroughly studied (e.g. Mackenzie and Hoover unpublished data), quantitative investigations of metals associated with terrestrial mass flux have yet to be undertaken. This study, motivated by a lack of well-constrained data, initiates the process by characterizing the evolving relationship between storm event mass transfer and metal fluxes, both natural and anthropogenic, as streams flow through a subtropical watershed.

The objectives of this study were: *(1)* to measure time-variant concentrations of trace metals in streams during storms, *(2)* to determine the relative importance of natural and anthropogenic sources of particulate metals, *(3)* to calculate loads for individual storms, and *(4)* to characterize temporal and spatial relationships between mass transport and metal input.

15.1.2
Study Area

The Ala Wai Canal Watershed lies landward of Waikiki in Honolulu, Oʻahu and encompasses Manoa and Palolo Valleys and its surrounding mountainous ridges (Fig. 15.1). Extending ~9 km from the ocean to the watershed divide with elevations ranging from 914 m to sea level, yearly rainfall varies from 50 cm along the coast in Waikiki to 450 cm on the ridges in the back of Manoa Valley (Giambelluca et al. 1986). This small subtropical watershed (48.7 km^2) is comprised of both urban (~55%) and conservation (~45%) regions with the latter harboring minimal agriculture. More than 200 000 residents occupy the watershed and vehicular traffic ranges upwards of 250 000 daily trips (Dashiell 1997).

Storm waters were collected at two stations in upper and lower segments of the Ala Wai Canal Watershed. Station Waikeakua (WK), named for a flashy mountain stream, is a sampling site situated on conservation lands in the upper reaches of Manoa Valley, upstream of urban activity (Fig. 15.1). Located ~2 km downstream from the watershed divide at an elevation of 90 m a.s.l., samples taken at station WK (a tributary to Manoa Stream) were hypothesized to most closely reflect "natural" conditions. Situated in a modified channel underlain by ancient reef limestone, station Kaimuki High School (KHS) is located ~8.5 km downstream from the watershed divide at an elevation of less than 1 m and anchors the lower urban watershed (Fig. 15.1). Station KHS is also positioned ~0.5 km below the confluence of Manoa and Palolo Streams (draining Manoa and Palolo Valleys, respectively) and ~200 m of the Ala Wai Canal Estuary. Samples collected from station KHS reflect material input from local urban surroundings as well as material carried downstream from conservation areas of the watershed.

15.2
Experimental

15.2.1
Sample Collection

Storm water samples were collected by ISCO® sequential samplers triggered by rising water level (stage) above a preset level. Samples were drawn through Teflon tubing into pre-cleaned 0.5-l high density polyethylene (HDPE) bottles at 30-minute intervals throughout the hydrograph of any storm with sufficient magnitude to exceed the trigger height. Automated sampling ceased when stage decreased below the preset trigger level or sample bottles were filled. Due to the unique and variable nature of local rainstorms (flashy to extended) and, at times, the exhaustion of available sample bottles, sample coverage varied for each event (Table 15.1). During the period of 19 October 1999 to 3 November 2000, automated collection yielded paired samples at stations WK and KHS for nine storm events in the Ala Wai Canal Watershed (Table 15.1).

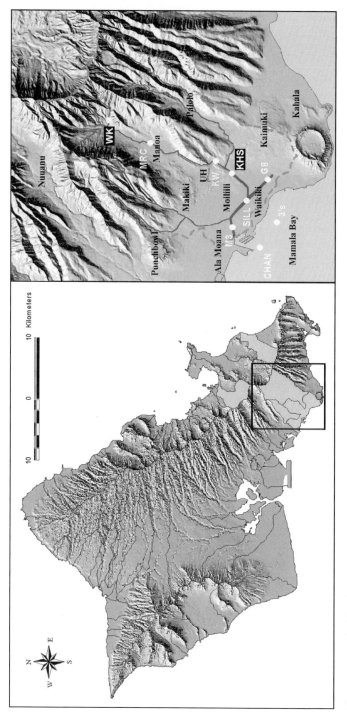

Fig. 15.1. Map of O'ahu (*left panel*) and the Ala Wai Canal Watershed (*right panel*) with sampling stations indicated

Table 15.1.
Sample coverage at sampling stations in the upper (Waikeakua-WK) and lower (Kaimuki High School-KHS) Ala Wai Canal Watershed, Oʻahu, Hawaiʻi

Storm event	Number of samples collected		
	WK	KHS	Total
19 Oct 1999	5	7	12
02 Dec 1999	10	7	17
20 Aug 2000	15	11	26
27 Aug 2000	4	5	9
28 Aug 2000	4	16	20
05 Sep 2000	6	5	11
28 Oct 2000	3	4	7
29 Oct 2000	6	10	16
03 Nov 2000	4	8	12
Total	57	73	130

15.2.2
Sample Processing and Analysis

Adapted from techniques of Gélinas et al. (1998), storm water samples were filtered to separate dissolved and particulate phases, followed by leaching of the solid phase in acid solution. Samples were filtered through a 0.22 μm membrane and collected particles were dried to a constant weight to determine the mass of suspended particulate matter (SPM). The filtrate was acidified and retained for determination of dissolved metal concentrations. Dried SPM and membranes were placed in high-pressure Teflon digestion vessels and treated with 2 ml H_2O_2 (30%, certified ACS) for organic matter decomposition. Particulate samples were then leached with 3.5 ml 15N HNO_3 using a CEM Model MDS-2100 microwave. Undissolved SPM remaining after leaching was removed by a second filtration through a 0.22 μm membrane, and the soluble leached fraction was diluted to a known mass with quartz-distilled 0.3N HNO_3.

Leaching of SPM yields distinct advantages over total particulate dissolution. Acid leaching releases to solution adsorbed components and easily solubilized metal oxides, leaving behind the resistant detrital fraction. Because aquatic organisms are typically only exposed to this easily released fraction, the toxicity of SPM should be primarily assessed from the concentrations of potential toxicants present in the leachable fraction (Gélinas et al. 1998). It should be noted that the 15 N HNO_3 leach used in this study represents a "worst case environmental scenario" in which all possible labile constituents are leached and only the most refractory constituents remain in the solid phase (Lorentzen and Kingston 1996; Gélinas pers. comm.).

Appropriate dilutions of leached fractions were analyzed for a suite of trace elements, including Pb, Zn, Cu, Ba, Co, As, Ni, V, and Cr, by inductively coupled plasma mass spectrometry (ICP-MS) using a VG Model PQ-2S (e.g. Wen et al. 1997). Analysis of a similar suite of dissolved trace elements in acidified storm water filtrate was carried out by flow injection analysis (FIA) in line with ICP-MS (De Carlo and Resing 1998). A series of aqueous multi-element standards produced by dilutions of a commercial stock solution (VG

ICPMS-2A) were used for instrument calibration, while analytical drift correction was achieved by monitoring the signal intensity of internal standards spanning the elemental mass range. Analysis of several standard reference materials (particulate analysis: USGS T-129 and T-155 water samples, NRC MESS-1 marine sediment; dissolved analysis: NRC CASS-3 and SLRS-3 water samples) in conjunction with storm samples provided quality control. Principal components analysis (PCA) was applied to the combined particulate metal concentrations of samples collected at both the conservation and urban stations to determine statistical relationships between quantitative variables.

15.3
Results and Discussion

15.3.1
Enrichment Factors

To better delineate variations in particulate metal concentrations in the Ala Wai Canal Watershed, enrichment factors were calculated for Pb, Zn, Cu, Ba, Co, As, Ni and Cr with respect to particulate V. V was chosen as a basis for normalization because, as will be demonstrated, concentrations at both sampling stations were relatively similar. Based on average metal concentrations at stations WK and KHS (Table 15.2), enrichment factors (Table 15.3) were calculated using the following formula:

$$\frac{C\text{Metal X at KHS} / C\text{V at KHS}}{C\text{Metal X at WK} / C\text{V at WK}}$$

Enrichment factors greater than one indicate urban enhancement, while factors less than one indicate enrichment in conservation lands. Enrichment factors near one denote relatively similar concentrations throughout the watershed.

15.3.2
Mineralogical Input

Concentrations of solid phase Ni, V, and Cr displayed overlapping values throughout the Ala Wai Canal Watershed (Table 15.2 and Fig. 15.2a–c). Not surprisingly, Table 15.2 and Fig. 15.2a–c also reveal that the variability in concentrations of particulate Ni, V, and Cr was comparable in upper (Ni RSD = 10%, V RSD = 17%, Cr RSD = 16%) and lower (Ni RSD = 12%, V RSD = 15%, Cr RSD = 10%) segments of the watershed. (Relative standard deviation (RSD) refers to $\pm 1\sigma$.) While the range of concentrations for these metals was relatively similar at both stations, mean concentrations of Ni, V, and Cr at station WK were slightly higher than those at station KHS (Table 15.2 and Fig. 15.2a–c).

The narrow yet scattered range of Ni, V, and Cr concentrations (Table 15.2 and Fig. 15.2a–c) during storms suggests a watershed distribution that is predominantly a function of mineralogical control. As shown in Table 15.3, enrichment factors (normalized to V) near one for particulate Ni (mean EF = 1.03) and Cr (0.94) indicate a relatively uniform primary source of these metals. With Ni, V, and Cr concentrations in Koʻolau basalt as high as 1 121 mg kg^{-1}, 277 mg kg^{-1}, and 894 mg kg^{-1}, respectively (Frey et al. 1994) (Table 15.4),

Chapter 15 · Storm-Driven Variability of Particulate Metal Concentrations in Streams   159

Table 15.2. Mean particulate metal concentrations (mg kg^{-1}) at stations *WK* and *KHS*. Also shown are 9-storm and 8-storm (excluding 19 October 1999 storm) means

Storm event	Pb		Zn		Cu		Ba		Co		As		Ni		V		Cr	
	WK	KHS	WK	KHS	WK	KHS	WK	KHS	WK	KHS	WK	KHS	WK	KHS	WK	KHS	WK	KHS
19 Oct 1999	20	311	170	1008	164	391	124	300	50	66	62	22	364	328	314	263	493	379
02 Dec 1999	28	51	169	256	155	172	133	221	47	52	63	46	324	322	278	241	507	384
20 Aug 2000	23	94	227	397	143	206	114	217	40	59	76	31	349	330	270	256	450	413
27 Aug 2000	17	140	157	494	143	261	107	258	43	69	61	31	309	312	268	280	406	385
28 Aug 2000	19	134	172	463	156	224	125	257	48	61	53	27	315	276	313	261	466	362
05 Sep 2000	20	134	159	389	155	229	120	258	45	82	76	35	326	334	302	313	438	435
28 Oct 2000	28	124	225	357	154	202	132	225	59	62	83	31	329	326	311	262	493	417
29 Oct 2000	23	34	236	244	173	185	162	230	62	67	51	12	350	271	300	218	468	394
03 Nov 2000	24	48	175	249	157	176	142	219	55	60	54	17	392	349	273	237	510	409
9-storm mean	23	114	193	424	154	223	127	242	48	63	66	27	340	310	287	255	469	392
8-storm mean	23	93	195	362	153	206	127	236	47	63	66	27	338	308	284	254	467	394
Max	43	612	298	1273	191	488	180	351	74	101	155	76	440	415	403	360	775	491
Min	15	24	130	174	128	135	95	176	32	41	19	4	266	218	151	123	360	277
SD	6	101	41	251	13	73	19	37	10	10	28	14	34	38	50	39	73	41
RSD (%)	27	89	21	59	8	32	15	15	20	15	42	53	10	12	17	15	16	10

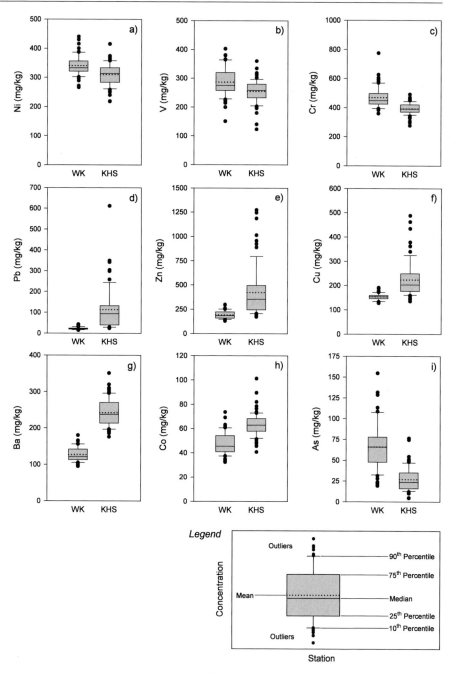

Fig. 15.2. Particulate concentrations (mg kg^{-1}) of **a** Ni, **b** V, **c** Cr, **d** Pb, **e** Zn, **f** Cu, **g** Ba, **h** Co, and **i** As at stations WK and KHS

Table 15.3. Mean enrichment factors for particulate metals (normalized to V)

Storm event	Pb	Zn	Cu	Ba	Co	As	Ni	V	Cr
19 Oct 1999	18.54	7.07	2.84	2.88	1.58	0.42	1.07	1.00	0.92
02 Dec 1999	2.10	1.75	1.28	1.91	1.28	0.84	1.15	1.00	0.87
20 Aug 2000	4.30	1.84	1.52	2.00	1.55	0.43	1.00	1.00	0.97
27 Aug 2000	7.87	3.01	1.74	2.31	1.53	0.49	0.96	1.00	0.91
28 Aug 2000	8.47	3.23	1.72	2.47	1.54	0.61	1.05	1.00	0.93
05 Sep 2000	6.46	2.36	1.42	2.07	1.77	0.44	0.99	1.00	0.96
28 Oct 2000	5.26	1.88	1.56	2.02	1.25	0.44	1.18	1.00	1.00
29 Oct 2000	2.04	1.42	1.47	1.96	1.50	0.32	1.07	1.00	1.16
03 Nov 2000	2.30	1.63	1.29	1.77	1.25	0.36	1.02	1.00	0.92
9-storm mean	5.58	2.47	1.63	2.14	1.49	0.46	1.03	1.00	0.94
Max	18.54	7.07	2.84	2.88	1.77	0.84	1.18	1.00	1.16
Min	2.04	1.42	1.28	1.77	1.25	0.32	0.96	1.00	0.87
SD	5.17	1.76	0.48	0.34	0.18	0.16	0.07	0.00	0.08
RSD	93	71	29	16	12	34	7	0	9

the physical and chemical weathering of basalt likely represents the most important watershed source for these metals. Similar scatter in Ni, V, and Cr concentrations in conservation and urban areas (Table 15.2 and Fig. 15.2a–c) is also consistent with a relatively homogeneous source of these elements throughout the watershed.

Higher mean concentrations of Ni, V, and Cr in SPM collected in conservation lands (Table 15.2 and Fig. 15.2a–c) may be the result of varying physiographic conditions throughout the watershed. Because the steep topography and fast currents in the upper watershed maintain larger particles in suspension than the slower currents present in the lower watershed, particles collected at station WK contain coarse grains of unweathered primary volcanic mineral particles, e.g. olivine, pyroxene. Finer particles carried by slower currents in the lower sections of the watershed include more clay minerals, which contain slightly less Ni, V, and Cr.

15.3.3
Anthropogenic Input

During storm events in the Ala Wai Canal Watershed, urban concentrations of particulate Pb, Zn, Cu, Ba, and Co displayed values exceeding those observed in SPM from conservation areas (Table 15.2 and Fig. 15.2d–h). Exhibiting mean enrichment factors greater than 1, Pb (mean EF = 5.58) displayed the strongest urban enhancement in the lower watershed, followed in order by Zn (2.47), Ba (2.14), Cu (1.63), and Co (1.49) (Table 15.3). Marked "first flush" effects, in which the highest concentrations are observed prior to peaks in stream flow and suspended solids, were also apparent

Table 15.4. Metal concentrations (mg kg^{-1}) in a variety of relevant settings. Values are total concentrations unless stated otherwise (RDS = road-deposited sediment)

Settings	Pb	Zn	Cu	Ba	Co	As	Ni	V	Cr
Surficial Ala Wai canal core[a]	215	497	242	67.8	43.7	18.6	217	191	332
Manoa stream bed sediments[a]	177	315	197						
Agric./conserv. soils (Manoa)[a]	0 – 75	50 – 1400	100 – 200						
Urban soils (Manoa)[a]	75 – 900	200 – 1000	160 – 325						
Oahu soils[b]						10 – 45			
Manoa RDS[c]	210	703	359	323	70	8	442	287	513
McCully RDS[c]	338	706	431	333	62	10	376	254	475
Manoa RDS 0.5 M HCl leach[d]	101	386	93	109	14		54	17	20
Manoa RDS[d]	106	434	167	200	40	4	177	203	273
Manoa background soil-leach[d]	8	28	21	79	36		31	32	12
Manoa background soil[d]	13	132	114	105	58	13	303	253	507
Pololu soil core (Big Island)[e]	4 – 14	66 – 160	65 – 200		41 – 87		58 – 211	174 – 464	
Big Island soils[f]	1.5 – 17.5	80 – 200	20 – 200						
Koolau basalts[g]		93 – 158		34 – 149	37 – 88		79 – 1121	170 – 277	67 – 894

[a] De Carlo and Anthony 2001;
[b] De Carlo and Uehara, unpublished data;
[c] De Carlo, unpublished data;
[d] Sutherland and Tolosa 2000;
[e] Li 1996;
[f] Halbig et al. 1985;
[g] Frey et al. 1994.

for particulate Pb, Zn, and Cu (Fig. 15.3a–c) and to a lesser extent for Ba and Co (Fig. 15.3d–e). Variability in concentration for Pb, Zn, and Cu was much lower in samples collected in the upper watershed compared to lower watershed samples (Table 15.2 and Fig. 15.2d–f). This, however, was not the case for Ba and Co as concentrations for both metals displayed similar variability at stations WK and KHS (Table 15.2 and Fig. 15.2g–h).

15.3.3.1
Pb

Solid phase Pb displayed significant urban enrichment (mean EF = 5.58) in the Ala Wai Canal Watershed (Table 15.3). Representing near background values for Hawai'i (e.g. Li 1996), the mean particulate Pb concentration in samples collected at station WK in the upper watershed was 23 mg kg^{-1} ±27% (Table 15.2 and Fig. 15.2d). In contrast, SPM collected at station KHS showed an average Pb concentration of 114 mg kg^{-1} ±89% (Table 15.2 and Fig. 15.2d), exceeding the mean conservation value by a factor of 5.

SPM from the upper watershed consists of material primarily derived from conservation areas containing relatively low concentrations of Pb. Particulate concentrations of Pb at station WK (Table 15.2 and Fig. 15.2d) displayed values comparable to Pb concentrations in relatively uncontaminated Big Island soils (Halbig et al. 1995; Li 1996) and within the larger range of Pb concentrations (near 0 to 75 mg kg^{-1}) observed by De Carlo and Anthony (2002) in agricultural and conservation soils from Manoa and Palolo Valleys (Table 15.4). On the other hand, particulate matter in the lower watershed encompasses a much broader range of material, including eroded soils from both conservation lands and intermediary areas upstream of station KHS, as well as input of eroded Pb-rich urban soils (up to 900 mg kg^{-1}) and road-deposited sediments (210 to 338 mg kg^{-1}) (Sutherland and Tolosa 2000; De Carlo and Anthony 2002) (Table 15.4).

The high variability of particulate Pb concentrations in the lower watershed (Table 15.2 and Fig. 15.2d) is also consistent with anthropogenic input. At station KHS, elevated concentrations of particulate Pb during the early stages of storms (Fig. 15.3a) represent a "first flush" of easily mobilized material, such as road-deposited sediment (Sutherland and Tolosa 2000) or contaminated residual soils (De Carlo and Anthony 2002), containing large contributions of anthropogenic Pb. As the storm progresses and rising stream flow mobilizes larger volumes of terrestrial mass, the local supply of anthropogenic Pb is depleted and becomes diluted by relatively Pb-poor SPM originating from conservation areas of the watershed.

Although many metals can be enriched in maturing soils relative to the parent rock due to the presence of metal-scavenging aluminum and iron hydroxides (Hayes and Leckie 1986; Erel et al. 1990), this natural process is likely overshadowed by the prevalence of anthropogenic sources in urban areas. Despite the phasing out of leaded gasoline in the late 1970s and its final elimination in Hawai'i in 1989 (State of Hawai'i Department of Business and Economic Development – as referenced in De Carlo and Spencer 1997), aged watershed soils still contain significant burdens of Pb that represent "averaged" anthropogenic input over the past half-century (Spencer et al. 1995). In particular, roadside watershed soils are often incorporated into street runoff and stream flow during storm events (e.g. Geschwind 2000; Sutherland and Tolosa 2000;

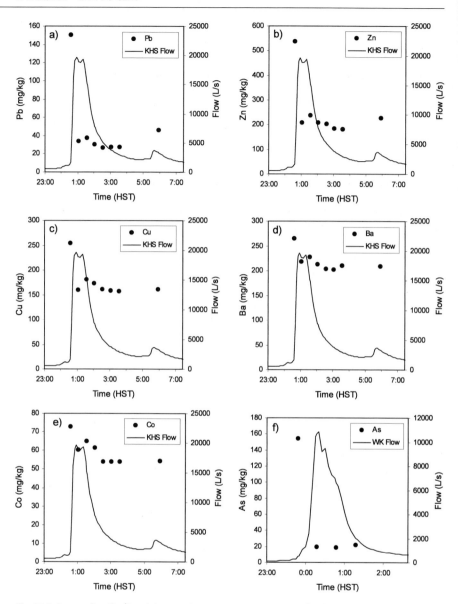

Fig. 15.3. Stream flow (l s⁻¹) and time-variant concentrations (mg kg⁻¹) of particulate **a** Pb, **b** Zn, **c** Cu, **d** Ba, **e** Co at station KHS and particulate **f** As at station WK during the 2–3 November 2000 storm

De Carlo and Anthony 2002). Another potentially important anthropogenic Pb source may stem from the use of lead-based paint especially in older residential neighborhoods, such as exists in Manoa and Palolo Valleys. Particulate matter derived from lead-based paint mixes with soils that can also be eroded and transported to streams by high intensity rainstorms.

15.3.3.2
Zn and Cu

Particulate concentrations of Zn and Cu in lower segments of the watershed also exhibited urban enrichment (mean Zn EF = 2.47, mean Cu EF = 1.63), though to a lesser extent than was observed for Pb (Table 15.3). The mean solid phase Zn concentration of 424 mg kg^{-1} ±59% observed at station KHS was 2.3 times greater than the mean concentration of 193 mg kg^{-1} ±21% in SPM from station WK (Table 15.2 and Fig. 15.2e). Average urban concentrations of particulate Cu (223 mg kg^{-1} ±32%) also displayed a value that was elevated with respect to the average concentration observed in conservation lands (154 mg kg^{-1} ±8%) (Table 15.2 and Fig. 15.2f).

As noted for Pb, the range in particulate concentrations of Zn and Cu was larger in urban segments of the watershed compared to conservation areas (Table 15.2 and Fig. 15.2e–f), consistent with pulsed urban input at station KHS. This variability during storm events can be largely explained by a "first flush" effect (Fig. 15.3b–c) in which the local flux of easily mobilized anthropogenic Zn and Cu diminishes as the storm progresses, and is subsequently diluted by large quantities of relatively Zn- and Cu-poor SPM originating upstream of station KHS.

Particulate concentrations of Zn and Cu from the upper watershed station (Table 15.2 and Fig. 15.2e–f) compare well with values for relatively pristine Big Island soils (Halbig et al. 1995; Li 1996) and agricultural and conservation soils from Manoa and Palolo Valleys (De Carlo and Anthony 2002) (Table 15.4), indicating little contamination of these metals in conservation lands. The incorporation of road-deposited sediments and urban soils at station KHS, however, provide a particularly potent urban source of Zn and Cu in lower segments of the Ala Wai Canal Watershed. Total particulate Zn and Cu concentrations in road-deposited sediments collected in Manoa Valley display values as high as 700 mg kg^{-1} and 430 mg kg^{-1}, respectively (De Carlo unpublished data) (Table 15.4). As shown in Tables 15.2 and 15.4, Zn and Cu concentrations in SPM collected at station KHS also fall within the range observed for urban Manoa soils (De Carlo and Anthony 2002). Owing to their presence in both brake pads (Armstrong 1994) and tires (Sadiq et al. 1989), increased particulate concentrations of Zn and Cu in the urban environment are likely characteristic of intense automotive use. Release of these metals from brake pad and tire wear and their mobilization into streams via road deposited sediment and roadside soils renders them of special concern in Hawai'i, largely because storm water is untreated and flows directly into water ways.

15.3.3.3
Ba and Co

While urban enrichments were also observed for solid phase Ba (mean EF = 2.14) and Co (1.49) (Table 15.3), intra-station variability for these metals was similar at stations WK (Ba RSD = 15%, Co RSD = 20%) and KHS (Ba RSD = 15%, Co RSD = 15%) (Table 15.2 and Fig. 15.2g–h). This contrasts with the much greater variability noted for particulate Pb, Zn, and Cu in the urban watershed relative to conservation areas (Table 15.2 and Fig. 15.2d–f). The comparable variability in Ba and Co concentrations at both watershed stations can be attributed to the dampened "first flush" effect for these metals at station KHS (Fig. 15.3d-e), suggesting a less significant input of Ba- and Co-rich SPM during the initial stages of storms and that sources of Ba and Co may be separate from those of Pb, Zn, and Cu.

Solid phase concentrations of Ba and Co reflect the different potential source materials in conservation lands and urban areas. In the upper Ala Wai Canal Watershed, concentrations of particulate Ba and Co (Table 15.2 and Fig. 15.2g-h) lie within the range for Koʻolau Basalts collected near Honolulu (Ba: 34–149 mg kg^{-1}, Co: 37–88 mg kg^{-1}) (Frey et al. 1994) (Table 15.4), suggesting a natural source for these metals at station WK. The higher mean urban concentrations of Ba (242 mg kg^{-1}) and Co (63 mg kg^{-1}) at station KHS (Table 15.2 and Fig. 15.2g-h), however, more closely resemble values observed by De Carlo (unpublished data) in road-deposited sediment (Table 15.4). Fossil fuel burning, vehicular exhausts, and its presence in a variety of alloys represent potential urban sources for Co (Smith and Carson 1981), while potential sources of Ba include fossil fuel and coal combustion (Miner 1969; Davis 1972) and the degradation of paint, bricks, tiles, glass, and rubber (Bodek et al. 1988; Venugopal and Luckey 1978).

15.3.4
Agricultural Input

Displaying an enrichment factor less than one (mean EF = 0.46) (Table 15.3), concentrations of As were substantially elevated in SPM collected in conservation lands relative to samples from the urban environment (Table 15.2 and Fig. 15.2i). As observed in Table 15.2 and Fig. 15.2i, the mean concentration of solid phase As in the upper segments of the watershed (66 mg kg^{-1} ±42%) exceeded the mean urban particulate As concentration (27 mg kg^{-1} ±53%) by a factor of 2.3. A strong "first flush" of As was also restricted to the upper watershed (Fig. 15.3f), yet the variability of As concentrations in SPM was similar at both stations WK and KHS (Table 15.2 and Fig. 15.2i).

Elevated particulate concentrations of As at the conservation station (Table 15.2 and Fig. 15.2i) may reflect the small extent of agricultural land use present in upper segments of the Ala Wai Canal Watershed. For example, De Carlo and Dollar (1997) observed significant As enrichment in Maui soils to which super-phosphate was applied. It is tempting to hypothesize that the application of such fertilizers by several flower farms located immediately upstream of station WK could contribute highly-mobile As to easily eroded surface soils, which can be subsequently transported during early stages of freshets. The strong "first flush" effect observed for particulate As concentrations in the upper watershed is consistent with this scenario (Fig. 15.3f).

Particulate As values observed at station WK (Table 15.2 and Fig. 15.2i) are defined by a mixture of sediment from conservation areas harboring low concentrations of As and a local agricultural input enriched in As. With little agricultural land use in the urban watershed, concentrations of As during storm events remain relatively low in the lower sections of the watershed (Table 15.2 and Fig. 15.2i). Urban concentrations of particulate As may simply reflect the seaward transport of upper watershed SPM, as evidenced by the comparable variability in As concentrations at both stations WK and KHS (Table 15.2 and Fig. 15.2i) and the general absence of a "first flush" at station KHS. Mean concentrations of As in SPM collected in the lower watershed (Table 15.2 and Fig. 15.2i) are slightly higher than As concentrations observed in Manoa Valley road-deposited sediments (10 mg kg^{-1}) and forested soils (2.3 mg kg^{-1}) from unimpacted reference areas (De Carlo et al. 2004), but within the range of 10 to 45 mg kg^{-1} observed in other O'ahu soils (De Carlo and Uehara unpublished data) (Table 15.4).

15.3.5
Principal Components Analysis

Principal components analysis (PCA) is used to examine statistical relationships between quantitative variables. These variables are grouped into several factors, which account for a significant portion of variance (i.e. eigenvalues) of the original variables. Extracted factors can, in turn, be interpreted based on the meaning of variables clumped within. Factor loads and scores are estimated for each variable and varimax rotation is used to maximize values in factor space, increasing the interpretability of each factor.

Applied to the particulate metal concentration data from all samples taken at both stations WK and KHS, PCA generated eigenvalues accounting for 77% of the total variance and separated quantitative variables into two principal factors. In Factor 1, solid phase Pb, Zn, Cu, Ba, and Co were associated with each other (Fig. 15.4). Elements with high Factor 1 loads are marked by anthropogenic enrichments in concentration in the lower urbanized watershed. Particulate As, Ni, V, and Cr were heavily weighted in Factor 2 (Fig. 15.4). Factor 2-associated metals indicate elements that do not display significant anthropogenic enrichment in the Ala Wai Canal Watershed. Concentrations of these elements derive largely from a basaltic source of olivine and pyroxene. Particulate As is most likely an exception, however, as it is the only element in Factor 2 that displays a strong upper watershed enrichment (Table 15.3). Factor analysis of solid phase metal concentration data from a multi-watershed study in Hawai'i positively correlates As and U in a third factor, which these authors interpret to represent elements with an agricultural source.

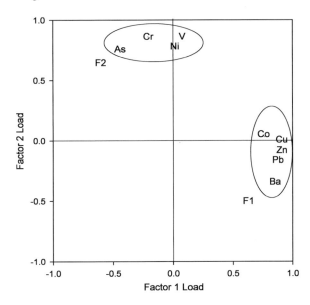

Fig. 15.4. Factor 1 and 2 loads for particulate metals in the Ala Wai Canal Watershed

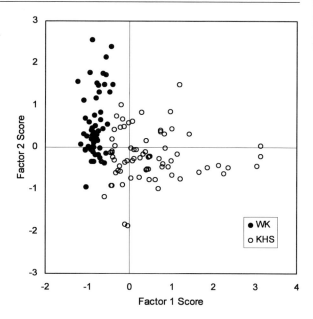

Fig. 15.5.
Factor 1 and 2 scores for samples collected at stations WK and KHS

PCA can also reveal distinct groupings of samples collected either in the upper or lower Ala Wai Canal Watershed. Figure 15.5 shows factor scores for Factor 1 plotted against Factor 2 for storm water samples from the conservation and urban sites. The most obvious feature is the general lack of overlap with respect to Factor 1 scores (Pb, Zn, Cu, Ba, Co) for samples collected at stations WK and KHS (Fig. 15.5). This separation reflects the enormous modifications in land use throughout the watershed, extending from relatively pristine areas at station WK to the highly urban environment of station KHS, and the resultant impact on the chemical composition of SPM during storms.

15.3.6
Urban Input during a "Kona" Storm

Although storm-derived particulate matter in the lower watershed generally consists of urban input diluted by downstream particulate transport, the "Kona" storm event of 19 October 1999 provided a scenario in which SPM collected in the lower watershed remained relatively unimpacted by upper watershed mass. Typical rainstorms in the Ala Wai Canal Watershed are generated by the orographic uplift of northeast trade winds and travel southwest from the mountains to the ocean. This results in a stream flow peak at the more northerly station WK that temporally precedes the peak in stream flow at station KHS (Fig. 15.6a). In contrast, "Kona" storms are generated by a large low-pressure system over the southern portion of the Hawaiian Islands and roughly travel north from the ocean to the mountains. As a result, peak stream flow at the urban station precedes that at the conservation station during "Kona" storms. During the 19 October 1999 "Kona" event, the peak in stream flow observed at station KHS preceded peak flow at station WK by ~30 min (Fig. 15.6b).

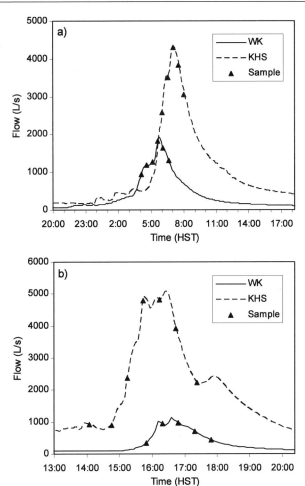

Fig. 15.6.
Stream flow (l s⁻¹) during **a** an orographic storm (5 September 2000) and **b** a "Kona" storm (19 October 1999) at stations WK and KHS. Also indicated are sample collection times

Samples collected in the lower segments of the Ala Wai Canal Watershed during the early stages of the 19 October 1999 "Kona" storm likely contained much lower amounts of conservation area-derived SPM than during a typical orographic storm. Thus, particulate concentrations of metals at station KHS should reflect a nearly undiluted urban metal input. Removal of the diluting effects of downstream transport of relatively uncontaminated SPM resulted in mean particulate concentrations of Pb, Zn, and Cu at station KHS that exceeded the 8-storm mean values (excluding the 19 October 1999 event) by factors of 3.3, 2.8, and 1.9, respectively (Table 15.2 and Fig. 15.7). Moreover, the highest storm-derived particulate concentrations of Pb (612 mg kg⁻¹), Zn (1 273 mg kg⁻¹), and Cu (488 mg kg⁻¹) were observed at the urban station during the 19 October 1999 storm (Table 15.2). This suggests that typical orographic storm flow transport of SPM from the upper watershed results in a roughly 3-fold dilution of solid phase Pb, Zn, and Cu in urban areas. Although elevated concentrations of Ba and Co were generally observed in the lower watershed, similarly large

Fig. 15.7.
Mean particulate metal concentrations (mg kg^{-1}) during "Kona" (19 October 1999) and orographic storms at station KHS

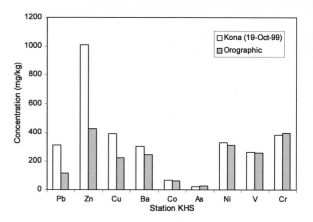

elevations in urban concentrations during the 19 October 1999 storm were not observed (Table 15.2 and Fig. 15.7). Concentrations of Ba and Co during the "Kona Storm" exhibited increases of only 28 and 4%, respectively, relative to their 8-storm mean values (Table 15.2 and Fig. 15.7), further suggesting that urban sources of Ba and Co may be different from those of Pb, Zn, and Cu. Displaying no urban enrichment, concentrations of As, Ni, V, and Cr at station KHS remained similar to their respective 8-storm mean values during the 19 October 1999 storm (Table 15.2 and Fig. 15.7).

15.3.7
Particulate and Dissolved Metal Loads

Loads were calculated for particulate and dissolved metals (Tables 15.5 and 15.6) during storms in the Ala Wai Canal Watershed. Because samples were generally collected in 30-minute intervals during storms, load calculations assumed that stream flow, SPM concentration, dissolved metal concentration, and particulate metal concentration remained constant 15 min before and after the time of sample collection. Interpolated values of dissolved and particulate metal concentrations were used when the sampling interval exceeded one hour. In this manner, a rough integration of each storm was undertaken and a metal load estimated.

Although metal loads varied widely between storm events due to large fluctuations in metal concentration, SPM concentration, and stream flow, the solid phase clearly dominated the transport of metals during storm events in the Ala Wai Canal Watershed (Tables 15.5 and 15.6). Storm water analysis of dissolved Pb, Zn, Cu, Co, Ni, and V revealed concentrations in the low µg kg^{-1} range, up to 6 orders of magnitude less than particulate metal concentrations of the same metals (Table 15.2). The observed partitioning of metals between the dissolved and solid phases was roughly consistent with experimentally determined K_d values (Irving 1998). The large disparity between dissolved and particulate metal concentrations, as well as the large amount of suspended sediment mobilized by storm flow, produced particulate metal loads that far exceeded loads in the dissolved phase (Tables 15.5 and 15.6). Dissolved loads at stations WK and KHS constituted less than 1% of their respective total (particulate plus dissolved) loads (Tables 15.5 and 15.6).

Table 15.5. Suspended particulate matter (SPM) loads (kg storm^{-1}) and particulate metal loads (g storm^{-1}) at stations **a** Waikeakua (*WK*) and **b** Kaimuki High School (*KHS*). *SD*: standard deviation, *RSD*: relative standard deviation

Storm event	Pb	Zn	Cu	Ba	Co	As	Ni	V	Cr	SPM
a Waikeakua (WK)										
19 Oct 1999	62	559	559	388	157	193	1 185	1 015	1 596	3 057
02 Dec 1999	3 661	21 000	18 828	16 276	6 417	6 900	39 591	31 846	79 092	121 674
20 Aug 2000	342	2 767	2 143	1 626	2 143	1 124	4 913	4 098	6 568	14 221
27 Aug 2000	30	263	241	177	73	110	523	457	697	1 688
28 Aug 2000	341	2 784	2 540	1 980	2 540	968	5 074	5 276	7 852	15 718
05 Sep 2000	56	453	444	344	444	216	935	862	1 257	2 856
28 Oct 2000	382	3 007	2 133	1 733	2 133	1 186	4 404	4 618	7 040	13 287
29 Oct 2000	3 458	33 890	28 369	25 077	28 369	5 612	52 195	36 844	69 037	159 148
03 Nov 2000	1 850	16 055	15 622	14 272	5 421	2 034	35 868	16 863	47 629	99 654
Total	10 182	80 777	70 879	61 873	47 698	18 343	144 687	101 879	220 767	431 304
Average	1 131	8 975	7 875	6 875	5 300	2 038	16 076	11 320	24 530	47 923
SD	1 486	11 975	10 376	9 233	8 933	2 489	20 384	14 012	31 689	61 279
RSD (%)	131	133	132	134	169	122	127	124	129	128
b Kaimuki High School (KHS)										
19 Oct 1999	2 881	10 443	4 024	2 964	653	204	3 283	2 591	3 848	9 582
02 Dec 1999	6 081	38 957	29 654	40 371	9 002	10 205	59 851	46 649	71 668	210 755
20 Aug 2000	2 395	9 854	5 468	5 667	5 468	1 131	9 025	7 366	11 433	26 505
27 Aug 2000	1 013	3 548	1 835	1 751	473	203	2 138	1 902	2 624	6 641
28 Aug 2000	4 060	14 717	7 555	8 585	7 555	1 282	9 776	9 550	13 670	35 357
05 Sep 2000	746	2 232	1 271	1 444	1 271	194	1 860	1 763	2 443	5 474
28 Oct 2000	2 645	9 808	5 779	6 752	5 779	1 130	10 090	8 751	13 390	31 089
29 Oct 2000	17 965	124 833	99 549	111 489	99 549	3 160	135 652	78 462	202 038	514 439
03 Nov 2000	4 651	28 843	22 231	29 286	8 160	3 738	50 500	33 140	60 045	133 063
Total	42 438	243 234	177 364	208 310	137 908	21 247	282 176	190 175	381 160	972 905
Average	4 715	27 026	19 707	23 146	15 323	2 361	31 353	21 131	42 351	108 101
SD	5 248	38 561	31 484	35 814	31 755	3 206	44 700	26 570	65 137	167 588
RSD (%)	111	143	160	155	207	136	143	126	154	155

Table 15.6.
Dissolved metal loads (g storm^{-1}) at stations **a** WK and **b** KHS. *SD*: standard deviation, *RSD*: relative standard deviation

Storm event	Pb	Zn	Cu	Co	Ni	V	Cd
a Waikeakua (WK)							
20 Aug 2000	0.60	17	55	1.21	21	37	0.84
28 Aug 2000	0.29	11	31	0.70	38	22	0.41
05 Sep 2000	0.13	6	17	0.34	7	13	0.09
28 Oct 2000	0.08	6	20	0.37	7	11	0.12
29 Oct 2000	0.68	35	77	19	30	86	0.58
Total	1.77	75	200	22	103	169	2.04
Average	0.35	15	40	4	21	34	0.41
SD	0.27	12	26	8	14	31	0.32
RSD (%)	76	80	64	190	69	92	78
b Kaimuki High School (KHS)							
20 Aug 2000	3	65	206	3	68	125	0.18
28 Aug 2000	9	181	302	6	103	196	0.72
05 Sep 2000	2	23	64	2	36	49	0.08
28 Oct 2000	1	13	57	9	24	35	0.09
29 Oct 2000	3	70	275	134	99	190	0.56
Total	17	352	905	154	330	595	1.63
Average	3	70	181	31	66	119	0.33
SD	3	67	115	58	36	76	0.30
RSD (%)	91	95	64	188	54	64	91

Dominant geographic sources of individual elements during storms events can be elucidated by the comparison of particulate metal loads derived from conservation lands (above WK) and urban areas (below WK). Loads calculated at station WK represent conservation land-derived material. Because particulate matter in the lower watershed typically reflects both urban particulate input and the downstream transport of SPM, loads calculated at station KHS are assumed to reflect the total load for the Ala Wai Canal Watershed. Thus, metal loads from areas downstream of station WK are determined by taking the difference between the total load at station KHS and the conservation land load at station WK. Elevated loads in the lower segments of the watershed (Fig. 15.8) suggest an urban source for solid phase Pb, Zn, Cu, Ba, and Co. Reflecting anthropogenic enrichment, urban loads for these metals accounted for between 60 and 75% of their respective total loads (Fig. 15.8). With 86% of its total load derived from areas upstream of station WK (Fig. 15.8), particulate As displayed a considerable upper watershed source presumably associated with agricultural input. Loads of solid phase Ni, V, and Cr indicated roughly equal metal fluxes in both the upper and lower portions of the watershed (Fig. 15.8).

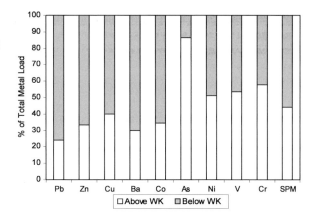

Fig. 15.8.
Percent of total particulate metal and SPM loads coming from areas above and below station WK. Values are based on cumulative 9-storm loads

15.4 Conclusions

This study has quantified concentrations of Pb, Zn, Cu, Ba, Co, As, Ni, V, and Cr during nine storm events in the Ala Wai Canal Watershed. Due to the extremely low concentration of metals in the dissolved phase and the large quantity of SPM transported by storm flow, a comparison of particulate and dissolved loads confirmed the expected dominance of the solid phase as a carrier of metals across the land-ocean divide. Dissolved metals comprised less than 1% of the total metal load (particulate and dissolved) in the Ala Wai Canal Watershed.

Reflecting urban land use, solid phase concentrations of Pb, Zn, Cu, Ba, and Co in storm-derived SPM exhibited elevated concentrations at the lower watershed station compared to values observed at the conservation site. Normalized to particulate V concentrations (due to its relative invariance throughout the watershed), Pb displayed the largest mean urban enrichment, followed in order by Zn, Ba, Cu, and Co. Moreover, particulate loads of these metals showed the presence of an urban flux roughly two times greater than that from conservation lands. Anthropogenic sources of Pb, Zn, Cu, Ba, and Co include automotive and industrial-related activities, although sources of Ba and Co are likely separate from those of Pb, Zn, and Cu.

In contrast, particulate As displayed an upper watershed enrichment in the Ala Wai Canal Watershed. Estimations of particulate As loads also indicated a much higher metal flux originating from conservation lands relative to areas below station WK. The upper watershed input of solid phase As is believed to be associated with the agricultural use of super-phosphate fertilizer by several flower farms upstream of the conservation site.

Displaying a relative spatial invariance throughout the Ala Wai Canal Watershed, concentrations of Ni, V, and Cr may be primarily controlled by the natural mineralogical release from the weathering of basalt. Solid phase loads of these metals exhibited a relative balance between input from areas in the upper and lower watershed.

Applied to particulate metal concentrations from both stations WK and KHS, principal components analysis (PCA) generated eigenvalues explaining 77% of the total variance and separated variables into two distinct factors. Factor 1 elements (Pb, Zn, Cu, Ba, and Co) are interpreted to represent metals with an urban anthropogenic source, whereas Factor 2 includes elements (As, Ni, V, and Cr) whose concentrations are not enhanced in the urban setting.

Sampling of a "Kona" storm, generated by an offshore low-pressure system, permitted the collection of urban input relatively undiluted by SPM from conservation areas. Higher concentrations of Pb, Zn, and Cu during the 19 October 1999 storm relative to the 8-storm mean (excluding the 19 October 1999 event) suggest that particulate material transported from the mountains during typical orographic storms dilutes the concentrations of these elements approximately 3-fold in urban sections of the watershed.

Acknowledgements

This work was supported in part by a grant from the National Oceanic and Atmospheric Administration, Project R/EL-15, which is sponsored by the University of Hawaii Sea Grant College Program, the US Environmental Protection Agency, the Hawai'i State Department of Health, and the US Geological Survey-National Institute for Water Research. We would also like to acknowledge Michael Tomlinson, Scott Narod, Norine Yeung, Vincent Todd, Julia Hubert, Chuck Fraley, and Sam Saylor for their invaluable contributions in the field and laboratory, as well as the gracious insights of Yuan-Hui (Telu) Li into the statistical treatment of the data. This is SOEST contribution 6435 and UH Sea Grant College Program contribution UNIHI-SEAGRANT-BC-00-01.

References

Armstrong LJ (1994) Contribution of heavy metals to storm water from automotive disc brake pad wear. Santa Clara Valley Non-Point Source Pollution Control Program, pp 26–36
Bodek I, Lyman WJ, Reehl WF (1988) Environmental inorganic chemistry: properties, processes, and estimation methods. Permagon Press, New York
Dashiell EP (1997) Ala Wai Canal Watershed water quality improvement project: management and implementation plan, vol. 1. City and County of Honolulu, State of Hawai'i
Davis WE (1972) National inventory of sources and emissions: barium, boron, copper, selenium, and zinc. US Environmental Protection Series, Washington DC
De Carlo EH, Anthony SS (2002) Spatial and temporal variability of trace element concentrations in an urban subtropical watershed, Honolulu, Hawai'i. Appl Geochem 17:475–492
De Carlo EH, Dollar S (1997) Assessment of suspended solids and particulate nutrient loading to surface runoff and the coastal ocean in the Honokowai Drainage Basin, Lahaina District, Maui, Hawai'i. West Maui Algal Bloom Study, Final report to Hawai'i State Department of Health
De Carlo EH, Resing JA (1998) Determination of picomolar concentrations of trace elements in high salinity fluids by FIA-ICP-MS. ICP Information Newsletter 23:82–83
De Carlo EH, Spencer KJ (1995) Records of lead and other heavy metal inputs to sediments of the Ala Wai Canal, O'ahu, Hawai'i. Pacific Science 49:471–491
De Carlo EH, Spencer KJ (1997) Retrospective analysis of anthropogenic inputs of lead and other metals to the Ala Wai Canal, Oahu, Hawaii. Applied Organometallic Chemistry 11(4):415–437
De Carlo EH, Tomlinson MS, Anthony SA (2004) Trace elements in streambed sediments of small subtropical streams on O'ahu, Hawai'i: results of the USGS NAWQA Program. Appl Geochem (in press)
Erel Y, Patterson CC, Scott MJ, Morgan JJ (1990) Transport of industrial lead in snow through soil to stream water and groundwater. Chem Geol 85:383–392

Frey FA, Garcia MO, Roden MF (1994) Geochemical characteristics of Ko'olau Volcano: implications of intershield geochemical differences among Hawaiian volcanoes. Geochim Cosmochim Acta 58:1441–1462

Gélinas Y, Barnes RM, Florian D, Schmit JP (1998) Acid leaching from environmental particles: expressing results as a concentration within a leachable fraction. Environ Sci Technol 32:3622–3627

Geschwind L (2000) A field and laboratory evaluation of the Fossil Filter in reducing heavy metal non-point source pollution from street runoff. Unpublished senior thesis, Global Environmental Science Program, University of Hawai'i at Manoa, Honolulu

Giambelluca TW, Nullet MA, Schroeder TA (1986) Rainfall atlas of Hawai'i. Department of Land and Natural Resources, State of Hawai'i

Halbig JB, Barnard WM, Johnston SE, Butts RA, Bartlett SA (1985) A baseline study of soil geochemistry in selected areas on the island of Hawai'i. Department of Planning and Economic Development, State of Hawai'i

Hayes KF, Leckie JO (1986) Mechanism of lead ion adsorption at the goethite-water interface. In: Davis JA, Hayes KF (eds) Geochemical processes at mineral interfaces. ACS, pp 114–141

Hollibaugh JT (ed) (1996) San Francisco Bay: the Ecosystem. AAAS, Pacific Division, San Francisco

Hoover D (2002) Fluvial nutrient fluxes to O'ahu coastal waters: land use, storm runoff, and impacts on coastal ecosystems. Dissertation, Department of Oceanography, University of Hawai'i at Manoa, Honolulu

Irving MM (1998) An investigation of reactions among heavy metals and naturally occurring minerals in synthetic and natural stream waters. Dissertation, Department of Chemistry, University of Hawai'i at Manoa, Honolulu

Juvik SP, Juvik JO (eds) (1998) Atlas of Hawai'i. University of Hawai'i Press, Honolulu, Hawai'i

Kanehiro Y, Sherman GD (1967) Distribution of total and 0.1 normal hydrochloric acid-extractable zinc in Hawaiian soil profiles. Soil Sci Society of American Proceedings 31(3): 394–399

Li H (1996) Baseline study of heavy metal concentrations and human activity impact analysis in the Hawaiian environment. Thesis, Department of Oceanography, University of Hawai'i at Manoa, Honolulu

Lorentzen EML, Kingston HM (1996) Comparison of microwave-assisted and conventional leaching using EPA method 3050B. Anal Chem 68:4316–4320

Luoma SN (1989) Can we determine the biological availability of sediment-bound trace elements? Hydrobiologia 176/177:379–396

Miner S (1969) Air pollution aspects of barium and its compounds. Litton Systems, Inc., Bethesda

Rivera-Duarte I, Flegal AR (1994) Benthic lead fluxes in San Francisco Bay, California, USA. Geochim Cosmochim Acta 58:3307–3313

Sadiq M, Alam I, El-Mubarek A, Al-Mohdhar HM (1989) Preliminary evaluation of metal pollution from wear of auto tires. Bull Environ Contam Toxicol 42:743–748

Sansalone JJ, Buchberger SG (1997) Partitioning and first flush of metals in urban roadway storm water. J Environ Engrg 134–143

Smith IC, Carson BL (eds) (1981) Trace metals in the environment, vol. 6: cobalt. Ann Arbor Science, Ann Arbor

Spencer KJ, De Carlo EH, McMurtry GM (1995) Isotopic clues to the sources of natural and anthropogenic lead in sediments and soils from O'ahu, Hawai'i. Pacific Science 49:492–510

Sposito G (1984) The surface chemistry of soils. Oxford University Press, New York

Sutherland RA, Tolosa CA (2000) Multi-element analysis of road-deposited sediment in an urban drainage basin, Honolulu, Hawai'i. Environ Pollut 10:483–495

Tinsley IJ (1979) Chemical concepts in pollutant behavior. John Wiley & Sons, New York

Tomlinson MS, De Carlo EH (2001) Investigations of Waimanalo and Kaneohe Streams. Final report to US-Environmental Protection Agency and Hawai'i State Department of Health, 34 pp

Venugopal B, Luckey TD (1978) Metal toxicity in mammals. Plenum Press, New York

Wen XY, De Carlo EH, Li YH (1997) Interelement relationships in ferromanganese crusts from the central Pacific Ocean: their implications for crust genesis. Mar Geol 136:277–297

Part II

Chapter 16

A Model for Predicting Heavy Metal Concentrations in Soils

L. Frolova · A. Zakirov · T. Koroleva

Abstract

In the absence of 'a priori' information about the probable distribution of heavy metals in soils, it is relevant to carry out multiple parameter analyses of soil concentrations and to take into account a variety of factors. A method was developed to group existing soil data from the Predvolgie region of the Tatarstan Republic to forecast concentrations of heavy metals in similar soil environments. The predicted concentration of heavy metals in soils is obtained by applying a model that combines soil data using iterative statistical procedures of functional clustering and fuzzy sets. Currently, researchers do not have enough data from experimental and field research to construct adequate maps of soil pollution and estimates of the ecological state of the environment. Our proposed method permits a preliminary solution of these problems.

Key words: soils, heavy metals, prediction model, functional clustering, fuzzy sets

16.1 Introduction

Accumulation of heavy metals in the atmosphere, soils, and waters of natural and modified landscapes is a common problem for many countries. Soils can be considered as an integrated indicator of long-term pollution processes in an environment reflecting the quality of soil, air and water resources. The influence of heavy metals on the environment has increased with the increase of economic and industrial activities in society, thus knowledge of the principles of spatial distribution of heavy metals has become increasingly important.

Because soils are rather complex systems, it is useful to estimate several parameters in soils and to take into account a variety of factors in the absence of 'a priori' information on the probable distribution of heavy metals in soils.

Modern research is characterized by new approaches in the theory of image recognition, methods of finite mathematics, and criteria of multivariate statistics that have influenced analytical methods to determine the spatial distribution of concentrations of heavy metals in soils. There have been two main approaches; a deterministic approach (Webster 1972; Stroganova et al. 1998), and a statistical approach (Moiseenkov et al. 1995; Alexeyenko et al. 1995; Norris 1972). The first provides numerical classification of soil taxa, and the later reveals functional dependencies between various soil parameters and the concentration of heavy metals. Most authors suggest that soil parameters are strongly related to heavy metals in soils, although some positive relationships are rather general. We also try to find the correlation of soil properties with Ni concentration using a standard method of multiple regression and we did not find a functional dependence, as well as some local researchers. Thus, it is necessary to continue to search for other forms of dependence and to develop methods for their detection.

Two main points were considered in developing a method for predicting concentrations of heavy metals. Statistical calculations should be based on a minimum amount of data. Collecting soil data is expensive and usually there are not many replicate samples. For statistical stability of the conclusions it is necessary to take into account that only a small number of samples are available and that finite population corrections are usually necessary (Frolova and Zakirov 2002).

Soil science literature points out that soil is a complex object in terms of data collection. Structural complexity, process dynamics, and uncertainty of parameter values describing a condition, or state, of these processes are also integral properties of soils. Because of the lack of mutually exclusive data classes, several authors have found it effective to apply fuzzy set theory in an analysis of dynamic conditions of some biological systems (Zadeh 1988; Frolova and Zakirov 1997).

16.2
Experimental

16.2.1.
Method of Grouping Soil Data

The prediction of heavy metal concentrations in soils is estimated with a method that combines and evaluates existing soil and environmental data and uses iterative statistical procedures of functional clustering and fuzzy sets. The method is illustrated with soil data for the total content of nickel.

A 10×10 km sampling grid was overlain on a general soil map of the region and physical and chemical analyses were made for about 130 point samples. Parameters for the heavy metal predictions consisted of selected physical and chemical data including concentrations of soluble and total Ni, Pb, Cu, Zn, Cr, and Mn; and qualitative estimates of ecological conditions and total heavy metal emissions to the atmosphere obtained from maps prepared by the Institute of Ecology of Natural Landscapes, Tatarstan Academy of Sciences. Values of heavy metals are missing for some of the point sample sites, for example, there were only 96 values of Ni.

The general task is to allocate in the multivariate space of the parameters, those clusters that are homogeneous in the sense that their variability is characterized by the *mean* and *variance* of a normal distribution. In this multivariate space we used six parameters: a qualitative estimate of ecological condition (rank); sum of emissions to atmosphere (rank); humus content (%); sum of exchangeable bases (meq/100 g); silt (1–50 μm) content (%); and heavy metal concentration (mg kg^{-1}). These six parameters can be combined into 146 ways ranging from a single group of the 6 parameters, (1,2,3,4,5,6) all the way through six groups each containing only one parameter (1)(2)(3)(4)(5)(6). Within the 6×96 matrix for Ni (6 parameters at 96 locations of soil tests) the data are normalized and all values are dimensionless. The *mean* of any parameter for all tests is zero and the *variance* of a parameter for all tests is one.

To obtain statistical stability each cluster must have at least a minimum data set described by a *mean* and *variance* of a normal distribution. An initial estimate of the cluster size is obtained by iteratively calculating a cluster size, N_c, according to Eq. 16.1:

$$N_c = \left(\frac{t_{n0}^2 S_{n0}^2}{\delta^2} \right) + 1 \qquad (16.1)$$

The right hand side of the equation is the 'point of statistical stability' where t is the value of Student's distribution, S^2 is the variance of the subsample, n_0 is the size of estimated subsample, δ is the half-width confidence interval with a user-designated value of measurement error: we commonly use 20% of the sample *mean*.

An initial subsample size n_0 is assumed, say 2, and N_c is calculated. If $N_c > n_0$ the subsample size n_0 is increased by 1 and N_c calculated again. These step increments are carried out until the subsample size n_0 equals or exceeds N_c at which point the subsample is statistically stable. Because most subsamples have groups with more than one parameter, the overall cluster size will be equal to a maximum of the cluster sizes for all parameters included in the group.

Fig. 16.1. Optimum combination of parameters

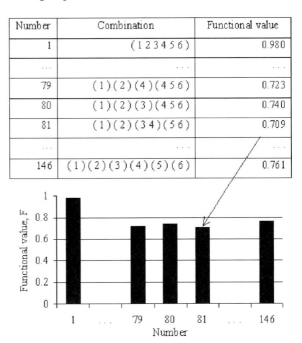

Number	Combination	Functional value
1	(1 2 3 4 5 6)	0.980
...	...	...
79	(1)(2)(4)(4 5 6)	0.723
80	(1)(2)(3)(4 5 6)	0.740
81	(1)(2)(3 4)(5 6)	0.709
...	...	...
146	(1)(2)(3)(4)(5)(6)	0.761

Table 16.1. Cluster number, sample size n_0, mean and variance of parameters for the optimal combination

Cluster	n_0	Mean	Variance	Mean	Variance
		Ecological condition			
1.1	21	2.24[a]	0.18		
1.2	19	2.53	0.56		
1.3	22	2.77	0.99		
1.4	19	2.95	1.31		
1.5	16	3.27	0.86		
		Sum of emissions to atmosphere			
2.1	48	2.00	0.00		
2.2	34	1.96	0.04		
		Humus plus		Sum of exchangeable bases	
3.1	24	3.78	3.90	24.1	42.2
3.2	24	4.03	4.09	33.5	89.5
3.3	27	5.04	5.75	34.1	153.6
		Silt plus		Ni (total)	
4.1	23	19.4	27.7	35.8	75.9
4.2	26	23.1	35.5	46.2	370.6
4.3	27	24.8	48.5	42.7	168.6

[a] The values for *mean* and *variance* have been rounded for this presentation.

For all stabilized clusters a functional value, F, is calculated from Eq. 16.2:

$$F = \frac{1}{N} \sum_{p=1}^{k_1} \sum_{q=1}^{k_2} \sum_{\substack{j \in G_p \\ i \in R_{qp}}} (x_{ij} - \bar{x}_{qp}^j)^2 \tag{16.2}$$

where N is the number of values in data matrix; p is the index of group in combination; k_1 is the number of groups in combination; q is the index of cluster for group p; k_2 is the number of clusters for group p; j is the index of parameter in group p; G_p is the set of parameters in group p; i is the index of test in cluster q; R_{qp} is the set of values in cluster q of group p; x_{ij} is the value of parameter j of the test i; and $\bar{x}_{qp}^j$ is the mean of values of parameter j of cluster q.

The distribution of functional values is random in shape over all combinations of parameters and the minimum value is where the overall *variance* is the least and indicates the combination that has the optimal separation of the parameters (Fig. 16.1).

In our example for total nickel the minimum functional value is 0.7098 associated with the parameter combination of (1)(2)(34)(56). The optimal combination for total nickel has a total of 13 clusters (Table 16.1); five for ecological conditions, two for sum of emissions, and three each for the humus plus sum of bases group, and the silt plus Ni (total) group.

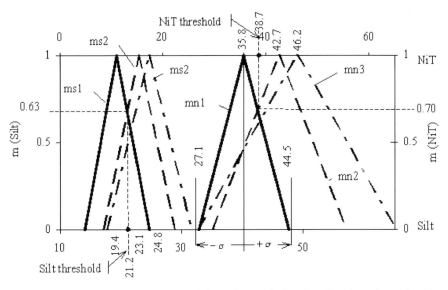

Fig. 16.2. Thresholds and correspondence of fuzzy classes of silt and total Ni through membership function, *m*, for the optimal parameter combination

16.2.2
Analysis of Clusters

The initial grouping of soil data provides statistically stable groups characterized by their *means* and *variances*. The final choice of an optimum data cluster is the combination of parameters whose overall *functional value* is minimal.

The optimal combination has a binary group of silt and total nickel, consequently this correspondence in three clusters can be used to predict the heavy metal based on the correspondence with the silt fraction. A simple regression of total Ni on silt has a regression coefficient of 0.23, thus that relationship is not useful for prediction.

Clusters of the silt fraction and total nickel can be diagrammed as a set of fuzzy classes (Fig. 16.2) that illustrates the location of thresholds and their corresponding membership functions *m*(Silt) and *m*(NiT) for clusters 4.1 (ms1, mn1), 4.2 (ms2, mn2), and 4.3 (ms3, mn3). The *mean* of each class can be placed as an apex at a membership function value of 1.0 and on two *x*-axes (Silt, NiT) the base is the *mean* plus and minus the *standard deviation* of the class. Thus for the first silt cluster the base line is from 14.1–24.7% silt value and the apex is located above 19.4% at the membership function value of 1.0. The strongly overlapping 2nd and 3rd clusters are constructed in a similar manner.

The threshold separating the silt clusters 1 and 2 occurs at *m*(Silt) = 0.63 and silt value = 21.2%. The minor threshold between clusters 2 and 3 is not significant for the current predictions.

The threshold separating the total Ni clusters occurs at about *m*(NiT) = 0.70 and as can be seen cluster 3 has a wider variance than cluster 2. If the value of silt is less than 21.2% the predicted interval of the total Ni concentration is between 27.1–44.5 mg kg^{-1} soil;

if it is more than 21.2% the range is 26.9–65.5 mg kg^{-1} soil. The predicted concentrations of other heavy metals can be calculated by the same methodology using Windows software for a PC developed by the authors.

16.3
Results and Discussion

Simple statistical regressions did not reveal significant relationships between soil parameters and heavy metals in soils of the Predvolgie region. Consequently we searched for tools that might detect other forms of dependence. We developed a method that enabled us to detect numeric correspondence between soil parameters and heavy metals using combinational procedures and fuzzy set concepts.

We prepared a matrix from existing environmental and soil data for a large area. The parameter data were normalized and evaluated to obtain stabilized clusters and statistically determine the optimum combination of the parameters.

From an optimum combination that contained clusters of heavy metal and other soil parameters, the normalized *means* and *variances* of the parameters in that cluster enabled us to construct fuzzy classes. For the overlapping classes of a given parameter, thresholds can be determined and these provide an initial stratification. A numeric correspondence between parameters, for example silt and total nickel, is obtained through the membership function of the fuzzy classes. This is a form of dependence that was undetected by usual statistical procedures.

Meaningful correspondence for heavy metals was detected as follows:

- Ni (total and soluble) and the silt fraction
- Cu (total and soluble) and the humus
- Mn (total) and the sum of exchangeable bases
- Mn (soluble) and the silt fraction
- Pb (total) and the humus
- Pb (soluble) and the silt fraction
- Zn (total) and the sum of exchangeable bases
- Zn (soluble) and the silt fraction
- Cr (soluble) and the silt fraction

The clusters for total Cr were not stable statistically thus no predictions for this heavy metal were possible.

For the parameters included in a cluster, such as silt and total nickel in Fig. 16.2, it is possible to estimate the value of any parameter in the cluster using predicted interval.

In the Predvolgie region agricultural lands have received more lime and fertilizers than natural lands, consequently their surface soils have higher contents of exchangeable bases. Therefore, in addition to predicting intervals of heavy metals from the correspondence with silt or humus fractions, it is commonly possible to predict results for soils having anthropogenic influence. Relationships and predictions are given in Table 16.2. As noted previously, the predicted interval for total Ni is 35.8 ±8.7 mg Ni(total)/kg soil when the amount of silt in the surface soil sample is less than or equal to 21.2%.

Table 16.2. Predicted intervals of heavy metal concentrations via threshold values of soil parameters

Physical and chemical parameter	Threshold value	Heavy metal (total (T) / soluble (S))	Predicted interval, (mg / kg soil)
Silt	≤21.2%	NiT	35.8 ±8.7
	≥21.2%		46.2 ±19.3
Silt	≤21.2%	NiS	11.7 ±3.2
	≥21.2%		14.1 ±5.3
Humus	≤4.2%	CuT	21.0 ±4.0
	≥4.2%		23.5 ±5.7
Humus	≤3.9%	CuS	5.3 ±1.1
	≥3.9%		6.0 ±1.6
Sum of exchangeable bases	≤30 meq / 100 gm	MnT	128.7 ±30.0
	≥30 meq / 100 gm		141.5 ±55.0
Silt	≤22.5%	MnS	137.4 ±32.0
	≥22.5%		153.0 ±63.4
Humus	≤4.1%	PbT	11.4 ±2.1
	≥4.1%		13.1 ±2.6
Silt	≤23.4%	PbS	2.1 ±1.4
	≥23.4%		2.9 ±1.8
Sum of exchangeable bases	≤28.5 meq / 100 gm	ZnT	25.5 ±5.0
	≥28.5 meq / 100 gm		29.2 ±7.6
Silt	≤23.0%	ZnS	5.6 ±1.6
	≥23.0%		7.2 ±2.0
Silt	≤22.0%	CrS	5.6 ±5.5

We interpret the results (Table 16.2) as characterizing the current "background" level of heavy metal concentrations in soils of the Predvolgie region of Tatarstan Republic. This model was applied to predict heavy metal concentrations in soils of two regions of Tatarstan Republic, Apastovsky and Kaybitsky. Predicted interval concentrations of Ni total based on correspondence with measured Silt are provided in Table 16.3. At present the model does not predict Ni intervals for measured silt values greater than 31.7%.

The model was used to calculate the predicted interval concentrations of other heavy metals based on correspondence with measured soil parameters the same way. At present the predictions are for rather large overlapping intervals of a parameter, however we are excited about reducing the acceptable error of measurement and refining the numerical dependence detected in this model.

In a preliminary assessment of this model of parameter dependence, we note that there was about 80% coincidence of predicted intervals and measured values.

Table 16.3. Application of the model

Silt (%)		Ni total (mg kg^{-1})		Coincidence
Measured	Threshold	Predicted interval	Measured	
Apastovsky				
23.5	>21.2	46.2 ±19.3	60.0	Yes
24.6	>21.2	46.2 ±19.3	36.0	Yes
40.9	>21.2	46.2 ±19.3	80.0	No
30.2	>21.2	46.2 ±19.3	90.0	No
23.8	>21.2	46.2 ±19.3	30.0	Yes
27.3	>21.2	46.2 ±19.3	41.0	Yes
Kaybitsky				
20.0	≤21.2	35.8 ±8.7	30.0	Yes
38.7	>21.2	46.2 ±19.3	65.5	Yes
23.3	>21.2	46.2 ±19.3	55.0	Yes
24.9	>21.2	46.2 ±19.3	38.0	Yes
28.4	>21.2	46.2 ±19.3	35.0	Yes
22.2	>21.2	46.2 ±19.3	28.0	Yes
29.2	>21.2	46.2 ±19.3	30.0	Yes
28.5	>21.2	46.2 ±19.3	88.0	No
31.0	>21.2	46.2 ±19.3	38.0	Yes
28.5	>21.2	46.2 ±19.3	35.0	Yes

16.4 Conclusions

When usual methods do not reveal clear relationships between soil properties and concentrations of heavy metals in soils, other techniques should be explored. This was the situation in part of the Tatarstan Republic where limited background information was available. We worked on a model to provide numeric correspondence between soil parameters and heavy metals. It was achieved with procedures of combinational theory and interpretations of fuzzy set classes which are illustrated in the paper. We are not aware that this type of model has been used by other researchers.

A preliminary assessment of the model using the thresholds calculated for silt and the predicted intervals for Ni indicate about an 80% coincidence between predicted and measured values for the Predvolgie region of the Tatarstan Republic. Because silt can readily be estimated by competent field pedologists we believe our model will permit a rapid and relatively inexpensive inventory of heavy metals in our region. Such preliminary information can then be used to plan a systematic assessment. Refinements of this model are logical next steps and we welcome your comments and suggestions.

Acknowledgements

We thank Dr. Boris Grigoryan, Associate Professor of Soil Department of Kazan State University and Head of the Biogeochemistry Laboratory of the Institute of Ecology of Natural Landscapes, Tatarstan Academy of Sciences for permission to use their data to develop and test this model. We thank Richard W. Arnold for his critical review, consultation, and patient assistance in preparing this English version of our manuscript.

References

Alexeyenko VA, Serikov VN, Alexeyenko LP (1995) Metals in agricultural geochemical landscapes of the southern part of European Russia. In: Tikhonov AN, et al. (eds) Geography. Moscow Univ Publ House, Moscow, pp 248–256

Frolova L, Zakirov A (1997) Decision-making using fuzzy sets for strategy selection of environment restoration. J Environ Radioecol Appl Ecology 3:26–35

Frolova LL, Zakirov AG (2002) Definition of sample size of heavy metal concentrations in various soils. In: W Lianxiang, W Deyi, T Xiaoning, N Jing (ed) Sustainable utilization of global soil and water resources. Proceedings of 12th International conference of ISCO, held in Beijing, 26–31 May 2002. Tsinghua University Press, Beijing, 2:33–37

Moiseenkov OV, Kasimov NS, Batoian VV (1995) Ecogeochemistry and methods of computer cartography. In: Tikhonov AN, et al. (eds) Geography. Moscow Univ Publ House, Moscow, pp 239–247

Norris J (1972) The application of multivariate analysis to soil studies: III soil variation. J Soil Sci 23:62–75

Stroganova MN, Myagkova AD, Prokofieva TV, Skvortsova TV (1998) Soils of Moscow and urban environment. PAIMS, Moscow

Webster R, Burrough P (1972) Computer-based soil mapping of small areas from sample data. Multivariate classification and ordination. J Soil Sci 2:210–221

Zadeh L (1988) Fuzzy logic. J Computer 21:83–92

Part II

Chapter 17

Phytoremediation of Thallium Contaminated Soils by *Brassicaceae*

H. Al-Najar · R. Schulz · V. Römheld

Abstract

In order to assess the efficacy of phytoremediation, the thallium (Tl)-hyperaccumulator plants kale, *Brassica oleracea acephala* L. cv. Winterbor F1, and candytuft, *Iberis intermedia* Guers., were grown in several Tl-contaminated soils. These soils differed in their total Tl concentration (1.4 to 153 mg Tl kg^{-1} soil), the origin of pollution (anthropogenic vs. geogenic), as well as Tl binding forms, which were characterized by a sequential extraction. In soils with geogenic Tl the percentage of easily accessible fractions was relatively low amounting to 5% of total Tl. In contrast, in soils with anthropogenic Tl pollution the pool of easily accessible Tl was large amounting to 23% of total Tl in the soil polluted by deposition from cement plant. As a consequence, the partition ratio of shoot Tl concentration vs. total soil Tl concentration ranged from 1 to 12 for soils contaminated by geogenic and anthropogenic sources, respectively. In general, there was no relationship between Tl uptake by the hyperaccumulator plants and the total Tl concentration of the soils. The plant uptake of Tl depended on the capacity of the soils to replenish the soil solution with Tl as well as the replenishment from less accessible binding forms.

Key words: anthropogenic, binding forms, geogenic, phytoextraction, sequential extraction

17.1
Introduction

Plant species capable of accumulating large concentrations of trace metals such as zinc (Zn), cadmium (Cd), nickel (Ni) and lead (Pb) in their shoots are generally referred to as hyperaccumulator plants (Brooks et al. 1977). Consequently, they have raised strong interest for phytoremediation of metal-contaminated soils (Baker and Brooks 1989; Baker et al. 1991; Whiting et al. 2001). In contrast to Tl, the toxic metals Cd, Zn, Cu and Ni have received considerable attention in phytoremediation research (Knight et al. 1997; Robinson et al. 1997; Yang et al. 1998). In literature Tl in plants has been studied mostly from a human toxicological point of view. Only few studies have considered the potential of hyperaccumulator plants to remediate Tl-polluted soils (Leblanc et al. 1999).

Thallium is a group IIIa element together with indium (In), gallium (Ga), aluminum (Al) and boron (B). It is highly reactive, readily soluble in acids and forms monovalent (thallous) and trivalent (thallic) salts, the latter being less stable (Kazantzis 1986). Tl encountered in the soil may have two origins: geogenic or anthropogenic. Natural Tl is derived from the geochemical composition of rocks (Tremel et al. 1997), but may be greatly exceeded by anthropogenic inputs as a reuslt of industrial emissions or mining activities. Tl has an abundance of about 0.7 µg g^{-1} in the crust of the earth (Green 1972). In the soil it is bound in different binding forms (fractions) or pools, which differ in their plant availability. In the order of decreasing availability, these fractions are: (1) water-soluble and mobile; (2) exchangeable; (3) bound to Mn oxides; (4) bound to organic matter; (5) bound to amorphous and poorly crystalline Fe oxides; (6) bound to crystalline Fe oxides; (7) bound in the soil matrix or in the primary minerals (Zeien and Brümmer 1989). By studying the Tl binding forms in a soil, the bioavailability of different fractions may be assessed in addition to the potential of plants to mobilize Tl from the less accessible fractions. Kale and candytuft have previously been identified as hyperaccumulator plants for Tl (Kurz et al. 1999; Leblanc et al. 1999).

The success of phytoremediation mainly depends on the concentration of toxic metals in the easy accessible binding forms in the soil (Gerritse et al. 1983; Brümmer et al. 1986; McGrath et al. 1997). However, the availability of heavy metals in the soil depends on many factors like soil pH (Gieslinski et al. 1996), clay mineral content (Herms and Brümmer 1984), Mn oxides concentration and organic matter content in the soil (King 1988), but also on plant factors.

The objectives of the current study were to determine the amount of Tl that could be taken up by kale and candytuft from anthropogenic and geogenic origins of pollution. Moreover, the relationship between plant uptake and Tl concentration in the soil in addition to the ability of different soils to replenish the easy accessible pool was investigated. These questions are essential for assessing the feasibility of phytoremediation of Tl from different origins of pollution.

17.2
Experimental

17.2.1
Soil Extraction and Preparation of Soils for Uptake Experiments

Soil solution was extracted by incubating 100 g soil of 20% moisture content for 24 h, followed by 3 500 rpm centrifugation for 30 min. and then filtration (Brümmer et al. 1986). Sequential Tl extraction was carried out according to Zeien and Brümmer (1989) as shown in Table 17.1.

After each of the extraction steps 2 to 6, the samples were washed with the respective preceding extraction solutions (Table 17.1). The washing solutions were combined with the preceding extract. Increasing fraction numbers denote increasing binding strength. However, it is not possible to extract exactly a specific metal binding form from soil using a particular extraction solution (Wilcke et al. 1999).

The soils contained Tl in different concentrations and from different origins of pollution (geogenic and anthropogenic). Two different sources of anthropogenic Tl pollution were used: *(1)* one soil near a cement plant in Leimen, Germany, denominated (L1); and *(2)* 6 soils polluted as a result of mining activities with different mobile and total concentrations of Tl, Cd and Zn from St. Laurent le Minier, Gard, South France, denominated SLM1-6, and three different soils with geogenic origin of Tl pollution from Schömberg, SW-Germany, denominated SW1-3.

As shown in Table 17.2, the pH of soils L1, SLM1-6 and SW1 was greater than 7.2, while SW2 and SW3 soils had a pH of less than 6. The soil from Leimen (L1) was a silty loam, the soils SLM1-6 had a sandy texture, while the soils SW1-3 had a loamy texture.

Table 17.1. Tl fractions as characterized by defined extraction solutions and shaking time according to Zeien and Brümmer (1989)

Fraction		Extraction solution	Shaking time
F1	Mobile	25 ml of 1 M NH_4NO_3	24 h
F2	Easily exchangeable	50 ml of 1 M NH_4Oac; pH 6, adjusted with 50% acetic acid	24 h
F3	Bound to Mn oxides	50 ml of 0.1 M NH_2OH-HCl + 1 M NH_4Oac; pH 6, adjusted with HCl	30 min
F4	Bound to organic matter	50 ml of 0.025 M NH_4-EDTA; pH 4.6, adjusted with NH_4OH solution	90 min
F5	Bound to amorphous and poorly crystalline Fe oxides	50 ml of 0.2 M NH_4-oxalate; pH 3.25, adjusted with NH_4OH solution	4 h
F6	Bound to crystalline Fe oxides	50 ml of 0.1 M ascorbic acid in 0.2 M NH_4-oxalate; pH 3.25, adjusted with NH_4OH solution	30 min in water of 96 °C
F7	Mainly bound to silicates	Aqua regia (3 parts HCl + 1 part HNO_3)	–

For the pot experiments the soils were treated as follows: water content was adjusted to 20 and 30% w/w for the anthropogenic and geogenic polluted soils, respectively. To supply sufficient nutrients to the plants 400 mg N kg^{-1} soil was added as NH_4NO_3, and 156 mg K kg^{-1} soil as well as 124 mg P kg^{-1} soil was added as KH_2PO_4. SW-soils were treated with 2 g Sedipur AF kg^{-1} soil (BASF, Ludwigshafen) as soil conditioner (due to heteropolar characteristics), while Leimen soil (L1) was treated with 1 g Sedipur AF kg^{-1} soil. The soils were filled in cylindrical polyethylene pots with a depth of 19 cm and a diameter of 9 cm for the uptake experiment with kale plants, and pots with a depth of 13 cm and a diameter of 7 cm for the experiment with candytuft. The experiment was conducted in a climate chamber with 450 µmol $m^{-2} s^{-1}$ photon flux using 4 replicates.

17.2.2
Plant Preparation and Analysis

The seeds of kale and candytuft were germinated for one and two weeks, respectively, in a peat-sand mixture, before 3 plants of kale and candytuft were transferred to each pot. After 20 d kale was harvested, while candytuft was harvested after 30 d. The shoots of both plants were weighted after drying at 60 °C for 4 d. Plant material was digested by microwave (MLS1200; MLS Ltd., Leutkirch, Germany) with a mixture of HNO_3 (65%) and H_2O_2 (30%) following the standard VDLUFA method (VDLUFA 1996). Tl, Cd and Zn concentrations of the digests were determined by inductively coupled plasma-mass spectrometry (ICP-MS) (Elan 6000, Perkin Elmer).

In order to correlate plant uptake with the total Tl concentration in the different soils, a plant-soil partition coefficient (PC) was calculated by dividing the Tl concentration in the plant shoot dry matter by the total concentration of the soil (Tremel et al. 1997).

Table 17.2. Some physical and chemical characteristics of the used anthropogenic and geogenic contaminated soils from Leimen, Germany (*L1*), St. Laurent le Minier, Gard, Southern France (*SLM1–SLM6*) and Schömberg, SW-Germany (*SW1–SW3*)

Soil Nr.	Origin of pollution	Soil texture	pH	Tl Mobile (µg/kg soil)	Tl Total (mg/kg soil)	Cd Mobile (µg/kg soil)	Cd Total (mg/kg soil)	Zn Mobile (µg/kg soil)	Zn Total (mg/kg soil)
L1[a]	Anthropogenic	Silt loam	7.2	106	1.4	20	0.1	70	46
SLM1	Anthropogenic	Sandy	7.3–7.8	628	49.1	1 500	190.0	169 000	49 200
SLM2[a]	Anthropogenic	Sandy	7.3–7.8	73	9.6	420	21.0	33 600	7 500
SLM3	Anthropogenic	Sandy	7.3–7.8	600	153.0	525	70.1	97 500	23 400
SLM4	Anthropogenic	Sandy	7.3–7.8	270	13.3	1 825	49.2	93 250	13 600
SLM5	Anthropogenic	Sandy	7.3–7.8	310	14.2	985	40.3	97 750	14 800
SLM6	Anthropogenic	Sandy	7.3–7.8	410	17.6	710	154.0	117 750	50 300
SW1[a]	Geogenic	Loamy	7.2	12	1.6	1	0.8	53	129
SW2[a]	Geogenic	Loamy	5.2	41	3.8	53	0.7	735	164
SW3	Geogenic	Loamy	5.4	112	2.7	70	1.0	870	188

[a] Soils selected for detailed soil extraction.

17.3
Results and Discussion

17.3.1
Heavy Metal Concentration of Various Soils

Soils with different textures and different origins of pollution were used to investigate the relation between Tl fractionation in the soils and the ability of hyperaccumulator plants to accumulate Tl in their shoots.

As shown in Table 17.2, Leimen soil (L1) contained a high concentration of Tl in the mobile fraction amounting to 106 µg kg^{-1} soil, while the geogenic soils (SW1 and 2) and mining soil SLM2 showed only moderate Tl concentrations in this fraction: they amounted to 73, 12 and 41 µg kg^{-1} soil for the soils SLM2, SW1 and SW2, respectively.

Total Cd and Zn concentrations of L1 and SW1 soils were very low in comparison to soil SLM2 (Table 17.2). Also the mobile Cd and Zn in soil SLM2 showed high concentrations and amounted to 420 µg Cd and 33 600 µg Zn kg^{-1} soil. This high Cd and Zn concentration in SLM soils is a result of metal mining activities. The amounts of mobile vs. total concentration of Tl, Cd and Zn of each soil was indicative for the origin of pollution. It is commonly considered that mining places are sources of pollution to specific mined elements, but the analysis indicated that Tl was also a major pollutant in this area. The distribution of Tl either horizontal or vertical in the soil profile indicate the origin of pollution. Anthropogenic pollutions mostly occur on the top layers, while Tl could be found in deep soil profiles as indication of geogenic pollution (Tremel et al. 1997). Heavy metals pollution form mining origins depends mainly on the parent material of the soil.

17.3.1.1
Effect of Pollution Origin on Tl Binding Forms

The two soils (L1 and SLM2) with anthropogenic Tl sources (emission and mining) and another two soils (SW1 and 2) with geogenic Tl origin were sequentially extracted in order to investigate the availability of Tl to plants. The different extent of the mobile fractions according to origin of pollution is expected to affect strongly the heavy metal availability.

As shown in Table 17.3, the concentration of Tl in the soil solution of each soil was different and accounted for 1.2, 0.35, 0.05 and 0.07 µg l^{-1} in the case of L1, SLM2, SW1 and SW2, respectively. High variations regarding the distribution of the various fractions in the different soils were observed. The anthropogenically contaminated soil L1 had a 3–10 times higher Tl concentration in the mobile fraction (F1) compared to SW1 and SW2. Likewise, fraction F1 of soil SLM2 was roughly 2–3 times higher than the respective fraction of SW1 and SW2 (Table 17.3). The soil polluted as a result of mining activities (SLM2) showed extremely high amounts of Tl bound to Mn and crystalline Fe oxides (fractions 3 and 6) compared to the other soils.

The fraction 1 to 4 had been indicated as "easily accessible" by previous studies, while the fractions F5 to F7 had been denominated as "less accessible" (Symeonides and McRae 1977). As shown in Table 17.3, the soils with Tl of geogenic origin had a

Table 17.3. Tl concentration in soil solution (µg l^{-1}) and in different Tl fractions (µg/kg soil) in different soils (description: see Table 17.2). F_1 (mobile), F_2 (easily exchangeable), F_3 (bound to Mn oxides), F_4 (bound to organic matter), F_5 (bound to amorphous and poorly crystalline Fe oxides), F_6 (bound to crystalline Fe oxides) and F_7 (residual fraction)

Soil	Soil solution	Easily accessible					F5	F6	F7	Total
		F1	F2	F3	F4	Sum				
L1	1.20 ±0.20	106 ±4	65 ±4	100 ±11	32 ±3	303	120 ±12	134 ±4	761 ±20	1318
SLM2	0.35 ±0.04	73 ±3	59 ±3	1273 ±55	71 ±4	1476	52 ±1	5044 ±126	3029 ±92	9601
SW1	0.05 ±0.01	12 ±1	18 ±2	45 ±5	12 ±1	87	10 ±1	135 ±3	1361 ±120	1593
SW2	0.07 ±0.00	41 ±5	39 ±3	43 ±6	23 ±5	146	37 ±4	111 ±3	3478 ±175	3772

Table 17.4. Yield and concentration of Tl, Cd and Zn in the shoots of kale and candytuft cultivated in different soils (description: see Table 17.2) and Tl shoot-soil partition coefficient (*PC*) of kale in brackets (calculation of *PC*: Tl conc. in shoot D.M. / total Tl conc. in soil) (*n.d.*: not detected)

Soil	Yield (g D.M.)		Tl (mg/kg D.M.)		Cd (mg/kg D.M.)		Zn (mg/kg D.M.)	
	Kale	Candytuft	Kale	Candytuft	Kale	Candytuft	Kale	Candytuft
L1	2.28 ±0.21	0.35 ±0.04	16.8 (12.0)	22.5	3.1 ±0.2	0.52	44.5	19
SLM1	0.28 ±0.01	<0.1	39.8	n.d.	6.4	n.d.	975.9	n.d.
SLM2	0.42 ±0.03	0.13 ±0.05	10.5 (1.1)	7.3	5.1	1.3	826.3	251
SLM3	0.23 ±0.03	<0.1	32.4	n.d.	4.2	n.d.	420.8	n.d.
SLM4	0.34 ±0.04	<0.1	32.5	n.d.	9.9	n.d.	1027.8	n.d.
SLM5	0.29 ±0.03	<0.1	42.4	n.d.	4.3	n.d.	688.0	n.d.
SLM6	0.30 ±0.03	<0.1	42.3	n.d.	5.6	n.d.	921.5	n.d.
SW1	1.88 ±0.22	0.32 ±0.04	2.7 (1.7)	3.5	1.0	0.2	26.4	24
SW2	2.31 ±0.16	0.36 ±0.06	3.6 (0.9)	9.3	1.7	0.7	39.7	28
SW3	2.65 ±0.27	0.33 ±0.04	19.9 (7.4)	23.7	1.5	0.7	35.3	26

relatively low concentration of "easy accessible" Tl (87 and 146 µg Tl kg^{-1} soil for SW1 and SW2, respectively), while the soils with anthropogenic pollution L1 and SLM2 had a relatively high pool of "easy accessible" Tl (303 and 1 476 µg Tl kg^{-1} soil).

The "easily accessible" fractions in L1, SW1 and SW2 soils represented 21%, 6% and 10% of the "easily accessible" fractions of SLM2 soil. The total Tl concentrations of the soils were also different: 1 318, 9 601, 1 593 and 3 772 µg Tl kg^{-1} soil for soils L1, SLM2, SW1 and SW2, respectively. Tl from anthropogenic sources usually is considered to be more accessible than that of geochemical origin, but some fractions of Tl from geochemical origin may become mobile due to weathering or other chemical changes like decreased soil pH (Markidis and Amberger 1989).

In conclusion, the soils with anthropogenic origins of Tl pollution are characterized by high easily accessible pools as shown by soils L1 and SLM2. Therefore the risk of food contamination should be taken into account on such soils. In contrast, the pool of easily accessible Tl was very limited on soils with geogenic origin of Tl. The availability of Tl from geogenic origin may be relevant in the case of certain physical and chemical changes of the soil characteristics, i.e. low pH.

17.3.2
Limitations for the Success of Tl Phytoextraction

The plant uptake of a specific heavy metal depends mainly on two factors, plant genotypical characteristics and the chemistry of a specific element in the soil. The following questions should be addressed:

1. Is Tl phytoextraction by kale and candytuft (*Brassicaceae*) feasible for different origins of pollution (geogenic and anthropogenic), especially when the soil is highly polluted with other heavy metals such as Cd and Zn?
2. What are the possible origins of soil Tl pollution which have to be considered for phytoextraction? Is Tl uptake restricted from the 'easily accessible' fractions in the soil? Hence, what is the fate of the 'less accessible' fractions of different origins of pollution?

17.3.2.1
Plant Tolerance to high Heavy Metals Pollution

As shown in Table 17.4, kale and candytuft showed strong growth depressions in the soils SLM1-6. This growth depressions were most probably caused by high concentrations of mobile Zn ranging from 34 to 169 mg kg^{-1} soil (Table 17.2). These high Zn concentrations in the soils polluted as a result of mining activities were in an average of 1 000 times higher than in soils L1 and SW1. High Zn and Cd concentrations inhibit iron uptake (Cakmak et al. 1996). Therefore, the leaves of some plants on a SLM soil were sprayed with iron (10^{-2} M Fe^{3+} citrate), which visibly decreased the toxicity symptoms, but did not improve the growth within the time of the experiment.

Kale and candytuft were harvested after 20 and 30 d of cultivation, respectively, and the shoot tissues analyzed for Cd, Zn, and Tl. The concentrations of Zn in the shoots of kale and candytuft grown in the SLM1-6 soils were on average 10–30 times higher than those of the same plants cultivated in the SW1-3 soils (Table 17.4). The extremely high mobile Cd and Zn concentration in the soils SLM1-6 identifies another problematic issue of phytoremediation. The Tl hyperaccumulator plants (kale and candytuft) could not tolerate the high Cd and Zn concentrations in the soil. Therefore the hyperaccumulation process failed completely in the case of these soils polluted as a result of mining activities.

The plants grown in the soils L1 and SW1-3 showed no toxicity symptoms. There were no significant differences of yield between the kale plants in L1 and SW1-3 soils with an average of 2.3 g DM pot^{-1}. After 30 d the growth of candytuft plants was still very limited. Its yield amounted to 0.35, 0.32, 0.36 and 0.33 g DM pot^{-1} for soils L1, SW1, SW2 and SW3, respectively. The success and future improvement of phytoremediation mainly depends on the selection of appropriate candidate plants showing most desirable and exploitable growth characteristics in addition to appropriate physiology, morphology and adaptability to agronomic practices (Baker et al. 2000). Due to the relatively high uptake and translocation rate of Tl within kale and candytuft in a nutrient solution experiment (Kurz et al. 1999; Leblanc et al. 1999), they were identified as Tl hyperaccumulator plants and used in this study regardless to the differences in the yield. This classification is in accordance with the definition of hyperaccumulator plants with high uptake and translocation of Cd and Zn from roots to shoots (Rascio 1977; Baker 1981; Homer et al. 1991).

Therefore, to optimize phytoremediation both the uptake potential of the plants and their yield must be taken into consideration in addition to the binding form of the heavy metals in the soil (availability) (Baker 1981). Selection trails are needed to identify the fastest growing (largest biomass potential and greatest nutrient responses) and most effectively metal-accumulating genotypes. However, such a combination may not be possible and a compromise between extreme hyperaccumulation and low biomass (or vice versa) may be acceptable (Baker et al. 2000). Taking the yield as well as the concentration of Tl in the shoots into account, the Tl hyperaccumulator kale proved to be feasible for Tl phytoremediation under the prerequisite that other toxic metals are present only in moderate quantities.

17.3.2.2
The Effects of Tl Binding Forms on Plant Uptake

As shown in Table 17.4, Tl concentrations in the shoots of kale and candytuft were quite different depending on the different origins of pollution of the various soils. Tl uptake by kale and candytuft from the anthropogenic soil L1 was nearly 6 times higher than the uptake from geogenic soil SW1. On soil L1 Tl concentration of kale and candytuft amounted to 17 and 23 mg Tl kg^{-1} DM (DM: dry matter), compared with 3 and 4 mg Tl kg^{-1} DM taken up from soil SW1, respectively. Comparing the mobile and easy accessible fractions of the soils will emphasize the role of Tl availability to the plant (Table 17.3). The uptake of Tl was directly related to its concentration in the mobile fraction F1 as well as in the easy accessible fractions (ΣF1–F4), except for the case of the soil polluted by mining activities SLM2. Plants in this soil showed strong toxicity symptoms due to high levels of Zn (and Cd). As shown in Table 17.3, the easily accessible fractions of SW2 were two times higher than in SW1, causing a greater uptake of Tl by kale and candytuft from SW2.

Using the shoot–soil partition coefficient (PC) both factors concerning the hyperaccumulator plant and the soil are addressed. The PC indicates the ability of the plant to accumulate Tl and in addition to evaluate the potential of the soil to replenish Tl concentration in the soil solution from the different binding forms in the polluted soil.

As shown in Table 17.4, the PC value for soil L1 was the highest of all soils cultivated with kale amounting to a value of 12. In the case of the other soils (SLM2, SW1-3) the PC did not exceed 2. The high PC value which was obtained for soil L1 may not only be related to the high Tl concentration in the soil solution (1.2, 0.35, 0.05 and 0.07 µg Tl l^{-1} for soils L1, SLM2, SW1 and SW2, respectively) and the mobilization of Tl from easily accessible fractions, but also to the potential of the less accessible pool to replenish the easily accessible pool. In a rhizobox experiment with soil L1, both the easily accessible and the less accessible pools were depleted by kale plants.

The uptake from geogenic soils SW1-2 was extremely restricted and even the depletion of the easily accessible pool was very small (Al-Najar et al. 2003). In conclusion, for different Tl-polluted soils the uptake of Tl mainly depends on the Tl concentration in the easily accessible pool and the ability of the soil to replenish the soil solution from other fractions.

Acknowledgement

The authors wish to thank Mrs. M. Ruckwied for ICP-MS analyses and Dr. Arno Kaschl (UFZ-Center of Environmental Research Industrial and Mining Landscapes, Leipzig, Germany) for the constructive ideas and the correction of the manuscript.

References

Al-Najar H, Schulz R, Römheld V (2003) Plant availability of anthropogenic thallium origin in the rhizosphere of hyperaccumulator plants: A key factor for assessment of phytoextraction. Plant Soil 249(1): 97–105

Baker AJM (1981) Accumulation and excluders – strategies in response of plants to heavy metals. J Plant Nutr 3:643–654

Baker AJM, Brooks RR (1989) Terrestrial higher plants which hyperaccumulate metalic elements – a review of their distribution, ecology and phytochemistry. Biorecov 1:81–126

Baker AJM, Reeves RD, McGrath SP (1991) In-situ decontamination of heavy metal polluted soil using crops of metal-accumulation plants – a feasibility study. In: Hinchee RE, Olfenbuttel RF (eds) In-situ bioreclamtion. Butterworth-Heineman, Boston, pp 600–605

Baker AJM, McGrath SP, Reeves RD, Smith JAC (2000) Metal hyperaccumulator plants: a review of the ecology and physiology of a biological resources for phytoremediation of metal-polluted soils. In: Terry N, Bannelos GS (eds) Phytoremediation of contaminated soil and water. CRC press LLC, pp 85–107

Brooks RR, Lee J, Reeves RD, Jaffre' T (1977) Detection of nickeliferous rocks by analysis of herbarium specimens of indicator plants. J Geochem Explorat 7:49–57

Brümmer GW, Gerth J, Herms U (1986) Heavy metal species, mobility and availability in soils. Z Pflanzenernaehr Bodenkd 149:382–398

Cakmak I, Sari N, Marschner H, Kaloyci M, Yilmaz A, Eker S, Bülüt K (1996) Dry matter production and distribution of zinc in bread and durum wheat genotypes differing in zinc deficiency. Plant Soil 180:173–181

Gerritse RG, Van Driel W, Smilde KW, Van Luit B (1983) Uptake of heavy metals by crops in relation to their concentration in the solution. Plant Soil 75:393–404

Gieslinski G, Van Rees KCJ, Huang PM, Kozak LM, Rostad HPW, Knotgt DR (1996) Cadmium uptake and bioaccumulation in selected cultivars of durum wheat and flax as affected by soil type. Plant Soil 182:115–124

Green J (1972) Elements: planetary abundance and distribution. In: Fairbridge R (ed) Encyclopedia of geochemistry and environmental sciences. Van Nostrand Reinhold, New York, pp 268–300

Herms U, Brümmer G (1984) Solubility and retention of heavy metals in soils. Z Pflanzenernaehr Bodenkd 147:400–424

Homer FA, Morrison RS, Brooks RR, Clemens J, Reeves RD (1991) Comparative studies of nickel, cobalt and copper uptake by some nickel hyperaccumulators of the genus *Alyssum*. Plant Soil 138:195–205

Kazantzis G (1986) Thallium. In: Editor? (ed) Handbook on the toxicology of metals. Elsevier Science Publishers B.V., pp 549–567

King LD (1988) Effect of selected soil properties on cadmium content of tobacco. J Environ Qual 17:251–255

Knight B, Zhao FJ, McGrath SP, Shen ZG (1997) Zinc and cadmium uptake by the hyperaccumulator *Thlaspi caerulescens* in contaminated soils and its effects on concentration and chemical speciation of metals in soil solution. Plant Soil 197:71–78

Kurz H, Schulz R, Römheld V (1999) Selection of cultivars to reduce the concentration of cadmium and thallium in food and fodder plant. J Plant Nutr Soil Sci 162:323–328

Leblanc M, Robinson BH, Petit D, Deram A, Brooks RR (1999) The phytomining and environmental significance of hyperaccumulation of thallium by *Iberis intermedia* from southern France. Econ Geol 94:109–114

Markidis H, Amberger A (1989) Uptake of thallium from cement factory dust in pot trials with green rape, bush beans and ryegrass. Landwirtsch Forsch 42:324–332

McGrath SP, Shen ZG, Zhao FJ (1997) Heavy metal uptake and chemical changes in the rhizosphere of *Thlaspi caerulescens* and *Thalaspi ochroleucum* grown in contaminated soils. Plant Soil 188:153–159

Rascio N (1977) Metal accumulation by some plants growing on zinc mine deposits. Oikos 29:250–253

Robinson BH, Chaiarucci A, Brooks RR, Petit D, Kirkman JH, Gregg PEH, De Dominics V (1997) The nickel hyperaccumulator plant *Alyssum bertolonii* as a potential agent for phytoremediation and phytomining of nickel. J Geochem Explorat 59:75–86

Symeonides C, McRae SG (1977) The assessment of easily accessible cadmium in soils. J Environ Qual 6:120–123

Tremel A, Masson P, Sterckeman T, Baize D, Mench M (1997) Thallium in French agrosystems, thallium content in arable soils. Environ Pollut 96:293–302

VDLUFA (1996) Umweltanalytik, Methodenbuch Bd. VII, 1. Auflage. VDLUFA-Verlag Darmstadt

Whiting SN, Leake JR, McGrath SP, Baker AJM (2001) Zinc accumulation by *Thlaspi caerulescens* from soils with different Zn availability: a pot study. Plant Soil 236:11–18

Wilcke W, Kretzschmar S, Bundt M, Zech W (1999) Metal concentrations in aggregate interiors, exteriors, whole aggregates, and bulk of Costa Rican soils. Soil Sci Soc Am J 63:1244–1249

Yang X, Shi WY, Fu QX, Yang MJ, He HC (1998) Copper-hyperaccumulators of Chinese native plants characteristics and possible use for phytoremediation. In: El Bassam N, Behl PK, Prochnow B (ed) Sustainable agriculture for food, energy and industry. James & James (science publishers) Ltd., pp 484–489

Zeien H, Brümmer GW (1989) Chemische Extraktion zur Bestimmung von Schwermetallbindungsformen in Böden. Mitteilgn Dtsch Bodenkund Gesellsch 59/I:505–510

Chapter 18

Mercury Recovery from Soils by Phytoremediation

L. Rodriguez · F. J. Lopez-Bellido · A. Carnicer · F. Recreo · A. Tallos · J. M. Monteagudo

Abstract

Due to a low environmental impact, phytoremediation could become an attractive method to remove heavy metals from contaminated soils. The work reported here describes preliminary results for the recovery of mercury from contaminated soils by phytoextraction. This process involves the removal of mercury by plants followed by combustion of the plant biomass to recover the extracted mercury. Three agricultural crop plants, barley, wheat and yellow lupin, were tested in a field experiment. Mean Hg content of the soil were 29.17 µg g^{-1} for the 0–10 cm horizon and 20.32 µg g^{-1} for 10–40 cm horizon with less than 2% of the total Hg being bioavailable. The results of a field experiment showed that all the crops extracted mercury with Hg concentration reaching up to 0.479 µg g^{-1} for wheat. The low Hg uptake by the plants was attributed to the low availability of the mercury in the soils. The best Hg phytoextraction yield was obtained for barley reaching up to 719 mg ha^{-1}. Further efforts are being made to improve the fraction of bioavailable mercury (by means of solubilization agents) and the biomass crop yields.

Key words: remediation, contamination of soils, mercury, phytoextraction

18.1 Introduction

Mercury contamination of the environment is a very serious problem that has to be faced not only in the highly industrialized regions but also in developing countries. Mercury attracts attention and concern due to its toxicity, mobility, bioaccumulation, methylation processes and long-rate transport in the atmosphere (Boening 2000). As a chemical element, mercury cannot be created or destroyed. However, mercury can cycle in the environment as part of both natural and human activities. Measured data and modelling results indicate that the amount of mercury mobilized by and released from the biosphere has increased since the beginning of the industrial age. The global mercury cycle includes its emission into the atmosphere from a variety of sources, e.g. mining and industrial activities, its dispersion and transport in the air, its deposition on the earth and its storage and transfer between the land, water and air. As a result of its high volatility, elemental mercury (Hg^0) can easily be released into the atmosphere both from ore deposits and associated mining and refining operations. This mercury may, in turn, be deposited on the soils by different ways. The mercuric ion bound to airborne particles and in a gaseous form is capable of deposition both by precipitation and in absence of it (dry deposit). In contrast, elemental mercury vapour has a strong tendency to remain airborne (US EPA 1997).

As any other metal, mercury may occur in the soil in various forms: dissolved as free ion or soluble complex; non-specifically adsorbed by binding mainly due to electrostatic forces; specifically adsorbed by strong binding due to covalent or coordinative forces; chelated, e.g. bound to organic substances; and precipitated, e.g. as sulphide, carbonate, hydroxide and phosphate. There are three soluble forms of Hg in the soil environment. The most reduced is Hg^0 metal, the other two forms being ionic: the mercurous ion Hg_2^{2+} and, the mercuric ion, Hg^{2+}, in oxidizing conditions especially at low pH. Hg(I) ion is not stable under environmental conditions since it dismutates into Hg^0 and Hg(II) (Schuster 1991). A second potential route for the conversion of mercury in the soil is methylation to methyl or dimethyl mercury by anaerobic bacteria (Leyva and Luszczewsky 1998; Boening 2000). This last process is particularly relevant since methyl mercury is the most toxic mercurial form.

Mercury-contaminated soils are notoriously hard to remediate. Current reclamation technologies involve soil excavation and either landfilling or soil washing, followed by physical or chemical extraction of the contaminant. These engineering methods are expensive (Salt et al. 1995). Phytoremediation is emerging as a cost-effective alternative. In addition, because this method is performed in situ, there is no need for soil excavation and transport.

Phytoremediation is the use of plants to remediate toxic chemicals found in contaminated soil, sludge, sediment, ground water, surface water and wastewater. This technique uses a variety of plant biological processes (Cunningham et al. 1995; Pivetz 2001). Phytoremediation encompasses a number of different methods that can lead to contaminant degradation, e.g. rhizodegradation and phytodegradation; removal either by accumulation, e.g. phytoextraction and rhizofiltration, or dissipation (phytovolatilization); and immobilization, e.g. phytostabilization and hydraulic control. Phytoremediation requires the understanding of the extraction mechanisms, the appropriate selection of plants, and the knowledge of the factors enabling plant growth. Phytoremediation offers a lower cost method for soil remediation. Extracted pollutants may also be recycled. Lastly, phytoremediation is the most ecological cleanup technology for contaminated soils and is also known as "green remediation".

Phytoextraction involves the uptake of a contaminant by roots followed by the subsequent accumulation into plant organs. It must be noted that roots usually accumulate more contaminant than shoots and leaves; therefore the plant must have the ability to translocate the contaminant from roots to shoots at high rates. The process generally finishes by harvest and ultimate disposal of the plant biomass. Phytoextraction is mainly applied to heavy metals, e.g. Cd, Co, Cu, Hg, Ni, Cd, Pb, Zn (Nellessen and Fletcher 1993; Qian et al. 1999; Del Rio 2000; Kayser et al. 2000), and radionuclides, e.g. ^{90}Sr, ^{137}Cs, ^{234}U, ^{238}U (Pivetz 2001). The success of the phytoextraction technique depends upon the identification of suitable plant species that hyperaccumulate heavy metals and produce large amounts of biomass using established crop production and management practices.

The aim of our investigation was focused on the removal of the mercury dispersed in the soils from the Almadén area (Ciudad Real, Spain), a historical mercury-mining centre, by phytoremediation techniques. The investigation involved several steps: soil characterization; determination and change of the soil properties implied by the mobility and bioavailability of mercury; phytoextraction experiments with selected herbaceous plants both in the laboratory and in the field; and biomass treatment, for the recovery of the extracted mercury, by combustion in a fluidised-bed furnace. We describe here the preliminary results obtained in the first year of investigation in the field.

18.2
Materials and Methods

18.2.1
Site Description

The experimental site is located approximately 7 km southeast of Almadén, Ciudad Real (Spain). This village is located in the central part of the Iberian Peninsula, 300 km southwest of Madrid. The climate in Almadén is under Mediterranean conditions, with an average annual rainfall less than 500 mm. Since the beginning of mercury mining activity more than 2 000 years ago the soils of the Almadén area have exhibited high mercury loadings, with average values in the range 20–50 $\mu g\ g^{-1}$.

18.2.2
Phytoextraction Experiment Conditions

Three cultivated plants were tested: *Hordeum vulgare* (barley), *Triticum aestivum* (wheat) and *Lupinus luteus* (yellow lupin). The crops were grown on plots of 8 × 10 m, with four replicates per species. Plants will be cultivated on the same plots in three consecutive years. In the first year, the sowing was performed by hand in February and crops were harvested in July. Fertilizer, chemical treatments or irrigation were not administered during the growth period.

18.2.3
Soil Characterization

Prior to the start of the field experiments, an exhaustive physical and chemical characterization of the soil was carried out. Four representative profiles of the experimental plot were taken, bulked on site, placed into plastic bags, and immediately transported to the laboratory. Profiles were divided in two horizons: 0–10 cm and 10–40 cm of depth. Physical characterization included the following parameters: texture, structure, granulometry, moisture and density (real and apparent). The chemical parameters analysed for the soil fraction <2 mm were: total mercury, pH (in H_2O and KCl), conductivity, organic matter content, total cation exchange capacity (CEC) and carbonate, sulphate and nitrate contents. Total mercury was analysed by atomic absorption spectroscopy (AAS) after microwave acid digestion of the samples. Digestion was performed with a mixture of concentrated HNO_3, HCl, $HClO_4$ and HF (4/10/5/1, v/v/v/v). Sulphate and nitrate contents were determined using a Hach DR/2010 spectrometer with standard methods. The remaining parameters were determined according to standard methods certified by the Ministerio de Agricultura, Pesca y Alimentación of Spain (MAPA).

In addition, a sequential extraction procedure proposed by DiGiulio and Ryan (1987) was applied in order to evaluate geochemical partitioning of mercury in the soil. This procedure consisted of several extractions and digestions, which led to six fractions of mercury in the soil: water soluble (extracted with deionised water), exchangeable (extracted with ammonium acetate), bonded to fulvic acids (extracted with an ammonia solution), bonded to humic acids (extracted with an ammonia solution and further precipitated by

HCl), sulforganic (extracted by digestion with a mixture of HNO_3 and H_2O_2) and residual (obtained by digestion with a mixture of H_2SO_4 and HNO_3). The sum of water soluble and exchangeable fractions, respectively, is the so-called bioavailable fraction of mercury. AAS was used for the analysis of the mercury concentrations in the extract solutions.

The presence of organic forms of mercury was investigated in several soil and plant samples. The Hg speciation analyses were carried out by the Analytical Chemistry Department of the Universidad del País Vasco (Spain).

18.2.4
Plant Analysis

Total dry-matter of the crops were harvested by cutting the plants with scissors and/or a sickle at a height of approximately 5–10 cm above ground. Representative sampling of the all subplots was performed in order to determine biomass production and mercury extraction yield. Harvested plants were finely crushed, weighed (to calculate biomass production) and homogenized prior to analysis of the mercury loading. Plant samples (typically 0.5 g) were microwave digested in a mixture of concentrated H_2O_2, HNO_3 and HF (1/5/1, v/v/v). Total mercury loadings of the digested samples were determined by inductively coupled plasma-atomic emission spectrometry (ICP-AES).

18.3
Results and Discussion

18.3.1
Mercury in Soils

Mercury concentration and pH values for the soil samples are summarized in Table 18.1. It can be seen that the soil from the experimental plot showed that it was calcareous with pH values ranging from 7.3 to 8.1. The total mercury concentration values of the soil samples were similar, ranging from 19.62 to 32.40 µg g^{-1} with averages of 29.17 µg g^{-1} for the 0–10 cm horizon and 20.32 µg g^{-1} for 10–40 cm horizon. The most important finding was the sharp decrease in the Hg concentration in the soil with depth, indicating that most of the Hg comes from atmospheric deposition as a result of the previous mining activity in the Almadén area (Ferrara et al. 1998).

On the other hand, the Hg sequential extraction showed that the soluble and exchangeable fractions of mercury (so-called bioavailable Hg) in the soil were less than 2% of the total Hg loading (Table 18.1). It was observed that mercury in the soil samples mainly appeared in the class "bonded to humic acids" (71–92% of total Hg) and in the "sulforganic" fraction (up to 17% of total Hg). These results are in agreement with the widely reported relationship between organic matter and mercury (Schuster 1991; Rodriguez Martín-Doimeadios et al. 2000).

The presence of organic forms of mercury in the soil was investigated. All the samples analysed showed methylmercury concentrations below 50 ng g^{-1}, which represents less than 0.2% of the total mercury content. Other organic forms of mercury, e.g. dimethylmercury, ethylmethylmercury and diethylmercury, were not detected in any sample (detection limit ≈1 ng g^{-1}). Since methylmercury is the most toxic form of mercury it can be concluded that the toxicity risk for biological systems is very low.

Table 18.1. Soil characterization: pH and mercury concentration

Sample	Depth (cm)	pH	Hg$_{Total}$ (µg g^{-1})	Bioavailable Hg ($\mu g\ g^{-1}$)	% of Hg$_{Total}$
1.1	0 – 10	7.6	28.78	0.60	2.07
1.2	10 – 40	8.0	19.71	0.26	1.34
2.1	0 – 10	8.1	32.40	0.25	0.78
2.2	10 – 35	8.1	23.90	0.36	1.51
3.1	0 – 10	7.3	26.85	0.29	1.08
3.2	10 – 40	7.9	18.03	0.57	3.17
4.1	0 – 10	7.3	28.63	0.15	0.51
4.2	10 – 40	7.9	19.62	0.15	0.79
Mean 1[a]	0 – 10	7.6	29.17	0.32	1.11
STD 1[a]	0 – 10	0.3	2.33	0.19	0.68
Mean 2[b]	10 – 40	8.0	20.32	0.34	1.70
STD 2[b]	10 – 40	0.1	2.51	0.18	1.03

[a] Mean and standard deviation of 0–10 cm horizon samples.
[b] Mean and standard deviation of 10–40 cm horizon samples.

18.3.2
Plant Selection

Three annual crops were selected for use in the mercury phytoextraction experiments: *Triticum aestivum* (wheat), *Hordeum vulgare* (barley) and *Lupinus luteus* (yellow lupin). The criteria used for plant selection were: adaptability to the soil and climate of the Almadén area; cheapness and ease of farming operations; and bibliographic precedents for heavy metal phytoextraction. Thus, wheat and barley are common crops in Almadén and have therefore demonstrated their suitability to the soil and climate of this area. As far as yellow lupin is concerned, it is a common crop in the southwest of Spain. Moreover, the three plant species used have been widely cited in the scientific literature in the context of uptake of heavy metals by plants (Nellessen and Fletcher 1993). Although several plant species have been reported specifically for mercury extraction, e.g. *Chlonis barbata*, *Cynodon dactylon* and *Ciperus rotundus* (Martín et al. 1998), their use is not recommended in the Almadén area.

18.3.3
Field Experiments Results

A simple factorial design with four replicates per species was used in the field experiments. Phytoextraction results for the four replicates are shown in Fig. 18.1. The three crops tested did show a capacity for mercury extraction although mercury uptake values were low. Mean values of Hg loading for wheat, barley and lupin were 0.479, 0.339 and 0.196 µg g^{-1}, respectively. These Hg concentration values represent less than 3% of the mean Hg concentration in the soil. However, Hg concentration for wheat

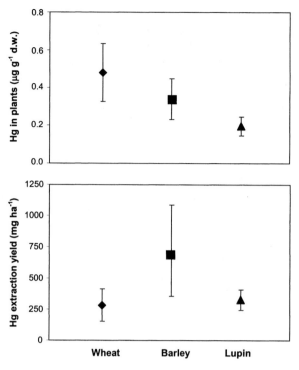

Fig. 18.1.
Phytoextraction of Hg by plants: mercury concentration in plants and phytoextraction yields (dry weight biomass production for lupin was estimated)

and barley are higher than the so-called bioavailable mercury concentration of the soil samples studied, e.g. 0.32 and 0.34 µg g^{-1} for the 0–10 and 10–40 cm horizons, respectively (Table 18.1). This fact suggests that the low uptake of mercury by the plants would be related to the low bioavailability of the Hg in the soil. It therefore seems necessary to carry out a detailed study of the factors that control the bioavailability of mercury in the soils and ascertain how to modify this parameter with different solubilization agents.

On the other hand, the presence of organic forms of mercury in the plants was investigated. Methylmercury was exclusively detected in all the plant samples analysed. It was present in concentrations ranging between 4.8–35 ng g^{-1}, enough low to discard any toxicity hazard for the chain food.

Biomass production differed greatly both between the different crops tested and different experimental plots. Mean values of dry matter biomass production for wheat and barley were 579 and 2 151 kg ha^{-1}, respectively. Growth of yellow lupin was insufficient for the measurement of biomass production. Growth of wheat was also very low, with biomass production lower than that normally observed for the same species in that zone (≈3 500 kg ha^{-1}). Barley exhibited the best biomass yield but it was again lower than those usually achieved in the Almadén area (≈4 600 kg ha^{-1}). This poor growth could be attributed to several factors: *(i)* sowing was performed somewhat late (in February rather than the usual November–December); *(ii)* the complete absence of rainfall in the last two months; and, *(iii)* the absence of any treatment with fertilizer. These aspects will be taken into account in subsequent years of experimentation. The differences existing in biomass production between different plots of the same crop could be caused by differences in some physicochemical properties of the plot soil or in moisture conditions (e.g. ground slope, field capacity).

Mercury phytoextraction yields can be calculated as the product of biomass production and Hg concentration in the plants. As a result of the factors described above, Hg extraction yields were also low (Fig. 18.1). Owing to its high biomass production, the best yield was obtained for barley with an average value of 719 mg ha^{-1}. Mean values of Hg extraction yield for wheat and lupin were 283 and 326 mg ha^{-1} (the latter value was calculated assuming a typical biomass production value of 1 667 kg ha^{-1}). At first, these yields could seem very low only taking into account the high mercury loading of the soil. Nevertheless, it must be thought that the mercury extracted by the plants practically accounts for the bioavailable – and thus the most toxic – fraction of soil. Moreover, it has to be considered that the investigation is in its initial stages and that these are preliminary results. We believe that mercury uptake can be significantly enhanced by addressing two aspects of the investigation: increasing the amount of bioavailable mercury by means of solubilization agents; and improving the biomass production by using more efficient farming techniques.

18.4
Conclusion

Phytoremediation techniques were investigated for the removal of mercury from soils. The concentration of Hg in soils was around 20–30 µg g^{-1} with less than 2% of the total Hg being available based upon sequential extraction. The decrease of mean Hg concentration from 29.17 µg g^{-1} at 0–10 cm horizon to 20.32 µg g^{-1} at 10–40 cm horizon demonstrated the anthropogenic origin of the mercury in the soil. Three crops were tested in the field experiments: *Triticum aestivum* (wheat), *Hordeum vulgare* (barley) and *Lupinus luteus* (yellow lupin). Preliminary results show that all crops extracted mercury, with Hg plant concentration reaching up 0.479 µg g^{-1} in wheat. The mercury concentration in the plants accounted for less than 3% of mercury concentration in the soil. However, the Hg concentrations in the plants were similar or even higher than that of the bioavailable Hg in the soils. Mercury extraction yields reached up to 719 mg ha^{-1} for barley. This is a significant amount if is taking into account that the extracted mercury represents the most available/toxic fraction. Further efforts are being made to improve the fraction of bioavailable mercury (by means of solubilization agents) and the biomass crop yields.

Acknowledgements

The investigation was carried out in co-operation with MAYASA (Minas de Almadén y Arrayanes S.A.) and CIEMAT (Centro de Investigaciones Energéticas Medioambientales y Tecnológicas, Ministry of Science and Technology, Spain). Special thanks are given to Victoriano Alcalde-Moraño for his assistance in both the field experiments and the soil analysis. Financial support from MCYT (Dirección General de Investigación, Project 1FD97-2033, Ministry of Science and Technology, Spain) and The European Union (FEDER) is gratefully acknowledged.

References

Boening DW (2000) Ecological effects, transport, and fate of mercury: a general review. Chemosphere 40:1335–1351

Cunningham SD, Berti WR, Huang JW (1995) Phytoremediation of contaminated soils. Trends Biotechn 13:393–397

Del Rio M, Font R, Fernández-Martínez J, Domínguez J, de Haro A (2000) Field trials of *Brassica carinata* and *Brassica juncea* in polluted soils of the Guadiamar River area. Fresenius Environ Bull 9:328–332

DiGiulio RT, Ryan EA (1987) Mercury in soils, sediments, and clams from a North-Carolina peatland. Water Air Soil Poll 33:205–219

Ferrara R, Maserti BE, Andersson M (1998) Atmospheric mercury concentrations and fluxes in the Almadén District (Spain). Atmos Environ 32(22):3897–3904

Kayser A, Wenger K, Keller A, Attinger W, Felix HR, Gupta SK, Schulin R (2000) Enhancement of phytoextraction of Zn, Cd and Cu from calcareous soil: the use of NTA and sulfur amendments. Environ Sci Technol 34:1778–1783

Leyva R, Luszczewsky A (1998) El mercurio en el medio ambiente. Ingeniería Química 344:179–183

Martín Gil FJ, Ramos Sánchez MC, Martín Gil J (1998) Fitoenmienda de medios contaminados. Residuos 34:68–70

Nellessen JE, Fletcher JS (1993) Assessment of published literature on the uptake, accumulation and translocation of heavy metals by vascular plants. Chemosphere 27:1669–1680

Pivetz BE (2001) Phytoremediation of contaminated soil and ground water at hazardous waste sites. Technology Information Office. United States Environmental Protection Agency. EPA/540/S-01/500

Qian J-H, Zayed A, Zhu Y-L, Yu M, Terry N (1999) Phytoaccumulation of trace elements by wetland plants: III. Uptake and accumulation of ten trace elements by twelve plant species. J Environ Qual 28:1448–1455

Rodríguez Martín-Doimeadios RC, Wasserman JC, García Bermejo LF (2000) Chemical availability of mercury in stream sediments from the Almadén area, Spain. J Environ Monit 2:360–366

Salt DE, Blaylock M, Kumar PBAN, Dushenhov V, Ensley BD, Chet I, Raskin I (1995) Phytoremediation: a novel strategy for the removal of toxic metals from the environment using plants. Biotechnology 13:468–474

Schuster E (1991) The behavior of mercury in the soil with special emphasis on complexation and adsorption processes – A review of the literature. Water Air Soil Poll 56:667–680

US EPA (United States Environmental Protection Agency) (1997) Mercury study report to Congress. EPA-452/R-97-003

Chapter 19

Effect of Cadmium and Humic Acids on Metal Accumulation in Plants

F. Baraud · T. W.-M. Fan · R. M. Higashi

Abstract

The natural ability of plants to accumulate, exclude or stabilize elements could be exploited to remediate soils contaminated with metals. To implement this alternative technology termed phytoremediation, it is crucial to better understand the various processes controlling metal mobilization or immobilization, uptake, and sequestration by the plants. Metal chelation is recognized as a vital biological process that regulates metal solubility, bioavailability, and internal storage in plants. Natural ligands, e.g. soil humates, root exudates components, or synthetic chelators, i.e. ethylenediaminetretraacetic acid or EDTA, can interact in a yet-to-be-defined way to influence metal uptake and sequestration by plants. Here, we investigated the interactive effect of Cd and soil humates on metal acquisition and translocation in wheat plants. Metal contents in tissues and root exudates composition were determined, using X-ray fluorescence for metals and gas chromatography-mass spectrometry (GC-MS) and nuclear magnetic resonance (NMR) for organic exudates. Cd inhibited biomass production, from –55% to –65% on tissue dry weight basis, and greatly reduced root exudation, of about –84% by dry weight. Cd treatment also resulted in a substantial co-accumulation of transition metals (Fe, Ni, Cu, Zn) and Cd in wheat roots. Moreover, co-treatment with humates alleviated some of the Cd effect showing biomass inhibition reduced by about 10% for the tissues and 17% for the exudates, while accumulation of some metals (Zn, Cu, Ni, Cd) in the root was enhanced. Thus, under Cd treatment, with or without humate, the enhanced accumulation of metals was not mediated via root exudation. This is contrary to the exudate-mediated Fe acquisition under Fe deficiency. The mechanism for this phenomenon is being sought.

Key words: humic acids, cadmium, root exudates, metal ion ligands, phytoremediation

19.1 Introduction

Toxic metals are released into the biosphere through various anthropogenic activities including mining operations, industrial and domestic waste discharges, combustion of fuels, and agricultural operations. Upon accumulation in organisms, many of these metals induce toxic damage, thus threatening both human and ecosystem health (Alloway 1990). To address metal contamination issues, both engineering and in-situ bioremediation are used, with the latter approach generally expected to be more economical for very large-scale remediation efforts. This is particularly the case for plant-based remediation, termed phytoremediation. Furthermore, basic understanding of the mechanisms of phytoremediation is indispensable for assessing biological risk of contamination in already-vegetated sites, or in situations where the goal of remediation is incompatible with physical-chemical treatments, e.g. where the goal is to protect the habitat or the agricultural viability of the soil.

There are three main approaches that can be utilized in phytoremediation of metals: phytoextraction, phytovolatilization, and phytostabilization (Cunningham and Ow 1996). Plants have evolved an extensive ability to mobilize or absorb elements from their environment. In particular, plant roots can acquire micronutrients which are present at very low available concentrations in soils (Marschner et al. 1986). This ability is being exploited for the phytoextraction approach. Plants can also transform metals and metalloids, e.g. Se, into volatile forms for dissipation, as in the case of the phytovolatilization approach (Terry and Banuelos 2000). Moreover, metal availability in soil can be altered by exudation of organic ligands by plants (e.g. Basu et al. 1997). Metal availability can also be modified by the transformation products of plant litters, e.g. humates, which is a major sink for metals in soils. These processes can be utilized in both phytostabilization and phytoextraction approaches. Regardless of the approach, metal ion ligands either directly produced by plants or resulting from transformation of plant matter is expected to play significant roles in phytoremediation. Before any commercial use of this alternative clean-up technology can be realized, it is crucial to acquire a better understanding of these ligands, from their chemical nature to their function in metal bioavailability in soil or sediment matrices. As mentioned earlier, this basic understanding of phytoextraction, phytovolatilization, and phyto-stabilization is also needed for an accurate, pre-treatment assessment of biological risk from contamination at vegetated sites.

It is now generally recognized that plants have evolved several strategies to mobilize metals and enhance their uptake. Along with microorganisms, they can manipulate the chemistry of the rhizosphere, through acidification, electron transfer, and release of ligands via root exudation (Marschner 1986; Laurie and Manthey 1994; Barona and Romero 1996). In particular, metal chelation with various ligands is recognized to be an important factor in regulating metal solubility, bioavailability, and internal storage in plants (Laurie and Manthey 1994; Salt et al. 1998). Biogenic chelating agents present in the soil solution are expected to increase metal solubility by forming highly soluble metal complexes which prevents metal ions from precipitation and sorption (Salt et al. 1995a). The metal chelates can also enhance metal mobility and bioavailability. The ligands that are expected to be present in soil solutions include: *(1)* natural ligands such as microbial chelating agents (e.g. bacteriosiderophores), ligands released in plant root exudates (e.g. phytosiderophores), and humic substances; and *(2)* synthetic organic ligands (most notably EDTA) released from agriculture, industrial and urban activities. These agents are deployed to modulate metal ion solubility and availability.

Both soil humic substances and root exudate ligands are among the most abundant organic chelators in soils, yet least understood, apparently due to the complex nature of the constituents which vary widely in sizes, polarity, and functional groups. The best known ligands in root exudates are the phytosiderophores released by gramineous plants in response to micronutrient deficiency. For instance, an increase in mugineic and avenic acid exudation by gramineous roots have been reported to facilitate Fe and Zn acquisition under Fe and Zn deficiency (Crowley and Gries 1994). There is also evidence that these siderophores might be involved in Cu and Mn uptake (Romheld 1991). Many other ligands have also been found in root exudates (Fan et al. 1997, 2000a), but their involvement in metal acquisition remains unclear. Moreover, phytosiderophores have been assumed to mediate the acquisition of contaminant metals, but whether this is the case remains to be seen. A broad understanding of the exudate chemistry in conjunction with its relation to metal acquisition by plants should help clarify these uncertainties.

Humic substances (HS) are a complex mixture of organometallic macromolecules, arising from the biotic and abiotic degradation of plant and animal residues in the environment. They are generally classified into humic acids, fulvic acids, and humin according to their solubility in acid and basic solutions (Hayes et al. 1989). It has long been recognized that humic substances have many beneficial effects on plant growth and crop productivity (MacCarthy et al. 1990; Chen and Aviad 1990; Chen et al. 1994). For example, they influence nutrient uptake, nitrogen metabolism, enzymes activities, and membrane permeability in plants (MacCarthy et al. 1990; Chen and Aviad 1990; Chen et al. 1994). In addition, HS have been shown to protect plants from toxic effects of excess metals and to alleviate toxic effect of Cd, Zn and Cu (Strickland et al. 1979; Kinnersley 1993). Despite the abundance of descriptive effects, little is known about the chemical mechanism(s) by which HS influence these biological activities.

In this study, we investigated the interactive effect of Cd (a priority metal pollutant) and soil humic acids on plant metal uptake. We also examined the impact of Cd and/or HS on the composition of root exudates, so that ligands that mediate the metal uptake may be revealed.

19.2
Experimental

19.2.1
Plant Growth

Seeds of Chinese Spring wheat were surface sterilized in 10% NaOCl (v/v in water) for 30 min, rinsed extensively with doubly deionized water and germinated on filter paper moistened with $CaSO_4$ at 5×10^{-4} M for 3 d at room temperature (first 24 h in complete darkness). The seedlings were transplanted onto dark polyethylene cups (4 seedlings per cup). Three cups each were positioned on the top of 4 black 9-l polyethylene containers containing one-half strength Hoagland solution as modified from Epstein (Epstein 1972) (composition in mg l^{-1}: K: 118; N: 113; Ca: 80; Na: 23; S: 16; Mg: 12; Si: 14; P: 1.55; Fe: 1.396; Cl: 0.888; B: 0.137; Zn: 0.065; Mn: 0.055; Cu: 0.016; Ni: 0.029; Mo: 0.024). The plants were aerated and maintained in a growth chamber (Sanyo) with 16/8 h light/dark periods, day/night temperature of 25/19 °C and constant humidity at 70%.

On day 10, soil humate and/or Cd was added to the nutrient solution in 3 of the 4 containers. Cd was added as $CdSO_4$ to the medium to reach a final concentration of 5 mg l^{-1} of Cd, in the 'Cd' treatment and 'HS + Cd' treatment. The soil humic acids used were isolated from a freeze-dried soil obtained from Chikugo prefecture, Japan as previously described (Higashi et al. 1998). The extraction and Tiron-treatment (to remove exchangeable metals ions) were performed according to the procedures previously described (Fan et al. 2000b). HS was added to get a final concentration of 1 mg l^{-1} in the 'HS' treatment and 'HS + Cd' treatment. Throughout the experiment, the solutions were maintained at pH 6.0 by daily adjustment with 1 M KOH or HCl. On day 20, roots exudates were collected as described previously (Fan et al. 2001). Immediately following exudate collection, the plants were harvested, rinsed thoroughly with doubly deionized water and separated into roots and shoots. The tissues were freeze-dried, the dry weight determined and the final material pulverized to <5 μm particles before storage at –70 °C.

19.2.2
Determination of Metals Contents

Total shoot and root metals contents were determined by energy dispersive X-ray fluorescence spectrometry (ED-XRF, JVAR Inc. EX3600 spectrometer). The analytical conditions were as follows: for Mn, Fe, Ni, Cu and Zn: high voltage 15 kV, Mo 0.05 mm filter, live time 1 000 s and for Cd: high voltage 45 kV, Mo 0.1 mm filter, live time 500 s. The dry ground tissues (approximately 20 mg per sample) along with a set of tomato leaf standards were pressed into 7 mm diameter pellets for X-ray fluorescence (XRF) measurements using an analytical press (Wilmad Inc.). The quantitative analysis was performed by interpolation from standard curves established with the tomato leaf standards. These standards were prepared by spiking tomato leaf tissues with appropriate multielemental standard solutions. Selected standards and wheat tissues were also analyzed by inductively coupled plasma-atomic emission spectroscopy (ICP-AES) after $HClO_4/HNO_3$ digestion. The ICP-AES results, as well as analysis of reference material (#1573a from the National Institute of Standards and Technology) by the XRF method confirmed the accuracy of the XRF method for roots and shoots analysis.

19.2.3
Root Exudates Composition

The lyophilized exudates were redissolved in 1 ml of doubly deionized water and centrifuged to remove particulates. An aliquot was lyophilized and then silylated with 1/1 (v/v) acetonitrile:N-methyl-N-[*tert*-butyldimethylsilyl]trifluoroacetamide (MTBSTFA) under sonication at 60 °C for 2–3 h and left overnight at room temperature. The silyl derivates were directly analyzed by gas chromatography-mass spectrometry (GC-MS) as previously described (Fan et al. 1997). The remaining solution was passed through a Chelex 100 cation exchange resin column to remove paramagnetic ions prior to 1-D and 2-D nuclear magnetic resonance (NMR), as described previously (Fan et al. 1997). Various organic and amino acids were identified and quantified from their GC-MS response calibrated against those of known standards. In addition, acetate, glycinebetaine (GB) and 2'-deoxymugineic acid (2'-DMA) were analyzed by NMR, as described previously (Fan et al. 1997; Fan et al. 2001).

19.3
Results and Discussion

We grew wheat on nutrient solutions containing Cd and/or humic substances (HS), in order to study Cd and HS influence on metal uptake. HS effect was then determined by comparing 'HS' treatment to '*Control*' (nutrient solution without Cd nor HS) and 'HS + Cd' to 'Cd' treatment. Cd influence was evaluated by comparison of 'Cd' treatment to '*Control*' and 'HS + Cd' to 'HS' treatment.

Table 19.1.
Dry weight (g) of roots and shoots as a function of Cd and humic substances (HS) treatments (triplicate samples). Standard deviation in *italic* (3 replicates). %-Values represent dry weight changes relative to:
a) 'Control'; b) 'HS'; c) 'Cd' values

	Control	Cd	HS	HS + Cd
Roots	0.61 *(0.33)*	0.21 *(0.02)* −66%[a]	0.71 *(0.02)* +16%[a]	0.25 *(0.03)* +18%[c] −65%[b]
Shoots	1.92 *(0.20)*	0.85 *(0.14)* −56%[a]	2.21 *(0.34)* +15%[a]	0.97 *(0.07)* +14%[c] −56%[b]

19.3.1
Effect of Cd and Humic Substances on Biomass Production

Toxic symptoms were visible in wheat plants exposed to Cd, including reduction of the root growth and leaf expansion as well as chlorosis. Both root and shoot dry weight decreased (Table 19.1), regardless of the presence of HS. This reduction was more pronounced for roots (−65%) than shoots (−55%). HS treatment ('HS') resulted in a small enhancement in root and shoot growth, relative to the 'Control', i.e. an increase of about +16% for the roots and +15% for the shoots dry weight (Table 19.1). A similar growth enhancement by HS was observed when Cd was also present ('HS + Cd' vs. 'Cd'), i.e. about +18 and +14%, respectively for roots and shoots.

Cd is known to be highly toxic to humans, animals, and plants. In plants, Cd causes a number of toxic symptoms, including growth reduction, inhibition of photosynthesis, perturbation of enzymatic activities (Reddy and Prasad 1992; Salt et al. 1995b). The observed large reduction in root and shoot growth was in agreement with our previous studies (Fan et al. 2000a; Shenker et al. 2001) and known toxic effects of Cd. The enhancement of biomass production by HS is consistent with the reported effects of HS (e.g. Chen and Aviad 1990). The small increase observed could be due to the low HS concentration (1 mg l^{-1}) employed for the study. More interestingly, this relative enhancement was not eliminated by the presence of Cd (Table 19.1). This positive influence suggests that HS alleviates the toxic action of Cd, as has been reported for plants grown in soils (Strickland et al. 1979).

19.3.2
Effect of Cd and Humic Substances on Root Exudation

Cd treatments ('Cd' and 'HS + Cd') caused a large decrease (about 84%) in the total root exudation, relative to 'Control' and 'HS', while HS plus treatments led to a small increase (21 to 27%) in the dry weight of the exudates relative to HS minus treatments ('HS', 'HS + Cd' vs. 'Control', 'Cd'). Some of this effect was attributable to the influence on root biomass. To correct for this, the exudate dry weight was normalized against root biomass (relative exudation), as shown in Fig. 19.1. A substantial effect of Cd was still observed, as relative root exudation was reduced by about 50%. In contrast, the HS effect on relative exudation was attenuated to only 4 to 8%, which was insignificant.

The large inhibitory effect of Cd on root exudation was also evident in our previous study (Fan et al. 2000a). This effect may be related to the ability of wheat plants to modulate metal uptake under toxic conditions.

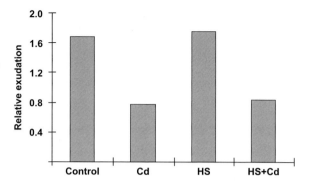

Fig. 19.1. Effect of Cd and humic substances (HS) on relative root exudation (relative exudation = exudate dry weigth / root dry weight)

Table 19.2. Effect of Cd and humic substances (HS) on root metal content (in mg kg^{-1}) and % changes. Standard deviation in *italic* (3 replicates). %-Values represent concentration changes relative to: (a) 'Control'; (b) 'HS'; (c) 'Cd' values. *n.d.*: not detected

	Mn	Fe	Ni	Cu	Zn	Cd
Control	35.9 (14.2)	425.2 (141.9)	66.8 (9.7)	40.3 (2.8)	90.7 (4.8)	n.d.
Cd	44.2 (10.5) +23%[a]	978.7 (152.1) +130%[a]	118.0 (15.4) +77%[a]	160.2 (20.1) +298%[a]	294.6 (25.2) +225%[a]	4567.1 (569.9)
HS	35.8 (8.4) 0%[a]	322.0 (36.9) −24%[a]	78.3 (5.8) +17%[a]	41.7 (5.3) +4%[a]	80.0 (8.3) −12%[a]	n.d.
HS + Cd	48.8 (11.9) +36%[b] +10%[c]	978.9 (83.5) +204%[b] 0%[c]	134.2 (9.2) +71%[b] +14%[c]	142.6 (3.7) +242%[b] −11%[c]	338.7 (10.6) +323%[b] +15%[c]	4942.2 (591.2) +8%[c]

19.3.3
Effect of Cd and Humic Substances on Tissue Metal Profiles

Cd accumulated in the root to very high levels, reaching 4567 ±570 µg g^{-1} for 'Cd' treatment, and 4942 ±591 µg g^{-1} for 'HS + Cd' treatment (Table 19.2). Large increases of other metal concentrations were also observed in the Cd-treated roots. The most significant variations (% change in Table 19.2) were for Zn (+225% for 'Cd' and +323% for 'Cd + HS'), Cu (+298% and +242%) and Fe (+130% and +204%). Note that the presence of HS slightly enhanced the Cd-induced accumulation of Mn, Ni, Zn, and Cd in the root. In contrast, for Cd-treated shoots a decreasing trend was observed for most of the transition metals, except for Ni and Zn (Table 19.3). 'HS + Cd' treatment resulted in a small increase (+30%) of Ni while the Zn concentration increased by 56–79% when Cd was present. It is also interesting to note that for both roots and shoots, HS alone treatment ('HS') tended to attenuate the accumulation of transition metals, except for Ni in the root (Table 19.2).

Thus, Cd exposure resulted in enhanced uptake for most transition metals into the root. However, except for Zn, the excess metals in the root did not appear to be translocated into the shoot (see section below). There also appeared to be an interaction of HS with Cd in enhancing the uptake of transition metals into the root.

Table 19.3. Shoot metal content: concentrations (in mg kg^{-1} or ppmw) and variation (%) Standard deviation in *italic* (3 replicates).%-Values represent concentration changes relative to: (a) 'Control'; (b) 'HS'; (c) 'Cd' values. *n.d.*: not detected. HS: humic substances

	Mn	Fe	Ni	Cu	Zn	Cd
Control	71.9 *(5.4)*	189.6 *(19)*	16.2 *(5.4)*	9.4 *(3.0)*	70.6 *(15.3)*	n.d.
Cd	51.8 *(10.2)* −28%^{a}	140.7 *(23.4)* −26%^{a}	14.7 *(5.2)* −9%^{a}	6.8 *(3.8)* −28%^{a}	110.4 *(8.3)* +56%^{a}	113.4 *(43.6)*
HS	52.0 *(9.1)* −28%^{a}	164.2 *(33.8)* −13%^{a}	12.5 *(8.1)* −23%^{a}	7.5 *(2.7)* −20%^{a}	48.6 *(14.8)* −31%^{a}	n.d.
HS + Cd	44.5 *(9.6)* −14%^{b} −14%^{c}	127.4 *(41.5)* −22%^{b} −10%^{c}	16.2 *(5.7)* +30%^{b} +10%^{c}	5.9 *(7.0)* −21%^{b} −13%^{c}	86.9 *(17.8)* +79%^{b} −21%^{c}	83.6 *(17.8)* −26%^{c}

Table 19.4. Root/shoot concentration ratio ($C_{R/S}$) for the various metals. *n.d.*: not detected. HS: humic substances

	Mn	Fe	Ni	Cu	Zn	Cd
Control	0.50	2.24	4.12	4.30	1.29	n.d.
Cd	0.85	6.95	8.01	23.66	2.67	40.29
HS	0.69	1.96	6.25	5.58	1.65	n.d.
HS + Cd	1.10	7.68	8.27	24.08	3.90	59.11

19.3.4
Effect of Cd and Humic Substances on Root/Shoot Concentration Ratio ($C_{R/S}$)

To better visualize the effect on metal translocation from roots to shoots, $C_{R/S}$ values were calculated, as shown in Table 19.4. Increasing values should indicate metal accumulation in the roots and restrained movement to the shoot.

Except for Mn, the $C_{R/S}$ values for all other metals were much greater than 1. Mn had the lowest values of $C_{R/S}$ for all treatments while Cd showed the highest values, i.e. about 2 to 100 times greater than those for the other metals. In addition, the $C_{R/S}$ value of 'HS + Cd' was the highest among the four treatments for all metals.

The large $C_{R/S}$ values of all metals but Mn is consistent with the knowledge that plants retain heavy metals in their roots (Reddy and Prasad 1992; Sauerbeck 1991). The $C_{R/S}$ values of Mn and Cd also suggest that Mn was the most "translocatable" while Cd was the most retained metal by the root. Moreover, the higher $C_{R/S}$ values of transition metals with 'HS + Cd' cotreatment than with either treatment alone suggest a synergistic effect of Cd and HS on metal retention in the root. This retention could be mediated through interaction with metal ion ligands such as phytochelatins. This is consistent with a much higher accumulation of phytochelatins in wheat roots than shoots observed in our previous study (Fan et al. 2000a) and with the ability of phytochelatins to sequester Cd in the vacuole of root cells (Dushenkov et al. 1995; Hart et al. 1998).

19.3.5
Effect of Cd and Humic Substances on Root Exudate Profile

To investigate whether root exudation could underlie the effect of HS or Cd on metal uptake into the root, a broad screen of root exudate for metal ion ligands was performed using nuclear magnetic resonance (NMR) and gas chromatography-mass spectrometry (GC-MS). The most abundant components detected were lactate, alanine, GAB (γ-aminobutyrate), malate, acetate, glycinebetaine (GB) and 2'-deoxy-mugineic acid (2'-DMA). The treatment-dependent profiles of these exudate components are shown in Fig. 19.2. Qualitatively, the exudation profiles for the four treatments were similar, e.g. lactate, acetate, and 2'-DMA remained most abundant. Quantitatively, the exudation of most components for 'Cd' and 'HS + Cd' treatments were distinctly lower than those for 'Control' and 'HS' treatments. In particular, exudation of these components dropped to near detectable limits under 'Cd' treatment. However, HS cotreatment alleviated somewhat this reduction for acetate and 2'-DMA.

These results are consistent with the notion that Cd inhibits root exudation and that Cd-induced metal sequestration into the root is not mediated via these small molecular weight ligands including the phytosiderophore 2'-DMA. This is in contrast to the important role of 2'-DMA in Fe acquisition under Fe deficiency (Marschner et al. 1986). However, the deinhibition of acetate and 2'-DMA exudation under 'HS + Cd' treatment may play a role in the small enhancement of metal sequestration into roots observed for this treatment as compared to the Cd treatment alone (cf. Table 19.2). This lack of dependence of metal accumulation on 2'-DMA appears to extend to several transition metal ions, as summarized in Fig. 19.3. This result indicates that, when Cd is present, the mechanism of accumulation of these transition metal ions – including Fe – is different from that under Fe deficiency and not dependent on low molecular weight organic exudates.

19.4
Conclusion

In conclusion, Cd inhibited the growth of wheat plants and root exudation while enhancing the accumulation of most transition metals in the root, in addition to its own accumulation to 4–5 g kg^{-1} levels. The majority of the excess metals remained in the root except for Zn which was in part translocated to the shoot. The presence of soil HS,

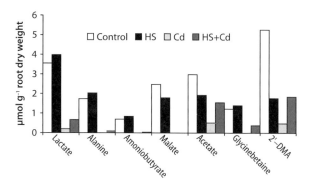

Fig. 19.2.
Effect of Cd and humic substances (HS) on root exudate profiles

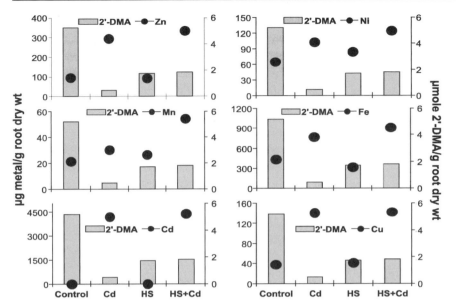

Fig. 19.3. 2'-DMA exudation vs. metals accumulated in wheat roots. *Gray bar:* 2'-DMA concentration in exudates (µmol g^{-1} of root dry weight, *right ordinate*); *Black circles:* metal concentration in the root tissue (µg g^{-1} root dry weight, *left ordinate*)

to a small extent, alleviated the inhibitory effect of Cd on growth and exudation while enhancing Cd accumulation and attendant accumulation of transition metal in the root. HS and Cd also had a synergistic effect on retaining transition metals and Cd in the root. Moreover, the Cd-induced accumulation of metals in the root was not correlated with the exudation of any small metal ion ligand including the phytosiderophore 2'-DMA. This indicates that the excess metal sequestration induced by Cd was not directly mediated by the exudation of the low molecular weight ligands.

References

Alloway BJ (1990) Heavy metals in soils. Wiley and Sons Ltd., New York
Barona A, Romero F (1996) Study of metal accumulation in wild plants using principal component analysis and ionic impulsions. Toxic Environ Chem 54:75–86
Basu U, McDonald-Stephens JL, Archambault DJ, Good AG, Briggs KG, Taing A, Taylor GJ (1997) Genetic and physiological analysis of doubled-haploid, aluminium-resistant lines of wheat provide evidence for the involvement of a 23 kD, root exudate polypeptide in mediating resistance. Plant Soil 196:283–288
Chen Y, Aviad T (1990) Effects of humic substances on plant growth. In: McCarthy P, et al. (eds) Humic substances in soil and crop sciences; selected readings. American Society of Agronomy and Soil Science Society of America, pp 161
Chen Y, Magen H, Riov J (1994) Humic substances originating from rapidly decomposing organic matter: properties and effects on plant growth. In: Senesi Miano (ed) Humic substances in the global environment and implications on human health. Elsevier Science B.V., pp 427–443
Crowley DE, Gries D (1994) Modelling of iron bioavailability in the rhizosphere. In: Mantley JA, Crowley DE, Luster DG (ed) Biochemistry of metal micronutrients in the rhizosphere. CRC Press Inc Lewis Publishers, pp 199–223

Cunningham SD, Ow DW (1996) Promises and prospects of phytoremediation. Plant Physiol 110:715–719

Dushenkov V, Nanda Kumar PBA, Motto H, Raskin I (1995) Rhizofiltration: the use of plants to remove heavy metals from aqueous streams. Environ Sci Technol 29(5):1239–1245

Epstein E (1972) Mineral nutrition of plants: principles and perspectives. Wiley, New York

Fan TW-M, Lane AN, Pedler J, Crowley D, Higashi RM (1997) Comprehensive analysis of organic ligands in whole root exudates using nuclear magnetic resonance and gas chromatography-mass spectrometry. Analytical Biochemistry 251:57–68

Fan TW-M, Baraud F, Higashi RM (2000a) Genotypic influence on metal ion mobilization and sequestration via metal ion ligand production by wheat. In: Eller PG, Heineman WR (ed) Nuclear site remediation. American Chemical Society Symposium Series, Washington DC, pp 417–431

Fan TW-M, Higashi RM, Lanes AN (2000b) Chemical characterization of a chelator-treated soil humate by solution-state multinuclear two-dimensional NMR with FTIR and pyrolysis-GCMS. Environ Sci Technol 34:57–68

Fan TW-M, Lane AN, Shenker M, Bartley JP, Crowley D, Higashi RM (2001) Comprehensive chemical profiling of gramineous plant root exudates using high-resolution NMR and MS. Phytochemistry 57:209–221

Hart JJ, Welch RM, Norvell WA, Sullivan LA, Kochian LV (1998) Characterization of cadmium binding, uptake and translocation in intact seedlings of bread and durum wheat cultivars. Plant Physiol 116:1413–1420

Hayes MHB, MacCarthy P, Malcolm RL, Swift RS (1989) Humic substances II: in search of structure. Wiley, Chichester, New York

Higashi RM, Fan TW-M, Lane AN (1998) Association of desferrioxamine with humic substances and interaction with cadmium(II) as studied by pyrolysis-gas chromatography – mass spectrometry and nuclear resonance spectroscopy. Analyst 123:911–918

Kinnersley AM (1993) The role of phytochelates in plant growth and productivity. Plant growth regulation 12:207–218

Laurie SH, Manthey A (1994) The chemistry and role of metal ion chelation in plant uptake processes. In: Mantley JA, Crowley DE, Luster DG (ed) Biochemistry of metal micronutrients in the rhizosphere. CRC Press Inc. Lewis Publishers, pp 27–62

Marschner H (1986) Mineral nutrition in higher plants. A PRESS, London

Marschner H, Roemheld V, Kissel M (1986) Different strategies in higher plants in mobilization and uptake of iron. J Plant Nutr 9:695–714

MacCarthy P, Clapp CE, Malcolm R, Bloom PR (1990) Humic substances in soil and crop sciences: selected readings. Proceedings of a symposium cosponsored by the International Humic Substances Society, Chicago, Illinois, 2 December 1985. American Society of Agronomy: Soil Science Society of America, Madison, Wis., USA

Reddy GN, Prasad MNV (1992) Characterization of cadmium binding protein from *Scenedesmus quadricauda* and Cd toxicity reversal by phytochelatin constituting amino acids and citrate. J Plant Physiol 140:156–162

Romheld V (1991) The role of phytosiderophores in acquisition of iron and other micronutriments in gramineous species: an ecological approach. Plant Soil 130:127–134

Salt DE, et al. (1995a) Phytoremediation: a novel strategy for the removal of toxic metals from the environments using plants. Biotechnology 13:468–474

Salt DE, et al. (1995b) Mechanisms of cadmium mobility and accumulation in Indian mustard. Plant Physiol 109:1427–1433

Salt DE, Smith RD, Raskin I (1998) Phytoremediation. Annu Rev Plant Physiol Plant Mol Biol 49:643–668

Sauerbeck DR (1991) Plant, element and soil properties governing uptake and availability of heavy metals derived from sewage sludge. Water Air Soil Poll 57–58:227–237

Shenker M, Fan TWM, Crowley DE (2001) Phytosiderophores influence on cadmium mobilization and uptake by wheat and barley plants. J Environ Qual 30:2091–2098

Strickland RC, Chaney WR, Lamoreaux RJ (1979) Organic matter influences phytotoxicity of cadmium soybeans. Plant Soil 52:393–402

Terry N, Banuelos GS (2000) Phytoremediation of contaminated soil and water. Lewis Publishers, Boca Raton

Chapter 20

Selection of Microorganisms for Bioremediation of Agricultural Soils Contaminated by Cadmium

D. Bagot · T. Lebeau · K. Jezequel · B. Fabre

Abstract

Accumulation of toxic metals in agricultural soils is an issue of health concern because toxic metals may be transferred to plants and food. The reclamation of metal-polluted soils, e.g. by using chemical extractants, is usually difficult and incomplete. An alternative option is to try to decrease the transfer of heavy metals from soil to plant by inoculating the soil with microorganisms selected for their ability to biosorb heavy metals. Here, we isolated one fungus, two actinomycetes and two bacteria from soil contaminated by cadmium. These microbes were found to grow in the presence of high cadmium level in laboratory conditions. The fungus and the actinomycetes have the best growth capacity in the soil extract in presence of 10 mg Cd l^{-1} which is representative of their ability to soil colonization. The bacillus, despite a low resistance to high cadmium concentrations, is very efficient in presence of 1 mg Cd l^{-1}. The percentage of cadmium biosorbed in the medium (up to 50% in presence of 10 mg Cd l^{-1}) and the specific biosorption (80 to 140 mg Cd g^{-1} of biomass) led to determine that one actinomycete and the bacillus are the most efficient microorganisms that will be used to later bioremediation experiments in soil microcosms.

Key words: cadmium, bioremediation, microorganisms, heavy metals biosorption

20.1 Introduction

The use of fertilisers and amendments in agriculture like sewage sludge or manure has led to an enrichment of toxic contaminants in soils because these emissions may contain high concentrations of heavy metals: high cadmium concentrations in phosphorous fertilisers for example. Heavy metals accumulate in the soils (Hamon et al. 1998). They may further concentrate in roots, stems and leaves (Mench et al. 1989). The main consequence of this bioaccumulation is that the crop production may become unfit both for animals and human consumption.

It's possible to remove heavy metals from wastewater by using microorganisms (Kapoor et al. 1998; Costley and Wallis 2000). Agricultural soils are characterised by low cadmium levels on wide areas. Soils cleaning up is difficult using conventional treatments (precipitation, ion exchange …) for technical and economical reasons. Phytoremediation could be employed (Krenlampi et al. 2000), but phytoremediation takes several years, during which no food crop is possible. Therefore an alternative to the cleaning up is the pollutant immobilisation in the soil to avoid its transfer to plants or groundwater. Heavy metals adsorption on mineral or organic amendments has been exploited (Bailey et al. 1999) but cadmium leakage has been observed due to pH change or soil temperature and humidity variations even in the presence of these amendments.

In order to immobilise Cd in the soil, we propose an inoculation with viable microorganisms able to accumulate Cd. Their ability to multiplicate in this environment is essential. Here, the toxicity of the Cd for the microflora is an issue (Giller et al. 1997; Nies 1999). Some species can indeed disappear while resistant strains can proliferate (Roane and Kellog 1996). Microorganisms resistance can be related to the metabolic paths and to the presence of metal-binding proteins or peptides (Cooksey 1994; Inouhe et al. 1996; Silver and Phung 1996; Brown et al. 1998). It is also dependent on the nature of the medium (Gimmler et al. 2001).

Screening of cadmium resistant microorganisms has been realised by some authors (Boularbah et al. 1992; Gabriel et al. 1996; Amoroso et al. 1998) in order to determine the ability of these strains to biosorb cadmium. Nevertheless the resistance and the accumulation of heavy metals have always been measured on synthetic media adapted to the selected microorganisms.

The aim of this work was to select microorganisms that will be used to inoculate soils with the goal to immobilise the cadmium present inside. For this reason we hoped to identify the ones really adapted to soil bioremediation. We wanted microorganisms able to colonise the contaminated soils. Therefore we used a liquid soil extract medium to study the microorganisms growth in conditions which are close to soil environment. So we made a comparison between synthetic media and soil extract medium with or without cadmium. We then studied the accumulation capacity of the microorganisms that is representative of the bioremediation efficiency of the selected strains.

20.2
Experimental

20.2.1
Microbial Strains and Maintenance

Five microorganisms were tested. The *Bacillus* Zan-044 (99.5% similar to *Bacillus simplex*), and the two *Streptomyces* R25 and R27, were respectively isolated by Valentine et al. (1996) and Amoroso et al. (1998) from sediments which were contaminated with cadmium among other metals. The basidiomycete *Fomitopsis pinicola* CCBAS 535 (Gabriel et al. 1996) is common in all forests and municipal parks in Europe. *Pseudomonas aeruginosa* PU21 was isolated from hospital sewage by Jacoby (1986). These microorganisms were stored at 4 °C on agar slants before they were used.

20.2.2
Culture Media

We used the liquid synthetic medium Peptone, Tryptone, Yeast extract, Glucose medium (PTYG (g l^{-1}): peptone, 0.5; tryptone, 0.5; yeast extract, 1; glucose, 1; $MgSO_4 \cdot 7H_2O$, 0.6; $CaCl_2 \cdot 2H_2O$, 0.07) (Valentine et al. 1996) for the *Bacillus* Zan-044; Minimal Medium (MM (g l^{-1}) L-Asparagine, 0.5; K_2HPO_4, 0.5; $MgSO_4 \cdot 7H_2O$, 0.2; $FeSO_4 \cdot 7H_2O$, 0.01; Glucose, 10) (Amoroso et al. 1998) for the *Streptomyces* R25 and R27; Liquid Broth (LB (g l^{-1}) tryptone, 10; yeast extract, 5; NaCl, 5) for *Pseudomonas* PU 21 and a Malt Extract medium (30 g l^{-1}) for *F. pinicola*. We also used a liquid soil extract for the cultivation of the five strains. Loamy, slightly acidic soil samples were taken from an experimental agronomic site at Aspach-le-

Bas (Haut-Rhin, France) in the arable horizon of the soil (0–30 cm depth). The soil was then mixed with the same weight of water (running water) and autoclaved at 130 °C during 1 h. After filtration, the pH and the cadmium concentration of the soil extract were respectively adjusted before sterilisation at 5, 6 or 7 (with HCl 1N or NaOH 1N) and at 1 or 10 mg Cd l^{-1} of medium (Cd was used in $Cd(NO_3)_2$, $4H_2O$ form), and the filtrate was autoclaved at 115 °C during 0.5 h. The chemical characteristics of the soil extract are: pH 5.6, Total sugar, 0.5–3 mg l^{-1}, Cl^-, 6.61 mg l^{-1} SO_4^{2-}, 10.24 mg l^{-1}, NO_3^-, 1.83 mg l^{-1}, PO_4^{2-}, 0 mg l^{-1}, Cd, 0 mg l^{-1}, Cu, 0.22 mg l^{-1}, Zn, 0.09 mg l^{-1}, Pb, 0.12 mg l^{-1}.

20.2.3
Cultivation Conditions

Microbial cells were precultured at 25 °C for about 2–3 d for the *Bacillus* Zan-044 and the *Pseudomonas* PU 21, 5 d for the *Streptomyces* R25 and R27 and 10 d for *Fomitopsis pinicola*. For the cultures, each Erlenmeyer was previously washed with 30% v/v aqueous HNO_3, rinsed out with distilled water and was filled with 50 ml of medium. We used the synthetic liquid medium or the soil extract for growth kinetics with free cells.

Each Erlenmeyer was inoculated with the same quantity of cell inoculum. The inoculum size corresponded to 0.04 g l^{-1} for the *Bacillus* Zan-044, the *P. aeruginosa* and the *Streptomyces* R25 and R27, and to 0.2 g l^{-1} for *F. pinicola*. The preculture of *F. pinicola* was centrifuged at 2000 g and 0.2 ml of the cell suspension was inoculated into the culture medium after supernatant elimination while a pellet of the *Streptomyces* preculture was inoculated in each Erlenmeyer. Erlenmeyer glasses were incubated at 25 °C and stirred at 175 rpm and triplicates were realised.

20.2.4
Growth

The growth kinetics were followed by estimating the dry weight of the biomass. For the *Streptomyces* and the fungus, the medium was filtered through a microporous membrane (HAWP filter from Millipore, Freehold, NJ; mean pore size, 0.45 µm) and weighed after drying at 105 °C for 24 h. The cell concentration (dry weight) of bacterial suspensions was determined by measuring the optical density (OD) of the samples at 600 nm and following calibration (dry weight (g l^{-1}) = 0.4 × OD) according to Valentine et al. (1996). Growth rates, µ (h^{-1}) were calculated using the relation = $\ln 2 / \rho$ where ρ (h) is the generation time estimated during the exponential stage of the growth kinetic.

20.2.5
Cadmium Analysis

The measurement of the cadmium biosorbed by microorganisms was achieved at the time corresponding to the stationary stage of the growth kinetics i.e. 3 d for the bacteria, 5 d for the *Streptomyces* R25 and R27 and 10 d for *F. pinicola*. The residual cadmium concentrations were measured using Inductively Coupled Plasma – Atomic Emission Spectroscopy (ICP-AES) (JY ULTIMA; Jobin Yvon HORIBA Group, Kyoto, Japan) after the filtration of the culture medium described above. The detection limit for cadmium is 0.09 mg l^{-1}.

20.3 Results and Discussion

20.3.1 Microorganisms Growth with Synthetic Media

We studied first the resistance to cadmium of the microorganisms cultivated with synthetic media. This allowed us to evaluate the range of cadmium concentrations where they were able to multiplicate. We investigated also the cadmium concentration tolerated for the growth of microorganisms according to the pH. Table 20.1 shows the maximal biomass production and the growth rates of the five strains.

Whatever the microorganism, the pH had little effect on the maximal biomass except for *Streptomyces* R25 (biomass was lower at pH 5) and *P. aeruginosa* (maximal biomass at pH 6). *Bacillus* Zan-044 showed a maximal growth rate at pH 7 though the biomass produced was not simultaneously enhanced.

The most sensitive strains to cadmium were *Streptomyces* R25 and *Bacillus* Zan-044 whose growth rates were affected by cadmium concentrations. In all conditions *F. pinicola* and *P. aeruginosa* provided the highest biomass amounts despite the lowest growth rate of *F. pinicola*.

Except for *P. aeruginosa* the microorganisms were very sensitive to toxicity of 10 mg Cd l^{-1}. With synthetic media the experiments allowed us to evaluate the potential of each strain we studied. The most resistant one was the *Pseudomonas aeruginosa* PU21 which is in accordance with Chang et al. (1997). *Fomitopsis pinicola* had also a

Table 20.1. Maximal biomass production (X in g l^{-1}) and growth rates (μ in h^{-1}) of the microorganisms in synthetic media

Microorganism	pH	Cadmium concentration in the medium					
		0 mg l^{-1}		1 mg l^{-1}		10 mg l^{-1}	
		X	μ	X	μ	X	μ
Fomitopsis pinicola	5	1.312	(0.028)	1.116	(0.019)	0.096	(0.000)
	6	0.964	(0.026)	2.026	(0.027)	0.098	(0.000)
	7	0.782	(0.022)	0.996	(0.019)	0.092	(0.000)
Streptomyces R 25	5	0.262	(0.076)	0.354	(0.029)	0.060	(0.000)
	6	0.498	(0.029)	0.054	(0.000)	0.066	(0.000)
	7	0.576	(0.019)	0.058	(0.000)	0.070	(0.000)
Streptomyces R 27	5	0.094	(0.037)	0.098	(0.003)	0.094	(0.035)
	6	0.149	(0.016)	0.071	(0.003)	0.070	(0.002)
	7	0.163	(0.037)	0.071	(0.031)	0.079	(0.033)
Bacillus Zan-044	5	0.406	(0.120)	0.009	(0.110)	0.002	(0.009)
	6	0.290	(0.180)	0.230	(0.160)	0.009	(0.000)
	7	0.380	(0.270)	0.286	(0.240)	0.016	(0.013)
Pseudomonas PU 21	5	0.872	(0.140)	0.676	(0.110)	0.676	(0.100)
	6	1.344	(0.160)	1.250	(0.160)	0.958	(0.110)
	7	1.094	(0.150)	0.920	(0.120)	1.070	(0.090)

good potential despite its low growth rate. Both *Streptomyces* and *Bacillus* Zan-044 showed under these conditions a moderate behaviour.

20.3.2
Microorganisms Growth with Soil Extract Medium

In comparison to synthetic media, the biomass production of the five microorganisms was lower, due to the medium poverty (Table 20.2), except for *Streptomyces* R27. *P. aeruginosa* was unable to grow with the soil extract medium even in absence of cadmium. Considering the biomass production or the growth rate the optimal pH was 6 for the four strains able to develop with soil extract, except for *Bacillus* (pH 7). In these conditions the microorganisms were less sensitive to cadmium toxicity.

The growth rates were little affected until 1 mg Cd l^{-1}. The *Streptomyces* R25 growth rate was more highly affected by 10 mg Cd l^{-1} than R27. *F. pinicola* was the microorganism which presented the less sensitivity to cadmium whatever the concentration or the pH. The results showed that pH 6 was favourable for the development of all the strains studied. Soil pH is frequently close to this value.

The culture results were representative for the ability of the microorganisms to soil colonisation which is one parameter of strain selection for bioremediation. Whatever the cadmium concentration and the pH, *F. pinicola* was the most efficient when cultivated with soil extract medium. Both *Streptomyces* and the *Bacillus* Zan-044 presented little differences in the ability to develop under these conditions.

Table 20.2. Maximal biomass production (X in g l^{-1}) and growth rates (μ in h^{-1}) of the microorganisms in soil extract medium

Microorganism	pH	Cadmium concentration in the medium					
		0 mg l^{-1}		1 mg l^{-1}		10 mg l^{-1}	
		X	μ	X	μ	X	μ
Fomitopsis pinicola	5	0.053	(0.017)	0.386	(0.038)	0.338	(0.024)
	6	0.910	(0.043)	0.826	(0.038)	0.929	(0.025)
	7	0.586	(0.024)	0.572	(0.017)	0.410	(0.007)
Streptomyces R 25	5	0.128	(0.025)	0.142	(0.019)	0.052	(0.000)
	6	0.164	(0.050)	0.136	(0.033)	0.052	(0.000)
	7	0.160	(0.035)	0.150	(0.033)	0.058	(0.005)
Streptomyces R 27	5	0.132	(0.014)	0.212	(0.008)	0.250	(0.008)
	6	0.262	(0.027)	0.194	(0.019)	0.116	(0.007)
	7	0.366	(0.020)	0.366	(0.012)	0.288	(0.006)
Bacillus Zan-044	5	0.083	(0.016)	0.078	(0.009)	0.080	(0.016)
	6	0.263	(0.042)	0.182	(0.019)	0.066	(0.085)
	7	0.326	(0.053)	0.216	(0.049)	0.100	(0.096)
Pseudomonas PU 21	5	0.064	(0.000)	0.056	(0.000)	0.057	(0.000)
	6	0.068	(0.000)	0.069	(0.000)	0.060	(0.000)
	7	0.064	(0.000)	0.067	(0.000)	0.079	(0.000)

20.3.3
Cadmium Biosorption with Synthetic Media or Soil Extract

The cadmium biosorption is representative of the bioremediation potentiality of the microorganisms selected to inoculate contaminated soils. With synthetic medium (Table 20.3) *P. aeruginosa* is the most efficient microorganism to immobilise cadmium since it biosorbed 90 to 99% of the cadmium present in the medium whatever the concentration. With 1 mg Cd l^{-1} both *Streptomyces* and the *Bacillus* were equivalent with 30% Cd biosorbed. *Bacillus* did not biosorbed cadmium in the presence of 10 mg Cd l^{-1} while the two *Streptomyces* biosorbed approximately 40%. Both *Streptomyces* presented the highest specific biosorption while *Bacillus* and *Pseudomonas* had the lowest.

With 1 mg Cd l^{-1} in the soil extract medium (Table 20.4), *F. pinicola*, *Streptomyces* R25 and *Bacillus* Zan-044 were the microorganisms which biosorbed the major part of the cadmium present in the medium (41 to 64%). The maximum biosorption of cadmium was close to 15% for *Streptomyces* R27 and *P. aeruginosa*. With 10 mg Cd l^{-1} concentration in the medium, *Bacillus* biosorbed only 0.35% of the cadmium because of its low biomass production. *F. pinicola* and *Streptomyces* R25 were the most efficient

Table 20.3. Cadmium biosorption by the microorganisms in synthetic media. *a* Cd biosorbed from the medium (mg l^{-1}). *b* Cd specific biosorption of the microorganisms (mg g^{-1}). *c* Percentage of Cd present in the medium biosorbed by the microorganisms

Microorganism	Cadmium concentration in the medium					
	1 mg l^{-1}			10 mg l^{-1}		
	a	b	c	a	b	c
Streptomyces R 25	0.376	(7.231)	38	4.373	(66.255)	44
Streptomyces R 27	0.288	(4.046)	29	3.541	(45.982)	35
Bacillus Zan-044	0.306	(1.330)	31	0.000	(0.000)	0
Pseudomonas PU 21	0.906	(0.725)	91	9.894	(10.328)	99

Table 20.4. Cadmium biosorption by the microorganisms in soil extract medium. *a* Cd biosorbed from the medium (mg l^{-1}). *b* Cd specific biosorption of the microorganisms (mg g^{-1}). *c* Percentage of Cd present in the medium biosorbed by the microorganisms

Microorganism	Cadmium concentration in the medium					
	1 mg l^{-1}			10 mg l^{-1}		
	a	b	c	a	b	c
Fomitopsis pinicola	0.409	0.722	41	4.388	3.407	44
Streptomyces R 25	0.641	4.868	64	5.024	86.621	50
Streptomyces R 27	0.153	0.794	15	0.772	6.308	8
Bacillus Zan-044	0.575	1.089	58	0.035	142.191	0
Pseudomonas PU 21	0.139	12.437	14	1.895	135.083	19

with a cadmium biosorption close to 50%. *P. aeruginosa* had a cadmium fixation of 19% without any biomass production in this medium.

These results showed that the cadmium biosorption by *F. pinicola* was essentially due to the magnitude of its biomass production while *Streptomyces* R 25 one was due to its higher specific biosorption. In spite of their lack of growth *Bacillus* and *Pseudomonas* had the highest specific biosorption.

The experiments of cadmium biosorption showed that *Bacillus* Zan-044 presented efficiency only with 1 mg Cd l^{-1}. *Bacillus* with 10 mg Cd l^{-1} and *P. aeruginosa* specific biosorptions had no significance due to the lack of growth. *F. pinicola* and *Streptomyces* R25 had a good efficiency in presence of low or high cadmium concentrations. *Streptomyces* R25 seemed to be particularly interesting, having a good efficiency in cadmium biosorption as well as a high specific biosorption while potentiality of *F. pinicola* was essentially related to its biomass production.

20.4
Conclusion

From the strains we tested for soil bioremediation the best adapted to development as to cadmium biosorption in soil extract were *Fomitopsis pinicola* and *Streptomyces* R 25 for high cadmium levels and *Bacillus* Zan-044 for low cadmium levels.

The most significant parameters for the evaluation of the ability of microorganisms to soil bioremediation appeared to be the biomass production coupled to a high specific biosorption level. These parameters were in relation to the multiplication of the inoculated microorganisms in the soil. The most influent factors on microorganisms development are the pH, the organic and chemical characteristics of the soil and the competitive relations with the microorganisms of the soil. The microorganisms we considered the most efficient were little sensitive to pH modifications between 5 and 7 and presented a good adaptation to the soil extract which was a poor medium.

To approach better the study of soil microcosm the next stage would be to implement the soil inoculation technique and to control the colonisation of soil samples by the selected microorganisms. Studies would be developed with immobilised microorganisms, to improve their competitiveness against the microorganisms of soils.

Acknowledgements

We are grateful to Dr Pascaline Dury for her councils and her competence in scientific English for the correction of this article.

References

Amoroso MJ, Castro GR, Carlino FJ, Romero NC, Hill RT, Oliver G (1998) Screening of heavy metal-tolerant actinomycetes isolated from the Sali River. J Gen Appl Microbiol 44:129–132

Bailey SE, Olin TJ, Bricka RM, Adrian DD (1999) A review of potentially low-cost sorbents for heavy metals. Water Res 33:2469–2479

Boularbah A, Morel JL, Bitton G, Guckert A (1992) Cadmium biosorption and toxicity to six cadmium-resistant Gram-positive bacteria isolated from soil. Environ Toxicol Water Qual 7:237–246

Brown NL, Lloyd JR, Jakeman K, Hobman JL, Bontidean I, Mattiasson B, Csregi E (1998) Heavy metal resistance genes and proteins in bacteria and their application. Biochem Soc Transactions 26:662–665

Chang JS, Law R, Chang CC (1997) Biosorption of lead, copper and cadmium by biomass of *Pseudomonas aeruginosa* PU 21. Water Res 31(7):1651–1658

Cooksey DA (1994) Molecular mechanisms of copper resistance and accumulation in bacteria. FEMS Microbiol Rev 14:381–386

Costley SC, Wallis FM (2000) Effect of flow rate on heavy metal accumulation by rotating biological contactor (RBC) biofilms. J Ind Microbiol Biot 24:244–250

Gabriel J, Vosahlo J, Baldrian P (1996) Biosorption of cadmium to mycelial pellets of wood-rotting fungi. Biotechnol Lett 10:345–348

Giller KE, Witter E, McGrath SP (1998) Toxicity of heavy metals to microorganisms and microbial processes in agricultural soils: a review. Soil Biol Biochem 30(10/11):1389–1414

Gimmler H, de Jesus J, Greiser A (2001) Heavy metal resistance of the extreme acidotolerant filamentous fungus *Bispora* sp. Microb Ecol 42:87–98

Hamon RE, McLaughlin MJ, Naidu R, Correll R (1998) Long-term changes in cadmium bioavaibility in soil. Environ Sci Technol 32:3699–3703

Inouhe M, Sumiyoshi M, Tohoyama H, Joho M (1996) Resistance to cadmium ions and formation of a cadmium-binding complex in various wild-type yeasts. Plant Cell Physiol 37:341–346

Jacoby GA (1986) Resistance plasmids in *Pseudomonas aeruginosa*. In: Gunsalus IC, Sokatch JR, Ornston LN (eds) The bacteria, vol X, chapter 17. Academic Press, Orlando, FL, pp 497–514

Kapoor A, Viraraghavan T, Cullimore DR (1999) Removal of heavy metals using the fungus *Aspergillus niger*. Biores Technol 70:95–104

Krenlampi S, Schat H, Vangronveld J, Verkleij JAC, van der Lelie D, Mergeay M, Tervahauta AI (2000) Genetic engineering in the improvement of plants for phytoremediation of metal polluted soils. Environ Pollut 107:225–231

Mench M, Tancogne J, Gomez A, Juste C (1989) Cadmium bioavailability to *Nicotiana tabacum* L., *Nicotiana rustica* L., and *Zea mays* L. grown in soil amended or not amended with cadmium nitrate. Biol Fert Soils 8:48

Nies DH (1999) Microbial heavy-metal resistance. Appl Microbiol Biotechnol 51:730–750

Roane TM, Kellog ST (1996) Characterisation of bacterial communities in heavy metal contaminated soils. Can J Microbiol 42:593–603

Silver S, Phung Le T (1996) Bacterial heavy metal resistance: new surprises. Ann Rev Microbiol 50:753789

Valentine NB, Bolton H, Kingsley MT, Drake GR, Balkwill DL, Plymale AE (1996) Biosorption of cadmium, cobalt, nickel, and strontium by a *Bacillus simplex* strain isolated from the vadose zone. J Ind Microbiol 16:189–196

Chapter 21

Electrodialytic Remediation of Heavy Metal Polluted Soil

L. M. Ottosen · I. V. Cristensen · A. J. Pedersen · A. Villumsen

Abstract

This article presents the historical background, the principle and case studies of electrodialytic remediation of heavy metal polluted soil. Remediation of fine grained soils polluted by heavy metals is problematic for most traditional methods such as pump and treat technologies and soil washing methods. On the other hand, electrochemical soil remediation methods are particularly suited for the fine grained soils. These methods are based on an applied electric current as cleaning agent. The current tends to pass the soil through the finest pores, i.e. next for the smallest soil particles, where the heavy metals are mainly adsorbed.

The heavy metals are mobile in the electric field as ions in solution, only. This means that the heavy metals must be desorbed during the process. For some soils it is necessary to add an enhancement solution to aid this desorption, whereas for other soils addition of water is sufficient to the remediation to occur.

Electrodialytic soil remediation was applied at laboratory scale for different soil and pollution combination. Soils with a low carbonate content and polluted with Cu, Pb and Zn are shown remediated by the electrodialytic method without use of enhancement solution to improve the mobility of the metal ions. It was shown possible to remove 97% Cu in one experiment and 87% Pb and 69% Zn in another experiment. During the electrodialytic process the soil becomes acidified, thus desorbing these metal ions. Further a carbonate soil polluted by Cu can be remediated using ammonia as enhancement solution. Indeed, Cu forms mobile complexes with ammonia. Thus the soil can be remediated at high pH, at which the carbonates are not dissolved. It was shown that addition of ammonia improved the remediation from nothing removed from the soil totally without ammonia to 73% Cu removed after addition of ammonia. In the case of wood preservation sites where the soil is polluted by Cu and As, ammonia can also be used as enhancement solution, since As is mobile in the alkaline environment. In an experiment 70% As and 53% Cu was removed. Nonetheless, for wood preservation sites polluted with Cu, Cr and As ammonia is not sufficient. Here ammonium citrate showed potential, since the citrate part is mobilizing Cr. Experimentally 33% Cr, 65% Cu and 66% As was removed from the soil at the same time after addition of ammonia citrate. Thus, by carefully choosing the enhancement solution for each case, electrodialytic remediation is possible, even of soils with a fine fraction ranging from 33% to 63% per weight as the experimental soils here.

Key words: heavy metals, soil pollution, electrochemical remediation

21.1
Introduction

Heavy metal polluted soils pose a risk to water quality and to health of humans at normal site use, and thus remediation of these sites are highly relevant. Remediation of heavy metal polluted fine grained soils is difficult or even impossible by more traditional methods like pump and treat technologies or soil washing methods. Electrochemical remediation methods, on the other hand, are very effective in fine grained soils, e.g. (Ho et al. 1995) or (Reed and Berg 1994) and this paper deals with such method.

Electrochemical soil remediation methods are based on the transport processes that occur when an electric current is passed through a soil. The electric current tends to pass the soil in the micropores due to the lower resistance here, and this means that the current is acting exactly where the heavy metals are mainly adsorbed in the soil. Thus the electrochemical methods are especially suitable for fine-grained soils.

The transport mechanisms of major importance to the electrochemical soil remediation methods are electro-migration and electro-osmosis (e.g. Acar and Alshawabkeh 1993). In the soil the current is carried by ions, and the anions will be transported toward the anode and the cations toward the cathode. This transport of ions in the electric field is electro-migration. Electro-osmosis is movement of pore water in the soil in the applied electric field. The negative charge on the surface of most soil particles causes an accumulation of positively charged cations near the surface in the diffuse electric double layer. Under the action of the electric field these cations will give a net flow of ions in the direction of the cathode, and by this water is forced to move towards the cathode.

During the electrochemical soil remediation processes the heavy metals are concentrated in electrolyte solutions in which the electrodes are placed. It is possible to recover the heavy metals from these solutions by different techniques. In the case of e.g. Cu, Zn, Pb and Cd electro-precipitation to metallic form can be done.

In Denmark a special concept for electrochemical soil remediation has been developed, in which the electro-kinetic method is combined with the process of electrodialysis. Ion exchange membranes are used to separate the soil and the solutions where the heavy metals are concentrated. This means that current will not be wasted in carrying harmless ions from one electrode compartment to the other, but the current will be used for carrying ions out from the soil only. The electrodialytic soil remediation method has shown its feasibility in laboratory scale to remediate different soil and pollution types of which examples are given in the present paper.

21.2
History of Electrochemical Soil Remediation Methods

During the 1930s the effects of applying an electric potential to a soil were first used. Puri and Anand (1936) tested the value of applying an electric potential across a soil for removing sodium ions in order to improve the soil quality. Casagrande (1948) was the pioneer who started to use the electro-osmotic effect for dewatering fine grained soils. He used the technique to stabilize earth masses where the classical methods had failed.

Years later Bruch and Lewis (1973) proposed to extend the usefulness of the electro-kinetic phenomenon to being a tool in ground water pollution control in soils of low hydraulic permeability. They based the proposal on theoretical considerations. In 1980 the movement of heavy metals in a soil in an applied electric field was first reported by Segall et al. (1980), who tried to dewater a dredged material disposal site and its embankment foundations in order to increase the capacity of the disposal site. In the liquid samples from the process they found various heavy metals. Even though this was just noted in the actual work, this finding may have given different researchers the idea of using an applied electric field for soil remediation, because in the end of the 1980s more teams started to develop electrochemical remediation methods. The majority of the published research conducted in the late 1980s and early 1990s originated from research teams from Geokinetics (Lageman 1989), Lousiana State University (Acar et al. 1999), Lehigh University (Pamucku et al. 1990) and Massachusetts Institute of Technology (Probstein and Hicks 1993).

The number of papers published in the late 1980s in the field of electrochemical remediation was less than 20, but during the 1990s more than 400 papers have been published. Among the scientists and companies working in the field a network has been established, and in this network a common reference list is made (Ottosen and Hansen 1999). The numbers given above about the publications are taken from this list. In 2001 this network constituted of 85 teams mainly from Europe and Northern America, but also teams from Asia, Southern America and Australia have joined the network. The methods where an electric field is applied to a soil to remove chemical species are variably called e.g. electrokinetic remediation, electro-reclamation, electrokinetic extraction, electrokinetic soil processing, electrochemical decontamination or electrodialytic remediation. Most of the electrochemical remediation methods are now in the state of field implementation. Many important and promising results have been obtained from laboratory work and a few full scale remediation actions have been performed with success, too.

21.3 Electrodialytic Remediation

21.3.1 Principle

Electrodialytic remediation is based on removal of polluting heavy metals from the soil matrix by electro-migration (Fig. 21.1). The heavy metals are concentrated in electrolyte solutions from where they can be recovered for reuse by different techniques such as electro-precipitation.

Figure 21.1 shows the basic principle of a laboratory cell for electrodialytic remediation. The polluted soil is placed in the central compartment and the soil is separated from the electrolyte solutions with ion exchange membranes. Between the soil and the anolyte an anion exchange membrane (AN) is placed. This membrane prevents cations from the anolyte from electro-migrating into the soil, but the electro-

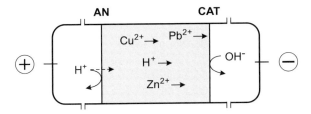

Fig. 21.1. Principle of electrodialytic soil remediation. The polluted soil is placed in the central compartment and within the applied electric field the ions in the soil move in accordance with their charge and are concentrated in the processing solutions (AN = anion exchange membrane, CAT = cation exchange membrane)

migrating anions in the soil can be removed into the anolyte. At the other end of the soil a cation exchange membrane (CAT) is placed and this membrane prevents anions in electro-migrating from the catholyte into the soil, but the electro-migrating cations from the soil can enter the catholyte. This use of membranes limits the number of ions that can electro-migrate in the soil, because it prevents new ions from entering from the electrode compartments. This means that the transport number for the heavy metals in the soil is higher than in a cell, where passive membranes are used as barriers between soil and electrolyte solutions, everything else equal.

The electrolyte solutions in the electrode compartments have other functions than being the media where the heavy metals are concentrated. Since electrodes used are inert electrodes, electrolysis of water occurs at both anode and cathode, i.e. production of O_2 gas and H^+ ions at the anode and production of H_2 gas and OH^- ions at the cathode. If the electrodes were placed directly in the soil, the electrode processes would easily be inhibited from the gas production, since the gasses could not be transported away from the electrode surface. The decisive factor for success in applying the electrodialytic remediation method to a soil is the ability to desorb the polluting heavy metals during the process, since the heavy metals are mobile by electro-migration as ions in the soil solution only.

When the electric DC current is passed through the electrodialytic soil remediation cell, the soil is acidified from the anode towards the cathode. This is due to water splitting ($H_2O \longrightarrow H^+ + OH^-$) at the anion exchange membrane, which is a catalyst of this process. The H^+ ions produced by the water splitting will electro-migrate into the soil and result in the acidic front (Ottosen et al. 2000a). The acidic environment in the soil causes different heavy metals as Cu, Zn, Pb and Cd to desorb and thus these heavy metals are mobile in the applied electric field. Thus some soils can be remediated by the electrodialytic method after addition of tap water, only, to a total water content of about 15–20% per weight. Meanwhile it is beneficial or even necessary to add an enhancement solution to the soil instead of water prior to the remediation for some soil types and pollution combinations in order to ensure the desorption of the heavy metals. Following examples are given with soils that can be remediated without any use of enhancement solution, and soils where enhancement solutions are beneficial or necessary.

The enhancement solutions shall aid desorption of the polluting heavy metals into an ionic form or into charged species so that they are mobile in the electric field. Furthermore the enhancement solution itself must be a harmless compound to the ecosystem, or degradable into harmless compounds, so that it will not form a new pollution problem.

21.3.2
Remediation of a Cu-Polluted Weathered Soil

In the following sections examples of different remediation experiments are given. The concentration of the elements in the soil was measured after pre-treatment of the soil as described in Danish Standard 259: 1.0 g of dry soil and 20.0 ml (1:1) HNO_3 were heated at 200 kPa (120 °C) for 30 min. The liquid was separated from the solid particles by vacuum filtration through a nuclepore filter and diluted to 100 ml. The elements were measured by atomic absorption spectrometry (AAS). The units used in this paper are mg kg^{-1} dry matter.

The figures for evaluating the electrodialytic method in this paper are shown as concentration profiles of the heavy metals in the soil. After each experiment the soil has been segmented into slices from anode towards cathode. The concentration of heavy metals was measured in each slice (minimum a double measurement) and these measurements are the basis for the figures.

Remediation of a Danish Cu polluted soil is the first example. The soil was sampled at a wood preservation site that was polluted during 40 years of wood preservation, which took place until 1976. The soil was taken in an area of the site where the pollution was so intensive that no vegetation could grow. The soil was weathered topsoil sampled at the depth of 0 to 15 cm. The carbonate content was less than 0.1%, the organic content was 3.6% and the fine fraction (<63 µm) was 33%. The actual soil may have been polluted with As, too, but this element was not included in this first investigation.

Three different experiments, which differed only in duration, were performed with this soil. The principle of the laboratory cell was as shown in Fig. 21.1. The length of the soil compartment was 15 cm and the internal diameter was 8 cm. The current density was kept constant at 10 mA, corresponding to 0.2 mA cm^{-2}. Figure 21.2 shows the Cu concentration profiles at the end of the experiments.

From Fig. 21.2 it is clearly seen that Cu is removed from the soil during the process. From the first 10 cm 96% Cu was removed during 70 d. After 12 d it is only the first slice closest to the anode that is remediated, and after 22 d two slices are remediated. In these two experiments an accumulation of Cu is seen in the slices next to the remediated ones. This picture is pH dependent. In the remediated slices pH was about 3 and in the slices with accumulation of Cu, pH was about 5.8 as initially. The remediation rate is thus linked to the rate of the developing acidic front. After 70 d the Cu concentration is below 500 mg kg^{-1} all through the soil and thus the soil is remediated to a class, where it can be used for certain purposes in Denmark. Initially the Cu concentration was too high for that and the soil could go to a waste dump, only.

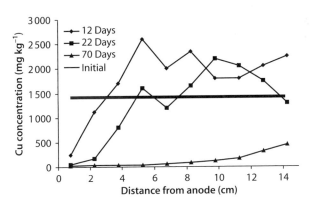

Fig. 21.2. Laboratory-scale electrodialytic remediation of a Cu polluted soil without use of enhancement solution. Cu concentration profiles in the soil from three different experiments, which varied in duration only are shown. The soil is acidified from the anode end towards the cathode and the mobility of Cu follows the acidic front. Thus Cu first mobilized in the acidified soil is precipitation in the part of the soil where the acidic front has not reached yet. Finally (here after 70 d) all soil volume is acidified and Cu removed. Modified from Ottosen et al. (1997)

21.3.3
Remediation of a Cu-Polluted Calcareous Soil

In calcareous soils the buffering capacity against the acidification is high and the acidic front develops very slowly during the treatment and thus the remediation rate is slow, too. Experiment A in Fig. 21.3 is an example of this taken from Ottosen et al. (1998). The actual soil is polluted from wire production and it has a carbonate content of about 11%. The soil was highly polluted with Cu and the initial concentration was 20 000 mg Cu kg^{-1}. A part of the polluting Cu is found precipitated as malachite in the actual soil, which has been shown by X-ray diffraction (XRD) (Kliem 2000).

In experiment A the soil with its natural moisture was added a small amount of 1 M HCl prior to the remediation in order to overcome a part of the buffering capacity. In the experimental cell the soil compartment was 10 cm long and the internal diameter was 4 cm. The current was 6.5 mA (0.5 mA cm^{-2}) and the duration of the experiment was 24 d. From Fig. 21.3 it can be seen, that the copper concentration decreased in the slice closest to the anode, only, and the concentration reached here was 3 600 mg Cu kg^{-1}, thus still high. To remediate this soil using the developing acidic front to mobilize the Cu would take very long.

Ammonia could be a good enhancement solution for this soil. Copper forms stable cationic complexes with ammonia following the equilibrium:

$$Cu^{2+} + 4NH_3 \longleftrightarrow (Cu(NH_3)_4)^{2+}$$

Ammonia (2.5% NH$_3$) was tested as enhancement solution for the actual soil (experiment B in Fig. 21.3) in an experiment where the experimental cell was the same as in experiment A. The current in experiment B was 8.2 m (0.68 mA cm^{-2}) but the duration was five days shorter and thus the number of charges moved in the two experiments were the same. In experiment B the Cu concentration was decreased in the whole soil volume. Only in the slice closest to the cathode the low level was not obtained, but it would probably have been if the current had been passed for a longer time. The reduction of Cu in the rest of the soil was 93%. pH in the soil after the remediation was still above 9.

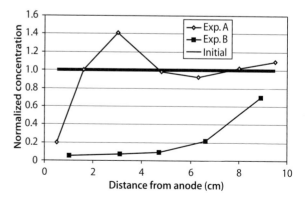

Fig. 21.3. Laboratory-scale electrodialytic remediation of a calcareous soil polluted by Cu. Shown is normalized Cu concentration profiles in the soil after experiment A and B without and with addition of ammonia, respectively. Modified from Ottosen et al. (1998)

From these two experiments there is no doubt that the addition of ammonia enhanced the remediation of this actual soil, furthermore the voltage was much lower after addition of ammonia and thus the energy consumption was much lower, too (Ottosen et al. 1998)

21.3.3
Remediation of Soil Polluted with Cu and As

For wood preservation a combination of As and Cu salts was used in Denmark in the period from about 1936 to 1954 and wood preservation sites polluted during this period is thus polluted with these two elements. The example here is a soil sampled at the same wood preservation site as the soil from the first example of this paper. The pollution level was 900 mg As kg^{-1} and 830 mg Cu kg^{-1}. An experiment where water was added to this soil is described in Ottosen et al. (2000b), and it was shown that As was little mobile, only, probably because As is present in uncharged species in the acidic environment created during the process. Meanwhile As is expected to be mobile under alkaline conditions and in Ottosen (1995) it was shown that As was mobilized by addition of NaOH to the soil, but in this experiment Cu was immobile. To mobilize both elements at the same time ammonia can be used as enhancement solution. As is mobile because of the high pH, and Cu forms $(Cu(NH_3)_4)^{2+}$ as described above.

Figure 21.4 show the normalized concentration profiles for an experiment with the actual soil. The experiment was performed in a cell that differed slightly from the principle in Fig. 21.1. The main difference is that between the soil and the electrolyte solution in the anode end is placed filter paper in stead of the anion exchange membrane. This is done for two reasons. At first the acidic front from the anion exchange membrane is unwanted since the remediation must proceed at high pH, and secondly ammonia can be supplied to the soil through the filter paper during the process and the ammonia can pass through the soil by electro-osmosis. More information about the actual set up can be found in Ottosen et al. (2000b). The soil compartment was 15 cm long and had an internal diameter of 8 cm. The current was 45 mA for 15 d and 50 mA for 27 d (corresponding to 0.9 and 1.0 mA cm^{-2}) and the maximum voltage during the experiment was 39.2 V. Prior to the remediation the soil was added 2.5% ammonia to a water content of 16%.

Fig. 21.4. Laboratory-scale electrodialytic remediation of a soil polluted by Cu and As. Shown is concentration profiles of As and Cu in the soil after a remediation experiment where 2.5% ammonia was used as enhancement solution. Modified from (Ottosen et al. 2000b)

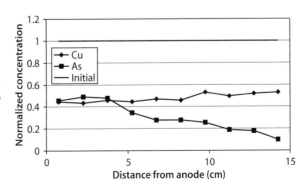

From Fig. 21.4 it is seen that As and Cu was removed from the soil at the same time. 70% As and 54% Cu was removed. The Cu concentration was stabilized at a level of about 400 mg kg^{-1}. From the profile it seems unlikely that the Cu concentration will reach the same low level of 20 mg kg^{-1} as it was the case for the soil from the same site that was remediated without enhancement solution (the first example of this paper), but the level reached here is still less than the 500 mg kg^{-1} that was the goal. It seems as if the As removal rate is slower than the one for Cu since the profile has not levelled out as the Cu profile. Probably more As would have been removed if the experiment had proceeded longer. The lowest As concentration reached was 90 mg kg^{-1} corresponding to a 90% removal.

21.3.4
Remediation of Soil Polluted with Cu, As and Cr

In the period from 1954 until 1993 a combination of Cu, Cr and As salts was used for the wood preservation process. The soil for the next experiment was sampled at a Danish wood preservation site where wood was preserved in the period 1960 to 1981. The soil was sampled in an area naked from vegetation due to the high pollution level: 8780 mg Cu kg^{-1}, 8420 mg Cr kg^{-1} and 14000 mg As kg^{-1}. The soil had a carbonate content of 0.5%, an organic content of 9.7% and the fine fraction was 49%.

It was shown (Ottosen 2001) that ammonia cannot be used as enhancement solution for soils polluted with Cr. Cr can form charged complexes with ammonium but the adsorption of Cr to the soil particles or precipitation of e.g. Cr(OH)$_3$ is too strong for the ammonium to mobilize Cr. Ammonia citrate was then tested as enhancement solution for the actual soil. Next to the advantages from the ammonia part of the enhancement solution, already described, the citric part can form negatively charged complexes with Cr(III). Negatively charged complexes can be expected formed with Cu, too.

An experiment was performed with 1 M ammonium citrate, with pH adjusted to 8 by ammonia, as enhancement solution. The cell had an internal diameter of 8 cm and the length of the soil compartment was 5 cm. The cell was constructed similar to the cell described in the former experiment, i.e. with filter paper next to the soil in the anode side. The cell was applied a current of 40 mA (0.8 mA cm^{-2}) for 35 d. The voltage was less than 16 all through the experiment, and for the first 27 d the voltage was about 5 V. In Fig. 21.5 the resulting profiles are shown.

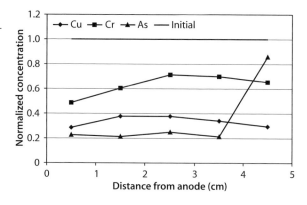

Fig. 21.5. Normalized concentration profiles of Cu, Cr and As from an experiment where ammonium citrate was used as enhancement solution (Ottosen 2001)

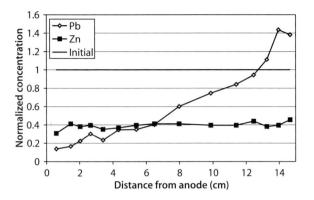

Fig. 21.6.
Normalized concentration profiles of Pb and Zn. No enhancement solution was used in this experiment

In this 30 d experiment 33% Cr, 65% Cu and 66% As was removed from the soil, i.e. ammonium citrate can be used to mobilize these three pollutants at the same time. For this experiment no optimisation was done on the concentration or pH of the ammonium citrate. An optimisation is necessary in order to increase the removal rate of Cr, but this experiment showed that the enhancement had good potential.

21.3.5
Soil Polluted with Pb and Zn

The last example here is remediation of a soil sample polluted from the production of different electronic devises. The soil was sampled in the depth of 10 to 50 cm and the main pollutant was Pb (940 mg kg^{-1}) but the concentration of Zn (340 mg kg^{-1}) was increased, too. The carbonate fraction of the soil was 0.5%, organic matter 4.1% and the fine fraction was 63%.

An electrodialytic remediation experiment was made in a cell with an internal diameter of 8 cm and a 15 cm long soil compartment. A constant current of 20 mA (0.4 mA cm^{-2}) was applied to the cell and the voltage varied between 10 V and 80 V.

Figure 21.6 shows that both Pb and Zn were removed during the experiment. Totally 39% Pb and 61% Zn was removed. Zn is mobile at a higher pH than Pb in the applied electric field (Ottosen et al. 2001) and this is the reason why no accumulation if Zn is seen in the slices closest to the cathode as it is the case of Pb. Closest to the anode pH was 2.8 and in the slices where Pb accumulated but Zn was removed the pH was about 3.7 to 3.9. The Zn concentration reached a stable level of about 130 mg kg^{-1} all through the soil. The Pb concentration was 130 mg kg^{-1} in the slice closest to the anode.

21.4
Conclusions

The examples given above showed that electrodialytic remediation of different soil types and pollution combinations is possible, but a key factor is the choice of a good enhancement solution in several cases.

It is not necessary to use enhancement solutions for soils of low buffering capacity polluted with heavy metals that will readily desorb under acidic conditions as e.g. Cu,

Pb and Zn, since an acidic front develops from anode towards cathode in the electrodialytic system and mobilizes the metals. Meanwhile in soils with a high buffering capacity the acidic front develops very slowly and it is beneficial to add an enhancement solution here. For instance, calcareous soils have a high buffering capacity. Acid could be added prior to the application of current and by this mobilizing the heavy metals, but the acid would also dissolve the calcium carbonates in the soil meaning that there would be a lot of mobile Ca^{2+} ions, too. This would increase the energy demand for the action because a lot of current is wasted in carrying these harmless ions out from the soil. Instead an enhancement solution which mobilizes the heavy metals at high pH should be chosen. In the case of Cu this could be ammonia, since charged amine complexes are formed with Cu at a high pH where the carbonates will not dissolve.

For remediation of soils polluted from wood preservation the choice of enhancement solution depends on whether the soil is polluted with Cu, Cr and As or only Cu and As. In the latter case ammonia can be used as enhancement solution because As is mobile due to the high pH and Cu forms mobile amine complexes. When Cr is present, too, ammonia cannot be used as enhancement solution, because Cr is not mobilized by the addition of ammonia. Instead ammonium citrate can be used in this case, since Cr forms mobile complexes with the citrate.

One of the major advantages of the electrodialytic method is that it is possible to remediate fine grained soils by the method. Examples of remediation of soils with a fine fraction (<63 μm) from 33% to 63% were shown.

Acknowledgements

AS Bioteknisk Jordrens SOILREM is acknowledged for financial support and Rune Dyre Jespersen is thanked for interesting collaboration.

References

Acar YB, Alshawabkeh A (1993) Principles of electrokinetic remediation. Environ Sci Technol 27(13): 2638–2647

Acar YB, Gale RJ, Hamed J, Putnam G, Wong R (1999) Electrochemical procesing of soils: theory of pH gradient development by diffusion, migration and linear convection. J Environ Sci Health A25(6):687–714

Bruch JC, Lewis RW (1973) Electroosmosis and the contamination of underground fluids. J Environ Plan Poll Cont 1(4):50–56

Casagrande L (1948) Electro-osmosis in soils. Geotechnique 1:159–177

Ho AV, Sheridan PW, Athmer CJ, Heitkamp MA, Brackin JM, Weber D, Brodsky PH (1995) Integrated in-situ remediation technology: the Lasagna process. Environ Sci Technol 29(10):2528–2534

Kliem BK (2000) Bonding of heavy metals in soil. Ph.D Thesis, Technical University of Denmark, pp 95–102

Lageman R (1989) Theory and praxis of electro-reclamation. NATO/CCMS Study. Demonstration of Remedial Action Technologies for Contaminated Land and Groundwater. Copenhagen, Denmark, 8–9 May 1989

Ottosen LM (1995) Electrokinetic remediation. Application to soils polluted from wood preservation. Ph.D. Thesis. Department of Geology and Geotechnical Engineering, Technical University of Denmark

Ottosen LM (2001) Elektrodialytisk rensning af jord fra træimprægneringsgrunde, Miljøprojekt 626 Miljøstyrelsen (Danish EPA) (in Danish)

Ottosen LM, Hansen HK (1999) Electrochemical soil remediation network. In: Proceedings from 2nd Symposium. Heavy Metals in the Environment and Electromigration Applied to Soil Remediation, Lyngby, Denmark, 7-9 July 1999, pp 35-40

Ottosen LM, Hansen HK, Laursen S, Villumsen A (1997) Electodialytic remediation of soil polluted from wood preservation industry. Environ Sci Technol 31:1711-1715

Ottosen LM, Hansen HK, Hansen L, Kliem BK, Bech-Nielsen G, Pettersen B, Villumsen A (1998) Electrodialytic soil remediation – improved conditions and acceleration of the process by addition of desorbing agents to the soil. Contaminated Soil '98. Proc. Sixth Int. Conference on Contaminated Soil. 17-21 May 1998, Edinbourgh, UK (Thomas Telford, London, UK) pp 471-478

Ottosen LM, Hansen HK, Hansen CB (2000a) Water splitting at ion-exchange membranes and potential differences in soil during electrodialytic soil remediation. J Appl Electrochem 30:1199-1207

Ottosen LM, Hansen HK, Bech-Nielsen G, Villumsen A (2000b) Electrodialytic remediation of an arsenic and copper polluted soil – continuous addition of ammonia during the process. Environ Technol 21:1421-1428

Ottosen LM, Hansen HK, Ribeiro AB, Villumsen A (2001) Removal of Cu, Pb and Zn in an applied electric field in calcareous and non-calcareous soils. J Hazard Mat B85:291-199

Pamukcu S, Khan LI, Fang HY (1990) Zinc detoxification of soils by electro-osmosis. Transportation Research Record 1288:41-45

Probstein RF, Hicks RE (1993) Removal of contaminants from soils by electric fields. Science 260:498-503

Puri AN, Anand B (1936) Reclamation of alcali soils by electrodialysis. Soil Sci 9(12):345-360

Reed BE, Berg MT (1994) In-situ electrokinetic remediation of a lead contaminated soil: (I) a theoretical overview. Hazardous and Industrial Wastes 26th Mid-Atlantic Industrial Waste Conference Papers, pp 480-487

Segall BA, O'Bannon CE, Matthias JA (1980) Electro-osmosis chemistry and water quality. ASCE. J Geotech Eng Div 106(GT10):1148-1152

Part II

Chapter 22

Electrodialytic Removal of Cu, Cr and As from Treated Wood

I. V. Cristensen · L. M. Ottosen · A. B. Ribeiro · A. Villumsen

Abstract

The service life of wood treated with Chromated Copper Arsenate (CCA) may be 30 years or even more due to the strong fixation of CCA in wood. The strong fixation also means that when the wood is removed from service, a large proportion of the copper, chromium and arsenic is still present and will enter into the waste stream unless actions are taken to prevent this. While the use of CCA is regulated in many countries, the handling of the waste wood is often not. The amount of treated wood being removed from service is expected to increase dramatically over the next few decades.

A method for safe handling of the waste wood and reuse of the wood resources, the contains – energy and metals, – would be environmentally beneficial. Here we tested electrodialytic remediation as a remediation method. Preliminary results show that more than 90% copper and approximately 85% chromium and arsenic are removed from the wood.

When the method will be optimised, it is expected that close to 100% of the metals will be removed during remediation. Afterwards the metals can be recovered and possibly reused in new wood preservatives. The wood chips can be reused or burned as it no longer contains metals.

Key words: wood, CCA, chromated-copper-arsenate, electrodialytic remediation

22.1 Introduction

The wood preservative Chromated Copper Arsenate (CCA) has been used worldwide since the 1950s. It is accepted as one of the most effective treatments for the protection of wood against fungi, insects and marine borers (Eaton and Hale 1993). The strong fixation of CCA in wood can give the wood a service life of 30 years or more. In recent years, a concern for the environmental impact of arsenic in preserved wood has led to restrictions on the use of CCA in some countries. For instance, in Denmark the use of arsenic, and thereby CCA, for wood preservation has been prohibited since 1993. In Sweden, the National Chemicals Inspectorate introduced restrictions on the use of chromium and arsenic containing preservatives in 1994, so that CCA may no longer be used above ground level with a few exceptions. This has resulted in a dramatically decrease in the use of CCA in Sweden. Before 1994 the CCAs had approximately 85% of the domestic market for treated sawn timber. In 1994 it was reduced to approximately 40% of this market (Jermer 2000). In Norway the wood preservation industry has entered a voluntary agreement with the environmental authorities, and the use of arsenic and chromium in wood preservation will be phased out by October 2002. Currently the European Union is proposing to limit the use of CCA-treated timber to industrial cooling towers, railway sleepers and electricity and telephone poles. If approved, this could become in affect by the middle of 2004. In the U.S.A. CCA will not be allowed for residential use, e.g. play-structures and picnic tables, by 2004.

While more and more restrictions on the production and use of CCA-treated timber are being proposed, only few countries have legislation on the handling of the waste wood. In some countries incineration is recommended, while others recommend land filling of the wood.

Denmark is one of the few countries with legislation on treated waste wood. Incineration of CCA-treated wood is banned, and deposition is the only alternative until new methods has been introduced that ensure reuse of the resources of the wood as energy and metals. When such a method is introduced the wood should be collected and treated separately (Miljø og Energiministeriet 1999). In Sweden only "new" treated waste wood, that has not yet left the preservation plant, is considered. It is recommended that this waste wood is to be burned only in approved incineration plants, and the ash is to be deposited in special landfills. When it comes to treated waste wood that has been removed from service, there are no regulations or recommendations.

One of the reasons given for the lack of regulation is the problems of identifying treated waste wood. Treated wood has a characteristic green colour, but after several years in service the wood may appear grey, and look similar to untreated weathered wood. One method for identifying treated weathered waste wood may be from information about the location or use of the wood in service; another could be visual identification using colour reagents like Chromazurol S, which will stain the wood blue when smeared on wood containing copper. X-ray fluorescence (XRF) is easier and faster in use than a colour reagent, but also far more expensive. In seconds it can tell the concentration of several metals in the wood, simply by pressing the instrument onto the piece of wood in question.

The amount of treated waste wood is expected to increase rapidly in the next years. In Denmark it was estimated that the amount of treated wood removed from service would increase from 17 000 t in 1992 to 100 000 t a year by 2010. Similar trends can be

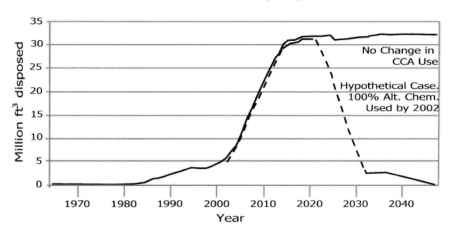

Fig. 22.1. Long-term disposal forecast for CCA-treated wood waste in Florida (From Solo-Gabriele and Townsend 2000). *CCA:* Cu, Cr, As

seen in other countries. In Fig. 22.1 a long-term disposal forecast for Florida predict that the amount of CCA-treated waste wood will increase from below 5 million ft^3 in the 1990s to 30 million ft^3 in 2015 (Solo-Gabriele and Townsend 2000).

The restrictions on the use of CCA will of course lead to a reduction in CCA-treated waste wood, but due to the long service life of CCA-treated wood, the expected decrease in waste will only become apparent much later (see Fig. 22.1). A method for safe handling of the waste wood and preferable reuse of the resources the wood contains (energy and metals) would be environmentally beneficial. Electrodialytic remediation could be such a method.

22.2
Electrodialytic Remediation

Electrodialytic remediation is a method developed and patented at the Technical University of Denmark for cleaning soils polluted with heavy metals. The method uses a direct electric current as cleaning agent, and combines it with the use of ion exchange membranes to separate electrolytes and soil (Hansen et al. 1999; Ottosen et al. 1997).

Figure 22.2 shows the electrodialytic cell in principle.

The laboratory cell consists of three compartments: an anode compartment (I), a cathode compartment (III) and a middle compartment (II) with the wood chips (Fig. 22.2). The catholyte is separated from the middle compartment by a cation exchange membrane, a membrane that only allows positive ions – cations – to pass. The anolyte is separated from the middle compartment by an anion exchange membrane, which allows only negative ions – anions – to pass.

When an electric potential is applied to the electrodes, the current in the cell will be carried by ions in the solutions in the compartments, and the ions will move in accordance to their charge. Cationic species will move towards the cathode and anionic species will move towards the anode. When ion exchange membranes are used, no current carrying ions can pass from the electrode compartments into the middle compartment due to the ion exchange membranes, while ions can be transported from the middle compartment into the electrode compartments. In this system the current is thus prevented in carrying highly mobile ions from one electrode compartment through the middle compartment into the other electrode compartment. Furthermore competition between these highly mobile ions from the electrode compartments and the ions in the middle compartment is avoided.

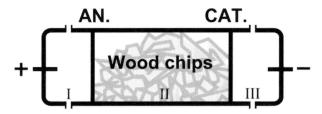

Fig. 22.2. Schematic presentation of an electrodialytic cell. Compartment I and III is the anode and cathode compartment respectively, compartment II contains the wood chips. *AN:* anionic exchange membrane. *CAT:* Cationic exchange membrane

Good results have been obtained using this method for soil remediation and subsequently the method has been tested on other materials, e.g. fly ash, sludge and most recently CCA-treated waste wood. The first remediation experiments on CCA-treated waste wood were performed with sawdust. The results were very promising with 93% Cu, 95% Cr and more than 99% As, removed at the end of the 30 d experiment (Ribeiro et al. 2000).

This study presents results obtained by electrodialytic remediation of CCA-treated wood, using wood chips for the experiments. Using woodchips instead of sawdust, less processing of the treated wood is required. Thereby the workers doing this will not be exposed to airborne particles to the same degree, and a safer work environment will be created.

In order to move in the electric field, the metals need to be present as ions. In most cases it is necessary to add an assisting agent to facilitate this process. In the case of CCA-treated waste wood, oxalic acid has proven effective. It forms negatively charged complexes with copper and chromium that will move in an electric field. Oxalic acid was used in the sawdust experiments (Ribeiro et al. 2000) and is used in the present experiment.

22.3
Experimental

Woodchips from the outermost 3 cm of an out of service electricity pole were used. The pole is a *Picea abies* L., treated with 12 kg m^{-3} sapwood "K33" (CCA-B) in 1962 by Collstrop A/S. The pole was removed from service in 1999.

Contents of copper, chromium was determined using atomic absorption spectrophotometry in flame (AAS). Arsenic was determined using inductively coupled plasma mass spectrometry (ICP-MS). Wood samples were pre-treated using microwave assisted pressurised digestion in concentrated HNO_3.

The average metal concentration in the wood were 426 mg kg^{-1} for Cu, 837 mg kg^{-1} for Cr and 589 mg kg^{-1} for As.

The electrodialytic cell used, resembles the cell in Fig. 22.2. In this experiment 32.16 g wood chips were used. Prior to the remediation the wood chips were treated with oxalic acid (2.5%) and placed in the middle compartment (II); an acrylic container (5 cm long, 8 cm i.d.). Platinized electrodes obtained from Bergsoe Anti Corrosion were placed in the electrode compartments (I and III) and ion exchange membranes from IONICS were placed according to Fig. 22.2. Oxalic acid was circulated using 1 l in each electrode compartment. 2.5% oxalic acid was used in the anode compartment and saturated oxalic acid (100 g l^{-1}) was used in the cathode compartment. The duration of the experiment was 7 d. The DC current was kept constant at 40 mA. This resulted in a voltage drop between 2.5 V and 3.2 V indicating low power consumption.

The first three days of the experiment a 15 ml sample was removed from both electrolytes and replaced with the same amount of 2.5% oxalic acid. pH and concentrations of Cu and Cr were determined in the samples.

At the end of the experiment, contents of Cu, Cr and As were determined in the wood chips, the membranes, the electrolytes and on the electrodes.

22.4
Results and Discussion

Wood treated with CCA was subjected to electrodialytic remediation treatment for 7 d. At the end of the experiment pH was approximately 1 in the electrolytes, and at no time during the experiment was pH higher than 2.

The distribution of metals after remediation are given in Table 22.1. The results show that more than 90% copper and approximately 85% of both chromium and arsenic were removed from the wood during the remediation.

22.4.1
Arsenic

83% of the initial arsenic has been removed. As seen in Table 22.1, the distribution after remediation shows that arsenic is primarily moved towards the anode, but is also found in the liquid from the middle compartment and to some degree in the catholyte. At pH 2, the dominating arsenic species is H_3AsO_4, but above pH 2.2 $H_2AsO_4^-$ dominates. The fact that pH in this experiment is in the same range, makes it probable that these species will be dominating in the anolyte and middle compartment respectively. The fact that some arsenic is found in the catholyte indicates that also cation species is present. According to Ribeiro et al. (2000) AsO^+, $As(OH)^+$, or in even more acid solutions As^{3+}, may exist and these species will move toward the cathode.

22.4.2
Copper

94% of the initial copper has been removed during remediation (Table 22.1). Almost all the removed copper is found in the anolyte. This is in agreement with the fact that oxalic acid and copper forms anionic chelates, most likely $Cu(Ox)_2^{2-}$.

22.4.3
Chromium

86% of the chromium has been removed. Almost all chromium is removed in the anode direction, most of the chromium was at the end of the experiment found in connection with the anionic exchange membrane. The anionic chromium species is

Table 22.1. Distribution of copper, chromium and arsenic after remediation. "Wood" refers to residual amount of metal in the wood

	Cu (%)	Cr (%)	As (%)
Wood	6	14	16
Catholyte	2	5	12
Anolyte	86	13	50
Middle compartment	2	4	20
Anion membrane	4	61	1
Cation membrane	0	2	0

probably $Cr(Ox)_3^{3-}$. There is also a theoretical possibility that chromium(VI) is present (as CrO_4^{2-} or $HCrO_4^-$), but since Cr(VI) is reduced to Cr(III) during the preservation, and given the fact that both wood and oxalic acid may react as an reducing agent, the dominating anionic chromium specie is presumed to be $Cr(Ox)_3^{3-}$.

22.4.4
Duration of Experiment

Figure 22.3 shows that the large majority of the metals are located in the electrolytes within the first two days. After that the amount of metals only increases slightly. This would suggest that the remediation could be finished sooner than the 7 d used in this experiment.

These results were obtained in a system with only a small degree of optimisation. Future work will be further optimisation of the process with regard to concentration of oxalic acid, current, duration of the experiments and size of wood chips used in the experiments. Also the possibilities for reuse of both the metals and the wood will be investigated.

The metals are proposed reused in the preservation industry after some adjustments. If arsenic is no longer allowed in wood preservation, another use or safe disposal is needed. Until now only a few countries has a total ban of arsenic in wood preservation.

As the wood no longer contains copper, chromium and arsenic, the wood may be burned and thereby the energy source will be utilized. In Fig. 22.4 a life cycle for the wood metals are proposed.

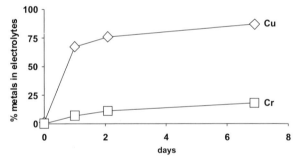

Fig. 22.3. Percentage of copper and chromium in the electrolytes as a function of remediation time. Total recovered metal at the end of the experiment is set as 100%

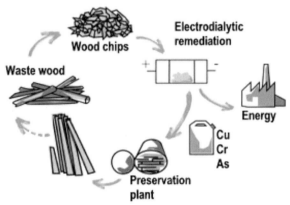

Fig. 22.4. Proposed life cycle for wood treated with CCA (where CCA is still allowed). *CCA:* Cu, Cr, As

22.5 Conclusions

In the near future the amounts of treated waste wood will increase dramatically. This wood contains large amounts of copper, chromium and arsenic. By using electrodialytic remediation it is possible to remove the metals from the wood. In the present study, 94% Cu, 86% Cr and 84% As were removed. The method creates no new waste products.

The laboratory scale experiment described above showed that the method may be an environmentally safe method for handling treated wood waste. Still optimisation is needed in order to reach a removal rate close to 100% for all the metals.

Future works also include a remediation experiment in larger scale in order to optimize e.g. the size of wood chips and the energy consumption. At present, pilot scale experiments to remediate up to 2 m^3 CCA-treated wood is being done at the Technical University of Denmark.

Acknowledgement

The support from EU's LIFE programme and Nordic Industrial Fond is acknowledged.

References

Eaton RA, Hale MDC (1993) Wood – decay, pests and protection. Chapman & Hall, UK
Hansen HK, et al. (1999) Electrodialytic remediation of soil polluted with heavy metals, key parameters for optimization of the process. Trans Ichem E Part A 77:218–222
Jermer J (2000) Wood preservation in sweden. American Wood Preservers' Association, paper prepared for the Ninety Sixth Annual Meeting 7–10 May 2000
Miljø og Energiministeriet (1999) Affald 21, Regeringens affaldsplan 1998–2004 (Waste 21). ISBN: 87-7909-297-7 (In Danish)
Ottosen LM, et al. (1997) Electrodialytic remediation of soil polluted with copper from wood preservation industry. Environ Sci Technol 98:1711–1715
Riberio AB, et al. (2000) Electrodialytic removal of Cu, Cr, and As from chromated copper arsenate-treated timber waste. Environ Sci Technol 34:784–788
Solo-Gabriele H, Townsend T (2000) Report #00-03. Florida Center for Solid and Hazardous Waste Management

Part II

Chapter 23

Treatment of Wastewater Contaminated by Mercury by Adsorption on the Crandallite Mineral

J. M. Monteagudo · J. M. Frades · M. A. Alonso · L. Rodriguez · R. Schwab · P. Higueras

Abstract

The present study has been undertaken to investigate a process that might remove inorganic mercury from mine waste water streams by using a compound of the crandallite type. In this work, an artificial amorphous crandallite, $Ca_{0.5}Sr_{0.5}Al_3(OH)_6(HPO_4)(PO_4)$, was synthesized in our laboratory and studied for the separation, removal and recovery of mercury from mercurial wastewaters. Since this compound exhibits an extremely wide range of ionic substitutions, Ca and Sr were interchanged with mercury. As a result, the mercury content of the waste water, ranging initially from 70 to 90 mg l^{-1}, was reduced to less than 0.1 mg l^{-1}. The process has been studied under batch conditions. The crandallite has been shown to have a high capacity for the absorption of mercury from mercuric nitrate solutions. The exchange capacity values of crandallite range from 0.90–1.50 meq g^{-1}. The equilibrium and kinetic behaviour was also studied.

Key words: mercury, immobilization, crandallite, wastewaters

23.1 Introduction

Over the years mercury has been recognized as having serious impacts on human health and the environment. This recognition has led to numerous studies that deal with the properties of various mercury forms, the development of methods to quantify and speciate the forms, fate and transport, toxicology studies, and the development of site remediation and decontamination technologies (Stepan et al. 1995; Higueras et al. 2001; Monteagudo et al. 2003).

Different mercury forms may occur at a contaminated site. These compounds may be transformed from one species to another under changing environmental conditions. Decontamination of different mercury forms may require different techniques. Because of its high vapor pressure, metallic mercury disperses relatively quickly into the atmosphere and, with suitable air movement, is taken up by plants and animals. Therefore, from the viewpoints of environmental chemistry, geochemistry, marine biology, and limnology, it is timely to set up a rapid, simple, sensitive and accurate method for the removal of mercury from water.

The most important starting material for mercury extraction is mercury-sulfide, HgS (cinnabar). The ore-bearing deposits are porous sedimentary rocks such as sandstone, bituminous shale and Silurian quartzite that contain mercury sulfide. The ore is extracted in several mines that generate mercury metallurgy waste waters containing mercury levels ranging from 70 to 90 mg l^{-1} (Higueras et al. 2001; Monteagudo and Ortiz 2000).

Mercury is widely used because of its diverse properties. In very small quantities, mercury conducts electricity, responds to temperature and pressure changes, and forms alloys with almost all other metals. Mercury serves an important role as a process or product ingredient in several industrial sectors. In the electrical industry, mercury is used in components such as fluorescent lamps, wiring devices and switches, e.g. thermostats, and mercuric oxide batteries. Mercury is also used in navigational devices, temperature and pressure sensors. It is also a component of dental amalgams used in repairing dental caries. In addition to specific products, mercury is used in numerous industrial processes. The largest quantity of mercury used in manufacturing in the U.S. is the production of chlorine and caustic soda by mercury cell chlor-alkali plants. Other processes include amalgamation, use in nuclear reactors, wood proccessing (as an antifugal agent), use as solvent for reactive and precious metals, and use as catalyst. Mercury compounds are also frequently added as a preservative to many pharmaceutical products (USEPA 1997).

For example, the majority of plants used in chlor-alkali electrolysis employ liquid mercury cathodes, resulting in waste water containing 10% mercury or more (Chall 1991).

Inorganic mercury (Hg(II)) contained in waste waters can be transformed into methylmercury (CH$_3$Hg) by the action of the bacteria present in the water. Methylmercury is the most toxic form that can enter the food chain. Methylmercury is more easily absorbed by fish and other aquatic fauna, either directly through the gills or by ingestion of contaminated aquatic plants and animals. The most widespread mercury-related health problem among humans involves the consumption of water fauna, such as fish, that have been contaminated with methylmercury (Stepan et al. 1995). By this means, it is necessary to remove the inorganic mercury as soon as possible from the waste waters produced in all processes involved mercury.

Among the techniques which can be used for the reclamation of mercury, we can mention the adsorption on a substrate such as the activated carbon (Dean et al. 1972; Sen and De 1987), the filtration and the ultrafiltration (Cheremisinof and Habib 1972), the metal reduction (Richard and Brookman 1975), the membranes processes (Draxler et al. 1988; Maes 1989), the ions exchange resins (Law 1971; Monteagudo and Ortiz 2000), and the biological reduction (Raskin et al. 1994; Dushenkov et al. 1995).

In this paper, we propose a new adsorbent-immobilizator which is available, inexpensive and has a very good capacity of fixing mercury. The choice of the crandallite type compound is due mainly to its chemical composition, great specific surface, high porosity and high capacity. The compounds of the crandallite type can be applied for immobilization of radioactive fission products and toxic heavy metals from natural and artificial sources. Ions like Hg^{2+} may enter mixed crystals and lower their concentration in solution. The Sr-Ca-induced crandallite works as a geochemical barrier or immobilizator, filtering ions out of solution (Schawb et al. 1990, 1993). This paper was undertaken as a part of an industrial scale project whose objective was to investigate a method for removing mercury at levels ranging from 70 to 90 mg l^{-1} from mine waste water streams by use a crandallitic compound. The Hg-crandallite is fed back to the plant of mercury metallurgy.

23.2
Experimental

23.2.1
Preparation of Amorphous Precursor Ca/Sr-Crandallite

The reaction of synthesis was defined as

$$3Al(OH)_3 + 2H_3(PO_4) + 1/2SrCO_3 + 1/2CaCO_3$$
$$\longrightarrow Ca_{0.5}Sr_{0.5}Al_3(OH)_6(HPO_4)(PO_4) + CO_2 + 4H_2O \quad (23.1)$$

The reaction was carried out in 1-dm³ flasks, magnetically stirred at 60 °C, 700 rpm and environmental pressure. The synthesis time was also studied from 5 to 30 d. Then the powder was filtrated, dried (40–45 °C), milled and analysed. The phase purity was checked via chemical and thermal analysis of the investigated compound, according to a previous report (Schawb et al. 1984).

23.2.2
Equilibrium Experiments

The equilibrium experiments were carried out in 1 dm³ flasks hermetically sealed and magnetically stirred, submerged in a temperature controlled thermostatic bath. 500 cm³ of mercuric nitrate $(Hg(NO_3)_2)$ solution was added into each of several flasks. Solution and crandallite were maintained at fixed temperature under vigorous stirring until the equilibrium was achieved. After that, the mixtures were filtered to remove the crandallitic compound and the filtrate was analyzed for mercury content by atomic emission spectrophotometry inductively coupled plasma (IPC-AES, JOBIN YVON JY48P) for mercury concentration between 0.15 mg l⁻¹ and 125 mg l⁻¹, or by atomic emission spectrophotometry advanced mercury analyzer (AMA 254) for mercury concentration below 0.15 mg l⁻¹. The crandallite phase composition was determined by mass balance from initial and equilibrium compositions of the aqueous phase, according to Eq. 23.2:

$$n^* = \frac{V}{W}(C_0 - C^*) \quad (23.2)$$

where C_0 and C^* are the initial concentration and equilibrium concentration of inorganic mercury in the liquid phase (meq dm⁻³ solution), respectively. n^* denotes the crandallite phase equilibrium concentration of mercury (meq g⁻¹ dry crandallite). V and W are the volume of solution and weight of dry crandallite (dm³ and g), respectively.

23.2.3
Batch Kinetic Studies

A set of comparative experiments were carried out increasing progressively the agitation rate in order to ensure that film mass transfer resistance was negligible. The evolution of concentration with time, until constant concentration, was obtained measuring the conductivity of the solution in a conductivity cell. Signals from the Crison 2201 conductimeter were monitored by a computer by means of a data adquisition program developed in our Department.

23.3
Results and Discussion

23.3.1
Equilibrium Studies

Since one of the controlling factors governing the use of ion exchangers is the equilibrium distribution of ions between the solids and solution phases which can be achieved in any given system, ion exchange equilibrium was studied. To fit the experimental equilibrium data the Langmuir, Prausnitz and Freundlinch equations were used:

Langmuir equation: $n^* = \dfrac{a_1 C^*}{a_2 + C^*}$ (23.3)

Freundlich equation: $n^* = b_1 (C^*)^{b_2}$ (23.4)

Prausnitz equation: $n^* = \dfrac{1}{\dfrac{1}{d_1 C^*} + \dfrac{1}{d_2 (C^*)^{d_3}}}$ (23.5)

Figure 23.1 shows the experimental equilibrium isotherm for the adsorption of Hg^{2+} by crandallite from mercurial wastewater, where the Langmuir equation (Eq. 23.3) was used to fit the equilibrium experimental data. The results show clearly that the equilibrium is very favourable for crandallite.

Table 23.1 lists the parameters of Eq. 23.3, 23.4 and 23.5, as well as the average deviation (s). They were determined by fitting the experimental data with the Langmuir,

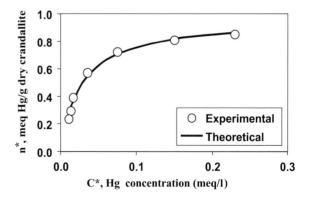

Fig. 23.1. Crandallite phase equilibrium concentration of mercury, n^* (meq g^{-1} dry crandallite) vs. equilibrium mercury concentration in the liquid phase, C^* (meq l^{-1} solution). Approximation to Langmuir equation

Table 23.1.
Experimental parameters of equations with average deviation ($s*$) calculated as followed:
$$s^* = \sum_{i=1}^{m}(n^*_{exp} - n^*_{eq})^2 / (m-2)$$

Parameter	Langmuir	Freundlich	Prausnitz
a_1	0.96611		
a_2	0.02737		
b_1		1.50653	
b_2		0.33622	
d_1			30.38039
d_2			0.82734
d_3			-0.10034
s^*	7.54×10^{-2}	6.36×10^{-1}	4.8×10^{-2}

Prausnitz and Freundlinch equations using a non-linear regression method. As it can be seen in Table 23.1, the data are correlated quite well by the equations.

The maximum asymptotic solid-phase solute concentration (meq g^{-1} dry crandallite), n^∞, and the equilibrium constant, K_I, were determined fitting the experimental data to the Langmuir equation, using a non-linear regression method, defined as follows:

$$n^* = \frac{n^\infty K_I C^*}{C_0 + (K_I - 1)C^*} \tag{23.6}$$

where n^∞ is the maximum asymptotic solid-phase solute concentration (meq g^{-1} dry crandallite), K_I is the equilibrium constant, and C_0, n^* and C^*, as defined above, are the initial solute concentration in solution, and the equilibrium solid and liquid phases solute concentrations, respectively. The sorption parameters obtained of Eq. 23.6 were: equilibrium constant (K_I) = 30.13 and saturation capacity (n^∞) = 0.93 meq g^{-1}, being the average deviation (s) 7.54×10^{-4}%. For all experiments, the K_I value is greater than 30 indicating replacement of the crandallite ions Ca^{2+} and Sr^{2+} with Hg^{2+} metal ions. This indicates that crandallites have higher affinity for Hg^{2+} than for Ca^{2+} and Sr^{2+} ions, which allows the ease the elimination of these wastewaters.

The values of the equilibrium constants, K_I, confirm that the isotherms are very favourable for crandallite. The following ion exchange reaction can occur with mercury in its oxidation state Hg^{2+}:

$$2\,Ca_{0.5}Sr_{0.5}Al_3(OH)_6(HPO_4)(PO_4) + 2Hg^{2+}$$
$$\longleftrightarrow 2HgAl_3(OH)_6(HPO_4)(PO_4) + Ca^{2+} + Sr^{2+} \tag{23.7}$$

Crandallite crystallizes in the alunite crystal lattice. Because of its open structure, the cations Ca^{2+} and Sr^{2+} are replaced by mercury; this element entering into the crystal network thus becomes immobilized. The crandallite used in this study show high selective binding of Hg^{2+} over Ca^{2+} and Sr^{2+}. It may be that the longer spacer provides more spatial possibility for the ligands to form complex with mercury ions.

23.3.2
Kinetic Studies

The experimental kinetic results are presented in Fig. 23.2. The amount of metal ion absorbed increases with time till the maximum capacity of the crandallite is attained. The plot shows that the amount of metal ion sorbed increase with time. When it ceases changing, the adsorption value was taken as the adsorption capacity of the crandallite. From the standpoint of an industrial application of this process, the discontinuous kinetic curves obtained are sufficient to provide the essential information required for process design. However, batch kinetic data were also accurate and extensive enough to allow a more detailed theoretical analysis to explain the different behaviour of the process studied.

Since the ion exchange rate of mercury did not change with the agitation rate, the ion exchange rate must be controlled by the intraparticle diffusion rate. Therefore, the ion exchange rates of the batchwise tests were analyzed by the squared driving force model (Sheng and Chang 1996; Urano and Tachikawa 1991), which leads to the following equation:

$$\log \frac{1}{1-F^2} = \frac{4\pi D_{eff} t}{d_{pw}^2} \qquad (23.8)$$

where D_{eff} (cm^2 s^{-1}) is the intraparticle effective diffusivity, d_{pw} (cm) the mean resin wet diameter, t (s) the time passed from the beginning of the experiment and $F = [(C_0 - C)/(C_0 - C_\infty)]$ represents the fractional uptake of mercury. C and C_∞ are the concentration of mercury in solution at any given time and at the end of the experiment, respectively. The results illustrate that the fractional uptake increases with time till the maximum value is attained.

The model is valid since the kinetic data for the crandallite are correlated quite well by the Eq. 23.8 being obtained an excellent fit between the experimental and theoretical data, where the squared correlation coefficient was 0.998. From results obtained in the equilibrium and kinetic studies, it may be stated that crandallite could be used for the purification of mercury metallurgy wastewater.

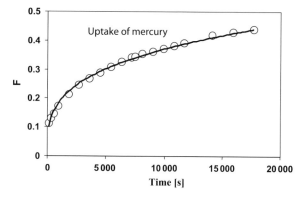

Fig. 23.2. Fractional uptake of mercury ($F = [(C_0 - C)/(C_0 - C_\infty)]$) vs. reaction time (s) ($C_0 = 80$ mg l^{-1}, $T = 20 \pm 0.1$ °C)

23.4
Conclusion

Crandallite has been shown to have a high adsorbing capacity of Hg^{2+} from mercurial waste waters (mercuric nitrate solutions). The separation factors of this crandallitic compound of mercury over Ca^{2+} and Sr^{2+} are very high. Crandallite crystallizes in the alunite crystal lattice. Because of its open structure, the cations Ca^{2+} and Sr^{2+} are replaced by mercury; this element entering into the crystal network thus becomes immobilized.

This process is capable of reducing mercury concentration in effluent to less than 0.1 mg l^{-1} for inlet mercury concentration in waste water averaging 80 mg l^{-1}. The method developed to serve the needs of the mercury metallurgy industry, it is applicable to most mercury bearing streams found in other industries waste water containing up to 100 mg l^{-1}.

References

Chall SP (1991) In Ullmann's encyclopedia of industrial chemistry, vol A15. pp 97–105
Cheremisinoff PN, Habib YH (1972) Cadmium, lead, mercury: a plenary account for water pollution: removal techniques. Water and Sewage Works 7:46–51
Dean JG, Bosqui FL, Lanouette KH (1972) Removing heavy metals from water. Environ Sci Technol 6:518–522
Draxler J, Fuerst W, Marr R (1988) Separation of metals species by emulsion liquid membranes. J Membrane Sci 38:281
Dushenkov V, Motto H, Raskin I (1995) Rhizofiltration: the use of plants to remove heavy metals from aqueous streams. Environ Sci Technol 29:1239–1245
Higueras P, Urbina M, Biester H, Lorenzo S (2001) Mercury environmental constraints in the Almadén mining district. RMZ – Materials and Geoenvironment 48:195–200
Law SL (1971) Methylmercury and inorganic mercury collection by a selective chelatin resin. Science 174:285–287
Maes M (1989) Membrane techniques, useful processess. Eau Ind Nuissances 33:133
Monteagudo JM, Ortíz MJ (2000) Removal of inorganic mercury from mine waste water by ion exchange. J Chem Technol Biotechnol 75:767–772
Monteagudo JM, Durán A, Carmona M, Schwab R, Higueras P (2003) Elimination of inorganic mercury from waste waters using crandallite type compounds. J Chem Technol Biotechnol 78:1–7
Raskin I, Nanda A, Dushenkov S, Saly DE (1994) Bioconcentration of heavy metals by plants. Curr Opin Biotechnol 5:285–290
Richard MD, Brookman G (1975) The removal of mercury from industrial waste waters by metal reduction. Engineering Bulletin Purdue University II:713–720
Schawb RG, Herold H (1984) Equilibrium of phosphates of crandallite type. Fortschritte der Mineralogie 62:230–241
Schawb RG, Herold H, Gótz C, de Oliveira NP (1990) Compounds of the crandallite-type: synthesis and properties of pure goyacite, corceixite and plumbogumnite. N Jb Miner Mh 3:113–126
Schwab RG, Goetz C, Herold H, de Oliveira NP (1993) Compounds of the crandallite type: thermodynamic properties of Ca-, Sr-, Ba-, Pb-, La-, Ce-, to Gd-phosphates and arsenates. Neues Jahrbuch für Mineralogie – Monatshefte 12:551–568
Sen AK, De AK (1987) Adsorption of mercury(II) by coal fly ash. Water Res 8:885–893
Sheng HL, Chang LW (1996) Ammonia removal from aqueous solution by ion exchange. Ind Eng Chem Res 35:553–558

Stepan DJ, Fraley RH, Charlton DS (1995) Remediation of mercury-contaminated soils: development and testing of technologies. Gas Research Institute Topical Report, GRI-94/0402

Urano K, Tachikawa H (1991) Process development for removal and recovery of phosphorus from wastewater by a new adsorbent. 1. Preparation method and adsorption capability of a new adsorbent. Ind Eng Chem Res 30:1893–1896

USEPA (1997) Mercury study report to congress. EAP-452/R-97-01

Chapter 24

Low Cost Materials for Metal Uptake from Aqueous Solutions

N. Fiol · J. Serarols · J. Poch · M. Martínez · N. Miralles · I. Villaescusa

Abstract

In this work the ability of some vegetable wastes from industrial processes such as cork and yohimbe bark, grape stalks and olive pits, to remove metal ions from aqueous solutions has been investigated. The influence of pH, sodium chloride and metal concentration on Ni(II) and Cu(II) uptake was studied. Metal uptake showed in all the cases a pH-dependent profile. Maximum sorption was found at an initial pH around 5.0–6.0. In some cases an increase of sodium chloride concentration induced a decrease in metal removal. Adsorption isotherms at the optimum pH were expressed by the non-competitive Langmuir adsorption model. When comparing the four materials, yohimbe bark waste was found to be the most efficient adsorbent for both metals studied.

Key words: low-cost adsorbents, nickel, copper, metal removal, sorption

24.1 Introduction

Conventional methods for removing metals from industrial effluents include chemical precipitation, coagulation, solvent extraction, electrolysis, membrane separation, ion exchange and adsorption (Patterson 1977). Considering the harmful effects of heavy metals, it is necessary to remove them from liquid wastes at least to a limit accepted by law. The sorption process is one of the few alternatives available for the removal of heavy metals at low concentrations from industrial efluents. Activated carbon, activated alumina or polymer resins which are non-regenerable and expensive materials, are the sorbents usually used for this purpose.

The high prices and regeneration cost of these materials limits their large-scale use for the removal of metals, and has encouraged researchers to look for low cost sorbing materials (Kratochvil and Volesky 1998; Bailey et al. 1999). Relatively recently, biological materials such as algae, bacteria, fungi and yeast (Mattuschka and Straube 1993; Pagnanelli et al. 2000) or certain waste products from industrial or agricultural operations, have also been recognised as new cheap sorbents for the removal of toxic metals. For instance, studies to assess the ability of crab shell (An et al. 2001), peat (McKay and Porter 1997), sunflowers stalks (Sun and Shi 1998), pine bark (Al-Asheh and Duvnajak 1998) or seafood processing waste sludge (Lee and Davis 2001) for the decontamination of metal-containing effluents, found these adsorbents to be moderately effective. The potential use of these residues as adsorbent materials present two advantages, its reuse and its low cost. They can be disposed off without expensive regeneration and represent a cheap alternative to conventional sorbents. Apart from these advantages the sorption process offers the possibility of effective metal concentration on the material.

The sorption of metals by this kind of materials might be attributed to their proteins, carbohydrates and phenolic compounds which have carboxyl, hydroxyl, sulphate, phosphate and amino groups that can bind metal ions (Madrid and Camara 1997). Different mechanisms may be involved in metal binding such as ion-exchange, chelation, complexation and surface adsorption depending on material and metal used as sorbent and sorbate, respectively (Brown et al. 2000).

In this work the ability of four different industrial wastes (cork, yohimbe bark, grape stalks and olive pits) to sorb metal ions has been investigated. Batch experiments at room temperature were designed to study the influence of pH, sodium chloride and metal concentration on the sorption processes.

24.2
Experimental

24.2.1
Reagents and Solutions

The wastes used in this work were generated after different industrial production processes: grape stalks from wine production, olive pits from olive oil extraction, cork bark from wine cork production and yohimbe bark from pharmaceutical alkaloid extraction. All these materials were washed, dried, ground, then sieved before their use. The particle size used was in the range of 1.0 to 1.5 mm.

Metal solutions were prepared by dissolving defined amounts of $NiCl_2 \cdot 2H_2O$ and $CuCl_2 \cdot 2H_2O$, in distilled water. NaOH and HCl were used for pH adjustment and NaCl was used as the ionic medium. These reagents were analytical grade and were purchased from Panreac (Barcelona, Spain). Metal standard solutions of 1 000 mg dm^{-3} purchased from Carlo Erba (Milano, Italy) were used for atomic absorption calibrations.

24.2.2
Procedure

The uptake of Cu(II) and Ni(II) was carried out by batch experiments at 25 °C. A fixed mass of 0.1 g of dry biomaterial was put into contact with 10 cm^{-3} of different aqueous metal solutions and shaken in a rotatory mixer (Cenco Instrument) at 25 rpm until equilibrium was reached. Then, samples were filtered through a 0.45 µm cellulose filter paper (Millipore Corporation). After filtration the metal concentration of the aqueous filtrate was determined by atomic absorption spectrometry using a Varian Absorption Spectrometer, Model 1275. The amount of metal removed by the biomaterial was calculated by a mass balance. Initial metal concentration was kept constant at 10 mg dm^{-3} (1.57×10^{-4} mol dm^{-3} for Cu and 1.70×10^{-4} mol dm^{-3} for Ni) when the influence of contact time, pH and sodium chloride concentration was investigated. Metal concentrations within the range 5–100 mg dm^{-3} (7.87×10^{-5}–1.57×10^{-3} mol dm^{-3} for Cu and 5.51×10^{-5}–1.70×10^{-3} mol dm^{-3} for Ni) were used to study the influence of initial metal concentration and to obtain the Langmuir isotherm. In all experiments the initial and equilibrium pH were measured using a Crison Model Digilab 517 pHmeter. The initial pH of the solution was tested within the range of pH 1.0 to pH 7.0. Attention was paid to avoid metal solid hydroxide precipitation (Baes et al. 1976). When the initial pH of metal solutions was adjusted to the desired value no efforts were made to maintain the solution pH while copper or nickel was being sorbed. To study the influence of sodium chloride concentration on metal ion removal, the NaCl concentration was varied from 0.1–2.0 mol dm^{-3}. Each test was carried out in duplicate and the average results are presented in this paper.

24.3
Results and Discussion

24.3.1
Equilibrium Contact Time

In order to ascertain the contact time that was necessary to achieve the equilibrium state, characterised by unchanging sorbate concentration in the solution, simple preliminary sorption-kinetic experiments were performed stirring the same amount of dry solid with 10 cm^3 of metal solution in different tubes. Samples were drawn at predetermined intervals of time for analysis. Initial metal concentration was 10 mg dm^{-3}: 1.57×10^{-4} mol dm^{-3} for Cu and 1.70×10^{-4} mol dm^{-3} for Ni.

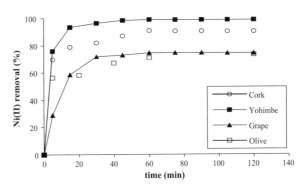

Fig. 24.1.
Ni(II) removal from aqueous solutions by sorption on various biomaterials. Total metal concentration 10 mg dm^{-3} (1.70×10^{-4} mol dm^{-3})

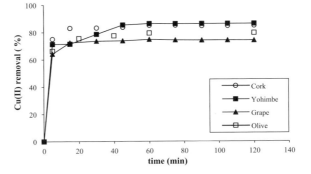

Fig. 24.2.
Cu(II) removal from aqueous solution by sorption on various biomaterials. Total metal concentration 10 mg dm^{-3} (1.57×10^{-4} mol dm^{-3})

The adsorbed metal concentrations were obtained from the difference between initial and final metal concentration. The percent removal was calculated as:

$$\%R = [(C_i - C_{eq})/C_i] \times 100$$

where C_i and C_{eq} are the initial and final metal concentration in solution respectively.

The results corresponding to the kinetics of Ni(II) and Cu(II) removal by the four studied materials are presented in Fig. 24.1 and 24.2, respectively.

As can be seen in both figures, metal sorption by the different biomaterials was quite rapid and reached a plateau after about 60 min. The rapid kinetic has significant practical importance, as they will facilitate the use of small sorbent volumes to ensure efficiency and economy. Based on the results obtained in Fig. 24.1 and 24.2, a shaking time of two hours was used in all further sorption experiments to ensure equilibrium.

24.3.2
Effect of pH on Metal Removal

In general, metal uptake by biomaterials has been reported to be strongly dependent on pH (Vecchio et al. 1998; Seco et al. 1997). Thus, it was of great importance to investigate the effect of pH on metal removal. For this purpose a set of experiments varying the initial pH within the range of pH 1.0 to pH 7.0 was carried out. The percentage of nickel and copper removal vs. equilibrium pH is presented in Fig. 24.3 and 24.4, respectively.

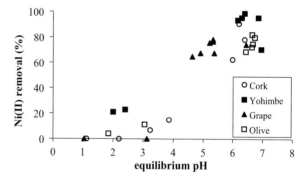

Fig. 24.3. Ni(II) uptake as a function of equilibrium pH for different sorbents. Total metal concentration 10 mg dm^{-3} (1.70 × 10^{-4} mol dm^{-3})

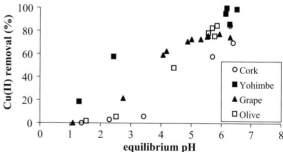

Fig. 24.4. Cu(II) uptake as a function of equilibrium pH for different sorbents. Total metal concentration 10 mg dm^{-3} (1.57 × 10^{-4} mol dm^{-3})

As it can be seen, metal removal increased with pH solution and a plateau was reached at around pH 5.5–6.0. The same trend has also been reported in the removal of these ions by some other materials (Seco et al. 1997). It can be also observed that low metal adsorption was found at low pH. This fact can be explained by the competition between protons and metal cations for the surface sites. When increasing the pH there is a decrease of positive surface charge of the biomaterial that results in a lower coulombic repulsion of sorbed metal and the surface. This effect seems to be more important in the case of nickel. The maximum uptake was in all cases at equilibrium pH values around 6.0–7.0 that corresponded to initial pHs around 5.0–6.0. Other authors found similar equilibrium pH values for the same metal sorption (Lee and Yang 1997; Yu and Kaewsarn 1999; Villaescusa et al. 2000). Taking into account these results, further experiments were carried out at initial pH values around 6.0 without any pH adjustment.

24.3.3
Effect of Sodium Chloride Concentration

Several possible mechanisms for metal sorption have been proposed but ion exchange can be one of the principal mechanism of interaction between metal and sorbent (Kratochvil and Volesky 1998). In this context, the influence of NaCl concentration on the sorption process was studied. In Table 24.1 results corresponding to Ni(II) and Cu(II) uptake by different biomaterials in different NaCl solutions are shown.

In general a significant influence of sodium chloride concentration on metal uptake for all the studied systems was observed. The results in Table 24.1 demonstrate that the presence of NaCl in solution induces in general a decrease in metal uptake. Only when yohimbe bark was used as adsorbent a slight effect was observed. For the other materials this effect was more important in the case of nickel uptake. The decrease in sorption efficiency observed can be due to different phenomena. Sorption is sensitive to the change in ionic strength if electrostatic attraction is a significant mechanism for metal removal. Some authors explain the reduction of metal removal percentage by the presence of competing Na^+ ions for metal binding (Lee

Table 24.1. Influence of NaCl concentration on metal uptake (%) by cork, yohimbe, grape stalks and olive pits at initial pH = 5.5–6.0. Total metal concentration 10 mg dm^{-3} (1.70 × 10^{-4} mol dm^{-3} for Ni and 1.57 × 10^{-4} mol dm^{-3} for Cu)

[NaCl] (mol dm^{-3})	Cork		Yohimbe bark		Grape stalks		Olive pits	
	Ni(II)	Cu(II)	Ni(II)	Cu(II)	Ni(II)	Cu(II)	Ni(II)	Cu(II)
0.0	90.8	84.9	98.9	100.0	74.6	74.7	78.1	79.7
0.1	7.7	57.7	93.0	92.4	20.5	44.4	14.0	22.7
0.5	0.0	20.6	88.0	92.2	6.2	29.5	4.6	17.3
1.0	0.0	18.7	82.6	91.6	0.0	16.2	1.2	10.5
2.0	0.0	16.6	80.2	90.0	0.0	0.0	0.0	0.0

and Yang 1997). Nevertheless, this reduction in metal removal can also be explained on the basis of the different ionic species present in solution at different chloride concentration. Most of divalent metals form chloro-complexes in presence of chloride ions. At zero or low chloride concentration only free cations are present in the solution whereas at higher chloride concentration neutral or anionic species predominate in the medium. Thus, the different species in solution may be sorbed or not depending on the material.

24.3.4
Langmuir Isotherm

The equilibrium isotherms were determined for each material and each metal separately. In order to optimise the design of a sorption system to remove these metal ions it is important to establish the most appropriate correlation for the equilibrium curves for each system. The most used isotherm equation for modelling equilibrium data is the Langmuir equation (McKay and Porter 1997) that, for diluted solutions may be represented by:

$$q = \frac{q_{max} b C_{eq}}{1 + b C_{eq}}$$

where q is the specific uptake (mol g^{-1} dry solid) and C_{eq} the metal concentration (mol dm^{-3}). The constant q_{max} is the maximum sorbate uptake per unit weight of sorbent and b is the Langmuir constant related to energy of sorption which reflects quantitatively the affinity between the sorbent and the sorbate.

The experimental data fitted satisfactorily the Langmuir sorption models. The values of q_{max} and b for the different systems (sorbent-metal) are presented in Table 24.2. If we

Table 24.2. Langmuir parameters for nickel and cooper in different sorbent materials

Metal ion	Material	q_{max} (mol g^{-1})	b (dm^3 mol^{-1})	R^2	Source
Nickel	Pine bark	1.07×10^{-4}	1.27×10^3		Al-Asheh and Duvnjak 1998
	Yohimbe bark	1.50×10^{-4}	6.16×10^4	0.993	This work
	Cork	7.00×10^{-5}	3.20×10^4	0.994	This work
	Grape stalks	7.85×10^{-5}	2.52×10^4	0.998	This work
	Olive pits	4.03×10^{-5}	1.58×10^3	0.997	This work
Copper	Pine bark	1.49×10^{-4}	5.50×10^3		Al-Asheh and Duvnjak 1998
	Peat	2.80×10^{-4}	6.35×10^3		McKay and Porter 1997
	Yohimbe bark	1.50×10^{-4}	1.60×10^4	0.999	This work
	Cork	4.70×10^{-5}	1.10×10^4	0.992	This work
	Grape stalks	8.83×10^{-5}	1.83×10^4	0.997	This work
	Olive pits	3.73×10^{-5}	1.53×10^3	0.974	This work

consider the four materials used in this work separately, a similar maximum uptake for nickel and copper was found for all of them. When comparing the q_{max} values the relative capacities were in the order yohimbe > grape stalks > cork > olive pits. From all the studied materials yohimbe showed the highest sorption capacity, 1.50×10^{-4} mol g^{-1}, similar to that found in literature for pine bark (Al-Asheh and Duvnjak 1998).

24.4
Conclusion

In summary the following conclusions can be drawn:

- Based on the results obtained the four materials tested can be used as sorbing material for Ni(II) and Cu(II) removal from aqueous solution.
- Metal sorption is pH-dependent and maximum sorption for both metals was found to occur at initial pHs around 5–6.
- The presence of high sodium chloride concentration significantly reduces metal removal.

Therefore, our results demonstrate the potential utility of vegetable wastes from industrial processes for the treatment of wastewater containing heavy metals. Finally, we would like to remark that our results have a double implication: *(i)* the re-use of an industrial waste that is abundant and usually incinerated and *(ii)* the elimination, recovery or concentration of heavy metals from wastewater.

Acknowledgements

This work has been supported by Ministerio de Ciencia y Tecnología, Spain, project PB97-0655.

References

Al-Asheh S, Duvnjak Z (1998) Binary metal sorption by pine bark: study of equilibria and mechanisms. Sep Sci Technol 33:1303–1329
An HK, Park BY, Kim DS (2001) Crab shell for the removal of heavy metals from aqueous solutions. Water Res 35:3551–3556
Baes CF, Mesmer RE (1976) The hydrolysis of cations. Wiley, New York, USA
Bailey SE, Olin TJ, Bricka RM, Adrian DD (1999) A review of potentially low-cost sorbents for heavy metals. Water Res 33(11):2469–2479
Brown PA, Gill SA, Allen SJ (2000) Metal removal from wastewater using peat. Water Res 34:3907–3916
Kratochvil D, Volesky B (1998) Advances in the biosorption of heavy metals. Trends Biotechnol 16:291–300
Lee SM, Davis AP (2001) Removal of Cu(II) and Cd(II) from aqueous solution by seafood processing waste sludge. Water Res 35:534–540
Lee S, Yang J (1997) Removal of copper in solution by apple wastes. Sep Sci Technol 32:1371–1387
Madrid Y, Camara C (1997) Biological substrates for metal preconcentration and speciation. Trends Anal Chem 16:36–44
Mattuschka B, Straube G (1993) Biosorption of metals by a waste biomass. J Chem Technol Biotechnol 58:57–63

McKay G, Porter JF (1997) Equilibrium parameters for the sorption of copper, cadmium and zinc ions onto peat. J Chem Technol Biotechnol 69:309–320

Pagnanelli F, Petrangeli PM, Toro L, Trifoni M, Veglio F (2000) Biosorption of metal ions on *Arthrobacter* sp: biomass characterization and biosorption modelling. Environ Sci Technol 34:2773–2778

Patterson JW (1977) Waste water treatment. Science Publishers, New York, USA

Seco A, Marzal P, Gabaldón C, Ferrer J (1997) Adsorption of heavy metals from aqueous solution onto activated carbon in single Cu and Ni systems and in binary Cu-Ni, Cu-Cd and Cu-Zn systems. J Chem Technol Biotechnol 70:23–30

Sun G, Shi W (1998) Sunflowers stalks as adsorbents for the removal of metal ions from wastewater. Ind Chem Res 37:1324–1328

Vecchio A, Finoli C, DiSimine D, Androni V (1998) Heavy metal biosorption by bacterial cells. Fresenius J Anal Chem 361:338–342

Villaescusa I, Martínez M, Miralles N (2000) Heavy metals uptake from aqueous solution by cork and yohimbe bark wastes. J Chem Technol Biotechnol 75:1–5

Yu Q, Kaewsarn P (1999) Binary adsorption of copper(II) and cadmium(II) from aqueous solutions by biomass of marine alga durvillaea potatorum. Sep Sci Technol 34(8):1595–1605

Chapter 25

Removal of Copper(II) and Cadmium(II) from Water Using Roasted Coffee Beans

M. Minamisawa · S. Nakajima · H. Minamisawa · S. Yoshida · N. Takai

Abstract

The adsorption behavior of heavy metals on arabica and robusta roasted coffee beans was investigated. To adsorb heavy metals, the coffee beans residues were suspended in aqueous solutions containing Cu(II) or Cd(II). Then the amount of heavy metal remaining in the solution was measured by atomic absorption spectrometry. The results show that the adsorption percentage of the heavy metal ions were above 90% for all coffee beans examined. Further, the adsorption capacities of Cu(II) and Cd(II) ions onto blend coffee were about 2.0 mg g^{-1}. This adsorption capacity is similar to that of zeolite, activated carbon and chitosan; and is higher than that of chitin and cerite. Blend coffee was thus found to be a good adsorbent for the removal of heavy metals from wastewater.

Key words: roasted coffee beans; removal of Cu(II) and Cd (II)

25.1 Introduction

The occurrence of toxic metals in the water environment has been known to cause severe health problems to animals and human beings (WHO 1971). The removal of heavy metals from river water, lake water and wastewater is a crucial issue of major health concern. Several methods have been proposed for the removal of heavy metals, e.g. ion exchange, filtration, coagulation and adsorption. Ion-exchange resin, membrane filter, hafunium hydroxide, activated carbon (Huang and Blankenship 1984), chelating resin and porous polymer employed are effective, but high cost materials (Patterson and Minear 1975; Panday et al. 1985).

Recently, the use of cheap agricultural wastes such as rice straw (Larsen and Schierup 1981), bark (Randall et al. 1974), Japanese green tea (Kimura and Yamashita et al. 1985), wool and coconut husks as adsorbents have been highlighted for metal removal from wastewater. Minamisawa et al. (1999) used chitin and chitosan for the adsorption of some metals such as Cu, Co, Au and Mn ions. Chitin is universally present in the exoskeletons of arthropods and manufactured in large scale from crab and shrimp shell wastes. Chitin and chitosan are nontoxic, readily biodegradable, and hence environmentally acceptable. Orhan and Buyukbungor (1993) and Macchi et al. (1986) have reported that turkish coffee, exhausted coffee, nut and walnut shells were useful for the heavy metals removal. However, these agricultural materials have to be chemically treated prior to use as adsorbent and hence the water treatment by use of these materials is allowed to be costly process.

In the previous paper (Minamisawa et al. 2002), we have demonstrated that Cu(II) and Cd(II) were almost removed from aqueous solution by use of roasted coffee beans with simple pretreatments of washing with water and drying. Thus, the roasted coffee beans are very useful for the removal of heavy metal as a new low cost material. The aim of present work is to elucidate the effect of the type of coffee and the roasting degrees of coffee beans on the adsorption behavior of heavy metals. The adsorption behavior of coffee beans is also compared with that of common adsorbents such as zeolite, activated carbon, chitosan, chitin and cerite.

25.2
Experimental

25.2.1
Materials

The four coffee beans, arabica species of Brazil, Columbia, Guatemala and Indonesia robusta coffees, were treated at five roasting temperature and time as follws; light roast (190~215 °C for 10 min), medium roast (190~215 °C for 15 min), city roast (200~230 °C for 15 min), full city roast (200~240 °C for 15 min), French roast (200~250 °C for 18~20 min). The preparation procedures of coffee beans have been described in a previous paper (Minamisawa et al. 2002). Coffee beans, chitine powder from KATOKICHI Ltd., Japan; KIMITSU chitosan Grade-F powder from Kimitu Chemical Industries, Japan; cerite from Tanabeshoko co., Japan; natural zeolite from Tochigi-pre., Japan; and activated carbon from Wako Pure Chemicals Co., Japan were employed as adsorbents. Cu(II) and Cd(II) solutions were prepared by the dilution of copper and cadmium standard solutions for atomic absorption spectrometry (Cu(II) or Cd(II) 1 g l^{-1} in HNO_3 0.1 mol l^{-1}) from Wako Pure Chemicals Co., Japan. All other reagents were of analytical and extra-pure reagent grade.

25.2.2
Adsorption Experiments

The adsorption experiments were carried out by a batch method. 2.5 g of coffee bean, chitin, chitosan, activated carbon, zeolite or cerite, was added to a 500 ml of sample solution containing Cu(II) or Cd(II) 5 mg l^{-1} as nitrate, which was adjusted to pH 6.5–6.7 with diluted ammonia water. The suspension was stirred for 180 min by use of a magnetic stirrer, and separated with a membrane filter. The adsorption amounts of Cu(II) and Cd(II) onto the adsorbents were determined by measuring the concentration of metals in the resulting filtrate on a SAS 7500 Seiko Instrument atomic absorption spectrometer.

25.3
Results and Discussion

25.3.1
Effect of Roasting Degrees of Coffee Beans

The effects of the roasting degrees of coffee beans on adsorption behavior of heavy metals were investigated. Arabica and robusta coffee beans were treated in 5 roasting degrees of light, medium, city, full city and French. The colors of light and me-

Fig. 25.1.
Adsorption ratio of Cu(II) and Cd(II) ions on the Guatemala coffee residue as a function of time

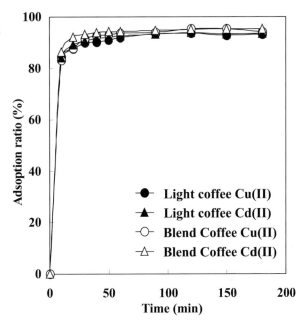

dium (American style) roasts are light brown and chestnut, respectively. While the city is the standard roast, called the New York style and the full city is deeper roast than city. French is the deepest grade roasted at the highest temperature, called the French style.

Figure 25.1 depicts time courses for the adsorption of Cu(II) and Cd(II) ions on the light roasted coffee and the blend roasted coffee (mixture of 5 degrees-roasted beans) of Guatemala coffee beans. The adsorption ratios of Cu(II) and Cd(II) ions increase rapidly and reached over 80% after 10 min, and then equilibrium is established in approximately 40 min. From Fig. 25.1, it is apparent that the metal adsorption on the coffee beans progressed very rapidly. A similar adsorption behavior of these metals was observed for coffee beans with other roasted degrees. The adsorptions ratios of Cu(II) and Cd(II) ions at 180 min were 92.6–95.5% for 5 roasting degrees and consequently the adsorption of heavy metal is hardly affected by roasting degrees of the coffee bean.

25.3.2
Adsorption Capacities of Coffee Beans

The adsorption capacities (Q), the amount (mg) of adsorbed metal per the weight (g) of roasted coffee beans, were determined from the Eq. 25.1 (Minamisawa et al. 2002).

$$Q = (C_0 - C) / W \qquad (25.1)$$

where C_0 and C are initial and final metal concentrations of solution (mg/500 ml), respectively, and W is the amount (2.5 g/500 ml) of coffee bean suspended. The adsorp-

Table 25.1. Adsorption capacities (mg g^{-1} at 180 min after adsorption experiments) of Cu(II) and Cd(II) ions for the various coffees

Coffee	Light	Medium	City	Full	French
Cu(II) ions adsorbed (mg g^{-1})					
Brazil	1.88	2	1.82	2	1.86
Columbia	1.82	1.9	1.9	1.88	1.82
Guatemala	1.92	1.7	1.78	1.86	1.86
Indonesia	1.82	1.82	1.82	1.96	1.88
Cd(II) ions adsorbed (mg g^{-1})					
Brazil	1.84	1.94	1.9	1.88	1.86
Columbia	1.82	1.82	1.84	1.82	1.94
Guatemala	1.94	1.9	1.88	1.92	1.98
Indonesia	1.88	1.88	1.9	1.9	1.96

tion capacities (mg g^{-1}) of Cu(II) and Cd(II) ions on several roasted coffee beans are summarized in Table 25.1. The adsorption capacities of coffee beans treated in five roasting degrees were 1.82–2.0 mg g^{-1} for Cu(II) and Cd(II). The values indicate that the heavy metals were concentrated about 400 times in the coffee bean from the aqueous solution containing 5 mg l^{-1}. If the adsorption experiments are carried out in aqueous metal solution of higher concentration than 5 mg l^{-1} employed in the present work, the capacity will become much greater than above values. The large capacity seems to give the adsorption ratio above 90%, moreover regardless of the kinds of beans and these roasting conditions. From these results, the adsorption of heavy metals was found to be hardly affected by the differences of not only the roasting degree but the a kind of coffee beans. Since the coffee bean residues are discarded as a blend form, these finding suggest that the waste coffee can be utilized as a new low cost adsorbent for the removal of toxic heavy metals.

25.3.3
Comparison of Various Absorbents

The adsorption of metals by a blend coffee was compared with the following common absorbents: chitin, chitosan, cerite, zeolite and activated carbon. Figure 25.2 and 25.3 depict the concentration of Cu(II) and Cd(II) in the suspension as a function of time during adsorption experiments, respectively. As shown in Fig. 25.2 and 25.3, the metals in the solution are removed above 80% within 10 min by zeolite, activated carbon and chitosan. The adsorption capability of blend coffee is comparable to that of these adsorbents. The adsorption ratios of zeolite, activated carbon, blend coffee and chitosan were greater than that of cerite and chitin, especially, for blend coffee, zeolite, and chitosan were above 94% up to 30 min. The equilibrium time was less than 40 min expect for cerite and chitin. From Fig. 25.3, the adsorption of Cd(II) ions onto blend coffee exceeded 10% than that of activated carbon.

Fig. 25.2.
Concentration change of Cu(II) ion in aqueous solution with time for adsorption on various adsorbents

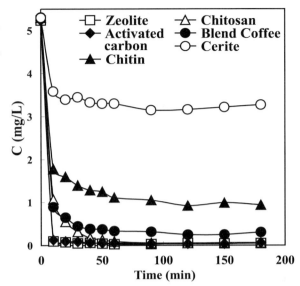

Fig. 25.3.
Concentration change of Cd(II) ion in aqueous solution with time for adsorption on the various adsorbents

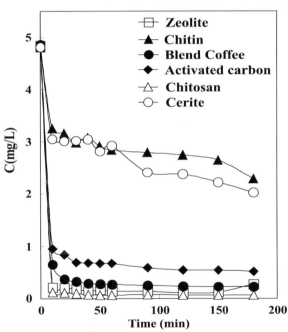

In the previous paper we have described that the adsorption activities of zeolite, activated carbon and chitosan are due to hydroxyl (OH), amino (NH_2) and carboxyl (COOH) groups (Minamisawa 1999). From the infrared absorption spectra (not shown) the coffee bean also was found to have COOH, hydrogen bond of OH, NH_2, and phenolic OH groups. These groups are derived from components of coffee bean such as cellulose, protein, and product including nitrogen. In view of this, the heavy metal ions is assumed to be incorporated into the core through active groups such as OH, COOH and NH_2 as well as zeolite, activated carbon and chitosan.

25.3.4
Adsorbent Capacity of Various Adsorbents

As shown in Table 25.2, the adsorption capacities of Cu(II) and Cd(II) ions onto blend coffee, zeolite, activated carbon and chitosan were from 1.83 to 2.08 mg g^{-1}. The blend coffee has a comparable loading capacity of Cd(II) to activated carbon and zeolite. The loading capacity of chitin and cerite was considerably less than that of other materials. However activated carbon and zeolite have high cost, though these materials are utilized in different fields such as water and wastewater engineering, chemical and metallurgical engineering, analytical chemistry (Ferro-Garcia et al. 1988).

On the other hand, the large amounts of coffee beans used are discarded as a mixed waste of the different type. Therefore, it is advantageous that the adsorption capability of the heavy metal in aqueous solution is independent of coffee type and its roast condition. In addition, it is interesting in view of low cost processing that the high adsorbed amount of Cu(II) and Cd(II) ions were obtained by the convenient pretreatment of washing with water and following drying. The high metal collection capability of coffee waste is promising to developing a novel adsorbent.

25.4
Conclusions

The coffee beans residue after extraction with hot water was found to have a high adsorption capability of heavy metals from aqueous solution. The cadmium(II) and copper(II) ions were removed very rapidly from aqueous solution containing these metals. Adsorption capacity of coffee beans is comparable to that of activated carbon and zeolite and the adsorption behavior was hardly affected by kinds of coffee beans and the roasting degrees. From these results, a great potentially of the coffee beans residue discarded as blend form was demonstrated as a convenient and low-cost adsorbent of heavy metals. The heavy metal removal technique using the coffee bean would be effective method for the economic treatment of wastewater.

Table 25.2. Adsorption capacities of Cu(II) and Cd(II) ions for the various adsorbents (mg g^{-1} at 180 min after adsorption experiments)

Adsorbent	Cu(II) ions	Cd(II) ions
Blend cofee	2.01	1.85
Zeolite	2.08	1.83
Activated carbone	2.07	1.74
Chitin	1.75	0.84
Chitosan	2.11	1.91
Cerite	0.81	1.03

References

Ferro-Garcia MA, Rivera-Utrilla J, Rodriguez-Gordillo J, Bautista-Toledo I (1988) Carbon 26(3):363-373
Huang CP, Blankenship DW (1984) Water Res 18(1)37-46
Kimura M, Yamashita H, et al. (1985) Bunseki Kagaku 35:400
Larsen VJ, Schierup H (1981) J Eviron Qual 10:188
Macchi G, Maroni D, Tiravarthi G (1986) Environ Technol Lett 7:431-444
Minamisawa H, Arai N, Okutani T (1999a) Anal Chim Acta 378:279-285
Minamisawa H, Arai N, Okutani T (1999b) Anal Chim Acta 398:289-296
Minamisawa M, Nakajima S, Mitue Y, Yoshida S, Takai N (2002) Nihon Kagaku Kaishi
Orhan Y, Buyukbungor H (1993) Water Sci Technol 28(2):247-225
Panday KK, Prased G, Singh VN (1985) Water Res 19:869-873
Patterson JW, Minear RA (1975) Heavy Metals in the Aquatic Environmental. In: Krenkel PA (ed) Pergamon Press, Oxford, pp 261-276
Randall JM, Reuter FW, Garrett V, Waiss Jr. AC (1974) Forest Proc J 24:80
WHO (1971) World Health Organization international standards for drinking water. WHO, Genova

Part III
Organic Pollutants

Part III

Chapter 26

Bioremediation for the Decolorization of Textile Dyes – A Review

A. Kandelbauer · G. M. Guebitz

Abstract

Textile dyeing effluents containing recalcitrant dyes are polluting waters due to their color and by the formation of toxic or carcinogenic intermediates such as aromatic amines from azo dyes. Since conventional treatment systems based on chemical or physical methods are quite expensive and consume high amounts of chemicals and energy, alternative biotechnologies for this purpose have recently been studied. A number of anaerobic and aerobic processes have been developed at laboratory scale to treat dyestuff. Some industrial pilot scale plants have even been set up. Additionally, biosorption shows very promising results for decolorizing textile effluents. In this contribution, we review fundamental and applied aspects of biological treatment of textile dyes.

Key words: biodegradation, bioremediation, textile effluents, dye decolorization, white-rot fungi, bacteria, mixed cultures

26.1 Introduction

In textile dyeing considerable amounts of dyestuff, e.g. up to 30% of reactive dyes, are lost and discharged with the effluents. Therefore, elimination of dyes from textile dyeing effluents currently represents a major ecological concern. Due to their high brilliance, low concentrations of dyes are highly visible and therefore, undesired in industrial effluents. Depending on the process used and on national regulations, the limits of dye concentration in rivers, about 1 ppm in the UK, would require a reduction of the dye concentration by up to 98% (Pierce 1994). The chemical structures of dye molecules are designed to resist fading on exposure to light or chemical attack and they prove to be quite resistant towards microbial degradation. For instance, azo dyes, which amount to around 60% of textile dyes display strongly adverse effects on growth of methanogenic bacterial cultures (Hu and Wu 2001). This toxicity may be due mainly to the azo functional group itself rather than to the products of reductive cleavage (Razo-Flores et al. 1997a) although aromatic amines are commonly known to be potential carcinogens (Benigni et al. 2000). The chromophoric group of azo dyes, the azo bond, can be cleaved anaerobically by bacteria in the human intestinal microflora (Rafii et al. 1990). A wide variety of environmental microorganisms and even helminths have been shown to degrade dyes (Chung and Stevens 1993). Nevertheless, dyes cause severe problems when released into municipal waste water treating plants since microbial cultures in conventional treating plants may be massively damaged by azo dyes and are not able to satisfyingly decolorize them. More than 100 liters of water are currently consumed in the textile finishing industry for the processing of 1 kg of textiles (Hillenbrand 1999). Thus, there is a strong demand for new recycling technologies to reduce this enormous water consumption. Microbial or enzymatic dye degradation would allow reuse of the water. Recently, it has been shown that enzymatically decolorized textile effluents can be used for the preparation of dyeing baths (Abadulla et al. 2000). Due to their high specificity, enzymes only attack the dye molecules while valuable dyeing additives or fibers are kept intact and can be reused. Both microorganisms and isolated enzymes have a high potential for the treatment of process effluents in the textile industry to allow their reuse, like it has also been shown in the pulp and paper industry (Zhang et al. 2000). Here, we will report the main aspects of biological treatment of textile dyes.

26.2
General

26.2.1
Modes of Bioremediation

The term bioremediation covers a wide variety of processes that use natural resources to control pollution problems caused by xenobiotics. Xenobiotics are characterized as compounds foreign to specific ecological systems which are often of anthropogenic origin and display high persistence in the environment. They may consist of aromatic ring systems substituted by electron-withdrawing groups like azo, nitro, or halogens (Knackmus 1996). To decrease toxicity levels induced by xenobiotics, several remediation techniques have been used: "microbial degradation" using microorganisms such as bacteria and fungi; "phytoremediation" by plants which involves several biological mechanisms; and "enzyme remediation" using specific enzymes to degrade pollutants. Different modes of bioremediation of colored effluents are summarized in Fig. 26.1.

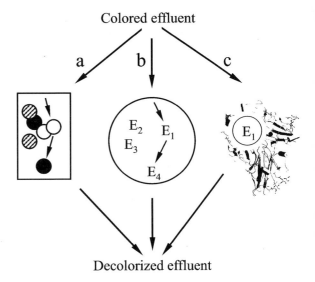

Fig. 26.1.
Different modes of dye bioremediation. Decolorization using *(a)* mixed cultures, *(b)* isolated organisms, and *(c)* isolated enzymes

For decolorization, reactors containing *(a)* mixed cultures, *(b)* isolated organisms, or *(c)* isolated enzymes can be used. With mixed cultures *(a)* one species may be involved in cleavage of the chromophoric group (*white circles*). Another species may further biotransform the modified dye (*black circles*) whereas others (represented by the *lined circle*) are not involved in bioremediation at all but may stabilize the overall ecosystem. Similarly, with isolated organisms *(b)* only few of the expressed enzymes (E_1 to E_4) are directly involved in dye biotransformation. In enzyme remediation *(c)* they may as such be used after separation from the biomass. Their action may depend on the presence of other substances like cofactors, cosubstrates or mediators.

The colored effluent may be pumped through a bioreactor containing cultures of one or more isolated microorganisms or containing mixed populations. In a mixed culture, where a consortium of different species is present, dye decolorization may be the result of the synergistic action of various microorganisms. One organism may be able to cause a biotransformation of the dye, which consequently renders it more accessible to another organism that otherwise is not able to attack this dye (Nigam et al. 1996a,b). In this way, the decolorization could mutually depend on the presence of several microorganisms and on their synergistic action. Alternatively a single microorganism may in fact be able to decolorize the solution by somehow modifying the chromophore but without concomitant complete degradation. Such decolorization yields metabolic endproducts that may be toxic, as shown for anaerobic reduction of azo dyes (Keck et al. 1997). If this unwanted metabolite can be accepted as a nutrient source by another organism, detoxification thereby could be achieved. Thus, the complete degradation of a xenobiotic leading to carbon dioxide, ammonia, and water may turn out only to be achievable within mixed populations. Such mineralization is the safest way to assure that no potentially harmful and unrecognized intermediate degradation products are released into the environment. In general, mixed populations usually exhibit higher stabilities towards environmental stress caused by changes in effluent characteristics like temperature, pH or composition.

Microbial cells can be fixed in reactors by different means of immobilization, depending on the reactor type. Fluidized-bed reactors contain freely mobile pellets covered with layers of immobilized biomass, while packed-bed reactors contain organisms that are fixed onto a suitable support material (Zhang et al. 1999). For dye degradation, bacterial cells may be immobilized on carriers like mineral material, seashells, or nylon (Nigam and Marchant 1995). Calcium alginate (Kudlich et al. 1996) and granular sludge (Tan et al. 1999; Shen et al. 1996) were used as support for mixed cultures. Bacterial cells have been immobilized on activated carbon to allow simultaneous adsorption of non-biodegradable matter and oxidation of biodegradable contaminants regenerating activated carbon in one single reactor (Walker and Weatherly 1999). Another process uses the flocculation of *Pseudomonas* sp. cells with aluminum sulfate for enhanced biodegradation of synthetic dyes (Tse and Yu 1997).

The living cell itself may be looked at as a decolorization reactor en miniature. Decolorization on the one hand may simply be the result of physical retention of the dye on the biomass by means of ion exchange or adsorption on the cell outer surface area. Since no metabolic transformation of the dye molecule is necessarily involved in this process, the contaminated biomass has to be disposed on landfills or treated in a second step (see Sect. 26.3.3).

> **Box 26.1.** Important oxidative enzymes used for dye decolorization
>
> $$\text{Dye} + \text{H}_2\text{O}_2 \xrightarrow{\text{Peroxidase}} \text{Oxidized dye} + \text{H}_2\text{O}$$
> $$\text{Dye} + \text{O}_2 \xrightarrow{\text{Laccase}} \text{Oxidized dye} + \text{H}_2\text{O}$$
> $$\text{Dye} + \text{O}_2 \xrightarrow{\text{Monooxygenase}} \text{Hydroxylated dye} + \text{H}_2\text{O}$$
> $$\text{Dye} + \text{O}_2 \xrightarrow{\text{Dioxygenase}} \text{Bishydroxylated dye}$$

Biochemical transformation of the dye on the other hand may either occur outside the cell if the enzymes are excreted into the medium or inside the cell, provided that the dye is readily transported into the cell, demonstrating the impact of its bioavailability. Again, one single enzyme or a whole group of enzymes may be responsible for decolorization, and low-molecular weight compounds like cofactors, co-substrates or mediators may indispensably be involved as well.

A number of microorganisms have been found to be able to decolorize textile dyes including bacteria, fungi, and yeasts (Banat et al. 1996; Martins et al. 1999). In general, any organism that produces the enzymes listed in Box 26.1 is a likely candidate for dye degradation. Preferentially, suitable organisms should excrete the active enzymes into the medium. Otherwise transport into the cells may be limiting for bioelimination. Another important requirement for an organism is its resistance against toxic effects of dyes and other substances present in the effluent. Therefore, in cases where the target molecule or additives inhibit growth, isolated enzyme systems may be preferred. This may happen especially at high concentrations of dyestuff as many studies have reported decreasing of decolorizing rates of microorganism with increasing dye concentrations above certain levels. In short, it has to be dealt with the following major problems: dyehouse effluents are complex mixtures containing high loads of additives (salts, detergents, dispergents, metals) and may strongly vary depending on the production charge. Dyehouse effluents often display extreme pHs and high temperatures. Dyes display a wide structural variety and thus do possess very different chemical and physical properties. They are designed to resist very harsh conditions. Thus, biological systems have to be designed that work under such conditions and still effectively not only decolorize but preferably completely degrade dyestuff. Due to these requirements there is currently no simple solution by just applying a bioremediation technique. Extensive reviews of general methods for decolorization of textile effluents have been reported (Hao et al. 2000; Slokar and Marechal 1998; Robinson et al. 2001b). Although combinations of chemical and microbial treatment have been described (Ledakowicz et al. 2001; Kunz et al. 2001; Pulgarin et al. 1999; Donlagic and Levec 1998; Van der Bruggen et al. 2001) in the following we will focus on bioremediation.

26.2.2
Enzymes Involved in Bioremediation

Textile dyes by themselves represent a huge diversity of chemical compounds. There are several possibilities to classify dyes. By chemical structure, dyes are characterized

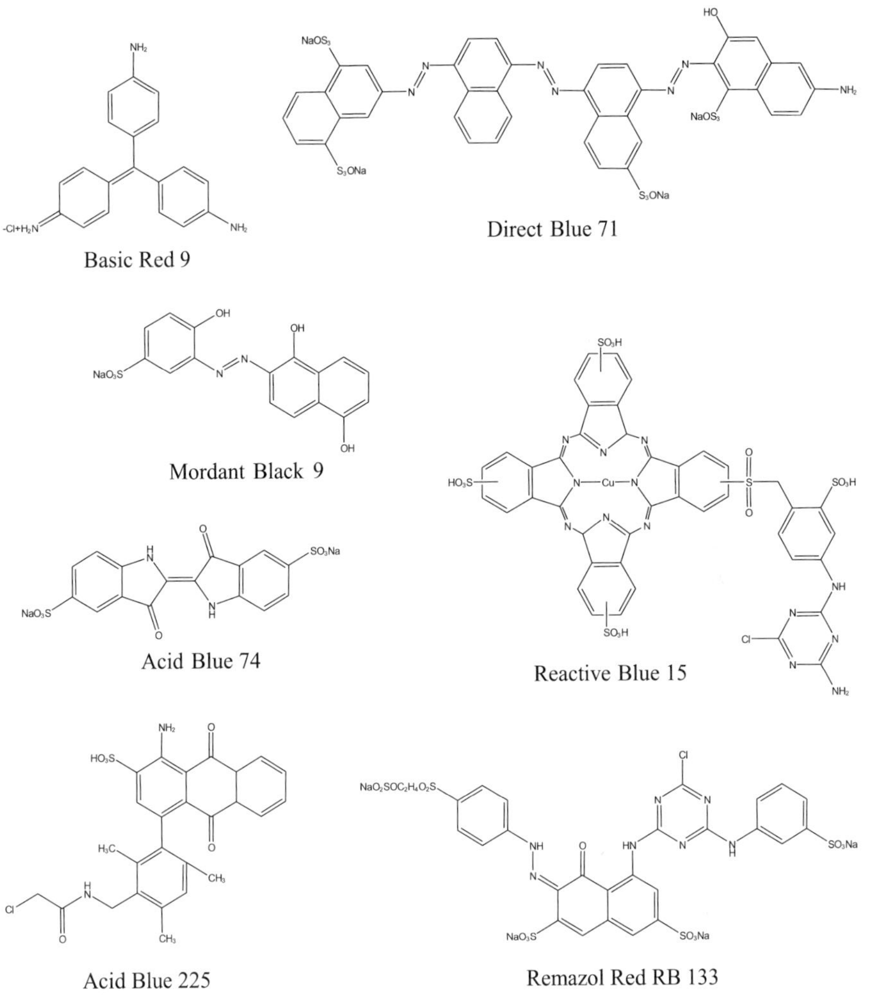

Fig. 26.2. Examples of common commercial dyes

by their chromophore. Several examples for different types of dye molecules are depicted in Fig. 26.2. Typical dyes are complex aromatic or heteroaromatic compounds that either contain azo bonds, or feature indigoid, triaryl methane, anthrachinoid, or phtalocyanoid carbon skeletons. They are substituted with various electron withdrawing or electron donating groups like hydroxy, amino, nitro, halogens or sulfonate. However, dyes are also classified according to their dyeing properties, e.g. disperse, reactive, and direct for applicational purposes. The dyeing properties in turn reflect solubility and chemical reactivity towards the fabric to be dyed.

Although dye molecules display a high structural variety, they are degraded by only few different enzymes. These biocatalysts have one common mechanistic feature. They are all redox-active molecules and thus, exhibit relatively wide substrate specificities.

There are reviews summarizing mechanistic and applied aspects of oxidative enzymes in the degradation of xenobiotics (Duran and Esposito 2000; Mester and Tien 2000). The most important dye degrading types of enzymes are listed in Box 26.1.

Extracellular oxidative enzymes such as laccases and peroxidases are typically produced by fungi, one of their natural functions being the degradation of lignin, a complex aromatic matrix in wood. For mechanistic discussions of their oxidative action on azo dyes, see Chivukula and Renganathan (1995). Intracellular mono- and dioxygenases are ubiquitously present in living organisms. They cause the breakdown of aromatic rings via incorporation of oxygen atoms (biohydroxylation) and subsequent cleavage of the ring system resulting in carboxylic acids, which are further used in metabolism (Smith 1990; Berry et al. 1987; Commandeur and Parsons 1990). Reductive enzymes like cofactor-dependent oxidoreductases or cytochrome P450 reductases may unspecifically transform dyes in the course of secondary metabolic pathways (Kudlich et al. 1997).

Laccases have the advantage of just requiring molecular oxygen as a co-substrate. Similarly, the use of peroxidases only depends on the availability of hydrogen peroxide as second substrate. Those enzymes may thus be promising candidates for enzyme remediation. In contrast, the application of reductases or oxidases requiring cofactors like NAD(H), NADP(H), or FAD(H) which are extremely expensive compounds is economically not feasible. Most decolorizations in connection with such enzymes usually take place in whole cell applications.

26.2.3
Enzymatic Cleavage of Dyes

In the following sections, we discuss some mechanistic aspects of microbial treatment of dye classes, for which degradation pathways have been suggested.

Azo dyes. Most mechanistic studies have focussed on azo dyes since they comprise the largest class of textile dyes. Azo dyes usually do not occur in nature. Thus, it is not surprising that cleavage of azo bonds under anaerobic conditions in general is referred to be due to unspecific reduction processes mediated by redox-active compounds like quinone-type substances (Keck et al. 1997; Kudlich et al. 1997), biochemical cofactors like NADH (Nam and Renganathan 2000) or reduced inorganic compounds like Fe^{2+} (Nerud et al. 2001) or H_2S (Yoo 2002) which are formed by certain strictly anaerobic bacteria as metabolic end products. However, some enzymes with pronounced specificities have been discovered (Zimmermann et al. 1984).

According to Chivukula and Renganathan (1995), the mechanism of laccase mediated azo-dye decomposition proceeds first via two sequential electron abstractions. This is followed by an attack of the nucleophile water on the resulting resonance stabilized cation. Subsequently, breakdown of the dye molecule concomitantly to the release of one proton and one molecule N_2 takes place yielding chinoid aromatics and transient hydroperoxides, respectively (Chivukula and Renganathan 1995). The main function of the laccase-catalyst thus consists of oxidatively rendering the azo dye more susceptible to hydrolysis and nitrogen is eliminated in molecular form.

The same group also described the mechanism of peroxidase action on azo dyes (Spadaro and Renganathan 1995; see also Goszcynski et al. 1994). Examples for sug-

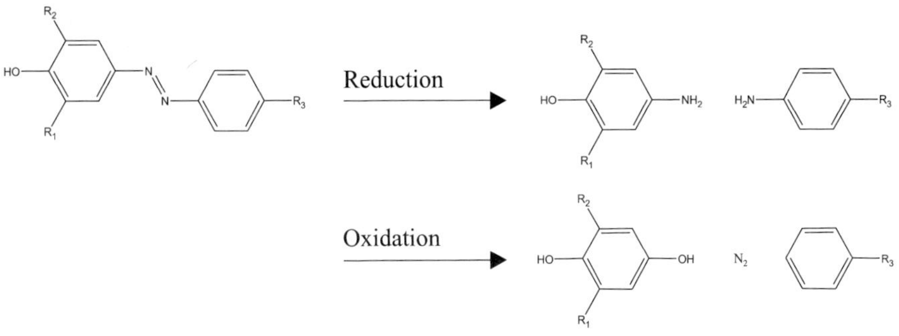

Fig. 26.3. Suggested decolorization products of azo bond cleavage via **a** reduction (Keck et al. 1997) and **b** oxidation (Chivukula and Renganathan 1995; Spadaro and Renganathan 1995)

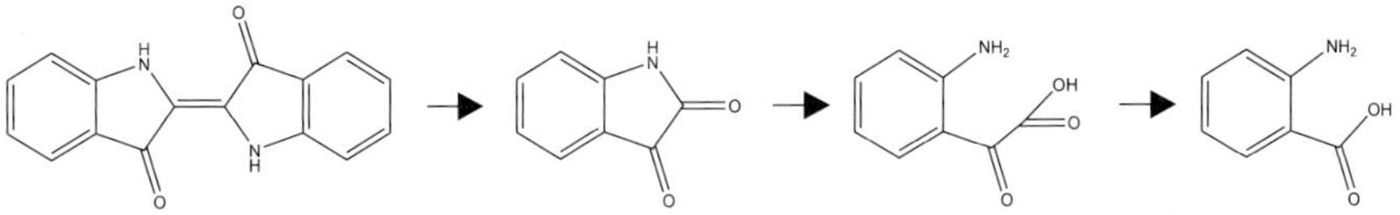

Fig. 26.4. Suggested oxidative degradation pathway for indigoid dyes (Campos et al. 2001)

gested azo-dye cleavage are shown in Fig. 26.3. For a comprehensive discussion of the microbial removal of azo dyes see the outstanding review of Stolz (Stolz 2001) which includes a thorough discussion of mechanisms as well as application aspects.

Indigoid dyes. Indigo (1), the most important dye in the manufacturing of blue jeans was demonstrated to be cleaved under laccase catalyzed electron transfer to give isatin (2) and upon further decarboxylation anthranilic acid (3) as the final stable oxidation product (Fig. 26.4). It was suggested that the degradation might proceed via dehydroindigo as a reaction intermediate. Again, the function of laccase may consist of increasing the susceptibility of the dye towards hydrolytic attack by water (Campos et al. 2001). Similarly, in the peroxidase catalyzed decolorization of Indigo Carmine, isatin sulfonic acid is formed, although a stable red oxidation product was observed when a manganese dependent peroxidase from *Phanerochaete chrysosporium* was employed (Podgornik et al. 2001). The authors suggested that this red product was a dimeric condensation product of Indigo Carmine which has not been formed with lignin peroxidase as the catalyst.

Triphenyl methane dyes. The application of organisms able to degrade triphenyl methane derivatives has recently been reviewed by Azmi and co-workers (Azmi et al. 1998). This group of dyes has proved to be especially recalcitrant towards biodegradation. Typical representatives like Methyl Violet or Crystal Violet were shown to have adverse effects on the respiration of activated sludge bacteria and strongly inhibited cell growth (Ogawa et al. 1988). Gentian Violet, another triphenylmethane dye and very effective in controlling fungal growth is extensively in use as a medical substance (Willian et al. 1978).

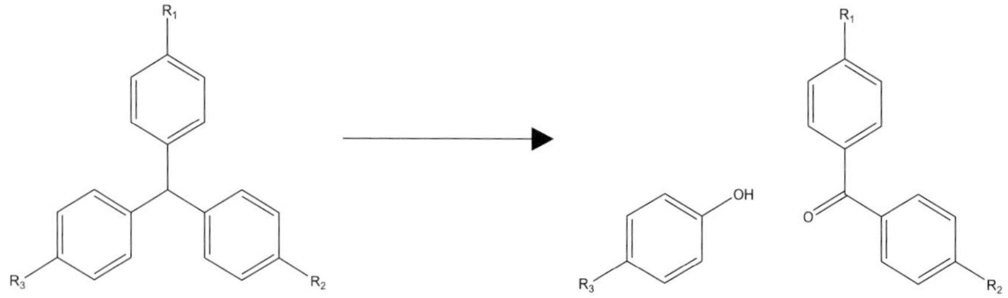

Fig. 26.5. Decolorization products suggested for triarylmethane dyes (Yatome et al. 1993)

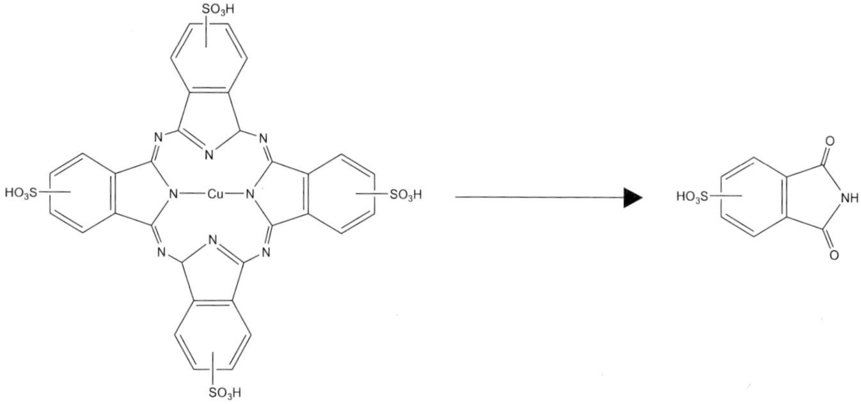

Fig. 26.6. Decolorization of phtalocyanide type dyes (Heinfling-Weidtmann et al. 2001)

Using thin-layer chromatographic methods, the decolorization of the triphenylmethane dye Crystal Violet was shown to yield Michler's ketone as a metabolic dead end product (Yatome et al. 1993) (Fig. 26.5). Different enzymes exhibit different substrate specificities. While a laccase treatment of Malachite Green, Crystal Violet, and Bromophenol Blue produced respectively 100%, 20%, and 98% of decolorization (Pointing and Vrijmoed 2000), an analogous experiment using a peroxidase yielded 46%, 74%, and 98%, respectively (Shin and Kim 1998).

Phtalocyanine dyes. The degradation of phtalocyanine dyes was recently described (Fig. 26.6) (Heinfling-Weidtmann et al. 2001; Reemtsma and Jakobs 2001). The action of the white-rot fungus *Bjerkandera adusta* on Reactive Blue 15 and Reactive Blue 38 was studied using HPLC and the reaction products were analyzed via electrospray mass spectrometry. Sulphophtalimides were identified as the main products.

Anthrachinoid dyes. Degradation of anthraquinoid dyes is assumed to proceed mainly via general aromatic metabolism pathways utilizing a variety of mono- and dioxygenases, respectively. The anthraquinonic dye Acid Green 27 has recently been demonstrated to serve as a laccase substrate (Wong and Yu 1999).

In general, the rate of dye degradation strongly correlates with the dye's electrochemical half-wave redox potential (Bragger et al. 1997) although structural features may play an important role as well (Pasti-Grigsby et al. 1992; Kulla et al. 1983; Xu 1996). Electron donating methyl and methoxy substituents seemed to enhance laccase activity while electron withdrawing chloro, fluoro and nitro substituents inhibited oxidation of azophenols and other substituted phenols and phenol analogs by fungal laccases (Chivukula and Renganathan 1995; Xu 1996).

26.2.4
Sources of Microorganisms

Microorganisms for dye decolorization may be obtained simply by isolation of existing dye degrading cultures from environmental samples (e.g. textile effluents), by adaptation of promising strains to conditions present in textile effluents or by construction of suitable organisms employing genetic methods.

26.2.4.1
Isolation and Adaptation of Naturally Occurring Microorganisms

In general, enrichment of microorganisms with special effectiveness in dye digestion via natural adaptation occurs at any site where these xenobiotics are present in amounts above average. Such sites may for example be natural ecosystems by chance long-term exposed to textile effluents or sewage treatment plants near textile mills. Usually, one does not try to isolate such naturally evolved strains but simply benefits from there presence, especially if they occur directly in a municipal sewage plant. Isolation of dye-degrading bacterial strains usually is a tedious and time-consuming task (Zimmermann et al. 1984; Nigam et al. 1996b). The enrichment of bacteria under chemostat conditions capable of growing on dye molecules as only carbon source – if successful at all – has been reported to take very long periods, from several months up to more than a year (Zimmermann et al. 1984). However, since dye degradation is mainly accomplished via secondary metabolic routes, this nutritional restriction is not principally needed. In the following paragraph we will briefly mention some important cultures that have been obtained by adaptation methods.

Thermophilic bacteria, selected by adaptation from a textile effluent have been shown to decolorize textile dyes at temperatures up to 60 °C (Banat et al. 1997). Although not completely identified yet, one isolate of this mixed culture resembled members of *Corynebacterium* and was able to decolorize commercial azo, diazo, reactive and disperse dyes. However, synergistic decolorization by various species has been observed (Nigam et al. 1996a,b).

Combined anaerobic and aerobic microbial treatments have been suggested to complete the degradation of azo dyes (O'Neill et al. 2000; Bortone 1995; Haug et al. 1991). The formed aromatic amines are generally not further degraded without oxygen. However, a methanogenic consortium was recently found to detoxify aromatic amines formed during the prior azo reduction step, thus completely mineralizing azo dyes under strict anaerobic conditions. This mixed population was grown on the amines as the sole N-sources (Razo-Flores et al. 1997b) and the azo-dye azodisalicylate was continuously degraded in a bioreactor for more than 100 d.

Complete mineralisation of an azo compound by an isolated aerobic bacterial strain has also been successful. Via continuous adaptation of *Hydrogenophaga palleronii* S1, a strain was developed growing on 4-carboxy-4'-sulfoazobenzene as the sole carbon and energy source. The detection of sulfanilate by high performance liquid chromatography (HPLC) revealed that indeed the reduction of the azo bond had occurred. This organism furthermore metabolized the resulting amines and complete mineralisation was achieved (Blümel et al. 1998). 4,4'-Dicarboxyazobenzene was cleaved as well in presence of oxygen whereas various other hydroxy substituted azo compounds were not accepted as substrates.

A major advantage of biological methods in contrast to physico-chemical processes lies in the continuous self-optimization of the decolorization system by evolutionary processes. Different isoenzymes of a laccase were shown to be expressed differently by the fungus *Pycnoporus cinnabarinus* in dependence of the reactor cycle state. This provides evidence for continuous adaptation of the fungus in response towards the environmental stress and evolutionary pressure caused by exposure to dyehouse effluent conditions (Schliephake and Lonergan 1996).

In this context it should be mentioned that the expression of enzymes involved in dye degradation can be significantly enhanced by medium components in the course of cell growth. It has been shown that heavy metal ions like Cu^{2+} and Cd^{2+} enhance laccase activity of a *Pleurotus ostreatus* (Baldrian and Gabriel 2002) and Cu^{2+} of various *Trametes* species (Galhaup and Haltrich 2001). Xenobiotics like xylidine, aniline or 9-fluorenone (Mougin et al. 2001) or various aromatic alcohols (Arora and Gill 2001) as well have been demonstrated to increase laccase expression of some fungi. An interesting discussion of the metabolic implications of environmental stress taking place in fungi is presented by Crowe and Olsson (2001).

Induction effects are of special interest for enzyme remediation technologies. The enzyme is produced off-site in high yields by addition of inducers to the culture medium. With on-site growing cell technologies such optimized environments hardly can be realized thus resulting in less-effective enzyme expression by the used organism.

26.2.4.2
Genetic Engineering of Dye Degrading Organisms

In the course of natural adaptation, organisms degrading xenobiotics evolve more or less naturally (Gottschalk and Knackmus 1993) or under controlled laboratory conditions (Zimmermann et al. 1984). Thus, well-directed optimized hybrid strains may as well be obtained directly via genetic engineering. By cloning and transferring genes encoding for dye degrading enzymes organisms could be designed that combine the abilities of mixed cultures within one single species (Knackmus 1996). A number of genes conferring the ability of dye decolorizing have been identified (Dabbs 1998; Heiss et al. 1992). Chang and co-workers have reported successful decolorization of an azo dye using *Escherichia coli* carrying the azoreductase gene from a wild-type *Pseudomonas luteola* (Chang et al. 2000; Chang and Kuo 2000). This approach could become a useful alternative for shortening down the extended time-periods otherwise needed to adapt appropriate cultures and isolated strains, respectively. Furthermore, introduction of heavy-metal resistance into dye-degrading organisms may help to solve the problem of environmental toxicity of such ions present in high amounts in textile bioremediation sites. Although in conventional sewage plants the use of a genetically highly modified organism seems to be unrealistic, plasmids with a broad host range of replication and metal-resistance expression could easily be introduced into the bacterial community. The presence of heavy metals should cause the plasmids to be stably maintained in the bacterial population (Nies 1999).

26.3
Bioremediation

26.3.1
Dye Degradation with Bacteria

A wide variety of bacteria including *Proteus* sp., *Enterococcus* sp., *Streptococcus faecalis*, *Bacillus subtilis*, *Bacillus cereus*, *Pseudomonas* spp. (Bumpus 1995), and even helminths (Chung and Stevens 1993) have been shown to reduce azo bonds of textile dyes. Azoreduction may be stimulated by addition of certain co-substrates, which are metabolized and which serve as reduction equivalents. Depending on the type of organism employed, different additional nutrients may thus be supplied as co-substrates. Whereas the addition of glucose has been shown to significantly improve the decolorization of Mordant Yellow 3 by an anaerobic bacterial consortium (Haug et al. 1991), glucose inhibited decolorization of Reactive Red 22 by a *Pseudomonas luteola* strain (Chang and Lin 2000). In the latter case, the amount of yeast extract added could be correlated quantitatively to the achieved extent of decolorization. Mixtures of low molecular weight carboxylic acids like acetate, propionate, and butyrate have also been employed to improve the decolorization of azo-disalicylate (Razo-Flores et al. 1997a) while others have utilized more complex co-substrates like tapioka starch as a supplemental carbon source (Chinvetkitvanich et al. 2000).

Aromatic amines in general are not further metabolized under strictly anoxic conditions. Only one report has been published so far of complete mineralization of a certain azo dye under anaerobic conditions (Razo-Flores et al. 1997b). Aromatic amines were shown to be formed in the human intestines (Rafii et al. 1990; Bragger et al. 1997) and are regarded as potent carcinogens (Benigni et al. 2000). They thus provide substantial risks for human health (Zimmermann et al. 1984). To be eliminated yet another aerobic step in biologic treatment has to subsequently take place. This is why currently the most promising technologies for azo-dye degradation are based on a combination of anaerobic and aerobic processes using mixed cultures in various designs of reactors (Kapdan and Kargi 2002; Coughlin et al. 2002; Kapdan et al. 2000a–c; O'Neill et al. 2000; Rajaguru et al. 2000; Tan et al. 1999).

Aerobic sequencing batch reactors using either mixed cultures from activated sludge units (Panswad et al. 2001) or enrichment cultures of glycogen- and polyphosphate accumulating organisms (Lourenco et al. 2001) have recently been described. During exposition of methanogenic granular sludge to oxygen, addition of ethanol has been shown to stimulate the respiration of facultative aerobic microorganisms present in the outer spheres of the colonized material, thereby preventing penetration of oxygen to strictly anaerobes. Consequently, azo dyes are reduced by methanogenic colonies in anaerobic microniches within the inner spheres of the material. Thus, aerated anaerobic/aerobic reactors with the ability of full azo-dye mineralization can be constructed (Tan et al. 1999; Field et al. 1995). For examples of recently described bacterial systems for treatment of textile dyestuff, see also Stolz (2001) and McMullan et al. (2001). Attempts have been made as well to implement anaerobic/aerobic treatment plants in textile industry (Sarsour et al. 2001; Krull et al. 2000).

Not only aromatic amines but dyes as well can be oxidized by bacteria (Greaves et al. 2001). Oxidative attack mediated by peroxidases has been observed with soil bacteria like a lignin-degrading *Streptomyces* sp. (Pasti-Grigsby et al. 1992). An azo-dye-degrading extracellular peroxidase is also released by a *Flavobacterium* sp. (Cao et al. 1993). Several bacteria such as *B. subtilis*, *Pseudomonas pseudomallei* and different *Corynebacterium*, *Mycobacterium*, and *Rhodococcus* species have been found to degrade triphenylmethane dyes as well. Thus, in principle dye degradation is possible by using solely aerobic cultures.

26.3.2
Dye Degradation with Fungi

White-rot fungi are able to degrade complex substrates like lignin via oxidative radical pathways. They can also degrade textile dyes due to the unspecific nature of their lignin degrading enzymatic system. The enzymes responsible for this action are peroxidases and laccases. The fact that these enzymes are excreted by the fungi makes these organisms especially interesting for bioremediation. A huge number of scientific papers showing decolorization by fungi have been recently published (Fu and Viraraghavan 2001). During typical experiments dyestuff is added to either a more or less purified enzyme solution, culture filtrate or fermentation broth still containing living organism. In other studies, it was tried to optimize the cultivation of some fungi with respect to dye decolorization (Robinson et al. 2001a; Tekere et al. 2001; Bakshi et al. 1999). This kind of data gives important information about the potential of microorganisms both in terms of potential in treating dye contaminated (model) waste water and in resistance towards dye toxicity under more or less native conditions.

Several white rot fungi are known to degrade the various types of dyes: *Phanerochaete chrysosporium* (Martins et al. 2001; Tatarko and Bumpus 1998), *Irpex lacteus* (Novotny et al. 2001), *Coriolus versicolor* (Swamy and Ramsey 1999a,b), *Pleurotus ostreatus* (Shin and Kim 1998), *Pycnoporus sanguineus* (Pointing and Vrijmoed 2000), *Pycnoporus cinnabarinus* (Schliephake et al. 2000), *Phlebia tremellosa* (Kirby et al. 2000), *Geotrichum candidum* (Kim et al. 1995), *Trametes hirsuta* (Abadulla et al. 2000), or *Neurospora crassa* (Corso et al. 1981). Recently, a new strain from the genus *Penicillium* has been shown to degrade various polymeric dyes (Zheng et al. 1999). Fungal systems (*Trametes versicolor*, *Pleurotus ostreatus*, *Phanerochaete chrysosporium*, *Piptoporus betulinus*, *Laetiporus sulphureus* and several *Cyathus* species) have been described in literature to degrade triphenylmethane dyes (Azmi et al. 1998). In general, *Phanerochaete chrysosporium* seems to be the most extensively investigated fungus for dye decolorization working on dyes of all classes. For an informative compendium of recent literature describing fungi able to decolorize dyes, see Fu and Viraraghavan (2001).

26.3.3
Biosorption

A very prominent method for removing color from effluents is physical adsorption of colored substances on various materials like sawdust (Khattri and Singh 1999), charcoals, activated carbon, clays, soils, diatomaceous earth, activated sludge, compost, living plant communities, synthetic polymers, or inorganic salt coagulants (Slokar and Marechal 1998). The corresponding process by using biomass commonly is referred to as biosorption.

Color removal via biosorption is usually achieved by adsorption on fungal mycelia, either with or without concomitant biodegradation, as in the case of *Aspergillus foetidus* (Sumathi and Manju 2000). Fungal cells may either be used as growing cells or in form of dead biomass (Fu and Viraraghavan 2001), although decolorization with active biomass is greater most probably due to parallel digestion (Aretxaga et al. 2001). Azo dyes have been shown to bind effectively onto the mycelium of *Aspergillus niger* resulting in extensive color removal higher than 95% (Sumathi and Manju 2000). A stationary culture of this fungus was also used to decolorize a complex wastewater from a textile company by an airlift bioreactor over a relatively wide pH range. Between pH 3 and 7 there was 100% decolorization, with pH 12 still about 60%. The process does not seem to be limited to a certain type of dye. Acid, basic, direct, reactive, and disperse dyes are reported to be cleared out of solution within a couple of hours (Assadi and Jahangiri 2001). Pellets consisting of activated carbon and mycelium of *Trametes versicolor* were used for textile dye decolorization (Zhang and Yu 2000). Combining biodegradation with adsorption, high decolorization rates could be achieved like it was also reported for a system using bacteria and carbon black as a carrier material (Walker and Weatherly 1999).

Recently, biosorption on agricultural residues was suggested by Nigam and coworkers as a first step prior to microbial treatment to concentrate dyes (Robinson et al. 2001b). Wheat straw and corncobs were shown to remove up to 70% of color from a simulated effluent containing 500 mg l^{-1} of dyestuff (Nigam et al. 2000). Since dye contents of typical effluents may lie well below this value (Kalliala and Talvenmaa 2000; Pierce 1994; Correia et al. 1994), this first step may prove to be successful in decolorizing the effluent. Subsequently, for complete mineralization of dyes solid state fermentation can be performed on the dried adsorbent using white rot fungi. Although spectrophotometric quantification was not feasible, color removal was visually evident and neither adsorbed dye molecules nor their breakdown products had any significant adverse effects on the growth of both, *Phanerochaete chrysosporium* and *Trametes versicolor* (Nigam et al. 2000).

Following a similar strategy, cellulosic anion exchange resins based on quaternized lignocellulose were used for adsorption of acid azo dyes from wastewaters, the underlying mechanism being ion exchange. Since in this case the adsorbent is not an agricultural waste but an expensive ion exchange column, the adsorbent has to be regenerated. This can be done by chemical means using bisulfite-mediated borohydride reduction of azo groups (Laszlo 1997) or biologically via anaerobic bacterial treatment of the adsorbed azo-dye molecules (Laszlo 2000). For this it is not even necessary to bring the column in direct contact to the reducing bacteria. Interestingly, by physically separating the bacterium *Burkholderia cepacia* from the dye using a dialysis tube, anthrachinone-2-sulfonate was shown to mediate the transfer of reducing equivalents from bacteria to adsorbent-bound dye. Consequently, regeneration of an anion exchanger bed was achieved with medium from a separate anaerobic reactor via a low molecular weight electron shuttle (Laszlo 2000).

26.3.4
Enzyme Remediation

Enzyme reactors display one major advantage over whole cell systems because of the distinct separation of different technological problems which otherwise would be intertwined with each other. The production of an enzyme is provided by specialized fermentation technologies. Enzyme expression of a suitable organism can be optimized separately by exploring induction events or using genetic engineering. Downstreaming of the enzyme and furthermore, the preparation of the biocatalyst (immobilization protocols, stabilized enzyme cocktails) are separated from the actual site of problem. Namely, the ready-made biocatalyst is produced separately and delivered for implementation. By using whole cell systems, all these different fields are to be dealt with at the same time, which drastically increases the complexity of the problem.

Enzyme reactors could prove useful for some special applications, e.g. for treatment of partial streams within the plant, flows of relatively constant/known composition. Little has been done on this field yet although for some related applications useful enzyme systems have already been designed and successfully tested in industry, e.g. immobilized catalases for hydrogen peroxide removal at high pHs and temperatures (Paar et al. 2001).

In terms of enzyme remediation of textile dyes, laccases seem to be the most promising enzymes. Laccases have been shown to decolorize a wide range of industrial dyes (Rodriguez et al. 1999; Reyes et al. 1999). In the presence of redox mediators this range even could be extended (Reyes et al. 1999; Soares et al. 2001a,b) or decolorization events of degradable dyes could be significantly enhanced (Abadulla et al. 2000). Low molecular weight compounds like 2,2'-azino-bis-(3-ethylbenzothiazoline-6-sulfonic acid) (ABTS) may also be necessary to mediate the actual electron transfer steps of laccases (Wong and Yu 1999).

Similarly, peroxidases addition of veratryl alcohol was shown to positively in-fluence the decolorization of azo and anthraquinone dyes catalyzed by lignin peroxidase. However, this effect may either be attributed to the protection of the enzyme of being inactivated by hydrogen peroxide or to the completion of the oxidation-reduction cycle of the lignin peroxidase rather than to just redox-mediation (Young and Yu 1997).

For technical applications, enzymes have to be immobilized. Immobilization of fungal laccases on various carrier materials such as activated carbon (Davis and Burns 1992), agarose (Reyes et al. 1999), Eupergit C (D'Annibale et al. 2000), sepharose (Milstein et al. 1993), and porosity glass (Rogalski et al. 1995; Rogalski et al. 1999) has been shown to increase stabilities of the enzyme at high pH and tolerance to elevated temperatures and to make the enzyme less vulnerable to inhibitors such as Cu-chelators. After treatment with immobilized enzymes, the decolorized dyeing effluents could be recycled within the dyeing process. This is not possible with effluents treated with microorganism since they require additional components to support growth. Both these substances added to the effluent and compounds secreted by microorganisms can cause problems in recycling of effluents (Abadulla et al. 2000).

Immobilized laccases have been shown to efficiently decolorize and even detoxify textile dyes (Abadulla et al. 2000). Reactors containing such preparations could be run in ten repeated decolorizations for about 15 h with high residual activity, retaining 85% of its initial activity. However, experiments with an authentic textile effluent but otherwise same conditions caused loss in laccase activity resulting in 14% retained activity. The authors investigated all known components of the effluent like salts, soap, and dispersant and their mixtures for laccase inactivation but none of these was detrimental to the enzyme (Reyes et al. 1999).

Laccases have been found as well in bacteria (Diamantidis et al. 2000) and were shown to be involved in the pigment formation with some bacterial spores (Solano et al. 2001). Since they are assumed to be widespread also among this class of microorganisms (Alexandre and Zhulin 2000), screening for oxidative enzymes expressed by thermoalkalophilic organisms seems to be of special interest with respect to their potential applicability at elevated temperatures. Since organisms of this kind are generally more difficult to cultivate, such enzymes could be cloned, genetically transferred and expressed by organisms for which production methods already have been well established (Kruus et al. 2001).

26.4
Conclusions

A number of strategies have been developed for the enzymatic and microbial treatment of textile dyeing effluents. The technology of choice definitely depends on the composition of the effluents and potential reuse processes. Enzymatic processes are very promising for the decolorization of dyeing effluents for reuse in dyeing. Provided that immobilized enzymes are used to avoid addition of proteins (enzymes), acceptable color differences between fabrics dyed in dyeing baths prepared with water and with enzymatically decolorized dyeing effluents were achieved. Furthermore, this process resulted in savings of dyeing additives, which are not attacked by enzymes. Although the cost for enzyme is decreasing due to new production technologies including genetic methods, the major drawback of enzymatic processes is the limited range of dyes that are decolorized by one single enzyme. These problems could be circumvented in the future by using enzyme cocktails or adjusting the dyeing protocols to those dyes, which are susceptible to enzymatic decolorization.

Microbial processes allow complete mineralization and detoxification of textile dyes and other pollutants contained in textile effluents. Although both isolated bacteria and fungi have been shown to decolorize surprisingly wide ranges of different dyes, the potential of mixed populations should not be neglected especially due to their higher stability against changes of the environment. Aerobic or anaerobic microbial treatment and combined systems are mainly used when dyeing effluents are treated together with other textile effluents of varying composition and reduction of e.g. chemical oxygen demand (COD) and toxicity is equally important as decolorization. Bioadsorption has as well been used to concentrate dyes from dilute effluents prior to biodegradation.

References

Abadulla E, Tzanov T, Costa S, Robra KH, Cavaco-Paulo A, Gübitz GM (2000) Decolorization and detoxification of textile dyes with a laccase from *Trametes hirsuta*. Appl Environ Microbiol 66:3357–3362
Alexandre G, Zhulin IB (2000) Laccases are widespread in bacteria. Tibtech 18:41–42
Aretxaga A, Romero S, Sarr M, Vicent T (2001) Adsorption step in the biological degradation of a textile dye. Biotechnol Prog 17:664–668
Arora DS, Gill PK (2001) Effects of various media and supplements on laccase production by some white-rot fungi. Biores Technol 77:89–91
Assadi MM, Jahangiri MR (2001) Textile wastewater treatment by *Aspergillus niger*. Desalination 141:1–6
Azmi W, Sani RK, Banerjee UC (1998) Biodegradation of triphenylmethane dyes. Enzyme Microb Technol 22:185–191
Bakshi DK, Gupta KG, Sharma P (1999) Enhanced biodecolorization of synthetic textile dye effluent by *Phanerochaete chrysosporium* under improved culture conditions. World J Microbiol Biotechnol 15:507–509
Baldrian P, Gabriel J (2002) Copper and cadmium increase laccase activity in *Pleurotus ostreatus*. FEMS Microbiol Lett 206:69–74
Banat IM, Nigam P, Singh D, Marchant R (1996) Microbial decolorization of textile dye containing effluents a review. Biores Technol 58:217–227
Banat IM, Nigam P, McMullan G, Marchant R (1997) The isolation of thermophilic bacterial cultures capable of textile dyes decolorization. Environ Int 23:547–551
Benigni R, Giuliani A, Franke R, Gruska A (2000) Quantitative structure-activity relationships of mutagenic and carcinogenic aromatic amines. Chem Rev 100:3697–3714
Berry DF, Francis AJ, Bollag JM (1987) Microbial metabolism of homocyclic and heterocyclic aromatic compounds under anaerobic conditions. Microbiol Rev 51:43–59
Blümel S, Contzen M, Lutz M, Stolz A, Knackmuss HJ (1998) Isolation of a bacterial strain with the ability to utilize the sulfonated azo compound 4-carboxy-4'-sulfoazobenzene as the sole source of carbon and energy. Appl Environ Microbiol 64:2315–2317
Bortone G (1995) Effects of an anaerobic zone in a textile wastewater treatment plant. Water Sci Technol 32:133–140
Bragger JL, Lloyd AW, Soozandehfar SH, Bloomfield SF, Marriott C, Martin GP (1997) Investigations into the azo reducing activity of a common colonic microorganism. Int J Pharm 157:61–71
Bumpus JA (1995) Microbial degradation of azo dyes. Prog Ind Microbiol 32:157–176
Campos R, Kandelbauer A, Robra KH, Cavaco-Paulo A, Gübitz GM (2001) Indigo degradation with purified laccases from *Trametes hirsuta* and *Sclerotium rolfsii*. J Biotechnol 89:131–139
Cao W, Mahadevan B, Crawford DL, Crawford RL (1993) Characterization of an extracellular azo-dye-oxidizing peroxidase from *Flavobacterium* sp. ATCC 39723. Enzyme Microb Technol 15:810–817
Chang JS, Kuo TS (2000) Kinetics of bacterial decolorization of azo dye with *Escherichia coli* NO_3. Biores Technol 75:107–111
Chang JS, Lin YC (2000) Fed batch bioreactor strategies for microbial decolorization of azo dye using *Pseudomonas luteola* strain. Biotechnol Prog 16:979–985
Chang JS, Kuo TS, Chao YP, Ho JY, Lin PJ (2000) Azo-dye decolorization with a mutant *Escherichia coli* strain. Biotechnol Lett 22:807–812
Chinvetkitvanich S, Tuntoolvest M, Panswad T (2000) Anaerobic decolorization of reactive dyebath effluents by a two-stage UASB system with tapioca as a co-substrate. Water Res 34:2223–2232
Chivukula M, Renganathan V (1995) Phenolic azo-dye oxidation by laccase from *Pyricularia oryzae*. Appl Environ Microbiol 61:4374–4377
Chung KT, Stevens Jr. SE (1993) Degradation of azo dyes by environmental microorganisms and helminths. Environ Toxicol Chem 12:2121–2132
Commandeur ICM, Parsons JR (1990) Degradation of halogenated aromatic compounds. Biodeg 1:207–220
Correia VM, Stephenson T, Judd S (1994) Characterisation of textile wastewaters – a review. Environ Technol 15:917–929
Corso CR, de Angelis DF, de Oliveira JE, Kiyan C (1981) Interaction between the diazo dye Vermelho Reanil P8B, and *Neurospora crassa* strain 74A. Eur J Appl Microbiol Biotechnol 13:64–66

Coughlin MF, Kinkle BK, Bishop PL (2002) Degradation of Acid Orange 7 in an aerobic biofilm. Chemosphere 46:11–19

Crowe JD, Olsson S (2001) Induction of laccase activity in *Rhizoctania solani* by antagonistic *Pseudomonas fluorescens* strains and a range of chemical treatments. Appl Environ Microbiol 67:2088–2094

Dabbs ER (1998) Cloning of genes that have environmental and clinical importance from rhodococci and related bacteria. Antonie van Leeuwenhoek 74:155–168

D'Annibale A, Stazi SR, Vinciguerra V, Sermanni GG (2000) Oxirane-immobilized *Lentinula edodes* laccase: stability and phenolics removal efficiency in olive mill wastewater. J Biotechnol 77:265–273

Davis S, Burns RG (1992) Decolorization of phenolic effluents by soluble and immobilized phenol oxidases. Appl Microbiol Biotechnol 37:474–479

Diamantidis G, Effosse A, Potier P, Bally R (2000) Purification and characterization of the first bacterial laccase in the rhizospheric bacterium *Azospirilium lipoferum*. Soil Biol Biochem 32:919–927

Donlagic J, Levec J (1998) Does the catalytic wet oxidation yield products more amenable to biodegradation. Appl Cat B Environ 17:L1–L5

Duran N, Esposito E (2000) Potential applications of oxidative enzymes and phenoloxidase-like compounds in wastewater and soil treatment: a review. Appl Cat B Environ 28:83–99

Field JA, Stams AJM, Kato M, Schraa G (1995) Enhanced biodegradation of aromatic pollutants in cocultures of anaerobic and aerobic bacterial consortia. Antonie van Leeuwenhoek 67:47–77

Fu Y, Viraraghavan T (2001) Fungal decolorization of dye wastewaters: a review. Biores Technol 79:251–262

Galhaup C, Haltrich D (2001) Enhanced formation of laccase activity by the white-rot fungus *Trametes pubescens* in the presence of copper. Appl Environ Biotechnol 56:225–232

Goszcynski S, Paszczynski A, Pasti-Grigsby MB, Crawford RL, Crawford DL (1994) New pathway for degradation of sulfonated azo dyes by microbial peroxidases of *Phanerochaete chrysosporium* and *Streptomyces chromofuscus*. J Bacteriol 176:1339–1347

Gottschalk G, Knackmuss HJ (1993) Bacteria and the biodegradation of chemicals: achieved naturally, by combination, or by construction. Angew Chem Int Ed 32:1398–1408

Greaves AJ, Churchley JH, Hutchings MG, Phillips DAS, Taylor JA (2001) A chemometric approach to understanding the bioelimination of anionic, water-soluble dyes by a biomass using empirical and semi-empirical molecular descriptors. Water Res 35:1225–1239

Hao OJ, Kim H, Chiang PC (2000) Decolorization of wastewater. Crit Rev Env Sci Tec 30:449–505

Haug W, Schmidt A, Nörtemann B, Hempel DC, Stolz A, Knackmuss HJ (1991) Mineralization of the sulfonated azo dye mordant yellow 3 by 6-aminonaphtalene-2-sulfonate-degrading bacterial consortium. Appl Environ Microbiol 57:3144–3149

Heinfling-Weidtmann A, Reemtsma T, Storm T, Szewzyk U (2001) Sulfophtalimide as major metabolite formed from sulfonated phtalocyanine dyes by the white-rot fungus *Bjerkandera adusta*. FEMS Microbiol Lett 203:179–183

Heiss GS, Gowan B, Danns ER (1992) Cloning of DNA from a *Rhodococcus* strain conferring the ability to decolorize sulfonated azo dyes. FEMS Microbiol Lett 99:221–226

Hillenbrand T (1999) Die Abwassersituation in der deutschen Papier-, Textil- und Lederindustrie. Gwf Wasser Abwasser 14:267–273

Hu TL, Wu SC (2001) Assessment of the effect of azo dye RP2B on the growth of a nitrogen fixing cyanobacterium *Anabaena* sp. Biores Technol 77:93–95

Kalliala E, Talvenmaa P (2000) Environmental profile of textile wet processing in Finland. J Clean Prod 8:143–154

Kapdan IK, Kargi F (2002) Simultaneous biodegradation and adsorption of textile dye stuff in an activated sludge unit. Process Biochem 37:973–981

Kapdan IK, Kargi F, McMullan G, Marchant R (2000a) Biological decolorization of textile dyestuff by *Coriolus versicolor* in a packed column reactor. Environ Technol 21:231–238

Kapdan IK, Kargi F, McMullan G, Marchant R (2000b) Effect of environmental conditions on biological decolorization of textile dyestuff by *C. versicolor*. Enzyme Microb Technol 26:381–387

Kapdan IKA, McMullan G, Marchant R (2000c) Decolorization of dyestuffs by a mixed bacterial consortium. Biotechnol Lett 22:1179–1181

Keck AA, Klein J, Kudlich M, Stolz A, Knackmuss HJ, Mattes R (1997) Reduction of azo dyes by redox mediators originating in the naphtalenesulfonic acid degradation pathway of *Sphingomonas* sp. Strain BN6. Appl Environ Microbiol 63:3684–3690

Khattri SD, Singh MK (1999) Colour removal from synthetic dye wastewater using a bioadsorbent. Water Air Soil Poll 120:283–294

Kim SJ, Ishikawa K, Hirai M, Shoda M (1995) Characteristics of a newly isolated fungus, *Geotrichum candidum* Dec 1, which decolorizes various dyes. J Ferm Bioeng 79:601–607

Kirby N, Marchant R, McMullan G (2000) Decolourisation of synthetic textile dyes by *Phlebia tremellosa*. FEMS Microbiol Lett 188:93–96

Knackmus HJ (1996) Basic knowledge and perspectives of bioelimination of xenobiotic compounds. J Biotechnol 51:287–295

Krull R, Hempel DCA, Alfter P (2000) Konzept zur technischen Umsetzung einer zweistufigen anoxischen und aeroben Textilabwasserbehandlung. Chem Ing Tech 72:1113–1114

Kruus K, Kiiskinen LL, Raettoe M, Viikari L, Saloheimo M (2001) Novel laccase enzyme and the gene encoding the enzyme. WO 01/92498 A1. PCT Int Appl 1–59

Kudlich M, Bishop E, Knackmuss HJ, Stolz A (1996) Simultaneous anaerobic and aerobic degradation of the sulfonated azo dye Mordant Yellow 3 by immobilized cells from naphtalenesulfonate-degrading mixed culture. Appl Microbiol Biotechnol 46:597–603

Kudlich M, Keck A, Klein J, Stolz A (1997) Localization of the enzyme system involved in anaerobic reduction of azo dyes by *Sphingomonas* sp. Strain BN6 and effect of artificial redox mediators on the rate of azo-dye reduction. Appl Environ Microbiol 63:3691–3694

Kulla HG, Klausener F, Meyer U, Lüdeke B, Leisinger T (1983) Interference of aromatic sulfo groups in the microbial degradation of the azo dyes Orange I and Orange II. Arch Microbiol 135:1–7

Kunz A, Reginatto V, Durán N (2001) Combined treatment of textile effluent using the sequence *Phanerochaete chrysosporium* – ozone. Chemosphere 44:281–287

Laszlo JA (1997) Regeneration of dye-saturated quaternized cellulose by bisulfite-mediated borohydride reduction of dye azo groups: an improved process for decolorization of texile waste waters. Environ Sci Technol 31:3647–3653

Laszlo JA (2000) Regeneration of azo-dye-saturated cellulosic anion exchange resin by *Burkholderia cepacia* anaerobic dye reduction. Environ Sci Technol 34:167–172

Ledakowicz S, Solecka M, Zylla R (2001) Biodegradation, decolourisation and detoxification of textile wastewater enhanced by advanced oxidation processes. J Biotechnol 89:175–184

Lourenco ND, Novais JM, Pinheiro HM (2001) Effect of some operational parameters on textile dye biodegradation in a sequential batch reactor. J Biotechnol 89:163–174

Martins MAM, Cardoso MH, Queiroz MJ, Ramalho MT, Campos AMO (1999) Biodegradation of azo dyes by the yeast *Candida zeylanoides* in batch aerated cultures. Chemosphere 38:2455–2460

Martins MAM, Cardoso Ferreira IC, Santos IM, Queiroz MJ, Lima N (2001) Biodegradation of bioaccessible textile dyes by *Phanerochaete chrysosporium*. J Biotechnol 89:91–98

McMullan G, Meehan C, Conneely A, Kirby N, Robinson T, Nigam P, Banat IM, Marchant R, Smyth WF (2001) Microbial decolourization and degradation of textile dyes. Appl Microbiol Biotechnol 56:81–87

Mester T, Tien M (2000) Oxidative mechanism of ligninolytic enzymes involved in the degradation of environmental pollutants. Int Biodeter Biodegr 46:51–59

Milstein O, Huettermann A, Majcherczyk A, Schulze K, Fruend R, Luedermann HD (1993) Transformation of lignin-related compounds with laccase in organic solvents. J Biotechnol 30:37–47

Mougin C, Kollmann A, Jolivalt C (2002) Enhanced production of laccase in the fungus *Trametes versicolor* by the addition of xenobiotics. Biotechnol Lett 24:139–142

Nam S, Renganathan V (2000) Non-enzymatic reduction of azo dyes by NADH. Chemosphere 40:351–357

Nerud F, Baldrian P, Gabriel J, Ogbeifun D (2001) Decolorization of synthetic dyes by the Fenton reagent and the Cu/pyridine/H_2O_2 system. Chemosphere 44:957–961

Nies DH (1999) Microbial heavy-metal resistance. Appl Microbiol Biotechnol 51:730–750

Nigam P, Marchant R (1995) Selection of a substratum for composing biofilm system of a textile-effluent decolourizing bacteria. Biotechnol Lett 17:993–996

Nigam P, McMullan G, Banat IM, Marchant R (1996a) Decolourisation of effluent from the textile industry by a microbial consortium. Biotechnol Lett 18:117–120

Nigam P, Banat IM, Singh D, Marchant R (1996b) Microbial process for the decolorization of textile effluent containing azo, diazo and reactive dyes. Process Biochem 31:435–442

Nigam P, Armour G, Banat IM, Singh D, Marchant R (2000) Physical removal of textile dyes from effluents and solid-state fermentation of dye-adsorbed agricultural residues. Biores Technol 72:219-226

Novotny C, Rawal B, Manish B, Patel M, Sasek V, Molitoris HP (2001) Capacity of *Irpex lacteus* and *Pleurotus ostreatus* for decolorization of chemically different dyes. J Biotechnol 89:113-122

O'Neill C, Lopez A, Esteves S, Hawkes F, Hawkes DL, Wilcox S (2000) Azo-dye degradation in an anaerobic-aerobic treatment system operating on simulated textile effluent. Appl Biochem Biotech 53:249-254

Ogawa T, Shibata M, Yatome C, Idaaka E (1988) Growth inhibition of *Bacillus subtilis* by basic dyes. Bull Environ Contam Toxicol 40:545-552

Paar A, Costa S, Tzanov T, Gudelj M, Robra K-H, Cavaco-Paulo A, Gübitz GM (2001) Thermoalkalistable catalases from newly isolated *Bacillus* sp. for the treatment and recycling of textile bleaching effluents. J Biotechnol 89:147-154

Panswad T, Iamsamer K, Anotai J (2001) Decolorization of azo-reactive dye by polyphosphate- and glycogen-accumulating organisms in an anaerobic-aerobic sequencing batch reactor. Biores Technol 76:151-159

Pasti-Grigsby MB, Pasczcynski A, Gosczcynski S, Crawford DL, Crawford RL (1992) Influence of aromatic substitution patterns on azo-dye degradability by *Streptomyces* spp. And *Phanerochaete chrysosporium*. Appl Environ Microbiol 58:3605-3613

Pierce J (1994) Colour in textile effluents – the origins of the problem. J Soc Dyers Col 110:131-134

Podgornik H, Poljansek I, Perdih A (2001) Transformation of Indigo Carmine by *Phanerochaete chrysosporium* ligninolytic enzymes. Enzyme Microb Technol 29:166-172

Pointing SB, Vrijmoed LLP (2000) Decolorization of azo and triphenylmethane dyes by *Pycnoporus sanguineus* producing laccase as the sole phenoloxidase. World J Microbiol Biotechnol 16:317-318

Pulgarin C, Invernizzi M, Parra S, Sarria V, Polania R, Püringer P (1999) Strategy for the coupling of photochemical and biological flow reactors useful in mineralization of biorecalcitrant industrial pollutants. Catal Today 54:341-352

Rafii F, Franklin W, Cerniglia CE (1990) Azoreductase activity of anaerobic bacteria isolated from human intestinal microflora. Appl Environ Microbiol 56:2146-2151

Rajaguru P, Kalaiselvi K, Palanivel M, Subburam V (2000) Biodegradation of azo dyes in a sequential anaerobic – aerobic system. Appl Microbiol Biotechnol 54:268-273

Razo-Flores E, Donlon B, Lettinga G, Field JA (1997a) Biotransformation and biodegradation of N-substituted aromatics in methanogenic granular sludge. FEMS Microbiol Rev 20:525-538

Razo-Flores E, Luijten M, Donlon B, Lettinga G, Field JA (1997b) Complete biodegradation of the azo-dye azodisalicylate under anaerobic conditions. Environ Sci Technol 31:2098-2103

Reemtsma T, Jakobs J (2001) Concerted chemical and microbial degradation of sulphophtalimides formed from sulfophtalocyanine dyes by white-rot fungi. Environ Sci Technol 35:4655-4659

Reyes P, Pickard MA, Vazquez-Duhalt R (1999) Hydroxybenzotriazole increases the range of textile dyes decolorised by immobilised laccase. Biotechnol Lett 21:875-880

Robinson T, Chandran B, Nigam P (2001a) Studies on the production of enzymes by white-rot fungi for the decolourisation of textile dyes. Enzyme Microb Technol 29:575-579

Robinson T, McMullan G, Marchant R, Nigam P (2001b) Remediation of dyes in textile effluent: a critical review on current treatment technologies with a proposed alternative. Biores Technol 77:247-255

Rodriguez E, Pickard MA, Vazquez-Duhalt R (1999) Industrial dye decolorization by laccases from ligninolytic fungi. Current Microbiology 38:27-32

Rogalski J, Jozwik E, Hatakka A, Leonowicz A (1995) Immobilization of laccase from *Phlebia radiata* on controlled porosity glass. J Mol Catal A Chem 95:99-108

Rogalski J, Dawidowicz AL, Jozwik E, Leonowicz A (1999) Immobilization of laccase from *Cerrena unicolor* on controlled porosity glass. J Mol Catal B Enzym 6:29-39

Sarsour J, Janitza J, Göhr F (2001) Biologischer Abbau farbstoffhaltiger Abwässer. Wasser/Abwassertechnik 44-46

Schliephake K, Lonergan GT (1996) Laccase variations during dye decolourisation in a 200L packed-bed reactor. Biotechnol Lett 18:881-886

Schliephake K, Mainwaring DE, Lonergan GT, Jones IK, Baker WL (2000) Transformation and degradation of the disazo dye Chicago Sky Blue by a purified laccase from *Pycnoporus cinnabarinus*. Enzyme Microb Technol 27:100-107

Shen CF, Miguez CB, Borque D, Groleau D, Guiol SR (1996) Methanotroph and methanogen coupling in granular biofilm under oxygen-limited conditions. Biotechnol Lett 18:495–500

Shin KS, Kim CJ (1998) Decolorization of artificial dyes by peroxidase from the white-rot fungus, *Pleurotus ostreatus*. Biotechnol Lett 20:569–572

Slokar YM, Marechal AML (1998) Methods of decoloration of textile wastewaters. Dyes Pigments 37:335–356

Smith MR (1990) The biodegradation of aromatic hydrocarbons by bacteria. Biodeg 1:191–206

Soares GM, Costa-Ferreira M, Pessoa d'A (2001a) Decolorization of an anthraquinone-type dye using a laccase formulation. Biores Technol 79:171–177

Soares GM, Pessoa d'A, Costa-Ferreira M (2001b) Use of laccase together with redox mediators to decolourize Remazol Brilliant Blue R. J Biotechnol 89:123–129

Solano F, Eljo P, López-Serrano D, Fernández E, Sanchez-Amat A (2001) Dimethoxyphenol oxidase activity of different microbial blue multicopper Proteins. FEMS Microbiol Lett 204:175–181

Spadaro JT, Renganathan V (1995) Peroxidase-catalyzed oxidation of azo dyes: mechanism of Disperse Yellow 3 degradation. Arch Biochem Biophys 312:301–307

Stolz A (2001) Basic and applied aspects in the microbial degradation of azo dyes. Appl Microbiol Biotechnol 56:69–80

Sumathi S, Manju BS (2000) Uptake of reactive textile dyes by *Aspergillus foetidus*. Enzyme Microb Technol 27:347–355

Swamy J, Ramsey JA (1999a) Effects of glucose and NH_4^+-concentrations on sequential dye decolorization by *Trametes versicolor*. Enzyme Microb Technol 25:278–284

Swamy J, Ramsey JA (1999b) The evaluation of white-rot fungi in the decolorization of textile dyes. Enzyme Microb Technol 24:130–137

Tan CG, Lettinga G, Field JA (1999) Reduction of the azo dye Mordant Orange 1 by methanogenic granular sludge exposed to oxygen. Biores Technol 67:35–42

Tatarko M, Bumpus JA (1998) Biodegradation of Congo Red by *Phanerochaete chrysosporium*. Water Res 32:1713–1717

Tekere M, Mswaka AYA, Read JS (2001) Growth, dye degradation and ligninolytic activity studies on Zimbabwean white-rot fungi. Enzyme Microb Technol 28:420–426

Tse SW, Yu J (1997) Flocculation of *Pseudomonas* with aluminum sulfate for enhanced biodegradation of synthetic dyes. Biotechnol Tech 11:479–482

Van der Bruggen B, De Vreese I, Vandecasteele C (2001) Water reclamation in the textile industry: nanofiltration of dye baths for wool dyeing. Ind Eng Chem Res 40:3973–3978

Walker GM, Weatherly LR (1999) Biological activated carbon treatment of industrial wastewater in stirred tank reactors. Chem Eng J 75:201–206

Willian AU, Pathak S, Chery J, Hsu TC (1978) Cytogenic toxicity of Gentian Violet and Crystal Violet on mammalian cells in vitro. Mutat Res 58:269–276

Wong Y, Yu J (1999) Laccase-catalyzed decolorization of synthetic dyes. Water Res 33:3512–3520

Xu F (1996) Oxidation of phenols, anilines, and benzenethiols by fungal laccases: correlation between activity and redox potentials as well as halide inhibition. Biochem 35:7608–7614

Yatome C, Yamada S, Ogawa T, Matsui M (1993) Degradation of Crystal Violet by *Nocardia corallina*. Appl Microbiol Biotechnol 38:565–569

Yoo ES (2002) Kinetics of chemical decolorization of the azo dye C.I. Reactive Orange 96 by sulfide. Chemosphere 47:925–931

Young L, Yu J (1997) Ligninase-catalysed decolorization of synthetic dyes. Water Res 31:1187–1193

Zhang F, Yu J (2000) Decolourisation of Acid Violet 7 with complex pellets of white rot fungus and activated carbon. Bioprocess Eng 23:295–301

Zhang TC, Fu YC, Bishop PL, Kupferle M, FitzGerald S, Jiang HH, Harmer C (1995) Transport and biodegradation of toxic organics in biofilms. J Hazard Mat 41:267–285

Zhang X, Stebbing DW, Saddler JN, Beatson RP, Kruus K (2000) Enzyme treatments of the dissolved and colloidal substances present in mill white water and the effects on the resulting paper properties. J Wood Chem Technol 20:321–335

Zheng Z, Levin RE, Pinkham JL, Shetty K (1999) Decolorization of polymeric dyes by a novel *Penicillium* isolate. Process Biochem 34:31–37

Zimmermann T, Gasser F, Kulla HG, Leisinger T (1984) Comparison of two bacterial azoreductases acquired during adaptation to growth on azo dyes. Arch Microbiol 138:37–43

Chapter 27

Degradation of the Indigo Carmine Dye by an Anaerobic Mixed Population

G. Fischer-Colbrie · J. Maier · K. H. Robra · G. M. Guebitz

Abstract

An anaerobic mixed population was found to be able to grow on acetate and Indigo carmine as sole carbon sources. After eighteen days of incubation, the dye was completely decolorized. Degradation products of indigo carmine monitored by high performance liquid chromatography coupled to ultraviolet/visible detector (HPLC-UV/VIS) were not detected after 25 d of incubation. Investigations on the degradation pathway of the anaerobic mixed population have been studied. To our best knowledge, it is the first time that an anaerobic mixed population growing on acetate, a dye bath additive in the textile industry, is able to mineralize indigo carmine.

Key words: anaerobic mixed population, textile dyes, indigo carmine

27.1 Introduction

Indigo carmine is a very widespread textile dyestuff used for dyeing of cotton and wool fabrics. It is a very well known model dyestuff in experimental textile chemistry as well. As a general environmental problem, most of the dyestuff used in a bath dyeing process is discharged with the process water. It has been estimated that about 15% of the dyes used in textile industry ends into waste water and is not recycled (Reyes et al. 1995; Wong and Yu 1999). Without a treatment, these substances are hazardous for the receiving water bodies since they inhibit the entry of light into the rivers and can be toxic or generate toxic cleavage products (Pierce 1994; Tratnyek et al. 1994).

The resulting problems are manifold: large amounts of dyestuff are lost and an expensive and sometimes not even a very successful treatment of the waste water has to be applied in order to provide clean water either for a reuse in the dyeing process or for the receiving water bodies.

In the last few years, several biological processes for textile effluent treatment have been developed. Previously, we have shown that immobilized laccases can efficiently decolorize textile dyeing effluents. However, drawbacks of this approach are high cost for enzymes and their limited stabilities. Another possible way to get rid of the color in the dye house effluent is the dye degradation by anaerobic mixed populations (Delée et al. 1998; Razo-Flores et al. 1999). Recent studies have shown that decolorization under anaerobic conditions is often more effective and faster than with an aerobic microbial consortium. Under anaerobic conditions, the reduction of one or more central double bonds of the chromophore within the respiratory chain of the bacteria has been suggested as the mechanism of decolorization.

However, toxic cleavage products may be recalcitrant to an attack by anaerobic bacteria and therefore remain in the wastewater streams. Although reduction is a very simple way of decolorization, it is generally not carried out by aerobic bacteria, which use oxygen as their only terminal electron acceptor. In contrast, anaerobic bacteria are in general unable to degrade further the aromatic system of the dye molecule after an initial cleavage. Since such findings have been obtained very often, the most effective way to degrade the dyestuffs in the dye house effluents seems to be an anaerobic-aerobic sequential treatment (O'Neill et al. 2000). However, this combined treatment is quite expensive and sometimes additional substrates have to be supplied to support the growth of microorganisms.

For the first time we show in this paper that after an adaptive period, a dye degrading mixed anaerobic population is able to use acetate, which is present in the dye house effluents (Shore 1995), as a sole carbon source.

27.2
Experimental

27.2.1
Culture Conditions

The anaerobic mixed population was grown in a medium with sodium acetate as the only carbon source. The medium consisted of sodium acetate (3 g l^{-1}), KH$_2$PO$_4$ (3 g l^{-1}), NH$_4$Cl (0.15 g l^{-1}), MgCl$_2 \cdot$ 7H$_2$O, CaCl$_2 \cdot$ 2H$_2$O, and FeCl$_3 \cdot$ 6H$_2$O, Merck. The first inoculum was taken from a long term fed batch fermentation of an anaerobic mixed population growing on acetate in the presence of several textile dyes. The pH was adjusted to 7.0–7.2 and the temperature was held at 38 °C. The bacterial growth took place under an atmosphere of nitrogen.

27.2.2
Indigo Carmine Degradation

Fermentations were carried out in 200-ml-, 1-l-, and 10-l-vessels as indicated below. The initial Indigo carmine (Sigma) concentration was 150 mg l^{-1} (300.9 mol l^{-1}) and the inoculum for each experiment was taken from a fermentation at the logarithmic growth phase. For each experiment a control was carried out with 1 ml l^{-1} thiomersal as a strong growth inhibitor for microorganisms in order to determine the non-microbial degradation of Indigo carmine at 38 °C.

27.2.3
Analysis

Ultraviolet/visible (UV/VIS) spectroscopy: the absorbance of indigo carmine was measured on a Hitachi U-2001 Spectrophotometer at the absorbance maximum of 611 nm.

High performance liquid chromatography coupled to UV/VIS detector (HPLC-UV/VIS): the samples for the HPLC-UV/VIS measurements were prepared as follows: after removing cells by centrifugation, proteins were precipitated by an equal volume of acetone. After evaporating the acetone under reduced pressure, ultra-filtration was carried out with a membrane's cut-off of 500 Dalton to get rid of macromolecules. This filtrate was concentrated to a defined volume appropriate for the HPLC analysis. The determination and quantification of the cleavage products of indigo carmine were carried out on a Kontron HPLC. The method chosen was according to Kraak and Huber (1974). Two solvents served as eluents: Solvent A was a 50 mmolar phosphate buffer at pH 7 with 5 mm of tri-butyl-ammonium-hydrogen-sulfate (TBAHS, obtained from Sigma) and the solvent B was a methanol water (9:1) mixture (methanol, HPLC-grade, Merck). The column used was a RP-18, 125 × 4 mm. Measurements were carried out at room temperature, the eluent flow was constant at 0.5 ml min^{-1}. For optimal elution a linear gradient was run from 100% solvent A to 100% solvent B within the 5th and 25th min of a 30-min run. Peaks were detected at two different wavelengths simultaneously, one in the UV range (245 nm) and one near the absorbance maximum of Indigo carmine at 600 nm.

27.3
Results and Discussion

27.3.1
Decolorization of Several Textile Dyes

An anaerobic mixed population was found to decolorize Indigo carmine and to degrade further the cleavage products metabolizing them. Additionally, this mixed population decolorized several other textile dyes such as azo dyes, anthra-chinonic dyes or triaryl-methane dyes.

Anaerobic degradation of various azo dyes yielded the following decolorization intensities (%): solophenyl blue (98.4), diamond black PV (82.4), remazol orange (>99%).

One anthra-chinonic dye tested was cibachron marone, which was decolorized to an extend of 62.4%. Furthermore, the triaryl-methane dye terasil pink was decolorized to >99%.

27.3.2
Degradation of Indigo Carmine

27.3.2.1
Spectrophotometrical Studies

The degradation of the textile dye indigo carmine was studied in more detail. Experiments were carried out in 1-l vessels, which were held at 37 °C in a water bath under nitrogen atmosphere. Every 24 h, one sample was taken and the absorbance was measured at 611 nm. According to the spectro-photometric measurements, decolorization of indigo carmine was complete after 18 d. It has to be mentioned that non-microbial degradation under the same conditions in the absence of growing bacteria was about 10% for the same period (see Fig. 27.1).

Comparing the graphs of the abiotic control and the biodegradation, it can be seen that the dye has been decolorized due to the microbial growth.

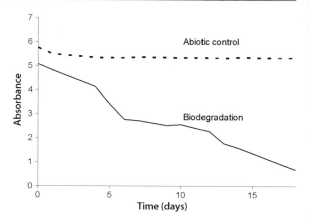

Fig. 27.1.
Decolorization of indigo carmine during 18 d. *Dotted line:* control without growing organisms. *Black line:* sample with the active population

27.3.2.2
HPLC/UV-VIS Studies

The degradation pathway of Indigo carmine by the mixed population was studied in detail using a 10-l fed batch reactor. Every 48 h, samples were prepared and analyzed as described in 2.3 Analysis. According to the results of the HPLC experiments, a mechanism of degradation of Indigo carmine by the anaerobic mixed population can be suggested (Fig. 27.2). The stability of the Indigo carmine anion has been previously discussed by Russel and Konoka (1967). As reference substances Indigo carmine, isatin-5-sulfonate and 4-amino-3-carboxybenzenesulfate (prepared in our lab from isatin-5-sulfonate) were used and peak detection could be carried out at two different wavelengths to verify the proposed pathway. A mixture of the three reference substances could be separated easily by the HPLC method mentioned above. The retention times were 5.09 min, 8.11 min, and 10.87 min for and 4-amino-3-carboxybenzenesulfate, isatin-5-sulfonate, and Indigo carmine, respectively.

According to the HPLC measurements, the Indigo carmine concentration decreased to 61.3% after eight days of incubation and to a level of 6% after 18 d (Fig. 27.3). Concerning the concentration profile of isatin-5-sulfonate, one can see that this substance appears while the concentration of Indigo carmine decreases. The cleavage product 4-amino-3-carboxybenzenesulfate did appear to a very small extend indicating that this product was obviously transported into the cells and therefore no longer detectable in the fermentation broth.

In a fermentation sample taken after 25 d of incubation with indigo carmine, which has been concentrated 1:100 of the original volume, no degradation products could be detected. In this case, only traces of unknown UV active molecules appeared on the chromatogram. At this point of the fermentation, a sample of the biomass was taken and analyzed. After breaking the cell walls and removing the fragments and the nuclei, proteins were separated as described above. After ultra-filtration, no characteristic peak of any Indigo carmine cleavage product could be detected. Therefore, it can be emphasized that Indigo carmine has been degraded to non-aromatic compounds, which were most likely further metabolized by the anaerobic mixed consortium.

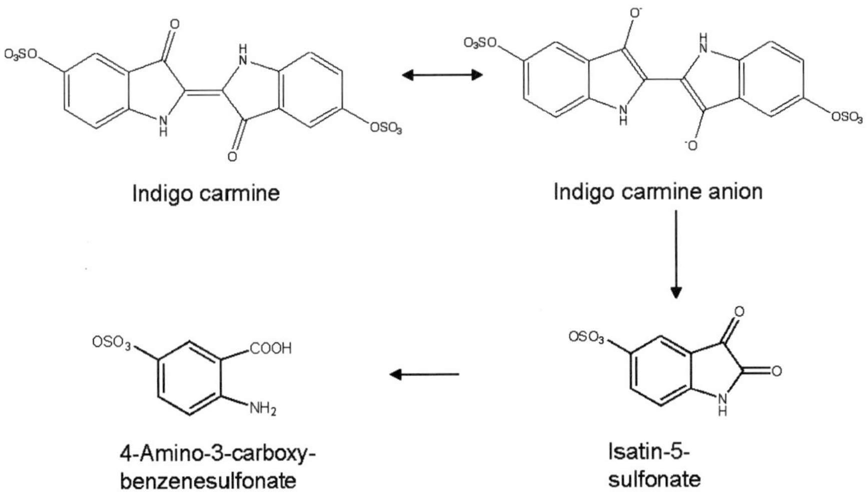

Chapter 27 · Degradation of the Indigo Carmine Dye by an Anaerobic Mixed Population

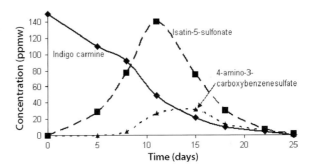

Fig. 27.2. Proposed mechanism of the degradation pathway of indigo carmine: the molecule undergoes an initial reversible reaction yielding the indigo carmine anion, which appears yellow. The cleavage of the central bond gives isatin-5-sufonate, which is further decarboxylized to 4-amino-3-carboxybenzenesulfonate

Fig. 27.3. Concentration profile of indigo carmine, isatin-5-sulfonate and 4-amino-3-carboxybenzenesulfonate

27.4 Conclusion

The population shows high potential in decolorization of textile dyes regardless their molecular structure. The mineralization of the dye stuff indigo carmine was not complete until an incubation period of 25 d. Taking into account that growth under anaerobic conditions is far slower than under aerobic conditions for the same carbon source, the degradation time still seems long. For the mechanism of the degradation it can be supposed that the first step, namely the cleavage of indigo carmine to isatin-5-sulfonate, is carried out by a non-specific reduction somehow connected to the bacterial metabolism. It was observed that after one day of incubation the fermentation broth had turned from dark blue to yellow. Since after aeration the blue color returned, it has to be assumed that this change in the color is due to the reversible reaction between the indigo carmine molecule and the indigo carmine anion, res. its hydroxylated form. This effect could not be observed in the replicate without the active population. The next step, the decarboxylation of the isatin-5-sulfonate, is supposed to be enzyme-catalyzed, since the chemical decarboxylation requires drastic conditions such as high temperatures and high pH. Finally, most likely the cleavage products of indigo carmine were metabolized by the anaerobic mixed population. What follows now is the breakdown of the aromatic system. As it can be seen from Fig. 27.2, no detectable aromatic degradation products were left either in the extra-cellular environment or in the cytosol of the bacteria.

Thus, this system has a high potential for the treatment of textile dying effluents especially since the mixed population can use effluent components (acetate) as carbon source to support growth.

References

Delée W, O'Neill C, Hawkes FR, Pinheiro HM (1998) Anaerobic treatment of textile effluents: a review. J Chem Technol Biotechnol 73:323–335

Kraak JC, Huber JF (1974) Separation of acidic compounds by high pressure liquid-liquid chromatography involving ion-pair formation. J Chromatogr 102:333–351

O'Neill C, Lopez A, Esteves S, Hawkes FR, Hawkes DL, Wilcox S (2000) Azo dye degradation in an anaerobic-aerobic treatment system operating on simulated textile effluent. Appl Microbiol Biotechnol 53:249–254

Pierce J (1994) Colour in textile effluents – the origins of the problem. JSCD 110:131–134

Razo-Flores E, Smulders P, Prenafeta-Boldú F (1999) Treatment of anthranilic acid in an anaerobic expanded granular sludge bed reactor at low concentrations. Water Sci Technol 40:187–194

Reyes P, Pickard MA, Vazquez-Duhalt R (1995) Hydroxybenzotriazole increases the range of textile dyes decolorized by immobilized laccase. Biotechnol Lett 21:875–880

Russel G, Konoka R (1967) Radical anions derived from indigo and bibenzimidazole. J Org Chem 32:234–236

Shore J (1995) The dyeing of cellulosic fibres. John Shore, Soc. of Dyers and Colorists, pp 1ff, 189ff, 321ff

Tratnyek PG, Elovitz MS, Colverson P (1994) Photoeffects of textile dyes in wastewaters: sensizitation of singlet oxygen formation, oxidation of phenols and toxicity to bacteria. Environ Toxicol Chem 13:27–33

Wong Y, Yu J (1999) Laccase-catalyzed decolorization of synthetic dyes. Water Res 33:3512–3520

Chapter 28

Biodegradation of Benzothiazoles by *Rhodococcus* Bacteria Monitored by ^{1}H Nuclear Magnetic Resonance (NMR)

N. Haroune · P. Besse · B. Combourieu · M. Sancelme · H. De Wever · A. M. Delort

Abstract

The biodegradation of benzothiazole, 2-hydroxybenzothiazole and 2-aminobenzothiazole by two strains of *Rhodococcus* was monitored by high performance liquid chromatography (HPLC) and by in-situ ^{1}H Nuclear Magnetic Resonance (NMR), which is directly performed on culture media, without prior purification. A common biodegradative pathway is evidenced; the benzothiazole compounds were biotransformed into hydroxylated derivatives. The chemical structure of these metabolites was determined by a long range ^{1}H-^{15}N heteronuclear shift correlation without any previous ^{15}N enrichment of the starting xenobiotic.

Key words: ^{1}H NMR, ^{15}N NMR, benzothiazole, degradation, *Rhodococcus*

28.1 Introduction

Benzothiazoles are a group of xenobiotics containing a benzene ring fused with a thiazole ring (Fig. 28.1). They are manufactured worldwide for a variety of applications. They are used as fungicides in lumber and leather production (Reemtsma et al. 1995), as herbicides (Wegler and Eue 1977; Hartley and Kidd 1987), as antialgal agents (Bujdakova et al. 1994), as slimicides in the paper and pulp industry (Meding et al. 1993) and as chemotherapeutics (Bujdakova et al. 1993). These applications clearly indicate that benzothiazoles have a wide spectrum of biological activities. Nonetheless, their main use is in rubber manufacture as catalysts in the vulcanisation process. Released from rubber products or from benzothiazole production plants, benzothiazoles have been detected in industrial wastewater, but also in various environmental

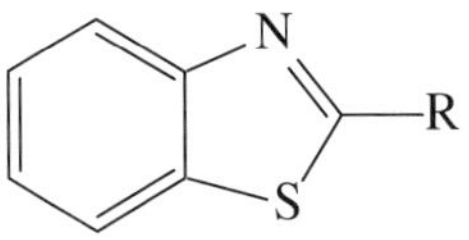

Substituent R	H	OH	NH_2	SO_3H
Name	benzothiazole	2-hydroxy-benzothiazole	2-amino-benzothiazole	benzothiazole-2-sulphonate
Code	BT	OBT	ABT	$BTSO_3$

Fig. 28.1. Structural formula of 2-substituted benzothiazoles

media (Fiehn et al. 1994; Reemtsma et al. 1995) and are of concern for aquatic environment due to their limited biodegradability and potential toxicity (Gold et al. 1993; De Wever and Verachtert 1997).

Actually only few bacterial isolates have been shown to degrade benzothiazoles as pure culture (De Wever et al. 2001). Gaja and Knapp (1997) described a *Rhodococcus* strain PA growing on benzothiazole (BT) as sole source of carbon, nitrogen and energy, and strain TA growing on 2-aminobenzothiazole (ABT). Two isolated strains, *Rhodococcus erythropolis* (BTS1) and *Rhodococcus rhodochrous* (OBT18), were shown to degrade benzothiazole (BT) and 2-hydroxybenzothiazole (OBT) (De Wever et al. 1997, 1998; Besse et al. 2001). Benzothiazole-2-sulfonate ($BTSO_3$) is also degraded by *R. erythropolis* (De Wever et al. 1998) and 2-aminobenzothiazole by *R. rhodochrous* (Haroune et al. 2001).

^{1}H Nuclear Magnetic Resonance (NMR) spectroscopy is a powerful tool to determine chemical structures, and has been used for example to study the biodegradation of xenobiotics by microorganisms (Delort and Combourieu 2000, 2001; Brecker and Ribbons 2000). In particular, in-situ ^{1}H NMR, directly performed on culture media, at natural abundance, allows to monitor biodegradation kinetics. We have used NMR successfully to establish the biodegradative pathway of morpholine, thiomorpholine and piperidine by strains of *Mycobacterium* (Combourieu et al. 1998a,b, 2000; Besse et al. 1998; Poupin et al. 1998). Further, Gradient Heteronuclear Multiple-Bond Correlation (^{1}H-^{15}N GHMBC) experiments at natural abundance have been done recently to study organic compounds containing N atoms, taking advantage of the valuable information contained in ^{1}H-^{15}N scalar couplings. This type of experiments was made possible at natural abundance since middle of the 1990s when spectrometers were equipped routinely with gradients. More details are available in a recent and very interesting review about the application of this technique to the determination of natural products structure, particularly of alkaloids (Martin and Hadden 2000).

Here, we present a detailed study of the biodegradative pathway of benzo-thiazole (BT), 2-hydroxybenzothiazole (OBT) and 2-aminobenzothiazole (ABT) by *Rhodococcus erythropolis* BTS1 and *Rhodococcus rhodochrous* OBT18 using ^{1}H Nuclear Magnetic Resonance (NMR) spectroscopy. Kinetics of biodegradation were monitored by in-situ ^{1}H NMR. Further precise determination of benzothiazole metabolites structure was made by using long-range ^{1}H-^{15}N heteronuclear shift correlation (Besse et al. 2001; Haroune et al. 2001).

28.2
Experimental

Chemicals: benzothiazole (BT), 2-hydroxybenzothiazole (OBT) and 2-aminobenzothiazole (ABT) (Aldrich). Tetradeuterated sodium trimethylsilylpropionate ($TSPd_4$) (Eurisotop, Saint Aubin, France).

Growth conditions: Rhodococcus erythropolis and *Rhodococcus rhodochrous* were grown in 100-ml Trypticase-soy broth (bioMerieux, Marcy l'Etoile, France) in 500-ml Erlenmeyer flasks incubated at 30 °C and 200 rpm. The cells were harvested after 20 h of culture.

Incubation with xenobiotic compounds: cells were centrifugated 15 min at 8 000 g at 5 °C. The pellets were washed twice with phosphate buffer: (K_2HPO_4: 1 g l^{-1}; KH_2PO_4: 1 g l^{-1}; $FeCl_3$: 4 mg l^{-1}, $MgSO_4 \cdot 7H_2O$: 40 mg l^{-1} – pH 6.7) and then resuspended in this buffer: 5 g wet cells in 100 ml buffer. The residual cells were incubated with 3 mM benzothiazole, 3 mM hydroxybenzothiazole or 0.5 mM aminobenzothiazole in 500-ml Erlenmeyer flasks at 30 °C under stirring (200 rpm). Controls consisted of preparations incubated under the same conditions without substrate or cells. Samples (1 ml) were taken every 30 min directly in the culture medium, then centrifugated 5 min at 12 000 g, then prepared for HPLC analysis or ^1H NMR analysis.

High performance liquid chromatography (HPLC): analyses were performed using a Waters 600E chromatograph fitted with a reversed-phase column (Interchrom Nucleosil C18, 5 µm, 250 × 4.6 mm – Interchim) at 20 °C. Mobile phase: acetonitrile/water 20/80, v/v, 1 ml min^{-1}, Waters 486 UV detector (295 nm).

Thin layer chromatography (TLC): analyses were performed using SiO_2 thin layers, the eluent was ethyl acetate/chloroform 40/60 (v/v). *Rf* values were 0.34 (2-aminobenzothiazole) and 0.11 (metabolite of 2-aminobenzothiazole).

^1H-Nuclear Magnetic Resonance (NMR): spectra were recorded on a Bruker Avance DSX300 spectrometer at 300.13 MHz at 298 K using 5 mm-diameter tubes. Water was suppressed by the classical double pulsed field gradient echo sequence: WATERGATE. 64 scans were collected: relaxation delay, 5 s; acquisition time, 3.64 s; spectral window of 3 420 Hz; 32 000 data points. A 0.3 Hz line broadening was applied before Fourier transformation and a baseline correction was performed on spectra before integration with Bruker software. $TSPd_4$ was used as internal reference for chemical shift (0 ppm) and quantification. The method for quantification of the metabolites is described elsewhere (Combourieu et al. 1998a).

^1H-^{15}N Gradient Heteronuclear Mutiple Bond Correlation (GHMBC): experiments were performed at 298 ±0.2 K on a Bruker Avance DSX300 spectrometer operating at 300.13 MHz and 30.41 MHz for ^1H and ^{15}N respectively. A 5-mm triple tuned ^1H-^{13}C-^{15}N probe equipped with a z-gradient coil was used. ^1H and ^{15}N 90° pulse lengths were 7.5 and 27 µs, respectively. No low-pass J-filter was used. Delay to allow $^nJ_{NH}$ correlations was set to 80 or 140 ms. Typically, 1 024 data points with 32 scans for each of 128 t_1 increments were acquired with spectral widths of 3 600 Hz in F2 and 6 100 Hz in F1. The required acquisition time was 285 ms. Zero-filling to 512 points and phase-shifted sine window function in t_1 and sine-squared window function in t_2 were applied prior to 2D Fourier transformation. A recycle delay of 1.5 s was used. The ^{15}N chemical shifts are reported negative upfield from CH_3NO_2 resonance (external reference, 0 ppm).

Isolation of the unknown metabolites: R. erythropolis (45 g wet cells) was incubated in 900 ml phosphate buffer with 3 mM hydroxybenzothiazole (9 Erlenmeyer flasks) while R. rhodochrous (60 g) was incubated with 1 mM aminobenzothiazole in 1 400 ml phosphate buffer (14 Erlenmeyer flasks). The biodegradation kinetics were monitored by HPLC in order to stop the experiment at the maximum metabolite concentration. In the case of OBT, after 2 h of incubation (OBT-metabolite concentration = 0.5 mM), the reaction mixture was centrifuged and the supernatant was extracted with ethyl acetate for 24 h. The organic layer was dried on $MgSO_4$, concentrated *under vacuum* then purified over a silica gel column (eluent: ethylacetate/chloroform 30/70, v/v). The pure metabolite was obtained as a pale yellow solid (*mp* = 219–221 °C). In the case of aminobenzothiazole, the same protocol was followed after 72 h of incubation (metabolite concentration = 0.4 mM). The pure metabolite was obtained as a pale yellow solid (*mp* = 250–252 °C). In order to determine the structure of these compounds, they were analyzed by mass spectrometry (Electronic Impact) on a Hewlett Packard MS 5989B spectrometer.

28.3
Results and Discussion

28.3.1
Biodegradation of 2-Aminobenzothiazole

The biodegradation of 0.5 mM 2-aminobenzothiazole by resting cells of *R. rhodochrous* was monitored by in-situ ^1H NMR. Briefly, after incubation of the bacteria in the presence of the pollutant, samples were taken at regular intervals; after centrifugation, the supernatant was directly analyzed by ^1H NMR. Spectra recorded after 1 h, 25 h and 94 h of incubation are shown in Fig. 28.2. ^1H NMR spectra showed the typical signals of the parent molecule 2-aminobenzothiazole at 7.17 (triplet), 7.37 (triplet), 7.47 (doublet), and 7.71 ppm (doublet). Then, after 1 h of incubation, a first metabolite was detected as observed by new peaks growing in the 6–8 ppm aromatic region: a doublet of doublets at 6.89 ppm and two doublets at 7.22 and 7.34 ppm.

The kinetic of biodegradation was also monitored by HPLC: only one peak could be detected on HPLC chromatogram. Its retention time (6 min) was shorter than that

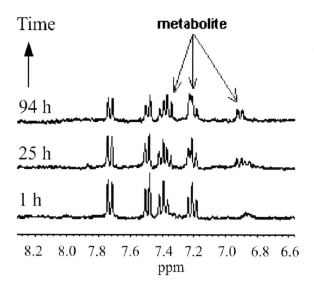

Fig. 28.2. Biodegradation of 2-aminobenzothiazole (0.5 mM) by *Rhodococcus rhodochrous* bacteria followed by in-situ ^1H NMR spectra taken after 1 h, 25 h and 94 h of incubation. Note the decrease of the xenobiotic peaks and the increase of peaks corresponding to the metabolite

of 2-aminobenzothiazole (24 min), indicating a higher polarity of this compound (data not shown). With this strain, the degradation rate was only 50% after 94 h, thus indicating a probable toxicity of the metabolite and/or the xenobiotic itself. With *Rhodococcus erythropolis*, the degradation rate of aminobenzothiazole was even slower. Only 30% of aminobenzothiazole disappeared within 100 h of incubation. The same polar metabolite was observed by HPLC.

28.3.2
Biodegradation of Benzothiazole

In-situ ^{1}H NMR spectra collected during the biodegradation of benzothiazole (3 mM) by resting cells of *Rhodococcus erythropolis* are presented in Fig. 28.3. While the signals of benzothiazole at 7.57 ppm (triplet), 7.65 (triplet), 8.14 (2 × doublet), 9.28 (singlet) decreased with time, new signals grew in the 7.0–7.6 aromatic region: two triplets at 7.24 and 7.38 ppm, and two doublets at 7.29 and 7.58 ppm. They were assigned to 2-hydroxybenzothiazole by spiking the pure 2-hydroxybenzothiazole into the NMR tube. This identification was also confirmed by HPLC co-elution.

Further, HPLC analysis revealed that once benzothiazole was fully transformed, a second metabolite showed up at a retention time of 7 min, which is shorter than that of 2-hydroxybenzothiazole (25 min), thus indicating a higher polarity. Using ^{1}H NMR, we observed also the decrease of 2-hydroxybenzothiazole peaks. However, no signal corresponding to the new metabolite observed by HPLC could be detected due to the lower sensitivity of NMR. With this strain, benzothiazole (BT) was quantitatively transformed into 2-hydroxybenzothiazole (OBT), which was then converted to an unknown more polar metabolite. This unknown metabolite was identified as 2,6-dihydroxybenzothiazole as shown in Sect. 28.4. Its highest concentration reached 0.5 mM after 2 h, it was then degraded in its turn within 6 h of incubation. The same unknown and more polar metabolite was obtained during the degradation of OBT (3 mM) by *R. erythropolis* (data not shown).

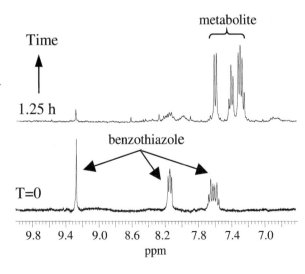

Fig. 28.3.
Biodegradation of benzothiazole (3 mM) by *Rhodococcus erythropolis* bacteria followed by in-situ ^{1}H NMR. Note the striking decrease of benzothiazole peaks after 1.25 h, and the increase of those of the metabolite 2-hydroxybenzothiazole

During the degradation of benzothiazole (BT), the successive formation of two metabolites was observed. The first one was identified as 2-hydroxybenzothiazole (OBT), the second one, more polar, was only observed by HPLC.

The kinetics of BT and OBT biodegradation observed with *R. rhodochrous* showed a different behavior. No OBT was detected during the BT biodegradation, whatever the analytical method used. Only the unknown metabolite was observed (data not shown).

In the two following sections (28.3.3 and 28.3.4) more sophisticated 2D NMR experiments (2D: two-dimensional) combined to mass spectrometry experiments were used to identify the unknown metabolites obtained during the degradation of 2-aminobenzothiazole, benzothiazole and 2-hydroxybenzothiazole. Because ^{1}H-^{15}N NMR experiments are less sensitive (^{15}N nucleus is difficult to detect), these unknown metabolites were produced in larger amounts and purified for analysis.

28.3.3
Characterization of the 2-Aminobenzothiazole Metabolite

In order to identify the unknown metabolite of 2-aminobenzothiazole, *R. rhodochrous* (60 g) was incubated in 1 400 ml phosphate buffer with a 1 mM solution of this compound. The biodegradation kinetic was monitored by HPLC in order to stop the experiment at the maximum concentration in metabolite (72 h). Extraction of the supernatant overnight with ethyl acetate, followed by purification over a silica gel column yielded the pure metabolite as a pale yellow solid. We checked that the isolated product corresponded to the unknown metabolite observed previously, by co-elution with a sample taken during the degradation of aminobenzothiazole, and also by thin layer chromatography (TLC). The purified product was first analyzed by ^{1}H NMR in CD_3OD. Only three signals are visible in the aromatic region, each one corresponding to an equivalent number of protons (Fig. 28.4, top trace). So, a substituent was present on the aromatic ring. By analyzing the coupling constants, we deduced that this substituent was either on position 5 or 6 (Fig. 28.6). However, the assignment of protons 4 and 7 was not possible because of the difficulty of knowing the respective electronic effects of the nitrogen and the sulfur atoms on these protons.

The chemical ionisation (CI, CH_4) mass spectrum of the isolated compound presented a peak with a mass/charge ratio of 167 ($[M + H]^+$), differing from the molecular weight of 2-aminobenzothiazole by 16 mass units. Note also that peaks at *m/z* 195 ($[M + C_2H_5]^+$) and 207 ($[M + C_3H_5]^+$) confirmed the quasi-molecular ion. Therefore the metabolite formed during biodegradation is an hydroxylated derivative.

To assign unambiguously the position of the hydroxyl group on the benzene ring of the metabolite, long-range ^{1}H-^{15}N heteronuclear shift correlation, and more precisely gradient heteronuclear mutiple bond correlation (GHMBC), was done. In the scheme of the pulse program a variable delay is included between the two first 90° ^{15}N pulses (for details see Martin and Hadden 2000). This evolution period allows the system to be converted into zero and double quantum coherence and is proportional to the coupling constant J between N and H. Thus this delay must be optimized in order to select ^{2}J, ^{3}J or ^{4}J couplings. In our case, delay to allow ^{n}J$_{NH}$ correlations was set to 80 and 140 ms in order to select assumed ^{3}J$_{NH}$ (6.2 Hz) and ^{4}J$_{NH}$ (3.5 Hz) respectively. As chemical shift is concerned, we used CH_3NO_2 as external reference (0 ppm) since there is not a single chemical shift referencing. The chemical shifts are reported negative upfield from CH_3NO_2 resonance.

Fig. 28.4.
^1H-^{15}N GHMBC spectrum of 2-amino-6-hydroxybenzothiazole recorded in CD_3OD; the experiment was set up for assumed ^4J (delay = 140 ms) long range coupling. The spots visible on the 2D spectrum correspond to the correlations between N3 and H4 ($^3J_{N3H4}$) and between N3 and H5 ($^4J_{N3H5}$). The presence of a proton on the 5 position of the benzene ring demonstrates that the hydroxyl group is on the 6 position. GHMBC: gradient heteronuclear multiple bond correlation. 2D: two-dimensional

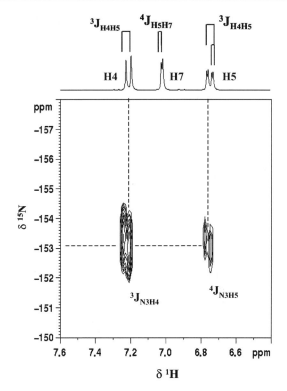

The ^1H-^{15}N gradient heteronuclear multiple bond correlation (GHMBC) recorded with an 80 ms evolution period showed only one correlation between the endocyclic $^{15}N3$ (−153.1 ppm) and the doublet at 7.24 ppm (data not shown). Thus, this proton was assigned to H4, since no other ^3J was observed in this compound. In Fig. 28.4, the delay fixed at 140 ms (while other conditions remained the same) allowed the observation of a ^4J coupling between nitrogen and the doublet of doublets resonating at 6.76 ppm. This proton was assigned to H5 because of its 8.5 Hz (^3J) coupling constant with H4. No doubt exists concerning the assignment of the substituent in position 6. Moreover, the $^{15}N_3$ chemical shifts of both metabolite and 2-aminobenzothiazole were very close, −153.1 and −153.5 ppm respectively, thus indicating no modification of the oxidation state of this atom. All these data show that the unknown metabolite is 2-amino-6-hydroxy-benzothiazole, corresponding to a hydroxylation in position 6 of aminobenzothiazole (Fig. 28.6).

28.3.4
Characterization of Benzothiazole and 2-Hydroxybenzothiazole Metabolites

In the case of the unknown metabolite obtained during benzothiazole and 2-hydroxybenzothiazole degradation, *R. erythropolis* was chosen for the quantitative assay, this strain giving the highest concentration in metabolite. *R. erythropolis* (45 g wet cells) was incubated in 900 ml phosphate buffer with 3 mM hydroxybenzothiazole. The biodegradation kinetic was monitored by HPLC in order to stop the experiment at the maximum concentration in metabolite (0.5 mM, 2 h). Extraction of the supernatant with ethyl acetate, followed by purification over a silica gel column yielded the pure metabolite as a pale yellow solid. A detailed structural analysis of this compound was performed as previously described for 2-aminobenzothiazole derivative. ^{1}H NMR spectrum in CD$_3$OD combined with mass spectrum clearly indicated that the metabolite formed was a hydroxylated derivative of 2-hydroxybenzothiazole.

Finally the position of the hydroxyl group on the benzene ring of the metabolite was assigned unambiguously by ^{1}H-^{15}N GHMBC experiments. The delay to allow ^{n}J$_{NH}$ correlations was set to 80 and 140 ms in order to select assumed ^{3}J$_{NH}$ (6.2 Hz) and ^{4}J$_{NH}$ (3.5 Hz) respectively. In ^{1}H-^{15}N GHMBC recorded with an 80 ms evolution period only one correlation could be observed with the doublet at 6.97 ppm. Thus, this proton was assigned to H4, since no other ^{3}J can be observed in this compound (not shown). In Fig. 28.5, the delay fixed at 140 ms (while other conditions remained the same) allowed the observation of a ^{4}J coupling between nitrogen and the doublet of doublets resonating at 6.73 ppm. This proton was assigned to H5 because of its 8.5 Hz (^{3}J) coupling constant with H4. These data cleary show that the unknown metabolite is 2,6-dihydroxybenzothiazole (diOBT), corresponding to a hydroxylation in position 6 of OBT (Fig. 28.6).

The same experiments were carried out starting from a quantitative assay with R. rhodochrous and led to an identical structure for the metabolite, observed during the biodegradation of BT and OBT with this strain.

Fig. 28.5.
^{1}H-^{15}N GHMBC spectrum of 2,6-dihydroxybenzothiazole recorded in CD$_3$OD; the experiment was set up for assumed ^{4}J (delay = 140 ms) long range coupling. Note the correlations between N3 and the three benzenic protons (H4, H5 and H7) and the absence of correlation between N3 and H6 showing the presence of a substituent in the 6 position. GHMBC: gradient heteronuclear multiple bond correlation

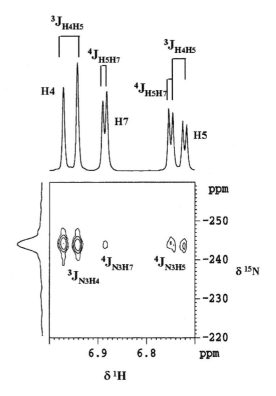

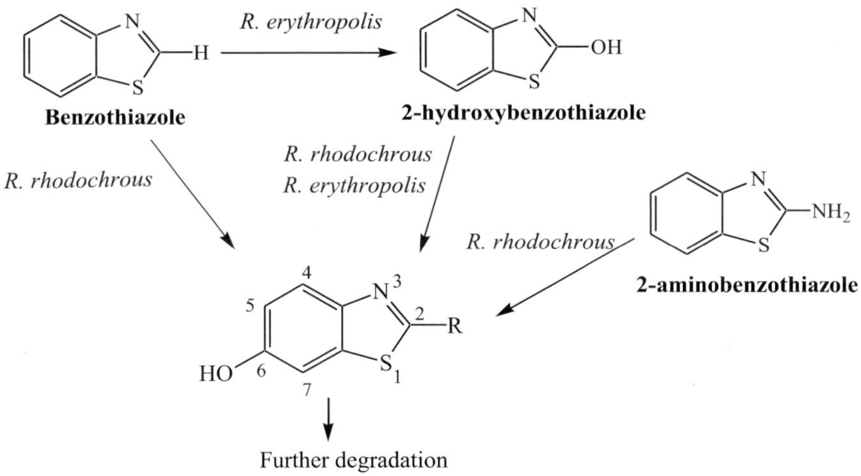

Fig. 28.6. Biodegradative pathways of 2-aminobenzothiazole, benzothiazole and 2-hydroxybenzothiazole with two strains of bacteria: *Rhodococcus rhodochrous* and *Rhodococcus erythropolis*

28.4 Conclusion

To date, only a few studies have been reported on the fate of organic pollutants in the environment using NMR spectroscopy. The recent development of ^{1}H NMR in the environmental field is quite promising. As shown in this paper, various techniques can be combined: first in-situ 1D NMR (1D: one-dimensional) can be used to monitor in real time microbial degradation of organic pollutants in liquid media, without any purification of the sample; second 2D NMR experiments are powerful tools to characterize the structure of unknown metabolites. Using this approach, we have studied the biodegradative pathway of three benzothiazoles, including benzothiazole, 2-hydroxybenzothiazole and 2-aminobenzothiazole by two *Rhodococcus* bacterial strains. Hydroxylation of benzothiazole derivatives on the aromatic ring in position 6 was shown to be a common step of this pathway (Fig. 28.6). We have recently shown that this hydroxylation process is operating in other microbial strains, namely *Rhodococcus pyrinidovorans* PA (degradation of benzothiazole and 2-hydroxybenzothiazole) (Haroune et al. 2002) and *Aspergillus niger* (degradation of methabenzthiazuron) (Malouki et al. 2003). We are currently studying the further steps corresponding to the cleavage of the benzyl ring of benzothiazoles which finally lead to mineralization of these compounds.

References

Besse P, Combourieu B, Poupin P, Sancelme M, Truffaut N, Veschambre H, Delort AM (1998) Degradation of morpholine and thiomorpholine by an environmental *Mycobacterium* involves a cytochrome P450. Direct evidence of the intermediates by in-situ ^{1}H-NMR. J Mol Biocatal B Enz 5:403–409

Besse P, Combourieu B, Boyse G, Sancelme M, De Wever H, Delort AM (2001) Long-range ^{1}H-^{15}N heteronuclear shift correlation at natural abundance: a tool to study benzothiazole biodegradation by two *Rhodococcus* strains. Appl Environ Microbiol 67:1412–1417

Brecker L, Ribbons DW (2000) Biotransformations monitored in situ by proton nuclear magnetic resonance spectroscopy. Trends Biochem 18:197–202

Bujdakova H, Kuchta T, Sidoova E, Gvozdjakova A (1993) Anti-*Candida* activity of four antifungal benzothiazoles. FEMS Microbiol Lett 112:329–334

Bujdakova H, Kralova K, Sidoova E (1994) Antifungal activity of 3-(2-alkylthio-6-benzothiazolyl-aminomethyl)-2-benzothiazolinethiones in vitro. Pharmazie 49:375–376

Combourieu B, Besse P, Sancelme M, Veschambre H, Delort AM, Poupin P, Truffaut N (1998a) Morpholine degradation pathway of *Mycobacterium aurum* MO1: direct evidence of intermediates by in-situ ^{1}H nuclear magnetic resonance. Appl Environ Microbiol 64:153–158

Combourieu B, Poupin P, Besse P, Sancelme M, Veschambre H, Truffaut N, Delort AM (1998b) Thiomorpholine and morpholine oxidation by a cytochrome P450 in *Mycobacterium aurum* MO1. Evidence of the intermediates by in-situ ^{1}H NMR. Biodeg 9:433–442

Combourieu B, Besse P, Sancelme M, Godin JP, Monteil A, Veschambre H, Delort AM (2000) Common degradative pathways of morpholine, thiomorpholine and piperidine by *Mycobacterium aurum* MO1: evidence from ^{1}H-nuclear magnetic resonance and ionspray mass spectrometry performed directly on the incubation medium. Appl Environ Microbiol 66:3187–3193

De Wever H, Verachtert H (1997) Biodegradation and toxicity of benzothiazoles. Water Res 31:2673–2684

De Wever H, de Cort S, Noots I, Verachtert H (1997) Isolation and characterization of *Rhodococcus rhodochrous* for the degradation of the wastewater component 2-hydroxybenzothiazole. Appl Microbiol Biotechnol 47:458–461

De Wever H, Vereecken K, Stolz A, Verachtert H (1998) Initial transformations in the biodegradation of benzothiazoles by *Rhodococcus* isolates. Appl Environ Microbiol 64:3270–3274

De Wever H, Besse P, Verachtert H (2001) Microbial transformations of 2-substituted benzothiazoles. Appl Microbiol Biotechnol 57:620–625

Delort AM, Combourieu B (2000) Microbial degradation of xenobiotics. In: Barbotin JN, Portais JC (eds) NMR in microbiology: theory and applications. Horizon Scientific, UK, pp 411–430

Delort AM, Combourieu B (2001) In-situ ^{1}H-NMR study of the biodegradation of xenobiotics: application to heterocyclic compounds. J Ind Microbiol Biot 26:2–8

Fiehn O, Reemtsma T, Jekel M (1994) Extraction and analysis of various benzothiazoles from industrial wastewater. Analyt Chim Acta 295:297–305

Gaja MA, Knapp S (1997) The microbial degradation of benzothiazoles. J Appl Microbiol 83:327–334

Gold LS, Slone TH, Stern BR, Bernstein L (1993) Comparison of target organs of carcinogenicity for mutagenic and non-mutagenic chemicals. Mutat Res 286:75–100

Haroune N, Combourieu B, Besse P, Sancelme M, Delort AM (2001) ^{1}H NMR: a tool to study the fate of pollutants in the environment. CR Acad Sci Paris, Chimie/Chemistry 4:759–763

Haroune N, Combourieu B, Besse P, Sancelme M, Reemtsma T, Kloepfer A, Diab A, Knapp JS, Baumberg S, Delort A-M (2002) Benzothiazole degradation by *Rhodococcus pyridinovorans* strain PA: evidence of a catechol 1,2-dioxygenase activity. Appl Environ Microbiol 68:6114–6120

Hartley D, Kidd H (1987) The agrochemical handbook. The Royal Society of Chemistry, Nottingham

Malouki MA, Giry G, Besse P, Combourieu B, Sancelme M, Bonnemoy F, Richard C, Delort AM (2003) Sequential bio- and phototransformation of herbicide methabenzthiazuron in water. Environ Toxicol Chem 22:2013–2019

Martin GE, Hadden CE (2000) Long-range ^{1}H-^{15}N heteronuclear shift correlation at natural abundance. J Nat Prod 63:543–585

Meding B, Toren K, Karlberg AT, Hagberg S, Wass K (1993) Evaluation of skin symptoms among workers at a Swedish paper mill. Am J Ind Med 23:721–728

Poupin P, Truffaut N, Combourieu B, Besse P, Sancelme M, Veschambre H, Delort AM (1998) Degradation of morpholine by an environmental strain of *Mycobacterium* involves a cytochrome P450. Appl Environ Microbiol 64:159–165

Reemtsma T, Fiehn O, Kalnowski G, Jekel M (1995) Microbial transformations and biological effects of fungicide-derived benzothiazoles determined in industrial wastewater. Environ Sci Technol 29:478–485

Wegler R, Eue L (1977) Chemie der Pflanzenschutz- und Schädlingsbekämpfungsmittel. 5. herbizide. Springer-Verlag, Berlin Heidelberg New York, pp 752

Chapter 29

Biotransformation of Nonylphenol Surfactants in Soils Amended with Contaminated Sewage Sludges

J. Dubroca · A. Brault · A. Kollmann · I. Touton · C. Jolivalt · L. Kerhoas · C. Mougin

Abstract

The biotransformation of nonylphenol was investigated in an agricultural soil treated with a mixture of ^{14}C-labelled and unlabelled surfactant. It was then studied in soil samples amended with sludges spiked with the mixture of chemicals. Nonylphenol amount in all samples of soil and soil/sludge mixtures was 40 mg kg^{-1}. In the soil free of sludge, the half-life of nonylphenol was found to be 4 d. When the soil was amended with sludge from the city of Ambares, France, it was about 16 d. In the soil amended with sludge from Plaisir, a 8-day lag phase was observed before the transformation starts, and nonylphenol half-life exceeded 16 d. In each case, nonylphenol transformation resulted in mineralization as well as stabilization of the chemical as bound residues within the soil. Further, some strains of filamentous fungi were isolated from the soil/sludge mixtures and identified to belong to the *Mucor* and *Fusarium* species. Most of them were able to efficiently transform nonylphenol in liquid cultures. In addition, the ligninolytic basidiomycete *Trametes versicolor* was able to catalyze partly the conversion of nonylphenol into carbon dioxide. Laccases purified from *T. versicolor* cultures are enzymes involved in nonylphenol oxidative coupling leading to oligomerization.

Key words: sewage sludge, soil, alkylphenols, fungi, enzymes, biotransformation, oxidative coupling

29.1
Introduction

The use of sewage sludge to fertilize agricultural land has been recognized in the past decades to be an economic and environmentally acceptable method for municipal sludge disposal. However, these sludges contain, in addition to nutrients, numerous heavy metals and organic contaminants, which can represent serious risks to human health and the environment. Thus, at the end of sewage treatment, sludges contain many substances which are not fully degraded. Among these compounds there are variable amounts of organic endocrine disrupters, whose presence in the environment is currently an increasing concern.

Alkylphenol ethoxylates are non-ionic surfactants used in domestic, industrial and agricultural applications. In wastewater treatment plants, they are partly degraded to nonylphenols, which can amount to about 80% of the ethoxylates. Nonylphenols are released in the environment as a mixture of 18 isomers (nonylphenol), including the well-known 4-*n*-nonylphenol (4 nNP) (Fig. 29.1). They comprise both hydrophobic branched nonyl groups and a hydrophilic moiety (Ahel et al. 1994a,b). The mixture exhibits a very high water solubility (5 000 mg l^{-1}) and high log$P$ value (5.92).

Nonylphenols represent an environmental hazard to the environment because they induce several toxic effects on wildlife and humans. They are strong endocrine disrupters,

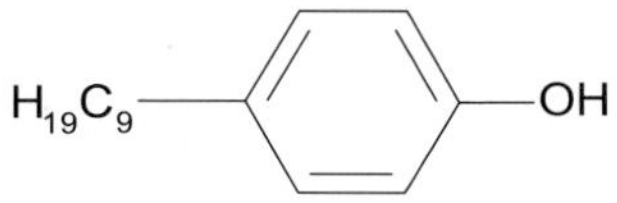

Fig. 29.1.
Structure of the 4-*n*-nonylphenol. Isomers have branched nonyl groups

$H_{19}C_9$—⟨phenyl ring⟩—OH

as well as inducers of breast cancer in women and of prostate in men (Routledge and Sumpter 1996). They also contribute to the feminization of male fish in sewage outflows (White et al. 1994). Endocrine disrupters have been designed to interfere with plant-*Rhizobium* signalling and nitrogen-fixing symbiosis (Fox et al. 2001). They induce also developmental abnormalities among freshwater sponges (Hill et al. 2002). Finally, they are responsible for morphological defects and for multiple physiological effects, partly due to uncoupled respiration in filamentous fungi and yeasts (Karley et al. 1997).

Following sewage treatment, low amounts of nonylphenols (less than 10 µg l^{-1}) are discharged in water (Ahel et al. 1994b), and both the acute and chronic toxicity of these compounds to aquatic organisms are well known. By contrast, high amounts of nonylphenol accumulate in sludge (more than 1 g kg^{-1}, Ahel et al. 1994a). To date, knowledge of their fate and impact is very scarce. This raises two questions of great concern: *(1)* what is the fate of nonylphenols in soils amended with sludges, and *(2)* what is their impact on soil (micro)organisms?

Our objectives are to assess the ecotoxicological risk induced by soil amendment with sewage sludges from urban origin, containing organic pollutants. We are currently involved in exposure assessment of soil organisms to chemicals. In this paper, we investigate the biotransformation of nonylphenol in an agricultural soil, amended or not with nonylphenols-containing sludges. Fungal biotransformation of the pollutants is also reported, as well as transformation by enzymes purified from fungi.

29.2
Experimental

29.2.1
Chemicals

4-*n*-Nonylphenol (4nNP) was obtained from Lancaster, whereas technical nonylphenols (nonylphenol, a mixture of structural isomers of 4nNP through branched nonyl groups) were obtained from Fluka. Other chemicals were available from Sigma. U-ring-^{14}C-labelled 4 NP (2 GBq mmol^{-1}, radiochemical purity >99%) was a generous gift of Dr. J.-P. Cravedi (INRA, Toulouse).

29.2.2
Characteristics of Soil and Sludges

We investigated the biotransformation of nonylphenol in a silt loam soil, composed of 25.5% sand, 55.0% silt and 19.5% clay (% refers to weight %). Its organic matter content was 1.65%. Soil pH was 8.1 and the cationic exchange capacity was 10.2 meq 100 g^{-1}. The soil, collected in the 10–20 cm layer from an experimental field in Versailles, was sieved (2 mm) and used immediately.

Table 29.1.
Origins and physico-chemical properties of the sludges used in the study. PAH: polycyclic aromatic hydrocarbons; PCB: polychlorobiphenyls. % refers to weight %

Property	Ambares (33)	Plaisir (78)
Dry matter (%)	75.71	33.12
Organic matter of dry matter (%)	51.41	45.20
Total nitrogen of dry matter (%)	3.97	2.66
pH water	7.8	12.5
Nonylphenol content (mg/kg d.w.)	200.0	506.0
PAH content[a] (mg/kg d.w.)	0.51	0.43
PCB content[b] (mg/kg d.w.)	0.089	0.071
Metal ion content[c] (mg/kg d.w.)	8880.71	113.86

[a] Fluoranthene, benzo[b]fluoranthene, benzo[a]pyrene.
[b] PCBs 28, 52, 101, 118, 138, 153, 180.
[c] Fe, Mn.

We used two sludges exhibiting different characteristics (Table 29.1). The sludge from Ambares was composed of both urban (90 000 equivalent inhabitants) and industrial wastewaters. That from Plaisir mostly contained urban wastewater (42 000 equivalent inhabitants).

29.2.3
Biotransformation of Nonylphenol in Soil and Soil/Sludge Mixtures

Soil samples (27 g dry soil) were incubated in 150-ml Erlenmeyer flasks, in darkness at 25 °C, for 96 d. Water was added to obtain a 80% soil moisture holding capacity. The moisture content was kept constant during the experiments by adding sterile water every week. The soil was supplemented with unlabelled and labelled (9 kBq) nonylphenol to give a final concentration of 40 mg kg^{-1}. The Erlenmeyer flasks were sealed with cotton plugs and incubated in 1-l sealed flasks with vials containing 10 ml of 1N NaOH (to trap CO_2) and 10 ml water (to keep moisture constant), according to Mougin et al. (1997).

We used the same protocol for incubations with soil/sludge mixtures. Sludges containing endogenous nonylphenol were spiked with unlabelled and labelled chemical. After addition of the chemicals, the solvent (acetone) was allowed to evaporate for 30 min, and nonylphenol was sorbed to the sludges for 24 h at 4 °C under nitrogen atmosphere to avoid any biotransformation. 2 g dry sludge samples were then mixed with 25 g dry soil for incubations. The final amount of nonylphenol was 40 mg kg^{-1} in all soil/sludge samples.

29.2.4
Soil and Soil/Sludge Mixtures Analysis

Soil samples were mixed with 1 g hyflosupercel (diatomaceous silica for filtration) and 20 ml water, then nonylphenols were extracted by adding 100 ml acetone-water (80/20, v/v) and shaking for 60 min. The liquid and solid phases were separated by filtration on a Büchner funnel, and the solid fraction was extracted a second time. The extracts were pooled. The radioactivity was measured by counting 500 µl aliquots of the extracts by liquid scintillation.

The extracts obtained from soil/sludge mixtures were purified prior to high performance liquid chromatography (HPLC) analysis. 5 ml aliquots were first evaporated to complete dryness and dissolved in 1 ml hexane. The solution was applied onto a glass column packed with 5 g Florisil (activated with 2% water), then eluted. The first 30 ml fraction of hexane was discarded. The radioactive compounds were then eluted by 90 ml dichloromethane. Finally, the solvent was concentrated to complete dryness and dissolved in 20 ml acetonitrile for HPLC analysis. Non-extractable radioactivity in the soil and soil/sludge samples was determined by combustion in a model 307 oxidizer (Packard Instrument, Rungis, France). $^{14}CO_2$ production was quantified by counting 500 µl aliquots of NaOH solution by liquid scintillation.

HPLC analysis was performed by injecting 100 µl of the organic extracts onto an analytical column TSK ODS-80TM (25 cm × 4.6 mm i.d., Varian, les Ulis, France) set at 30 °C. A Varian 9010 pump delivered the mobile phase consisting of a mixture of acetonitrile-water-H_3PO_4 (50/50/0.05, v/v/v) at a rate of 1 ml min^{-1}. It was increased to 70%v acetonitrile in 1 min and maintained at this value during 22 min. Another linear increase to 100%v acetonitrile occurred in 2 min, and then followed by a stationary phase of 5 min and a return to initial conditions. Radioactivity and A_{224} of the column eluate were monitored. Our analytical protocol allowed a high ^{14}C recovery amounting to 98.0% of initial radioactivity.

29.2.5
Isolation of Fungi from Soil/Sludge Mixtures and Biodegradation Tests

Fusarium and *Mucor* strains were isolated from soil/sludge mixtures by layering soil aggregates onto plates containing malt (20 g l^{-1}), agar (15 g l^{-1}) and yeast extract (1 g l^{-1}), supplemented with a mixture of chloramphenicol, streptomycin, penicillin G and chlortetracycline, each at 50 mg l^{-1}. The plates were incubated for one week at 25 °C. Then, according to their morphology, fungal strains were picked, individually cultured, and identified. Fungal strains from our collection were also taken into account.

Fungal biotransformation of nonylphenol was then assessed by culturing the strains at 25 °C in Erlenmeyers flasks containing 10 ml liquid media as described earlier (Mougin et al. 2000). Glycerol and maltose were used as carbon sources for *P. chrysosporium* and all the other strains, respectively. Nonylphenol was added at 11 mg l^{-1} after 3 d of growth.

A detailed study of the biotransformation of 4nNP was achieved in *T. versicolor* cultures using a chemical solution spiked with 5 kBq labelled chemical. It was immediately added to the culture without the 3 d of pregrowing phase. Liquid and solid phases were analyzed separately.

29.2.6
Analysis of Fungal Cultures

The remaining nonylphenol was extracted three-times from the biomass with 20 ml hot methanol (60 °C). After cooling, the alcoholic fractions were reduced under vacuum, pooled with the medium, and the aqueous solution was extracted twice with 35 ml diethyl ether. Then, the solvent was evaporated and the compounds were dissolved in 1 ml methanol for HPLC analysis. Analysis was performed as described for soil experiments.

29.2.7
Enzymatic Assays with Purified Laccases

Laccases (*para*-diphenol:dioxygen oxidoreductases, EC 1.10.3.2) were purified according to published procedures (Jolivalt et al. 1999). They were then incubated in aerated 0.1 M citrate/phosphate buffer (1 unit enzyme in 1 ml buffer, pH ranging from 4 to 7) under stirring at 30 °C, in the presence of 5 mg l^{-1} 4nNP. Aliquots (100 µl) of the incubation media were then directly analyzed by HPLC using the procedures described above.

29.2.8
Identification of Nonylphenol Metabolites

We determined the chemical structure of the metabolites formed by purified enzymes using a mass-spectrometer Nermag R30-10C. The spectra were obtained by Desorption Chemical Ionization with ammonia (NH_3) as a reactant gas. The spectrometer was set in positive mode.

29.3
Results and Discussion

29.3.1
Biotransformation of Nonylphenol in an Agricultural Soil Free of Sludge

We studied the ability of the endogenous microflora to transform nonylphenol in an agricultural soil. Soil samples were incubated in darkness at 25 °C for 96 d in the presence of 40 mg kg^{-1} nonylphenol (a mixture of unlabelled and labelled chemicals). Our results showed that the greatest part of the ^{14}C contained in nonylphenol was associated to the solid fraction of the soil, and about 40.0% of initial ^{14}C content was measured following soil combustion after 96 d (Fig. 29.2). In addition, 31.4% of the labelled compounds were transformed to carbon dioxide during the same period. The radioactivity present in the organic extract was 30.7% at the end of the experiment. HPLC analysis showed a decrease of the nonylphenol content with time in the extract. They were the main components of the extract during the first week of incubation, and became negligible after 32 d of incubation. Residual nonylphenol was assumed to be the sum of the ^{14}C non extractable from the soil and of the surfactant remaining in the organic extracts analyzed by HPLC. Thus, we calculated a half-life of 4 d for nonylphenols in our incubation conditions. The value corresponded to the time required to transform (by stabilization or degradation) 50% of initial amounts of nonylphenol. The remaining fraction of the radioactivity in the extract was attributed to polar unidentified transformation products. Our results show that nonylphenol is rapidly transformed by the endogenous microflora of the aerated soil, in agreement with the previous study of Topp and Starratt (2000).

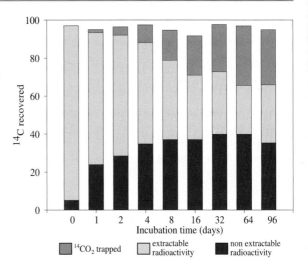

Fig. 29.2.
Distribution of labelled nonylphenol during the incubations with the soil free of sludge. Note the increase of both $^{14}CO_2$ produced and non extractable radioactivity due to biological activity. On the other hand, extractable radioactivity decreased with incubation time

Table 29.2. Distribution of labelled nonyphenol during incubation of soil/sludge mixtures. Note that the transformation kinetics and pathways of nonylphenols differ when the chemicals are in the soil free of sludge or brought by sludges from two origins

Soil/sludge mixture	% of initial radioactivity					
	Bound		Extracted		Mineralized	
	8 days	96 days	8 days	96 days	8 days	96 days
Soil from Versailles	37.2	40.0	41.9	30.7	15.9	31.4
Soil/sludge from Ambares	26.9	21.5	20.5	5.2	32.9	54.6
Soil/sludge from Plaisir	12.7	44.9	81.1	7.6	0.0	36.1

29.3.2
Biotransformation of Nonylphenol in Soils Amended with Sewage Sludges

This experiment was intended to study the transformation of nonylphenol in soils amended with sludges. Nonylphenol was brought by sludges in order to mimic the upper layer of an agricultural soil amended with contaminated sludges. The biotransformation of the chemical is supposed to be mediated by soil microflora. Mixtures were incubated in darkness at 25 °C for 96 d in the presence of 40 mg kg^{-1} of a mixture of unlabelled and labelled nonylphenol. In the soil amended with the sludge from Ambares, the amount of ^{14}C bound to soil materials represented 21.5% of the initial radioactivity after 96 d of incubation (Table 29.2). At the opposite, the total $^{14}CO_2$ produced in the incubation vials during the same period amounted to 54.6% of the initial radioactivity. On the other hand, extractable radioactivity was drastically reduced over the incubation period, representing only 5.2% of the initial radioactivity after 96 d. Our results indicated that the kinetics of the transformation of nonylphenol brought by the sludge from Ambares and this observed in the soil free of sludge were different, as its half-life was calculated to be 16 d in the presence of the sludge. The amendment mainly increased the mineralization of nonylphenol whereas its stabilization in soil was decreased. In addition, extracted radioactivity was low in soil/sludge samples. These results suggested that the sludge from Ambares also modified the biotransformation pathways of nonylphenol.

Results of nonylphenol incubation were also different in the soil amended with the sludge from Plaisir. Indeed, a 8-day lag phase occurred, during which no transformation could be observed (data not shown). The depressive effects of the sludge on biodegradation can be attributed to several factors: *(1)* toxicity of the sludges, because of the presence of numerous organic contaminants and high amounts of metal ions, thus requiring the selection of resistant microorganisms, *(2)* a high biological oxygen demand, due to the high organic matter content of the sludges, *(3)* an increase of nonylphenol adsorption in the mixtures by comparison with the soil alone. Nevertheless, relatively few studies have been published, which address the effects of sludge-bound chemicals on microbial processes in soils (Gejlsberg et al. 2001).

After the lag phase, mineralization started rapidly to reach 36.1% of the initial radioactivity after 96 d, a level quite identical to that noticed in the soil incubated alone. Following a similar pattern, non-extractable radioactivity represented 44.9% of the initial radioactivity. Then, extractable radioactivity was also reduced to 7.6% at the end of the experiment. Analysing the organic extract by HPLC allowed estimating a half-life of nonylphenol between 16 and 20 d in these conditions. These results clearly demonstrated that the nature of the sludge affect nonylphenol biotransformation in the soil, resulting in an increase of the half-life of the chemical.

29.3.3
Biotransformation of Nonylphenol by Filamentous Fungi

The aim of that study was to evidence the ability of fungi to transform nonylphenol in liquid cultures. Half-lives of the chemicals is based on their residual content in the cultures with respect to initial amounts (Table 29.3). Degradation of the 4nNP with half-lives lower than 2 d proceeds in general more rapidly than that of nonylphenol (>2 days). However, the two white-rot fungi, *T. versicolor* and *P. chrysosporium*, transformed nonylphenol more extensively than 4nNP. In addition, *Fusarium* strains, isolated from soils amended with sludges from both origins, were poor degraders of isoNPs. By contrast, *Mucor* strains, only present in the sludge from Ambares, and *Fusarium*, were able to efficiently transform the mixture. The results demonstrated that the extent of nonylphenol transformation varied according to the fungal taxonomic groups. White-rot fungi have been described for many years as efficient tools for organic pollutant breakdown, because of the secretion of exocellular oxidases (Pointing 2001). By contrast, no data are available concerning the enzymatic systems from strains belonging to *Mucor* and *Cunninghamella* genera able to transform phenolic compounds.

Table 29.3.
Transformation of nonylphenol compounds in fungal liquid cultures. Half-lives have been calculated on the basis of residual content of the chemicals in the cultures, with respect to initial amounts

Strains	Half-life (days)	
	4-n-nonylphenol	nonylphenol
Soil amended with sludge from Ambares		
Mucor racemosus	2.0	3.0
Fusarium oxysporum a. solani	2.0	>8.0
Mucor hiemalis	1.5	5.0
Soil amended with sludge from Plaisir		
Fusarium solani	1.0	>8.0
Fusarium sp.	1.5	>8.0
Fusarium oxysporum	2.0	6.0
Strains from our collection		
Trametes versicolor	1.0	<1.0
Phanerochaete chrysosporium	6.0	3.0
Cunninghamella elegans	1.0	2.0
Cunninghamella echinulata	1.0	2.0

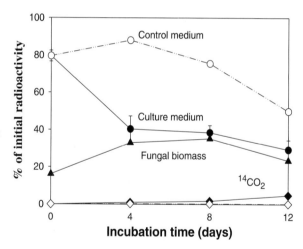

Fig. 29.3.
(^{14}C)nonylphenol mass-balance in *T. versicolor* cultures (*black symbols*) and non-inoculated controls (*white symbols*). Note the decrease of radioactivity in the control cultures due to adsorption on glass, and its partitioning between medium and biomass in fungal cultures

29.3.4
Biotransformation of 4-n-Nonylphenol by the Fungus *T. versicolor*

The biotransformation of 4-n-nonylphenol isomer has been investigated in liquid cultures of *T. versicolor*, previously shown as the most efficient alkylphenol degrader. In non-inoculated controls, ^{14}C amounts in the medium were 50% lower during the 12 d of incubation, due to adsorption on glass. No mineralization occurred. During the same period, radioactivity partitioned between the medium and the biomass in the inoculated cultures amounted to 29.2 and 23.4% at the end of the experiment, respectively. 4nNP was partly converted by the fungus into labelled carbon dioxide (6%).

Using HPLC, we failed to detect any radioactive peaks corresponding to transformation products in analyzing the organic extract obtained from the medium, although it contained significant amounts of ^{14}C. Direct injection of culture medium into the HPLC system showed that the radioactivity was retained in the analytical column (data not shown). These results suggest that a high-molecular weight compound could be formed by the fungus, and that its ability to mineralize nonylphenol is very reduced. For that reason, such a biodegradation process observed in soils may be rather due to the catabolic activity of bacteria.

29.3.5
Biotransformation of 4-*n*-Nonylphenol by Laccases Purified from *T. versicolor* Cultures

Laccases are the main exocellular enzymes produced by *T. versicolor*. This experiment addresses their ability to transform phenolic compounds such as nonylphenol. For that purpose, purified enzymes have been incubated in the presence of 5 mg l^{-1} 4nNP. The concentrations of 4nNP were dramatically reduced in the incubation media in the absence of laccase, and the residual chemical represented only 34% of its theoretical value after only 30 s of incubation (Fig. 29.4). This apparent loss was due to a strong adsorption of 4nNP on the glass of the incubation vessels. The disappearance of 4nNP was increased when laccase was added to the media, and more than 90.0% of the chemical were transformed after 5 min of incubation. The reaction was pH-dependant. Higher 4nNP disappearances were observed at acidic pH (4 and 5). Most of 4nNP (80%) was transformed after 5 min at pH 4, 60.0% at pH 5 and only 30% at pH 6. The reaction was totally inhibited at pH 7.

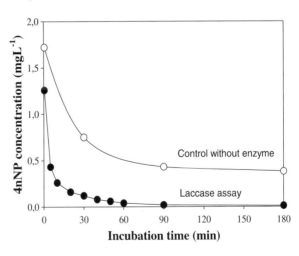

Fig. 29.4.
Transformation of 4-*n*-nonylphenol by purified laccases. Note the more rapid decrease of nonylphenol content in the buffer containing the enzymes than in the control medium

No soluble compounds corresponding to transformation products could be detected after analysis of the medium by HPLC. However, a brown precipitate accumulated on the bottom of the incubation vials, with respect to time. It was not soluble in alcohols, was only slightly solubilized by classical organic solvents, such as dichloromethane, ethyl acetate, or hexane.

29.3.6
Identification of Nonylphenol Metabolite

We attempted to identify the metabolites formed by incubations of 4-*n*-nonylphenol with purified laccases by mass-spectrometry. The spectrum of the fraction solubilized in organic solvents showed ions at *m/z* 220, 438 + 456, 656 + 674, 874 + 892, 1111 + 1115 (Fig. 29.5). The first ion (220) resulted from the fragmentation of larger molecules, and corresponded to the molecular weight (MW) of 4nNP. The doublets, all exhibiting a difference of 18 amu due to the loss of water, corresponded to M$^+$ ions and MNH$_4^+$ adducts. They were attributed to oligomerized 4nNP as dimeric (MW = 438), trimeric (MW = 656), tetrameric (MW = 874) and pentameric (MW = 1092) compounds. For the last compound, the discrepancy between theoretical and experimental data was due to the calibration range of the spectrometer, which was not optimized for high mass values. Additional experiments are under progress to confirm these results using NMR and infra-red spectrometry.

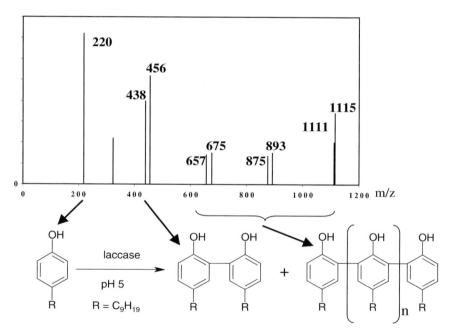

Fig. 29.5. Mass-spectrum and possible structure of oligomers formed by incubation of 4-nonylphenol with fungal laccase

29.4 Conclusions

Nonylphenol is rapidly transformed in aerated soils by mineralization and stabilization (formation of bound residues). The degradation rates are somewhat decreased when the soil is amended with sewage sludges containing similar amounts of nonylphenol. The strongest slowing effect was observed in the presence of the sludge from Plaisir. The filamentous fungi isolated from the soil/sludge mixtures and from our collection transformed nonylphenol with different efficiencies. Yet, *T. versicolor* (basidiomycete) and *Mucor* strains (zygomycetes) transformed the chemical, whereas *Fusarium* strains (deuteromycetes or ascomycetes) were poor degraders. Fungal laccases, produced mainly by white-rot basidiomycetes such as *T. versicolor*, are enzymes which catalyze the oxidative coupling of nonylphenol to produce oligomers. The reaction illustrates one of the mechanisms involved in the stabilization of xenobiotics in soils. Nonylphenol mineralization is mainly attributed to bacterial catabolic activity, whereas fungi are rather involved in the chemical stabilization in the soil through oxidative coupling. Although sludges modify the adsorption of nonylphenol onto soil, our results obtained using soil/sludge mixtures clearly establish that the exposure of soil organisms to the surfactant can be increased by slowing down its biodegradation. This result emphasizes the need to assess the effects of nonylphenol on soil microorganisms by further investigations.

References

Ahel M, Giger W, Koch M (1994a) Behaviour of alkylphenol polyethoxylate surfactants in the aquatic environment – I. Occurrence and transformation in sewage treatment. Water Res 28:1131–1142

Ahel M, Giger W, Schaffner C (1994b) Behaviour of alkylphenol polyethoxylate surfactants in the aquatic environment – II. Occurrence and transformation in rivers. Water Res 28:1143–1152

Fox JE, Starcevic M, Kow KY, Burow ME, McLachlan JA (2001) Endocrine disrupters and flavonoid signal. Nature 413:128–129

Gejlsberg B, Klinge C, Samsoe-Petersen L, Madsen T (2001) Toxicity of linear alkylbenzene sulfonates and nonylphenol in sludge-amended soil. Environ Toxicol Chem 20:2709–2716

Hill M, Stabile C, Steffen LK, Hill A (2002) Toxic effects of endocrine distupters on freshwater sponges: common developmental abnormalities. Environ Pollut 117:295–300

Jolivalt C, Raynal A, Caminade E, Kokel B, Le Goffic F, Mougin C (1999) Transformation of N',N'-dimethyl-N-(hydroxyphenyl)ureas by laccase from the white rot fungus *Trametes versicolor*. Appl Microbiol Biotechnol 51:676–681

Karley AJ, Powell SI, Davies JM (1997) Effect of nonylphenol on growth of *Neurospora crassa* and *Candida albicans*. Appl Environ Microbiol 63:1312–1317

Mougin C, Pericaud C, Dubroca J, Asther M (1997) Enhanced mineralization of lindane in soils supplemented with the white rot basidiomycete *Phanerochaete chrysosporium*. Soil Biol Biochem 29:1321–1324

Mougin C, Boyer F-D, Caminade E, Rama R (2000) Cleavage of the diketonitrile derivative of the herbicide isoxaflutole by extracellular fungal oxidases. J Agric Food Chem 48:4529–4534

Pointing SB (2001) Feasibility of bioremediation by white-rot fungi. Appl Microbiol Biotechnol 57:20–33

Routledge EJ, Sumpter JP (1996) Estrogenic activity of surfactants and some of their degradation products assessed using a recombinant yeast screen. Environ Toxicol Chem 15:214–248

Topp E, Starratt (2000) Rapid mineralization of the endocrine-disrupting chemical 4-nonylphenol in soil. Environ Toxicol Chem 19:313–318

White R, Jobling S, Hoare SA, Sumpter JP, Parker PG (1994) Environmentally persistent alkylphenolic compounds are estrogenic. Endocrinology 135:175–182

Part III

Chapter 30

Quantification of in-situ Trichloroethene Dilution versus Biodegradation Using a Novel Chloride Concentration Technique

C. Walecka-Hutchison · J. L. Walworth

Abstract

The objective of this study was to evaluate the effectiveness of in-situ trichloroethene (TCE) bioremediation, and to determine whether the observed decrease in TCE concentrations was attributable to biological degradation versus abiotic processes. An enhanced in-situ TCE bioremediation project in which groundwater amended with microbe stimulating compounds was injected into the contaminated subsurface was analyzed. Dilution, attributed to mixing between the injected clean and contaminated waters, was calculated using a modified groundwater mixing equation and chloride concentrations of the waters at various times in the study. Over the course of the trial, spatially averaged TCE concentrations within the aquifer decreased by 41%. The chloride calculations suggested that a 29% reduction may be attributable to dilution, and that only a 12% decrease in concentrations may be attributable to biological degradation.

Key words: trichloroethene, biodegradation, groundwater mixing, chloride, dilution

30.1 Introduction

The widespread use of chlorinated ethenes as organic solvents, and their subsequent improper disposal practices, has resulted in extensive global soil and groundwater contamination. The most frequently detected organic groundwater pollutant in the United States is trichloroethene (TCE). It has been estimated that between 9 and 34% of the country's drinking water supply sources have been contaminated with this suspected carcinogen (Agency for Toxic Substances and Diseases Registry 1989). Conventional means of solvent contaminated aquifer restoration not only entail great expense, but often merely transfer the contaminants to another medium. In-situ bioremediation is a more favorable alternative as it transforms TCE to stable, non-toxic end products (carbon dioxide, chloride, and water (Little et al. 1988)) without bringing groundwater to the surface, thus potentially reducing remediation costs.

30.2 Review on Trichloroethene Biodegradation

TCE biodegradation occurs in both aerobic and anaerobic environments. Anaerobically, TCE undergoes reductive dechlorination. This process involves the removal of a chlorine atom from chlorinated ethenes followed by its replacement with a hydrogen atom. The TCE becomes biotransformed to 1,2-dichloroethene (predominantly *cis*-1,2-DCE, with low quantities of *trans*-1,2-DCE), vinyl chloride, and eventually ethene (Ensley 1991). The rates of reductive dehalogenations are higher for highly chlorinated compounds, resulting in the persistence of the less chlorinated daughter products under reducing environmental conditions (Vogel et al. 1987). The potential persistence of vinyl chloride as the end product of this pathway is of concern as the compound is a known human carcinogen.

Transformation rates of chlorinated solvents under aerobic conditions are highest for the least chlorinated species (Vogel et al. 1987). Consequently, highly chlorinated compounds like perchloroethene (PCE) are resistant to oxidative processes. Aerobic TCE degradation is a cometabolic or fortuitous transformation due to broad specificity of microbial enzyme systems (McCarty and Semprini 1994). The microorganism requires a primary substrate (electron donor) for growth, but due to the broad enzyme specificity, the microorganism can also degrade the chlorinated solvent. No energy is gained by the microorganism from the transformation of the solvent.

The initial step of the aerobic TCE biodegradation pathway is the formation of TCE epoxide, which can be biotransformed for example by methanotrophic bacteria, (organisms that oxidize methane for energy and growth) to dichloroacetic and glyoxilic acid. The acids can be further broken down to carbon dioxide, chloride, and water by heterotrophic bacteria (Little et al. 1988). Although the TCE epoxide is toxic and carcinogenic, its 12-second half-life makes it an intermediate product of lesser importance (Oldenhuis et al. 1989). Therefore, the aerobic pathway is frequently preferred for site remediation.

Enzymes responsible for oxidation of TCE include methane monooxygenase (particulate and soluble), ammonia monooxygenase, toluene oxygenase (mono and dioxygenase), and propane monooxygenase. The cometabolic inducers of these oxygenases include: *(1)* methane, *(2)* ammonia, *(3)* toluene, phenol and *(4)* propane (Ensley 1991). These enzymes are produced by a variety of microorganisms, many of which experience contaminant toxicity if the TCE they co-oxidize is encountered at concentrations above 6 000 µg l^{-1} (Broholm et al. 1990). Additionally, the metabolites or by-products of the aerobic degradation pathway have shown toxic effects toward the oxygenase enzyme systems. Methane monooxygenase has shown complete inactivation (was no longer able to oxidize methane for energy and growth and/or transform TCE) after 200 transformations of TCE (Ensley 1991). Although no oxygenase is resistant to these toxic effects, the toluene monooxygenase enzyme is more resistant than the toluene dioxygenase (Ensley 1991).

The biodegradation efficacy is frequently measured via observed reductions in aqueous contaminant concentrations, and further supported by an increase in degradation product concentrations, CO_2 evolution, microbial enumeration, and/or utilization of amended electron acceptors and nutrients. However, contaminant disappearance does not distinguish biodegradation from other abiotic processes such as dispersion, sorption, diffusion or volatilization. The quantification of biodegradation relies on a firm understanding of the physical and chemical processes controlling the fate of the contaminant in question (Rittmann et al. 1994). However, accurate mass balances of contaminant abiotic processes at field sites are often unknown and/or unattainable. Therefore, proving whether contaminant removal is due to biotic or abiotic processes is difficult and generally only qualitative (Madsen 1991). The assessment of aerobic in-situ TCE biodegradation has proved to be especially challenging. Unlike its anaerobic counterpart, the aerobic pathway results in short-lived, and consequently difficult to detect, intermediate products. Additionally, due to its cometabolic and aerobic nature, primary substrates, oxygen and frequently nutrients need to be supplemented. This results in the added and generally unaccounted for abiotic component of dilution resulting from recharge of the amended water.

In conjunction with the measured change in TCE concentrations, first-order rate estimations have been used to evaluate biodegradation. Sorenson et al. (2000) compared three first-order rate methods and concluded that the graphical extraction method (Ellis 1996) as well as the Buscheck and Alcantar (1995) method resulted in overestimation of degradation rates as they did not adequately account for dispersion. The third method normalized TCE concentrations to the concentrations of internal plume tracers (co-contaminants tritium and tetrachloromethane) and consequently allowed degradation to be distinguished from dispersion.

Carbon fractionation has also been used in the attempt to substantiate in-situ biodegradation of organic pollutants. Suchomel and Long (1990) used the $^{13}C/^{12}C$ signature in soil CO_2 to prove aerobic TCE biodegradation. They found higher soil CO_2 concentrations with lower $\delta^{13}C$ values at TCE contaminated vs. uncontaminated sites. Furthermore, these values asymptotically approached the $\delta^{13}C$ values of the TCE. They attributed the signature to the different carbon isotopic composition of the contaminant and/or the fractionation occurring during the microbial degradation process. Lollar et al. (2001) used carbon isotope ratios ($^{13}C/^{12}C$) to quantify TCE biodegradation and found that during the anaerobic degradation of TCE the light (^{12}C) vs. heavy isotope (^{13}C) bonds are preferentially degraded, resulting in isotopic enrichment of the residual contaminant in ^{13}C.

The purpose of this study was to evaluate the effectiveness of aerobic in-situ TCE bioremediation. An enhanced field scale aerobic TCE biodegradation project located in Tucson, Arizona, was evaluated. The objective was to determine whether the observed decreases in TCE concentrations were attributable to biological degradation vs. other physical/chemical processes. Because the site was amended with a substrate (electron donor), oxygen (electron acceptor), and nutrients, dilution attributed to mixing between clean and contaminated water was a contributing factor in reducing the contaminant concentrations. The use of aqueous chloride concentrations, in conjunction with a modified groundwater mixing equation, allowed the quantification of dilution and differentiation between biotic and abiotic losses.

30.3
Experimental

30.3.1
Site Background

Air Force Plant number 44 (AFP#44) is located in southern Arizona, 15 miles south of the metropolitan Tucson area. It is owned by the United States Air Force and operated by Raytheon Missile Systems. The plant was constructed in 1951 and has been continuously utilized for the production of a variety of missile and weapon systems

(Hargis+Associates, Inc. 1996). During the 1950s, 1960s, and 1970s the facility utilized chlorinated solvents, primarily chlorinated aliphatic hydrocarbons, as degreasing agents. Disposal of these industrial wastes has consequently led to the contamination of soil and groundwater at this site. Groundwater contamination was detected in 1981, and in 1982 the United States Environmental Protection Agency listed AFP#44 as one of six project areas of the Tucson International Airport Area Superfund Site. The predominant pollutants of the site are TCE and 1,1-dichloroethene (1,1-DCE).

30.3.2
Site Hydrogeology

AFP#44 hydrogeology is broken down into the following units: the unsaturated zone, the shallow groundwater zone, the upper zone of the regional aquifer, the aquitard and the lower zone of the regional aquifer. The shallow groundwater zone is encountered at a depth of 18 to 26 m below ground surface and extends to 37 m. It consists of saturated, unconsolidated, dense, reddish clay to sandy clay with some very thin discontinuous lenses of fine sand (Hargis+Associates, Inc. 1996, 1997). This fine-grained unit is found above the transmissive upper zone of the regional aquifer thereby functioning as a confining unit.

The heterogenous sediments comprising the shallow groundwater zone exhibit relatively low hydraulic conductivity. The estimated average bulk hydraulic conductivity is 1.8×10^{-4} cm s^{-1} (GRC 1993). The general groundwater flow is in the north and northwest direction with an average linear groundwater velocity of 1.2 cm d^{-1} (data not shown). The total dissolved mass of TCE and 1,1-DCE in the shallow groundwater zone is believed to be 222 and 40 kg respectively (Hargis+Associates, Inc. 1996). TCE is the primary groundwater contaminant with concentrations ranging from 4.6 to 2 100 µg l^{-1} and a geometric mean of 40 µg l^{-1} (Fig. 30.1).

The shallow groundwater zone does not yield sufficient quantities of water to be a true aquifer. However, it slowly drains into the underlying upper zone of the regional aquifer and thus serves as a continuing source of groundwater contamination. The approximate contaminant contribution to the regional aquifer is 6 µg l^{-1} and 1 µg l^{-1} TCE and 1,1-DCE respectively (Hargis+Associates, Inc. 1996). Therefore, the primary objective in remediating the shallow groundwater zone was to minimize the impact of its contaminants on water quality of the underlying regional aquifer which serves as the as the primary source of water for the City of Tucson. The segment of the transmissive regional aquifer underlying AFP#44 contains the bulk of the TCE contamination of the site. Consequently, it has been undergoing containment/remediation via pump-and-treat technology since 1987 in the hope of returning it to drinking water quality (5 µg l^{-1} TCE).

30.3.3
Shallow Groundwater Zone Remediation

Bench scale studies performed to determine whether aerobic TCE biodegradation was feasible in the shallow groundwater zone demonstrated a 50% removal of the contaminant by microbial populations indigenous to the site (WCC 1996a). Additionally, both toluene dioxygenase and soluble methane monooxygenase were identified as present

Chapter 30 · Quantification of in-situ Trichloroethene Dilution versus Biodegradation

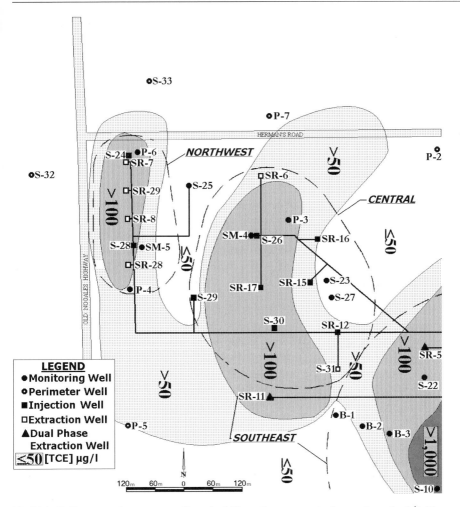

Fig. 30.1. Shallow groundwater zone wells and trichloroethene concentration contours (µg l^{-1}). The contamination ranges from 4.6 to 2 100 µg l^{-1}, being highest in the southeast portion of the region

in the shallow groundwater zone via gene probe analysis (WCC 1996b). Of the two major forms, the soluble methane monooxygenase demonstrates a broader substrate range relative to its particulate methane monooxygenase counterpart (Oldenhuis et al. 1989).

Field scale enhanced in-situ bioremediation began in April of 1997 and lasted through May of 1998 in the northwest and central portions of the shallow groundwater zone (TCE concentrations ranging from 47–170 µg l^{-1}, Fig. 30.1). The higher contamination of the southeast region of the shallow groundwater zone warranted remediation via dual-phase extraction. This technology extracts both groundwater and soil vapor from a single extraction well.

For the first three months of the enhanced bioremediation project, only "clean" groundwater was injected into the subsurface. Due to the limitations in organic carbon, oxygen, and inorganic nutrients (nitrogen and phosphorous) at the site, amendment injection began in July and consisted of hydrogen peroxide as an oxygen source, diammonium phosphate as both nitrogen and phosphorous source, and methanol as a potential cometabolic inducer and/or a carbon source (WCC 1998). The injection of the hydrogen peroxide and methanol was pulsed: hydrogen peroxide was amended for 24 out of each 36 h at a concentration of 38 mg l^{-1}, followed by methanol for the remaining 12 out of 36 h at a concentration of 75 mg l^{-1}. Diammonium phosphate was delivered continuously at 15 mg l^{-1}. During the lifespan of the project the total volume of water recharged into and extracted from the shallow groundwater zone was 3.3×10^7 and 1.5×10^7 l respectively (Montgomery and Associates 1998).

Monitoring wells within the shallow groundwater zone (Fig. 30.1) were sampled on a quarterly basis before as well as throughout the duration of the bioremediation project to determine changes in contaminant concentrations and in water level elevations. Heterotrophic plate counts as well as changes in chloride and dissolved oxygen concentrations were also evaluated on a quarterly basis during the lifespan of the bioremediation project. These data were assessed to determine the extent of microbial activity and its contribution to the changes in the contaminant concentrations.

30.3.4
Chloride Data Dilution Factors

Mixing between groundwaters of different recharge origins, different aquifers, and different flow systems have been quantified using isotopic methods (Clark and Fritz 1997). Simple linear algebra and $\delta^{18}O$ or $\delta^{2}H$ have been used to quantify mixing between two distinct groundwaters thereby depicting a portion of groundwater "A" in a mixture of "A" and "B" as follows: $\delta_{sample} = X\delta_A + (1-X)\delta_B$ ^{18}O and ^2H preserve the mixing ratio as they are non-reactive in nature. However the relationship does not indicate where the mixing occurs.

The availability and temporal variability of chloride concentrations within the monitoring wells of the shallow groundwater zone, coupled with the conservative nature of chloride, and its known concentration of 21 mg l^{-1} in the amended injected water, allowed the use of well chloride concentrations in determining dilution attributed to mixing between the clean injected and contaminated waters. A modified version of the groundwater mixing equation described above was used as follows:

$$C_c = C_a X + C_b(1-X)$$

where C_c was the final detected chloride concentration in each monitoring well, C_a was the injected amended clean water chloride concentration, C_b was the initial well chloride concentration before the onset of the bioremediation project, and X was the volume fraction of water ($V_a/(V_a + V_b)$, where V_a is the volume of injected clean water, and V_b the volume of the initial contaminated water). Solving the equation for X provides the percentage of the total volume of water within each well attributable to the injected clean water. The amount of contaminant removal attributable to dilution within the well could further be determined by dividing the contaminant concentration by a dilution factor, which can be calculated as: $1/(1-X)$.

30.4 Results and Discussion

Contaminant concentrations generally decreased throughout the shallow groundwater zone following the enhanced bioremediation technology (Table 30.1). Similar trends of contaminant reductions were observed within the monitoring wells and ranged from 26% to 87% and 55% to 90% in trichloroethene (TCE) and 1,1-dichloroethene (1,1-DCE) concentrations, respectively. The slightly greater observed decrease in the 1,1-DCE concentrations is not surprising as biological oxidations occur at faster rates in less chlorinated contaminants. The increased concentrations of both contaminants in well P-6 are unique and most likely attributable to contaminant desorption and diffusion from a low hydraulic conductivity layer which impedes flow and amendment delivery into this portion of the heterogeneous site (data not shown).

During the 13 months of operation, the in-situ bioremediation technology recharged 3.3×10^7 l of water into the shallow groundwater zone. The total volume of water in the shallow groundwater zone, excluding the southeast portion was calculated to be 1.6×10^8 l (Hargis+Associates, Inc. 1996). Therefore, approximately 0.21 pore volumes of clean water were injected into the sediments of this site. The chloride dilution calculations demonstrated that during the time span of the bioremediation project the groundwater in the shallow groundwater zone was diluted 1 to 100 times (Table 30.2).

Table 30.1. Shallow groundwater zone monitoring well contaminant concentrations before and after the bioremediation project

Well No.	Trichloroethene				1,1-Dichloroethene			
	Before (μg l^{-1})	After (μg l^{-1})	Decline (%)		Before (μg l^{-1})	After (μg l^{-1})	Decline (%)	
SM-04D	480	140	71		88	25	72	
P-3	225	30	87		12	2	83	
P-4	77	35	55		18	8	55	
SM-05D	100	74	26		42	4	90	
P-6	109	155	–		3	11	–	

Table 30.2. Shallow groundwater zone monitoring well dilution factors. TCE: trichloroethene, DCE: dichloroethene

Well No.	Water (%)		Dilution	Dilution loss (μg l^{-1})		Biotic loss (μg l^{-1})	
	Clean	Contaminated		TCE	1,1-DCE	TCE	1,1-DCE
SM-04D	51	49	1:2	240	44	100	19
P-3	99	1	1:100	223	12	–	–
P-4	93	7	1:14	72	17	–	–
SM-05	15	85	1:1.2	17	7	9	31
P-6	2	98	1:1.0	0	0	–	–

The greatest dilution (1:100) and largest reduction of TCE levels (87% removal) occurred in monitoring well P-3 (Tables 30.1 and 30.2). 99% of the total water in this well consisted of clean recharged water, and just 1% ambient groundwater. At the other extreme, only 2% of the water within monitoring well P-6 could be attributed to injected clean water. As previously discussed, a low hydraulic conductivity layer most likely impeded flow into this portion of the heterogeneous site.

The chloride-derived dilution factors of each shallow groundwater zone monitoring well were further used to quantify the change in contaminant concentrations attributable to dilution. For example, dividing the baseline TCE and 1,1-DCE concentrations of well SM-04D (480 µg l^{-1} and 88 µg l^{-1} respectively) by its dilution factor of 2, suggests that decreases in contaminant concentrations to 240 µg l^{-1} and 44 µg l^{-1} respectively resulted from dilution. However, the TCE and 1,1-DCE concentrations observed at the end of the bioremediation study were lower than due to dilution alone: 140 and 25 µg l^{-1}, respectively (Table 30.1 and 30.3). The additional removal of 100 µg l^{-1} of

Table 30.3. Shallow groundwater zone monitoring well trichloroethene (µg l^{-1}), 1,1-dichloroethene (µg l^{-1}), chloride concentrations (mg l^{-1}), and heterotrophic plate counts (CFU ml^{-1}) with respect to time. *TCE*: trichloroethene, *DCE*: dichloroethene, *HPC*: heterotrophic plate counts

Well No.	Param.	Feb '97	Jun '97	Jul '97	Oct '97	Jan '98	Apr '98
SM-04D	TCE	480	226	–	110	160	140
	DCE	88	41	–	16	15	25
	HPC	–	1.93×10^4	1.19×10^3	5.20×10^4	1.20×10^4	5.70×10^5
	Cl⁻	100	138	132	70	53	60
P-3	TCE	225	201	–	54	76	30
	DCE	12	11	–	3.3	1.4	2
	HPC	–	1.47×10^3	7.3×10^2	4.10×10^3	2.70×10^3	1.20×10^3
	Cl⁻	110	101	125	30	22	22
P-4	TCE	77	38	–	47	65	35
	DCE	18	16	–	18	5.8	8
	HPC	–	1.20×10^3	4.25×10^2	1.50×10^4	2.20×10^4	5.80×10^3
	Cl⁻	148	31	26	33	27	30
SM-5D	TCE	100	38	–	–	74	–
	DCE	42	10	–	–	4.2	–
	HPC	–	9.90×10^3	–	–	1.47×10^4	2.20×10^3
	Cl⁻	55	40	–	–	35	50
P-6	TCE	109	–	–	150	200	155
	DCE	3	–	–	7.4	5.8	11
	HPC	–	–	2.38×10^3	6.90×10^3	1.10×10^4	7.90×10^3
	Cl⁻	770	–	876	750	815	786

TCE and 19 µg l^{-1} of 1,1-DCE could likely be attributed to biodegradation (Table 30.2). Microbial enumerations for this well ranged from 1.2×10^3 to 5.7×10^5 colony forming units ml^{-1} throughout the duration of the project (Table 30.3). Between February and June of 1997, chloride concentrations increased from 100 to 138 mg l^{-1} (Table 30.3). These observations are indicative of biological activity and thus supportive of microbial degradation.

Although contaminant concentrations decreased in monitoring wells P-3 and P-4 throughout the bioremediation project, the final detected concentrations were higher than anticipated (Table 30.1 and 30.2). After subtracting the calculated losses attributed to dilution, the expected TCE concentrations in these two wells at the end of the study were 2 and 5 µg l^{-1}, respectively (Table 30.2). However, the measured concentrations in the two wells were 30 and 35 µg l^{-1} of TCE respectively (Table 30.1 and 30.3). Because these wells demonstrated high recharge (dilution factors of 100 and 14 respectively) it is possible that additional aqueous contaminants were transported into these portions of the heterogeneous aquifer during infiltration. Desorption and diffusion may also have been contributing processes as the majority of the shallow groundwater zone contamination is encountered within fine-grained sediment. Microbial activity was observed in both wells, and a noticeable increase in chloride concentrations (101 to 125 mg l^{-1}, Table 30.3) was observed in well P-3. Therefore, biodegradation most likely occurred within these areas but is probably underestimated due to the unknown contaminant mass balance within the aquifer.

The shallow groundwater zone monitoring wells showed an overall decrease in contaminant concentrations following the bioremediation project. However, the specific contaminant diminutions and corresponding dilution factors were variable due to the heterogeneity of the site (Table 30.2). Therefore, to determine the overall effectiveness of the enhanced in-situ TCE bioremediation technology, geometric mean values of contaminant concentrations as well as water level elevations with respect to time were calculated to represent the shallow groundwater zone as a whole. The geometric mean dilution factor could further be used to quantify the overall shallow groundwater zone reduction in TCE concentrations attributable to dilution. The spatially averaged TCE concentration had decreased by a total of 41% by the end of the bioremediation project (from 105 to 62 µg l^{-1}, Fig. 30.2). However a 29% reduction (down to 75 µg l^{-1} TCE) may be attributed to dilution based on chloride data representative of the site (geometric mean dilution factor of 1.4). Therefore, only a 12% reduction of TCE may be attributed to biological degradation at this site.

Microbial activities in conjunction with increased aqueous chloride concentrations observed in the monitoring wells throughout the project were indicative of biological transformation. However, the unaccounted for increase in contaminant concentrations attributable to the mass transfer processes may have resulted in underestimating biological removal. The calculated removal attributed to dilution is also conservative, as it does not account for the addition of chloride to the groundwater during the biodegradative process.

The bioremediation operations ceased in May of 1998 when it became apparent that a contaminated shipment of methanol (obtained between February and April) and its subsequent subsurface injection resulted in perchloroethene (PCE) contami-

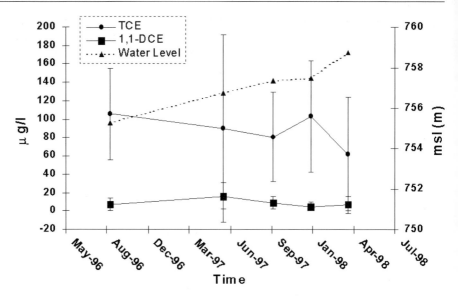

Fig. 30.2. Mean monitoring well contaminant concentrations (µg l^{-1}) and water level elevations (shown in m above mean sea level) representing shallow groundwater zone as a whole. *Error bars* represent contaminant spatial distribution. *TCE*: trichloroethene, *DCE*: dichloroethene

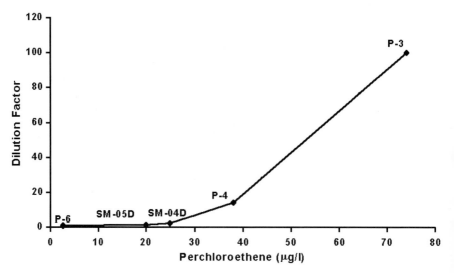

Fig. 30.3. Shallow groundwater zone monitoring wells with corresponding dilution factors and perchloroethene (PCE) concentrations

nation of the shallow groundwater zone. The varying PCE concentrations detected within the monitoring wells of the site corresponded well with the dilution factors calculated by this method. Highest PCE concentrations and dilution factors were seen in well P-3, with lowest values of both observed in well P-6 (Fig. 30.3). The dilution factors therefore appeared to be well representative of the heterogeneities of the site. The accidental injection of this contaminant further substantiated the use of this method in quantifying TCE dilution at heterogeneous sites.

30.5
Conclusions

Biodegradation efficacy is determined by an observed reduction in aqueous contaminant concentrations, and further supported by an increase in degradation product concentrations, CO_2 evolution, microbial enumerations and/or utilization of amended electron acceptors and nutrients. However, contaminant disappearance does not distinguish biodegradation from abiotic processes such as dispersion, dilution, sorption, diffusion or volatilization. An understanding of the physical and chemical processes that control the fate of a contaminant is therefore imperative in determining the effectiveness of biodegradation projects. However, accurate mass balances of contaminant abiotic processes at field sites are often unknown and/or unattainable. Proving whether contaminant removal is due to biotic or abiotic processes is therefore difficult and generally only qualitative. Detection of specific degradative intermediates or metabolites can substantiate biological degradation. In the anaerobic TCE degradative pathway, compounds like 1,2-DCE and vinyl chloride are easily detectable. However, the preferred aerobic pathway results in very short-lived and therefore undetectable metabolites (epoxides).

The cometabolic nature of aerobic TCE biodegradation often requires supplementation of primary substrates for the growth and energy needs of the degrading bacteria. The injection of water amended with these substrates (as well as electron acceptors and nutrients necessary for degradation) displaces and dilutes the contaminant concentrations. Enhanced bioremediation field studies conducted without a tracer in the injection package are therefore generally equivocal because of the inability to distinguish between dilution and biodegradation.

This study determined the overall effectiveness of an enhanced aerobic TCE bioremediation field project by quantifying dilution attributed to mixing between the injected clean and contaminated waters. A modified groundwater mixing equation and inherent chloride concentrations were used. Because chloride is a by-product of aerobic TCE biodegradation, the dilution factors obtained via this method can be used to substantiate bioremediation without the added time and cost of tracer studies usually necessary for this purpose. Although the calculated dilutions do not indicate where the mixing occurred, they appeared well representative of the heterogeneities encountered within this site as supported by the variable and well-corresponded PCE concentrations resulting from the contaminants accidental injection into the groundwater. During the 13 months of operating the in-situ bioremediation technology, approximately 0.21 pore volumes of clean water were injected into the sediments of this site. Spatially averaged TCE concentrations decreased by 41% by the end of the project. The chloride calculations indicated that a 29% reduction was attributable to dilution, suggesting that no more than a 12% decrease in concentrations could be attributable to biological degradation. Whereas a cursory examination of the data from this site suggests that enhanced bioremediation was a success, this more thorough evaluation shows that this cleanup method produced limited results.

References

Agency for Toxic Substances and Disease Registry (1989) Toxicological profile for trichloroethylene. U.S. Department of Health and Human Services, U.S. Public Health Service, Atlanta

Broholm K, Jensen BK, Christensen TH, Olsen L (1990) Toxicity of 1,1,1-trichloroethane and trichloroethene on mixed culture of methane-oxidizing bacteria. Appl Environ Microbiol 56:2488–2493

Buscheck TE, Alcantar CM (1995) Regression techniques and analytical solutions to demonstrate intrinsic bioremediation. In: Hinchee (ed) Intrinsic bioremediation. Battelle Press, Columbus, pp 109–116

Clark I, Fritz P (1997) Environmental isotopes in hydrology. Lewis, New York, pp 104–108

Ellis DE (1996) Intrinsic remediation in the industrial marketplace. In: U.S. Environmental Protection Agency (ed) Symposium on natural attenuation of chlorinated organics in groundwater. Office of Research and Development, U.S. Environmental Protection Agency, Washington, DC, EPA/540/R-96-509, pp 120–123

Ensley BD (1991) Biochemical diversity of trichloroethylene metabolism. Annu Rev Microbiol 45:283–99

Groundwater Resources Consultants, Inc. (GRC) (1993) Drilling, construction and testing of perched zone extraction wells S-24 through S-31. U.S. Air Force Plant No. 44, Tucson, Arizona

Hargis+Associates, Inc. (1996) Final draft shallow groundwater zone removal action engineering evaluation/cost analysis. United States Air Force Plant No. 44, Hughes Missile Systems Company, Tucson, Arizona

Hargis+Associates, Inc. (1997) Final draft shallow groundwater zone removal action implementation report. United States Air Force Plant No. 44, Hughes Missile Systems Company, Tucson, Arizona

Little CD, Palumbo AV, Herbes SE, Lidstrom ME, Tyndall RL, Gilmer PJ (1988) Trichloroethylene biodegradation by methane-oxidizing bacterium. Appl Environ Microbiol 54(4):951–956

Lollar BS, Slater GF, Sleep B, Witt M, Klecka GM, Harkness M, Spivack J (2001) Stable carbon isotope evidence for intrinsic bioremediation of tetrachloroethene and trichloroethene at area 6, Dover Air Force Base. Environ Sci Technol 35:261–269

Madsen EL (1991) Determining in-situ biodegradation: facts and challenges. Environ Sci Technol 25(10):1663–1673

McCarty PL, Semprini L (1994) Groundwater treatment for chlorinated solvents. In: Kerr RS (ed) Handbook of bioremediation. Lewis, Boca Raton, pp 87–116

Montgomery and Associates, Inc. (1998) Evaluation of shallow groundwater zone bioremediation project. U.S. Air Force Plant No. 44, Tucson, Arizona

Oldenhuis R, Vink RLJ, Janssen DB, Witholt B (1989) Degradation of chlorinated aliphatic hydrocarbons by *Methylosinus trichosporium* OB3b expressing soluble methane monooxygenase. Appl Environ Microbiol 55:2819–2826

Rittman BE, Seagren E, Wrenn BA, Valocchi AJ, Chittaranjan R, Lutgarde R (1994) In-situ bioremediation, 2nd edn. Noyes, New Jersey, pp 38–45, 123–132

Sorenson KS, Peterson LN, Hinchee RE, Ely RL (2000) An evaluation of aerobic trichloroethene attenuation using first-order rate estimation. Bioremediation Journal 4:337–357

Suchomel K, Long A (1990) Production and transport of carbon dioxide in a contaminated vadose zone: a stable and radioactive carbon isotope study. Environ Sci Technol 24:1824–1831

Vogel TM, Criddle CS, McCarty PL (1987) Transformation of halogenated aliphatic compounds. Environ Sci Technol 21:722–736

Woodward-Clyde Consultants (WCC) (1996a) Bioremediation treatability study: final draft shallow groundwater zone removal action engineering evaluation/cost analysis. United State Air Force Plant No. 44, HMSC, Tucson, Arizona

Woodward-Clyde Consultants (WCC) (1996b) Soluble methane monooxygenase study results, shallow groundwater zone. U.S. Air force Plant No. 44, Tucson, Arizona

Woodward-Clyde Consultants (WCC) (1998) Draft report shallow groundwater zone bioremediation system operation report March 1997 through March 1998. United States Air Force Plant No. 44, Raytheon Tucson, Arizona

Chapter 31

Anthropogenic Organic Contaminants Incorporated into the Non-Extractable Particulate Matter of Riverine Sediments from the Teltow Canal (Berlin)

J. Schwarzbauer · M. Ricking · B. Gieren · R. Keller · R. Littke

Abstract

Anthropogenic activities induce significant alterations of the macromolecular organic matter (MOM) in riverine systems mainly by emission of pollutants and their subsequent incorporation into geopolymers (bound residues). We have characterized the non-extractable residues of highly polluted riverine sediments (Spree River, Teltow Canal, Germany) in order to investigate the occurrence, alteration and distribution of several organic xenobiotics in situ, e.g. plasticizers, pesticides, and metabolites brominated and chlorinated aromatics, fragrances, technical additives and nitro compounds. Therefore this study intended a comprehensive characterization of riverine MOM combining different analytical techniques (pyrolytic analyses and chemical degradation techniques), in order to provide information concerning the incorporation mechanism and the mode of binding of a variety of organic pollutants with different chemical properties.

31.1 Introduction

In urban and industrial regions the organic matter in riverine sediments is highly controlled by the anthropogenic input due to enhanced emissions of organic contaminants and other pollutants. Not only the qualitative and quantitative composition of the extractable fraction but also of the non-extractable organic matter is affected by anthropogenic contributions. In contrast to the numerous investigations dealing with the occurrence and fate of low molecular weight pollutants in water and particulate matter of riverine systems only a few studies were carried out in order to analyse the alteration within the non-extractable fraction. This alteration via anthropogenic pollution was observed for both the macromolecular substances and the associated low molecular weight compounds. The anthropogenic contribution to the macromolecular organic matter of riverine systems can be generally attributed to three different processes:

- Low molecular weight substances can be strongly incorporated into bio- or geopolymers, e.g. humic substances.
- Natural polymers can be altered by technical processes, e.g. chlorination of drinking water or bleaching processes of paper, and released into the aquatic environment.
- Xenobiotic polymers can also be emitted into the aquatic environment, e.g. polysiloxanes.

Different kinds of anthropogenic compounds and their occurrence within the non-extractable matter are reported in a couple of studies. Most of the investigations are related to the occurrence and fate of associated low molecular contaminants, the so called "bound residues", and are published within the last 30 years (e.g. Li and Felbeck 1972; Kaufman et al. 1976; Liechtenstein et al. 1977; Wheeler et al. 1979; Liechtenstein 1980; Khan 1982; Boul et al. 1994; Lichtfouse 1997; Houot et al. 1997; Northcott and Jones 2000). Earlier investigations dealed especially with the occurrence and fate of bound pesticides in soils. After introducing the pesticides and their metabolites into soils, sediments or waters a weak association of a significant proportion to geopolymers was pointed out leading to insufficient re-extraction rates of the observed compounds by means of regular solvent extraction procedures. Recent studies confirmed this environmental behaviour in soils for a couple of organic contaminants such as atrazine, 2,2',5,5'-tetrachlorobiphenyl, 3,4-dichloroaniline, naphthalene and chlorinated phenols (Hsu and Bartha 1976; Palm and Lammi 1995; Barriuso and Houot 1996; Kan et al. 1997). The quota of the bound fraction ranged between 25 and 90%. The linkage to the geopolymers covers a wide diversity of modes ranging from weaker interactions like adsorption or van der Waals forces up to strong ionic interactions and covalent bonds. The mechanism of incorporation depends on a variety of chemical and physico-chemical properties and conditions, e.g. functional groups within the molecules, pH and redox potential as well as charactersistics of the geopolymer (e.g. Ziechmann 1972; Parris 1980; Senesi 1992; Piccolo et al. 1992; Schulten and Leinweber 1996; Pignatello and Xing 1996; Luthy et al. 1997; Klaus et al. 1998; Nanny 1999; Weber et al. 2001).

The occurrence of bound residues is not only restricted to soils but also to aquatic sediments and suspended matter. Similar associations and assimilations of low molecular weight compounds into aquatic geopolymers, aggregates or organo-mineral complexes were observed (e.g. Murphy et al. 1990; Chin and Gschwend 1992; Buffle and Leppard 1995; Klaus et al. 1998; Zwiener et al. 1999). An important aspect especially in riverine systems is the mobilisation of pollutants associated with colloids, suspended particulate matter or dissolved geopolymers and the subsequent enhanced spatial distribution (s.a. McCarthy and Zachara 1989; Johnson and Amy 1995). Modified transport processes in the presence of aquatic geopolymers are reported for pyrene and the pesticides amitrol, terbutylazine as well as pendimethaline (Huber et al. 1992; Herbert et al. 1993; Oesterreich et al. 1999).

Not only the mode of association but also the strength of bonding and the reversibility of the incorporation process are important for the microbial and abiotic degradation as well as toxicological aspects. Due to a limited bioavailability the toxicological effects of bound organic pollutants are generally reduced in comparison to the free substances (e.g. Lichtfouse 1997). Several investigations demonstrated a decrease of toxicity with progressive aging after application of pollutants to soils without a change of the absolute concentrations (Liechtenstein et al. 1977; Robertson and Alexander 1998). An increasing toxicity for substances (e.g. 2,4-dichlorophenol) in the presence of geopolymers was hardly reported (Steinberg et al. 1992).

In addition, the mode and rate of degradation processes of individual substances can be modified by the incorporation into geopolymers and the subsequent decrease of bioavailability. Frequently a higher stability or even persistence was observed for bound contaminants (e.g. Perdue and Wolfe 1982). Considering geochemical and biogeochemical cycles, the conservation of natural organic compounds in association with geopolymers is well investigated (Tissot and Welte 1984; Engel and Macko 1993). For bound residues not only a higher persistence but also modified degradation or transformation pathways were observed, e.g. for amitrole (Oesterreich et al. 1999). Jensen-Korte et al. (1987) reported an enhanced photolytic degradation of normally persistent pesticides in an aquatic environment induced by the addition of humic substances. Modified transformation processes were also observed in the case of an incorporation of metabolites into geopolmyers. Variations in the degradation pathway via the transfer of metabolites into the non-extractable matter has also to be assumed in the case of PCB and PAH. Richnow et al. (1994) and Michaelis et al. (1995) demonstrated an incorporation of chlorinated benzoic acids and polycyclic aromatic acids, wellknown metabolites of PCB and PAH, by means of covalent bonds. An association of mainly hydroxylated and/or dealkylated metabolites was formerly reported for atrazine in soils (Capriel et al. 1985).

Next to the alteration of non-extractable organic matter by bound residues, several anthropogenic activities also modify directly organic geopolymers. Following discharge these modified polymers affect mainly the aquatic environment. The most important technical modification of biopolymers and aquatic geopolymers are chlorination processes. Hence, in recent studies several chlorinated lignin-, humin- and cellulose-derived macromolecules were identified in the aquatic environment, mainly in Scandinavia (Dahlmann et al. 1993; Hyötyläinen et al. 1998a,b; Miikki et al. 1999), but also in Portugal, Germany and the Netherlands (Bultermann et al. 1997; Santos and Duarte 1998).

In addition to the input of modified natural macromolecules a further mode of anthropogenic alteration of the non-extractable organic matter is the emission of technical macromolecular products. Only very few investigations were reported concerning these emissions of xenobiotic polymers. Examples are the investigations by Fabbri et al. (1998a), as well as Requejo et al. (1985), characterising the input of polystyrene into the coastal and riverine aquatic environment via characteristic pyrolysis products as low molecular weight markers.

Considering the nature of the non-extractable organic matter the analytical methods used for the chemical characterisation and quantification can be divided generally into two different kinds of approaches. Non-destructive methods include IR-, UV/VIS-, NMR- and ESR-spectroscopy, also in combination with liquid chromatography or size exclusion chromatography (e.g. Burns et al. 1973; Senesi et al. 1987; Schlautmann and Morgan 1993; Fabbri et al. 1998b; Nanny 1999; Zwiener et al. 1999). Destructive analytical approaches transfer the bound or macromolecular fraction into low molecular weight compounds either by pyrolysis or by chemical degradation. Using pyrolysis different procedures were applied including on-line and off-line methods, with or without derivatisation of the products and subsequent gas chromatographic or gas chromatographic-mass spectrometric analysis (e.g. Horsfield et al. 1989; Schulten and Leineweber 1996; Stankiewicz et al. 1998; Asperger et al. 1999; Mongenot et al. 1999). Additionally, appropriate chemical degradation procedures allow the selective release of components depending on the mode of incorporation or binding. Common degradation reactions include acidic hydrolysis (e.g. Grasset and Ambles 1998), ether and ester cleavages using boron tribromide or boron trichloride (e.g. Richnow et al. 1994) or oxidation e.g. with CuO or RuO_4 (e.g. Hatcher et al. 1993; Hyötyläinen et al. 1998a). The degradation products were usually extracted, fractionated and analysed by traditional LC, GC or GC-MS. Using a sequential degradation approach with an increasing order of reactivity, a differentiation of individual modes of binding can be indicated for associated substances.

On the contrary, the modes of incorporation are also investigated on a laboratory scale by spiking appropriate samples with labelled (^{14}C, ^{13}C, D) or non-labelled model compounds. Following, these artificial bound residues were released and characterized by selective analytical methods as described earlier (e.g. Hatcher et al. 1993; Richnow et al. 1998; Guthrie et al. 1999). Recent reviews on the analytical topics are given by Northcott and Jones (2000), as well as Kögel-Knabner (2000).

In summary, anthropogenic activities cause significant alterations of the macromolecular organic matter (MOM) in riverine and lacustrine systems mainly by emission of pollutants and their subsequent incorporation into geopolymers (bound residues).

In the presented study we have characterized highly polluted MOM of riverine sediments (Spree River, Teltow Canal, Germany) in order to investigate the occurrence, alteration and distribution of several organic xenobiotics in situ. Thus, these investigations intend a comprehensive characterization of riverine MOM via combining different analytical techniques (pyrolytic analyses and chemical degradation techniques) in order to provide information concerning the incorporation mechanisms and the mode of binding for a variety of organic pollutants with different chemical properties.

31.2
Material and Methods

31.2.1
Samples

Four sediment samples were taken in 1998 and 1999 from three locations at the Teltow Canal in Berlin, as indicated in Fig. 31.1. In addition to the surface sediment samples T2 and T3 taken by means of a 4L Ekman-Birge grab sampler a short sediment core T1 was obtained by using a tube coring device. The sediment core was subdivided into an upper part T1(u) (0–3 cm) and a lower part T1(l) (3–10 cm), that represents an older accumulation time between 1980 and 1990.

31.2.2
Pre-Extraction

All sediment samples were pre-extracted first with 40 ml of methanol or butanol and subsequently with solutions of hexane/acetone (1/1, v/v) by means of shaking for 24 h in the dark and ultrasonication at 2×450 W (Badelin, Berlin, FRG) in a water bath for 30 min. Each extraction step was followed by centrifugation at 1 800 g and the combined extract was reduced in volume by rotary evaporation.

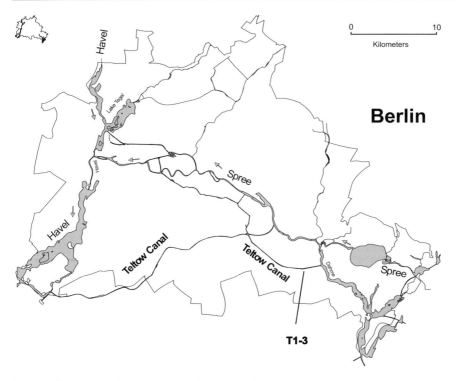

Fig. 31.1. Sampling location at the Teltow Canal, Berlin (Germany)

31.2.3
Chemical Degradation

The chemical degradation steps were carried out in two different modes. In a first set aliquots of the preextracted samples were treated seperately with KOH/MeOH, BBr_3 and RuO_4. Following, in a second step the pre-extracted and saponified residues were treated once more with BBr_3 or RuO_4. In Fig. 31.2 the flow scheme of the analytical procedure is given.

31.2.3.1
Hydrolysis

Aliquots of 150 to 500 mg of the pre-extracted samples were placed in 8 ml vials and 120 mg of KOH dissolved in a mixture of 0.2 ml pre-extracted water and 8 ml of methanol were added. Following, the closed vials were heated for 24 h at 105 °C. After cooling 3 ml of pre-extracted water was added and the solutions were filtered using Whatman glass fiber filters (0.7 µm pore diameter). The mixture was acidifed to pH 3–5 by addition of approx. 1 ml of a 10% hydrochloric acid solution. Subsequently the solution was extracted three times with 5 ml of dichloromethane. The combined organic lay-

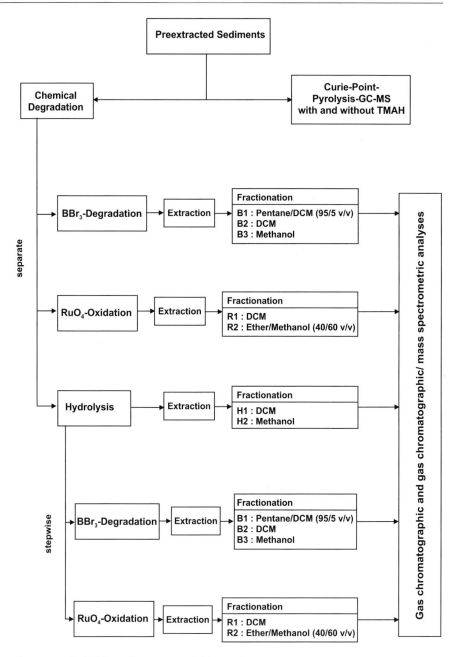

Fig. 31.2. Analytical flow scheme. *DCM:* dichloromethane

ers were dried with anhydrous granulated sodium sulphate and concentrated to a volume of approx. 0.5 ml.

The crude extracts were separated into two fractions by column chromatography (Baker, 2 g silica gel 40 μm) using dichloromethane (fraction 1) and methanol (fraction 2) as the eluent, respectively. Prior to analysis, 50 μl of an internal standard containing 90 ng μl^{-1} d_{34}-hexadecane in n-hexane were added to each fraction, and the volume was reduced to 200 μl by rotary evaporation at room temperature.

31.2.3.2
BBr_3-Treatment

To approx. 150 mg of pre-extracted or pre-extracted/saponified samples 5 ml of a 1.0 M boron tribromide solution in dichloromethane was added and the flasks were closed. Following, ultrasonication in a water bath for 2 h was performed. After 24 h of stirring at room temperature a second ultrasonic step was applied. Subsequently 2 ml of diethylether were added, the supernatant was decanted and filtered using a Whatman glass fiber filter (0.7 μm pore diameter). The solid residue was washed twice with diethylether and the combined organic solutions were added after filtration to the filtered reaction mixture. The combined organic layers were washed twice with 5 ml of pre-extracted water and dried with anhydrous granulated sodium sulphate. Prior to fractionation the solutions were concentrated to a volume of approx. 0.5 ml by rotary evaporation.

The crude extracts were separated into three fractions by column chromatography (Baker, 2 g silica gel 40 μm) using the following mixtures as the eluent: Fraction 1: n-pentane/dichloromethane 95/5, v/v; fraction 2: dichloromethane; fraction 3: methanol. Prior to analysis, 50 μl of an internal standard containing 90 ng μl^{-1} d_{34}-hexadecane in n-hexane were added to each fraction, and the volume was reduced to 100 μl by rotary evaporation at room temperature.

31.2.3.3
RuO_4-Oxidation

A mixture of 8 ml tetrachloromethane, 8 ml acetone and 1 ml of pre-extracted water was added to aliquots of 20 to 170 mg of pre-extracted or pre-extracted/saponified samples. In addition, 1500 mg sodium perjodate and 10 mg of ruthenium(IV)oxide were added and the reaction mixture was stirred for 4 h in darkness. The reaction was stopped by addition of 50 μl of methanol and 2 drops of concentrated sulphuric acid. The liquid phase was separated by decantation and the residue was washed twice with tetrachloromethane. The combined organic layers were collected in a separatory funnel and washed with 5 ml of pre-extracted water as well as 1 ml of a sodium thiosulfate solution. All water layers were combined and re-extracted five times with 10 ml of diethylether. The diethylether solution was added to the organic solution, the extract was dried with anhydrous granulated sodium sulphate and concentrated to a volume of 0.5 ml prior to fractionation.

The crude extracts were separated into two fractions by column chromatography (Baker, 2 g silica gel 40 μm) using the following mixtures as the eluent: Fraction 1: dichloromethane, fraction 2: diethylether/methanol (40/60, v/v). Prior to analysis, 50 μl of an internal standard containing 90 ng μl^{-1} d_{34}-hexadecane in n-hexane was added to each fraction, and the volume was reduced to 100 μl by rotary evaporation at room temperature.

31.2.4
Pyrolysis-Gas Chromatography, Pyrolysis-Gas Chromatography-Mass Spectrometry

Pyrolysis-gas chromatography was performed using a Horizon Curie-Point Pyrolator with a pyrolsis temperature of 610 °C held for 10 s. The interface was heated to 300 °C and the capillary column was directly inserted into the pyrolysis chamber. The gas chromatographic separation was carried out on a GC 4100 gas chromatograph (Carlo Erba, Milano, I) equipped with a 25 m × 0.25 mm i.d. × 0.25 µm film SE52 fused silica capillary column (CS Chromatographie Service, Langerwehe, FRG). For Py-GC/MS analyses the same Pyrolysator device was linked to a HP5890 gas chromatograph (Hewlett Packard, Palo Alto, USA) which was equipped with a 30 m × 0.25 mm i.d. × 0.25 µm film BPX5 fused silica capillary column (SGE, Weiterstadt, FRG). Chromatographic conditions were: 1 µl split/splitless injection at 60 °C, splitless time 60 s, 3 min hold, then programmed at 3 °C min^{-1} to 300 °C, helium carrier gas velocity was 40 cm s^{-1}. The mass spectrometric detection was carried out on a Finnigan MAT 8222 mass spectrometer (Finnigan, Bremen, FRG) which was operated at a resolution of 1 000 in electron impact ionization mode (EI$^+$, 70 eV) with a source temperature of 200 °C scanning from 35 to 700 amu at a rate of 1 s decade^{-1} with an inter-scan time of 0.1 s.

31.2.5
Gas Chromatographic Analysis

Gas chromatographic analysis was carried out on a GC8000 gas chromatograph (Fisons Instruments, Wiesbaden, FRG) equipped with a 25 m × 0.25 mm i.d. × 0.25 µm film SE54 fused silica capillary column (CS Chromatographie Service, Langerwehe, FRG). The end of the capillary column was splitted to lead the eluate separately to a flame ionization detector (FID) and an electron capture detector (ECD) for a simultaneous detection of the analytes. Chromatographic conditions were: 1 µl split/splitless injection at 60 °C, splitless time 60 s, 3 min hold, then programmed at 3 °C min^{-1} to 300 °C, hydrogen carrier gas velocity was 25 cm s^{-1}.

Acidic compounds in the polar fractions were methylated prior to analysis by addition of 5 µl of a TMSH solution to aliquots of 5 µl of the extracts. The mixture was ultrasonicated for 5 min and the volume was reduced to approx. 2 µl. The total solution was injected into the gas chromatograph.

31.2.6
Gas Chromatographic-Mass Spectrometric Analysis

GC/MS analyses were performed on a Trace MS mass spectrometer (Thermoquest, Egelsbach, FRG) linked to a Mega Series 5140 gas chromatograph (Carlo Erba, Milano, I) which was equipped with a 30 m × 0.25 mm i.d. × 0.25 µm film BPX5 fused silica capillary column (SGE, Weiterstadt, FRG). Chromatographic conditions were: 1 µl split/splitless injection at 60 °C, splitless time 60 s, 3 min hold, then programmed at 3 °C min^{-1} to 300 °C, helium carrier gas velocity was approx. 40 cm s^{-1}.

The mass spectrometer was operated in electron impact ionization mode (EI$^+$, 70 eV) with a source temperature of 200 °C scanning from 35 to 700 amu at a rate of 0.5 s decade^{-1} with an inter-scan time of 0.1 s.

31.2.7
Identification of Organic Compounds

The identification of individual compounds was based on comparison of EI$^+$-mass spectra with those of reference compounds, mass spectral data bases (NIST/EPA/NIH Mass Spectral Library NIST98, Wiley/NBS Registry of Mass Spectral Data, 4th Edn., electronic versions) and gas chromatographic retention times, published elution patterns or retention indices.

31.3
Results

31.3.1
Pyrolysis

The main compounds yielded by flash pyrolysis are transformation or degradation products of biogenic precursors. Next to the amino acid glycine (*1*) most of the identified compounds are structurally related to carbohydrates, amino acids and condensed molecules of both components resulting from Maillard reactions. Examples include furfural (*2*), methylfurfural (*3*) and pyrrol-2-carboxaldehyde (*4*) (see Fig. 31.3).

Apart from these obviously biogenic compounds only very few definite anthropogenic substances were identified. Within the group of xenobiotics 2,4'- and 4,4'-dichlorophenylmethane, DDT-derived metabolites, were most abundant. A pyrolytic conversion of related DDT metabolites to DDM cannot be excluded. Thus the occurrence of DDM in the pyrogram is only suggestive for DDT group metabolites in general.

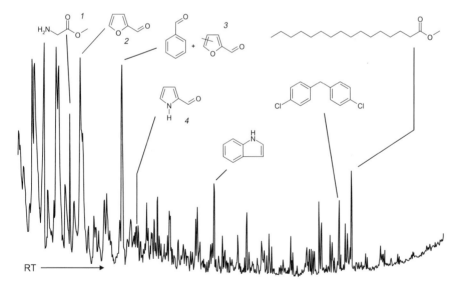

Fig. 31.3. Curie-point pyrolysis gas chromatogram of the non-extractable residue of a Teltow Canal sediment (Curie-point temperature 610 °C). *RT:* rentention time

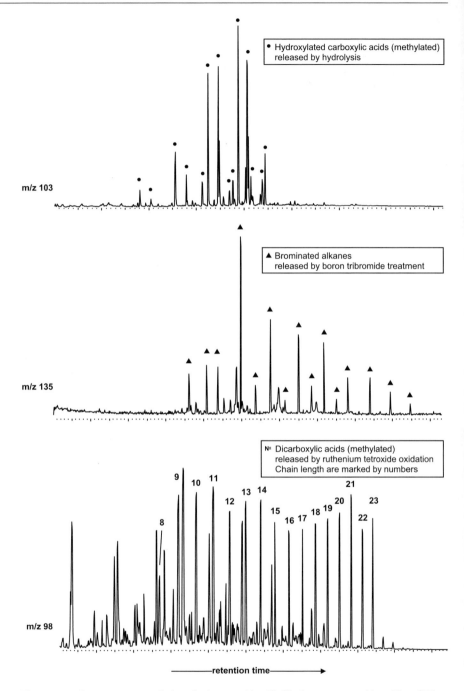

Fig. 31.4. Ion chromatogramms of selected substances identified in the non-extractable residue of Teltow Canal sediments reflecting the chemical composition of the natural macromolecular organic matter

31.3.2
Chemical Degradation

Depending on the degradation method a variety of biogenic compounds was identified reflecting the natural contribution to the MOM. Dominating groups of substances are illustrated in Fig. 31.4.

Hydrolysis revealed mainly fatty acids, fatty alcohols and long chain *n*-amides accompanied by unsaturated and branched isomers. In addition, steroid alcohols appeared in high amounts, but noteworthy the oxidized analogeous, the steroid ketones, e.g. cholestanone or coprostanone, were not present. Therefore a covalent association of the hydroxylated isomers by ester bondings is evident.

Boron tribromide cleavage released long chain carboxylic and dicarboxylic acids including their hydroxylated isomers as well as brominated alkanes, reflecting long chain aliphatic units of the naturally occuring macromolecules. The aromatic moieties were represented by brominated benzenes and alkylated homologues. Also most of the substances released after ruthenium tetroxide oxidation can be attributed to the aromatic proportion within the bio-/geopolymers. Examples include methoxylated and alkylated benzenes, methoxylated phenoles and phenylalkyl carboxylic acids.

Additionally, in all degradation product mixtures numerous anthropogenic compounds were identified. Variations of the chemical composition within the anthropogenic proportion of the reaction mixtures were observed depending either on the degradation mode or on the sample. The following presentation of results, subdivided and arranged by the type of degradation reaction, focuses on the occurrence of the anthropogenic contaminants. All identified substances are summarized in Table 31.1 with a semi-quantitative estimation of the relative concentrations.

31.3.2.1
Hydrolysis

Hydrolysis revealed numerous anthropogenic compounds that can be attributed to different technical applications or widespread domestic usages. Within the group of plasticizers the phthalates, which are well-known and ubiquiteous pollutants, were most abundant. Also the plasticizers tributylphosphate and 2,4,4-trimethylpentan-1,3-diol-diisobutyrate occurred in minor concentrations mainly in sample T1. Due to their molecular structures a non-covalent association of these compounds to the geopolymers has to be assumed.

In addition a significant contribution of technical additives to the non-extractable organic matter was observed. Mainly the isopropyl ester of palmitinoic acid, used as additive in cosmetics and washing agents, and compounds including a 2-ethylhexyl moiety (2-ethylhexanol, hexanedioic acid 2-ethylhexylester (5)) were identified. Both, the *iso*-propyl and the 2-ethylhexyl moieties (see Fig. 31.5), are very probably of anthropogenic origin due to the absence or rarely occurrence of similar molecular substructures within the biogenic compounds.

Furthermore as a result of industrial emissions the Teltow Canal sediments are highly polluted by the pesticides DDT and methoxychlor, accompanied by several metabolites (Schwarzbauer et al. 2001). Accordingly, numerous DDT-related com-

Table 31.1. Organic contaminants identified in the non-extractable organic matter of Teltow Canal sediments after chemical degradation procedures. o; +; ++; +++ = very low to high concentration; (me) = identified as methyl ethers; (m) = identified as methyl esters; ( ) = not all isomers detected

Compounds	BBr₃ treatment			RuO₄ oxidation				Hydrolysis				BBr₃ treatment after hydrolysis				RuO₄ oxidation after hydrolysis			
	T1 (u)	T2	T3	T1 (u)	T1 (l)	T2	T3	T1 (u)	T1 (l)	T2	T3	T1 (u)	T1 (l)	T2	T3	T1 (u)	T1 (l)	T2	T3
Technical additives, solvents																			
N-Methylpyrrolidone	o																		
N-Ethylaniline																+		o	+
Benzophenone				+	o												+		
Benzylbenzoat				+															
Dibutylmaleat				o															
Butylpalmitat				o															
Isopropyldodecanoate				+															
Isopropylpalmitate				+			+	++	++	+	+	++				+	++	o	+
Hexandioic acid di-iso-propyl ester	+			+	+			++	+			+	+			+	+	o	
2-Ethylhexanol	+	o		+	+				o				+	++	o				
Hexandioic acid 2-ethylhexyl ester (me)	++									o	+							o	
Bisphenol A (me)	+										o							o	
Ionol											o								o
Di-tert-butylchinone							+												
Detergent residues, fragrances																			
Linear alkylbenzenes				o		o	o				o								
Galaxolide	o						+	o	o										
Tonalide	o						o		o										

Table 31.1. Continued

Compounds	BBr₃ treatment				RuO₄ oxidation				Hydrolysis				BBr₃ treatment after hydrolysis				RuO₄ oxidation after hydrolysis			
	T1 (u)	T1 (l)	T2	T3	T1 (u)	T1 (l)	T2	T3	T1 (u)	T1 (l)	T2	T3	T1 (u)	T1 (l)	T2	T3	T1 (u)	T1 (l)	T2	T3
Plasticizers																				
Alkylsulfonic acid phenyl esters	o				o		o	o												
Tri-n-butylphosphate	++				+	o	++		o	o										
2,4,4-Trimethylpentan-1,3-diol-diisobutyrate	+	o			++				+	+	+	o								
Di-n-butylphthalate	++		+		++				+	+	+		+	+	+	++	+	++	++	++
Di-iso-butyl-phthalate	++				++				+	+	+	+	+	+	++	+	+	++	+	++
Bis(2-ethylhexyl)-phthalate	+	++			+								++	++	++	++	+	++	++	++
Tris(2-ethylhexyl)-trimellitate					+															
Phthalic acid 2-ethyl-hexyl monoester⁽ᵐ⁾	++	++	++	++	+	+	+	+	++	+	++	+		++	++	++	+	++	+	++
Phthalic acid⁽ᵐ⁾	++	++	++	++	+	+	++	++	++	++	++	+	++	++	++	++	+	++	+	++
Nitrogen containing compounds																				
4-Nitrobenzoic acid⁽ᵐ⁾					++	++														
4-Aminobenzoic acid⁽ᵐ⁾					+		+													
N-Methyl-4-aminobenzoic acid⁽ᵐ⁾					o															
2,4-Di-tert-butyl-6-nitrophenol					+		+													
2-tert-Butyl-4,6-dinitrophenol					+															
Pesticides and metabolites																				
Hexachlorocyclohexanes (α-, β-, γ-, δ-)																				
2,4'-DDD					+															

Table 31.1. Continued

Compounds	BBr₃ treatment				RuO₄ oxidation				Hydrolysis				BBr₃ treatment after hydrolysis				RuO₄ oxidation after hydrolysis			
	T1 (u)	T1 (l)	T2	T3	T1 (u)	T1 (l)	T2	T3	T1 (u)	T1 (l)	T2	T3	T1 (u)	T1 (l)	T2	T3	T1 (u)	T1 (l)	T2	T3
Pesticides and metabolites (continued)																				
4,4'-DDD	+	+			+	+	+													
4,4'-DDMS	‡	‡			o	+	‡													
2,4'-DDM	‡	‡			+	+	+													
4,4'DDM					‡		‡	‡												
2,4'-DDE			o						+	+	+	+								
4,4'-DDE			o	+					+	+	+	+	o	o						
2,4'-DDMU									o	o	o	o								
4,4'-DDMU									o	o	o	+								
2,4'-DDNU									+	+	o	+								
4,4'-DDNU									+	+	+	o								
4,4'DDEthane									o	+	o	+								
2,4'-DDA(m)	+	+							+	+	+	o								
4,4'-DDA(m)	+	+			‡	‡		‡	+	+	‡		+	+	‡	‡	+	‡	+	‡
4,4'-DDOH	‡	‡			+	‡	‡	‡	+	+	+	‡	+	+	‡	‡	+	‡	+	‡
2,4'-DBP	+	‡			+	‡	+		+	+	+									
4,4'-DBP	+	+			+												+	+		
2,4'-DDCN																	+	+	+	
4,4'DDCN												+								
4,4'-MDE									+	+		+					+	+		
4,4'Dimethoxybenzophenone																				

Table 31.1. *Continued*

Compounds	BBr₃ treatment				RuO₄ oxidation				Hydrolysis				BBr₃ treatment after hydrolysis				RuO₄ oxidation after hydrolysis			
	T1 (u)	T1 (l)	T2	T3	T1 (u)	T1 (l)	T2	T3	T1 (u)	T1 (l)	T2	T3	T1 (u)	T1 (l)	T2	T3	T1 (u)	T1 (l)	T2	T3
Pesticides and metabolites *(continued)*																				
2,2-Bis(4 dimethoxyphenyl)acetic acid[m]	++									+	o	++	++							
Chloropropylate						++														
Halogenated aromatics																				
Dichlorobenzene (1 isomer)		o																		
Trichlorobenzene (2 isomer)					(o)	o														
Tetrachlorobenzene (1 isomer)					o	o														
Pentachlorobenzene						o														
1-Chloronaphthalene									o		(o)	o								
Dichloronaphthalene (3 isomers)					+	o				(o)	o	+								
1-Bromonaphthalene					(o)	+			(o)	o	(o)	+								
Dibromonaphthalenes (3 isomers)					+	+			o		+	+								
PCB (Cl₄–Cl₆)											o				+					
4-Chlorobenzoic acid[m]			o	o		+				o					o	+				
2,4-Dichlorobenzoic acid[m]		+	++	o																
2,4-Dibromoaniline		o	++	o										o	+	+				
2,4,6-Tribromoaniline				o																
Bromophenol[me]														o	o	o				
Dibromophenol[me]		+												+	+	+				
Tribromophenol[me]		o												+	o	o				

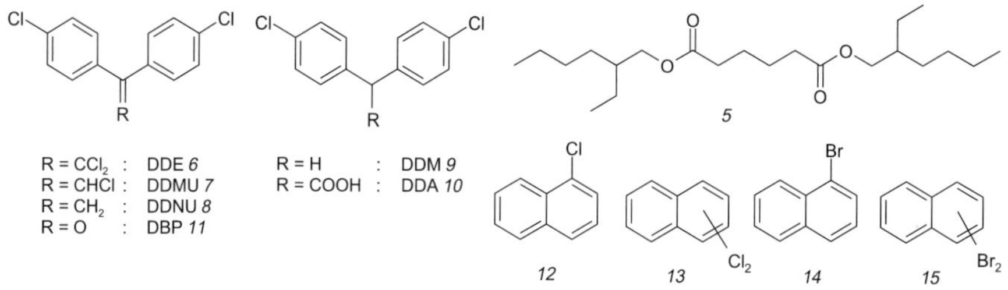

Fig. 31.5. Molecular structures of selected anthropogenic compounds released by hydrolysis

pounds were identified in the hydrolysis extracts including DDE (6), DDMU (7), DDNU (8) and DDM (9). Highest concentrations were observed for DDA (10) and DBP (11), the more polar degradation products of DDT.

A second group of specific xenobiotics in Teltow Canal sediments are halogenated aromatics. Several chlorinated and brominated mono- and diaromatic hydrocarbons were detected in high amounts within the extractable organic matter as reported previously (Schwarzbauer et al. 2001). The halogenated arenes identified in the hydrolysis extracts included mono- and dichlorinated naphthalenes (12) + (13), mono- and dibrominated naphthalenes (14) + (15), tetra- to hexachlorinated biphenyls (PCB) and 2,4,6-tribomoaniline. The peak pattern of the chlorinated naphthalenes was similar to the congener distribution in technical mixtures e.g. Halowax 1000 (Falandysz 1998).

31.3.2.2
BBr_3-Treatment

The treatment of the extracted residues with the Lewis acid boron tribromide revealed numerous compounds that were also detected in hydrolysis extracts (see Fig. 31.6). Examples including hexanedioic acid 2-ethylhexylester (5), 2,4,6-tribromoaniline (16) as well as galaxolide (17) and tonalide (18), persistent synthetic musk substitutes widespread used as fragrances in soaps, perfumes, detergents and other household cleaning products. In addition, low amounts of bisphenol A (19), used as plasticizer, fungicide and intermediate in polymer syntheses, were detected. For bisphenol A estrogenic activities were evident (Safe and Gaido 1998). Within the group of xenobiotics the DDT-related compounds became most abundant with DBP (11) as the main component.

Noteworthy, the boron tribromide treatment applied to saponified residues generated a slightly different pattern of compounds. As a result of the sequential degradation procedure additional compounds were identified and, furthermore, higher concentration of selected individual substances, detected in both BBr_3 extracts, were observed. In detail, DDA, hexanedioic acid di-*iso*-propyl ester and brominated naphthalenes were observed exclusively in the BBr_3 extract of the sequential procedure.

The origin of the brominated phenols is not obvious. As the most important result of boron tribromide treatment aliphatic and aromatic ethers are transformed to the corresponding alcohols and bromides. Hence, the occurrence of brominated phenols exclusively in the BBr_3 extracts indicated either a biogenic origin as a result of the cleavage of mono- to tetraalkoxylated aromatic substructures. Alternatively, with respect to the miss-

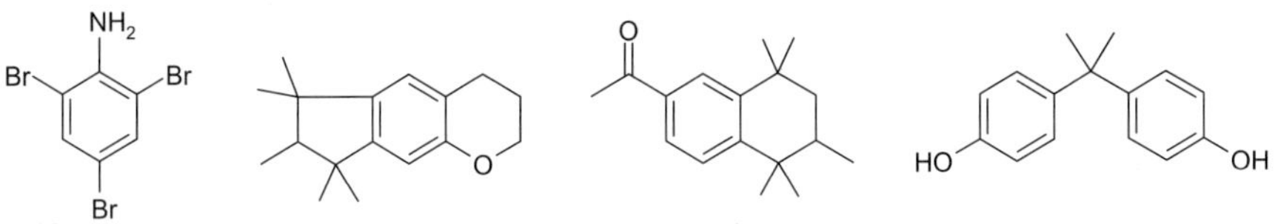

Fig. 31.6. Molecular structures of selected anthropogenic compounds released by boron tribromide teatment

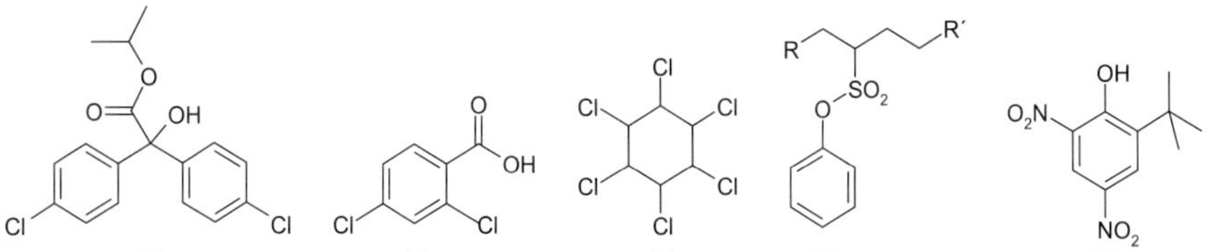

Fig. 31.7. Molecular structures of selected anthropogenic compounds released by ruthenium tetroxide oxidation

ing corresponding brominated catechols the chemical degradation released already brominated phenoxy moieties linked by covalent ether bondings to the geopolymers.

31.3.2.3
RuO_4-Oxidation

GC-MS analyses of the extracts obtained after ruthenium tetroxide oxidation revealed either compounds previously described as hydolysis or BBr_3 treatment products or novel compounds only occuring in RuO_4 extracts. Examples for the first group of anthropogenic contaminants are linear alkylbenzenes (LAB) with 11 to 13 carbon side chain length, bisphenol A (*19*), DDA, DBP, brominated naphthalenes and the pesticide chloropropylate (*20*) (see Fig. 31.7).

Numerous individual substances were detected only in RuO_4 extracts e.g. di- to pentachlorinated benzenes, 4-chlorobenzoic acid and 2,4-dichlorobenzoic acid (*21*), hexachlorocyclohexanes (α-, β-, γ- and δ-HCH) (*22*), a technical mixture obtained during the synthesis of lindane, and the plasticizers alkylsulfonic acid phenylesters (*23*). These plasticizers were recently identified in riverine sediments (Franke et al. 1998). Furthermore, nitro-substituted benzoic acid and alkylated phenols (*24*) were observed. The occurrence of aromatic nitro compounds as a result of the oxidation of anilines can be excluded due to the contemporary appearance of amino compounds, e.g. 4-aminobenzoic acid or N-ethylaniline. However, the origin as well as the emission pathway of these compounds is still unknown.

In contrast to the enhanced release of organic compounds by a sequential application of hydrolysis and BBr_3, the sequential procedure led in the case of RuO_4 to a minor portion of released organic compounds. Hence specific contaminants (e.g. halogenated arenes, nitro compounds) were not observed within the extracts of RuO_4 oxidation products applied to the saponified residues.

31.4 Discussion

The occurrence of anthropogenic substances in the degradation reaction mixtures has to be discussed either by their mode of incorporation, their modification due to incorporation and their appearance in comparison to the substances obtained by traditional extraction techniques. Hence in Table 31.2 the anthropogenic compounds identified in the extracts are summarized as previously published (Schwarzbauer et al. 2001; Ricking et al. 2003).

Table 31.2. Summary of selected anthropogenic contaminants identified in the extracts of the Teltow Canal sediments as published previously (Schwarzbauer et al. 2001; Ricking et al. 2003). Compounds not identified in the non-extractable residue after application of chemical or pyrolytic degradation procedures are given in *italics*

Technical additives, solvents	Halogenated aromatics
Dibenzylether	Chlorinated benzenes (Cl_1–Cl_6)
Benzophenone	*Chlorinated styrenes (Cl_3–Cl_8)*
Benzylbenzoate	Chlorinated naphthalenes (Cl_1–Cl_7)
Dimethyladipate	Brominated naphthalenes (Cl_1–Cl_2)
Fragrances, UV-protectors	Polychlorinated biphenyls (PCB, Cl_4–Cl_7)
4-Methoxycinnamic acid 2-ethylhexyl ester	4,4'-Dibromobenzophenone
Galaxolide	*Pentachloroanisole*
Tonalide	*2,4-Dichlorobenzaldehyde*
4-Oxoisophorone	2,4-Dichlorobenzoic acid[m]
Nitro compounds	**Pesticides and metabolites**
Nitrobenzene	Hexachlorocyclohexanes (α-, β-, γ-, δ-)
4-Ethylnitrobenzene	DDT
4-Nitrobenzoic acid[m]	DDD
Detergents related compounds	DDMS
Nonylphenols (10 isomers)	DDE
Linear alkylbenzenes (LAB)	DDMU
Plasticizers	DDCN
Tributylphosphate	DDA methyl ester
Tritolylphosphate	DBP
Alkylsulfonic acid phenyl/cresyl esters	*Methoxychlor*
Dimethylphthalate	MDD
Di-*n*-butylphthalate	MDE
Di-*iso*-butylphthalate	*Chlorfensone*
Bis(2-ethylhexyl)phthalate	*Bromopropylate*

Most of the compounds identified in the degraded non-extractable residues were formerly reported as constituents of the extractable fraction (Schwarzbauer et al. 2001; Ricking et al. 2003). Hence these compounds known as pesticides, technical additives or industrial agents represent the unaltered bound substances and reflect the incorporated proportion of organic pollutants introduced into the aquatic environment by anthropogenic emissions.

Considering their molecular structures and their frequent but not systematical occurrence in extracts of different selective degradation steps it has to be stated that the major portion of these substances was not associated by covalent linkages but by weaker interactions like adsorption or van der Waals forces. Thus the majority of compounds was released by destruction of the macromolecular matrix and not by a selective bond breaking. Consequently no dramatic variation within the spectra of contaminants was observed with respect to the mode of chemical degradation. However, in detail minor differences of the contamination pattern obtained by the various degradation methods were noted. Substances released by hydrolysis and RuO_4 oxidation covered a wider range of substances as compared to the group of compounds revealed by the BBr_3 degradation.

With respect to the sequential degradation procedures we observed two slightly different trends. In the case of RuO_4 oxidation an decreasing quantity of contaminants can be stated within the already saponified residues as compared to the formerly untreated oxidation products, despite the different selectivity and reactivity of the degradation agents. On the contrary, the BBr_3 treatment released an pattern of compounds unaffected by a former hydrolysis.

Both observations, *(i)* the similar quality of released bound contaminants in case of hydrolysis and RuO_4 oxidation and *(ii)* the decreased quantity of compounds revealed by RuO_4 oxidation after hydrolysis indicate that organic contaminants comparablely associated to the macromolecular organic matter were affected by both degradation methods in a very similar mode. The BBr_3 treatment released organic contaminants which are incorporated in a different way.

Furthermore a correlation between the concentration analysed in the extractable fraction and the appearance of individual substances within the degradation extracts was not observed. The relative concentration of various abundant extractable compounds decreased in the degradation product mixtures and fell partly below the detection limit, e.g. alkylsulfonic acid phenylesters, tritolylphosphates, and hexachlorocyclohexanes. On the contrary, a few contaminants with very low concentration in the extractable fraction were also identified in the bound fraction (e.g. bisphenol A, chloropropylate). Additionally, several compounds occurred in both the bound and the extractable fraction at higher concentration levels, e.g. chlorinated and brominated naphthalenes, phthalates, and DDT-group substances. With respect to the molecular structure and the relative concentrations these observations suggested no preference in the association of selected classes of compounds.

Most of the unaltered bound contaminants discussed were detected at a low to very low concentration level as compared to the degradation products of the natural organic components. Most abundant within the group of xenobiotics are the group of DDT-related compounds that were detected at elevated amounts. For the DDT metabolites a significant alteration has to be stated as compared to the DDT-related compounds detected in the extractable fraction. The main components in all degradation extracts were 4,4'-DBP, 4,4'-DDA and 4,4'-DDM. In addition, 4,4'-DDM was detected at rather high concentrations by pyrolytic analyses. The DDT metabolites DDMU, DDOH, DDMS occurred at minor concentrations, whereas DDD, DDE, DDCN and DDT itself were either not detected or at a significant lower level. This quantitative proportion of DDT metabolites was in contrast to the distribution observed in the extractable organic matter (Schwarzbauer et al. 2001). 4,4'-DDD was absolutely dominant in the extractable fraction according to the anaerobic degradation pathway of DDT (see: http://umbbd.ahc.umn.edu), whereas DDE, the main metabolite of aerobic microbial DDT-degradation, and the other metabolites occurred at significantly minor concentrations.

The alteration within the group of bound DDT-metabolites indicates either a different degradation pathway of incorporated DDT or the selective association of individual metabolites due to their different molecular structures. The significantly higher concentration of the DDT-related compounds as compared to unaltered bound contaminants suggests an enhanced incorporation of DDT or of its metabolites. A very similar phenomenon was observed for methoxychlor related compounds. Also the more polar metabolites 4,4'-dimethoxybenzophenone and 2,2-bis(4-dimethoxy-phenyl)acetic acid became most abundant within the bound organic fraction.

31.5
Summary

The aim of our investigation was to characterize the alteration of the non-extractable organic matter due to anthropogenic emissions and to elucidate the subsequent incorporation of the organic pollutants into riverine geopolymers. Hence we investigated the occurrence, alteration and distribution of several organic xenobiotics within the non-extractable organic matter of highly polluted riverine sediments (Spree River, Teltow Canal, Germany). We combined different analytical techniques (pyrolytic analyses and chemical degradation techniques) in order to provide information concerning the incorporation mechanisms and the mode of binding for a variety of organic pollutants with different chemical properties.

Briefly the following conclusions can be deduced from the results of the presented study considering the occurrence, molecular structure and semi-quantitative amounts of the identified anthropogenic contaminants:

- Most of the compounds identified in the degraded non-extractable residues represent the unaltered bound substances and reflect the incorporated proportion of organic pollutants introduced into the aquatic environment by anthropogenic emissions. The major portion of these substances was not associated by covalent linkages but by weaker interactions.
- Hydrolysis and RuO_4 oxidation affected the interactions of the associated substances with the macromolecular organic matter and the alteration of the macromolecular matrix on a very similar mode, despite the different selectivity and reactivity of the degradation agents. The BBr_3 treatment affected the incorporation of organic contaminants in a rather different way.

- The appearance of individual substances within the degradation extracts did not correlate with the concentrations determined in the extractable fraction. Considering the molecular structures of the contaminants and the corresponding chemical and physico-chemical properties a favoured association of selected classes of compounds cannot be assumed.
- The altered distribution of the bound DDT-metabolites and the significantly higher concentrations as compared to unaltered bound contaminants indicates either a different degradation pathway of incorporated DDT or the selective and enhanced association of individual metabolites due to their different molecular structures.

Acknowledgements

This work is financially supported by the Deutsche Forschungsgemeinschaft (DFG).

References

Asperger A, Engewald W, Fabian G (1999) Analytical characterization of natural waxes employing pyrolysis-gas chromatography-mass-spectrometry. J Anal Appl Pyrolysis 50:103–115
Barriuso E, Houot S (1996) Rapid mineralization of the s-trazine ring of atrazine in soils in relation to soil management. Soil Biol Biochem 28:1341–1348
Boul HL, Garnham MLG, Hucker D, Baird D, Alslable J (1994) Influence of agricultural practices on the levels of DDT and its residues in soil. Environ Sci Technol 28:1397–1402
Buffle J, Leppard GG (1995) Characterization of aquatic colloids and macromolecules: 1. Structure and behavior of colloidal material. Environ Sci Technol 29:2169–2175
Bulterman AJ, van Loon WMGM, Ghijsen RT, Brinkmann UAT, Huitema IM, de Groot B (1997) Highly selective determination of macromolecular chlorolignosulfonic acids in river and drinking water using Curie-point pyrolysis-gas chromatography-tandem mass spectrometry. Environ Sci Technol 31:1946–1952
Burns IG, Hayes MHB, Stacey M (1973) Spectroscopic studies on the mechanisms of adsorption of paraquat by humic acid and model compounds. Pestic Sci 4:201–209
Capriel P, Haisch A, Khan U (1985) Distribution and nature of bound (nonextractable) residues of atrazine in a mineral soil nine years after the herbicide application. J Agric Food Chem 33:587–589
Chin YP, Gschwend PM (1992) Partitioning of polycyclic aromatic hydrocarbons to marine porewater organic colloids. Environ Sci Technol 26:1621–1626
Dahlman O, Mörck R, Ljungquist P, Reimann A, Johansson C, Boren H, Grimvall A (1993) Chlorinated structural elements in high molecular weight organic matter from unpolluted waters and bleached-kraft mill effluents. Environ Sci Technol 27:1616–1620
Engel MH, Macko SA (1993) Organic geochemistry – principles and applications. Plenum, New York
Fabbri D, Trombini C, Vassura I (1998a) Analysis of polystyrene in polluted sediments by pyrolysis-gas chromatography-mass spectrometry. J Chromatogr Sci 36:600–604
Fabbri D, Mongardi M, Mintanari L, Galletti GC, Chiavari G, Scotti R (1998b) Comparison between CP/MAS ^{13}C-NMR and pyrolysis-GC/MS in the structural characterization of humins and humic acids of soil and sediments. Fresenius J Anal Chem 362:299–306
Falandysz J (1998) Polychlorinated naphthalenes: an environmental update. J Environ Pollut 101:77–90
Franke S, Schwarzbauer J, Francke W (1998) Arylesters of alkylsulfonic acids in sediments. Part 3 of organic compounds as contaminants of the Elbe River and its tributaries. Fresenius J Anla Chem 353:580–588
Grasset L, Ambles A (1998) Structure of humin and humic acid from an acid soil as revealed by phase transfer catalyzed hydrolysis. Org Geochem 29:881–891
Guthrie EA, Bortiatynsky JM, van Heemst JD, Richamn JE, Hardy KS, Kovach EM, Hatcher PG (1999) Determination of (^{13}C)pyrene sequestration in sediment microcosms using flash-pyrolysis-GC-MS and ^{13}C-NMR. Environ Sci Technol 33:119–125

Hatcher PG, Bortiatynsky JM, Minard RD, Dec J, Bolag J (1993) Use of high resolution ^{13}C-NMR to examine the enzymatic covalent binding of ^{13}C-labelled 2,4-dichlorphenol to humic substances. Environ Sci Technol 27:2098–2103

Herbert BE, Bertsch PM, Novak JM (1993) Pyrene sorption by water-soluble organic Carbon. Environ Sci Technol 27:398–403

Horsfield B, Disko U, Leistner F (1989) The microscale simulation of maturation: outline of a new technique and its potential applications. Geologische Rundschau 78:361–374

Houot S, Benoit P, Charnay MP, Barriuso E (1997) Experimental techniques to study the fate of organic pollutants in soils in relation to their interactions with soil organic constituents. Analusis Magazine 25:9–19

Hsu TS, Bartha R (1976) Hydrolysable and nonhydrolysable 3,4-dichloroaniline-humus complexes and their respective rates of biodegradation. J Agric Food Sci 24:118–222

Huber SA, Scheunert I, Dörfler U, Frimmel FH (1992) Zum Einfluss des gelösten organischen Kohlenstoffs (DOC) auf das Mobilitätsverhalten einiger Pestizide. Acta Hydrochim Hydrobiol 20:74–81

Hyötyläinen J, Knuutinen J, Malkavaara P, Siltala J (1998a) Pyrolysis-GC-MS and CuO-Oxidation-HPLC in the characterization of HMMs from sediments and surface waters downstream of a pulp mill. Chemosphere 36:291–314

Hyötyläinen J, Knuutinen J, Malkavaara P (1998b) Transport of high molecular mass lignin material in the receiving water system of a mechanical pulp mill. Chemosphere 36:577–587

Jensen-Korte U, Anderson C, Spitteller M (1987) Photodegradation of pesticides in the presence of humic substances. Sci Total Environ 62:335–340

Johnson WP, Amy GL (1995) Facilitated transport and enhanced desorption of polycyclic aromatic hydrocarbons by natural organic matter in aquifer sediments. Environ Sci Technol 29:807–817

Kan AT, Fu G, Hunter MA, Tomson MB (1997) Irreversible adsorption of naphthalene of naphthalene and tetrachlorobiphenyl to lula and surrogate sediments. Environ Sci Technol 31:2176–2185

Kaufman DD, Still GG, Paulson GD, Bandal SK (eds) (1976) Bound and conjugated pesticide residues. Amer Chem Soc Symp Ser 29

Khan SU (1982) Bound pesticide residues in soil and plants. Residue Reviews 84:1–25

Klaus U, Oesterreich T, Volk M, Spiteller M (1998) Interaction of aquatic dissolved organic matter (DOM) with amitrole: The nature of bound residues. Acta Hydrochim Hydrobiol 26:311–317

Kögel-Knabner I (2000) Analytical approaches for characterizing soil organic matter. Org Geochem 31:609–626

Li GC, Felbeck GT (1972) A study of the mechanism of atrazine adsorption by humic acid from muck soil. Soil Sci 113:430–433

Lichtfouse E (1997) Soil, a sponge for pollutants. Analusis 25:M16–M23

Liechtenstein EP (1980) Bound residues in soils and transfer of soil residues in crops. Residue Reviews 76:147–155

Liechtenstein EP, Katan J, Anderegg BN (1977) Binding of "persistent" and "nonpersistent" ^{14}C-labelled insecticides in an agricultural soil. J Agric Food Chem 25:43–47

Luthy RG, Aiken GR, Brusseau ML, Cunningham SD, Gschwend PM, Pignatello JJ, Reinhard M, Traina SJ, Weber WJ, Westall JC (1997) Sequestration of hydrophobic organic contaminants by geosorbents. Environ Sci Technol 31:3341–3347

McCarthy JF, Zachara JM (1989) Subsurface transport of contaminants: mobile colloids in the subsurface environment may alter the transport of contaminants. Environ Sci Technol 23:496–502

Michaelis W, Richnow HH, Seifert R (1995) Chemically bound chlorinated aromatics in humic substances. Naturwissenschaften 82:139–142

Miikki V, Hänninen K, Knuutinen J, Hyötyläinen J (1999) Pyrolysis of humic acids from digested and composted sewage sludge. Chemosphere 38:247–253

Mongenot T, Derenne S, Largeau C, Tribovillard NP, Lallier-Verges E, Dessort D, Connan J (1999) Spectroscopic, kinetic and pyrolytic studies of kerogen from the dark parallel laminae facies of the sulphur-rich Orbangnoux deposit (Upper Kimmeridgian, Jura). Org Geochem 30:39–56

Murphy EM, Zachara JM, Smith SC (1990) Influence of mineral-bound humic substances on the sorption of hydrophobic organic compunds. Environ Sci Technol 24:1507–1516

Nanny MA (1999) Deuterium NMR characterization of non-covalent interactions between monoaromatic compounds and fulvic acids. Org Geochem 30:901–909

Northcott GL, Jones KC (2000) Experimental approaches and analytical techniques for determining organic compounds bound residues in soil and sediment. Environ Pollut 108:19–43

Oesterreich T, Klaus U, Volk M, Neidhart B, Spiteller M (1999) Environmental fate of amitrole: influence of dissolved organic matter. Chemosphere 38:379–392

Palm H, Lammi R (1995) Fate of pulp mill organochlorines in the Gulf of Bothnia sediments. Environ Sci Technol 29:1722–1727

Parris GE (1980) Covalent binding of aromatic amines to humate. 1. Reactions with carbonyls and quinones. Environ Sci Technol 14:1099–1106

Perdue EM, Wolfe NL (1982) Modification of pollutant hydrolysis kinetics in the presence of humic substances. Environ Sci Technol 16:847–852

Piccolo A, Celano G, De Simone C (1992) Interactions of atrazine with humic substances of different origins and their hydrolysed products. Sci Total Environ 117/118:403–412

Pignatello JJ, Xing B (1996) Mechanisms of slow sorption of organic chemicals to natural particles. Environ Sci Technol 30:1–11

Requejo AG, Brown J, Boehm PD (1985) Thermal degradation products of non-volatile organic matter as indicators of anthropogenic inputs to estuarine and coastal sediments. In: Sigleo AC, Hattori A (ed) Marine and estuarine geochemistry. Lewis Publishers, Chelsea (Michigan), pp 81–96

Richnow HH, Seifert R, Hefter J, Kästner M, Mahro B, Michaelis W (1994) Metabolites of xenobiotica and mineral oil constituents linked to macromolecular matter in polluted environments. Advances in organic geochemistry 1993, Org Geochem 22:671–681

Richnow HH, Eschenbach A, Mahro B, Seifert R, Wehrung P, Albrecht P, Michaelis W (1998) The use of ^{13}C-labelled polycyclic aromatic hydrocarbons for the analysis of their transformation in soils. Chemosphere 36:2211–2244

Ricking M, Schwarzbauer J, Franke S (2003) Molecular markers of anthropogenic activity in sediments of the Havel and Spree Rivers (Germany). Wat Res 37:2607–2617

Robertson BK, Alexander M (1998) Sequestration of DDT and dieldrin in soil: disappearance of acute toxicity but not the compounds. Environ Toxicol Chem 17:1034–1038

Safe SH, Gaido K (1998) Phytoestrogens and anthropogenic estrogenic compounds. Environ Toxicol Chem 17:119–126

Santos EBH, Duarte AC (1998) The influence of pulp and paper mill effluents on the composition of the humic fraction of aquatic organic matter. Water Res 32:597–608

Schlautmann MA, Morgan JJ (1993) Binding of fluorescent hydrophobic organic probe by dissolved humic substances and organically-coated aluminium oxide surfaces. Environ Sci Technol 27:2523–2532

Schulten HR, Leinweber P (1996) Characterisation of humic and soil particles by analytical pyrolysis and computer modelling. J Anal Appl Pyrolysis 38:1–53

Schwarzbauer J, Ricking M, Franke S, Franke W (2001) Halogenated organic compounds in sediments of the Havel and Spree Rivers (Germany). Part 5 of organic compounds as contaminants of the Elbe River and its tributaries. Environ Sci Technol 35:4015–4025

Senesi N (1992) Binding mechanisms of pesticides to soil humic substances. Sci Total Environ 123/124:63–76

Senesi N, Testini C, Miano TM (1987) Interaction mechanism between HAs of different origin and nature and electron-donor herbicides – a comparative IR and electron-spin-resonance study. Org Geochem 11:25–30

Stankiewicz BA, van Bergen PF, Smith MB, Carter JF, Briggs DEG, Evershed RP (1998) Comparison of the analytical performance of filament and curie-point pyrolysis devices. J Anal Appl Pyrolysis 45:133–151

Steinberg CEW, Sturm A, Kelbel J, Kyu Lee S, Hertjorn N, Freitag D, Kettrup AA (1992) Changes of acute toxicity of organic chemicals to *Daphnia magna* in the presence of dissolved humic material (DHM). Acta Hydrochim Hydrobiol 20:326–332

Tissot BP, Welte DH (1984) Petroleum formation and occurence. Springer-Verlag, Berlin

Weber WJ, Lebouef EJ, Young TM, Huang W (2001) Contaminant interactions with geosorbent organic matter: insights drawn from polymer sciences. Water Res 35:853–868

Wheeler WB, Stratton GD, Twilley RR, Ou LT, Carlson DA, Davidson JM (1979) Trifluralin degradation and binding in soil. J Agric Food Chem 27:702–706

Ziechmann W (1972) Über die Elekronen-Donator und Acceptor-Eigenschaften von Huminstoffen. Geoderma 8:111–113

Zwiener C, Kumke MU, Abbt-Braun G, Frimmel FH (1999) Absorbed and bound residues in fulvic acid fractions of a contaminated groundwater – isolation, chromatographic and spectroscopic characterization. Acta Hydrochim Hydrobiol 27:208–213

Chapter 32

Behaviour of Dioxin in Pig Adipocytes

P. Irigaray · G. Rychen · C. Feidt · F. Laurent · L. Mejean

Abstract

Due to their lipophilic properties, dioxins can be integrated in the lipidic vacuole of adipocytes (fat cells). The aim of this study was to determine the kinetics of incorporation and release of ^{3}H-labelled palmitic acid and ^{14}C-labelled 2,3,7,8-TCDD in isolated adipocytes from pigs. The incorporation of 2,3,7,8-TCDD and palmitic acid was found to be concomitant in conditions of lipogenesis, under the effect of increasing quantities of insulin and in presence of glucose. The release of these two compounds was found to be dependant of a lipolytic agent (adrenalin). These results suggest the risk of an strong increase of 2,3,7,8-TCDD in blood induced by lipolysis for animals or humans previously exposed to this dioxin.

Key words: dioxin, 2,3,7,8-TCDD, adipocytes, lipogenesis, lipolysis

32.1
Introduction

Adipose cells (adipocytes) are known for their essential role in the regulation of the body fat mass by two main mechanisms. The first is fat burning, in order to supply energy for the organism. The second mechanism is the storage of fatty acids. In this respect, insulin favors fat retention. Further, adipocytes are also involved in other mechanisms. Recent studies suggest the role of the adipose tissue as a protective agent against 2,3,7,8-tetrachlorodibenzo-*p*-dioxin (2,3,7,8-TCDD) and other liposoluble contaminants (Lassiter and Hallam 1990; Geyer et al. 1990, 1993, 1994, 1997). Due to their lipophilic properties, they are mainly concentrated in the lipids of the food chain (especially in animal food products). Therefore, the daily consumption of contaminated food products can lead to accumulation of dioxins in human adipose tissue. Geyer et al. (1990) even suggested that dioxins integration within adipose tissue is easier than their release. Thus, our objective is to understand how dioxins are integrated in adipocytes and the way they are released.

32.2
Materials and Method

32.2.1
Chemicals

4-(2-hydroxyethyl)-1-piperazine ethanesulfonic acid (HEPES), fatty acids free bovine serum albumin (BSA), collagenase type II and the ^{3}H-labelled palmitic acid tritium of concentration 1 mCi ml^{-1} with a specific activity of 50 mCi mM^{-1} were obtained from Sigma. The

2,3,7,8-tetrachloro[U-^{14}C]dibenzo-p-dioxin in a solution of toluene (98% radioactivity pure) of concentration 0.042 mCi ml^{-1} with a specific activity of 45.8 mCi mM^{-1} was obtained from ChemSyn, ISOBIO. Ultima Gold liquid scintillation was obtained from Packard.

32.2.2
Cell Preparation

The adipose tissue used in these studies was the perirenal fat pad from Large-White pigs. These adipose tissue are known to have the most important lipogenic power (Mourot et al. 1999). Adipocytes were isolated from adipose tissue using Rodbell's method (Rodbell 1964) modified as followed. Fat tissue was washed in Krebs Ringer-HEPES buffer (KRH) composed of 118.7 mm NaCl, 4.8 mM KCl, 3.6 mM CaCl$_2$, 1.2 mM KH$_2$PO$_4$, 1.2 mM MgSO$_4$ and 25 mm HEPES, pH 7.4, 37 °C. 60 g samples were cut into small pieces then fitted in polypropylene beaker with 120 ml KRH containing 5 mM glucose, 4% BSA, and 0.5 mg ml^{-1} collagenase type II for 45 min in a shaking incubator at 37 °C (40 cycles min^{-1}). Cells were washed in a Krebs Ringer-bicarbonate buffer (KRB) composed of 118.7 mm NaCl, 4.8 mm KCl, 3.6 mM CaCl$_2$, 1.2 mM KH$_2$PO$_4$, 1.2 mM MgSO$_4$ and 25 mm carbonate acid sodium, pH 7.4 containing 5 mm glucose, 4% BSA and filtered through 200 µm nylon mesh, sequentially three times.

32.2.3
Lipogenesis Assays

20 ml samples of stock cells suspension were poured into polypropylene beaker (20% of cells, v/v) containing 0.252 µCi of ^{14}C-2,3,7,8-TCDD (0.006 µM) and 0.36 µCi of ^{3}H-palmitic acid (0.007 µM) for 30 ml of KRB containing 5 mm glucose, 4% BSA, and increased amount of insulin. The concentrations of insulin used were 30, 60 and 90 mm. Then the cells were incubated 1 h at 37 °C. The gas phase was 95% O$_2$, 5% CO$_2$. 1 ml cells samples were taken after 0, 5, 10, 15, 30, 45, 60 min. Medium's radioactivity was quantified by liquid scintillation (Ultima Gold) counting in a Packard Tricarb to determine the percent of radioactivity not incorporated in the cells. This experiment was performed in presence or absence of ^{3}H-palmitic acid.

32.2.4
Lipolysis Assays

25 ml of stock cells suspension were poured into polypropylene beaker (25% of cells, v/v) containing 0.252 µCi of ^{14}C-2,3,7,8-TCDD (0.006 µM) and 0.36 µCi of ^{3}H-palmitic acid (0.007 µM) for 30 ml of KRB containing 5 mm glucose, 4% BSA, and 90 nM insulin The cells were then incubated 1 h at 37 °C. The gas phase was 95% O$_2$, 5% CO$_2$. Immediately after incubation, the beaker was dipped into ice, 100% KRB (v/v) was added in order to decrease insulin concentration of a half. The cells (60% of cells, v/v) were then poured into polypropylene tubes containing lipolytic medium.

This new medium was composed of KRB containing 1 mM glucose, 4% BSA and increasing amounts of adrenalin. In order to prove that the release of 2,3,7,8-TCDD occurs only during lipolysis, two adrenalin concentrations have been used: 10^{-6} and

10^{-5} M. Then the cells were incubated at 37 °C. The gas phase was 95% O_2, 5% CO_2. Incubation was performed at different times: 15, 30, 45, 60, 75 and 90 min.

Immediately after incubation, tubes were plunged into ice. Supernatants which contained cells suspension, were punctured (0.8 ml and 10 ml of liquid scintillation were added to the medium in order to take off the 2,3,7,8-TCDD still remaining intimately bound to the plastic coat).

32.2.5
Experimental Design

Incorporation of 2,3,7,8-TCDD. This experiment was conducted using a single adipose tissue. For each time point and insulin doses, three repetitions were performed. The same procedure was accomplished in presence or absence of palmitic acid.

Release of 2,3,7,8-TCDD. This experiment was performed using three adipose tissue samples submitted to three different adrenalin concentrations to detect if the 2,3,7,8-TCDD release was dependent of the lipolytic state. Each measuring was performed in triplicate.

32.2.6
Calculations and statistical analysis

The simultaneous measurement of ^{14}C-2,3,7,8-TCDD and ^{3}H-palmitic acid kinetics was made by double radioactive labelling within the same sample. Two different markers were used: ^{3}H and ^{14}C, β radioactivity in both cases. Since β-maximum energies are of 18.6 keV for ^{3}H and 156 keV for ^{14}C, it is possible to obtain an excellent separation. The use of this proportioning requires the realization of a range standard in order to determine a possible quenching and especially to evaluate the output of the counting of the ^{14}C in the area of tritium and the area of the ^{14}C. The same steps were carried out for tritium. To evaluate the release of 2,3,7,8-TCDD, we used known quantities of labelled molecules present in the supernatant and in the medium as described:

$$\frac{[(\text{labelled molecules})_{\text{supernatant}} - (\text{labelled molecules})_{\text{medium}}]}{(\text{labelled molecules})_{\text{medium}}} 100$$

All data were analysed using an analysis of multiple linear regression (Systat 5.0.4).

32.3
Results and Discussion

32.3.1
Incorporation of 2,3,7,8-TCDD and Palmitic Acid in the Adipose Cells

2,3,7,8-TCDD and palmitic acid have been placed separately or together in an adipocytes suspension in presence of different insulin concentrations. Kinetics of incorporation of 2,3,7,8-TCDD in the presence of palmitic acid are reported in Fig. 32.1. Data are expressed as the percentage of the molecule quantity introduced at the be-

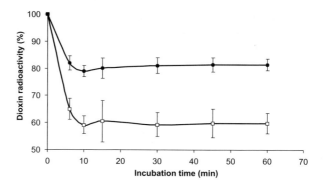

Fig. 32.1. Incorporation of dioxin in pig adipose cells. Kinetic of incorporation of the TCDD in presence (●) or in absence (□) of palmitic acid. This incorporation was realized during the first ten minutes. In presence of palmitic acid, the percentage of decrease of 2,3,7,8-TCDD as palmitic acid was 20% and in absence of this palmitic acid, incorporation of 2,3,7,8-TCDD was always realized during the first ten minutes but was twice higher than with palmitic acid (40%). Values are means with their standard deviations represented by vertical bars ($n = 9$)

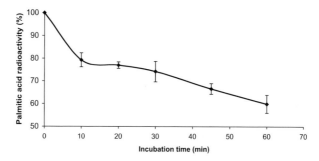

Fig. 32.2. Kinetic of incorporation of palmitic acid in pig adipose cells. Incorporation of palmitic acid was linear and increase with the time. Values are means with their standard deviations represented by vertical bars ($n = 9$)

ginning of the experiment. Kinetics of incorporation were similar for each insulin concentration. For this reason, a single chart has been presented.

Figure 32.1 shows also that incorporation kinetics of 2,3,7,8-TCDD and palmitic acid are similar when these two compounds are together in the medium. These two compounds seems to penetrate into adipocytes in the same way. This incorporation was realized during the first ten minutes. In presence of palmitic acid, the percentage of decrease of 2,3,7,8-TCDD as palmitic acid was 20%. The 2,3,7,8-TCDD entrance into adipocytes has also been observed without palmitic acid. And in absence of this palmitic acid, incorporation of 2,3,7,8-TCDD was always realized during the first ten minutes but was twice higher than with palmitic acid (40%). Incorporation of palmitic acid was linear and increase with the time (Fig. 32.2).

In fact, fatty acids entrance into adipocytes is not insulin-dependant while this requires the activation of a carrier in the case of glucose. However fatty acids storage takes place only in presence of glucose. But the entrance of the palmitic acid and

2,3,7,8-TCDD in adipocytes is realized quickly. We can infer that the 2,3,7,8-TCDD as well as the palmitic acid get within adipocytes without insulin. That is comprehensible by the hydrophobic property ($\log K_{OW}$ = 6.80) of the 2,3,7,8-TCDD. This incorporation takes place in the same way and in the same proportion in presence of the palmitic acid (20% for the two molecules). These results concerning lipogenesis can be related to those by Victor et al. (1985) who obtained a complete incorporation of the 2,3,7,8-TCDD within hepatocytes in suspension within 2 min.

32.3.2
Lipogenesis and Lipolysis: The Release of 2,3,7,8-TCDD

To study the release of 2,3,7,8-TCDD a lipogenesis was realized with 90 nM insulin and after 1 h, adipocytes were then poured into a lipolytic medium. Adrenalin concentration of 10^{-5} M resulted in a growing 2,3,7,8-TCDD and palmitic acid concentrations with time in the medium (Fig. 32.3). The same observations were made with 10^{-6} M of adrenalin (data not shown). Results suggest that release of 2,3,7,8-TCDD as well as that of palmitic acid were performed during lipolysis (p = 0.001) (Fig. 32.3). Figure 32.3 shows a different release for 2,3,7,8-TCDD (25%) than for palmitic acid (60%). Palmitic acid concentrations in the medium are twice higher despite equivalent incorporated amounts.

It has to be noticed that if incorporation of 2,3,7,8-TCDD (Fig. 32.1) takes place within adipocytes in spontaneous manner, the release of these compounds from adipocytes occur only in presence of adrenalin. 2,3,7,8-TCDD incorporation kinetics is similar to palmitic acid incorporation but release of these compounds is different. Release kinetics of 2,3,7,8-TCDD is more weak than release kinetics of palmitic acid. Our results seem to consolidate the hypothesis of an higher storage than release of 2,3,7,8-TCDD which suggest a protective role of the adipose tissue against this type of molecule.

Moreover, our results suggest the hypothesis that the release of 2,3,7,8-TCDD could be carried out after a strong reduction of triglycerides within the lipidic vacuole of the cell. 2,3,7,8-TCDD would behave like a fatty acid for the input in the cell but its mobi-

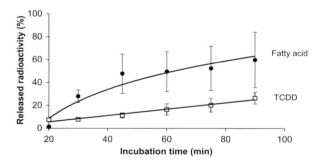

Fig. 32.3. Release of dioxin (2,3,7,8-TCDD) and palmitic acid from pig adipose cells. Kinetic of liberation of TCDD (□) and palmitic acid (●) during the time generated by 10^{-5} M of adrenalin. Note that Palmitic acid concentrations in the medium were twice higher than 2,3,7,8-TCDD despite equivalent incorporated amounts. Values are means with their standard deviations represented by vertical bars (n = 9)

lization seems not to follow the same process. At this time, our hypothesis is that the release of 2,3,7,8-TCDD is not dependent of the lipolytic agent but depend to the triglycerides concentration in the vacuole.

Concerning lipolysis, our work allows to confirm a hypothesis of Koppe (1995) who underled the fact that mobilization of fatty acids from adipocytes could involve the release of dioxins previously stored. In this studies, mobilization of the fatty acids since adipose tissue seems to be the cause of the release of dioxins.

32.4 Conclusion

We observed the fast entrance of 2,3,7,8-TCDD and palmitic acid in the adipocytes (10 min). TCDD incorporation is lower (20%) when these two molecules are present in the medium than without palmitic acid (40%). This difference could be explained by a phenomenon of saturation, which seems to be observed after ten minutes of incubation (Fig. 32.1 and 32.2). The release of 2,3,7,8-TCDD is only observed under the influence of the adrenalin. This release reaches approximately 60% for the palmitic acid and 25% for 2,3,7,8-TCDD at the end of 90 min (Fig. 32.3). 2,3,7,8-TCDD get inside adipose cells in spontaneous manner after ten minutes. But its release is not as easy as its incorporation. 20% of the incorporated 2,3,7,8-TCDD are released after 90 min of incubation with adrenalin against 60% for the palmitic acid. At this time the hypothesis according to which the liberation of the 2,3,7,8-TCDD would be achieved after a strong reduction only in triglycerides within the vacuole can be made.

It is therefore easier to understand the time of half-life (7 years) of this type of molecule within the human organism (Kedderis et al. 1993; Pegram et al. 1995; Fernandez-Salguero et al. 1996). Because if it's easy to grow bigger, it is on the contrary less easy to lose fat, and according to our results, it is likely to be the same for 2,3,7,8-TCDD. This reinforces adipose tissue in its protective role but poses a question about the becoming of these molecules accumulated in human fat in the case of brutal slimming.

References

Fernandez-Salguero PM, Hilbert DM, Rudikoff S, Ward JM, Gonzalez FJ (1996) Aryl-hydrocarbon receptor-deficient mice are resistant to 2,3,7,8-tetrachlorodibenzo-p-dioxin-induced toxicity. Toxicol Appl Pharmacol 140:173–179

Geyer HJ, Scheunert I, Karl R, Kettrup A, Korte F, Greim H, Rozman K (1990) Correlation between acute toxicity of 2,3,7,8-tetrachlorodibenzo-p-dioxin (2,3,7,8-TCDD) and total body fat content in mammals. Toxicology 65:97–107

Geyer HJ, Steinberg CEW, Scheunert I, Brügggemann R, Kettrup A, Rozman K (1993) A review of the relationship between toxicity (LC_{50}) of γ-hexachlorocyclohexane (γ-HCH, Lindane) and total lipid content of different fish species. Toxicology 83:169–179

Geyer HJ, Scheunert I, Brügggemann R, Matthies M, Steinberg CEW, Zitko V, Kettrup A, Garrison W (1994) The relevance of aquatic organisms' lipid content to the toxicity of lipophilic chemicals: toxicity of Lindane to different fish species. Ecotox and Env Saf 28:53–70

Geyer HJ, Schramm KW, Scheunert I, Shughart K, Buters J, Wurst W, Greim H, Kluge R, Steinberg CEW, Kettrup A, Madhukar B, Olson JR, Gallo MA (1997) Considerations on genetic and environmental factors that contribute to resistance or sensivity of mammals including humans to toxicity of 2,3,7,8-tetrachlorodibenzo-p-dioxin (2,3,7,8-TCDD) and related compounds. Ecotox and Env Saf 36:213–230

Kedderis LB, Mills JJ, Andersen ME, Birnbaum LS (1993) A physiologically based pharmacokinetic model for 2,3,7,8-tetrabromodibenzo-*p*-dioxin (TBDD) in the rat: tissue distribution and CYP1A induction. Toxicol Appl Pharmacol 121:87–98

Koppe JG (1995) Nutrition and breast-feeding. Eur J Obstet Gynecol Reprod Biol 61:73–78

Lassiter RR, Hallam TG (1990) Survival of the fattest: implication for acute effects of lipophilic chemicals on aquatic populations. Environ Toxicol Chem 9:585–595

Mourot J, Kouba M, Salvatori G (1999) Facteurs de variation de la lipogenèse dans les adipocytes et les tissus adipeux chez le porc. Prod Anim 12:311–318

Pegram RA, Diliberto JJ, Moore TC, Gao P, Birnbaum LS (1995) 2,3,7,8-tetrachlorodibenzo-*p*-dioxin (2,3,7,8-TCDD) distribution and cytochrome P4501A induction in young adult and senescent male mice. Toxicol Lett 76:119–126

Rodbell M (1964) The metabolism of isolated fat cells. Effects of hormones on glucose metabolism and lipolysis. J Biol Chem 239:375–380

Victor J, Wroblewski VJ, Olson JR (1985) Hepatic metabolism of 2,3,7,8-tetrachlorodibenzo-*p*-dioxin (2,3,7,8-TCDD) in the rat and guinea pig. Toxicol Appl Pharmacol 81:231–240

Part III

Chapter 33

Control of Halogenated By-Products During Surface Water Potabilisation

E. Chauveheid

Abstract

Organic and inorganic halogenated by-products are generally formed when chlorine is used for surface water potabilisation. The main halogenated by-products relevant for drinking water production are trihalomethanes and bromate anion. Strategies to control and reduce such by-products in drinking water are presented and discussed. Replacing chlorine with ozone and removing natural organic matter on granular activated carbon (GAC) filters reduces trihalomethanes. Acidic pH reduces the formation of bromate anion when ozone is used for disinfection. By combining these strategies and operating them at an industrial level for surface water treatment, drinking water with excellent bacteriological quality and low halogenated by-products is produced and supplied to the customer.

Key words: surface water, drinking water, halogenated by-products, chlorine, ozone, trihalomethanes, bromate, natural organic matter, activated carbon

33.1 Introduction

Given the more and more stringent quality standards, the production of drinking water from surface water resources needs a combination of physical and chemical treatments such as coagulation, filtration, oxidation and disinfection. To achieve good bacteriological quality, the treated water should be oxidised and disinfected with strong oxidants, such as chlorine (Cl_2 or $HClO$), chlorine dioxide (ClO_2) or ozone (O_3).

For many years, chlorine has been used for the oxidation and disinfection of surface water. While chlorine is quite effective for these purposes, its use leads to many chlorinated by-products such as trihalomethanes (CHX_3), first detected in treated water by J. Rook in 1974 (Rook 1974, 1976). As these compounds are carcinogenic, their concentration in water has been progressively reduced and regulated in many countries. In Europe, trihalomethanes were first regulated in 1980, but without any maximal concentration (European Directive 80/778). The latest European legislation for drinking water, published in 1998, regulates the sum of trihalomethanes (which represents the sum of 4 compounds) at 150 µg l^{-1}. This maximal admissible concentration will be reduced to 100 µg l^{-1} for 2008 (European Directive 98/83). The use of chlorine also leads to a large variety of other toxic organic by-products, which have been progressively identified but never regulated in Europe. This directive also stipulates that reduction of trihalomethanes in drinking water should not prevent efficient disinfection at the consumer tap. Hence, treatment processes must be designed to achieve good water disinfection and minimal trihalomethanes generation.

In order to reduce halogenated organic by-products, the massive use of chlorine was replaced by other oxidants, such as chlorine dioxide and ozone (Symons 1978). Chlorine dioxide has been shown to generate much less halogenated organic by-products than chlorine, especially trihalomethanes. Nevertheless, its use leads to the formation of significant amounts of chlorite (ClO_2^-), undesirable in water due to its toxicity. In presence of chlorine, chlorite is transformed into chlorate (ClO_3^-), which is recognised as less toxic. Finally, ozone has been used as an oxidant and disinfectant because it has a very strong oxidising power and is a chlorine free reagent. The use of ozone mainly leads to non-halogenated organic compounds (Langlais 1991), but in the presence of bromide (Br^-), some brominated trihalomethanes (CHX_3, see Fig. 33.3 for formulas) and brominated organic matter can be generated. Their concentrations are generally marginal compared to those obtained with chlorine. While ozone seems to be the ideal oxidant for water potabilisation, it has one major drawback: bromide will be oxidised to bromate (BrO_3^-), which is known as a highly carcinogenic compound (Croué et al. 1996). The latest European drinking water directive has set a maximal allowed concentration at 10 µg l^{-1} for bromate (European Directive 98/83). This concentration level is not easy to achieve at an industrial level given that efficient disinfection must be guaranteed simultaneously.

Due to European regulations and because of known toxicity of halogenated by-products, surface water potabilisation processes must be designed to achieve efficient disinfection and low halogenated by-products formation at the same time, meaning that a compromise must be found. This paper will briefly review the origin of main organic and inorganic halogenated by-products generated from surface water potabilisation and will present some strategies to improve the compromise between efficient disinfection and non-desirable halogenated by-products. The concepts will be illustrated with results obtained from an industrial water potabilisation facility and from monitoring these non-desirable halogenated by-products in the distribution network of the city Brussels.

33.2
Origin of Halogenated By-Products and Strategies to Reduce these Undesirable Compounds

Halogenated by-products formed during surface water potabilisation can be classified into organic and inorganic compounds. Beside chlorinated by-products, obtained from chlorine usage, brominated compounds can be generated when bromide is present in water, even at low concentration. Chlorine in water, in form of hypochlorous acid and hypochlorous anion, will oxidise bromide into hypobromous acid, as illustrated in Fig. 33.1.

Because bromide can easily be oxidised in presence of hypochlorous acid, a complete conversion into hypobromous acid will occur. At usual pH for surface water

$$H^+ + ClO^- \rightleftharpoons HClO \xrightarrow[-Cl^-]{+Br^-} HBrO$$

Fig. 33.1. Hypohalogenous species present during surface water treatment (between 7.5 and 9), hypochlorous acid ($pK_a = 7.35$) is mainly in its ionised form, while hypobromous acid ($pK_a = 8.8$) is the main species. As their acidic form is the most reactive to organic matter, brominated organic products will be generated in significant amounts even if bromide is at low concentration in water. Since both oxidants are generally present, mixed chlorinated and brominated organic by-products are generated.

33.2.1
Organic Halogenated By-Products

The presence of natural organic matter (NOM) is the source of organic halogenated by-products when chlorine is used (Norwood et al. 1983). NOM, characterised by total organic carbon, is usually in the range of 2 to 5 mg C l^{-1} for surface water, and can be as high as 10 mg l^{-1} for some waters. The generic term "natural organic matter" represents a large collection of diverse organic compounds originating essentially from microbial catalysed decomposition of plants. The main fraction of natural organic matter is made of polymeric compounds such as humic acids, fulvic acids, polypeptides and carbohydrates, together with smaller molecules like terpenes, acetogenins, fatty acids and many others. Non natural organic matter or anthropogenic pollutants, such as waste effluents, hydrocarbons and pesticides, are often found in surface waters. Nevertheless, NOM represents the major source of halogenated by-products when considering its concentration relative to other organic pollutants (see Fig. 33.2). The most current chlorination by-products are the following trihalomethanes: chloroform, bromodichloromethane, dibromochloromethane and bromoform (Rook 1974, 1976). Besides trihalomethanes, other soluble and low molecular weight halogenated molecules can be formed from further breakdown of oxidised natural organic matter, such as halogenated phenols (Rockwell 1978), halogenated acetonitrile (CHX$_2$CN) (Bieber and Trehy 1983), halogenated acetic acid (Bieber and Trehy 1983), chloral hydrate (CCl$_3$CHO) (Norwood et al. 1983; Meyer 1993), and even furanic compounds (see Fig. 33.3 for structure) (Kronberg 1988). The main origin for these chlorinated by-products is mainly ascribed to reactions of chlorine with meta-dihydroxybenzene sub-units found in the complex polymeric structure of humic acids. Other polymeric compounds found in natural organic matter have also been proposed as generating halogenated by-products.

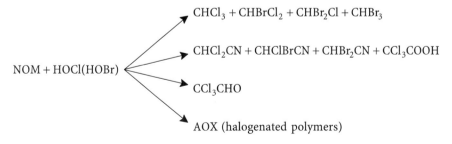

Fig. 33.2. Main halogenated by-products generated upon water chlorination. *NOM:* natural organic matter

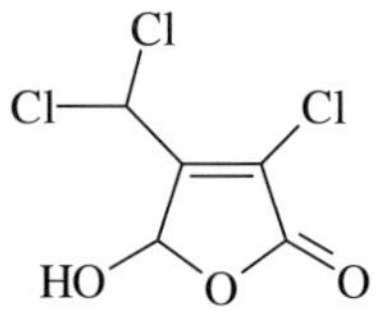

Fig. 33.3.
Main furanone derivative generated upon water chlorination

The natural organic matter can also be chlorinated without breakdown so that halogenated polyphenols and other halogenated polymers are generated, generally characterised as adsorbable organic halogenated compounds (AOX) (Glaze et al. 1980). New halogenated by-products are often characterised, such as halogenated furanone derivatives (Fig. 33.3), known to be highly carcinogenic (Kronberg 1988).

As trihalomethanes are generated from the reaction of natural organic matter with chlorine, reducing both natural organic matter before chlorination and the amount of chlorine used for oxidation and disinfection will reduce their presence. Natural organic matter can be reduced mostly through filtration on granular activated carbon. Most of these filters can work in a biological mode, meaning that natural polymeric carbon will be transformed into biomass and carbon dioxide. The activated carbon allows biofilm attachment and its development because of a huge contact surface available. At the same time, activated carbon allows physical adsorption of organic matter, mainly through hydrophobic interactions. Thus, adsorbed organic matter will be made available for the microorganisms in the biofilm. Oxidation of natural organic matter before biological filtration on activated carbon improves its removal because of enhanced bioavailability of organic compounds for the biofilm fixed on the granular activated carbon filtration material. This can be achieved with oxidants such as chlorine, chlorine dioxide or ozone. Nowadays, ozone is the preferred chemical for oxidation stages, because it is a powerful oxidant leading mainly to non-halogenated organic products. Moreover, ozone is quickly degraded in the presence of activated carbon, allowing biological filtration mode. So, from an industrial point of view, reduction of trihalomethanes can be achieved by using ozone oxidation instead of chlorine, and by combining an ozonation stage with granular activated carbon filtration, the ozonation being thus applied before the filtration.

33.2.2
Inorganic Halogenated By-Products

Inorganic halogenated by-products are fewer than organic derivatives. The main ones detected are chlorite, chlorate and bromate. Chlorite is obtained as a by-product of chlorine dioxide oxidation, and can be further oxidised to chlorate when chlorine is used as a disinfectant. Chlorite is undesirable in treated water, because of its toxicity, while chlorate is of less concern. To avoid chlorite formation, ozone can be used in conjunction with chlorine dioxide (Fig. 33.4) to oxidise chlorite into chlorate (Siddiqui 1996).

Chlorate is also present in concentrated hypochlorous acid used as a source of chlorine, since hypochlorous acid disproportionates into chlorate. By using chlorine gas, excessive chlorate concentrations can be avoided. If chlorine is used in conjunction with ozone (Fig. 33.5), hypochlorous acid is oxidised and transformed into chlorate (Siddiqui 1996).

Chapter 33 · Control of Halogenated By-Products During Surface Water Potabilisation

$$ClO_2 \xrightarrow{+1e^-} ClO_2^- \xrightarrow{O_3} ClO_3^- + O_2$$

Fig. 33.4. Oxidation of chlorine dioxide to chlorate in the presence of ozone

$$Cl_2 + H_2O \longrightarrow HClO \xrightarrow{O_3} ClO_3^-$$

Fig. 33.5. Oxidation of hypochlorous acid to chlorate with ozone

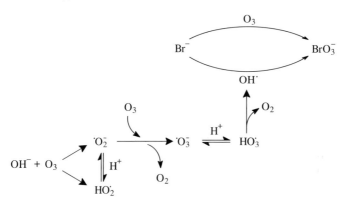

Fig. 33.6. Bromate generation upon water ozonation

Bromate is mainly obtained when ozone is used. In the presence of ozone, bromide is transformed into bromate according to molecular and free radical pathways, as illustrated in a simplified way in Fig. 33.6 (see Haag and Hoigné 1983 and von Gunten and Hoigné 1994 for more detailed mechanistic pathways). The radical pathway, initiated by hydroxyl radical (OH·), has been recognised as a significant pathway when radicals are not scavenged, while the molecular oxidation of bromide (direct reaction with molecular ozone) is dominant when radicals are efficiently scavenged (von Gunten and Hoigné 1994). The radical mechanism is initiated by the hydroxyl radical, generated from ozone catalysed decomposition (Staehelin and Hoigné 1985) as shown in Fig. 33.6.

Bromide oxidation to bromate is slowed down when the pH is more acidic (Croué et al. 1996; Kruithof et al. 1993) as ozone decomposition into hydroxyl radical is slower and the hypobromous acid, which is not oxidised by molecular ozone, is present in greater proportion. The ozone dosage and the reaction contact time should also be optimised, in order to control bromate formation and get efficient water disinfection at the same time. So, as for trihalomethanes, a compromise must be found between bromate formation and efficient disinfection.

Minor concentrations of bromate can be found in commercial concentrated hypochlorous solutions, often used for chlorination, since trace levels of bromide are oxidised to bromate in these concentrated solutions. The use of chlorine gas instead of hypochlorous solutions significantly reduces this source of bromate in final treated water. Reduction of natural bromide concentrations could be another interesting option to limit bromate formation but, unfortunately, bromide is not easily removed from water in a selective way.

33.3
Industrial Experience to Control Halogenated By-Products

The effective control of halogenated by-products during surface water potabilisation is presented for the Tailfer plant (15 km south of the city Namur, Belgium), located on the river Meuse, and its related distribution network in the city of Brussels (80 km away from the Tailfer plant). The Tailfer plant treats surface water from the river Meuse, with several treatment stages organised in the following order: screening, pre-ozonation, flocculation-coagulation (sulphuric acid, aluminium sulphate, silicate), decantation, filtration on double layer sand/granular activated carbon (GAC) filters, ozonation, ozone destruction with bisulphite, post-filtration on GAC, chlorination and pH adjustment. The discussion concerning control of halogenated by-products at an industrial scale is focused on trihalomethanes and bromate.

33.3.1
Control of Trihalomethanes

A general way to reduce trihalomethanes formation is to reduce chlorine consumption to a minimum. This goal is achieved by replacing chlorine with ozone, which can be used as an oxidant in a pre-oxidation stage (pre-ozonation) and as a disinfectant in an ozonation stage. Ozonation stages used in conjunction with natural organic matter removal stages, such as double layer filtration and granular activated carbon post-filtration, lead to low concentrations of trihalomethanes at the outlet of the Tailfer process. As ozone is not a persistent disinfectant, some chlorine should be added at the end of the treatment process, especially if the produced water has a long residence time prior to consumption. The reduction of the yearly maximal trihalomethanes concentration (representing the sum of the four compounds) is illustrated in Fig. 33.7 for the produced water from the Tailfer plant and in Fig. 33.9 for the water supplied to the surroundings of Brussels. As shown in Fig. 33.7, this yearly maximal concentration at the outlet of the Tailfer process has been reduced from 25 to 5 µg l^{-1} during the last decade. A significant reduction was observed in 1996 due to the transformation of sand filters into double layer sand/GAC filters, which increased natural organic matter removal. A seasonal effect on the concentration of trihalomethanes was observed because more chlorine was added at higher water temperature (summer), due to higher disappearance rates with higher water temperature.

Since 2001, the trihalomethanes concentrations are further reduced to concentrations as low as 3 µg l^{-1} and the seasonal effect has disappeared. This enhanced reduc-

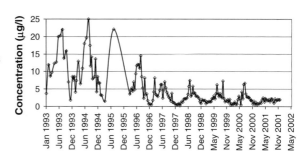

Fig. 33.7.
Total trihalomethanes concentrations in water produced from Tailfer process, from 1993 to 2002

tion of total trihalomethanes concentrations in produced water is ascribed to the implementation of a new filtration stage at the end of the process, following the double layer sand/GAC filters and ozone disinfection stage. This filtration stage, called post-filtration, is filled with granular activated carbon (3 m height) and is aimed at removing further natural organic matter from water. In fact, this stage is working as a biological reactor removing a significant fraction of the previously oxidised organic matter, as shown in Fig. 33.8. As a consequence, the total trihalomethanes concentrations found at the end of the treatment process are lower when chlorine is added as disinfectant. In 2001, the total organic carbon was less than 1 mg l^{-1}, while concentrations between 1 and 2 mg l^{-1} were observed before 2001 (see Fig. 33.8). As the post-filtration has not been operated at full capacity in 2001, an even more significant total organic carbon (TOC) reduction is expected for the future. So, it might be expected that the TOC could reach 0.5 mg l^{-1} to 0.7 mg l^{-1}, which would represent an 80% reduction of the natural organic carbon from its usual concentration in the raw surface water.

The water leaving the treatment process is supplied to 2 large reservoirs (60 000 m^3 each), called the Callois reservoirs and located 50 km away from the Tailfer plant (pipe distance). The generation of trihalomethanes during water supply to and storage in the Callois reservoirs is also reduced with the removal of organic matter achieved with the Tailfer process (Fig. 33.9).

During water supply from Tailfer plant to Callois reservoirs, two chlorine injections are realised on the way. One injection is made at a site called Mazy, located 20 km away from Tailfer (pipe distance), where chlorine level in water is adjusted to a pre-selected concentration from measurement of residual chlorine after injection of chlorine gas in water. This

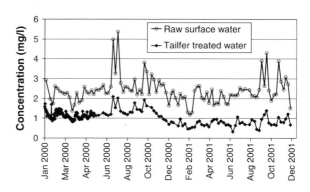

Fig. 33.8. Total organic matter concentrations in water produced from Tailfer process, from 2000 to 2002

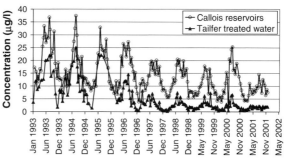

Fig. 33.9. Total trihalomethanes concentrations at the outlet of Callois reservoirs, from 1993 to 2002

pre-selected concentration at Mazy is chosen so that constant chlorine residual is achieved at the inlet of the Callois reservoirs (located 30 km away from Mazy). When looking at the pre-selected chlorine concentration at Mazy (Fig. 33.10), less chlorine has been used in 2001 to reach the target 0.10–0.15 mg l^{-1} residual chlorine at the inlet of Callois reservoirs, when compared to 2000. The difference is most spectacular during the summer.

This fact is a direct consequence of the newly installed and operated post-filtration stage at Tailfer plant. As seen in Fig. 33.10, the pre-selected chlorine level at Mazy is very close to the usually 0.10 to 0.15 mg l^{-1} pre-selected chlorine level at the outlet of the Callois reservoirs, needed to ensure efficient and continuous disinfection along the distribution network. As a matter of fact, less chlorine is added at Mazy and so fewer trihalomethanes are generated. Moreover, the improved stability of dissolved chlorine guaranties good bacteriological stability in the distribution network.

The water supplied from the Callois reservoirs to the northern part of Brussels (Callois is located in the southern part of the city) is chlorinated at the outlet of the reservoirs and once more on the way. The total trihalomethanes concentrations in the northern part of the distribution network are quite unchanged when compared to those observed at the Callois reservoirs (Fig. 33.11), while the pipe distance between both areas is approximately 35 km. Between 1998 and 2000, the maximal total trihalomethanes concentrations were in the range of 20 to 30 µg l^{-1} in the northern distribution area of Brussels. With the implementation of the post-filtration stage, operated since 2001, the total trihalomethanes concentrations have been further reduced to a maximum of 15 µg l^{-1} in this distribution area, which is well below the future European limit of 100 µg l^{-1}.

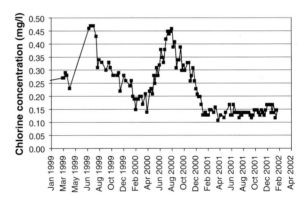

Fig. 33.10.
Pre-selected chlorine concentration at Mazy to get constant residual at Callois reservoirs, from 1999 to 2002

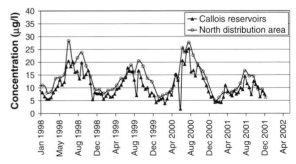

Fig. 33.11.
Total trihalomethanes concentrations at the outlet of Callois reservoirs and in the northern part of Brussels, from 1998 to 2002

Fig. 33.12.
Bromate concentrations in water produced from Tailfer process, from 1999 to 2002

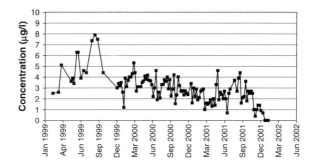

33.3.2
Control of Bromate

As mentioned earlier, bromate is a typical halogenated by-product generate from ozone application. The use of ozone in a pre-ozonation stage does not lead to bromate when usual bromide concentrations are present, because ozone reacts quickly with many organic (natural organic matter, algae) and inorganic compounds (dissolved metals, mineral colloids) present at higher concentrations in the raw surface water. During surface water treatment, the bromate formation is mainly observed at the ozone disinfection stage, where many contaminants have been removed with previous coagulation-decantation-filtration steps. The bromide anion is going through these coagulation-decantation-filtration stages and reacts then with ozone at the disinfection stage. At Tailfer plant, the raw water contains usually less than 30 µg l^{-1} of bromide. With such low natural concentrations of bromide, less than 10 µg l^{-1} bromate is generally obtained in the treated water. Before 2000, bromate concentrations varied between 3 and 8 µg l^{-1} in treated water. These concentration levels were achieved because the water still contained an ozone residual at the outlet of the ozonation stage that reacted further with bromide during storage of ozonated water in the Tailfer reservoirs. The contact time with ozone was estimated to approximately 30 to 60 min before ozone was completely consumed. Since 2000, a more efficient automated pH control was realised before the ozonation stage so that a stable pH between 7.2 and 7.4 was reached. Well-controlled and more acidic pH than that of raw water (varies from 7.7 to 8.4) slowed down the formation of bromate, as illustrated in Fig. 33.12.

As a consequence, bromate concentrations as high as 8 µg l^{-1} were achieved when ozonated water was stored in the Tailfer reservoirs. With implementation of post-filtration, ozone contact time has been reduced to less than 10 min, since ozone destruction with bisulphite is realised before the post-filtration stage. In these conditions, bromate concentrations at the outlet of the process are lower than 4 µg l^{-1} and no bromate is detected when water temperature is low, as illustrated for the beginning of year 2002.

33.4
Conclusions

Control of the most relevant halogenated by-products generated from a surface water potabilisation process is achieved by combining several treatment stages. The generation of the four trihalomethanes is reduced by using ozone for water oxidation and disinfection, and with the removal of organic matter by granular activated carbon filtration working in a biological mode. With such a design, total trihalomethanes concentrations can be limited to a maximum of 15 µg l^{-1} when the treated water is supplied as far as 80 km from the production site. At the same time, less chlorine is needed to achieve good bacteriological quality, and disinfection is improved since chlorine stability has been increased. For bromate, the concentration can be reduced to less than 4 µg l^{-1} by running ozonation at pH close to neutrality and reducing contact time after the ozonation stage. By lowering the pH before ozonation, ozone stability is significantly improved, leading to a slow bromate formation rate and improved disinfection efficiency. Thus, trihalomethanes and bromate by-products can be easily controlled at an industrial scale and comply with the stringent European drinking water regulation.

References

Bieber TI, Trehy ML (1983) Dihaloacetonitriles in chlorinated natural waters. In: Jolley (ed) Water chlorination: environmental impact and health effects, vol. 4(1). Ann Arbor Science, pp 85
Croué JP, Koudjonou BK, Legube B (1996) Parameters affecting the formation of bromate ion during ozonation. Ozone Sci Eng 18:1–18
European Directive 80/778/EEC, OJC L229/11 (15/7/1980)
European Directive 98/83/EEC, OJC L330/32 (5/12/1998)
Glaze WH, Saleh FY, Kinstley W (1980) Characterization of non-volatile halogenated compounds formed during water chlorination. In: Jolley (ed) Water chlorination: environmental impact and health effects, vol. 3. Ann Arbor Science, pp 99
Haag WR, Hoigné J (1983) Ozonation of bromide containing waters: kinetics of formation of hypobromous acid and bromate. Environ Sci Technol 17:261–267
Kronberg L (1988) Identification and quantification of the Ames mutagenic compound 3-chloro-4-(dichloromethyl)-5-hydroxy-2(5H)-furanone and of its geometric isomer (E)-2-chloro-4-(dichloromethyl)-4-oxobutenoic acid in chlorine treated humic water and drinking water extracts. Environ Sci Technol 22:1097–1103
Kruithof JC, Meijers RT, Schippers JC (1993) Formation, restriction of formation and removal of bromate. Water Supply 11(3/4):331–342
Langlais B (1991) Ozone in water treatment: application and engineering. Lewis Publishers
Meyer I (1993) Influence of biofilm on disinfection by-products in a distribution network. Water Supply 11(3/4):355–364
Norwood DL, Johnson JD, Christman RF (1983) Chlorination products from aquatic humic material at neutral pH. In: Jolley (ed) Water chlorination: environmental impact and health effects, vol. 4(1). Ann Arbor Science, pp 191
Rockwell AL, Larson RA (1978) Aqueous chlorination of some phenolic acids. In: Jolley (ed) Water chlorination: environmental impact and health effects, vol. 2. Ann Arbor Science, pp 67
Rook JJ (1974) Formation of haloforms during chlorination of natural water. J Water Treat Exam 23(2):234–243
Rook JJ (1976) Haloforms in drinking water. JAWWA 68(3):168–172
Siddiqui MS (1996) Chlorine-ozone interactions: formation of chlorate. Water Res 30(9):2160–2170
Staehelin J, Hoigné J (1985) Decomposition of ozone in water in the presence of organic solutes acting as promoters and inhibitors of radical chain reactions. Environ Sci Technol 19:1206–1213
Symons JM (1978) Ozone, chlorine dioxide and chloramines as alternatives to chlorine for disinfection of drinking water. In: Jolley (ed) Water chlorination: environmental impact and health effects, vol. 2. Ann Arbor Science, pp 555
von Gunten U, Hoigné J (1994) Bromate formation during ozonation of bromide containing waters: interaction of ozone and hydroxyl radical reactions. Environ Sci Technol 28:1234–1242

Chapter 34

Organic Pollutants in Airborne Particulates of Algiers City Area

N. Yassaa · B. Y. Meklati · A. Cecinato

Abstract

The concentrations of particle-bound organic compounds comprising n-alkanes, n-alkanoic acids, polycyclic aromatic hydrocarbons (PAH) and nitrated polycyclic aromatic hydrocarbons (NPAH) in ambient air of Algiers city area, were measured from May 1998 to February 1999. Motor vehicle were found to be the main source of airborne particles in downtown Algiers, while combustion and bacterial activity seemed to be responsible of the air pollution at the Oued Smar waste landfill. The in-situ generation of some NPAH seemed to contribute to air pollution, especially during summertime. In general, the wintertime concentrations of the organic pollutants in Algiers were similar to those measured in Europe, especially over the Mediterranean Basin. The chemical characterisation of organic compounds in smoke particulate matter emitted from fats and bitumes industries were also investigated and revealed specific distribution profiles of n-alkanes, n-alkanoic and n-alkenoic acids, and PAH.

Key words: organic aerosols; PAH; NPAH; air pollution; Algiers; waste landfill; fat factory, bituminous materials

34.1 Introduction

Organic compounds are the main constituents of the fine fraction of atmospheric aerosols in highly industrialised and urbanised areas (Schauer et al. 1999). There are several emission sources responsible for the presence of organic aerosols in the atmosphere. Among them, road traffic and industries play a key role in terms of both chemical balance of the atmosphere and health risk for humans. Aerosols have been also recognised as influencing the light balance of the Earth through absorption of solar radiation and, on the other hand, by scattering or reflecting it back to space (Charlson et al. 1992).

Whilst in the developed world both stationary and mobile sources of pollution are subjected to strict limitations nowadays, these still remain free from any control in developing countries. However, about 80% of human population is concentrated on these areas. People living in these countries often experience high ambient pollution because man-made activities are pooled in metropolis and megalopolis regions where any environmental policy is far from being adopted or, at least, applied in practice. Technologies capable of abating the release of toxic aerosols and gaseous effluents into the atmosphere are very limited and pollutants are simply spread out into the environment. Furthermore, the exposure of such strongly polluted airsheds to sunlight is likely to give rise to secondary pollution, which promotes annoyance and other epidemiological risks of population exposed (Finlayson-Pitts and Pitts 1997; Yassaa et al. 2001a).

Algeria, which covers an area of ca. 2 382 000 km^2 and looks on the Mediterranean Sea along its big coast at the Northern border (ca. 1 200 km broad), can play a main role in investigations dealing with potentially toxic organic substances occurring in the air of hot temperate, tropical and desert regions of the world. Studies in Algeria could provide a better insight on the pathways as well as spatial and temporal scales through which contaminants are spread out over the whole Mediterranean Basin. Algiers city was chosen as subject of investigation because the most industrialised and densely populated city of the country lies there. Indeed, with more than 3 millions inhabitants, Algiers concentrates more than 12% of the entire population of Algeria and belongs to the largest cities facing the Mediterranean coast.

During the two last decades, the suburbs of Algiers have experienced a high increase of motor vehicles, which caused a dramatic worsening of the air quality and related effects, e.g. high levels of gas mixtures and fine aerosols affecting visibility, human health, and material damage. Besides that, not far from Algiers city lies the open-air municipal landfill of Oued Smar, which is not yet subjected to any control by public authorities. Therefore, uncontrolled accumulation of wastes, degradation and exhausts of refuses coming from private houses, industries and hospitals contribute to the global pollution, thus affecting the air quality at both local and regional scale. In fact, smogs are often observed. These events occur mainly during the night or at sunrise, whenever the meteorological stability is well developed, concurrently with sweet winds blowing and strong thermal inversion dominating the area. This situation is very usual, and inhabitants and workers exposed to Algiers pollution experience increased adverse effects, in particular respiratory difficulties, irritation of eyes and throat. Children, old and sick persons suffer the main vulnerability to air pollution.

In recent years, algerian institutions have promoted field investigations aimed at assessing the pollution impact resulting from industries and power plants and the first data sets being acquired dealt with the chemical composition of gas and particle-bound organic components released from bitumes manufacture (Yassaa et al. 2001b) and fatty product factory (Yassaa et al. 2001c). The area of Bab-Ezzouar, which is nearby the industrial plants and waste landfill of Oued Smar and often lies downwind to the plant plumes, has been investigated in the outskirts of Algiers (Yassaa et al. 2001b). The increased number of respiratory symptoms and diseases recorded at Bab-Ezzouar in the last few years is suspected to depend upon the exposition of population to air pollutants associated with *fine* aerosol particles, which readily penetrate into the lungs. Nature and concentration levels of organic components at Bab-Ezzouar provide a good data-base for assessing both spatial and temporal pathways through contamination originated in the landfill and from bituminous product industry is spread out over the Algiers region. More recently, our concern was addressed to the biggest oil and fat production plant of Algeria, located in the regional domain of Algiers, in order to identify and quantify its organic emission and to draw information about its impact onto regional air quality and human exposure to particle-bound pollution (Yassaa et al. 2001c). This paper presents an overview of the aerial concentrations of *n*-alkanes, *n*-alkanoic acids, *n*-alkenoic acids, polycyclic aromatic hydrocarbons (PAH) and nitrated PAH present in the Algiers atmosphere and generated by traffic road, waste incineration, fats and bitumes industries on a compound by compound basis.

34.2
Experimental

34.2.1
Sampling Sites

Aerosols were collected in downtown Algiers, where the motor vehicle exhaust was recognised as the main source of both NO_x, volatile organic compounds (Yassaa et al. 2001d) and black carbon. The atmospheric aerosol samples were collected from May 1998 to February 1999 at about 200 m far from the sea, 5 m from a traffic rushing road and at about 3 m from the ground level. The Oued Smar municipal landfill was chosen as subject of our investigation because it is the largest one existing in Algeria. It is located at about 13 km from downtown Algiers, eastward from the city and also in the west part of the same industrial zone; it covers a surface of about 37.5 ha. The landfill represents an important emission source for a number of atmospheric pollutants comprising gaseous species compounds, e.g. volatile organics, SO_2, NO_x, CO, NH_3, as well as solids such as dusts, ashes, and soot. According to a recent estimation, 4 000 t d^{-1} of wastes are usually stored there. House wastes account for 1 600 t d^{-1} and the industrial ones, including toxic substances, for the remaining. All refuses are burnt in open air. Neither combustion parameters nor the release of exhaust are subjected to any control. Within our investigation, aerosol samples were collected from May 1998 to February 1999 at about 50 m from the municipal waste landfill and at 2 m above the ground.

A field campaign was also performed from March to May 2000 in the fatty manufacture plant, namely ENCG, acronym of *Entreprise Nationale des Corps Gras*, located in the center of Algiers, nearby the city harbour and at less than 20 m from the sea. It represents the biggest fatty production plant in Algeria. Although some improvements to the industrial processes have been recently undertaken, plant management is still under uncontrolled conditions.

Bab-Ezzouar University, i.e. Houari Boumediene University of Science and Technology, USTHB, was chosen as an other sampling location for the current study. It is located nearby the industrial region of Oued Smar and far from any traffic road. Thus, it represents an ideal site to evaluate the contribution of asphalt manufacture, which is the main industry in that area, to the organic particulate budget without any interference from autovehicular exhaust emissions. Aerosol samples were collected during August 1999 in the Bab-Ezzouar University campus on the rooftop of a building, 10 m from the ground level. Bab-Ezzouar University is located about 14 km far from downtown Algiers, south-east side to the city. With more than 25 000 students and an area of 105 ha, it represents among the largest universities in Africa. It lies at about 200 m from a big asphalt plant and at 1 km from the municipal waste landfill of Oued Smar.

34.2.2
Sampling

Air particles were collected on Teflon inert membrane by means of a medium-volume sampling apparatus (16.7 l min^{-1}) equipped with a size-selective inlet, suitable to enrich from air only particles smaller than 10 μm, and a volume counter measuring the air passed through the filter. Samplings started at 7 A.M. and lasted 24 h; samples collected in filters were wrapped in aluminium foils and stored at 4 °C until to be subjected to chemical determinations.

34.2.3
Sample Extraction and Clean-Up

Chemical determination of organic aerosols was performed by using a procedure described extensively elsewhere (Ciccioli et al. 1996; Yassaa et al. 2001a). Briefly, samples were spiked with a solution containing internal reference compounds for the analysis, i.e. 1-bromotetradecane, 1-bromoeicosane, perdeuterated 1-nitropyrene, phenanthrene-d_{10}, pyrene-d_{10}, chrysene-d_{12} and perylene-d_{12}. Then the organic content of aerosols was extracted by refluxing enriched filters in a soxhlet apparatus, using a dichloromethane-acetone mixture (4/1, v/v). The extract was spiked with an internal standard mixture and evaporated, then the residue was divided in two aliquots. Most of the extract (4/5 of the total) was dissolved in toluene and fractionated through a column chromatography on neutral alumina. Non-polar aliphatic compounds were eluted first with n-hexane, whilst the bulk of polynuclear aromatics were recovered by eluting dichloromethane through the column. The second eluate was further separated into three subfractions containing PAH, Nitro-PAH and more polar compounds, respectively, by means of normal-phase high performance liquid chromatography (HPLC), by using an eluent gradient from n-hexane to dichloromethane (the eluent program and further conditions are reported by Ciccioli et al. 1989, 1996). The second aliquot of sample extract (i.e. 1/5 of the total) was lead to react with boron trifluoride in excess of methanol to convert organic acids to their methyl ester analogues. A further elution through a silica column of the reacted material, run by using dichloromethane, allowed to clean alkanoic acid methyl esters from possible interference. All sample fractions were stored in the dark at 4 °C until to be analysed.

34.2.4
Chemical Analyses

n-Alkanes, esterified acids, PAH and nitrated PAH were determined by using a HP-5890-type gas chromatograph coupled with a HP-5970B mass spectrometric detector (GC-MS) operating in selected ion mass (SIM) mode (Hewlett Packard). The separations of the analytes were obtained through a 25 m long capillary column coated with a HP-5-type methylphenyl silicone stationary phase (i.d. = 0.2 mm, film thickness = 0.33 µm), provided by Hewlett Packard. The column temperature was maintained at 80 °C for 2 min, then programmed to 170 °C at 20 °C min^{-1} and held constant for 2 min; a second ramp (4 °C min^{-1}) heated column up to 280 °C and elution was completed at this temperature for 15 min. Mass spectrometer system was operated in electron impact mode (ion source energy = 70 eV) and GC-MS data were acquired by a dedicated software purchased from Hewlett-Packard. PAH were identified by recording molecular ion traces and fragments, isotopic and doubly charged ions ([M], [M−1]$^+$, [M+1]$^+$ and [M/2]$^+$, respectively). Molecular [247]$^+$ and four characteristic fragment ions (i.e. [217]$^+$, [201]$^+$, [200]$^+$, and [189]$^+$) were selected for detecting nitrated fluoranthenes and pyrenes, while [M]$^+$, [74]$^+$, [143]$^+$ and [227]$^+$ traces were selected for methylated acids.

34.3
Results and Discussion

34.3.1
n-Alkanes

Table 34.1 reports the mean concentrations of *n*-alkanes belonging to the C_{16}–C_{31} range, which we recorded in downtown Algiers, waste landfill of Oued Smar, fatty products manufacture (ENCG) and Bab-Ezzouar (USTHB). The *n*-alkane pattern observed in downtown Algiers was similar to that of motor vehicular exhaust emissions (Rogge 1993), allowing us to identify the main source of carbon particles there. Stocking and burning of several kinds of materials including biomass and refuses, i.e. plants, foodstuffs, animals and plastics, caused the huge presence of *n*-alkanes in the aerosols at the Oued Smar site. The air contamination consequent from anthropic activities appeared more severe at Oued Smar than in downtown Algiers.

The mean concentrations of *n*-alkanes in particulate organic matter emitted from the fatty manufacture plant appeared very high, compared to those recorded in urban area of Algiers and waste landfill of Oued Smar. Within the semi-volatile range, i.e. below C_{25}, *n*-alkanes present rather an even-to-odd preference, with the highest concentrations reached by C_{22} and C_{24} homologues. Such a distribution was observed for the first time, confirming that the smoke plumes emitted from the fatty manufacture give raise

Table 34.1.
Mean *n*-alkane concentrations (ng m^{-3}) in downtown Algiers, Oued Smar landfill, fatty manufacture and Bab-Ezzouar University

Compound	Downtown Algiers	Landfill	Fatty manufacture	Bab-Ezzouar University
n-C_{16}	2.5	10.4	11.5	3.2
n-C_{17}	7.7	36.8	24.6	15.7
n-C_{18}	7.3	24.7	43.5	18.1
n-C_{19}	5.7	24	47.2	12.1
n-C_{20}	6.9	26.3	64.0	11.4
n-C_{21}	6.6	19.2	55.2	8.9
n-C_{22}	8.7	19.5	80.8	10.6
n-C_{23}	7.35	17.1	83.8	9.1
n-C_{24}	6.5	18.4	94.9	9.1
n-C_{25}	4.1	12.8	65.2	7.6
n-C_{26}	2.8	7.9	62.1	7.4
n-C_{27}	2.9	9.1	55.1	5.7
n-C_{28}	1.5	6.4	52.4	4.6
n-C_{29}	2.4	4.3	52.3	4.4
n-C_{30}	0.6	2.4	50.4	2.9
n-C_{31}	0.7	1.8	41.5	3.0
Total	55	241.5	884.5	133.8

to a peculiar fingerprint. Indeed, while the presence of hydrocarbons with a long odd-numbered carbon chain has been explained as resulting from decarboxylation of biogenic even-numbered fatty acids, the same explanation for even-numbered compounds is less reliable. Thus: are they resulting from the complete reduction of even-numbered fatty acids? Otherwise, do they come out, at least in fat plants, from decarboxylation of odd-numbered fatty acids recognised among constituents of waxes and cutins of plant tissues and generated by the α-oxidation of even-carbon numbered congeners?

Particle-bound normal alkanes observed in Bab-Ezzouar can be regarded as belonging to the category of petroleum residues. The n-alkane concentrations peaked in the range of C_{17}–C_{22} with C_{17} and C_{18} as predominant congeners, which is consistent with a petrogenic input. This pattern was similar to that reported for the roofing tar fumes by Rogge et al. (1997), suggesting that they were originated from similar materials. Since during distillation and production steps the most volatile species were preferentially emitted, lower concentrations of highly-numbered n-alkanes were observed. In contrast, at urban Algiers, n-alkanes up to C_{18} were present over all in the gas phase, whilst congeners over C_{24} were preferably present in aerosols. The quite high concentration of C_{17}–C_{20} n-alkanes found at Bab-Ezzouar University was very likely caused by strong emission outcoming from the asphalt production factory. Even after dilution, at the sampling site some extent of semi-volatile n-alkanes appeared in particle phase as a result of the partition equilibrium between the gas and aerosol, developing when the plume vapours were cooled. Although wind transport of pollutants from Oued Smar landfill could be suspected as a secondary input of organic pollution to the sampling site, nevertheless, n-alkanes contents in air at Bab-Ezzouar were higher than at Oued Smar and different patterns were observed for both emitters. This fact confirms by far that the asphalt plant is the main source affecting the area.

34.3.2
Monocarboxylic Acids

Figure 34.1 shows the mean concentrations of n-alkanoic acids (C_{10}–C_{30}) recorded in downtown Algiers and waste landfill of Oued Smar. Total n-alkanoic acids reached 101 and 482 ng m^{-3} in Oued Smar in summer and winter, respectively, and 69 and 256 ng m^{-3} in downtown Algiers. All species were more abundant in Oued Smar than in downtown Algiers. However, the percent composition of the acid fractions was the same in the both sites. The most abundant components belonged to the C_{12}–C_{24} range and the absolute maxima were identified with C_{16} and C_{18}, depending on the time of the year.

A strong even-to-odd carbon number preference was also observed, which is peculiar of a natural origin. The concentration profile of n-alkanoic acids was useful for distinguishing the true contribution of vascular plant emission from the impact of the microbial activity, which is usual in urban aerosols. Homologues lighter than C_{20} have been related to microbial sources and cooking operations, whilst those exceeding C_{22} to vascular plant waxes (Simoneit 1978; Simoneit and Mazurek 1982; Rogge et al. 1993a). Thus, the huge presence of n-alkanoic acids up to C_{20} in downtown Algiers might be related to anthropogenic sources, mainly the emission from a fatty manufacturing plant sited not far from the sampling point. Their dramatic accumulation in winter strongly supports this insight.

Fig. 34.1.
Seasonal mean of the *n*-alkanoic acid concentrations (ng m^{-3}) in downtown Algiers (a) and Oued smar landfill (b), 1998–1999

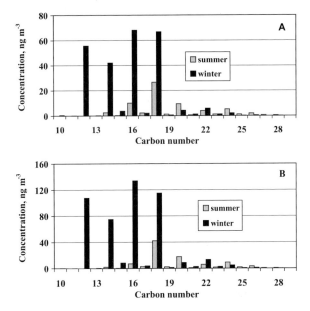

Soot collected at Oued Smar was richer of acids than that of downtown Algiers. The uncontrolled open-air combustion of several organic materials especially meat burning (foodstuffs and animals) is the major source contributing to the high levels of acids in that area.

Concentrations of *n*-alkanoic and *n*-alkenoic acids (C_9–C_{24}) recorded both in the fatty manufacture and Bab-Ezzouar University are reported in Table 34.2. Acids were always present at huge levels in the fatty manufacture, the maximum reaching 5 468 ng m^{-3}. They accounted for more than 90% of the organic aerosol composite, consistently with a fatty material composition, where acids reach 90% of total. These concentrations were never observed before in emissions or in urban atmospheres and exceeded also those reported for the most important known *n*-alkanoic acids sources, i.e. charbroiling and cooking operations (Rogge et al. 1991; Schauer et al. 1999). The distribution pattern of homologues among fatty acids showed a strong even-to-odd carbon number predominance. Palmitic and stearic acids, the most abundant ones, accounted respectively for 26% and 22% of the total atmospheric acids.

Although a number of unsaturated fatty acids have been recognised as present in the atmosphere, nevertheless only palmitoleic (i.e. *cis*-hexadecen-9-oic acid, $C_{16=1}$), oleic (*cis*-octadecen-9-oic acid, $C_{18=1}$) and cetoleic (*cis*-docosen-9-oic acid, $C_{22=1}$) acids were found in aerosols in our field experiment. Oleic acid was the most abundant one in all samples. Unsaturated *n*-fatty acids have been recognised as constituents of emission released by microbial sources and processing, degradation, and combustion of vegetable and animal materials (Rogge 1993). Among monounsaturated acids synthesised by living organisms, oleic acid is very usual and may be the most abundant one not only among its homologues, but among all acids also (Naudet 1996). Rogge et al. (1991) showed that meat food is an important source of *n*-alkenoic acids, mainly oleic and palmitoleic acids, in particular cooking with seed oils, margarins, or

Table 34.2.
Mean monthly n-alkanoic and n-alkenoic acids concentrations (ng m^{-3}) recorded in the fatty manufacture and Bab-Ezzouar University

Compound	Fatty manufacture	Bab-Ezzouar University
A_{10}	96	5.4
A_{11}	14.9	0.1
A_{12}	1207	54.0
A_{13}	1.9	0.6
A_{14}	1644	29.0
A_{15}	561	2.7
$A_{16=}$	65	–
A_{16}	3728	76.5
A_{17}	1.1	3.1
$A_{18=}$	668	n.ev.
A_{18}	3152	48.6
A_{19}	12.4	0.6
A_{20}	680	2.2
A_{21}	143	0.6
$A_{22=}$	374	n.ev.
A_{22}	1184	2.5
A_{23}	n.d.	0.7
A_{24}	447	2.1
Total	13979.3	228.7

animal fats. In contrast to plant waxes, seeds and seed oils, plant organelles, leaf cells, chloroplasts and pollen contain mainly palmitic and stearic, mono-unsaturated, di-unsaturated and poly-unsaturated fatty acids (Ching and Ching 1962; Hitchcock and Nichols 1971; Jamieson and Reid 1972; Laster and Valle 1971). Phytoplankton and bacteria also contain a number of unsaturated fatty acids (Hitchcock and Nichols 1971; Lechevalier 1977; Shaw 1974). In general, acid concentrations and homologue distributions measured at the fatty manufacture were consistent with the origin of the manufactured oils from vegetable materials and their huge presence in oil and soap composition. Thus the production of soaps appeared to be the most important source of fatty acids present in the atmosphere.

The concentrations for monocarboxylic n-alkanoic acids in Bab-Ezzouar samples were higher as compared to n-alkanes. A strong even-to-odd predominance was found. Reaching about 32% of the total identified congeners, hexadecanoic acid (C_{16}) was the most abundant acid. The huge abundance of n-alkanoic acids in the aerosols could be mainly explained by strong microbial activities related to high temperature and humidity typical of that time. The microbial degradation of several industrial refuses, yeast manufacture in particular, which were wasted into the river flowing close to Bab-Ezzouar University, was very likely the principal explanation of acids presence in the atmosphere, although it has been reported that hexadecanoic and octadecanoic acids can arise from fossil fuel burning (Rogge et al. 1993b). This hypothesis is supported by the finding that acid concentrations were higher than for n-alkanes, which is in disagreement with the experiment of Rogge et al. (1997), where acids were up to two orders of magnitude less abundant than the n-alkanes. Furthermore, the acid contents in air at Bab-Ezzouar University were higher not only than those measured in downtown Algiers during the whole year, but also those recorded in waste landfill during summer.

34.3.3
Polycyclic Aromatic Hydrocarbons (PAH)

Table 34.3 reports the aerial content of individual PAH in downtown Algiers, Oued Smar, fatty manufacture and Bab-Ezouar University. The most abundant PAH in urban Algiers was pyrene, whereas the air of Oued Smar was enriched in chrysene. PAH contents in the air were anytime higher at Oued Smar than in downtown Algiers. In Algiers city, PAH were at extents similar to those found at busy street sites in Europe (Cecinato 1999).

PAH concentrations recorded in fatty product industry appeared to be, at some extent, similar to those measured in urban Algiers, suggesting that they were originated from the same source: autovehicular exhaust emissions.

The most abundant PAH congeners in Bab-Ezzouar were fluoranthene and pyrene, accounting for about 25% of the PAH identified. Benzo[a]pyrene, which is typically monitored in roofing tar fumes (Rogge et al. 1997) and often used as indicator of the whole PAH carcinogenicity (WHO 1987), consisted about 2.7% of the total PAH amount consisted. A tentative source reconciliation was obtained by comparing both contents

Table 34.3. Mean PAH concentrations (ng m^{-3}) recorded in downtown Algiers, Oued Smar landfill, fatty manufacture and Bab-Ezzouar University

Compound	Algiers	Landfill	Fatty manufacture	Bab-Ezzouar University
Fluoranthene	3.4	9.5	3.9	9.6
Pyrene	4.4	8.2	3.3	9.3
Cyclopentapyrene	2.9	2.4	0.3	n.ev.
Benzo[a]anthracene	1.7	4.4	0.9	4.5
Chrysene	2.5	8.4	2.8	5.8
Benzofluoranthenes	3.1	6.7	2.7	8.8
Benzo[e]pyrene	1.5	2.5	1.1	4.0
Benzo[a]pyrene	1.3	2.1	0.8	2.0
Indeno[1,2,3-cd]pyrene	1.0	1.1	1.2	3.0
Dibenzo[a,h]anthracene	0.1	0.1	0.1	1.0
Benzo[g,h,i]perylene	2.3	1.5	2.0	3.5

and pattern of PAH with those recorded in urban Algiers. In general, the nature and contents of PAH well reproduce their petrogenic origin at both sites. Nevertheless, the lack of duty-traffic roads at Bab-Ezzouar, suggests that the release during the production of asphalt material is the most probable source of PAH.

34.3.4
Nitro-Polycyclic Aromatic Hydrocarbons (NPAH)

The mean concentrations of 2-nitrofluoranthene (0.26 and 0.16 ng m^{-3}), 1-nitropyrene (0.07 and 0.04 ng m^{-3}) and 2-nitropyrene (0.02 and 0.01 ng m^{-3}) were recorded respectively at Algiers urban site and waste landfill. The NPAH in Algiers reached levels similar to those recorded in urban areas of Europe (Marino et al. 2000). The in-situ generation of NPAH induced by photochemical processes seemed to affect the air quality of Algiers more than the direct emission. In fact, many investigations already outlined that diesel powered vehicles are strong emitters of 1-nitropyrene (Schuetzle et al. 1982; Ciccioli et al. 1989); by contrast, two-step nitrations of parent PAH induced by OH radical plus NO$_2$ or NO$_3$ develop 2-nitrofluoranthene and 2-nitropyrene (Arey et al. 1987; Pitts et al. 1985; Zielinska et al. 1989; Atkinson and Arey 1994), which are found in the air. Unlike other components, the air was richer in NPAH in downtown Algiers than at Oued Smar; that was probably caused by the scarce levels of precursors such as NO$_x$ at the waste landfill.

In Bab-Ezzouar, only 2-nitrofluoranthene (1.5 ng m^{-3}) was identified. This is consistent with the in-situ generation. The huge abundance of this species is surprising and of health concern with respect to its well known potential mutagenic activity that is more adverse than the parent PAH (Cecinato and Zagari 1997). It was found at extent higher (more than one order of magnitude) than that observed in urban Algiers. The possible explanation might be the fact that the parent PAH was abundant and the samples were collected during hot and sunny days experienced the region in August (average temperature around 35 °C) that promote the formation of a such toxic pollutant. Finally, the absence in Bab-Ezzouar of 1-nitropyrene recognised as emitted directly from diesel powered vehicles (Schuetzle et al. 1982; Ciccioli et al. 1989), supports the fact that the emissions from asphalt production may dominate by far over the auto-vehicular exhaust in the budget of saturated and polycyclic aromatic hydrocarbons.

34.4
Conclusion

The composition of both *n*-alkane and *n*-alkanoic acid fractions of aerosols seems to indicate that the release of organic species from Oued Smar landfill contributed to the air pollution in Algiers city, although the motor vehicle emission was the main source for higher PAH and NPAH. From the oils and the soaps plants (ENCG) emissions, the *n*-alkane distribution profile, which showed the even-to-odd carbon predominance similarly to the composition of unsaponifiable fraction of fats, could be regarded as a tracer for the industrial fatty product emissions. The concentration levels reached by organic pollutants in Algiers city during wintertime were similar to those measured in European towns, especially over the Mediterranean Basin.

References

Arey J, Zielinska B, Atkinson R, Winer AM (1987) Polycyclic aromatic hydrocarbons and nitroarene concentrations in ambient air during a winter time high-NO$_x$ episode in the Los Angeles basin. Atmos Environ 21:1437–1444

Atkinson R, Arey J (1994) Atmospheric chemistry of gas-phase polycyclic aromatic hydrocarbons: formation of atmospheric mutagens. J Environ Health Perspect 120:117–126

Cecinato A (1999) Atmospheric PAH in Italy: experience and concentration levels. Fresenius Environ Bull 8:586–594

Cecinato A, Zagari M (1997) Nitroarenes of photochemical origin: a possible source of risk to human health. J Environ Pathol Toixcol Oncol 16:93–99

Charlson RJ, Schwartz SE, Hales JM, Cess RD, Coakley JA, Hansen JE, Hofmann DJ (1992) Climate forcing by anthropogenic aerosols. Science 255:423–430

Ching TM, Ching KK (1962) Fatty acids in pollen of some coniferous species. Science 138:890–891

Ciccioli P, Cecinato A, Brancaleoni E, Liberti A (1989) Evaluation of nitrated polycyclic aromatic hydrocarbons in anthropogenic emission and air samples. Aerosol Science and Technology 10:296–310

Ciccioli P, Cecinato A, Brancaleoni E, Frattoni M, Zacchei P, Miguel AH, Vasconcellas CP (1996) Formation and transport of 2-nitrofluoranthene and 2-nitropyrene of photochemical origin in the troposphere. J Geophys Res 101:19567–19581

Finlayson-Pitts BJ, Pitts Jr. JN (1997) Tropospheric air pollution: ozone, airborne toxics, polycyclic aromatic hydrocarbons and particles. Science 276:1045–1052

Hitchcock C, Nichols BW (1971) The lipid and fatty acid composition of specific tissues. In: Hitchcock C, Nichols BW(ed) Plant lipid biochemistry. Academic Press, London, pp 59–80

Jamieson GR, Reid HH (1972) The leaf lipids of some conifer species. Phytochemistry 11:269–275

Laster JL, Valle R (1971) Organics associated with the outer surface of airborne urediospores. Environ Sci Technol 5:631–634

Lechevalier MF (1977) Lipids in bacterial taxonomist's view. CRC Critical Rev Microbiol 5:109–210

Marino F, Cecinato A, Siskos PA (2000) Nitro-PAH in ambient particulate matter in the atmosphere of Athens. Chemosphere 40:533–537

Naudet M (1996) Fatty acids. In: Karleskind A (ed) Oils & fats, vol. 1. Lavoisier, pp 67–80

Pitts Jr. JN, Sweetman JA, Zielinska B, Winer AM, Atkinson R (1985) Determination of 2-nitrofluoranthene and 2-nitropyrene in ambient particulate matter: evidence of atmospheric reactions. Atmos Environ 19:1601–1608

Rogge WF (1993) Molecular tracers for sources of atmospheric carbon particles: measurements and model predictions. Ph.D. thesis, California Institute of Technology, Pasadena

Rogge WF, Hildemann LM, Mazurek MA, Cass GR, Simoneit BRT (1991) Sources of fine organic aerosol. 1. Charbroilers and meat cooking operations. Environ Sci Technol 25:1112–1125

Rogge WF, Hildemann LM, Mazurek MA, Cass GR, Simoneit BRT (1993a) Sources of fine organic aerosol. 4. Particulate abrasion products from leaf surfaces of urban plants. Environ Sci Technol 27:2700–2711

Rogge WF, Mazurek MA, Hildemann LM, Cass GR (1993b) Quantification of urban organic aerosols at a molecular level: identification, abundance and seasonal variation. Atmos Environ 27:1309–1330

Rogge WF, Hildemann LM, Mazurek MA, Cass GR, Simoneit BRT (1997) Sources of fine organic aerosol. 7. Hot asphalt roofing tar pot fumes. Environ Sci Technol 31:2726–2730

Schauer JJ, Kleeman MJ, Cass GR, Simoneit BRT (1999) Measurement of emissions from air pollution sources. 1. C_1 through C_{29} organic compounds from meat charbroiling. Environ Sci Technol 33:1566–1577

Schuetzle D, Roley TL, Prater TJ, Harvey TM, Hunt DF (1982) Analysis of nitrated polycyclic aromatic hydrocarbons in diesel particulates. Anal Chem 54:265–271

Shaw N (1974) Lipid composition as a guide to the classification of bacteria. Adv Appl Microbiol 17:63–108

Simoneit BRT (1978) The organic chemistry of marine sediments. In: Riley JP, Chester R (eds) Chemical oceanography, 2nd edn., vol. 7. Academic Press, New York, pp 233–311

Simoneit BRT, Mazurek MA (1982) Organic matter of the troposphere-II. Natural background of biogenic lipid matter in aerosols over the rural Western United States. Atmos Environ 16:2139–2159

WHO (1987) Polynuclear aromatic hydrocarbons (PAH). In: WHO (ed) Air quality guidelines for Europe. WHO Regional Publications, European Series No. 23. World Health Organization, Geneva, pp 105–117

Yassaa N, Meklati BY, Cecinato A, Marino F (2001a) Organic aerosols in urban and waste landfill of Algiers metropolitan area: occurrence and sources. Environ Sci Technol 35:306–311

Yassaa N, Meklati BY, Cecinato A, Marino F (2001b) Chemical characteristics of aerosol in Bab-Ezzouar (Algiers). Contribution of bituminous product manufacture. Chemosphere 45:315–322

Yassaa N, Meklati BY, Cecinato A (2001c) Chemical characteristics of organic aerosols in Algiers city area: influence of a fat manufacture plant. Atmos Environ 35:6003–6013

Yassaa N, Meklati BY, Brancaleoni E, Frattoni M, Ciccioli P (2001d) Polar and non-polar volatile organic compounds (VOCs) in urban Algiers and saharian sites of Algeria. Atmos Environ 35:787–801

Zielinska B, Arey J, Winer AM, Atkinson R (1989) The nitroarenes of molecular weight 247 in ambient particulate samples collected in Southern California. Atmos Environ 23:223–229

Chapter 35

A Reactive Transport Model for Air Pollutants

K. Iinuma · Y. Satoh · S. Uchida

Abstract

A general solution of coupled diffusion-reaction equations governing the spatial and temporal evolution of an arbitrary number of inter-reacting air pollutants has been developed. Although the present reaction chemistry is just limited to gas-phase pseudo-first-order process, the solution is of use to determine the reaction rate constants of complex reaction paths consisting of successive-, reversible-, cyclic-, and concurrent-reaction. Transport parameters (diffusion and advection) of the air pollutants are also involved in this solution. Moreover, equilibrium concentrations (or number densities) of all the air pollutants can be determined using a dynamic equilibrium solution (steady state solution) derived from the balance condition between their transport and reaction processes. For an appropriate model analysis, we have examined a toluene chemistry model that comprises ten related chemicals with twenty-six reaction rate constants under normal urban area conditions.

Key words: air pollutants, reactive transport model, dynamic equilibrium, toluene

35.1 Introduction

The study on the atmospheric chemical transport models has a long history, and several mathematically well-founded models for reactive transport analysis of multiple chemicals are now available (Seinfeld and Pandis 1998). Unfortunately, however, most of the model analyses appear to be too numerical to deduce from the calculated results which parameters mainly affect the spatiotemporal behavior of the chemicals of interest. This may be due to the fact that the tractable and explicit formulation including the transport and reaction coefficients of all the chemicals is not yet available. Development of such formulation is possible when we limit the analysis to the pseudo-first-order reaction chemistry.

Transport and reaction of air pollutants in the urban atmosphere are complex. Their production, diffusion, advection, various kinds of reactions, and removal process are intertwined in a complex way. Of these processes, the oxidation reaction in air is exceedingly important, generating a variety of precursors for production of relatively stable air pollutants. Because of a large amount of oxygen molecules and perpetual production of OH radicals in air, most of the oxidation reactions basically belong to the pseudo-first-order process (Finlayson-Pitts and Pitts, Jr. 1999). For this kind of first-order system, the spatiotemporal behavior of the chemical concentrations can be treated by coupled linear partial differential equations including some constant parameters. By limiting the analysis to relatively small urban areas where transport and reaction parameters are almost constant, we can reduce it to a coupled diffusion-reaction equations. We have derived a general solution of one-dimensional coupled diffusion-reaction equations, using the Fourier transform technique and matrix algebra (Iinuma 1991). This integral-type solution with respect to space coordinates also leads us to an analytical form of the steady-state solution (dynamic equilibrium solution). The generalized equilibrium constant for multiple chemicals is easily derived from the equilibrium solution (Iinuma et al. 1993). All these formulae have a potential to be useful to search for the predominant reaction path and to evaluate the proper reaction rate.

Toluene is one of the simplest aromatic hydrocarbons, but its oxidation reactions with OH radicals as well as oxygen molecules in air certainly produce various kinds of intermediates and innumerable ring-cleavage products. Based upon a model of urban toluene chemistry developed mainly by Atkinson (1990), we have calculated the area distribution of the concentrations of 10 chemicals including 6 intermediates generated from toluene.

35.2
General Reactive Transport Model of Multiple Chemicals

In general, the inter-reacting system of N chemicals comprises $N(N-1)$ reaction paths, forming a complicated network-type reactions consisting of the forward, the reverse, and the branching process. Figure 35.1 shows this schematically. The concentrations (or the number densities) of the chemical species as a function of the elapsed time (t) and the space coordinate (z) are governed by the following coupled partial differential equations (coupled transport-reaction equations) (Eq. 35.1);

$$\frac{\partial X_j}{\partial t} = D_j \frac{\partial^2 X_j}{\partial z^2} - v_j \frac{\partial X_j}{\partial z} - \left(\sum_{k=1}^{n} \alpha_{kj} + \beta_j\right) X_j + \sum_{m=1}^{n} \alpha_{jm} X_m + F_j \quad (35.1)$$

where, $X_j(z, t)$ is the number density (concentration) of j-species ($j = 1, 2, 3, ..., n$), D_j is the diffusion coefficient, v_j is the advection velocity, α_{kj} is the reaction frequency from j- to k-species, β_j is the removal coefficient, and F_j is the source term.

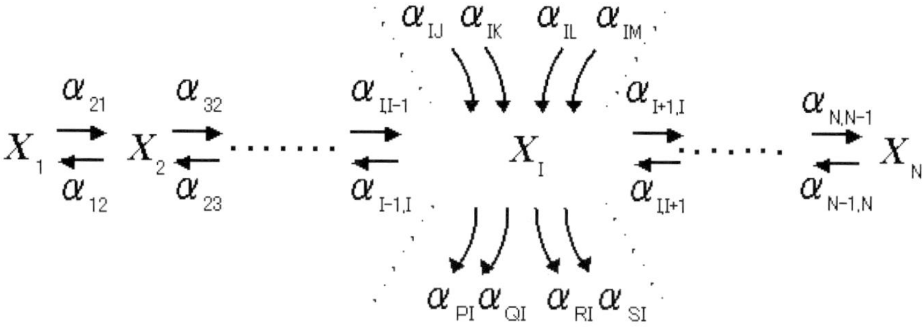

Fig. 35.1. A general network-type reaction system for N-chemicals. X_I is the number density of I-species and α_{IJ} is the reaction frequency from J-species to I-species; α_{IJ} can be calculated from the corresponding reaction rate

35.2.1
General Solution

Fourier transform of $X_j(z, t)$ with respect to z is expressed as $Y_j(\omega, t)$ (Eq. 35.2):

$$Y_j(\omega,t) = \int_{-\infty}^{\infty} X_j(z,t) \exp(-i\omega z) dz \qquad (35.2)$$

With the help of the eigenvalues (the roots), λ_p ($p = 1, 2, ..., n$), of the secular equation derived from the following matrix (Eq. 35.3);

$$(A) = \begin{pmatrix} T_1 & \alpha_{12} & \alpha_{13} & \cdots & \alpha_{1j} & \cdots & \alpha_{1n} \\ \alpha_{21} & T_2 & \alpha_{23} & \cdots & \alpha_{2j} & \cdots & \alpha_{2n} \\ \alpha_{31} & \alpha_{32} & \ddots & & \vdots & & \vdots \\ \vdots & \vdots & & \ddots & \vdots & & \vdots \\ \alpha_{j1} & \alpha_{j2} & \cdots & \cdots & T_j & \cdots & \alpha_{jn} \\ \vdots & \vdots & & & \vdots & \ddots & \vdots \\ \alpha_{n1} & \alpha_{n2} & \cdots & \cdots & \alpha_{nj} & \cdots & T_n \end{pmatrix} \qquad (35.3)$$

and the diagonal elements represented as (*i*: imaginary units)

$$T_j = -D_j \omega^2 - iv_j \omega + R_j - \beta_j, \qquad R_j = -\sum_{m=1}^{n} \alpha_{mi} \qquad (35.4)$$

we can represent the explicit form of $Y_j(\omega, t)$ as

$$Y_j(\omega,t) = \sum_{p=1}^{n} \frac{g_j(\lambda_p, t)}{f(\lambda_p)} \exp(\lambda_p t) \qquad (35.5)$$

For the explicit form of $g_j(\lambda_p, t)$ and $f(\lambda_p)$, please refer to the related references (Iinuma 1991; Iinuma et al. 1993).

Several appropriate numerical integration techniques are of use to carry out the inverse Fourier transform of Eq. 35.5. The solution derived as a function both of z and t enables us to visualize the space and time evolution of the concentrations of multiple chemicals from their initial stage to the final dynamic equilibrium. This interesting spatiotemporal evolution is the ubiquitous and common feature of reactive transport process in chemistry (references; vide supra).

35.2.2
Dynamic Equilibrium Solution

Our main concern in atmospheric chemistry may be the evaluation of the concentrations of multiple chemicals in equilibrium. Dynamic (not static) balance between generation and removal of chemicals realizes a chemical equilibrium. Using Eq. 35.5 which includes both the source term (F_j) and the removal term (β_j), we can calculate the equilibrium constant of two chemicals. It is noted that this equilibrium constant, in general, includes all the reaction rates involved in the system; this never be the ratio of only two reaction rates in it.

As a simple but useful realistic stationary source term, we examine the difference of two Heaviside-step-functions with constant emission rate,

$$F_j(z,t) = A_j[H(z+a_j) - H(z-a_j)] \tag{35.6}$$

where A_j is the constant emission rate of j-species and $2a_j$ is its emission area. From Eq. 35.6, we can derive the following dynamic equilibrium solution, $X_j(z, \infty)$

$$X_j(z,\infty) = \frac{1}{2\pi}\int_{-\infty}^{\infty} Y_j(\omega,\infty)\exp(i\omega z)d\omega \tag{35.7}$$

with

$$Y_j(\omega,\infty) = \frac{g_j(\omega)}{\lambda_1\lambda_2\lambda_3\cdots\lambda_n} \tag{35.8}$$

The denominator represents the product of all the eigenvalues of Eq. 35.3. The numerator $g_j(\omega)$ is expressed by the determinant of Eq. 35.3, which now includes the Fourier transformed source term (Eq. 35.6) in its j-th column (Iinuma et al. 1993).

The generalized equilibrium constant, $K_{eq}(j, k)$, for chemicals j and k is defined using Eq. 35.8 as

$$K_{eq}(j,k) = \frac{g_j(0)}{g_k(0)} = \frac{\delta(j)}{\delta(k)} \tag{35.9}$$

Equation 35.9 expresses the ratio of the concentrations of j-th ($\delta(j)$) and k-th ($\delta(k)$) chemicals in equilibrium. $K_{eq}(j, k)$, therefore, contains every reaction frequency (or corresponding reaction rate constant) of all the elementary reaction-paths shown in Fig. 35.1.

35.3
Application to Toluene Chemistry in Urban Areas

Analysis of toluene chemistry in urban area is suitable for present pseudo-first-order reaction model, because a large amount of oxygen molecules as well as the perpetual production of OH radicals effectively oxidize toluene in the atmosphere. The toluene chemistry basically proceeds by two routes of oxidation reactions (Warneck 1988). The oxidation and conversion pathways for present toluene chemistry are shown in Fig. 35.2. The first route is hydrogen atom abstraction from C-H bonds, going through three intermediates (benzyl radical, benzyl peroxy radical, and benzyl alkoxy radical)

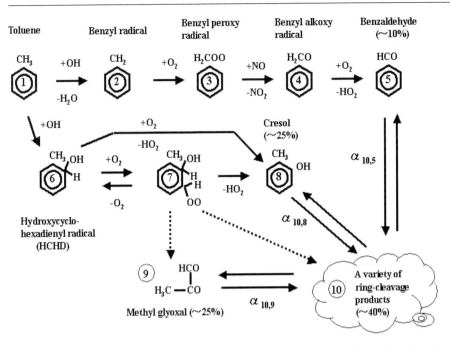

Fig. 35.2. Toluene chemistry in urban areas. The concentrations of stable fourchemicals estimated from field observation as well as laboratoryexperiments are: benzaldehyde (~10%), cresol (~25%), methyl glyoxal (~25%), and a variety of ring-cleavage products (~40%). The chemicals are numbered from 1 (toluene) to 10 (ring-cleavage products)

in order to reach benzaldehyde. The second route is OH radical addition to the aromatic ring to produce cresol and a variety of ring-cleavage products. This pathway is not yet fully understood, but hydroxycyclo-hexadienyl radical and peroxidic intermediates were experimentally observed (Kenley et al. 1978). The ratio of the equilibrium concentrations for four relatively stable chemicals was estimated to be

Benzaldehyde : Cresol : Methyl glyoxal : Ring-cleavage Products = 10 : 25 : 25 : 40 (35.10)

This chemical model consists at least of 26 reaction frequencies (reaction rate constants). However, to our knowledge, only six values were cited in the standard literatures (e.g. Atkinson 1990). So, the other twenty reaction frequencies need to be evaluated by substituting the concentration ratio Eq. 35.10 into Eq. 35.9 (Eq. 35.11),

$$\delta(5) : \delta(8) : \delta(9) : \delta(10) = 10 : 25 : 25 : 40 \qquad (35.11)$$

To calculate the area distribution of ten chemicals, their diffusion coefficients (D_j) and the advection velocities (v_j), in addition to these reaction rates, need to be introduced because these transport parameters are explicitly involved in Eq. 35.3 through Eq. 35.4.

Table 35.1. Reaction frequencies α_{ij} (h^{-1}) for toluene chemistry (Atkinson 1990). The suffix i and j are the same as those in Fig. 35.2. The assumed concentrations of OH, O$_2$, and NO are 1.0×10^6 cm^{-3}, 5.4×10^{18} cm^{-3}, and 1.6×10^9 cm^{-3}, respectively. The values in *itacics* are recommended by Atkinson. The estimated α_{ij} are $\alpha_{1,2}, \alpha_{1,6}, \alpha_{2,3}, \alpha_{3,2}, \alpha_{3,4}, \alpha_{,5}, \alpha_{5,10}, \alpha_{6,7}, \alpha_{6,8}, \alpha_{7,8}, \alpha_{7,9}, \alpha_{7,10}, \alpha_{8,7}, \alpha_{8,10}, \alpha_{9,7}, \alpha_{,10}, \alpha_{10,5}, \alpha_{10,7}, \alpha_{10,8},$ and $\alpha_{10,9}$. The residual 64 α_{ij} were assumed to be negligibly small, because the direct conversion of the corresponding chemical reaction path appears to be almost impossible under normal urban area conditions

i \ j	1	2	3	4	5	6	7	8	9	10
1	–	0	0	0	0	0	0	0	0	0
2	3.6E–3	–	0	0	0	0	0	0	0	0
3	0	1.9E+10	–	0	0	0	0	0	0	0
4	0	0	34.56	–	234	0	0	0	0	0
5	0	0	0	2.3E+7	–	0	0	0	0	0.02
6	*1.9E–2*	0	0	0	0	–	0	0	0	0
7	0	0	0	0	0	6.8E+7	–	100	10	10
8	0	0	0	0	0	2.3E+7	4.0E+06	–	0	0.02
9	0	0	0	0	0	0	5.0E+05	0	–	0.4
10	0	0	0	0	0.05	0	1.0E+06	0.2	1	–

While these parameters generally depend on the space coordinates as well as the elapsed time, we assume all of them to be constant by limiting the present analysis in relatively small urban areas. For toluene, benzaldehyde, cresol, and methyl glyoxal, these were estimated from the standard eddy diffusivity K_{yy} (Seinfeld and Pandis 1998) and the gas kinetic velocities at normal temperature. For other six labile intermediates, these were determined using their molecular weight. The magnitudes of D_j and v_j are ranging between 2.3 ~ 2.8 km^2 h^{-1} and 5.5 ~ 6.8 km h^{-1}, respectively.

Table 35.1 lists ninety reaction frequencies used for the present calculation. Of these reaction frequencies, the following six are obtainable; $\alpha_{2,1}, \alpha_{4,3}, \alpha_{5,4}, \alpha_{6,1}, \alpha_{7,6},$ and $\alpha_{8,6}$; we derived them from Atkinson's reaction rates (k_{ij}) using the number densities of three oxidants, OH, O$_2$, and NO, which are commonly accepted at STP condition in air; $\alpha_{ij} = k_{ij} \times$ (oxidants).

Evaluation of other eighty-four α_{ij} was carried out as follows; twenty of them (listed in the caption of Table 35.1) were evaluated using the equilibrium condition, Eq. 35.11. The residual sixty-four α_{ij} were set to be zero because of their unrealistic reaction paths under normal urban area conditions.

These reaction frequencies and the estimated transport parameters were introduced into Eq. 35.7 for calculation of area distribution of the 10 chemicals in equilibrium. Figure 35.3 shows the result in the area 0–80 km obtained from a simple numerical integration method. Two source parameters for emission of toluene in Eq. 35.6, a_1 and A_1, were set to be 1 km and the appropriate rate, respectively. The ratio of relative con-

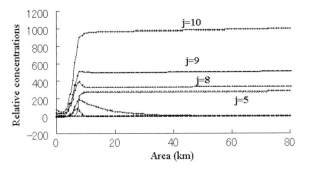

Fig. 35.3.
The calculated area distributions of 10 chemicals in equilibrium. $j = 5, 8, 9,$ and 10 are benzaldehyde, cresol, methyl glyoxal, and ring-cleavage products, respectively. Small concentrations of other chemicals are observed in the vicinity of source area

centration for benzaldehyde, cresol, methyl glyoxal, and ring-cleavage products determined from this profile was

$$\delta(5) : \delta(8) : \delta(9) : \delta(10) = 13 : 16 : 24 : 47 \qquad (35.12)$$

The agreement between this ratio and the recommended ratio (Eq. 35.11) is not yet satisfactory. Lack of accurate reaction data and some numerical instability occurred in the present calculations may cause this discrepancy. Much more sophisticated inverse Fourier-transform technique must be introduced for future analysis, together with the appropriate and accurate numerical method to calculate the determinants of larger matrices.

Three-dimensional reactive transport analysis of air pollutants is more interesting than the present one-dimensional analysis. This extension may need some consideration to the cross terms occurred in the original coupled diffusion-reaction equations. In usual atmospheric chemical processes, however, this kind of complexity may not play a critical role because of relatively small amount of candidates for air pollution. Therefore, three-dimensional analysis will possibly be carried out by combining three orthogonal one-dimensional solutions into three-dimensional solution (Overcamp 1990).

35.4
Conclusion

The analytical solution of a coupled transport/pseudo-first-order reaction equations was developed. The general and dynamic equilibrium solutions are of use for spatiotemporal evolution analysis of multiple chemicals in gas-phase. Relatively simple transport and reaction model of toluene chemistry comprising 10 chemicals in urban area was examined to evaluate twenty reaction frequencies using the dynamic equilibrium solution.

Acknowledgements

The authors would like to acknowledge H. Suto and T. Sato for skillful support of the preparation of this article.

References

Atkinson R (1990) Gas-phase tropospheric chemistry of organic compounds: a review. Atmos Environ 24A:1-41
Finlayson-Pitts BJ, Pitts Jr. JN (1999) Chemistry of the upper and lower atmosphere: theory, experiments, and applications. Academic Press, pp 130-138
Iinuma K (1991) Analysis of reactive ion transport in weakly ionized gas mixtures. Can J Chem 69: 1090-1099
Iinuma K, Sasaki N, Takebe M (1993) A general analysis of reactive ion transport in dynamic equilibrium. J Chem Phys 99:6907-6914
Kenley RA, Davenport JE, Hendry DG (1978) Hydroxyl radical reactions in the gas phase. Products and pathways for the reaction of OH with toluene. J Phys Chem 82:1095-1096
Overcamp TJ (1990) Diffusion models for transient release. J Appl Meteorology 29:1307-1312
Seinfeld JH, Pandis SN (1998) Atmospheric chemistry and physics: from air pollution to climate change. Wiley, New York, pp 938-943, pp 1193-1244
Warneck P (1988) Chemistry of the natural atmosphere. Academic Press, pp 266-267

Part IV
Polycyclic Aromatic Compounds

Part IV

Chapter 36

Analysis of High-Molecular-Weight Polycyclic Aromatic Hydrocarbons by Laser Desorption-Ionisation/Time-of-Flight Mass Spectrometry and Liquid Chromatography/Atmospheric Pressure Chemical Ionisation Mass Spectrometry

J. Čáslavský · P. Kotlaříková

Abstract

Laser desorption-ionisation/time-of-flight mass spectrometry (LDI/TOF MS) and liquid chromatography/atmospheric pressure chemical ionisation-ion trap mass spectrometry (HPLC/APCI-ITMS) were used for the analysis of polycyclic aromatic hydrocarbons with molecular weight exceeding 278 Da (HMW-PAHs) in air, water and soil samples from the contaminated area of DEZA chemical plant, Valašské Meziříčí, Czech Republic, and from its vicinity. Semipermeable membrane devices (SPMDs) were employed for passive sampling. LDI-TOF MS proved to be a suitable method for quick evaluation of the HMW-PAHs distribution; the presence of PAHs with molecular mass exceeding 500 Da in real samples was proved by this method. Identification and quantitation of individual PAHs was realised using LC/APCI-ITMS. LDI-TOF mass spectra and selected LC/APCI-MS profiles (m/z 303, 327 and 351) were used to confirm the source of contamination by high-molecular-weight PAHs in this area.

Key words: high-molecular-weight PAHs, LDI-TOF MS, HPLC/APCI-MS, atmospheric pressure chemical ionisation, SPMDs

36.1 Introduction

36.1.1 Polycyclic Aromatic Hydrocarbons

Polycyclic aromatic hydrocarbons (PAHs) represent a group of widespread organic contaminants, which are distributed throughout the environment as complex mixtures (e.g. May and Wise 1984). They arise mainly from incomplete combustion of organic matter and they surely came into existence a long time before the first human beings appeared on the Earth. These compounds have been accompanied mankind from his beginnings – their first important sources were probably the open fires and their concentrations in caves of early humans and later in the abodes of medieval people, but at this time the air was quite clear. The amount of PAHs emitted into the environment has dramatically increased in the second half of the 19th century as the result of industrial development. Several studies focused on the PAH levels in dated sediments offer evidence of an increasing level of these compounds in the latter part of 19th century (e.g. Wickstrom and Tolonen 1987; Barrick and Prahl 1987; Sanders et al. 1993).

Nowadays, the combustion of fossil fuels in power stations, in automobile engines and in residential heating systems represents the most important sources of PAHs (Menzie et al. 1992). They also enter the environment from fires, from fossil fuel discharges and from many industrial technologies like coke and iron production. In some regions, volcanic activity is also an important source. PAHs were the first compounds whose potential carcinogenic effects were discovered. Toxic, genotoxic and mutagenic properties of PAHs have been extensively studied to date (e.g. Harvey 1985; Fu et al. 1980; Busby et al. 1984; Busby et al. 1988; Kalf et al. 1997; McConkey et al. 1997; Monson et al. 1999).

Most studies dealing with the occurrence of PAHs in the environment are focused on compounds containing from 2 to 6 fused rings with molecular masses from 128 to 278 Da. High-molecular-weight PAHs (HMW-PAHs) with molecular masses exceeding 278 Da are analysed quite rarely, in spite of the fact that they are ubiquitous in various matrices like carbon black, coal tar and asphalt. They were found on urban air particulates (Marvin et al. 1999) and also in hydrotermal crude oils (Simoneit and Fetzer 1996). High-molecular-weight PAHs show significant biological activity; e.g. the PAH subfraction containing compounds with molecular masses from 302 to 352 Da exhibited positive mutagenic response accounting for approximately 25% of the total genotoxic response of the urban air particulate extracts (Marvin et al. 1999). The PAH fraction with 7 and more condensed rings from the flue gas condensate of the coal-fired residential furnaces represented 11.7 weight% of the total extract, but it was responsible for 54.7% of the carcinogenic potential (Grimmer et al. 1991). The mutagenic potency of dibenzo[a,l]pyrene was found to be more than 20 times higher in comparison with that of benzo[a]pyrene (Durant et al. 1999) and the carcinogenity of dibenzo[a,l]pyrene was evaluated as extremely strong (Cavalieri et al. 1989; Cavalieri et al. 1991).

The chromatographic separation of HMW-PAHs is a very difficult task because of the large number of possible isomers that rapidly increase with molecular weight. The most powerful separation technique – capillary gas chromatography – cannot be used for their analysis due to their very low volatility. Only a few successful applications of this method have been described, e.g. the separation of HMW-PAHs up to 472 Da on FSOT columns of 15–20 m length; column temperatures up to 400 °C, hydrogen carrier gas and on-column injection were necessary (Bemgard et al. 1993). Short capillary columns (5.5 m) were used for the separation of compounds from coronene (300 Da) to rubrene (532 Da) (Grob 1974); the resolving power of such columns was quite low.

The most suitable method for the analysis of polycyclic aromatic hydrocarbons is high performance liquid chromatography (HPLC) on reversed phases with UV and/ or fluorescence detection. The C_{18} polymeric stationary phases exhibit excellent selectivity for the separation of 16 PAHs on the EPA priority pollutant list, UV-VIS (ultraviolet-visible) detectors of diode-array type offer the possibility of eluted compounds identity confirmation by the comparison of their UV spectra with library, fluorescence detectors with programming of the excitation and emission wavelength during the chromatographic run are extremely sensitive and specific. Moreover, quantitative results obtained by liquid chromatography are usually more exact due to injection volume precision available with loop injectors in comparison with that of gas chromatographic syringe injectors. Less sample cleanup in comparison with gas chromatography is also usually required.

PAHs in environmental samples occur in very complex mixtures due to the large number of possible isomers and alkyl substituents. Standard HPLC columns do not possess the necessary resolving power for satisfactory separation of all components; to measure accurately individual PAHs in these mixtures either multi-dimensional HPLC approach, or highly selective detection methods are necessary. The multi-dimensional approach is based on the subsequent application of at least two chromatographic separations using different retention mechanisms. Wise et al. (1977) described the method employing normal phase HPLC on amino-propyl-silane stationary phase for the semi-preparative fractionation of complex PAH mixtures according to the number of condensed rings, and subsequently C_{18} reversed-phase HPLC separation of individual PAHs (Wise et al. 1977). A similar method was used for the fractionation of PAHs with molecular masses up to 400 Da from coal tar contaminated sediment (Marvin et al. 1995).

Mass spectrometric detection as the method with the highest and tuneable selectivity was applied at an early stage for the analysis of complex PAH mixtures, and various LC-MS interfaces have been employed. The moving belt (MB) was used for the characterisation of PAHs in coal-derived liquids as early as in 1977 (Dark et al. 1977). Subsequently this interface was applied for the analysis of PAHs in coal tar derived samples (Perreault et al. 1991). Compounds with molecular mass up to 580 Da were successfully ionised and EI (electronic impact) mass spectra were obtained, but the quantification of more volatile compounds, e.g. naphthalene, was rather difficult (Anacleto et al. 1995). The main drawbacks of this interface were its mechanical awkwardness and limitations on the mobile phase composition and flow rates. The particle beam (PB) showed no such restrictions and provided EI spectra (Pace and Betowski 1995), but the response for PAHs less than 200 Da and larger than 380 Da was very poor (Anacleto et al. 1995). Besides, the response of particle beam interface for some compounds was found to be non-linear. Electrospray ionisation (ESI) was able of forming molecular radical cations of PAHs, but success was strongly dependent on the correct choice of solvent, which was generally dichloromethane with 0.1% trifluoroacetic acid (Vanberkel et al. 1991; Vanberkel et al. 1992). This is of course incompatible with gradients used for PAH separation. The heated pneumatic nebulizer (HPN) with atmospheric pressure chemical ionisation (APCI) was finally evaluated as the most suitable interface for the analysis of PAH mixtures in LC/MS mode (Anacleto et al. 1995; Marvin et al. 1999).

36.1.2
Environmental Sampling of Organic Pollutants Using SPMDs

Semipermeable membrane devices (SPMDs) developed by Huckins et al. (1990) are passive samplers for selective sampling of semivolatile organic pollutants based on their selective permeation through synthetic membrane and accumulation in sequestered lipid. These devices – triolein-filled lay-flat low-density polyethylene tubes – were originally used as concentrators of non-polar organic compounds in aquatic environments (Huckins et al. 1990). They mimic the bioconcentration of these compounds in living organisms; although the composition of polymeric membranes and biomembranes is substantially different, the diffusion of some non-polar organics through them was found to be surprisingly similar (Johnson 1991). Shortly after the discovery of the ability of SPMDs to effectively uptake organic contaminants from air they were also successfully applied for air sampling (Petty et al. 1993). These devices are small, simple and inexpensive; their field manipulation (deployment, harvest and transport to laboratory) is straightforward and does not require highly experienced personnel. They take up pollutants during the whole exposition period, which is usually from several days to several weeks. Captured compounds are subsequently released by simple and effective dialysis.

The quantitative evaluation of sequestered compounds is more complicated. The simplest approach proposed by Huckins et al. (1993) is based on the presumption of the diffusion through the polyethylene membrane as the controlling step in the process of analyte transport from the water to the device sequestered lipid. Assuming constant temperature and negligible biofouling of the outer membrane surface the uptake rate F (g h^{-1}) could be expressed as:

$$F = \frac{DA(C_{MO} - C_{MI})}{Y} = k_p A(C_{MO} - C_{MI}) = k_w A(C_W - C_{WI}) = V_s \frac{dC_S}{dt} \qquad (36.1)$$

where D is diffusivity or permeability of the analyte in the membrane (m^2 h^{-1}), A is the membrane surface area (m^2), Y is the membrane thickness (m), k_p is the membrane mass transfer coefficient or D/Y (m h^{-1}), k_w is the mass transfer coefficient in the boundary water layer (m h^{-1}), V_s is the volume of the lipid (m^3), t is time (h), C_{MO} and C_{MI} are analyte concentrations (g m^{-3}) at the outer and inner surface, C_W and C_{WI} are analyte concentrations (g m^{-3}) in the bulk water and at the interface, and C_S is the analyte concentration in the lipid (g m^{-3}).

At non-equilibrium conditions, when the concentrations in water and lipid are far enough from equilibrium, the uptake of pollutants is approximately linear and their ambient concentrations can be evaluated using the simple equation:

$$C_W = \frac{C_{SPMD}}{R_s t} \qquad (36.2)$$

where C_W and C_{SPMD} are the concentrations of the target compound in water (ng l^{-1}) and in SPMD (ng SPMD^{-1}), respectively, t is time of deployment (d) and R_s is effective sampling rate (l d^{-1}), i.e. the volume of water extracted by the device per time unit.

SPMD sampling rates for various organic pollutants in water have been published – e.g. Huckins et al. 1994; Huckins et al. 1999; Sabaliunas and Sodergren 1997; Gustavson and Harkin 2000.

A similar approach was applied for air sampling and consequently SPMD air sampling rates for selected groups of organic pollutants are also available (e.g. Huckins et al. 1994; Ockenden et al. 1998).

36.1.3
Aim of the Study

The main goal of this study was to obtain the first piece of information about the occurrence and levels of high-molecular-weight-PAHs in the area of the DEZA chemical plant, Valašské Meziříčí, Czech Republic, and in its close vicinity. The passive sampling using SPMDs was the method of choice, because this method is safe and could be employed inside the chemical factory at workplaces with a fire risk. Laser desorption-ionisation/time-of flight mass spectrometry and liquid chromatography/atmospheric pressure chemical ionisation-ion trap mass spectrometry were selected as final analytical methods.

36.2
Experimental

36.2.1
Sampling Site

DEZA chemical plant (Valašské Meziříčí, Czech Republic) launched in 1963 is a well-known producer of black pitch, aromatic solvents (benzene, toluene, xylenes), PAHs (naphthalene, anthracene etc.), plasticizers and carbon black in the Czech Republic. Raw benzol and coal tar from the coke production in the Ostrava industrial region are the most important raw materials. The factory itself represents an important point source of PAHs and its area as well as its vicinity is heavily contaminated both by the raw materials and by the products.

Open air was sampled in three sampling locations (Fig. 36.1). The first one (A1) was the black pitch granulation tower, approx. 15 m in height, where the black pitch pellets are produced by dipping the thin stream of melted black pitch into water in the open air. Small particles of the black pitch are released during this process and transported to the vicinity. The second air sampling location (A2) was the roof of the biological cleaning facility in the western part of the factory, approx. 10 m above the surface; owing to the prevailing wind direction this should be the cleanest part of the factory. The third air-sampling location was situated on the border of the village of Mštěnovice (approx. 1 km from the factory).

Water was sampled on three sampling locations. The first one (W1) was inflow of the industrial wastewater to the lagoon, which represents the last step of the wastewater cleaning process. The second sampling location (W2) was the outflow from this lagoon, and the third one (W3) was situated in the Bečva River, approximately 200 m downstream the wastewater discharge.

Soil was sampled at three sampling locations. The first one (S1) was situated inside the factory on the small meadow among production halls, near the black pitch granulation tower. The second one (S2) was in the close vicinity of the factory (50 m from the factory fence) and the third one (S3) in the village of Mštěnovice. All locations are shown in Fig. 36.1.

36.2.2
Chemicals

In Table 36.1, all compounds selected for the study are given; Figure 36.2 shows their chemical structures. Coronene (CAS No. 191-07-1, purity 99%) and naphtho[2,3-a]pyrene (CAS No. 196-42-9, purity 99.5%) was obtained from Sigma-Aldrich Chemie (Schnelldorf, Germany), benzo[a]perylene (CAS No. 19-85-5, purity 99.0%), dibenzo[a,e]fluoranthene (CAS No. 5385-75-1, purity 99.4%), dibenzo[a,k]fluoranthene (CAS No. 84030-79-5, purity 99.8%), dibenzo[a,e]pyrene (CAS No. 192-65-4, purity 99.8%), dibenzo[a,h]pyrene (CAS No. 189-64-0, purity 99.8%), dibenzo[a,i]pyrene (CAS No. 189-55-9, purity 99.9%), dibenzo[a,l]pyrene (CAS No. 191-30-0, purity 99.8%), benzo[a]coronene (CAS No. 190-70-5, purity 99.8%) were supplied by Promochem (Wesel, Germany), and decacyclene (CAS No. 191-48-0, purity 99.3%) was obtained from Riedel-de-Haën (Seelze, Germany). Acetonitrile (HPLC super gradient grade) and water for gradient elution were purchased from Riedel-de-Haën. The other organic solvents used were for organic trace analysis.

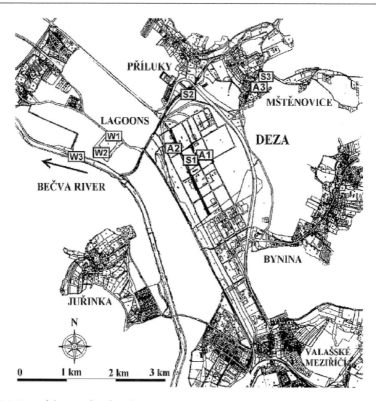

Fig. 36.1. Map of the sampling locations

Table 36.1. List of PAHs selected for this study

Compound	Abbreviation	Molecular mass	Relative mutagenic potency (Durant 1999)
Coronene	Cor	300	–
Benzo[a]perylene	BaPe	302	4.2
Dibenzo[a,e]fluoranthene	DB[ae]F	302	4.2
Dibenzo[a,k]fluoranthene	DB[ak]F	302	1.0
Dibenzo[a,e]pyrene	DB[ae]P	302	2.9
Dibenzo[a,h]pyrene	DB[ah]P	302	1.4
Dibenzo[a,i]pyrene	DB[ai]P	302	3.6
Dibenzo[a,l]pyrene	DB[al]P	302	24.0
Naphto[2,3-a]pyrene	N[2,3-a]P	302	9.3
Naphtho[a]perylene	NaPer	326	–
Benzo[a]coronene	BaCor	350	–
Decacyclene	Deca	450	–

Chapter 36 · Analysis of High-Molecular-Weight Polycyclic Aromatic Hydrocarbons

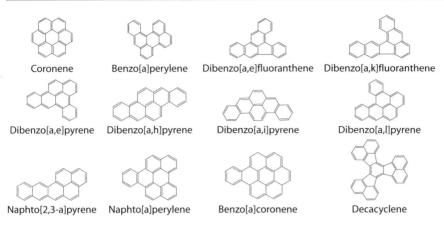

Fig. 36.2. Chemical structures of PAHs investigated

36.2.3
Field Methods

SPMDs were prepared following the procedure described in literature (e.g., Lebo et al. 1992). Briefly, the lay flat polyethylene tubing (low-density polyethylene without additives, width 28 mm, wall thickness 75 μm, Cope Plastics, Fargo, U.S.A.) was cut into 104-cm pieces, which were heat-sealed on one end. The tubes were cleaned by dialysis in n-hexane (48 h, solvent exchange after 24 h). The pre-cleaned tubes were filled with 1 ml of triolein (95% purity, Sigma-Aldrich) and after squeezing out the air and forming the smooth lipid layer the other end of the tubing was heat-sealed. SPMDs were then immediately closed in air-tight containers and stored in a deep-freezer at –18 °C. For the exposition in outdoor air and water the SPMDs were fixed in home-made exposition structures (Čáslavský et al. 2000). The devices were exposed for 4 weeks. After the exposition the SPMDs were transported to the laboratory in pre-cleaned paint cans on ice and stored in a deep freezer.

Soil samples were taken from a depth of 5 to 20 cm.

36.2.4
Extraction and Clean-Up

Exposed SPMDs were washed with iso-propanol to remove particles stuck to the surface. Dialysis in n-hexane (2 × 24 h) was used to release sequestered pollutants. A small amount of lipid released from SPMDs during dialysis was removed by size exclusion chromatography on Bio-Beads S-X3 200-400 mesh (Bio-Rad Laboratories) in 8 × 500 mm stainless-steel column (Tessek Ltd., Prague) with chloroform as a mobile phase at a flow rate of 0.6 ml min^{-1}. First 14 ml of the eluate containing lipidic compounds were discarded, following 11 ml containing PAHs were collected and the volume was reduced to 2 ml under a gentle stream of nitrogen. Soil samples were dried at normal temperature, ground and sieved. Organic pollutants were released by 8-hour Soxhlet extraction using dichloromethane. Silica gel (Kieselgel 60, Merck, BRD) activated 3 h at 250 °C (column dimensions: 1 × 10 cm) was employed for the separation of the pre-cleaned dialysates and soil extracts to aliphatic and aromatic fraction (elution with n-hexane and dichloromethane, respectively). Aromatic fraction was further separated by semi-preparative HPLC on amino-propyl-silica column (Separon SGX-NH$_2$, 8 × 250 mm, Tessek Ltd., Prague, Czech Republic) with 10% dichloromethane in n-hexane as mobile phase (flow rate 1.5 ml min^{-1}). PAHs with molecular weight exceeding 252 were isolated.

36.2.5
Laser Desorption-Ionisation/Time-of-Flight Mass Spectrometry

The Kompact MALDI IV mass spectrometer (Shimadzu-Kratos, Manchester, Great Britain) was used for the acquisition of Laser desorption-ionisation/time-of-flight mass spectra. 1 µl of the HMW-PAH fraction was deposited on the MALDI target and dried at laboratory temperature. PAH molecules were ionised by the UV-laser pulse (N$_2$ laser, 337 nm, pulse width 3 ns), accelerating voltage was set to 5.2 kV. Time-delayed extraction of ions, linear flight path and positive ion mode were used. Each spectrum represents an average from 50 laser shots along the sample spot.

LDI-TOF mass spectra of the HMW-PAH fraction from black pitch produced in the DEZA plant and soils S1, S2 and S3 shows Fig. 36.3.

36.2.6
Liquid Chromatography/Mass Spectrometry

LC/MS was performed on Esquire-LC instrument (Bruker Daltonics, Bremen, Germany). The HPLC used was Hewlett-Packard 1100 Series Liquid Chromatograph with binary gradient pump, vacuum solvent degasser, autosampler, column thermostat, and diode-array detector (Hewlett-Packard, Waldbronn, Germany). LC-PAH column 2.1 × 250 mm with Supelguard LC-18 column 2.1 × 20 mm (both Supelco) was used for the separation of HMW-PAHs, with a binary gradient using acetonitrile-dichloromethane (DCM) (0–10 min: 0% DCM, 25 min: 25% DCM, 35 min: 55% DCM, 40–50 min: 100% DCM, 55 min: 0% DCM). The flow rate was set to 0.3 ml min^{-1} with a column temperature of 35 °C. Ion trap mass spectrometer was connected to HPLC via the atmospheric pressure chemical ionisation interface. The conditions of APCI were as follows: The pressure of nebulizing gas (N$_2$) was 50 psi, the drying gas flow was 9 l min^{-1}. Drying temperature and APCI temperature were 350 and 500 °C, respectively. The scan range of ion trap was set to 250–520 amu, positive ions were detected. Results are summarised in Fig. 36.4 and 36.5 and in Table 36.2 and 36.3.

36.3
Results and Discussion

The aim of our study was to gain the first piece of information about the occurrence and levels of high-molecular-weight PAHs in the area of the DEZA Chemical Plant, Valašské Meziříčí, Czech Republic, and in its vicinity. To this purpose we used two methods: laser desorption-ionisation/time-of-flight mass spectrometry and liquid chromatography/atmospheric pressure chemical ionisation-mass spectrometry.

36.3.1
Laser Desorption-Ionisation/Time-of-Flight Mass Spectrometry

The ionisation of PAHs by UV light was utilised in the 1970s when photoionisation detectors (PID) became commercially available in gas chromatography (Driscol and Clarici 1976). PAHs were successfully ionised by photons having energies 9.5 or 10.2 eV. Currently, a special instrument for on-line analysis of PAHs in combustion products was introduced, based on laser photoionisation and time-of-flight mass spectrometry (Castaldi and Senkan 1998). We used a commercial MALDI-TOF instrument for the analysis of high-molecular-weight-PAHs. Samples were deposited directly on the MALDI target without any matrix and subsequently ionised by UV-laser pulse. In contrast to the results mentioned above, no response was obtained for PAHs with molecular mass lower than 202 – these compounds were probably volatilised from the MALDI target under vacuum in the ion source (Čáslavský et al. 2000). HMW-PAHs were readily ionised producing $[M]^+$ ions, but to our surprise also negative ions $[M]^-$ were observed. Their intensities were substantially lower than those of positive ions (unpublished results). Figure 36.3 shows LDI-TOF mass spectra of HMW-

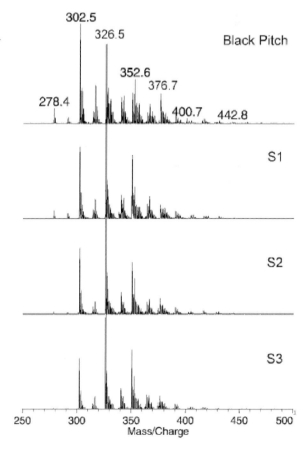

Fig. 36.3.
Laser desorption-ionisation/time-of-flight mass spectra of high-molecular-weight-PAHs from black pitch and soils S1, S2 and S3

PAHs from the black pitch produced in the DEZA factory and from soil samples. Compounds with molecular masses ranging from 278 up to almost 500 Da were found. The LDI-TOF mass spectrometry proved to be a very quick method; the analysis from the deposition of the sample on the target to the spectra acquisition usually takes only several minutes. The limits of detection depend on the compound properties; e.g. the signal to noise ratio of 10 was achieved for 100 pg of coronene on the MALDI target. Unfortunately, quantitative evaluation of the results obtained is more or less impossible because all compounds of the same molecular mass are summarised in one peak in the MALDI-TOF mass spectrum and the relative responses of individual compounds are quite different (Čáslavský et al. 2000).

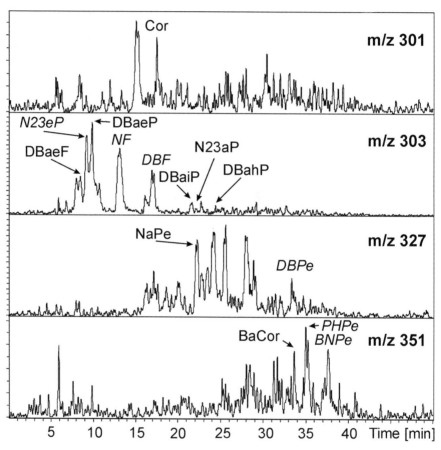

Fig. 36.4. Liquid chromatography/atmospheric pressure chemical ionisation mass spectrometry of HMW-PAHs from black pitch produced in DEZA factory. Compounds identified: *Cor:* coronene, *DBaeF:* dibenzo[a,e]fluoranthene, *N23eP:* naphtho[2,3-e]pyrene, *DBaeP:* dibezo[a,e]pyrene, *NF:* naphthofluoranthene, *DBF:* dibenzofluoranthene, *DBaiP:* dibenzo[a,i]pyrene, *N23aP:* naphtho[2,3-a]pyrene, *DBahP:* dibenzo[a,h]pyrene; *NaPe:* naphtho[1,2,3,4-ghi] perylene; *DBPe:* dibenzo[cd,lm]perylene; *BaCor:* benzo[a]coronene; *PHPe:* phenanthro[5,4,3,2-efgi]perylene; *BNPe:* benzo[pqr]naphtho[8,1,2-bcd]perylene. Compounds in italics were tentatively identified by comparison with published data

36.3.2
Liquid Chromatography/Atmospheric Pressure Chemical Ionisation-Mass Spectrometry

The ionisation of PAHs under atmospheric pressure chemical ionisation is more complicated; two competing mechanisms have been described. The first one is proton transfer from water clusters to form $[M + H]^+$, the second one is electron transfer to $[N_2]^{·+}$, $[O_2]^{·+}$ and possibly $[NO]^+$ to form $[M]^+$ ions (Anacleto et al. 1995). The relative importance of these two mechanisms varied with the partial pressure of water vapour within the APCI plasma. In contrast to these results we found in our experiments, where acetonitrile-dichloromenthane gradients were applied, that the proton transfer dominated if acetonitrile was present in mobile phase; electron transfer prevailed only in pure dichloromethane at the end of gradient run. Benzo[a]perylene was the only exception forming $[M+31]^+$ ion; this could be explained by the adduction of any of species present in APCI plasma (like HNO or CH_4N).

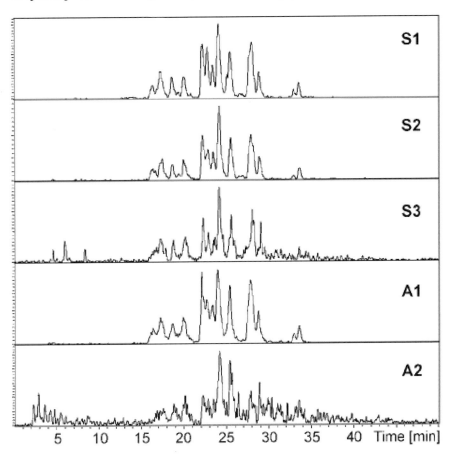

Fig. 36.5. Comparison of ion fragmentograms m/z 327 of soil and air samples

Figure 36.4 shows the *m/z* 301, 303, 327 and 351 traces from the LC/MS analysis of black pitch. Several compounds on these fragmentograms were identified by the comparison of the retention with standards, some other (with names written in italics) were tentatively identified by the comparison with published data (Marvin et al. 1999; Pace and Betowski 1995; Wise et al. 1993).

The identification of HMW-PAHs in environmental samples is a quite difficult task. These compounds were found in very complex mixtures in all samples taken in the area of the DEZA factory and in its close vicinity. The separation power of standard LC microbore column (2.1 × 250 mm) was insufficient which is clearly indicated by number of co-elutions on fragmento-grams presented in Fig. 36.4.

36.3.3
High-Molecular-Weight PAHs in the DEZA Factory and Neighbourhood

The origin of high-molecular-weight PAHs in the area studied could be deduced both from the comparison of PAH profiles obtained by laser desorption-ionisation/time-of-flight mass spectrometry (Fig. 36.3), and by the shape of fragmentograms from liquid chromatography/atmospheric pressure chemical ionisation-mass spectrometry (Fig. 36.5). The similarity of all profiles is the proof of contamination of the factory and its neighbourhood clearly shows the black pitch granulation technology as the main source of these compounds.

The amounts of HMW-PAHs sequestered by SPMDs from air (see Table 36.2) are surprisingly high, especially at sampling place A1, where SPMD was located in the

Table 36.2. Amounts of HMW-PAHs in SPMDs exposed in air and water. *n.a.*: not analysed, *n.d.*: not detected, *n.q.*: not quantified. Limit of quantitation: 0.8 ng SPMD^{-1}

Sampling Locality	Amounts of HMW – PAHs (ng SPMD^{-1})					
	A1	A2	A3	W1	W2	W3
Coronene	n.a.	98.4	n.d.	313.9	68.2	n.d.
Benzo[a]perylene	n.a.	19.8	2.3	9.2	8.3	1.2
Dibenzo-fluoranthenes [a,e+a,k]	n.a.	38.8	n.d.	229.2	117.9	n.d.
Dibenzo[a,e]pyrene	87 480	10.6	n.d.	280.1	186.4	n.d.
Dibenzo[a,h]pyrene	4 327	87.7	n.d.	16.4	7.9	n.d.
Dibenzo[a,i]pyrene	5 886	33.0	n.d.	n.d.	n.d.	n.d.
Dibenzo[a,l]pyrene	n.a.	5.5	n.d.	n.d.	n.d.	n.d.
Naphto[2,3-a]pyrene	n.q.	84.4	n.d.	n.d.	n.d.	n.d.
Naphtho[a]perylene	n.a.	1695	71.6	77.9	47.3	n.d.
Benzo[a]coronene	n.a.	1456	76.3	56.6	117.2	68.4
Decacyclene	n.a.	14.7	7.62	10.3	n.d.	6.0
Total	97 693	3 545	155.5	993.6	553.2	75.6

Table 36.3.
Concentrations of HMW-PAHs in soils (ng g^{-1}). *n.q.*: not quantified (due to coelution). Limit of quantification: 10 ng g^{-1}

Compound	Sampling place		
	S1	S2	S3
Benzo[a]perylene	1810	165	147
Dibenzo-fluoranthenes [a,e+a,k]	20060	4811	579
Dibenzo[a,e]pyrene	20890	6935	501
Dibenzo[a,h]pyrene	28840	n.q.	535
Dibenzo[a,i]pyrene	17160	3610	460
Naphtho[2,3-a]pyrene	12450	n.q.	335
Total	101210	15521	2557

area where an operator spends the main part of his working shift. The sequestered amounts of HMW-PAHs in air fall quite rapidly with increasing distance from the black pitch granulation tower.

HMW-PAHs were found in the wastewater from DEZA factory (Table 36.2). The comparison of their levels in samples W1 and W2 also shows the efficiency of 3-day retention in the lagoon, which is the last step of wastewater cleaning process before the discharge into the Bečva River. The amounts of HMW-PAHs decreased by a factor of 0.55. Similar values were found for the 16 priority pollutants PAHs at these sampling locations (Čáslavský et al. 2000). Almost no detectable amounts of HMW-PAHs were sequestered by SPMDs from the Bečva River water (sample W3).

HMW-PAHs were also found in all soil samples (Table 36.3). Similarly to air, their amounts decreased rapidly with increasing distance from the point source of contamination. This fact suggests that a long-range transport of these compounds could not be considered in this case.

36.4
Conclusions

The laser desorption-ionisation/time-of-flight mass spectrometry proved to be a useful tool for the quick evaluation of high-molecular-weight-PAH distribution. The main drawback of this method is a difficult quantification.

Liquid chromatography/atmospheric pressure chemical ionisation-mass spectrometry was successfully applied for the identification and quantification of individual high-molecular-weight PAHs. Main problems are caused by insufficient resolving power of standard micro-bore LC columns and also by the lack of standard compounds.

Polycyclic aromatic hydrocarbons with molecular weight exceeding 278 Da were found in all air, water and soil samples taken in the DEZA factory and in its vicinity. The amounts sequestered by SPMDs during exposition in air and water were substantial, as well as their concentrations in soils. The open-air technology of black pitch granulation in the DEZA chemical plant is evidently the source of contamination by these compounds in the area studied.

Acknowledgement

This work was supported by the grant No. 205/98/1265 by the Grant Agency of the Czech Republic. Authors thank to Dr. Miroslav Machala, PhD. from the Veterinary Research Institute Brno for furnishing of some high-molecular-weight-PAH standards.

References

Anacleto JF, Ramaley L, Benoit FM, Boyd RK, Quilliam MA (1995) Comparison of liquid-chromatography mass-spectrometry interfaces for the analysis of polycyclic aromatic-compounds. Anal Chem 67:4145–4154

Barrick RC, Prahl FG (1987) Hydrocarbon geochemistry of the puget sound region. 3. Polycyclic aromatic hydrocarbons in sediments. Estuar Coast Shelf Sci 25:175–191

Bemgard A, Colmsjo A, Lundmark B (1993) Gas-chromatographic analysis of high-molecular-weight polynuclear aromatic hydrocarbons. 2. Polycyclic aromatic hydrocarbons with relative molecular masses exceeding 328. J Chromatogr 630:287–295

Busby WF, Goldman ME, Newberne PM, Wogan GN (1984) Tumorigenicity of fluoranthene in a newborn mouse lung adenoma bioassay. Carcinogenesis 5:1311–1316

Busby WF, Stevens EK, Kellenbach ER, Cornelisse J, Lugtenburg J (1988) Dose-response relationships of the tumorigenicity of cyclopenta[cd]pyrene, benzo[a]pyrene and 6-nitrochrysene in a newborn mouse lung adenoma bioassay. Carcinogenesis 9:741–746

Čáslavský J, Zdráhal Z, Vytopilová M (2000) Application of SPMDs for PAH sampling in the DEZA chemical factory. Polycyc Aromat Comp 20:123–141

Castaldi MJ, Senkan SM (1998) Real-time, ultrasensitive monitoring of air toxics by laser photoionization time-of-flight mass spectrometry. J Air Waste Manag Assoc 48:77–81

Cavalieri EL, Rogan EG, Higginbotham S, Cremonesi P, Salmasi S (1989) Tumor-initiating activity in mouse skin and carcinogenicity in rat mammary-gland of dibenzo[a]pyrenes – the very potent environmental carcinogen dibenzo[a,l]pyrene. J Cancer Res Clin Oncol 115:67–72

Cavalieri EL, et al. (1991) Comparative dose-response tumorigenicity studies of dibenzo[a,L]pyrene versus 7,12-dimethylbenz[a]anthracene, benzo[a]pyrene and 2 dibenzo[a,L]pyrene dihydrodiols in mouse skin and rat mammary-gland. Carcinogenesis 12:1939–1944

Dark WA, McFadden WH, Bradford DL (1977) Fractionation of coal liquids by HPLC with structural characterization by LC-MS. J Chromatogr 15:454–460

Driscol JN, Clarici JB (1976) Ein neuer Photoionisationdetektor für die Gas-Chromatogaphie. Chromatographia 9:567–570

Durant JL, Lafleur AL, Busby WF, Donhoffner LL, Penman BW, Crespi CL (1999) Mutagenicity of $C_{24}H_{14}$PAH in human cells expressing CYP1A1. Mutat Res 446:1–14

Fu PP, Beland FA, Yang SK (1980) Cyclopenta-polycyclic aromatic hydrocarbons – potential carcinogens and mutagens. Carcinogenesis 1:725–727

Grimmer G, Brune H, Dettbarn G, Jacob J, Misfeld J, Mohr U, Naujack KW, Timm J, Wenzelhartung R (1991) Relevance of polycyclic aromatic hydrocarbons as environmental carcinogens. Fresenius J Anal Chem 339:792–795

Grob K (1974) Heutige Grenzen der Hochauflösenden Gas-Chromatographie. Chromatographia 7:94–98

Gustavson KE, Harkin JM (2000) Comparison of sampling techniques and evaluation of semipermeable membrane devices (SPMDs) for monitoring polynuclear aromatic hydrocarbons (PAHs) in groundwater. Environ Sci Technol 34:4445–4451

Harvey RG (1985) Polycyclic aromatic hydrocarbons and carcinogenesis. American Chemical Society, Washington

Huckins JN, Tubergen MW, Manuweera GK (1990) Semipermeable membrane devices containing model lipid: a new approach to monitoring the bioavailability of lipophilic contaminants and estimating their bioconcentration potential. Chemosphere 20:533–552

Huckins JN, Manuweera GK, Petty JD, Mackay D, Lebo JA (1993) Lipid-containing semipermeable membrane devices for monitoring organic contaminants in water. Environ Sci Technol 27: 2489–2496

Huckins JN, Petty, JD, Orazio CE, Zajicek JL, Gibson VL, Clark RC, Echols KR (1994) Semipermeable membrane device (SPMD) sampling rates for trace organic contaminants in air and water. Proc. of 15. Annual Meeting, Society of Environmental Toxicology and Chemistry, 30 October–3 November 1994, Denver, CO, Abstract MB01

Huckins JN, Petty JD, Orazio CE, Lebo JA, Clark RC, Gibson VL, Gala WR, Echols KR (1999) Determination of uptake kinetics (sampling rates) by lipid-containing semipermeable membrane devices (SPMDs) for polycyclic aromatic hyrocarbons (PAHs) in water. Environ Sci Technol 33:3918–3923

Johnson GD (1991) Hexane-filled dialysis bags for monitoring organic contaminants in water. Environ Sci Technol 25:1897–1903

Kalf DF, Crommentuijn T, Vandeplassche EJ (1997) Environmental-quality objectives for 10 polycyclic aromatic hydrocarbons (PAHs). Ecotox Environ Safety 36:89–97

Lebo JA, Zajicek JL, Huckins JN, Petty JD, Peterman PH (1992) Use of semipermeable membrane devices for in-situ monitoring of polycyclic aromatic hydrocarbons in aquatic environment. Chemosphere 25:697–718

Marvin C, Lundrigan J, McCarry B, Bryant D (1995) Determination and genotoxicity of high-molecular-mass polycyclic aromatic-hydrocarbons isolated from coal-tar-contaminated sediment. Environ Toxicol Chem 14:2059–2066

Marvin CH, Smith RW, Bryant DW, McCarry BE (1999) Analysis of high-molecular-mass polycyclic aromatic hydrocarbons in environmental samples using liquid chromatography-atmospheric pressure chemical ionization mass spectrometry. J Chromatogr A 863:13–24

May WE, Wise SA (1984) Liquid chromatographic determination of polycyclic aromatic hydrocarbons in air particulate extracts. Anal Chem 56:225–232

McConkey BJ, Duxbury CL, Dixon DG, Greenberg BM (1997) Toxicity of a PAH photooxidation product to the bacteria *Photobacterium phosphoreum* and the duckweed *Lemna gibba* – effects of phenanthrene and its primary photoproduct, phenanthrenequinone. Environ Toxicol Chem 16:892–899

Menzie CA, Potocki BB, Santodonato J (1992) Exposure to carcinogenic PAHs in the environment. Environ Sci Technol 26:1278–1284

Monson PD, Call DJ, Cox DA, Liber K, Ankley GT (1999) Photoinduced toxicity of fluoranthene to northern leopard frogs (*Rana pipiens*). Environ Toxicol Chem 18:308–312

Ockenden WA, Prest HF, Thomas GO, Sweetman AJ, Jones KC (1998) Passive air sampling of PCBs: field calculation of atmospheric sampling rates by triolein-containing semipermeable membrane devices. Environ Sci Technol 32:1538–1543

Pace CM, Betowski LD (1995) Measurement of high-molecular-weight polycyclic aromatic-hydrocarbons in soils by particle-beam high-performance liquid chromatography-mass spectrometry. J Am Soc Mass Spectrom 6:597–607

Perreault H, Ramaley L, Sim PG, Benoit FM (1991) Use of a moving-belt interface for online high-performance liquid chromatography mass-spectrometry characterization of polycyclic aromatic compounds of molecular-weight up to 580 Da in environmental samples. Rapid Commun Mass Spectrom 5:604–610

Petty JD, Huckins JN, Zajicek JL (1993) Application of semipermeable membrane devices (SPMDs) as passive air samplers. Chemosphere 27:1609–1624

Sabaliunas D, Sodergren A (1997) Use of semi-permeable membrane devices to monitor pollutants in water and assess their effects: A laboratory test and field verification. Environ Pollut 96:195–205

Sanders G, Jones KC, Hamiltontaylor J, Dorr H (1993) Concentrations and deposition fluxes of polynuclear aromatic hydrocarbons and heavy metals in the dated sediments of a rural English lake. Environ Toxicol Chem 12:1567–1581

Simoneit BRT, Fetzer JC (1996) High molecular weight polycyclic aromatic hydrocarbons in hydrothermal petroleums from the Gulf of California and Northeast Pacific Ocean. Org Geochem 24:1065–1077

Vanberkel GJ, McLuckey SA, Glish GL (1991) Preforming ions in solution via charge-transfer complexation for analysis by electrospray ionization mass spectrometry. Anal Chem 63:2064–2068

Vanberkel GJ, McLuckey SA, Glish GL (1992) Electrochemical origin of radical cations observed in electrospray ionization mass-spectra. Anal Chem 64:1586–1593

Wickstrom K, Tolonen K (1987) The history of airborne polycyclic aromatic-hydrocarbons (PAH) and perylene as recorded in dated lake-sediments. Water Air Soil Poll 32:155–175

Wise SA, Chesler SN, Hertz HS, Hilpert LR, May WE (1977) Chemically-bonded aminosilane stationary phase for the high-performance liquid chromatographic separation of polynuclear aromatic compounds. Anal Chem 49:2306–2310

Wise SA, Sander LC, May WE (1993) Determination of polycyclic aromatic hydrocarbons by liquid chromatography. J Chromatogr 642:329–349

Chapter 37

Atmospheric Polycyclic Aromatic Hydrocarbons (PAHs) in Two French Alpine Valleys

N. Marchand · J.-L. Besombes · P. Masclet · J.-L. Jaffrezo

Abstract

In Europe, Alpine valleys represent one of the most important crossroads for heavy traffic. The vehicle impact on air quality is not well-known due to a lack of data in valley systems. Besides a health toxicity concern, the study of atmospheric polycyclic aromatic hydrocarbons (PAHs) is of geo-chemical interest because they are emitted mainly by combustion processes. This class of compounds is also particularly interesting for the study of the potential impact of heavy duty traffic on air pollution. PAHs are therefore actually regarded as priority pollutants of our air environment. As part of the program "Pollution des Vallées Alpines" (POVA), we performed two sampling surveys of PAHs in two sensitive valleys: the valley of Chamonix and the valley of Maurienne. In each valley, two sites were instrumented for atmospheric PAH sampling, and for others pollutant monitoring such as NO_x, ozone, and particulate matter (PM10). The first sampling campaign was performed in summer 2000 and the second in winter 2001 both periods occurring during the corresponding closure of the Mont Blanc tunnel. During both seasons the total particulate PAH concentrations were higher in the valley of Chamonix despite the stop of international traffic through the Chamonix Valley. In summer, the average total PAH concentration was nearly twice as high in the Chamonix Valley (1.3 ng m^{-3}) than in Maurienne Valley (0.8 ng m^{-3}). In winter the difference between the two valleys is larger since the average PAH concentrations reached 48 ng m^{-3} and 18 ng m^{-3} in Chamonix and Maurienne Valley, respectively. In addition PAH total concentration reached very high levels (155 ng m^{-3}) in the valley of Chamonix especially during anticyclonic periods. This very sharp increase of the PAH concentrations can be connected to an increase of the emissions in winter.

Key words: polycyclic aromatic hydrocarbons; aerosol; air pollution; alpine valley; tunnel

37.1
Introduction

Polycyclic Aromatic Hydrocarbons (PAHs) are ubiquitous in the atmosphere, being present as volatile, semi-volatile and particulate pollutants. The importance of PAHs to air pollution chemistry and public health has been recognised since 1942, with the discovery that organic extracts of particles, collected from ambient air, can produce cancer in experiments with animals (Finlayson-Pitts and Pitts 1997). PAHs are now regarded as priority pollutants by both the United States Environmental Protection Agency and the European Community, since some PAHs have been classified as potential human carcinogens (IARC 1983; IARC 1987). Next to the public health concern, the study of atmospheric PAHs is of geo-chemical interest as they are emitted mainly by combustion processes such as diesel and gasoline exhaust (Rogge et al. 1993; Schauer et al. 1999; Oda et al. 2001), residential coal or wood combustion (Rogge et al. 1998; Schauer et al. 2001; Freeman and Cattell 1990), and biomass burning (Jenkins et al. 1996; Masclet et al. 1995). They are therefore good indicators of these emissions (Khalili et al. 1995; Li and Kamens 1993).

The results presented here were obtained during the program "Pollution des Vallées Alpines" (POVA). Alpine valleys represent unique transportation pathways in Europe, and a small number of corridors conveys the ever increasing traffic of heavy duty vehicles. This study benefits from an exceptional context when in March 1999, the accident in the Mont Blanc tunnel stopped international traffic through the Chamonix Valley. Consequently, most heavy duty traffic in the area must now pass through the Fréjus tunnel, in the Maurienne Valley. This traffic is now as large as that expected in 2010 (in 2001, 7 428 vehicles per day including 4 244 trucks). The general objectives of this program are the comparative study of atmospheric pollution and the modeling of atmospheric emissions and transport in these two French alpine valleys of Chamonix and Maurienne, before and after the reopening of the Mont Blanc tunnel to heavy duty traffic.

We present the results obtained for PAH concentrations from two intensive field campaigns that took place in summer and winter in each of these valleys. These results are discussed in terms of comparison of the concentrations between seasons and valleys, and supported by other measurements performed by other investigators.

37.2
Experimental

37.2.1
Sampling Campaigns

During this program, two 15-days campaigns were performed during the closure of the Mont Blanc tunnel. The first campaign took place from 14 until 21 August 2000 in the Chamonix Valley and 22–29 August in the Maurienne Valley. The second campaign took place from 16 until 22 January 2001 in the Chamonix Valley and 25–31 January in the Maurienne Valley. In each valley, five sampling sites were instrumented with various aerosol or gas samplers, including automatic analysers for NO_x, ozone, and PM10 (PM10: mass fraction of particles with aerodynamic diameter less than 10 µm). The locations of these sites are presented in Fig. 37.1. Atmospheric PAHs were collected at two sites per valley during winter: Le Clos de l'Ours and Les Houches for the Chamonix Valley, and Modane and Orelle for the Maurienne Valley. During summer, PAHs were collected at one site per valley: le Clos de l'Ours and Modane. The sites at le Clos de l'Ours and Modane are located in suburban areas, while the sites at Les Houches and Orelle are more rural, and located several hundred meters away from the main road of each valley.

37.2.2
Sampling and Analytical Procedures

The sampling and analytical procedures for the determination of PAH atmospheric concentrations are described in details by Besombes et al. (2001). Briefly, airborne particulates were collected by high volume samplers (50 $m^3 h^{-1}$, on average) on precleaned glass fibre filter (GF/F Whatmann filter, 210 × 270 mm). Sampling duration at all sites was 24 h during the summer campaign and 12 h during the winter campaign.

Fig. 37.1.
Map of the two alpine valleys, valley of Chamonix (*above*) and valley of Maurienne (*below*), and sampling sites during the program POVA. In **bold** and *underlined*, sites instrumented for PAH sampling

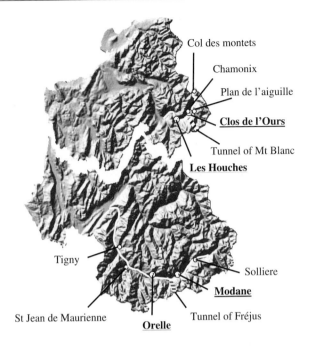

Before analysis, filters were stored at cold temperature (−4 °C) in aluminium sheets sealed in polyethylene bags.

Samples were soxhlet extracted for 3 h (BUCHI B-811) with a mixture of dichloromethane and cyclohexane (2/1, v/v). PAHs analyses are performed by reversed phase high performance liquid chromatography on a Vydac C18 column (length 25 cm, granulometry 5 µm and internal diameter 4.5 mm) with a ternary elution gradient. The elution program consisted of an initial methanol water acetonitrile mixture (75/10/15%, v/v/v) which was maintained for 5 min with a flow rate of 1 ml min^{-3} followed by a 15 min gradient to 100% acetonitrile. The eluant was then held at 100% acetonitrile for 10 min with a flow rate of 1.5 ml min^{-3}. All solvents were degassed using helium sparging to eliminate possible oxygen quenching during fluorescence. PAHs are identified and quantified by UV fluorescence at variable excitation and emission wavelengths. Seven pairs of wavelengths of excitation and emission (λ_{ex} and λ_{em}) are used to obtain the best sensitivity and selectivity. With this method, the following 12 PAHs are determined quantitatively: phenanthrene (PHE, λ_{ex} = 255 nm and λ_{em} = 410 nm), anthracene (ANT, λ_{ex} = 255 nm and λ_{em} = 410 nm), fluoranthene (FLA, λ_{ex} = 255 nm and λ_{em} = 410 nm), pyrene (PYR, λ_{ex} = 255 nm and λ_{em} = 410 nm), chrysene (CHR, λ_{ex} = 275 nm and λ_{em} = 380 nm), benzo[a]anthracene (BaA, λ_{ex} = 275 nm and λ_{em} = 380 nm); benzo[b]fluoranthene (BbF, λ_{ex} = 261 nm and λ_{em} = 415 nm), benzo[k]fluoranthene (BkF, λ_{ex} = 300 nm and λ_{em} = 430 nm), benzo[a]pyrene (BaP, λ_{ex} = 300 nm and λ_{em} = 430 nm); benzo[ghi]perylene (BghiP, λ_{ex} = 300 nm and λ_{em} = 415 nm); indenopyrene (IP, λ_{ex} = 300 nm and λ_{em} = 500 nm) and coronene (COR, λ_{ex} = 300 nm and λ_{em} = 444 nm). The overall analytical errors can be estimated between 7 and 34% depending on the PAH (Bresson et al. 1984).

37.3
Results and Discussion

37.3.1
Summer Campaign

Particulate PAH concentrations were investigated in two sampling sites in order to realise a comparative study of atmospheric pollution in the two alpine valleys. Figure 37.2 presents the daily total PAH concentrations for the sampling sites at Le Clos de l'Ours (Chamonix Valley) and Modane (Maurienne Valley). The average total concentrations obtained are 1.3 ng m^{-3} and 0.8 ng m^{-3} at the site Le Clos de l'Ours and at the site Modane, respectively. These PAH levels are close to those observed in urban areas during similar season (Kiss et al. 1998; Cecinato et al. 1999; Menichini et al. 1999). In addition, the total PAH concentration is nearly twice as high at Le Clos de l'Ours than at Modane, despite the stop of the international traffic through the Chamonix Valley.

The total PAH concentrations in the Chamonix Valley are rather similar throughout the week, except on Friday 18 August and Sunday 20 August when the maximum and minimum values are measured. On Friday, the concentration reaches 2.1 ng m^{-3} and then decreases on Sunday up to 0.8 ng m^{-3}. The sharp increase on Friday seems connected to an increase of vehicular emissions within the valley. Indeed, this day is also characterised by the higher concentration of NO$_x$ (Fig. 37.3) which is a good marker of the vehicular emission. At the same time, the higher relative concentrations of heavy PAHs (BghiP, IP, and COR) are observed at the site of Le Clos de l'Ours (Fig. 37.3). These compounds are well known to be emitted from vehicle exhaust (Miguel et al. 1998; Li and Kamens 1993). On Sunday, the lower activity near the sampling site can

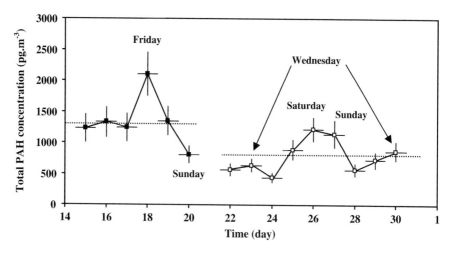

Fig. 37.2. Evolution of the total particulate PAH concentration vs. time during the summer campaign (16–30 August 2000), in *black square* for Le Clos de l'Ours (valley of Chamonix) and in open square for Modane (valley of Maurienne). The *dotted lines* depict the average total PAH concentration, 1.3 ng m^{-3} at Le Clos de l'Ours and 0.8 ng m^{-3} at Modane. Total PAH concentration is the sum of all individual particulate PAHs except PHE and ANT which are the most volatile ones

Fig. 37.3.
Evolution vs. time (15–20 August 2000) at Le Clos de l'Ours of: **a** NO$_x$ concentration and **b** relative concentration of heavy PAHs. Relative concentrations are calculated by the ratio of the sum of three PAHs (B(ghi)P + IP + COR) with the total PAH concentration

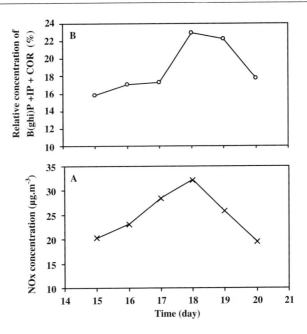

explain the concentration decrease, as it is usually observed for primary pollutants such as NO and NO$_x$.

Total concentrations show more variability in the Maurienne. The higher total PAH concentrations are obtained on Saturday and Sunday. However, these two days were also marked by a strong input of Saharan dust associated with anthropogenic inputs coming most probably from the heavily industrialised Torino area in Italy, as discussed in Colomb et al. (2002). We will not discuss it further. Except for Saturday and Sunday, the higher PAH concentrations are observed on Wednesdays (Fig. 37.2). This result could be in relation to the road traffic. Indeed, Wednesday is known to be the day of maximum heavy duty traffic in the valley of Maurienne (6 000 trucks on average). The comparison of this result with the other data obtained in the POVA program will be necessary to corroborate this hypothesis.

37.3.2
Winter Campaign

Figure 37.4 presents the total PAH concentrations for the two sites of the Chamonix Valley, for each of the 12 h sampling periods. The average concentrations at Le Clos de l'Ours and Les Houches are 48 and 18 ng m^{-3}, respectively. Therefore, the average concentration is 35 times higher during winter than during summer at the suburban site of Le Clos de l'Ours. This very sharp increase of the concentrations between the two seasons is partly connected to an increase in the emissions, particularly the occurrence of residential heating as a new source in winter. Indeed, PAH concentration at the more rural site of Les Houches is 3 times lower than at Le Clos de l'Ours, in agreement with the proximity of potential sources close to the latter site.

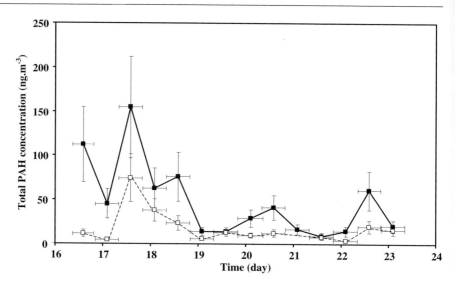

Fig. 37.4. Evolution of the total particulate PAH concentration vs. time during the winter campaign in the valley of Chamonix (16–22 January 2000): in *black square* and *continuous line* for Le Clos de l'Ours and in *open square* and *dotted line* for Les Houches. Total PAH concentration is the sum of all individual particulate PAHs except PHE and ANT which are the most volatile ones

The meteorological situation probably also has a fundamental impact on the PAH levels, with strong temperature inversions preventing the dispersion of the aerosols in this narrow valley. This is confirmed by the strong variability of the total PAH concentration during the sampling period. During the first part of the campaign from 16 to 18 January, anticyclonic conditions prevailed with low temperature (−4.5 °C on average) and low wind speed. High concentrations of PAHs are recorded at that time. In addition, the higher PAH levels observed during this period can also be linked with the shift of the gas/particles partitioning toward the particulate phase induced by the temperature decrease. Afterwards, from 19 to 22 January, different meteorological conditions occur with more disturbed weather, higher temperatures (−1.5 °C on average) and higher wind speed. These conditions allow a better dispersion of the particles leading to lower PAH concentrations.

The influence of meteorological conditions can also be illustrated by correlations between the total PAH concentration and PM10 concentrations (Fig. 37.5). From 16 to 18 January, correlations between PAHs and PM10 are high and similar at the two sites, since the correlation coefficients obtained are 0.97 and 0.98 at Le Clos de l'Ours and Les Houches, respectively. This result indicates the large scale nature of the processes involved during this period. Conversely, during the disturbed meteorological period, a good PAH/PM10 correlation is only observed at Le Clos de l'Ours, close to the larger sources of emissions (correlation coefficient 0.94). Such correlations between PAHs and PM10 are unusual, as PAH are generally associated with submicron aerosol (Aceves and Grimalt 1993; Pistikopoulos et al. 1990). This fraction of aerosol generally represents only a small fraction of the PM10 mass balance. Since PAHs are primary anthropogenic compounds, such correlations indicate that PM10 concentrations are essen-

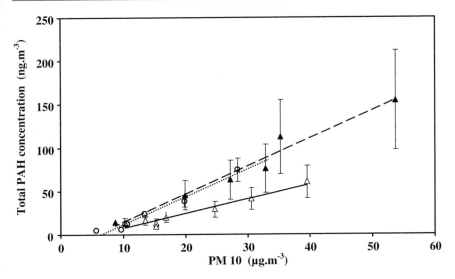

Fig. 37.5. Correlations between the total PAH concentration and PM 10 obtained in the Chamonix Valley in the two sampling site during the winter campaign: in *black triangle* and *large dotted line* at Le Clos de l'Ours and in *open circle* and *little dotted line* at Les Houches during 16–18 January, in *open triangle* and *continuous line* at le Clos de l'Ours during 19–22 January

tially influenced by primary anthropogenic emissions and that inputs of natural aerosol are probably marginal at that time. In addition, the two sites present the same variations of total PAH concentrations during the week (Fig. 37.4). These results also seem to indicate that atmospheric concentrations in the Chamonix Valley are largely influenced by purely local emission sources.

Our data show that during marked anticyclonic periods in winter, PAH concentrations can reach exceptionally high values (up to 150 ng m^{-3}) in the suburban area of Chamonix. Recently, the UK government Expert Panel on Air Quality Standard has set a UK air quality standard for BaP of 0.25 ng m^{-3} (Dimashki et al. 2001). Such a standard would be particularly challenging in the context of deep valleys, given that mean concentrations of particulate BaP were 2.5 ng m^{-3} on average during the week, with peak value at 4.5 ng m^{-3} for the second day of the campaign.

The average total PAH concentrations measured in the valley of Maurienne are 9.7 and 3.5 ng m^{-3} at Modane and Orelle, respectively. Like in the Chamonix Valley, average total concentrations are higher during winter than during summer but the difference between the two seasons is not as large. Also, the average concentration is 3 times lower at the more rural site of Orelle than at Modane. There are therefore similarities regarding total concentration differences between the sites of Chamonix and of Maurienne. However, PAH levels are 5 times more important in Chamonix than in Modane despite the differences in heavy duty traffic in the two valleys. An hypothesis to explain the higher concentrations in the Chamonix Valley with much lower traffic could be a better dispersion of pollutants within the Maurienne Valley, which is much wider and longer. This is supported by the lack of covariation of the PAH concentrations between the two sites of the Maurienne Valley that are only 13 km apart.

Finally, the average concentrations in BaP are 0.70 and 0.22 ng m^{-3} at Modane and Orelle, respectively, with a maximum of 1.05 ng m^{-3} at Modane during the first night. Such concentrations are much closer to the air quality standard for BaP than those observed in the Chamonix Valley.

37.4 Conclusion

The main result of our study is that during both winter and summer the total PAHs concentrations are still higher in the valley of Chamonix than in that of Maurienne, despite the closure of the Mont Blanc tunnel and the stop of international traffic through the Chamonix Valley. Furthermore, this tendency is more marked during winter, when strong temperature inversions close to the ground limit the pollutants dispersion. Regarding total concentration, these dynamic phenomena seem more important in the valley of Chamonix than in the valley of Maurienne. But we can not exclude the influence of different sources between summer and winter and between the two valleys. A detailed study of PAHs profiles and other species is currently in progress in order to improve our knowledge of the aerosol in the two valleys. Nevertheless, these preliminary results indicate an overwhelming importance of the geomorphology of the valleys for the dispersion of pollutants.

Acknowledgements

The program POVA is supported in France by the Région Rhône Alpes, ADEME (Agence de l'Environnement et de la Maîtrise de l'Energie), METL (Ministère de l'Equipement, des Transports et du Logement), and MEDD (Ministère de l'Ecologie et du Développement Durable). We would like to thanks Météo France for providing the meteorological data.

References

Aceves M, Grimalt JO (1993) Seasonally dependent size distributions of aliphatic and polycyclic aromatic hydrocarbons in urban aerosols from densely populated area. Environ Sci Technol 27:2896–2908

Besombes JL, Maître A, Patissier O, Marchand N, Chevron N., Stocklov M, Masclet P (2001) Particulate PAH observed in the surrounding of a municipal incinerator. Atmos Environ 35:6093–6104

Bresson MA, Beyne S, Masclet P, Mouvier G (1984) Optimisation des méthodes de prélèvement et d'analyse des HAP et de leurs dérivés azotés. Détermination de leur stabilité dans l'atmosphère. In: Resteli G, Angeletti G (ed) Physicochemical behaviour of atmospheric pollutants. Reidel, Dordreecht, pp 125

Cecinato A, Marino F, Di Filippo P, Lepore L, Possanzini M (1999) Distribution of n-alkanes, polynuclear aromatic hydrocarbons between the fine and coarse fractions of inhalable atmospheric particles. J Chromatogr A 846:255–264

Colomb A, Jacob V, Debionne J L, Aymoz G, Jaffrezo JL (2002) VOC's evolution during a Saharan dust episode in an alpine valley in August 2000. Fresenius Environ Bull 11:6–13

Dimashky M, Lee HL, Harrison RM, Harrad S (2001) Temporal trends, temperature dependence, and relative reactivity of atmospheric polycyclic aromatic hydrocarbons. Environ Sci Technol 35:2264–2267

Finlayson-Pitts BJ, Pitts Jr. JN (1997) Tropospheric air pollution: ozone, airbone toxics, polycyclic aromatic hydrocarbons, and particles. Science 276:1045–1052

Freeman DJ, Cattell FRC (1990) Woodburning as a source of atmospheric polycyclic aromatic hydrocarbons. Environ Sci Technol 24:1581–1585

IARC (1983) Polynuclear aromatic compounds, Part 1. Chemical, environmental and experimental datas. IARC Monographs on the Evaluation of Carcinogenic Risk of Chemicals to Humans, vol. 32. International Agency for Research on Cancer, Lyon

IARC (1987) Overall evaluations of carcinogenicity: an updating of IARC monographs. IARC monographs on the evaluation of carcinogenic risk of chemicals to humans, vol. 1–42 (suppl 7). International Agency for Research on Cancer, Lyon

Jenkins BM, Jones AD, Turn SQ, Williams RB (1996) Particle concentrations, gas-particle partionning, and species intercorrelations for polycyclic aromatic hydrocarbons (PAH) emitted during biomass burning. Atmos Environ 30:3825–3835

Kiss G, Varga-Puchony Z, Rohrbacher G, Hlavay J (1998) Distribution of polycyclic aromatic hydrocarbons on atmospheric aerosol particles of different sizes. Atmos Research 46:253–261

Li CK, Kamens RM (1993) The use of polycyclic aromatic hydrocarbons as sources signatures in receptor modeling. Atmos Environ 27A:523–532

Masclet P, Cachier H, Liousse C, Wortham H (1995) Emissions of PAH by savannah fires. J Atmos Chem 22:41–45

Menichini E, Monfredi F, Merli F (1999) The temporal variability of the profile of carcinogenic polycyclic aromatic hydrocarbons in urban air: a study in a medium traffic area in Rome, 1933–1998. Atmos Environ 33:3739–3750

Miguel AH, Kirchstetter TW, Harley RB, Hering RA (1998) On-road emissions of particulate polycyclic aromatic hydrocarbons and black carbon from gasoline and diesel vehicles. Environ Sci Technol 32:450–455

Oda J, Nomura S, Yasuhara A, Shibamoto T (2001) Mobile sources of atmospheric polycyclic aromatic hydrocarbons in a roadway tunnel. Atmos Environ 35:4819–4827

Pistikopoulos P, Wortham H, Gomes L, Masclet-Beyne S, Bon Nguyen E, Masclet P, Mouvier G (1990) Mechanisms of formation of particulate polycyclic aromatic hydrocarbons in the relation to the particle size distribution; effects on meso-scale transport. Atmos Environ 24A: 2573–2584

Rogge WF, Hidemann LM, Mazurek MA, Cass GR, Simoneit BRT (1993) Sources of fine organic aerosol. 2. Noncatalyst and catalyst equipped automobiles and heavy-duty diesel trucks. Environ Sci Technol 27:636–651

Rogge WF, Hidemann LM, Mazurek MA, Cass GR, Simoneit BRT (1998) Sources of fine organic aerosol. 9. Pine, oak, synthetic log combustion in residential fireplaces. Environ Sci Technol 32:13–22

Shauer JJ, Kleeman MJ, Cass GR, Simoneit BRT (1999) Measurement of emissions from air pollution sources. 3. C_1-C_{30} organic compounds from medium diesel trucks. Environ Sci Technol 33:1578–1587

Shauer JJ, Kleeman MJ, Cass GR, Simoneit BRT (2001) Measurement of emissions from air pollution sources. 3. C_1-C_{29} organic compounds from fireplace combustion of wood. Environ Sci Technol 35:1716–1728

Part IV

Chapter 38

Evaluation of the Risk of PAHs and Dioxins Transfer to Humans via the Dairy Ruminant

C. Feidt · S. Cavret · N. Grova · C. Laurent · G. Rychen

Abstract

To evaluate the risk of PAHs and dioxins transfer to humans several studies on dairy ruminant exposure and on intestinal absorption have been conducted. In order to assess PAHs feed-milk transfer, the transfer of three ^{14}C-labelled PAHs (^{14}C-phenanthrene, ^{14}C-pyrene, ^{14}C-benzo[a]pyrene) and ^{14}C-2,3,7,8-TCDD has been studied after a single oral ingestion (2.6×10^6 Bq) to lactating goats. Radioactivity associated to the labelled PAHs and 2,3,7,8-TCDD has been detected in milk seven hours after the molecules administration. Cumulated part of ingested radioactivity recovered in milk after five days have reached a similar level after 103 h for ^{14}C-phenanthrene and ^{14}C-pyrene (1.5% and 1.9% respectively). ^{14}C-benzo[a]pyrene did not appear significantly in milk (0.2% of ingested radioactivity). Transfer of ^{14}C-2,3,7,8-TCDD to milk was higher than for PAHs, reaching 7.8% of ingested radioactivity. The bioavailability of PAHs and dioxins was assessed using two animal models. In one hand, we have characterized the in-vitro transfer of PAHs and dioxin through intestinal barrier using Caco-2 cells cultivated on permeable filters. In the other hand, we have described the specific arterial apparition profile of the studied micropollutants in growing pig. The in-vitro experiment showed that ^{14}C-pyrene, ^{14}C-phenanthrene, ^{14}C-benzo[a]pyrene and ^{14}C-2,3,7,8-TCDD were able to cross the intestinal barrier. ^{14}C-phenanthrene was transported 1.1-, 1.8-, and 6.7-folds more than respectively ^{14}C-pyrene, ^{14}C-benzo[a]pyrene and ^{14}C-2,3,7,8-TCDD after 6 h exposure. Regarding arterial apparition profiles, ^{14}C arterial level from ^{14}C phenanthrene was about 3 and 10 times more elevated than ^{14}C level from ^{14}C-benzo[a]pyrene and ^{14}C-TCDD respectively. These results can be related to the intestinal barrier transfer profile of the same molecules. The results presented in this paper contribute to give new insights on evaluation of the risk of PAHs and dioxins transfer to humans via the dairy ruminant. They particularly clarify the way by which organic micropollutants are transferred to milk and to living organisms during digestion and absorption.

Key words: dairy ruminant, milk, PAH, dioxin, bioavailability

38.1 Introduction

Polycyclic aromatic hydrocarbons (PAHs) are potentially mutagenic compounds (IARC 1983) widely occuring in natural media such as soils, atmosphere, sediments and plants (Baek et al. 1991; Lorber et al. 1994; Yang et al. 1998). There are natural as well as anthropogenic sources of PAHs, such as vegetation fires, petroleum seepage and vehicle exhausts. The main processes of PAH formation are *(1)* incomplete combustion of organic matter, for example, fuel and wood burning, *(2)* the slow maturation of sedimentary organic matter to yield fossil fuels, e.g. coal and petroleum and *(3)* the rapid aromatisation of organic substances in modern sediments. Several PAHs are known to be potential human carcinogens.

PAHs are ubiquitous and occur at low levels of contamination of environmental media such as water, air, and soil. Their physical and chemical properties explain their migration through food chain with hydrophobic compartments, then their accumulation in lipids at the end of the chain (Fürst et al. 1990; Theelen et al. 1993; Madhavan and Naidu 1995; Fries 1995; Edulgee and Gair 1996; McLachlan 1997; Bosset et al. 1998; Roeder et al. 1998). Human exposure to PAHs occurs mainly by feeding except for smokers and specifically exposed workers. The understanding of PAH transfer pathways through the food chain is of major concern for food safety. Thus, there is a strong need to study the relationships between plant and food contamination and food PAHs bioavailability for humans. In order to evaluate the risk of organic micropollutants transfer to humans several studies have been undertaken either on dairy ruminant exposure or on intestinal absorption of these micropollutants. Indeed, it has been suggested that main human exposure originates from the atmosphere via the pathway air-feed-cow-milk-man (Fürst et al. 1990; McLachlan et al. 1990; Jödicke et al. 1992; Körner et al. 1993; McLachlan 1993; Pluim et al. 1993).

38.2
PAHs Exposure of the Dairy Ruminant

38.2.1
Influence of PAHs Emissions Sources on Milk Contamination

Since there is little data on PAHs contamination of milk (Dennis et al. 1983), our first objective was to assess PAHs concentration in milk sampled at different farms located near potential contaminating sources. Dairy farms were chosen in the east part of France according to 3 parameters: type of source, distance of the fields to the source (a perimeter of 4 km around the source) and cows fed with fodder produced in summer on the farm fields (winter ration composed with maize and grass silage, hay and complements). Three kinds of source were selected: stationary sources (4 farms located near a cement work), mobile sources (4 farms located by a motorway), combined sources (3 farms located by stationary sources: steelworks, cement works, municipal waste incinerator and motorway). Three control farms were chosen with fields more than 30 km away from any major source of contamination. For each dairy farm, a milk sample of 500 ml was collected in winter, directly in the milk tank (4 °C) and submitted to PAHs analysis as described by Grova et al. (2000).

Figure 38.1 shows PAHs concentration profiles in milk sampled in the different farms located by potential contaminating. Among the US-EPA list, only 8 PAHs have been detected: naphtalene, acenaphtylene, acenaphtene, fluorene, anthracene, fluoranthene, pyrene and benzo[a]anthracene with concentrations ranging from 0.1 to 16.2 ng g^{-1} milk fat. PAHs with more than 4 aromatic cycles were not detected. Except for benzo[a]anthracene, only non-mutagenic PAHs were detected (IARC 1983).

The detected PAHs present low molecular weight from 128.2 g mol^{-1} (naphtalene) to 278.35 g mol^{-1} (benzo[a]anthracene) and high volatility measured by vapour pressure from 1.04 to 2.71 10^{-5} Pa, compared to the less volatile compound which is 2.6×10^{-9} Pa (indenol[1,2,3-c,d]pyrene). In the atmosphere, PAHs are mainly combined with particles. The transport of these airborne particles depends on their size, their vola-

Chapter 38 · Evaluation of the Risk of PAHs and Dioxins Transfer to Humans via the Dairy Ruminant

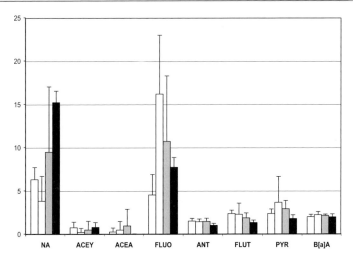

Fig. 38.1. Tank winter milk PAHs concentration (ng g^{-1} milk fat) at various location in Lorraine, France. Except for naphtalene and fluorene, there was no statistical difference between control and exposed farms

tilisation properties and meteorological conditions (Whitby et al. 1980). The properties of PAHs found in our study may explain their capacity to be transported over long distances and be deposited on fields far from any contaminating sources (i.e. control farms) and consequently on those located at less than 4 km from the source. This hypothesis is confirmed by Bryselbout et al. (2000) who demonstrated that light compounds with 4 or less aromatic cycles could be transported over a longer distance than heavy compounds with more than 4 aromatic cycles deposited rapidly after emissions and whose soil concentration massively decrease six meters away from motorway.

In our investigation (Fig. 38.1), naphtalene and fluorene show statistically higher concentrations ($p < 0.05$) than the other six PAHs. Further we observed no statistical correlation between contamination sources and PAH levels in milk, except for naphtalene and fluorene. Indeed, naphtalene is detected with a statistically higher level (15.2 ±1.3 ng g^{-1} milk fat) in milk sampled in farms located near combined contamination sources. Fluorene is also detected with statistically higher level (16.2 ±6.8 ng g^{-1} milk fat) in milk sampled in farms located by cement works. These results indicate that PAHs in milk show similar profiles whatever the sources of emissions are, except for naphtalene and fluorene.

PAHs milk profiles can be used to assess motorway source (acenaphtylene, fluorene and fluoranthene) and combined sources such as incinerator, cement works or steelworks (acenaphtylene, fluorene, fluoranthene) except for compounds with more than four aromatic cycles: chrysene, benzo[a]pyrene and indeno[1,2,3-cd]pyrene (Yang et al. 1998). Grova et al. (2000) observed that milk PAHs profiles are different in summer and winter. Only five compounds were detected: naphtalene, phenanthrene, anthracene, fluoranthene and pyrene. The three major PAHs were phenanthrene (10.9 ±4.5 ng g^{-1} milk fat), naphtalene (8.0 ±3.0 ng g^{-1} milk fat) and pyrene (7.0 ±4.8 ng g^{-1} milk fat). Several hypothesis could explain phenanthrene absence in winter milk and fluorene absence in summer milk: *(i)* season affect PAHs concentration (with 2 or 4 times higher concentrations in winter (Baek et al. 1991)), *(ii)* different sources emissions profiles along the year: quantities and characteristics of PAHs emitted from industrial stacks depend on several factors: the type of input, the manufacturing process, air pollution control devices (Yang et al. 1998), *(iii)* other contamination pathways like breathed air, depending on the type of source or soil ingestion with soil PAHs profiles different from the fodder profiles.

38.2.2
PAHs and Dioxin Feed-Milk Transfer in the Dairy Ruminant

In order to elucidate mechanisms ruling PAH transfer from feed to milk, we studied the transfer of [9 ^{14}C] phenanthrene, [4,5,9,10 ^{14}C] pyrene, [7,10 ^{14}C] benzo[a]pyrene and [U ^{14}C] 2,3,7,8-tetrachlorodibenzo-*p*-dioxin (TCDD) after a single oral ingestion of 2.6×10^6 Bq to lactating goats. 2,3,7,8-TCDD has been used as a bibliographic basis reference for the transfer of dioxins in dairy ruminants by Fries (1995). The three PAHs choice was made according to their different physical and chemical properties (Table 38.1). The goats received a single oral administration of PAHs or dioxin (2.5×10^6 Bq goat^{-1}) directly in the mouth with a sterile syringe and have been milked twice a day for 5 d. A blank sample was collected two hours before pollutants administration.

Figure 38.2 shows the food-milk transfer of ^{14}C-labelled PAHs and TCDD in milk. The data demonstrates the different behaviour of the pollutants during transfer. It should also be mentioned that only bulk radioactivity has been measured, not molecular compounds. Therefore, the radioactivity data shown on Fig. 38.2 refer to both pollutants and their immediate degradation products. Radioactivity associated to three PAHs and 2,3,7,8-TCDD has been detected in milk as early as the first milking, 7 h after the molecules administration. ^{14}C-benzo[a]pyrene was found in milk at very low concentration during the whole experimental period (Fig. 38.2). ^{14}C-pyrene and ^{14}C-phenanthrene concentrations in milk were quite similar with a later absorption peak for pyrene. ^{14}C-2,3,7,8-TCDD showed a higher transfer level with a concentration peak at 22 h of 48.8 Bq ml^{-1}. 2,3,7,8-TCDD behavior can be distinguished from that of PAHs by showing higher concentration in milk (Fig. 38.2). For example, 22 h after spiking, the milk of the goats that have ingested ^{14}C-2,3,7,8-TCDD showed five times more radioactivity than for ^{14}C-pyrene. At this stage, behaviour differences can not easily be related to their chemical and physical properties. The fast and high appearance of 2,3,7,8-TCDD in milk is in agreement with the work of Jones et al. (1989), showing that the incorporation of 2,3,7,8-TCDD in the secretary mammary gland was effective very quickly.

Table 38.1. Physical and chemical properties of the studied organic micropollutants (Mackay et al. 1991)

Compound	Fused benzene rings number	Lipophilicity Log K_{ow}	Water solubility (mg l^{-1})	Molecular weight (g mol^{-1})
2,3,7,8-TCDD	2	6.80	1.93×10^{-5}	321.98
Benzo[a]pyrene	5	6.31	3.80×10^{-3}	252.00
Pyrene	4	4.88	1.30×10^{-1}	202.30
Phenanthrene	3	4.50	1.21	178.20

Chapter 38 · Evaluation of the Risk of PAHs and Dioxins Transfer to Humans via the Dairy Ruminant

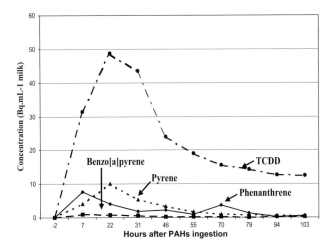

Fig. 38.2. Radioactivity in goat milk following a single oral ingestion of 2.5 × 10⁶ Bq ^{14}C-phenanthrene, ^{14}C-pyrene, ^{14}C-benzo[a]pyrene or ^{14}C-2,3,7,8-TCDD

Table 38.2. Cumulated part of ingested radioactivity recovered in milk after a single oral ingestion of 2.6 × 10⁶ Bq ^{14}C-compounds in lactating goats

Hours after ingestion	Phenanthrene (%)	2,3,7,8-TCDD (%)	Pyrene (%)	Benzo[a]pyrene (%)
−2	0.00	0.00	0.00	0.00
22	0.72	3.16	1.08	0.09
46	1.05	5.49	1.67	0.12
70	1.42	6.55	1.84	0.13
94	1.51	7.29	1.91	0.15
103	1.54	7.79	1.93	0.16

Cumulated parts of ingested radioactivity recovered in milk after five days are presented in Table 38.2. ^{14}C-phenanthrene and ^{14}C-pyrene have reached a similar level after 103 h (1.5% and 1.9% respectively). ^{14}C-benzo[a]pyrene did not appear significantly in milk (0.2% of ingested radioactivity). This behaviour could be explained by the size of benzo[a]pyrene (5 rings) compared to phenanthrene (3 rings) and pyrene (4 rings). The higher number of rings could indeed limit the transfer through ruminal, intestinal or mammary epithelial walls. This low transfer of benzo[a]pyrene has also been shown by West and Horton (1976) after a single oral administration of the ^{14}C-3-methylcholanthrene and the ^{14}C-benzo[a]pyrene blended with food in lactating ewes. An another explanation is that benzo[a]pyrene could be metabolised (Vetter et al. 1985). Further, transfer of 2,3,7,8-TCDD from oil to milk is much higher than for PAHs reaching 7.8% ingested radioactivity (Table 38.2). This could be explained by a higher lipophilicity of TCDD (K_{OW} 6.80) according to McLachlan's (1995) and by very low degradation of 2,3,7,8-TCDD (Wroblewski and Olson 1985).

38.3 Bioavailability of PAHs and Dioxins

38.3.1 Introduction

It is now well established that milk can be polluted by PAHs and dioxins via the ruminant feed. The next issue is to better understand the transfer processes from milk to humans. To this end, we have conducted several experiments regarding the transfer of dioxins and polycyclic aromatic hydrocarbons to blood through the intestinal barrier. Two animal models have been selected to carry out these studies. On one hand we have used human cells cultivated on permeable filters to assess the in-vitro transfer of PAHs and dioxins through intestinal barrier and on the other hand we have investigated the pollutant transfer from milk to pig blood because pigs are model animal for humans (Rowan et al. 1994; Pointillard et al. 1986).

38.3.2 Transfer of PAHs and Dioxins Through Intestinal Barrier Using Caco-2 Cell Line

Little is known about the factors governing intestinal absorption of these molecules in humans or animals (Vetter et al. 1985; Rahman and Barrowman 1986; Kadry et al. 1995; Van Schooten et al. 1997; Laurent et al. 2001). In recent studies, cultures of Caco-2 cell monolayers, isolated from a human colon carcinoma, have been used as a model system to study the intestinal uptake and transport processes of hydrophobic xenobiotics such as polychlorinated biphenyls (Dulfer et al. 1996; Dulfer et al. 1998). Differentiated post-confluent Caco-2 cells exhibit well-developed microvilli and, grown on semi-permeable supports, form tight monolayers with a polarised distribution of brush border enzymes (Trotter and Storch 1991). Thus, we used the Caco-2 cell line to study the uptake and transport of ^{14}C-labelled organic micropollutants: one dioxin (2,3,7,8-TCDD) and three PAHs (phenanthrene, pyrene and benzo[a]pyrene). Cell culture was accomplished as described by Cavret et al. (2003).

The percentage of radioactivity found in basal medium was considered as the absorbed part of the radioactivity brought on the other side of the cells (apical medium). As radioactivity was significantly observed in basal medium (Fig. 38.3), the studied micropollutants were able to cross the intestinal barrier.

Despite of this, transfer of the four molecules showed large differences. Apparition of radioactivity due to benzo[a]pyrene in the basal medium seemed initially very low (only 0.0 to 1% between 15 and 180 min). However, radioactivity detected increased significantly after 180 min and reached 5% at 360 min ($P < 0.001$). Pyrene showed a similar behaviour with a radioactivity detected that remained quite low between 15 and 180 min (0.2 to 2.6 between 15 and 180 min), and increased dramatically to 8.5% of apical radioactivity which crossed intestinal barrier ($P < 0.001$) (Fig. 38.3). ^{14}C-phenanthrene was significantly more present as soon as 90 min ($P < 0.001$). Finally, pyrene and phenanthrene appeared the most and the fastest absorbed compounds after a 6 h exposure. Radioactivity associated to ^{14}C-phenanthrene was transported into basal side 1.1-, 1.8- and 6.7-folds more than respectively ^{14}C-pyrene, ^{14}C-benzo[a]pyrene and ^{14}C-2,3,7,8-TCDD ($P < 0.001$).

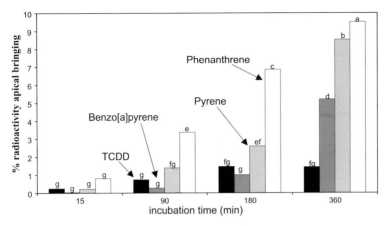

Fig. 38.3. Transfer of PAHs and dioxin through human cells. ^{14}C-radioactivity measured in basal medium (internal compartment) expressed as percentage of total radioactivity brought in apical medium vs. time of incubation (15–360 min) columns with a same letter do not differ significantly ($P > 0.05$)

Among the studied pollutants, the quantity of radioactivity measured in basal medium were inversely reliable to their lipophilicity and molecular weight. Phenanthrene, the less lipophilic ($\log K_{OW} = 4.5$) and the lightest molecule, had the highest transfer; whereas the other compounds saw their passage decreasing in rapidity and quantity as their lipophilicity and molecular weight increased.

To conclude, size and lipophilicity could explain the different transfer observed but not the delay needed. Previous works supported a PAHs metabolisation, particularly in liver, but also in intestinal cells (Bock et al. 1979; Vetter et al. 1985). Further work should be carried out to determine if transferred molecules are native molecules or their metabolites.

38.3.3
Milk-Arterial Transfer of PCDDs/Fs and PAHs

Since we demonstrated that organic micropollutants could be transferred through intestinal barrier using Caco-2 cell line, the next step was to characterise their specific apparition and delivery in arterial blood, which provides the different tissues and organs with nutrients. Thus, organic micropollutants bioavailability is linked to subsequent postprandial delivery in arterial blood. Pigs provide a valid model for studying digestion and absorption in humans (Pointillard et al. 1986; Rowan et al. 1994).

Two experiments were carried out in order to study the milk-arterial transfer of either a solution of native ^{12}C-PCDDs and PCDFs or ^{14}C-2,3,7,8-TCDD, ^{14}C-benzo[a]pyrene and ^{14}C-phenanthrene. The animal protocol as well as the experimental design and the analysis of PCDD/Fs or PAHs in arterial blood have been described by Rychen et al. (2002) and Laurent et al. (2002).

Figures 38.4 and 38.5 indicate the specific milk-arterial transfer profile of PCDD/Fs and PAHs. At time point 0 h, no traces of dioxins were detected. All studied PCDD/Fs were detected in arterial blood and presented a similar kinetic behaviour: arterial

Fig. 38.4.
PCDD/FFs plasma-fat concentrations following ingestion of 900 ml spiked milk with 17 dioxins by growing pigs

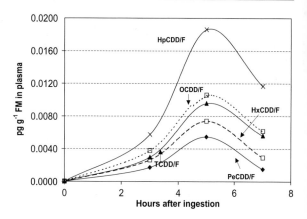

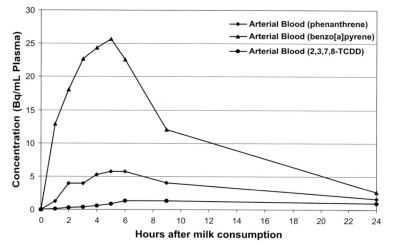

Fig. 38.5. Arterial kinetics of [^{14}C] after ingestion by the growing pig of 1000 ml milk spiked with ^{14}C-phenanthrene, ^{14}C-benzo[a]pyrene or ^{14}C-2,3,7,8-TCDD (mean value, $n = 2$)

concentrations of congeners increased from 3 h to 5 h after spiked milk ingestion and then decreased between 5 h and 7 h (Fig. 38.4). It is the same case for the labelled organic micropollutants: the radioactivity increased rapidly to a maximum about 4–6 h and decreased to reach background levels after 24 h (Fig. 38.5). These results suggest that dioxins and PAHs absorptions are connected with milk fat behaviour (Thomson et al. 1992; Dubois et al. 1996), and differ notably from absorption of glucose and of protein (Mahe et al. 1994; Rychen et al. 2002). Moreover, the organic micropollutants seemed to be rapidly absorbed by the tissues.

However, dioxins and HAPs absorptions appeared different. At time point 5 h, the transfer ratio "plasma fat/milk fat" of PCDD/Fs was usually found between 0.7 and 3% (16 dioxins) and appeared higher for 1,2,3,4,6,7,8-HpCDF (nearly 6%) (Table 38.3). These results indicate that all dioxins are transferred from milk fat to plasma fat at a similar level as 2,3,7,8-TCDD except for one furan "1,2,3,4,6,7,8-HpCDF" which appears in blood

Table 38.3. PCDD/Fs transfer ratio from milk fat to plasma fat 5 h ingestion of 900 ml spiked milk by growing pigs

Compound	Milk fat (pg/g MG)	Plasma fat (pg/g MG)	Ratio plasma/milk
2,3,7,8 TCDD	90 000	1 195.05	1.33
1,2,3,7,8 PeCDD	216 000	2 866.46	1.33
1,2,3,4,7,8 HxCDD	213 000	3 150.81	1.48
1,2,3,6,7,8 HxCDD	209 000	4 976.88	2.38
1,2,3,7,8,9 HxCDD	220 000	4 692.01	2.13
1,2,3,4,6,7,8 HpCDD	189 000	5 410.79	2.86
OCDD	427 000	8 943.71	2.09
2,3,7,8 TCDF	74 000	514.96	0.70
1,2,3,7,8 PeCDF	155 000	2 637.75	1.70
2,3,4,7,8 PeCDF	171 000	1 914.08	1.12
1,2,3,4,7,8 HxCDF	242 000	3 308.05	1.37
1,2,3,6,7,8 HxCDF	235 000	6 500.12	2.77
2,3,4,6,7,8 HxCDF	244 000	3 865.75	1.58
1,2,3,7,8,9 HxCDF	278 000	2 133.12	0.77
1,2,3,4,6,7,8 HpCDF	174 000	9 971.47	5.73
1,2,3,4,7,8,9 HpCDF	202 000	3 460.48	1.71
OCDF	385 000	6 524.94	1.69

fat in higher concentration than 2,3,7,8-TCDD. Thus, the level of dioxins in arterial plasma seems not related to solubility or lipophilicity properties of the molecules. Regarding ^{14}C-labelled 2,3,7,8-TCDD, benzo[a]pyrene and phenanthrene arterial kinetics, ^{14}C plasma level from ^{14}C-phenanthrene was about 3 and 10 times more elevated than ^{14}C level from ^{14}C-benzo[a]pyrene and from ^{14}C-2,3,7,8-TCDD, respectively (Fig. 38.5). These results can be related to the intestinal barrier transfer profile of the same molecules described in Section 38.3.2 (Fig. 38.3) and demonstrate that contrary to PCDD/Fs absorption, the PAHs transfer from milk to blood seems to be controlled by the chemical and physical properties of molecules (Table 38.1). At last, the PAHs transfer from milk to blood appears more elevated than the dioxins transfer.

To our knowledge, these studies present for the first time the apparition profile of ingested milk dioxins or PAHs in arterial blood. The measure of these events is of great physiological importance since it allows to precise the bioavailability of organic micropollutants and its transfer level from food to the organism. For all studied molecules highest concentration were found around 5 h after ingestion of spiked milk. These results suggest that PCDD/Fs and PAHs absorption is connected with milk fat absorption. But the dioxins and PAHs absorption was governed by different parameters: for PAHs, the chemical and physical properties (lipophilicity and solubility) seem to explain the blood transfer, whereas for dioxins, these parameters do not allow to predict the blood transfer. Finally, it is important to notice that the PAH transfer is higher than the PCDD/Fs one.

38.4 Conclusion

We have studied here the transfer of PAHs and dioxins through the food chain using three complementary approaches: *(1)* the study of PAH levels in milk at various farms, *(2)* the study of feed-milk transfer of ^{14}C-labelled pollutants, and *(3)* the study of milk-blood transfer of pollutants. Our investigation of PAH levels in milk at various locations shows that light molecules are present and ubiquitous although the heaviest (5 rings and more) are not detected in milk. Our experiments using ^{14}C-labelled compounds to transfer from milk to pig blood show that all studied molecules can be transferred to milk and to organism during digestion and absorption.

Further research work is now planned to precise and better characterize

- the different "feed-milk" transfer coefficients in ruminant and the part of metabolites or native molecules in milk;
- the transfer coefficient "milk-rat tissues" using a continuous exposure of contaminated feed;
- the selectivity of intestinal epithelial barrier focusing on PAH interaction with cytochrome P450.

Acknowledgements

This study was supported by French Ministeries of Agricultural and Research, Agence de l'Environnement et de la Maîtrise de l'Energie, Fédération Régionale des Coopératives Laitières and Conseil Régional de Lorraine.

References

Baek SO, Field RA, Goldstone ME, Kirk PW, Lester JN, Perru RA (1991) A review of atmospheric polycyclic aromatic hydrocarbons: source, fate and behaviour. Water Air Soil Poll 60:279–300

Bock KW, Clausbruch UCV, Winne D (1979) Absorption and metabolism of naphtalene and benzo[a]pyrene in the rat jejunum in situ. Med Biol 57:262–264

Bosset JO, Bütikofer U, Dafflon O, Koch H, Scheurer-Simonet L, Sieber R (1998) Teneur en hydrocarbures aromatiques polycycliques de fromages avec et sans flaveur de fumée. Sci Aliments 18:347–359

Bryselbout C, Henner P, Carsignol J, Lichtfouse E (2000) Polycyclic aromatic hydrocarbons in highway plants and soils. Evidence for a local distillation effect. Analysis 28:32–35

Cavret S, Laurent C, Feidt C, Laurent F, Rychen G (2003) Intestinal absorption of ^{14}C from ^{14}C-phenanthrene, ^{14}C-benzo[a]pyrene and ^{14}C-tetrachlorodibenzo-para-dioxin: approaches with Caco-2 cell line and with portal absorption measurements in growing pigs. Reproduction Nutrition Développement 43(2):145–154

Dennis MJ, Massey RC, McWeeny DJ, Knowles ME, Watson D (1983) Analysis of polycyclic aromatic hydrocarbons in the UK total diets. Food Chem Toxicol 21:569–574

Dubois C, Arnaud M, Férézou J, Beaumier G, Porugal H, Pauli AM, Bernard PM, Bécue T, Lafont H, Lairon D (1996) Postprandial appearance of dietary deuterated cholesterol in the chylomicron fraction and whole plasma in healthy subjects. Am J Clin Nutr 64:47–52

Dulfer WJ, Govers HAJ, Groten JP (1996) Effect of fatty acids and the aqueous diffusion barrier on the uptake and transport of polychlorinated biphenyls in Caco-2 cells. J Lipid Res 37:950–961

Dulfer WJ, Govers HAJ, Groten JP (1998) Kinetics and conductivity parameters of uptake and transport of polychlorinated biphenyls in the Caco-2 intestinal cell line model. Environ Toxicol Chem 17:493–501
Eduljee GH, Gair AJ (1996) Validation of a methodology for modelling PCDD and PCDF intake via the foodchain. Sci Total Environ 187:211–229
Fries GF (1995) A review of the significance of animal food products as potential pathways of human exposures to dioxins. J Anim Sci 73:1639–1650
Fürst P, Fürst C, Groebel W (1990) Levels of PCDDs and PCDFs in food-stuffs from the federal republic of Germany. Chemosphere 20:787–792
Grova N, Laurent C, Feidt C, Rychen G, Laurent F, Lichtfouse E (2000) Gas chromatography-mass spectrometry study of polycyclic aromatic hydrocarbons in grass and milk from urban and rural farms. Eur J Mass Spectr 6:457–460
Grova N, Feidt C, Laurent C, Rychen G (2003) ^{14}C milk, urine and faeces excretion kinetics in lactating goats after an oral administration of ^{14}C polycyclic aromatic hydrocarbons. Int Dairy Sci, (in press)
International Agency for Research on Cancer (1983) IARC Monographs on Evaluation of Polynuclear Aromatic Compounds, Part 1: Chemical, environmental, and experimental data. IARC Monograph Evaluation Risk Chem Human
Jödicke B, Ende M, Helge H, Neubert D (1992) Fecal excretion of PCDDs/PCDFs in a 3-month-old breast-fed infant. Chemosphere 25:1061–1065
Jones KC, Stratford JA, Waterhouse KS, Furlong ET, Giger W, Hites R, Schaffner C, Johnston AE (1989) Increases in the polynuclear aromatic hydrocarbon content of an agricultural soil over the last century. Environ Sci Technol 23:95–101
Kadry AM, Shoronski GA, Turkall RM, Abdel-Rahman MS (1995) Comparison between oral and dermal bioavailability of soil absorbed phenanthrene in female rats. Toxicol Lett 78:153–163
Körner W, Dawidowsky N, Hagenmaier H (1993) Fecal excretion of PCDDs and PCDFs in two breast-fed infants. Chemosphere 27:157–162
Laurent C, Feidt C, Lichtfouse E, Grova N, Laurent F, Rychen G (2001) Milk-blood transfer of ^{14}C-tagged polycyclic aromatic hydrocarbons (PAHs) in pigs. J Agric Food Chem 49:2493–2496
Laurent C, Feidt C, Grova N, Mpassi D, Lichtfouse E, Laurent F, Rychen G (2002) Portal absorption of ^{14}C after ingestion of spiked milk with ^{14}C-phenanthrene, ^{14}C-benzo[a]pyrene or ^{14}C-TCDD in growing pigs. Chemosphere 48:843–848
Lorber M, Cleverly D, Schaum J, Phillips L, Schweer G, Leighton T (1994) Development and validation of an air – to beef food chain model for dioxin – like compounds. Environ Sci Technol 15:39–65
Mackay D, Shin WY, Ma KC (1991) Illustrated handbook of physical chemical properties and environmental fate of organic chemicals, vol II. Lewis Publisher
Madhavan ND, Naidu KA (1995) Polycyclic aromatic hydrocarbons in placenta, maternal blood, umbelical cord blood and milk of Indian women. Hum Exp Toxicol 14:503–506
Mahe S, Roos N, Benamouzig R, Sick H, Baglieri A, Huneau JF, Tome D (1994) True exogenous and endogenous nitrogen fractions in the human jejunum after ingestion of small amounts of ^{15}N-labeled casein. J Nutr 124:548–555
McLachlan MS (1993) Digestive tract absorption of polychlorinated dibenzo-p-dioxins, dibenzofurans, and biphenyls in a nursing infant. Toxicol Appl Pharmacol 123:68–72
McLachlan MS (1995) Accumulation of PCDD/F in agricultural food chain. Organohalogen Compounds 26:105–108
McLachlan MS (1997) A simple model to predict accumulation of PCDD/Fs in an agricultural food chain. Chemosphere 34:1263–1276
McLachlan MS, Thoma H, Reissinger M, Hutzinger O (1990) PCDD/F in an agricultural food chain, Part 1: PCDD/F mass balance of a lactating cow. Chemosphere 20:1013–1020
Pluim HJ, Wever J, Koppe JG, Sikke vd JW, Olie K (1993) Intake and fecal excretion of PCDD/F in breast-fed infants at different ages. Chemosphere 26:1947–1952
Pointillart A, Cayron B, Gueguen L (1986) Utilisation du calcium et du phosphore et mineralisation osseuse chez le porc consommant du yaourt. Sci Aliments 6:15–30
Rahman A, Barrowman JA (1986) The influence of bile on the bioavailibility of polycyclic aromatic hydrocarbons from the rat intestine. Can J Physiol Pharmacol 64:1214–1218

Roeder RA, Garber MJ, Schelling GT (1998) Assessment of dioxins in foods from animal origins. J Anim Sci 76:142–151

Rowan AM, Moughan PJ, Wilson MN, Maher K, Tasman-Jones C (1994) Comparison of the ileal and faecal digestibility of dietary amino acids in adult humans and evaluation of the pig as a model for digestion studies in man. Brit J Nutr 71:29–42

Rychen G, Laurent C, Feidt C, Grova N, Lafargue PE, Hachimi A, Laurent F (2002) Milk-arterial plasma transfer of PCDDs and PCDFs in pigs. J Agric Food Chem 50:4640–4642

Theelen RMC, Liem AKD, Slob W, van Wijnen JH (1993) Intake of 2, 3, 7, 8 chlorine substituted dioxins, furans, and planar PCBs from food in the Netherlands: media and distribution. Chemosphere 27:1625–1635

Thomson ABR, Schoeller C, Keelan M, Smith L, Clandinin MT (1992) Lipid absorption: passing through the unstirred layers, brush-border membrane, and beyond. Can J Physiol Pharmacol 71:531–555

Trotter PJ, Storch J (1991) Fatty acid uptake and metabolism in a human intestinal cell line (Caco-2): comparison of apical and basolateral incubation. J Lipid Res 32:293–304

Van Schooten FJ, Moonen EJC, van der Wal L, Levels P, Kleinjans JCS (1997) Determination of polycyclic aromatic hydrocarbons (PAH) and their metabolites in blood, feces, and urine of rats orally exposed to PAH contaminated soils. Arch Environ Contam Toxicol 33:317–322

Vetter RD, Caray MC, Patton JS (1985) Coassimilation of dietary fat and benzo[a]pyrene in the small intestine: an absorption model using the killifish. J Lipid Res 26:428–434

West CE, Horton BJ (1976) Transfer of polycyclic hydrocarbons from diet to milk in rats, rabbits and sheep. Life Sci 19:1543–1552

Whitby KT, Sverdrup GM (1980) California aerosols: their physical and chemical characteristics. The character and origins of smog aerosols. In: Hidy GM, Mueller PK, Grosjean D, Appel BR, Wesolowski JJ (eds) Advances in environmental science and technology, vol. 9. John Wiley & Sons, Inc., New York, pp 477–517

Wroblewski VJ, Olson JR (1985) Hepatic metabolism of 2,3,7,8-tetrachlorodibenzo-p-dioxin (TCDD) in the rat and Guinea pig. Toxicol Appl Pharmacol 81:231–240

Yang HH, Lee WJ, Chen SJ, Lai SO (1998) PAH emission from various industrial stacks. J Hazard Mat 60:159–174

Chapter 39

Polycyclic Aromatic Hydrocarbons (PAHs) Removal during Anaerobic and Aerobic Sludge Treatments

E. Trably · D. Patureau · J.-P. Delgenes

Abstract

Polycyclic aromatic hydrocarbons (PAHs) are of particular interest because of their potential toxic and carcinogenic properties. Due to their low water solubility and their high affinity for organic matter, PAHs are easily concentrated in sewage sludge and may contribute to the contamination of agricultural soils by spreading. In this study, the behavior of 13 PAHs was assessed during anaerobic and aerobic mesophilic treatments of naturally PAH-contaminated sewage sludge. It was shown that abiotic losses were strictly limited to the light PAHs, e.g. fluorene, phenanthrene and anthracene. Under methanogenic conditions, PAH removal was about 50% whatever PAH molecular weight. More specifically, PAH removal was closely linked to solids reduction implying limitation by bioavailability. Under aerobic conditions, the aerated process enhanced PAH removal up to 90%. In contrast, the aerobic treatment is more efficient than the anaerobic treatment to remove PAHs from contaminated sludge by favoring the PAH diffusion. Moreover, the aerobic process was successful for sludge decontamination because outlet concentrations in dry weight were lower than actual French required values, for fluoranthene and benzo[b]fluoranthene.

Key words: adapted methanogenic ecosystems, continuous bioreactor, methanogenic and aerobic conditions, PAHs, sewage sludge

39.1
Introduction

39.1.1
Sewage Sludge Management and Treatment Processes

There is a growing concern about the management of sewage sludge, because environmental regulations have recently induced a fast growth of sludge production reaching more than 7.5 millions of tons of dry solids in 1998 in Europe. Sewage sludge management is henceforth a critical problem for the protection of the environment and for the administration of local communities, due to their huge volumes and their harmful composition.

Stabilization processes such as composting, anaerobic/aerobic digestion or lime stabilization can be used in order to reduce the volume, the fermentable power and the organic pollutant contents of sludge. Whatever treatment conditions, the final composition of stabilized sludge still presents high amounts of organic matter and nutritive elements such as total organic carbon, total nitrogen and total phosphorus compounds. Therefore, one of the best options to recycle sludge consists of their spreading on agricultural soils, to improve nutrient contents and to promote biological activity. Alternative options of treatment such as incineration and landfill disposal are more expensive and less sustainable. Moreover, if all produced sewage sludge were used on land, the total treated surface would represent only few percents (2–3%) of total agricultural soil area in Europe. However, due to their high organic content, sewage sludge concentrates many forms of recalcitrant pollutants such as heavy metals, pathogens and organic pollutants and sludge spreading may contribute to the contamination of agricultural soils.

39.1.2
Polycyclic Aromatic Hydrocarbons Persistence and Biodegradation

Polycyclic aromatic hydrocarbons are widely distributed in the environment and are known for their potential toxic and carcinogenic properties even at trace levels (Partanen and Boffetta 1994). Due to their high-hydrophobic properties, PAHs are easily adsorbed on hydrophobic organic surfaces such as particles in air, soil and sewage sludge. Three of the most suspected carcinogenic PAHs are the subject of concern for the French legislative procedure concerning sludge used on land: fluoranthene, benzo[b]fluoranthene and benzo[a]pyrene with level limits of 5 mg $kg_{d.w.}^{-1}$ (*d.w.*: dry weight), 2.5 mg $kg_{d.w.}^{-1}$ and 2 mg $kg_{d.w.}^{-1}$ respectively. Lower levels are anticipated in the frame of current European policies. Indeed, the European Commission has recently proposed a limit value of 6 mg $kg_{d.w.}^{-1}$ for the sum of 11 PAHs from acenaphthene to indeno[123cd]pyrene. In this context, the fate of the PAHs during anaerobic and aerobic treatments of naturally contaminated sludge is little known.

PAH biodegradation has been already widely studied under aerobic conditions in case of highly contaminated soils and sediments (g $kg_{d.w.}^{-1}$), but not yet, to our knowledge, in sewage sludge (Bouwer et al. 1997; Leduc et al. 1992; Mihelcic and Luthy 1988; Wild and Jones 1993). Many microorganisms are implicated in PAH biodegradation under aerobic conditions. The most common are: *Aeromonas* sp., *Micrococcus* sp. and *Pseudomonas* sp. (Wilson and Jones 1993). Some fungi can also participate in PAH biodegradation, especially the white rot fungi *Phanerochaete* sp. and *Bjerkandera* sp. (Kotterman et al. 1998). Aerobic PAH biodegradation follows two main mechanisms in complex media: the first involves the use of PAH as sole carbon and energy source for the growth of the microorganisms and for cellular maintenance. The second involves a co-metabolism with other carbon sources, meaning that PAHs are degraded but not used for cellular growth. This second mechanism is mainly occurring for the heaviest PAHs (four and five membered rings) (Wilson and Jones 1993).

By comparison, only little is known about PAHs biodegradation under strict anaerobic conditions. Mihelcic and Luthy (1988) reported, for the first time, the biodegradation of light-PAHs under denitrifying conditions. However, denitrifying anaerobic conditions remain less favorable for PAHs degradation than aerobic conditions (Wilson and Bouwer 1997). Recently, some studies reported PAHs degradation under sulfate-reducing conditions in the case of high contaminated marine sediments (Coates et al. 1996; Rockne and Strand 1998; Chang et al. 2001). Under methanogenic conditions, PAH biodegradation was recently described by Chang et al (2002) for light PAHs by addition to a synthetic PAH-contaminated soil of an enriched adapted culture coming from a long term highly polluted sediment. Trably et al (2003) also demonstrated significant methanogenic PAHs removal under continuous bireactors. In methanogenic batch reactors, PAH removal remained non significant (Kirk and Lester 1990). In all cases, PAHs biodegradation under anaerobic conditions seems more unfavorable than under aerobic conditions.

In this study, the behavior of 13 PAHs was determined under mesophilic anaerobic and aerobic digestion of naturally contaminated sewage sludge. Naturally PAH-contaminated sludge was used to represent the complex interactions between PAH and sludge matrix. Indeed, the added PAHs in already contaminated spiked soils are more rapidly degraded than the "older" PAHs, probably hardly linked to the matrix (Eggen and Majcherczyk 1998). The comparison of anaerobic and aerobic conditions was realized by calculation of PAH removal efficiencies into anaerobic (A series) and aerobic (B series) biological reactors, called R_A and R_B respectively. Sterilized control reactors were used to estimate abiotic losses during the process (volatilization, photolysis).

39.2
Experimental

39.2.1
Biological and Control Reactors

Four laboratory-scale continuous stirred tank reactors were realized to determine the behavior of 13 PAHs under mesophilic conditions. On one hand, two "control" reactors, respectively anaerobic (CR_A) and aerobic (CR_B) reactors, were chemically sterilized by addition of 6.6 g l^{-1} sodium azid (NaN3, Riedel de Haën) to inhibit bacterial activity. These control reactors were operated to assess PAH abiotic losses during the processes. On the other hand, two "biological" reactors were operated to assess the biologically-mediated removal of PAHs under anaerobic (R_A) and aerobic (R_B) conditions. The anaerobic reactors (R_A and CR_A) were initially inoculated with 5 liters each of anaerobic digested sludge. This sludge was sampled in the outlet of an industrial anaerobic digester located on a urban wastewater treatment plant (WWTP) contaminated by PAHs for more than 10 years. The aerobic reactors (R_B and CR_B) were initially inoculated with 5 liters of activated sludge sampled in the same PAHs-contaminated WWTP. In both cases, the starting inoculii corresponded to methanogenic or aerobic PAH-adapted ecosystems.

39.2.2
Operating Conditions

All reactors were well-mixed with a hydraulic retention time of about 20 ±1 d, a regulated mesophilic temperature of 35 ±1 °C, a daily organic load about 1.2 kg$_{COD}$ m^{-3} d^{-1} (COD: Chemical Oxygen Demand) and a reactional volume of 5 liters. The biogas outlets were cooled to avoid water losses during the process. The substrate feeding tank was changed once a week and was cooled to limit initial biodegradation. Magnetic stirring was used to agitate anaerobic reactors (250 ±2 rpm). Mechanic stirring at 250 ±2 rpm was used to mix aerobic reactors. The pH was not regulated but did not change significantly: 7.5 ±0.1 for anaerobic treatments and 7.1 ±0.1 for aerobic treatments. The reactors were fed with a mixture of primary and secondary sludge sampled in the PAH-contaminated WWTP mentioned above. The ranges of PAH concentrations in feeding sludge were (minimum, maximum in mg $kg_{d.w.}^{-1}$): fluorene (0.45, 0.59), phenanthrene (3.41, 4.49), anthracene (0.98, 1.1), fluoranthene (8.08, 11.25), pyrene (8.65, 11.22), benzo[a]anthracene (3.41, 3.66), chrysene (1.23, 1.54), benzo[b]fluoranthene (4.2, 4.63), benzo[k]fluoranthene (2.11, 2.43), benzo[a]pyrene (4.04, 4.08), dibenzo[ah]anthracene (0.66, 0.74), benzo[ghi]perylene (2.47, 2.73) and indeno[123cd]pyrene (3.38, 3.85). PAH levels were two times above the French maximum allowed concentrations for spreading on agricultural soils. The sum of the 13 studied PAHs reached more than 45 mg $kg_{d.w.}^{-1}$.

39.2.3
PAH Analysis – Sample Preparation

PAH analysis of sludge sample was conducted in three steps after centrifugation: *(1)* extraction from the liquid phase by solid phase extraction or SPE, *(2)* extraction from the solid phase by accelerated solvent extraction or ASE and *(3)* analysis of the extracts by reverse phase-high performance liquid chromatography with fluorimetric detection. All steps of the analytical method were previously tested and validated in the laboratory to obtain repeatability and reproducibility errors lower than 2% (two replicates of the same sludge sample). This internally validated method allowed to monitor PAH concentrations in laboratory-scale experimental reactors (Trably et al. 2004). The sludge samples corresponded to 2 d outlet collecting of each reactor (≥350 ml). The substrate feeding tank was changed once a week and an aliquot of 350 ml was taken. 300 ml of sludge sample (substrate or reactor outlet) were centrifuged 25 min at 20 000 g. Aqueous phases were stored in cool chamber (–20 °C) for further SPE. Solid pellets were ground with glass beads (diameter 4 mm) then dried 60 h at 40 °C in a ventilated oven. Dried samples were 2 mm-sieved and then stored at –20 °C.

39.2.4
PAHs Extraction from the Liquid Phase by SPE

PAHs extraction from the liquid phase was performed by SPE on PAH-affinity column (Supelco ENVI-18, 6 ml). The column was conditioned with 6 ml toluene/methanol (50/50, v/v), 6 ml methanol and 6 ml water successively. 200 ml of aqueous sample were first eluted 3 times under vacuum. Then PAHs were eluted with 2 ml toluene/methanol (50/50, v/v). The extract was then concentrated to dryness under nitrogen flow. Residues were dissolved in 2 ml of acetonitrile for further HPLC analysis.

39.2.5
PAHs Extraction from the Solid Phase by ASE

PAHs were extracted from dried sieved samples using the ASE-200 system (DIONEX). The extracting solvent was hexane/acetone (50/50, v/v). The conditions were: 120 °C, 100 bar, 2 extraction cycles, 5 min static time, 60% cell flush and 120 s purge time. The extracting cells were filled with 0.5 g dried sieved sample, 0.5 g alumina (SIGMA A-1522) and 1.5 g hydromatrix-celite (VARIAN). The 20 ml extract was then concentrated under nitrogen flow to dryness. Residues were dissolved in 5 ml of acetonitrile for further HPLC analysis.

39.2.6
Extracts Analysis by RP-HPLC-Fluorimetric Detection

The extracts analysis chain was composed of a multi-sample injector (WATERS-717-Plus), a solvent degasser (Waters inline Degasser), a peristaltic pump system (WATERS-600 controller) and a fluorimetric detector JASCO FP-1520. PAH separation was performed by reversed phase high performance liquid chromatography (RP-HPLC) on PAH column (BAKERBOND PAH16-Plus). The flow rate was fixed at 0.3 ml min^{-1} and elution temperature at 25 °C. The linear gradient elution (35 min) started after 5 min of elution with solvent mixture from 40% acetonitrile – 60% water to 100% acetonitrile. After 70 min, the column was rinsed with 40% acetonitrile – 60% water. Total analysis time was about 95 min. The fluorimetric PAH detecting program was optimized for each PAH and for each $\lambda_{\text{excitation wavelength}}$ and $\lambda_{\text{emission wavelength}}$, respectively: fluorene (266/312), phenanthrene (250/370), anthracene (250/400), fluoranthene (280/430), pyrene (320/404), benzo[a]anthracene (280/430), chrysene (268/384), benzo[b]fluoranthene (234/420), benzo[k]fluoranthene – benzo[a]pyrène (270/400), dibenzo[ah]anthracene – benzo[ghi]perylene (300/407) and indeno[123cd]pyrene (300/500).

39.2.7
Calculation Method of PAH Removal

The calculation of PAH removal was done by PAH mass balance between the inlet (substrate) and the outlet of each reactor at steady state (after 60–80 d of digestion). Mass balance losses corresponded to abiotic losses for the control reactors (CR$_A$ and CR$_B$) and to abiotic and biological PAH removal for the biologic reactors (R$_A$ and R$_B$). The total PAH concentration per volume of sludge (µg l^{-1}) was used for the calculation of the mass balance. As the PAH concentration in the liquid phase was non significant, it was neglected for the calculation of total PAH concentration. This one corresponds then to the product of the solid phase PAH concentration (mg kg$^{-1}_{\text{d.w.}}$) and the solid concentration (g$_{\text{d.w.}}$ l^{-1}).

39.3
Results and Discussion

39.3.1
Estimation of PAH Abiotic Losses in Anaerobic and Aerobic Control Reactors

In a first step, the abiotic losses due to the process, such as volatilization, photolysis or chemical oxidation, were estimated in the cases of the chemically sterilized control reactors. Under anaerobic conditions (CR$_A$ reactor), the abiotic losses were limited to the PAHs with less than three aromatic rings such as fluorene, phenanthrene and anthracene (Fig. 39.1). The light PAHs correspond to the most water soluble PAHs with the lowest melting point. Such properties favor volatilization in main part, chemical transformation or unspecific combination with organic matter in aqueous media. Our findings are in agreement with bioremediation experiments of contaminated soils

Fig. 39.1.
PAH abiotic losses under anaerobic (CR$_A$ reactor – *black boxes*) and aerobic (CR$_B$ reactor – *white boxes*) sterile conditions in function of the PAH molecular weight. Abiotic losses were calculated by using a mass balance between the inlet and the outlet of the control reactors at steady state. PAH abbreviations are explained in appendix

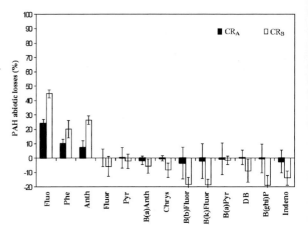

(Richnow et al. 1998). For the heavy PAHs (≥4 aromatic rings), the results showed that abiotic losses could be generally neglected under anaerobic conditions (Fig. 39.1). In this case, all PAHs injected to the system were recovered in the outlet of the anaerobic reactor. Wild and Jones (1993) reported also that PAH abiotic losses were not significant for PAH with more than 4 aromatic rings in contaminated soils bioremediation experiments thus confirming our results.

Under aerobic conditions, abiotic losses were approximately 2 times greater than under anaerobic conditions for the light PAHs (fluorene, phenanthrene and anthracene) (Fig. 39.1). This finding is probably due to the aerating system which induced more losses by either volatilization or oxidation with the organic matter. Confirming these results, Leduc et al. (1992) have already shown that abiotic losses of light PAHs represented the main losses in soils under aerating conditions. On the other hand, more heavy PAHs were recovered in the outlet of the aerobic reactor than in the substrate (up to 20% represented by negative losses in Fig. 39.1). This result suggests that, under aerobic conditions, the aerating system and the mechanical stirring increased the exchanges between the sludge matrix and the aqueous media. PAH diffusion was thus enhanced by inducing desorption of the heaviest PAHs from the non-extractable compartment. Despite the use of a highly efficient extraction method under high pressure and high temperature such as Accelerated Solvent Extraction, a significant part (20%) of PAH content still remained and can later be desorbed from the sludge matrix to the environment.

To resume, the aerobic process presented more abiotic losses than the anaerobic one for the light PAHs, but enhanced greatly the PAH diffusion and may be used to desorb the heavy PAHs from the non-extractable fraction.

39.3.2
PAH Removal by Anaerobic and Aerobic Biological Treatments

In the methanogenic anaerobic biological reactor (R$_A$), the calculated PAH removal efficiencies were about 50% (Fig. 39.2). The PAH removals were independent of the PAH molecular weight and of the wide range of PAH concentration (from 10 to

Fig. 39.2.
PAH removal under anaerobic (R_A reactor – *black boxes*) and aerobic (R_B reactor – *white boxes*) conditions. PAH removals were calculated by using a mass balance between the inlet and the outlet of the biological reactors at steady state. PAH abbreviations are explained in appendix

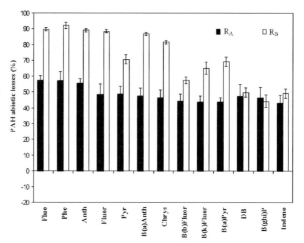

220 µg l^{-1}). The PAH removals seemed to be closely linked to the solids reduction rate of the process (about 53%). By comparison, the PAH removal efficiencies were significantly higher under aerobic conditions reaching up to 90% for the light PAHs and about 50% for the heaviest PAHs (Fig. 39.2). These findings are in agreement with previous studies which reported lower PAH degrading rates under nitrate reducing conditions than under aerobic conditions (Leduc et al. 1992; Mihelcic and Luthy 1988; Wilson and Bouwer 1997). Additionally, in case of the aerobic reactor, the number of aromatic rings and the PAH molecular weight had a significant negative influence on PAH removal efficiency. This result can be the consequence of a biological restriction or of a diffusion limitation. Others studies reported that PAH biological degradation was hardly slowed down with the increase of the molecular weight because of the high recalcitrance of the heavy PAHs to biodegradation (Sutherland et al. 1995). Moreover, in the case of the aerobic control reactor, heavy PAH diffusion was greatly enhanced. Thus, our results suggest that the significant decrease of PAH removal in function of the aromatic ring number was more the result of a biological limitation with lower efficiencies for the heavy PAHs. The aerobic conditions presented anyway better removal efficiencies than the anaerobic process, especially for the light PAHs.

39.3.3
Influence of Operating Conditions on PAH Diffusion

Due to their high hydrophobic properties, PAHs are mainly concentrated in the solid phase of the sludge samples (PAH concentrations in liquid phase were determined as non-significant). But the solids reduction rate reached about 53% in the anaerobic process vs. only 38% under aerobic conditions. In order to compare both processes independently of the solids reduction rates, a dimensionless factor has been calculated: the efficiency factor (Fig. 39.3). This factor corresponds to the ratio of PAH removal efficiency on the solid reduction rate. For instance, efficiency factors higher than 1 represent a decrease of PAH concentration in solids in the reactor effluent. On the opposite, efficiency factors lower than 1 represent an increase of the PAH concen-

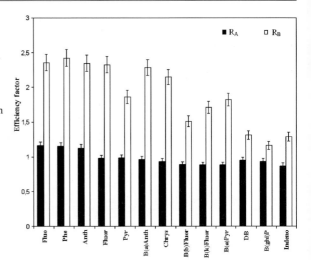

Fig. 39.3.
PAH efficiency factor of the anaerobic (R_A reactor – *black boxes*) and aerobic (R_B reactor – *white boxes*) reactors. The efficiency factor corresponds to the ratio of the PAH removal and the solids reduction rate. PAH abbreviations are explained in appendix

tration in solids during the process. Under anaerobic conditions, the solids reduction rate was about 53% and the PAH removal about 49% for all PAHs. Consequently, the factor efficiency was about 1 for all (Fig. 39.3). Thus, PAH removal in the anaerobic process seemed to be closely linked to the solids reduction rate and probably to be limited by the PAH bioavailability. Under aerobic conditions, solids reduction rate was significantly lower than under anaerobic digestion (38%). Additionally, the PAH removal efficiencies were higher for all PAHs in comparison with the anaerobic system (up to 90%). Therefore, the efficiency factors were generally better under aerobic conditions. The efficiency factors reached about 2.5 for the lightest PAHs under aerobic conditions (Fig. 39.3). These results confirmed the enhanced PAH diffusion conditions in the aerated system in accordance with the results obtained for the control reactor. In addition, as PAH removals decrease with the increase of the aromatic ring number, it can be assumed that they were limited in this case by the biological abilities.

39.3.4
PAH Levels in Untreated and Treated Sewage Sludges

Under anaerobic conditions, PAH removals and the solids reduction rate were of the same order of magnitude (50%). This result means that PAH concentration in solids did not decrease significantly during the process and therefore they were similar in the substrate (raw sewage sludge) and in the effluent of the anaerobic reactor (Table 39.1). Moreover, only the PAH concentrations in solids are the object of concern by the French legislation for sludge spreading on agricultural soils. Thus, the anaerobic process was not successful considering the contamination level of the treated sludge.

By comparison, PAH concentrations in solids decreased significantly under aerobic conditions due to the enhancement of the PAH diffusion (Table 39.1). The PAH levels in the aerobically treated sludge reached values close to the maximum levels required

Table 39.1. PAH concentrations (mg $kg_{d.w.}^{-1}$) in untreated and treated sewage sludge. Untreated sludge corresponds to the inlet (substrate) of the reactors and the treated sludge corresponds to the outlet of the anaerobic and aerobic biological reactors. Maximal relative error was of 6% for all results

	Fluo	Phe	Anth	Fluor	Pyr	B(a)A	Chrys	B(b)F	B(k)F	B(a)P	DB	B(ghi)P	Ind
French max. values	–	–	–	5	–	–	–	2.5	–	2	–	–	–
Untreated (substrate)	0.48	3.15	0.69	10.50	10.32	3.84	4.34	4.35	2.35	3.81	0.54	2.54	3.04
Treated (anaerobic)	0.50	3.49	0.90	11.20	10.86	3.92	4.70	5.22	2.67	4.41	0.79	2.98	4.37
Treated (aerobic)	0.09	0.45	0.126	1.64	3.26	0.80	1.30	2.40	1.34	2.24	0.48	2.50	2.52

for using sludge on land. Thus, the aerobic treatment of contaminated sludge is a successful process for the decontamination of naturally PAH-contaminated sewage sludge.

39.4 Conclusion

In this study, mesophilic anaerobic and aerobic treatments of PAH contaminated sludge were studied with laboratory-scale continuous bioreactors. It was shown that abiotic losses were only limited to the light PAHs under anaerobic conditions. The aerobic conditions enhanced significantly the abiotic losses for these light PAHs and the diffusion of the heavy PAHs from the non-extractable compartment. Under anaerobic conditions, PAH removals reached about 50% for all and were mainly due to the bacterial activity. In this case, PAH removal seemed to be hardly limited by the solids reduction rate and therefore by the PAH bioavailability. By comparison, the aerobic process was more efficient with PAH removals up to 90%, due to a PAH-diffusion enhancement of the aerating system. Heavy PAH removal seemed to be limited in this case by the bacterial degrading potential. Moreover, aerobic treatment of PAH contaminated sludge was successful with a significant decrease of the PAH concentration in solids which are concerned by the French legislation.

Appendix

PAH abbreviations: *Fluo:* fluorene; *Phe:* phenanthrene; *Anth:* anthracene; *Fluor:* fluoranthene; *Pyr:* pyrene; *B(a)A:* Benzo[a]anthracene; *Chrys:* chrysene; *B(b)F:* Benzo[b]fluoranthene; *B(k)F:* Benzo[k]fluoranthene; *B(a)P:* benzo[a]pyrene; *DB:* Dibenzo[ah]anthracene; *B(ghi)P:* Benzo[ghi]perylene; *Ind:* indeno[123cd]pyrene.

Acknowledgment

Agence de l'Environnement et de la Maîtrise de l'Energie (ADEME, Angers, France) is especially thanked for the grant to Eric Trably.

References

Bouwer EJ, Zhang W, Wilson LP, Durant ND (1997) Biotreatment of PAH-contaminated soils/sediments. Ann NY Acad Sci 829:103–117

Chang BV, Chang JS, Yuan SY (2001) Anaerobic degradation of phenanthrene in river sediment under nitrate-reducing conditions. Bull Environ Contam Toxicol 67:898–905

Chang BV, Shiung LC, Yuan SY (2002) Anaerobic biodegradation of polycyclic aromatic hydrocarbon in soil. Chemosphere 48:717–724

Coates JD, Anderson RT, Lovley DR (1996) Oxidation of polycyclic aromatic hydrocarbons under sulfate-reducing conditions. Appl Environ Microbiol 62:1099–1101

Eggen T, Majcherczyk A (1998) Removal of polycyclic aromatic hydrocarbons (PAH) in contaminated soil by white rot fungus *Pleurotus ostreatus*. Int Biodeter Biodegr 41:111–117

Kirk PW, Lester JN (1990) The fate of polycyclic aromatic hydrocarbons during sewage sludge digestion. Environ Technol 12:13–20

Kotterman MJJ, Vis EH, Field JA (1998) Successive mineralization and detoxification of benzo[a]pyrene by the white rot fungus *Bjerkandera* sp. strain BOS55 and indigenous microflora. Appl Environ Microbiol 64:2853–2858a

Leduc R, Samson R, Al-Bashir B, Al-Hawari J, Cseh T (1992) Biotic and abiotic disappearance of four PAH compounds from flooded soil under various redox conditions. Water Sci Technol 26:51–60

Mihelcic JR, Luthy RG (1988) Microbial degradation of acenaphthene and naphthalene under denitrification conditions in soil-water systems. Appl Environ Microbiol 54:1188–1198

Partanen T, Boffetta P (1994) Cancer risk in asphalt workers and roofers: review and meta-analysis of epidemiologic studies. Am J Ind Med 26:721–740

Richnow HH, Eschenbach A, Mahro B, Seifert R, Wehrung P, Albrecht P, Michaelis W (1998) The use of ^{13}C-labelled polycyclic aromatic hydrocarbons for the analysis of their transformation in soil. Chemosphere 36:2211–2224

Rockne KJ, Strand SE (1998) Biodegradation of bicyclic and polycyclic aromatic hydrocarbons in anaerobic enrichments. Environ Sci Technol 32:3962–3967

Sutherland JB, Rafii F, Khan AA, Cerniglia CE (1995) Mechanisms of polycyclic aromatic hydrocarbon degradation. In: Young LY, Cerniglia CE (eds) Microbial transformation and degradation of toxic organic chemicals. Wiley-Liss, inc., New-York

Trably E, Delgenes N, Patureau D, Delgenes JP (2004) Statiscal tools for the optimization of a highly reproducible method for the analysis of Polycyclic Aromatic Hydrocarbons in sludge samples. Int J Environ Anal Chem, (in press)

Wild SR, Jones KC (1993) Biological and abiotic losses of polynuclear aromatic hydrocarbons (PAHs) from soils freshly amended with sewage sludge. Environ Toxicol Chem 12:5–12

Wilson LP, Bouwer EJ (1997) Biodegradation of aromatic compounds under mixed oxygen/denitrifying conditions: a review. Ind Microbiol Biotechnol 18:116–130

Wilson SC, Jones KC (1993) Bioremediation of soil contaminated with polycyclic aromatic hydrocarbons (PAHs): a review. Environ Pollut 81:229–249

Chapter 40

Photodegradation of Pyrene on Solid Phase

A. Boscher · B. David · S. Guittonneau

Abstract

Photodegradation of 1% weight ratio pyrene on different model supports SiO_2, $CaCO_3$ and montmorillonite was investigated. The kinetic studies show that the pyrene pseudo half-lives depend on the nature of the support, and on the presence of water: on dry support, $t_{1/2}$ = 3 h on SiO_2, 3 h on $CaCO_3$, and 6 h on montmorillonite whereas in presence of water, $t_{1/2}$ = 9.9 h, 9.9 h, and 221 h, respectively. The irradiation of pyrene revealed degradation products such as 1-hydroxypyrene, 4,5-dihydro-4,5-dihydroxypyrene, 1,6 and/or 1,8-dihydroxypyrene, 1,6 and/or 1,8-pyrenequinone and 1,1'-bipyrene. Given *(i)* the byproducts identified, *(ii)* the inhibition of pyrene transfer $S_1 \longrightarrow T_1$ with $HgCl_2$ and *(iii)* an oxidation non photosensitized by rose bengal, a possible degradation pathway is proposed which involves the formation of a pyrene radical cation via the singlet state of pyrene.

Key words: pyrene, SiO_2, $CaCO_3$, montmorillonite, photodegradation, mechanism

40.1
Introduction

Pyrene and other polycyclic aromatic hydrocarbons (PAHs) are either of petrogenic origin as constituents of coals, petroleum and kerogen, or of pyroginic origin as a result of incomplete combustion processes (Prado et al. 1981; Wislocki et al. 1986). PAHs released into the air may be dispersed into water by rain or gravity, and owing to their high hydrophobicity, accumulate in soils and sediments (Liste and Alexander 1999; Jones et al. 1989). In the environment, PAHs, like pyrene, can be degraded by bacteria under aerobic conditions (Schneider et al. 1996). Nevertheless, this process is slow because of the anaerobic conditions which often occur on sediments and soils (Sharak Genthner et al. 1997). Their degradation can be achieved partially by photooxidation, which can occur in the environment, in the water phase and on solid substrates.

The photolysis rate of PAH adsorbed on solid phase is usually slow. The photolytic half-lives of pyrene and others PAHs adsorbed on different substrates such as alumina, silica, fly ash, and carbon black range between 17 and 1 000 h (Behimer and Hites 1988). Chemical and physical surface properties such as carbon content, surface area, particle porosity size and chemical functional groups have an effect on the photodegradation rate (Korfmacher et al. 1980; Yokley et al. 1986; Cope and Kalkwarf 1987; David and Boule 1993). However, one of the most important parameter in relation with the degradation process is the color of the support (Behimer and Hites 1988). In addition, when a photochemical reactivity is observed at the solid/air interface of an unactivated silica gel (Reyes et al. 2000), the main oxidized products are

1-hydroxypyrene, 1,6- and 1,8-pyrenequinone, and the formation of a pyrene radical cation as primary species is observed.

In comparison to the photolysis of adsorbed pyrene, the direct photolysis of pyrene dissolved in water is rapid. The half-lives of PAHs (Smith et al. 1979; Zepp and Schlotzhauer 1979) vary between several minutes for the most reactive PAH (e.g. $t_{1/2}$ = 10 min for 9-methylanthracene), 41 min for pyrene $t_{1/2}$, to several hours for the less reactive substances (e.g. $t_{1/2}$ = 70 h for naphthalene). The main photoproducts of pyrene degradation are also 1,6- and 1,8-pyrenequinones. The first step proposed in the literature for the photochemical oxidation of pyrene and other PAHs involves either an electron transfer from the excited singlet state of the PAH to molecular oxygen by a charge-transfer (Sigman et al. 1998), or the oxidation of the PAH by singlet oxygen formed during the irradiation (Barbas et al. 1996).

In this study, we investigate the influence of the support and water on the photoreaction course of pyrene sorbed on supports. Experiments were performed with the model supports SiO_2, $CaCO_3$ and montmorillonite, which are all representative of the main constituents of the soils and sediments (Sparks 1995). To complete data often given at 254 or 365 nm, we focused our attention on the photodegradation rate of pyrene under a solar-like irradiation and tried to confirm one of the mechanisms proposed in the literature.

40.2
Experimental

40.2.1
Chemicals

Pyrene (99%, Fluka), 1-hydroxypyrene (Aldrich), rose bengal (Aldrich), mercuric chloride (99.5%, Prolabo) and the different supports silica (SDS), calcite (Fischer Chemicals), montmorillonite (KSF, Aldrich) were used as purchased.

40.2.2
Sample Preparation

Pyrene adsorbed on dry support were prepared as following: 10 mg of pyrene were mixed mechanically with 1 g of various supports (SiO_2, $CaCO_3$ and montmorillonite). The mixture of 1% weight ratio was placed on petri dishes covered with a polymer film transparent in the close UV and visible region at λ > 240 nm. The surface of the sample irradiated was 28 cm^2. Samples were placed under the lamp at about 15 cm from the source. For each kinetic study, a run was made with 6 to 7 samples. At a desired time, a sample was removed from the irradiation chamber and analysed.

In presence of water, the previous pyrene preparations were put into 50 ml of deionised water. The samples were also covered with a film and put in the dark for two hours in order to reach the equilibrium of pyrene partition between water and solid phase. The surface of the sample irradiated was 15.9 cm^2 with a 4 cm water column above the support.

Saturated aqueous solution of pyrene was prepared in deionised water at 0.135 mg l^{-1} after 24 h of stirring. Prior to the usage, the solution was filtered twice on a 5 μm nitrate of cellulose filter.

40.2.3
Irradiation

Irradiations were performed in a solar simulator SUNTEST CPS+ Atlas equipped with a xenon lamp containing a UV filter ($\lambda > 290$ nm). The incident irradiance was 765 W m^{-2} and the temperature was maintained by air-cooling at 35 ±3 °C, inside the apparatus.

40.2.4
Extraction Procedure

After irradiation, the residual amount of pyrene and photoproducts were extracted from the solid phase with 10 ml of acetone in an ultrasonic bath (47 kHz) during 10 min. The mixture was then centrifuged and diluted prior to analysis by gas chromatography (GC) or liquid chromatography (LC). Under these conditions, the extraction recovery of pyrene from the support was approximately 90 ±7%.

40.2.5
Analysis

Analyses of pyrene were performed on a Waters 486 LC system (LC: liquid chromatography), using a C18 Nucleosil column (5 µm, 150 × 4.6 mm) with an acetonitrile/water mixture (85/15, v/v) as mobile phase. The UV detector was fixed at 242 nm. The photoproducts were identified by an Agilent 6890 GC/MS system (GC/MS: gas chromatography-mass spectrometry), equipped with a HP5 column (0.25 µm, 30 m × 0.32 mm) with a constant oven temperature of 190 °C and an injector temperature fixed at 250 °C.

Fluorescence emission spectra of pyrene solution and pyrne on solid supports were plotted at $\lambda_{exc.} = 336$ nm, on a Perkin Elmer LS5 spectrometer.

40.3
Results and Discussion

40.3.1
Kinetics of Pyrene Photodegradation on Sorbed Phase

The kinetics and the rate constants of the photodegradation of pyrene sorbed on SiO$_2$, CaCO$_3$ and montmorillonite were evaluated both with pyrene crystals and pyrene in aqueous solution. Experiments were carried out on both dry support and in presence of a water column above the support.

The data presented on Fig. 40.1 show an apparent zero order law for the pyrene degradation on dry SiO$_2$, CaCO$_3$, montmorillonite and pyrene crystals. The kinetic law of these photochemical reactions is in agreement with the fact that in all cases pyrene is present under its crystalline form at the surface of the support. The emission spectra of 1% weight ratio pyrene sorbed on SiO$_2$, CaCO$_3$ and montmorillonite is similar to that of pyrene crystals with a wide fluorescence band between 440–520 nm (excimer form) whereas, pyrene in dilute aqueous solution (0.101 mg l^{-1}) shows an emission band at 360–420 nm (molecular form).

Fig. 40.1.
Kinetic of pyrene conversion on solid supports, and degradation of the pyrene crystalline form

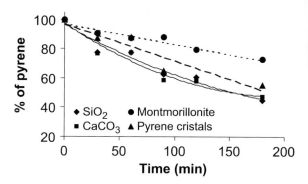

Table 40.1. Rate constant of pyrene photodegradation on dry supports and in presence of water, and intensity of light diffused at the solid surface

Solid support	Rate constant k_1 dry support (mol g^{-1} min^{-1})	Rate constant k_2 support in water (mol g^{-1} min^{-1})	k_1/k_2	$I_{em.}$ of light diffused (a.u) dry support $\lambda_{exc.}$ = 350 nm, $\lambda_{em.}$ = 350 nm
SiO$_2$	1.4 × 10^{-7}	4.3 × 10^{-8}	3.3	636
CaCO$_3$	1.4 × 10^{-7}	4.3 × 10^{-8}	3.3	688
Montmorillonite	0.7 × 10^{-7}	5.9 × 10^{-10}	36.8	186
Pyrene crystals	1.0 × 10^{-7} (mol min^{-1})	–	–	–

As compared to the photodegradation of pyrene crystals, the sorption of pyrene on montmorillonite decreases the degradation rate constant whereas, faster rate constants are obtained with SiO$_2$ and CaCO$_3$. In accordance with the literature data (Behimer and Hites 1988), these results can be related to the color of the different supports. A rapid pyrene degradation on SiO$_2$ and CaCO$_3$, two white supports, is observed while the degradation is slower on the grey montmorillonite (Table 40.1). A relation of the light reflectance at the solid surface and the photolytic degradation rate can be suggested. Results of light diffusion carried out with the fluorescence spectrometer at 350 nm support this assumption and are given in Table 40.1.

When the solid phase is in water, there is a partition of pyrene in the two phases, solid and water. Hence, after separation, the removal of pyrene was measured in both phases. Kinetics of pyrene still adsorbed on supports (not shown) follow a zero order law similar to the degradation kinetics observed on dry supports. Nevertheless, the rate constants are quite lower as mentioned in Table 40.1. The decrease of rate constants of pyrene on the support in water may be attributed to lower oxygen concentration in water as compared to the air, and also to a scattering of the light emitted by the lamp by solid particles suspended in the water column.

In the aqueous phase, pyrene is first solubilized until its concentration reaches the limit of solubility in water, about 0.10 mg l^{-1}. As indicated on Fig. 40.2, pyrene is then photochemically degraded in solution according to an initial first order law.

A residual concentration of pyrene still remains after 100 min of irradiation in water at about 0.02 mg l^{-1}. This can be interpreted as an equilibrium between the rate

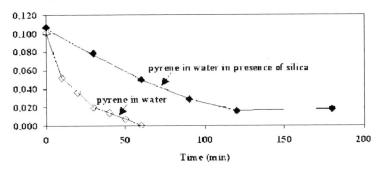

Fig. 40.2. Kinetics of pyrene degradation in water and in water in presence of silica

of pyrene desorption/solubilization from the support and the rate of photodegradation of pyrene in water. In water alone, the kinetic rate is faster ($k = 9.2 \times 10^{-4}$ s^{-1}) than it is in water in presence of silica. This can be explained by the previous phenomenon and partly because water is a transparent medium (transmission at 300 nm = 100%) compared to the aqueous phase, where silica slightly decreases the transmission of light (transmission at 300 nm = 97%).

An important parameter characterizing the environmental photostability of a compound is its half-life time. Calculations from kinetic data give the following half-life times for pyrene sorbed on dry supports with an incident irradiation of 765 W m^{-2}: 3 h on SiO_2, 3 h on $CaCO_3$, and 6 h on montmorillonite whereas, on wet support $t_{1/2}$ = 9.9 h, 9.9 h, and 221 h, respectively. In the aqueous phase and with the crystalline form, the half-lives of pyrene are respectively 0.3 h and 2.9 h. The photochemical activity of pyrene is enhanced in water (molecular state) in comparison to pyrene in the adsorbed state (crystalline form). This behavior can be explained by an inner filter effect as well as by a concentration effect.

40.3.2
Identification of Photoproducts

According to the kinetic studies and to the brown color developed at the surface of the supports after 1 h of irradiation, pyrene is degraded into further photoproducts. LC and GC/MS analyses revealed that the major photoproducts are more polar than pyrene. Five photoproducts have been identified with the help of their mass spectra in comparison with the literature data (Table 40.2). 1-Hydroxypyrene was identified by comparison with the retention times of a standard in LC, with its UV spectrum and its mass spectrum. In the absence of commercial compounds, the substitution position of carbonyl and hydroxyl functions (1,6- and/or 1,8-) was deducted by comparison of the experimental Rf values on TLC (cyclohexane/ethyl acetate 0.75/0.25) to the experimental LC chromatogram (Fig. 40.3) with data presented in the literature (Launaen et al. 1995; Mao et al. 1994) (TLC: thin layer chromatography). Except for 1,1'-bipyrene, the dimer of pyrene, all the other compounds are byproducts of photooxidation processes which involves O_2.

Table 40.2. Main pyrene photoproducts detected

Molecules detected	Molecular ion ($m\,z^{-1}$)	Mass spectral data
1-hydroxypyrene	218	189 (-COH), 163
4,5-dihydro-4,5-dihydroxypyrene	236	218, 205, 189, 176
1,6 and/or 1,8-dihydroxypyrene	234	205 (-COH), 176 (-2COH)
1,6 and/or 1,8-pyrenequinone	232	204 (-CO), 176
1,1'-bipyrene	402	356, 327, 291

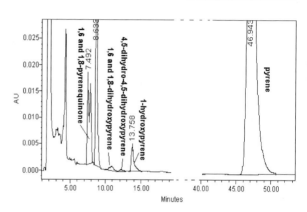

Fig. 40.3. Separation of photoproducts by liquid chromatography

40.3.3 Mechanism of Pyrene Photolysis

In order to assess the role of the singlet molecular oxygen O_2 ($^1\Delta_g$) in the photolytic degradation process at the solid surface, pyrene was irradiated in presence of rose bengal (RB), a well-known sensitizer, which leads to the formation of singlet oxygen. The irradiation was carried out between 450 and 570 nm affected RB but not pyrene. As described in the literature (Reyes et al. 2000), RB is coadsorbed with pyrene on the support, and after the photolytic generation of singlet oxygen (a,b), pyrene is oxidized in a final step (c).

$$RB + h\nu \longrightarrow RB^* \qquad (a)$$

$$RB^* + O_2 \longrightarrow RB + O_2^* \qquad (b)$$

$$O_2^* + \text{pyrene} \longrightarrow \text{oxidized products} \qquad (c)$$

Experiments carried out on SiO_2 in presence of RB and pyrene, have shown a very low degradation yield of pyrene, 7% after 4 h of irradiation. Therefore, singlet molecular oxygen does not play a significant role during the photodegradation.

In order to characterize the excited state of pyrene involved in the mechanism of pyrene phototransformation, $HgCl_2$ was used as a quencher of cross intersystem trans-

Box 40.1. Proposed mechanism of photodegradation of pyrene sorbed on SiO_2

$$\text{pyrene} \xrightarrow{h\nu} \text{pyrene}^* + e^{o-} \xrightarrow{O_2, SiO_2} O_2^{o-}, SiO_2^{o-} \xrightarrow{H_2O} \text{pyrene-OH} \cdots$$

(reaction scheme showing formation of hydroxypyrene, then quinone intermediates via H_2O / O_2 or SiO_2, leading to dihydroxypyrene products)

fer of pyrene $S_1 \longrightarrow T_1$ (Sigman et al. 1998). The increase of the photodegradation rate of pyrene in presence of $HgCl_2$ (91% of degraded pyrene with $HgCl_2$ on SiO_2 and 40% without $HgCl_2$ on SiO_2, after 4 h of irradiation) shows that pyrene is more easily degraded when the singlet state S_1 of pyrene is the only one state involved.

The mechanisms of pyrene photooxidation at the solid surface and in solution which are described in the literature, show that the pyrene radical cation is the first species formed (Sigman et al. 1998; Lawrence et al. 1995). Then, the photodegradation of pyrene proceeds predominantly by an electron transfer mechanism between pyrene and O_2 and/or between pyrene and support (Surapol and Thomas 1991; David and Boule 1993). Accordingly, the following oxidation pathway consistent with the photoproducts detected is suggested (Box 40.1).

40.4 Conclusion

Pyrene is phototransformed by the solar light at the surface of all the supports studied. The rate of degradation depends mainly on the intensity of light diffused at the surface of the support. Higher rates of degradation are observed with the white supports having the highest intensity of light diffused. In presence of water, the photolysis can occur in both water and solid phase but the rate of pyrene degradation is quite slowed down. A mechanism for the photooxidation of pyrene is proposed with the formation of a pyrene radical cation associated with an electronic transfer on the support or with O_2 molecular, which has to be confirmed by flash photolysis studies. This study will be continued in order to improve our knowledge about the photochemical behavior of pyrene on natural sediments.

Acknowledgements

The authors are grateful to CNRS for its financial support, Programme Environnement Vie et Société.

References

Barbas JT, Sigman ME, Dabestani R (1996) Photochemical of phenanthrene sorbed on silica gel. Environ Sci Technol 30:1776–1780

Behimer TD, Hites RA (1988) Photolysis of polycyclic aromatic hydrocarbons adsorbed on fly ash. Environ Sci Technol 22:1311–1319

Cope VW, Kalkwarf DR (1987) Photooxidation of selected polycyclic aromatic hydrocarbons and pyrenequinones coaled on glass surfaces. Environ Sci Technol 21:643–648

David B, Boule P (1993) Phototransformation of hydrophobic pollutants in aqueous suspension. Chemosphere 26:1617–1630

Jones KC, Stratford JA, Waterhouse KS, Furlong ED, Giger W, Hites RA, Schaffner C, Johnston AE (1989) Increase in the PAH content of an agricultural soil over the last century. Environ Sci Technol 23:95–101

Korfmacher WA, Wehry EL, Mamantov G, Natush DFS (1980) Resistance to photochemical decomposition of polycyclic aromatic hydrocarbons vapor-adsorbed on coal fly ash. Environ Sci Technol 14:1094–1098

Launen L, Pinto L, Wiebe C, Kieklmann E, Moore M (1995) The oxidation of pyrene and benzo[a]pyrene by nonbasidiomycete soil fungi. Can J Microbiol 41:477–488

Lawrence JF, Weakland S, Plummer EF, Busby WF, Lafleur AL (1995) Photoxidation of selected PAHs in aqueous organic media in the presence of Ti(IV)oxide. Int J Environ Anal Chem 60:113–122

Liste H, Alexander M (1999) Accumulation of phenanthrene and pyrene in rhizosphere soil. Chemosphere 40:11–14

Mao Y, Iu KK, Thomas JK (1994) Chemical reaction of pyrene and its chlorinated derivative on silica-alumina surfaces induced by ionizing radiation. Langmuir 10:709–716

Prado G, Westmoreland PR, Andon PH, Leary JA, Biemann K, Thilly WG, Longwell JP, Howard JB (1981) Formation of polycyclic aromatic hydrocarbons in premixed flames. Chemical analysis and mutagenicity. In: Cooke M, Dennis AJ (eds) Chemical analysis and biological fate: PAHs. Batelle, Ohio, pp 189–198

Reyes CA, Medina M, Crespo-Hernandez C, CedenoMZ, Arce R, Rosario O, Steffenson DM, Ivanov IN, Sigman ME, Dabestini R (2000) Photochemistry of pyrene on unactived and actived silica surfaces. Environ Sci Technol 34:415–421

Schneider J, Grosser R, Jayasimhulu K, Xue W, Warshawsky D (1996) Degradation of pyrene, benz[a]anthracene, and benzo[a]pyrene by *Mycobacterium* sp. strain RJGII-135, isolated from a former coal gasification site. Appl Environ Microbiol 62:13–19

Sharak Genthner BR, Townsend GT, Lantz SE, Mueller JG (1997) Persistence of PAH components of creosote under anaerobic enrichment conditions. Arch Environ Contam Toxicol 32:99–105

Sigman ME, Schuler PF, Ghosh MM, Dabestani RT (1998) Mechanism of pyrene photochemical oxidation in aqueous and surfactant solutions. Environ Sci Technol 32:3980–3985

Smith JH, Mabey WR, Bahonos N, Holt BR, Chou TW, Lee SS, Wenberger, DC Mill T (1979) Environmental pathways of selected chemicals in fresh water system. US Environmental Protection Agency, EPA-600/7-78-074

Sparks DL (1995) Inorganic soil components. In: Sparks DL (ed) Environmental Soil Chemistry. Academic Press, London, pp 23–38

Surapol P, Thomas K (1991) Reflectance spectroscopic studies of the cation radical and the triplet of pyrene on alumina. J Phys Chem 95:6990–6996

Wislocki PG, Bagan ES, Lu AYH (1986) Tumorigenicity of nitrated derivatives of pyrene, benzo[a]anthracene, chrysene and benzo[a]pyrene in the newborn mouse assay. Carcinogenesis 7:1317–1322

Yokley RA, Garrison AA, Wehri EL, Mamantov G (1986) Photochemical transformation of pyrene and benzo[a]pyrene vapor-deposited on eight coal stack ashes. Environ Sci Technol 20:86–90

Zepp RG, Schlotzhauer PF (1979) Photoreactivity of selected aromatic hydrocarbons in water. In: Jones PW, Leber P (ed) Polynuclear aromatic hydrocarbons. Ann Arbor Science Publishers, pp 141–158

Chapter 41

Degradation of Polycyclic Aromatic Hydrocarbons in Sewage Sludges by Fenton's Reagent

V. Flotron · C. Delteil · A. Bermond · V. Camel

Abstract

The aim of this study was to investigate the use of Fenton's reagent for the degradation of polycyclic aromatic hydrocarbons (PAHs) in sewage sludges. As Fenton's reagent generates hydroxyl radicals that further oxidise pollutants in the solution, our efforts focused on finding experimental conditions that would ensure the stability of aqueous PAH solutions. The use of Teflon recipients along with the addition of a co-solvent were required to limit adsorption effects. A non ionic surfactant (Brij-35) was found to be a more suitable co-solvent instead of using an organic solvent as it minimises hydroxyl radical consumption. However it was not efficient in desorbing PAHs contained in sludge samples after 24 h contact time, even at concentrations above its critical micellar concentration.

Key words: Fenton's reagent; oxidation; polycyclic aromatic hydrocarbons; remediation; sewage sludges

41.1 Introduction

Polynuclear aromatic hydrocarbons (PAHs) are ubiquitous pollutants. Urban areas are important sources of PAHs due to domestic fuel combustion, industrial emissions, car exhausts and natural background atmospheric deposition. As a consequence, rainwater and waste water draining from urban areas into the sewerage system contain relatively large amounts of PAHs. Due to their hydrophobicity, PAHs are primarily adsorbed onto biomass during activated sludge treatment of the waste waters. They remain partly undegraded due to their biological recalcitrance (Wild et al. 1990; Manoli and Samara 1999). When sewage sludges are used as soil amendment, the presence of PAHs is of great concern due to their mutagenic and carcinogenic potential (Santodonato et al. 1981; Santodonato 1997). In France, recent regulation imposes a maximum acceptable limit for three PAHs in sewage sludges: 5.0 mg kg^{-1} d.w. for fluoranthene, 2.5 mg kg^{-1} d.w. for benzo[b]fluoranthene and 2.0 mg kg^{-1} d.w. for benzo[a]pyrene (see chemical structures in Fig. 41.1). Due to the toxicity of these compounds, a treatment capable of partially or completely detoxifying the sludges would be of interest (Perez et al. 2001). Because of the very low biodegradability of these organic pollutants, a chemical oxidative treatment has to be considered. It could be included in currently used sludge treatments, preferentially before the usual dehydration step. As biodegradable oxidation by-products may be generated, it could be followed by a biological treatment for the removal of such compounds. The combination of chemical oxidation and biodegradation would offer the additional benefit of reducing the volume of final sludges obtained.

Fig. 41.1.
Chemical structure of the investigated PAHs. Reported solubilities: 265, 1.5 and 4 µg l^{-1} for fluoranthene, benzo[b]fluoranthene and benzo[a]pyrene respectively (Lopez Garcia et al. 1992; Martens and Frankenberger 1995)

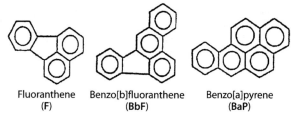

Fluoranthene (F)   Benzo[b]fluoranthene (BbF)   Benzo[a]pyrene (BaP)

Fenton's reagent (Fe(II)-H$_2$O$_2$) is very attractive for such treatment due to its moderate cost, simplicity of operation, and advanced oxidation potential as hydroxyl radicals OH˙ are formed according to Reaction 41.1 (Barb et al. 1951; Edwards and Curci 1992; Walling 1975; Wardman and Candeias 1996). Hydroxyl radicals are highly reactive species that have been shown to degrade many organic compounds, either by addition of an hydroxyl group (Reaction 41.2) or by hydrogen abstraction (Reaction 41.3). The presence of iron minerals in sludges may allow the Fenton oxidation to proceed without any addition of iron (Venkatadri and Peters 1993; Watts et al. 1999).

$$Fe^{2+} + H_2O_2 \longrightarrow Fe^{3+} + OH^- + OH˙ \tag{41.1}$$

$$R + OH˙ \longrightarrow ROH˙ \tag{41.2}$$

$$RH + OH˙ \longrightarrow R˙ + H_2O \tag{41.3}$$

This study was undertaken to investigate the feasibility of using Fenton's reagent to degrade PAHs in sewage sludge samples. This requires the control of three successive steps: *(1)* PAH desorption from the sludges, *(2)* stability of aqueous PAH solutions, *(3)* PAH oxidation by OH˙. Due to the high sorptive capacity and very low water solubility of PAHs, we studied the stability of PAH solutions, before investigating their oxidation by Fenton's reagent. An application to sludge samples was then performed.

41.2
Experimental

All experiments were conducted at room temperature and done in triplicate. The initial pH of the solutions was adjusted to 3 with H$_2$SO$_4$. Recipients were always covered with aluminium foil to avoid photolytic degradation of the PAHs. The percentage of PAHs adsorbed onto recipient walls was estimated by rinsing the recipient with acetonitrile during 2 h, and analysing this rinsing solvent.

41.2.1
Reagents and Chemicals

Stock solutions (10 mg l^{-1} in acetonitrile) of individual PAHs were obtained from CIL Cluzeau (purity: 97–99.7%). Analytical-reagent grade copper metal and nitric acid solution 68% were supplied by Prolabo, as well as hydrogen peroxide 30% by weight solution and HPLC grade methanol (MeOH), ethanol (EtOH), acetonitrile (ACN), *n*-hexane, dichloromethane and acetone. Iron(II) sulphate heptahydrate was supplied by Merck and the non ionic surfactant Brij-35 (polyoxyethylene lauryl ether) by Fluka. Deionised water (Milli-Q) was used. Stock solutions of Fe(II) and H_2O_2 were prepared daily at 0.2 mol l^{-1} and 0.4 mol l^{-1} respectively.

41.2.2
Sewage Sludge Samples

Sewage sludge samples were obtained from a municipal waste water plant near Paris. Upon reception at the laboratory, sludges were kept frozen to avoid any modification during storage. Before use, large aliquots (100 g) were taken, dried in an oven (40 °C, 24 h) before being homogenised with a mortar and sieved (at 2 mm). For the analysis of the PAH content, 1 g dried sample was extracted with 30 ml hexane-acetone (1/1, v/v) in a focused microwave-assisted extractor (Soxwave 100, Prolabo) at 30 W for 10 min. Extraction was performed in presence of 1 g activated copper bars to remove sulphur. Extracts were further filtered, concentrated to near 5 ml with a rotary evaporator, and finally to about 2 ml under a gentle stream of nitrogen. Clean-up was performed on disposable solid-phase extraction silica cartridges (Supelclean LC-Si, 1 g, Supelco) using a Visiprep vacuum manifold system (Supelco). Cartridges were conditioned with 5 ml n-hexane. PAHs were eluted using 4 ml n-hexane and 4 ml n-hexane/dichloromethane (1/1, v/v). After mixing both fractions, the solvent was completely evaporated under a gentle stream of nitrogen and the residue redissolved in 2 ml acetonitrile.

41.2.3
Liquid chromatography-Fluorescence Detection

A Varian 9010 pump equipped with a 20 µl loop connected to a Rheodyne injector 7125 was used, with a Thermo Separation Science FL3000 fluorimetric detector. Detection was performed at selected excitation and emission wavelengths: 230–410 nm for fluoranthene, 250–420 nm for benzo[b]fluoranthene and benzo[a]pyrene. A Supelco LC-PAH analytical column (250 × 4.6 mm i.d., 5 µm) was used with a pre-column. The flow-rate was 1.5 ml min^{-1} with the following gradient: 0–5 min 60% ACN/water, 5–30 min 60% ACN/water to 100% ACN, 30–45 min 100% ACN. External calibration was performed (in the range 25–200 µg l^{-1}) in different aqueous media.

41.2.4
Aqueous Solution Experiments

For PAH stability experiments, solutions containing the three PAHs (80 µg l^{-1}) were prepared in the different aqueous media tested. For Fenton oxidation experiments, solutions containing the three PAHs (80 µg l^{-1}) and Fe(II) (1.6 × 10^{-4} mol l^{-1}) were prepared. Fenton's reaction was started upon addition of H_2O_2 (3.2 × 10^{-4} mol l^{-1}). The initial [H_2O_2]/[Fe(II)] ratio was 2, with a large excess of reagents as the ratio [Fe(II)]/[Total PAHs] was around 155. This large excess was used to simulate further experimental conditions for Fenton sludge oxidation, taking into account the fact that the sludge organic matter will probably consume large amounts of OH·. The PAH content of solutions was followed over time by regularly withdrawing 1 ml, which was mixed with 0.25 ml methanol to quench radical reactions, before being neutralised (pH 6) with NaOH and then analysed.

41.2.5
Sludge Experiments

For Fenton oxidation experiments, 1 g sludge samples were put in contact with 5 ml aqueous solution (pH 3) containing Fe(II) (1.4×10^{-4} mol l^{-1}). Fenton's reaction was started upon addition of H$_2$O$_2$ (0.2 mol l^{-1}). The initial [H$_2$O$_2$]/[Fe(II)] ratio was 1 430. Stirring was ensured during all experiments. At the end of the chosen time, methanol (0.44 ml) was added to quench radicals. The mixture was then centrifuged, and the supernatant further centrifuged 15 min at 6 000 rpm. The solid residue was then analysed for its PAH content. For desorption experiments, 2 g sludge samples were put in contact with 20 ml aqueous solution (pH 3) of Brij-35 at three different concentrations in Teflon vessels. The slurry was mixed by head-over-tail shaking. After the desired time, the supernatant was centrifuged 15 min at 6 000 rpm. The solid residue was then analysed for its PAH content.

41.3
Results and Discussion

41.3.1
Stability of PAH Aqueous Solutions

PAH concentrations in standard aqueous solutions were followed over time. As shown in Fig. 41.2, a decrease was noted for all PAHs, especially for the more hydrophobic, benzo[b]fluoranthene and benzo[a]pyrene. This was due to adsorption of PAHs onto the glass walls of the recipient, as PAHs could be recovered after rinsing the recipient with acetonitrile. With polypropylene recipients the disappearance of the PAHs was even worse (data not shown) as already observed with polyethylene material (Lopez Garcia et al. 1992; Pinto et al. 1994). The use of Teflon recipients resulted in a reduction in PAH adsorption. However adsorption effects were still observed for long contact times (nearly 35 and 50% for benzo[a]pyrene and benzo[b]fluoranthene respec-

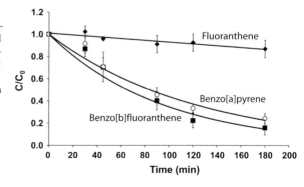

Fig. 41.2. Effect of time on the PAH concentrations in water contained in glass vessels. C_0: initial concentration, C: concentration at time t. Note that the decrease of the PAH concentrations can be explained by their adsorption onto glass surfaces

tively after 180 min), in agreement with a previous study reporting such adsorption effects in Teflon material (Lopez Garcia et al. 1992). In addition, the repeatability of the experiments was quite poor, probably due to the formation of colloids by hydrophobic interactions (Li and Lee 2001).

The addition of a co-solvent has been tested as a mean of increasing PAH solubility, thereby reducing their adsorption. So, the same experiments were conducted in aqueous solutions containing either an organic solvent (MeOH, EtOH and ACN) or a non ionic surfactant (Brij-35). Results are presented in Table 41.1. Methanol at 20% was insufficient to completely prevent adsorption, which is consistent with previous results reporting that 25% methanol was not sufficient for the high molecular weight PAHs (Brouwer et al. 1994). A higher percentage of methanol (i.e. 40%) was in fact necessary to avoid adsorption. On the other hand, due to its longer alkyl chain, 20% ethanol was almost sufficient. Acetonitrile, due to its π-π interactions with PAHs, was even more efficient. This complements a previous study that recommended to use acetonitrile-water 40:60 to avoid any adsorption effects in either borosilicate glass or Teflon recipients (Lopez Garcia et al. 1992). With all these organic solvents, repeatability was satisfactory as hydrophobic interactions were greatly reduced, leading to quite homogeneous solutions. Similar results were observed upon addition of the non ionic surfactant (Brij-35) at low concentrations (below its critical micellar concentration which is 10^{-4} mol l^{-1}), in agreement with previous studies (Brouwer et al. 1994; Lopez Garcia et al. 1992).

41.3.2
Fenton Degradation of PAHs in Aqueous Solutions

Since a co-solvent is added for avoiding adsorption effects over time, its effect in presence of Fenton's reagent needs to be investigated as it may compete with the PAHs in consuming hydroxyl radicals as indicated by Reactions 41.4 and 41.5.

$$PAH + OH^{\cdot} \longrightarrow PAH_{oxidised} \qquad (41.4)$$

$$CoS + OH^{\cdot} \longrightarrow CoS_{oxidised} \qquad (41.5)$$

It has been reported that reaction rate constants of $OH^{\cdot}$ with ethanol, methanol and acetonitrile are respectively 2.1×10^9, 1.2×10^9 and 6×10^6 mol^{-1} l s^{-1} (Edwards and Curci 1992). The low rate constant for acetonitrile is due to the presence of a strong electron attracting group in the molecule. In the same way, the presence of electron donating groups enhances reactivity. In the case of PAHs, the rate constant is assumed to be near 10^{10} mol^{-1} l s^{-1} due to the high electronic density of these polyaromatic molecules (Haag and Yao 1992). Even though the co-solvents used have lower reactivities than PAHs, they will compete for $OH^{\cdot}$ as they are present in large excess, compared to the very low PAH concentrations in the solutions.

The results of Fenton experiments conducted in different aqueous media are reported in Table 41.2. Solubilised PAHs were directly analysed by HPLC, while adsorbed PAHs were quantified after rinsing the recipient with acetonitrile and analysis of this rinsing solvent. Oxidised PAHs were estimated based on difference between the ini-

Table 41.1. Effect of the solvent composition on the recovery (%) of PAHs. PAHs were solubilised in different aqueous media stirred in Teflon recipients for 180 min. The concentration of solubilised PAH was analysed at the beginning of the experiments (C_0) and after 180 min (C). The recovery (%) is given as $100 \times C/C_0$ (RSD: relative standard deviation)

PAH	Aqueous medium													
	H_2O		H_2O/MeOH 80/20 v/v		H_2O/MeOH 60/40 v/v		H_2O/EtOH 80/20 v/v		H_2O/ACN 80/20 v/v		H_2O/Brij-35 10^{-5} M		H_2O/Brij-35 5×10^{-5} M	
	Mean value	RSD	Mean value	RSD	Mean value	RSD	Mean value	RSD	Mean value	RSD	Mean value	RSD	Mean value	RSD
Fluoranthene	90.5	9.9	100.9	17.7	105.4	3.3	98.1	1.3	101.4	3.6	87.2	3.1	94.9	2.9
Benzo[b]fluoranthene	47.4	25.9	60.4	1.5	100.0	1.2	94.5	4.5	99.8	4.2	78.8	18.2	98.1	3.7
Benzo[a]pyrene	62.6	16.5	72.1	4.4	99.5	0.9	92.0	5.0	100.1	3.5	86.4	8.6	87	10.2

Table 41.2. Effect of the solvent composition on the behaviour of PAHs in presence of Fenton's reagent. PAHs were solubilised in different aqueous media stirred in Teflon recipients for 180 min. The percentage of PAH solubilised was evaluated by analysing the solution. The percentage of PAH adsorbed was estimated after rinsing the recipient with acetonitrile and analysing this rinsing solvent. The percentage of PAH oxidised was determined by difference with the initial concentration

PAH	Aqueous medium								
	H_2O		$H_2O/MeOH$ (60/40 v/v)		$H_2O/EtOH$ (80/20 v/v)		$H_2O/Brij-35$ (10^{-5} M)	
	Mean value	RSD	Mean value	RSD	Mean value	RSD	Mean value	RSD
Solubilised								
Fluoranthene	62.6	3.2	101.4	1.6	109.2	14.1	75.4	2.9
Benzo[b]fluoranthene	11.9	111.9	98.0	4.3	98.9	7.1	92.7	8.4
Benzo[a]pyrene	6.0	87.3	78.1	0.9	81.9	2.0	14.1	4.0
Adsorbed								
Fluoranthene	6.7	86.0	0	0	0	0	2.4	16.7
Benzo[b]fluoranthene	74.3	10.8	4.6	3.1	6.7	0.4	13.9	8.6
Benzo[a]pyrene	8.9	88.6	6.4	8.5	8.8	0.2	0.2	175.0
Oxidised								
Fluoranthene	30.7	18.1	0	0	0.8	141.7	22.2	17.8
Benzo[b]fluoranthene	17.2	87.9	0	0	0	0	0	0
Benzo[a]pyrene	85.2	15.3	15.6	7.8	9.3	20.1	85.7	1.1

tial PAH content and the solubilised and adsorbed proportions. From these results it appears that the addition of an organic solvent prevents PAH oxidation due to strong competition (ACN was not studied in detail due to the production of a very odorous product in Fenton experiments). On the contrary, the presence of the non ionic surfactant in the solution has little influence, with degradation rates similar to those obtained in water. These results show that Brij-35 does not react rapidly with OH˙, so that it may be used in a PAH degradation medium to minimise adsorption effects. Nevertheless, future experiments should be performed to investigate the possible formation of by-products upon oxidation of the surfactant. In addition, experimental conditions will be optimised to ensure both PAH and surfactant advanced oxidation in order to avoid subsequent disposal of surfactant solutions, as obtained in the case of soil samples (Saxe et al. 2000).

It is interesting to note that PAHs do not react identically with the hydroxyl radicals, even though OH˙ is considered to be a non selective oxidative species. Whereas benzo[a]pyrene is readily degraded, the oxidation of fluoranthene is limited, while benzo[b]fluoranthene is hardly degraded. Similar discrepancies in PAH reactivity towards OH˙ have also been reported recently, with pyrene and benzo[a]pyrene having the highest reactivity (Kelley et al. 1991; Nam et al. 2001). The presence of a 5-carbon ring in the structure seems to inhibit the attack of OH˙ radicals. This is consistent with the mechanism of attack of OH˙ on aromatic compounds which is analogous to an electrophilic substitution (Anbar et al. 1966; Edwards and Curci 1992). In the case of the most readily oxidised compound (i.e. benzo[a]pyrene), comparison of the disappearance observed in the different media shows that similar efficiencies were obtained in pure water and Brij-35 solutions (see Fig. 41.3). An attempt was made to estimate the corresponding kinetic rate constants. Considering Reaction 41.4, the following equation describes the disappearance of benzo[a]pyrene:

$$-d[BaP]/dt = k[OH˙][BaP]$$

In the early stage of the experiments, a steady-state OH˙ concentration is assumed (the limiting step being Reaction 41.1), so that one can express the disappearance as a pseudo-first order reaction rate, leading to the following equation:

$$\mathrm{Ln}\,[BaP]/[BaP]_0 = -k_{exp} t$$

The curves of $\mathrm{Ln}\,C/C_0$ for that pollutant vs. reaction time were plotted. In presence of methanol or ethanol, a pseudo-first order reaction was observed during the first 10 min with a very low estimated kinetic constant (around 0.017 min^{-1}) due to the strong scavenging effect of the organic solvent present in the medium. In Brij-35 aqueous solution, a pseudo-first order reaction was observed during the first 10 min, with an approximate kinetic constant of 0.144 min^{-1}, compared with 0.185 min^{-1} found in pure water solutions. This result is in agreement with a previous study reporting a value of 0.183 min^{-1} in the first 30 min of Fenton's oxidation for the same pollutant (Lee and Hosomi 2001). The shorter pseudo-first order phase observed in our study may be due to the low initial concentration of the benzo[a]pyrene, which results in its rapid disappearance and in a non-steady state for OH˙ after a short period of time.

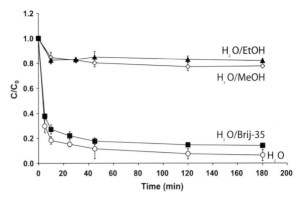

Fig. 41.3.
Effect of time on the benzo[a]pyrene concentration in different aqueous media containing Fenton's reagent (Teflon vessels). C_0: initial concentration, C: concentration at time t. Note that the disappearance of BaP can be explained by its OH· oxidation

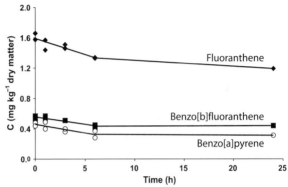

Fig. 41.4.
Effect of time on PAH concentrations of sludge samples submitted to Fenton's reagent

41.3.3
Application to Sewage Sludge Samples

Experiments were undertaken to study the behaviour of PAHs contained in sludge samples treated by Fenton's reagent. For that purpose, sludge samples were put in contact with an aqueous solution (pH 3) containing Fe(II) and H_2O_2, with continuous stirring. The PAH concentration of the sludge was then analysed over time. As indicated in Fig. 41.4, a decrease in PAH content of the sludge was observed in the first 6 h, which showed that partial desorption of the compounds occurred. However this desorption was quite limited due to the low water solubility of PAHs. In addition, as the performance of blank experiments (i.e. sludge slurry without Fenton's reagent) led to similar disappearance of PAHs in the sludge, we suspected that desorbed PAHs were not affected by Fenton's reagent under the tested conditions.

Additional experiments were conducted to enhance PAH desorption from the sludge. For that purpose, a co-solvent was added to the aqueous solution. As Brij-35 was previously found to be a suitable medium for Fenton oxidation of dissolved PAHs, we tested its efficiency in desorbing PAHs from sludge samples. No significant PAH desorption could be achieved up to 24 h whatever the Brij-35 concentrations tested (5×10^{-5}, 10^{-4} and 10^{-3} mol l^{-1}). These results were unexpected as non ionic surfactants have been reported to allow desorption of organic pollutants, such as naphthalene, from soil and sediment matrices (Pramauro et al. 1998). However, in our case the highly hydrophobic character of the sludge may have led to a much slower desorption of the PAHs and/or to the adsorption of the surfactant on the matrix. Therefore, longer extraction times, higher concentrations, and/or a more efficient surfactant need to be tested. In particular, the efficiency of biosurfactants will be studied.

41.4
Conclusions

This study shows that careful optimisation of experimental conditions is required to degrade PAH in sewage sludges with Fenton's reagent. In particular, the use of Teflon recipients and the addition of a co-solvent are required to limit adsorption effects. Ethanol and a non ionic surfactant (Brij-35) should be recommended for that purpose, the latter being more suitable due to a lower scavenging effect towards OH˙. We found different reactivity towards OH˙ for the three PAHs tested, benzo[a]pyrene being rapidly degraded while fluoranthene, and especially benzo[b]fluoranthene, were less degraded. Future experiments will therefore require optimisation of the Fenton conditions (especially the ratio H_2O_2/Fe(II)) to degrade all PAHs. In addition, experimental conditions should be found to enhance PAH desorption from the sludge. Finally, the effect of both the desorption and the oxidation steps on the mobility of metals contained in the sludge will be studied.

References

Anbar M, Mereystein D, Neta P (1966) The reactivity of aromatic compounds toward hydroxyl radicals. J Phys Chem 70:2660–2662

Barb WG, Baxendale JH, George P, Hargrave KR (1951) Reactions of ferrous and ferric ions with hydrogen peroxide. Part I: The ferrous ion reaction. Trans Faraday Soc 47:462–500

Brouwer ER, Hermans ANJ, Lingeman H, Brinkman UATh (1994) Determination of polycyclic aromatic hydrocarbons in surface water by column liquid chromatography with fluorescence detection, using an on-line micelle-mediated sample preparation. J Chromatogr A 669:45–57

Edwards JO, Curci R (1992) Fenton type activation and chemistry of hydroxyl radical. In: Strukul G (ed) Catalytic oxidations with hydrogen peroxide as oxidant. Kluwer Academic Publishers, The Netherlands, pp 97–151

Haag WR, Yao CCD (1992) Rate constants for reaction of hydroxyl radicals with several drinking water contaminants. Environ Sci Technol 26:1005–1013

Kelley RL, Gauger WK, Srivastava VJ (1991) Application of Fenton's reagent as a pretreatment step in biological degradation of polyaromatic hydrocarbons. Gas Oil Coal Environ Biotechnol 3:105–120

Lee BD, Hosomi M (2001) Fenton oxidation of ethanol-washed distillation-concentration benzo[a]pyrene: reaction product identification and biodegradability. Water Res 35:2314–2319

Li N, Lee HK (2001) Assessment of colloid formation and physical state distribution of trace polycyclic aromatic hydrocarbons in aqueous samples. Anal Chem 73:5201–5206

Lopez Garcia A, Blanco Gonzalez E, Garcia Alonso JI, Sanz-Medel A (1992) Potential of micelle-mediated procedures in the sample preparation steps for the determination of polynuclear aromatic hydrocarbons in waters. Anal Chim Acta 264:241–248

Manoli E, Samara C (1999) Occurrence and mass balance of polycyclic aromatic hydrocarbons in the Thessaloniki sewage treatment plant. J Environ Qual 28:176–187

Martens DA, Frankenberg WT (1995) Enhanced degradation of polycyclic aromatic hydrocarbons in soil treated with an advanced oxidative process – Fenton's reagent. J Soil Contamination 4:175–190

Nam K, Rodriguez W, Kukor JJ (2001) Enhanced degradation of polycyclic aromatic hydrocarbons by biodegradation combined with a modified Fenton reaction. Chemosphere 45:11–20

Perez S, Farré M, Garcia MJ, Barcelo D (2001) Occurrence of polycyclic aromatic hydrocarbons in sewage sludge and their contribution to its toxicity in the ToxAlert® 100 bioassay. Chemosphere 45:705–712

Pinto CG, Pavon JLP, Cordero BM (1994) Cloud point preconcentration and high-performance liquid chromatographic determination of polycyclic aromatic hydrocarbons with fluorescence detection. Anal Chem 66:874–881

Pramauro E, Prevot AB, Vincenti M, Gamberini R (1998) Photocatalytic degradation of naphthalene in aqueous TiO_2 dispersions: effect of nonionic surfactants. Chemosphere 36:1523–1542

Santodonato J (1997) Review of the estrogenic and antiestrogenic activity of polycyclic aromatic hydrocarbons: relationship to carcinogenicity. Chemosphere 34:835–848

Santodonato J, Howard PH, Basu DK (1981) Health and ecological assessment of polynuclear aromatic hydrocarbons. J Environ Pathol Toxicol 5:1–364

Saxe JK, Allen HE, Nicol GR (2000) Fenton oxidation of polycyclic aromatic hydrocarbons after surfactant-enhanced soil washing. Environ Engng Sci 17:233–244

Venkatadri R, Peters RW (1993) Chemical oxidation technologies: ultraviolet light/hydrogen peroxide, Fenton's reagent, and titanium dioxide-assisted photocatalysis. Haz Waste & Haz Mater 10:107–149

Walling C (1975) Fenton's reagent revisited. Acc Chem Res 8:125–131

Wardman P, Candeias LP (1996) Fenton chemistry: an introduction. Radiat Res 145:523–531

Watts RJ, Udell MD, Kong S, Leung SW (1999) Fenton-like soil remediation catalyzed by naturally occurring iron minerals. Environ Engng Sci 16:93–103

Wild SR, McGrath SP, Jones KC (1990) The polynuclear aromatic hydrocarbon (PAH) content of archived sewage sludges. Chemosphere 20:703–716

Part V
Pesticides

Part V

Chapter 42

Pesticide Mobility Studied by Nuclear Magnetic Resonance

B. Combourieu · J. Inacio · C. Taviot-Guého · C. Forano · A. M. Delort

Abstract

The adsorption-desorption mechanisms at the interface between organic and inorganic soil colloids influence the movement of pesticides, and in turn their bioavailability. Here we demonstrate the potential of high resolution magic angle spinning nuclear magnetic resonance (^{1}H HR-MAS) to assess in-situ interactions of pesticides at the solid aqueous interface. We used layered double hydroxides (LDH) also called anionic clays, due to their environmental relevance as soil models and cleaner materials for decontamination processes. We studied the adsorption behaviour of the pesticide 4-chloro-2-methylphenoxyacetic acid (MCPA) on anionic MgAl-clays. Our findings demonstrate that the mobile and immobile fractions of the pesticide can be unambiguously distinguished by ^{1}H high resolution magical angle spinning nuclear magnetic resonance (^{1}H-HR-MAS-NMR).

Key words: bioavailability, pesticides, MCPA, adsorption, layered double hydroxides, anionic clay, ^{1}H HR-MAS NMR

42.1 Introduction

The fate of pesticides in the environment depends greatly on adsorption–desorption mechanisms that take place at the interface between organic and inorganic soil colloids. These processes control transport and mobility of pesticides and in turn their bioavaibility for plant or microbial uptake and degradation.

Pesticide adsorption on cationic clays, the major mineral component in soils, has been widely studied (Bailey et al. 1968; Hermosin et al. 1993; Laird et al. 1994; Haderlein et al. 1996; Filomena et al. 1997; Sawhney and Singh 1997; Celis et al. 1998; Aguer et al. 2000; Cox et al. 2000). Anionic clays, on the other hand, are rare in nature but easy to prepare. The structure of these lamellar compounds is derived from that of brucite $Mg(OH)_2$ (Allman 1968). They are built by stacking of positive $[M^{2+}_{1-x}M^{3+}_x(OH)_2]^{x+}$ layers, separated by interlayer domains containing exchangeable hydrated anions $[X^{m-}_{x/m} \cdot nH_2O]$. These anionic exchangers are characterized by the general formula:

$$[M^{2+}_{1-x}M^{3+}_x(OH)_2]^{x+}[X^{m-}_{x/m} \cdot nH_2O]$$

abbreviated as $[M^{2+}M^{3+}X]$ (de Roy 1998; Vaccari 1999) and can be considered as the anti-type of 2:1 smectites. A wide range of composition can be obtained for these materials by changing the nature of the metal cations, the cation M^{II}/M^{III} ratios thus the layer charge density ($0.20 \leq x \leq 0.33$) as well as the nature of the intercalated anion X_m^-. Due to their high anion exchange capacities, they represent promising materials for the removal of toxic inorganic and organic anions from waste water. Few papers have reported the use of LDH as organic contaminant adsorbents, for instance, the adsorption of phenolic compounds (Hermosin et al. 1993; Ulibarri et al. 1995; Hermosin et al. 1996) or pesticides on organic modified clays (Celis et al. 1999; Villa et al. 1999). Sorption studies recently demonstrated that pesticides of the chlorophenoxy carboxylate family [2,4-dichlorophenoxyacetate (2.4-D), 2,4,5-trichlorophenoxyacetate (2.4.5-T), 4-chloro-2-methylphenoxyacetic acid (MCPA)] display very strong affinities to anionic clays. (Lakraïmi et al. 2000; Inacio et al. 2001).

Nuclear magnetic resonance (NMR) is a powerful method to assess interactions of organic xenobiotics with soil organic matter and minerals, e.g. clays. The various following approaches have been developed. First, liquid state nuclear magnetic resonance (NMR) has been used to study the interaction of pollutants with dissolved humic or fulvic acids (Nanny 1998, 1999; Herbert and Bertsch 1998; Bortiatynsky et al. 1998; Nanny and Maza 2001). Here, chemical shifts and longitudinal relaxation times T_1 of isotopically-tagged organic compounds (^{13}C, ^{15}N, ^{2}H, ^{19}F) were measured. Such studies yield information on non-covalent interactions, but they are limited to aqueous solutions.

Second, the global structure of soil organic matter in the solid state has been investigated by cross polarization magic angle spinning (^{13}C CP-MAS-NMR). Here, relationships between organic functions and adsorption values (K_{OC}) have been assessed (Ahmad et al. 2001; Xing and Chen 1999). In principle, these correlations could be used to predict xenobiotic sorption on different type of soils, however there are still indirect methods.

Third, direct analyses of dry parts of soils in presence of xenobiotics have been reported using solid-state CP-MAS NMR spectroscopy. For example, it was used to characterize the ^{13}C-atrazine residues and to follow the binding reactivity of organic matter (Benoit and Preston 2000). Also ^{15}N CP-MAS NMR allowed to study bound residues of 2,4,6-trinitrotoluene (^{15}N-TNT) in soils (Knicker et al. 1999; Bruns-Nagel et al. 2000). Reports on interactions of pollutants with clay fractions are scarce and focus on clay structure using ^{29}Si or ^{27}Al solid state NMR (Minear and Nanny 1998). These experiments, performed on dry samples, give an altered fingerprint because soil samples are usually highly hydrated outdoors. In particular they give few indication on the bioavailability of the pollutant.

Finally, recent experiments were carried out on hydrated matrices of whole soil (Simpson et al. 2001) or clay (Combourieu et al. 2001) by using liquid state ^{1}H high-resolution (HR) MAS NMR. Indeed, HR MAS NMR allows the characterization of inhomogeneous compounds with liquid-like dynamics (Piotto et al. 2001). In such systems with restricted or anisotropic motions, the effects of dipolar coupling, chemical shift anisotropy and inhomogeneous susceptibility on resonance line widths are not averaged to zero in static experiments but are dramatically decreased by magic angle spinning (MAS). Moreover the use of ^{1}H NMR avoids the problem of xenobiotic labeling.

The aim of the present work was first to assess the sorption capacity of MgAl anionic clays towards the pesticide 4-chloro-2-methylphenoxyacetic acid (MCPA), a phenoxyacetic acid widely spread for the treatment of grain, corn, sugar, cane and rice (Cluzeau 1996). The environmental relevance of this chemical is well established (Draoui et al. 1998). We looked at the influence of the chemical composition and the layer charge density of the adsorbents on the adsorption properties. In a second part, we present the differentiation of mobile vs. immobile fractions of the pesticide 4-chloro-2-methylphenoxyacetate (MCPA) using ^{1}H HR MAS NMR (Combourieu et al. 2001).

42.2
Experimental

42.2.1
Synthesis of [Mg$_R$AlCl]

The anionic clays were prepared with various cation molar ratios M^{II}/M^{III} (Mg/Al = R = 2, 3, 4), by coprecipitation at controlled pH (Miyata 1975). Typically, the [Mg$_3$AlCl] phase was prepared at 65 °C, under N$_2$ and vigorous magnetic stirring: about 50 ml of an aqueous solution of MgCl$_2$ · 6H$_2$O, 0.75 mol l^{-1} and AlCl$_3$ · 6H$_2$O, 0.25 mol l^{-1} was added dropwise to the reactor previously filled with 100 ml of deionized water. The pH was kept at 10.0 ±0.1 by simultaneous addition of 2 mol l^{-1} NaOH. The precipitate was left for 12 h at 65 °C in the mother solution for ageing, then separated by 3 repeated washing/centrifugation cycles and dried in air at room temperature. The results of the elemental analysis of the samples (Centre d'analyses de Vernaison CNRS, France) given in Table 42.1 are in agreement with the desired formulae; we note a general contamination by carbonate anions from air for all phases. The experimental Mg^{2+}/Al^{3+} molar ratio is close to the theoretical value, which will be used hereafter to identify the samples. The X-ray powder diffraction patterns of Mg$_R$AlCl phases, recorded on Siemens D501 diffractometer with copper $K\alpha$ radiation, are typical of pure anionic clays.

42.2.2
Adsorption of Pesticide on Clays

Adsorption isotherms and kinetic studies were carried out twice at 25 °C in open bottles. Typically, 10 mg of clay were dispersed 24 h in 25 ml of deionized/decarbonated water before herbicide addition. The pH was then adjusted to 7.0 using either HCl (0.1N) or NaOH (0.1N). The initial MCPA concentrations ranged from 0.05 to 2 mmol l^{-1} with a final volume of 50 ml (V). The suspensions were centrifuged. Concentrations were analyzed by ultraviolet (UV) spectroscopy using a Perkin Elmer lambda 2S spectrometer (279 nm). The amount of clay-adsorbed MCPA (Q, mmol g^{-1}) was calculated by difference between initial (C_i, mmol l^{-1}) and final (or equilibrium) concentrations (C_e, mmol l^{-1}) of the herbicide, per gram of clay adsorbent (m, g): $Q = (C_i - C_e) * V m^{-1}$. Then, the adsorption isotherms were obtained by plotting the amount of MCPA adsorbed (Q) vs. the adsorbate concentration (C_e) in the equilibrium solution.

Table 42.1. Elemental analysis of [Mg$_R$AlCl]

Compound	Formula
[Mg$_2$AlCl]	Mg$_{2.22}$Al(OH)$_{6.38}$Cl$_{0.96}$CO$_3^{2-}$$_{0.05}$, 1.88H$_2$O
[Mg$_3$AlCl]	Mg$_{3.09}$Al(OH)$_{7.51}$Cl$_{1.09}$CO$_3^{2-}$$_{0.29}$, 2.78H$_2$O
[Mg$_4$AlCl]	Mg$_{4.37}$Al(OH)$_{10.58}$Cl$_{1.00}$CO$_3^{2-}$$_{0.08}$, 1.88H$_2$O

42.2.3
Nuclear Magnetic Resonance (NMR)

42.2.3.1
^1H MAS and ^{13}C CP-MAS Experiments

The high-resolution solid state MAS spectra were obtained on a Bruker Avance DSX-300 spectrometer with a double bearing Magic Angle Spinning (MAS) probehead. For the high-power ^1H and ^{13}C MAS experiments, the samples were packed in 4 mm zirconia rotors fitted with Kel-F caps, and spun at the magic angle with a speed of 10 kHz at ambient temperature. The ^1H MAS spectra were recorded at 300.13 MHz with a single pulse experiment. ^{13}C cross polarisation (CP)-MAS spectra were performed at 75.46 MHz using a standard CP sequence. Hartmann-Hahn cross polarization from protons to carbon was carried out at a proton nutation frequency of 70 kHz over a contact time of 1.5 ms. The proton decoupling field strength was 68 kHz. A recycle delay of 3 s was used. Free induction decays (FID) were digitized into 3 K data points and Fourier-transformed after applying 10 Hz exponential line-broadening. The number of transients was fitted to reach optimal signal to noise ratios (typically 1 024 for ^{13}C experiments). Chemical shifts were referenced to an external reference taken from the carbonyl of glycine (176.03 ppm) (Aldrich) and from the singlet (0 ppm) of tetradeuterated sodium trimethylsilylpropionate (TSPd$_4$) (Eurisotop) for ^{13}C and ^1H respectively. These references were set before each spectrum.

42.2.3.2
^1H HR-MAS Experiments

To assess the mobility of adsorbed vs. intercalated MCPA in anionic clays, 40 mg samples were swollen with 80 µl D$_2$O. The sample volume was restricted to approx. 50 µl in the center of the rotor using a Teflon insert to increase the radio frequency field homogeneity. The ^1H HR MAS NMR spectra were acquired at 300.13 MHz with the same 4 mm MAS probehead using a single pulse experiment ($t90° = 6.2$ µs) with presaturation during relaxation delay (4 s) for water resonance suppression. The samples were spun in a speed range of 1.33 to 1.80 kHz. Chemical shifts are reported in ppm relative to TSPd$_4$ as external standard. The number of transients was typically 256 to achieve good signal to noise ratio.

42.3
Results and Discussion

42.3.1
MCPA Adsorption Isotherms for [Mg$_3$AlCl]

Preliminary adsorption experiments, reported in a previous paper (Inacio et al. 2001), were conducted in order to determine the optimal conditions for the adsorption of MCPA on anionic clays regarding the contact time and the pH value. The kinetic study showed that the adsorption equilibrium state is reached after a contact time of ca. 30–45 min. Hence, the pesticide adsorption on clays is very fast as already observed

Fig. 42.1.
Adsorption of the pesticide MCPA on anionic clays: **a** [Mg$_2$AlCl], **b** [Mg$_3$AlCl], **c** [Mg$_4$AlCl]

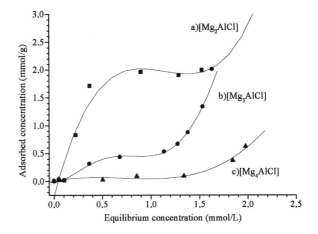

elsewhere with hydrotalcite-like compounds (Hermosin et al. 1993; Hermosin et al. 1996) and with Montmorillonite (Sannino et al. 1997). Furthermore, the adsorption capacity reaches a maximum in the pH range from 5.0 to 7.0, pH values commonly encountered in natural soils. Therefore, experiments were carried out at a pH value of 7.0 at which MCPA is in the anionic form (pK_a = 2.9).

Figure 42.1 displays the pesticide MCPA adsorption isotherms on [Mg$_4$AlCl], [Mg$_3$AlCl] and [Mg$_2$AlCl] phases. Three different parts within the adsorption isotherms of [Mg$_2$AlCl] and [Mg$_3$AlCl] can be distinguished. At low adsorbed concentration in the equilibrium solution C_e (below 0.5 mmol l^{-1}), a rapid increase of the amount of MCPA adsorbed Q is observed, then Q reaches a plateau the length of which depends on the Mg/Al ratio. At high C_e value, the amount of MCPA adsorbed increases again more rapidly than at low C_e value. These results suggest that the adsorption proceeds in two steps, first the saturation of the external sites at low equilibrium concentration while an interlayer process would occur at a high C_e value. For [Mg$_2$AlCl], the internal process begins at higher C_e values. Furthermore, the adsorbate-adsorbent affinity, evidenced by the slope at the origin, weakens while the Mg^{2+}/Al^{3+} molar ratio increases i.e. the layer charge density decreases.

The shape of the adsorption isotherm suggests information about the adsorbate-adsorbent interaction and Giles et al. (1960) have classified the adsorption isotherms into four main types. Here, the shape of the isotherms makes it difficult to classify them; they are hybrids of the L- and the S-type according to the above classification. The L-type isotherm shows a rapid adsorption reflecting a relative high affinity between adsorbate and the adsorbent. On the other hand, the S-type characterized first by a weak adsorption which then gradually increases suggests a weak surface interaction and competitive adsorbate-adsorbate interactions. Even [Mg$_3$AlCl] cannot be considered clearly as pure L-type because of the first part of the curve. [Mg$_2$AlCl] is similar to S-type but tends to L-type for C_e > 1.2 mmol l^{-1}. More difficult is the case of [Mg$_4$AlCl] which only adsorbs at C_e > 1.5. The shape from this point is clearly S-type and this isotherm assignment is supported by Freundlich 1/n values (Beck et al. 1993). These changes from S to L within the same isotherm are likely to be due to changes from external to interlayer adsorption process.

Table 42.2. Freundlich parameters for the pesticide MCPA adsorption on clays (Kf: adsorption capacity; 1/n: reciprocal of the affinity; and r correlation factor) and specific surface areas (S.A.) of original materials

Clay	K_f (mmol g^{-1})	$1/n$	r	S.A. (m^2 g^{-1})
[Mg$_2$AlCl]	1.6069	1.316	0.88	29
[Mg$_3$AlCl]	0.6665	1.4909	0.98	44
[Mg$_4$AlCl]	0.1145	2.702	0.90	37

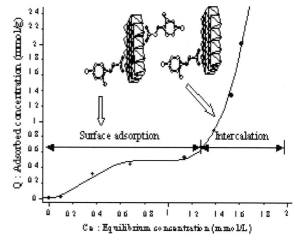

Fig. 42.2. MCPA adsorption isotherm for [Mg$_3$AlCl]

Several models have been developed to describe the adsorption phenomena. Here, owing to the shape of the isotherms, the sorption data were analyzed according to Freundlich equation: $Q = KC_e^{1/n}$ where Q is the amount of pesticides per unit weight of clay, C_e denotes the equilibrium concentration of the adsorbate and K_f and n are constants that give estimates of the adsorption capacity and intensity, respectively. This equation is used in a linearized form as follows: $\log Q = \log K + 1/n \cdot \log C_e$. The results of the fits are reported in Table 42.2. Under these conditions, n is represented by the reciprocal of the slope of the line.

The results of Table 42.2 show that the adsorption capacity increases with the layer charge density of the mineral support i.e. [Mg$_4$AlCl] < [Mg$_3$AlCl] < [Mg$_2$AlCl]. The pesticide adsorption maximum reached at the plateau (for $C_e = 1.0$ mmol l^{-1}) is 0.05, 0.50 and 1.90 mmol g^{-1} for [Mg$_4$AlCl], [Mg$_3$AlCl] and [Mg$_2$AlCl], respectively, which corresponds to 2%, 15% and 47% of the anion exchange capacity value (AEC). Hence, the adsorption must proceeds via an anion exchange mechanism.

The mechanism remains the same at high equilibrium concentration. The layer charge decrease resulting in a reduction in affinity would explain the fact that the interlayer adsorption is steeply delayed to higher concentrations (Cox et al. 1995). X-ray diffraction analysis shows an enlargement of the basal spacing ($d_{003} = 2.21$ nm) for [Mg$_3$AlCl], indicating that part of internal chloride anions were exchanged by MCPA. This high basal spacing results from the formation of intertwined double-layers of MCPA in the interlayer spaces, as reported by Prevot et al. (2001), with the anions hydrogen bonded to the hydroxylated surfaces in a perpendicular orientation (Fig. 42.2).

42.3.2
Nuclear Magnetic Resonance (NMR)

Solid state ^{1}H magic angle spinning (MAS) NMR experiments performed on both adsorbed and intercalated clays showed very broad signals and thus do not allow the differentiation of mobile vs. intercalated fractions of the pesticide MCPA. Further ^{13}C cross polarization (CP) MAS revealed only a slight difference of signal intensities for the two populations due to a less efficiency of cross polarization (CP) technique at the surface (data not shown). Therefore, in order to assess the contribution of adsorbed components, the samples (40 mg) were swollen with D$_2$O (80 μl).

Figure 42.3 (*top trace*) shows the ^{1}H spectrum of the adsorbed MCPA on clay hydrated with D$_2$O. All protons of MCPA are visible and chemical shifts are identical to those of the solution state NMR, thus indicating an efficient tumbling of the molecules at the surface. Note the presence of the ^{3}J coupling constant (9 Hz) of H$_5$ and H$_6$ aromatic protons. These sharp signals are characteristic of weakly bounded compounds. The spectrum of the intercalated pesticide (Fig. 42.3, *bottom trace*), recorded under the same conditions, displays only residual signals. In this sample the MCPA anions are tightly packed either inside the structure or at the surface because of a high packing of organic molecules which are not observed by HR MAS. These small signals could be explained by the release of a minor fraction of MCPA due to a slight hydrolysis of the matrix. This is confirmed by the presence of two broad peaks at 0.85 and 1.25 ppm, identified as protons of metallic hydroxides under colloidal form.

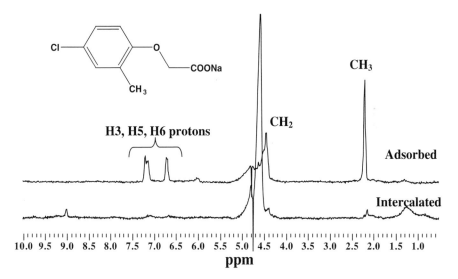

Fig. 42.3. Distinction of the adsorbed (*top trace*) and intercalated (*bottom trace*) fractions of the pesticide 4-chloro-2-methylphenoxyacetic acid (MCPA) on clays by ^{1}H high resolution magic angle spinning nuclear magnetic resonance (HR MAS NMR). When MCPA is absorbed at the surface, its mobility is high enough to give narrow and well resolved ^{1}H NMR signals. On the contrary when the pesticide is intercalated, very broad signals are recorded due to a restricted mobility

42.4
Conclusion

This communication demonstrates the potential of the ^{1}H HR MAS technique to study in-situ interactions of pesticides at the solid-aqueous interface of a solid matrix. This was made possible by using a well-characterized model of soil. We are now investigating the quantitative aspect of this technique in order to evaluate the bioavailable (mobile) fraction of a xenobiotic in such a solid matrix. The environmental interactions in a whole soil will also be investigated in the future by this technique since, even when a large amount of species are present, all the powerful and sensitive solution-state experiments such as total correlation spectroscopy (1D or 2D TOCSY) and nuclear Overhauser spectroscopy (NOESY) can be used.

Acknowledgements

We thank the CNRS (centre national de la recherche scientifique) for supporting this research programm (Programme Environnement Vie et Société).

References

Aguer JP, Hermosi MC, Calderon MJ, Cornejo J (2000) Fenuron sorption on homoionic natural and modified smectites. J Environ Sci Health B35(3):279–296
Ahmad R, Kookana RS, Alston AM, Skjemstad JO (2001) The nature of soil organic matter affects sorption of pesticides. 1. Relationships with carbon chemistry as determined by ^{13}C CPMAS NMR Spectroscopy. Environ Sci Technol 35:878–884
Allman R (1968) The crystal structure of pyroaurite. Acta Cryst B24:972–977
Bailey GW, White JL, Rithberg T (1968) Adsorption of organic herbicides by montmorillonite: role of pH and chemical character of adsorbate. Soil Sci Soc Amer Proc 32:222–234
Beck AJ, Johnston AEJ, Jones KC (1993) Movement of nonionic organic chemicals in agricultural soils. Crit Rev Environ Sci Technol 23:219–248
Benoit P, Preston CM (2000) Transformation and binding of ^{13}C- and ^{14}C-labelled atrazine in relation to straw decomposition in soil. Eur J Soil Sci 1:43–54
Bortiatynski JM, Hatcher PG, Minard RD (1998) The development of ^{13}C labeling and ^{13}C NMR spectroscopy technique to study interaction of pollutants with humic substances. In: Nanny MA, Minear RA, Leenheer JA (eds) Nuclear magnetic resonance in environmental chemistry. Oxford University Press, New-York, pp 26–49
Bruns-Nagel D, Knicker H, Drzyzga O, Butehorn U, Steinbach K, Gemsa D, Von Löw E (2000) Characterization of ^{15}N-TNT residues after anaerobic/aerobic treatment of soil/molasses mixtures by solid-state NMR spectroscopy. 2. Systematic investigation of whole soil and different humic fractions. Environ Sci Technol 34:1549–1556
Celis R, Cornejo J, Hermosin MC, Koskinen WC (1998) Sorption of atrazine and simazine by model association of soil colloids. Soil Sci Soc Am J 62:911–915
Celis R, Koskinen WC, Cecchi AM, Bresnahan GA, Carrisoza MJ, Ulibarri MA, Pavlovic I, Hermosin MC (1999) Sorption of the ionizable pesticide imazamox by organo-clays and organohydrotalcites. J Environ Sci Health B34:929–941
Cluzeau S (1996) Index phytosanitaire ACTA (ed.) Paris
Combourieu B, Inacio J, Delort AM, Forano C (2001) Differenciation of mobile and immobile pesticides on anionic clays by ^{1}H HR-MAS NMR spectroscopy. Chem Commun 2214–2215
Cox L, Hermosin MC, Cornejo J (1995) Adsorption mechanism of thiazafluron in mineral soil clay components. Eur J Soil Sci 46:431–438

Cox L, Celis R, Hermosin MC, Cornejo J (2000) Natural colloids to retard leaching of simazine and 2,4-D in soils. Environ Sci Technol 1:93–99

de Roy A (1998) Lamellar double hydroxides. Mol Cryst Liq Cryst 311:173–193

Draoui K, Denoyel R, Chgoura M, Rouquerol J (1998) Adsorption of pesticides on soil component, a microcalorimetric study. J Therm Anal 51:831–839

Filomena S, Violante A, Gianfreda L (1997) Adsorption-desorption of 2,4-D by hydroxy aluminium montmorillonite complexes. Pestic Sci 51:429–435

Giles CH, Mac Ewan TH, Nakhwa SN, Smith P (1960) Studies in adsorption. Part XI. A system of classification of solution adsorption isotherms and its use in diagnosis of adsorption mechanisms and in measurement specific surface of solids. J Chem Soc 3973–3993

Haderlein SB, Weissmahr KW, Schwazenbach RP (1996) Specific adsorption of nitroaromatic explosives and pesticides to clays minerals. Environ Sci Technol 30:612–622

Herbert BE, Bertsch PM (1998) A ^{19}F and ^{2}H NMR spectroscopic investigation of the interaction between nonionic organic contaminants and dissolved humic material. In: Nanny MA, Minear RA, Leenheer JA (eds) Nuclear magnetic resonance in environmental chemistry. Oxford University Press, New-York, pp 73–90

Hermosin MC, Pavlovic I, Ulibarri MA, Cornejo J (1993) Trichlorophenol adsorption on layered double hydroxide: a potential sorbent. J Environ Sci Health 9:1875–1888

Hermosin MC, Pavlovic I, Ulibarri MA, Cornejo J (1996) Hydrotalcite as sorbent for trinitrophenol sorption capacity and mechanism. Water Res 30:171–177

Inacio J, Taviot-Guého C, Forano C, Besse JP (2001) Adsorption of MCPA pesticide by MgAl-layered double hydroxides. Appl Clay Sci 18:255–264

Knicker H, Bruns-Nagel D, Drzyzga O, Von Löw E, Steinbach K (1999) Characterization of ^{15}N-TNT residues after anaerobic/aerobic treatment of soil/molasses mixtures by solid-state NMR spectroscopy. 1. determination and optimization of spectroscopic parameters. Environ Sci Technol 33:343–349

Laird DA, Yen PY, Koskinen WC, Steinheimer TR, Dowdy RH (1994) Sorption of atrazine on soil clay components. Environ Sci Technol 28:1054–1061

Lakraïmi M, Legrouri A, Barroug A, de Roy A, Besse JP (2000) Preparation of new stable hybrid material by chloride-2,4-dichlorophenoxyacetate ion exchanged into zinc-aluminium-chloride layered double hydroxides. J Mater Chem 10:1007–1011

Minear RA, Nanny MA (1998) Solution and condensed phase characterization. In: Nanny MA, Minear RA, Leenheer JA (eds) Nuclear magnetic resonance in environmental chemistry. Oxford University Press, New-York, pp 123–129

Miyata S (1975) The synthesis of hydrotalcite-like compounds and their structures and physico-chemical properties the systems MgAlNO$_3$, MgAlCl, MgAlClO$_4$, NiAlCl, ZnAlCl. Clays Clay Miner 23:369–375

Nanny MA (1998) Sorption process in the environment. In: Nanny MA, Minear RA, Leenheer JA (eds) Nuclear magnetic resonance in environmental chemistry. Oxford University Press, New-York, pp 19–25

Nanny MA (1999) Deuterium NMR characterization of noncovalent interactions between monoaromatic compounds and fulvic acids. Org Geochem 30:901–909

Nanny MA, Maza JP (2001) Noncovalent interactions between monoaromatic compounds and dissolved humic acids: A deuterium NMR T_1 relaxation study. Environ Sci Technol 35:379–384

Piotto M, Bourdonneau M, Furrer J, Bianco A, Raya J, Elbayed K (2001) Destruction of magnetization during TOCSY experiments performed under magic angle spinning: effect of radial B1 inhomogeneities. J Mag Reson 149:114–118

Prevot V, Forano C, Besse JP (2001) Hybrid derivatives of layered double hydroxides. Appl Clay Sci 18:3–15

Sannino F, Violante A, Grianfreda L (1997) Adsorption-desorption of 2,4-D by hydroxy aluminium montmorillonite complexes. Pestic Sci 51:429–435

Sawhney BL, Singh SS (1997) Sorption of atrazine by Al- and Ca-saturated smectite. Clays Clay Miner 45:333–338

Simpson AJ, Kingery WL, Shaw DR, Spraul M, Humfer E, Dvorsak P (2001) the application of ^{1}H HR-NMR spectroscopy for the study of structures and associations of organic components at the solid-aqueous interface of whole soil. Environ Sci Technol 35:3321–3325

Ulibarri MA, Pavlovic I, Hermosin MC, Cornejo J (1995) Hydrotalcite-like compounds as potential sorbents of phenol from water. Appl Clay Sci 10:131–145

Vaccari A (1999) Clays and catalysis: a promising future. Appl Clay Sci 14:161–198

Villa MV, Sanchez-Martin MJ, Sanchez-Camazano M (1999) Hydrotalcites and organo-hydrotalcites as sorbents for removing pesticides from water. J Environ Sci Health B34:509–525

Xing B, Chen Z (1999) Spectroscopic evidence for condensed domains in soil organic matter. Soil Sci 164:40–47

Chapter 43

Photo- and Biodegradation of Atrazine in the Presence of Soil Constituents

P. Besse · M. Sancelme · A. M. Delort · J. P. Aguer · B. Lavédrine · C. Richard
T. Alekseeva · C. Taviot-Guého · C. Forano · A. Kersanté · N. Josselin · F. Binet

Abstract

The photo- and biodegradation of atrazine in the presence of several soil constituents, in particular earthworm casts collected from agricultural soils, were investigated. Earthworm casts were found to favour the photodegradation of atrazine but did not affect its biodegradation, as well as commercially available humic acids. The organic matter content of casts, even rather poor, increased the photodegradation rate. In contrast, the lack of effect of earthworm casts on the biodegradability may be due to the absence of minerals with swelling layers.

Key words: atrazine, soil, earthworm casts, photodegradation, biodegradation

43.1 Introduction

The present study is part of an interdisciplinary research program, entitled Tranzat, that addresses a major and actual environmental concern related to agriculture, that is the diffuse contamination of soils and freshwater ecosystems by a wide range of xenobiotic chemicals used for crop protection. This work aims to determine to what extent and by which mechanisms both the soil biota and the soil microflora control the fate of herbicide at various spatial scales in the soil ranging from the molecular to the soil profile scale.

Atrazine was chosen in this program as an environmental contaminant model of priority status. As it is still nowadays a widely used herbicide in agriculture in Europe, it is frequently detected in freshwater ecosystems and its concentration often exceeds 10 to 100 fold the policy level of 0.1 µg l^{-1} for ground and surface waters. The part of atrazine reaching the soil undergoes a series of chemical and biological transformations. The abiotic degradation through solar light absorption occurs early at the soil surface. In soil, degradation of herbicide is mediated by microorganisms and, additionally, influenced by the interactions between herbicide and soil components.

Here we report our investigations on the photochemical and biological transformations of atrazine in the presence of several soil components. In particular, the effect of earthworm casts on these transformations were investigated.

43.2
Experimental

43.2.1
Sampling and Characterization of Earthworm Cast

Soil was collected in the 20 cm top of an agricultural soil located at Pleine-Fougère, 10 km south-west of the Mont Saint-Michel Bay (Britany, France) in spring 2001. Earthworms from the same location were extracted with formaldehyde. With respect to its great soil disturbance activity, the large anecic species *Aporrectodea giardi* was reared at the laboratory under 12 °C for 4 weeks with the fresh and undisturbed collected soil. The initial soil moisture was maintained and organic matter such as grass litter was added. After one week for worm adaptation, newly deposited casts were collected every day and then air-dried prior to be used as natural clay-humic complex for comparative experiments on atrazine degradation.

Earthworm cast composition: Mineralogical composition of earthworm casts was investigated with XRD using Philips X'Pert diffractometer with Cu-Kα radiation. The clay fraction for analysis was separated by sedimentation in distilled water after crumbling sample in a wet paste and parallel-oriented slides were prepared by the sedimentation from water suspension. Prior XRD examination clay was pretreated with 10% H_2O_2 solution on the boiling water-bath (3 times) to remove organic matter. Mg-saturated samples were prepared using 1N solution of $MgCl_2$. Specimens used in case of bulk samples and silt fraction were cuvets with sample powder. 1N Ammonium acetate extract (pH 4.8) (solid/liquid ratio 1/10) was used to determine the content of exchangeable cations – Ca, Mg, K and Na. Concentration of cations in extracts was determined by atomic absorption spectroscopic (AAS) techniques.

Castings are characterized by close to neutral pH value (7.55), coarse granulometrical composition (11.25% of particles with size < 2 µm), low exchangeable properties with almost similar Ca and Mg content in exchangeable complexes (3.00 and 3.70 mg-eq 100 g^{-1} respectively), and a very low organic content (1.87% for the fraction < 2 µm). Worm casts (bulk sample) is enriched in quartz. The other minerals presented were: several varieties of feldspars (albite, anortite, orthoclase, microcline), dioctahedral mica (muscovite), orthopyroxene, calcite, magnetite. The clay fraction (<2 µm) is enriched in muscovite, and contains also Fe-chlorite and kaolinite. The clay fraction contains also quartz and minor quantities of feldspars and calcite. Thus, the specific feature of the composition of the clay fraction is the absence of minerals with swelling layers (smectite type). The silt fraction (2–20 mm) is enriched in quartz, but contains visibly more layer silicates (the same set as clay fraction) in comparison with bulk sample.

43.2.2
Photolytic Experiments

In aqueous medium. Atrazine was solubilized at a level of 7.4×10^{-5} mol l^{-1} in pure water (Milli-Q, Millipore) or in neutral water containing humic substances (40 mg l^{-1}) under stirring. The solutions were exposed to solar light in capped Pyrex glass reactors (14 mm internal diameter) during the period April–May 2001 at Clermont-Ferrand

(46°N, 3°E). Aliquots of 0.5 ml were removed simultaneously from the two solutions and analysed by HPLC. Dark control experiments showed no loss of atrazine during the time required for the irradiations in pure water and solutions of humic substances.

On soil-sorbed phase. Atrazine was sorbed on solid supports (soil, worm casting, sand) using the following procedure: 3 g of solid support and 3 mg of atrazine were stirred in 20 ml of diethylether until the complete evaporation of the solvent. The solid-atrazine mixture was then dried in a dessicator for one hour. Two grams were sieved uniformly onto the surface of plates (8.5 mm internal diameter) recovered with a PVDF film and exposed to solar light. One gram was kept in the dark for control. Samples of 200 mg were taken at selected intervals, and diluted in 20 ml of dichloromethane and centrifuged at 6 000 rpm for 15 min to separate the solid residue and the supernatant. The organic phase was evaporated and the extract residue was solubilized in a methanol/water (50/50, v/v) mixture.

Analyses. Atrazine and photoproducts were analysed by HPLC using a Gilson chromatograph equiped with a UV-Visible detector and a conventional reverse phase column. The eluent was a mixture of methanol-water (60/40, v/v) and the detection wavelength was set to 263 nm. UV spectra were recorded on a Cary 3 (Varian) spectrophotometer. A 1 cm path quartz cell was used for all the experiments.

43.2.3
Biodegradation Experiments

Growth conditions. Several liquid growth media were tested in a first step: the "atrazine medium" described previously by Mandelbaum et al. (1993) containing 0.1% (wt/v) sodium citrate as the carbon source and atrazine as the only carbon source at a final concentration of 100 ppm (0.46 mM); Trypticase Soy broth (bioMérieux, Marcy l'Etoile, France); Nutrient Agar (bioMérieux) and LB medium (Difco). For the further studies, *Pseudomonas* sp. strain ADP was grown in 100-ml portions of Trypticase Soy broth in 500-ml Erlenmeyer flasks incubated at 30 °C at 200 rpm. The cells were harvested after 15 h of culture.

Atrazine metabolism by resting-cell culture. Cells were centrifuged at 9 000 g for 15 min at 5 °C. The pellet was washed a first time with NaCl solution (8 g l^{-1}) and then with mineral water (Volvic, France). Cells (55 mg dry weight) were resuspended in 10 ml of a 0.1 mM atrazine solution prepared previously (22 mg of atrazine in 1 l of Volvic water or phosphate buffer 50 mM, pH 6.7) and incubated in 50-ml Erlenmeyer flasks at 30 °C under agitation (200 rpm). The controls consisted of preparations incubated under the same conditions without substrate or cells. Samples (500 μl) were taken every 30 min directly from the culture medium. They were centrifuged at 12 000 g for 5 min and the supernatants were analysed by HPLC.

Incubation in the presence of a solid. To 10 ml of the 0.1 mM atrazine solution was added either 1 g of earthworm casting or various quantities of humic acids (40 mg, 120 mg or 1 g l^{-1}) (Aldrich). After 24 h of stirring, resting-cells of *Pseudomonas* sp. strain ADP (55 mg dry weight), obtained as previously described, were added. Control was obtained in the same conditions without cells.

Analyses. HPLC analyses were performed using a Waters 600E chromatograph fitted with a reversed-phase column (Merck C18; 5 µm, 150 × 4.6 mm). The mobile phase was a mixture acetonitrile/acetate buffer 50 mM pH 4.6 (30/70, v/v) with a flow rate of 1 ml min^{-1}. Waters 486 UV detector was set at 225 nm.

43.3 Results and Discussion

The aim of our study was to determine the physico-chemical mechanisms at the molecular scale involved in the triple interactions between the soil constituents, the atrazine molecule (sorption process) and a model bacteria *Pseudomonas* ADP, a well-known atrazine degrading bacteria (Mandelbaum et al. 1995). Therefore, we performed comparative experiments on the abiotic and biotic degradation of atrazine with various soil constituents and also with earthworm casts, biogenic soil structures that could be defined as particular natural clay-humic complex.

43.3.1
Photolytic Experiments

Kinetics of disappearance of atrazine irradiated in aqueous medium are shown in Fig. 43.1. In pure water, atrazine underwent a slow phototransformation. The decrease of atrazine concentration was faster when humic substances were added reaching about 50% of conversion after 40 d. Deisopropyl- and deethyl-atrazine were formed along with hydroxyatrazine. On dry solid supports, atrazine was also phototransformed (see Fig. 43.1). The reaction was faster on soil and earthworm casts than on sand. Similar consumption rates were obtained with these two latter supports. Again, deisopropyl- and deethyl-atrazine were found as main photoproducts (Fig. 43.2). Formation of hydroxyatrazine was not observed.

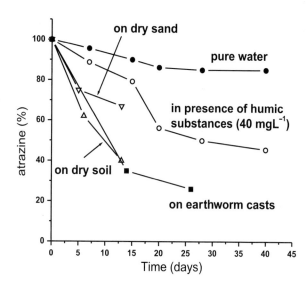

Fig. 43.1. Phototransformation of atrazine in solar light: compared decrease of atrazine concentration in various media. The percentage of atrazine was obtained by dividing the concentration measured in the irradiated samples by the initial concentration

Fig. 43.2.
HPLC chromatogram of extract obtained after irradiation of atrazine on earthworm casts (conversion extent = 50%) (*1*: deisopropylatrazine; *2*: deethylatrazine)

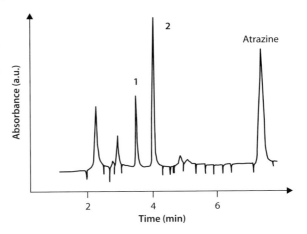

The phototransformation of atrazine has received large interest. The reaction has been studied in various experimental conditions i.e. in pure or natural water (Torrents et al. 1997; Minero et al. 1992; Curran et al. 1992), in water containing nitrate ions (Torrents et al. 1997), acetone (Burkhard and Guth 1976), soil organic matter (Minero et al. 1992) or riboflavine (Rejto et al. 1983). Studies on solid supports have been also reported (Curran et al. 1992; Konstantinou et al. 2001; Gong et al. 2001). The photolysis in pure water and in solar light was generally shown to be slow due to the fact that the absorption spectrum of atrazine overlaps poorly with the solar spectrum. On the other hand, reported results on the photosensitized reactions in homogeneous phase or on soils showed some discrepancies.

In a positive agreement with literature, we found that aqueous atrazine undergoes a slow photolysis on solar light. Our results show that humic substances increase the phototransformation rate. The action of humic substances is related to photochemistry and not to the formation of non-analysable bound residues since in the dark no or very low atrazine consumption was observed. The photoproducts observed are deisopropylatrazine, deethylatrazine and hydroxyatrazine. Dealkylation is likely to involve the abstraction of H atom from NH with the intermediary formation of an anilino radical. Dealkylation may occur further to oxidation and rearrangement of the radical. Alternatively, the anilino radical might add on humic substances itself yielding bound residue. Minero et al. (1992) who irradiated at $\lambda > 340$ nm found also that natural organic matter had an accelerating effect on the reaction, while Torrents et al. (1997) and Konstantinou et al. (2001) who irradiated at shorter wavelength ($\lambda > 290$ nm) found an inhibiting influence of humic substances. These differences show the importance of the irradiation conditions. In solar light or at $\lambda > 340$ nm, the direct photolysis of atrazine is a negligible reaction and the photoinductive effect of humic substances is measurable. In contrast, irradiations within the wavelength range 290–300 nm favor direct photolysis and make negligible photoinduced reactions.

We observed phototransformation on solid supports as Gong et al. (2001). Photolysis is slower on sand than on soil or earthworm casts. It may indicate that natural organic matter has a photoinductive effect. Alternatively minerals contained in soils may act as photocatalysts. From the analytical point of view our results are very different from those of Gong: we found again dealkylated products while he observed the formation of hydroxyatrazine.

43.3.2
Optimization of the Atrazine Biodegradation Conditions

In the 1990s, a number of bacterial strains, able to use atrazine as the sole source of nitrogen or carbon, were isolated: various *Pseudomonas* strains (Yanze-Kontchou and Gschwind 1994; Mandelbaum et al. 1995; Radosevich et al. 1995), *Agrobacterium radiobacter* J14a (Struthers et al. 1998), *Nocardioides* sp. (Topp et al. 2000), etc. Some of these strains are able to mineralise atrazine others catalyze an hydrolitic dechlorination reaction, producing hydroxyatrazine, or N-dealkylation reactions (Wackett et al. 2002). The bacterium, *Pseudomonas* sp. strain ADP was chosen as a model in this study as its atrazine metabolic pathway, as well as the different genes involved, were well-known and identified. However, several parameters needed to be optimised to carry out such experiments:

i. High biomass quantities were required to be able to monitor several assays in parallel. So it was not possible to use the "atrazine medium" described previously by Mandelbaum et al. (1993). Different growth media were tested: Trypcase soy broth (TS), Nutrient Broth (NB) and LB medium. The comparison of the OD_{575} (after dilution by ten: TS: 1.07; NB: 0.45; LB: 0.91) and the dry weight of cells obtained after 16 h of culture in a 100-ml portions of the different media (TS: 264 mg; NB: 105 mg; LB: 209 mg) showed that TS medium was the most appropriate for our further studies. The biodegradation of atrazine (0.1 mM) in 10 ml of 50 mM phosphate buffer (pH 6.7) was then carried out with cells grown on TS-medium (150 mg dry weight) in order to check if they still have their biodegradative abilities. Atrazine was completely degraded within 30 min and the only metabolite detected by HPLC was hydroxyatrazine as previously described (Mandelbaum et al. 1995; Bichat et al. 1999; Katz et al. 2001; Wackett et al. 2002).
ii. The medium in which the atrazine biodegradation was carried out, had to be neutral regarding the solid matrixes, the biodegradation or the interactions. Two incubations were made in parallel with a 0.1 mM atrazine solution prepared either in phosphate buffer 50 mM, pH 6.7 or in mineral water (Volvic, France, pH 7.1). In both cases, atrazine has completely disappeared within 30 min and the biodegradative pathway was not modified.
iii. The last point concerned the best conditions for the correct and accurate analysis of the potential effect of the solid on the biodegradation, that should be neither too rapid, nor too slow. Different cell concentrations were tested: 10, 30, 55 and 110 mg dry-weight cells in 10 ml of a 0.1 mm atrazine solution. The lower the concentration in cells is, the slower the biodegradation of atrazine gets. 23 h for complete degradation of atrazine seems to be a reasonable time to follow precisely the kinetics, so the cell concentration of 55 mg dry-weight cells was chosen.

43.3.3
Atrazine Biodegradation in the Presence of Humic Acids

As humic acids play an important role of photoinducer in the biotic process as previously shown, their effect was also tested on the atrazine biodegradation. After equilibration of the slurry for 24 h (10 ml of a 0.1 mM atrazine solution in the presence of

40 mg or 120 mg l^{-1} of humic acids), the experiments were initiated by inoculation with the bacterial resting-cells. Representative atrazine degradation time courses in the presence of the different concentrations in humic acids are shown on Fig. 43.3.

No significant adsorption of atrazine was observed during 24 h of pre-incubation, even with the highest concentration in humic acids (data not shown). No significant decrease of atrazine concentration was observed in the controls either, i.e. incubations carried out under the same conditions without cells (data not shown). The literature is full of examples showing the major role of organic matter in the sorption of herbicides, often directly related to their transport and dynamics in the soils (Bollag et al. 1992; Senesi 1992; Piccolo et al. 1998). Under the conditions used here, it is clear that binding of atrazine with humic substances was very limited and should not avoid degradation by reducing bioavailability. Meredith and Radosevich (1998) estimated that less than 1% of atrazine was associated with humic acids under similar conditions.

If the kinetics of atrazine degradation were relatively similar whatever the humic acid concentration, the time course evolution of the metabolite (hydroxyatrazine) changing with the humic acid concentration. However, it is difficult to draw clear conclusions. Comparing results obtained with 0.40 and 120 mg l^{-1} of humic acids, it seems that the higher the concentration in humic acids, the slower the rate of hydroxyatrazine

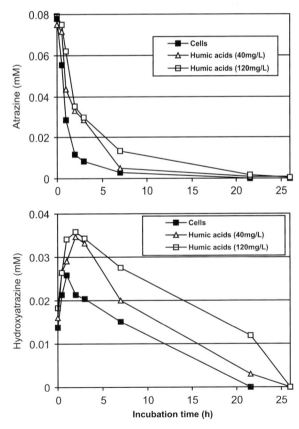

Fig. 43.3.
Concentration of atrazine (*up*) and its degradation product hydroxyatrazine (*down*) with time of incubation with *Pseudomonas* ADP in the absence or in the presence of humic acids

degradation gets. Martin-Neto et al. (2001) studied the interaction mechanisms between hydroxyatrazine and humic substances and showed that this metabolite readily forms electron-transfer complexes with humic substances. The existence of these complexes may explain a strong adsorption and its low bioavailability.

43.3.4
Atrazine Biodegradation in the Presence of Earthworm Casts

During the twenty four hours of pre-incubation of 1 g earthworm casts with 10 ml of a 0.1 mM solution of atrazine, its concentration did not show a significant decrease. The controls, corresponding to incubation without cells in the presence or in the absence of earthworm casts, indicated that atrazine is chemically stable in water and that its sorption on the castings used is negligible (Fig. 43.4). Resting-cells of *Pseudomonas* sp. strain ADP were then added to the (atrazine – earthworm casts) slurry. The kinetics were compared with those obtained with cells alone (Fig. 43.4).

Atrazine concentration decreased below detectable levels after 24 h of incubation. The lower concentration observed at time zero (0.08 mM instead of theoretical 0.1 mM) is due to the dilution corresponding to the addition of the bacterial cells suspension but could also be explained by the biosorption of cells. This phenomenon has been described for fungal strains (Benoit et al. 1998). The sole metabolite detected was hydroxyatrazine, compound that was not observed in the controls. Its formation was then only due to the biological activity. This metabolite was rapidly formed: its concentration was already of 10 µM after 5 min of incubation. After 24 h of incubation, hydroxyatrazine had almost completely disappeared. Therefore the presence of the earthworm casts in the incubation medium had no effect neither on the biodegradative rate of atrazine, nor on the kinetics of appearance and disappearance of its metabolite.

The absence of effect of the earthworm casts on the sorption or on the biodegradative kinetics of atrazine could be explained by their chemical composition (TC = 1.16%). Their poor content in organic matter and the absence of minerals with swelling layers (see experimental section), both considered as main actors in the interaction processes of pesticide–soil, could explain the results obtained. The initial low content of clays in the agricultural soil may have reduced the capability for the worms to concentrate it in casts. These results are in agreement with those obtained by Farenhorst and Bowman (2000), who have shown that the degree of atrazine sorption on castings increased with increasing organic carbon content.

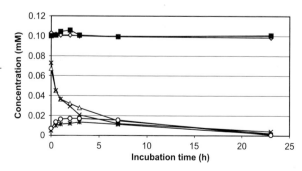

Fig. 43.4. Time courses of the concentration of atrazine and hydroxyatrazine in the absence (△ for atrazine and ○ for hydroxyatrazine) or in the presence of earthworm casts (× for atrazine and ✳ for hydroxyatrazine). Controls (without cells): atrazine (◇); atrazine + casts (■)

43.4
Conclusions

A detailed study of earthworm casts composition revealed that contents of minerals with swelling layers and organic matter are both low. In accordance, we found a negligible sorption of atrazine on earthworm casts. Regarding transformation processes, earthworm casts were found to favour the photodegradation of atrazine but did not affect its biodegradation. Since humic substances photoinduce the transformation of atrazine in homogeneous phase, the photoinductive effect of earthworm casts might be connected to their organic matter content although rather poor. The lack of enhancing effect of earthworm casts on the biodegradability may be due to the poor content in both minerals with swelling layers and organic matter. Studies on soil samples and worm casts collected from the riparian wetland close to the river and naturally rich in clay fraction and in organic matter are still in progress. They should permit a better insight into the role of each soil constituent in the photo and biodegradation of atrazine.

Acknowledgement

The authors want to thank Dr. Hans-Peter Buser from Syngenta Crop Protection AG (Basel, Switzerland) for having providing them atrazine and its metabolites.

References

Benoit P, Barriuso E, Calvet R (1998) Biosorption characterization of herbicide 2,4-D and atrazine and two chlorophenols on fungal mycelium. Chemosphere 37:1271–1282

Bichat F, Sims GK, Mulvaney RL (1999) Microbial utilization of heterocyclic nitrogen from atrazine. Soil Sci Soc Am J 63:100–110

Bollag JM, Myers CJ, Minard RD (1992) Biological and chemical interactions of pesticides with soil organic matter. Sci Total Environ 123/124:205–217

Burkhard N, Guth J (1976) Photodegradation of atrazine, atraton and ametryne in aqueous solution with acetone as photosensitiser. Pestic Sci 7:65–71

Curran W, Lox M, Liebl R, Simmons W (1992) Photolysis of imidazolinone herbicides in aqueous solution and on soil. Weed Science 40:143–148

Farenhorst A, Bowman BT (2000) Sorption of atrazine and metolachlor by earthworm surface castings and soil. J Environ Sci Health B35:157–173

Gong A, Ye C, Wang X, Lei Z, Liu J (2001) Dynamics and mechanism of ultraviolet photolysis of atrazine on soil surface. Pest Managem Sci 57:380–385

Katz I, Dosoretz CG, Mandelbaum RT, Green M (2001) Atrazine degradation under denitrifying conditions in continuous culture of *Pseudomonas* ADP. Water Res 35:3272–3275

Konstantinou I, Zarkadis A, Albanis T (2001) Photodegradation of selected herbicides in various natural waters and soils under different environmental conditions. J Environ Qual 30:121–130

Mandelbaum RT, Wackett LP, Allan DL (1993) Mineralisation of the s-triazine ring of atrazine by stable bacterial mixed cultures. Appl Environ Microbiol 59:1695–1701

Mandelbaum RT, Allan DL, Wackett LP (1995) Isolation and characterization of a *Pseudomonas* sp. that mineralised the s-triazine herbicide atrazine. Appl Environ Microbiol 61:1451–1457

Martin-Neto L, Traghetta DG, Vaz CM, Crestana S, Sposito G (2001) On the interaction mechanisms of atrazine and hydroxyatrazine with humic substances. J Environ Qual 30:520–525

Meredith CE, Radosevich M (1998) Bacterial degradation of homo- and heterocyclic aromatic compounds in the presence of soluble/colloidal humic acid. J Environ Sci Health B33:17–36

Minero C, Pramauro E, Pelizzetti E, Dolci M, Marchesini A (1992) Photosensitized transformations of atrazine under simulated sunlight in aqueous humic acid solution. Chemosphere 24:1597–1606

Piccolo A, Conte P, Scheunert I, Paci M (1998) Atrazine interactions with soil humic substances of different molecular structure. J Environ Qual 27:1324–1333

Radosevich M, Traina SJ, Hao Y, Tuovinen OH (1995) Degradation and mineralization of atrazine by a soil bacterial isolate. Appl Environ Microbiol 61:297–302

Rejto M, Saltzman S, Acher A, Muszat L (1983) Identification of sensitized photooxidation products of s-triazine herbicides in water. J Agric Food Chem 31:138–142

Senesi N (1992) Binding mechanisms of pesticides to soil humic substances. Sci Total Environ 123/124:63–76

Struthers JK, Jayachandran K, Moorman TB (1998) Biodegradation of atrazine by *Agrobacterium radiobacter* J14a and use of this strain in bioremediation of contaminated soil. Appl Environ Microbiol 64:3368–3375

Topp E, Mulbry WM, Zhu H, Nour SM, Cuppels D (2000) Characterization of s-triazine herbicide metabolism by a *Nocardioides* sp. isolated from agricultural soils. Appl Environ Microbiol 66:3134–3141

Torrents A, Anderson B, Bilboulian S, Johnson W, Hapeman C (1997) Atrazine photolysis: mechanistic investigations of direct and nitrate-mediated hydroxy radical processes and the influence of dissolved organic carbon from the Chesapeake Bay. Environ Sci Technol 31:1476–1482

Wackett LP, Sadowsky MJ, Martinez B, Shapir N (2002) Biodegradation of atrazine and related s-triazine compounds: from enzymes to field studies. Appl Environ Biotechnol 58:39–45

Yanze-Kontchou C, Gschwind N (1994) Mineralization of the herbicide atrazine as a carbon source by a *Pseudomonas* strain. Appl Environ Microbiol 60:4297–4302

Chapter 44

Behaviour of Imidacloprid in Fields. Toxicity for Honey Bees

J. M. Bonmatin · I. Moineau · R. Charvet · M. E. Colin · C. Fleche · E. R. Bengsch

Abstract

Following evidence for the intoxication of bees, the systemic insecticide imidacloprid was suspected from the mid nineties of having harmful effects. Recently, some studies have demonstrated that imidacloprid is toxic for the bees at sub-lethal doses. These doses are evaluated in the range between 1 and 20 µg kg^{-1}, or less. It appeared thus necessary to study the fate of imidacloprid in the environment at such low levels. Thus, we developed methods for the determination of low amounts, in the µg kg^{-1} range, of the insecticide imidacloprid in soils, plants and pollens using high pressure liquid chromatography – tandem mass spectrometry (LC/APCI/MS/MS). The extraction and separation methods were performed according to quality assurance criteria, good laboratory practices and the European Community's criteria applicable to banned substances (directive 96/23 EC). The linear concentration range of application was 1–50 µg kg^{-1} of imidacloprid, with a relative standard deviation of 2.9% at 1 µg kg^{-1}. The limit of detection and quantification are respectively LOD = 0.1 µg kg^{-1} and LOQ = 1 µg kg^{-1} and are suited to the sub-lethal dose range. This technique allows the unambiguous identification and quantification of imidacloprid. The results show the remanence of the insecticide in soils, its ascent into plants during flowering and its bioavailability in pollens.

Key words: imidacloprid, insecticide, Gaucho®, analysis, soils, plants, pollens, bees

44.1 Introduction

Various insecticides to protect crops against insects have been used over the last 40 years. Most insecticides were applied by spraying in large quantities, thus inducing pollution of air, soils and waters. In the 1990s, new insecticides were sold and announced as being efficient. Their implementation allowed the reduction in use of large quantities of pesticides and thus, to reduce pollution. Gaucho® is one of these new insecticides often used as a seed-dressing and imidacloprid is its active compound.

Imidacloprid is a systemic chloro-nicotinyl insecticide (Placke and Weber 1993). This propriety allows pesticide to rise into the sap and its distribution in the plant. It is used for soils, seeds and foliar applications for the control of sucking insects, including rice hoppers, aphids, thrips, whiteflies, termites, turf insects, soil insects and some beetles. It is most commonly used on rice, cereal, maize, sunflowers, potatoes and vegetables. It is especially systemic when used as a seed or soil treatment (Nauen et al. 1998). The active chemical works by interfering with the transmission of stimuli in the insect's nervous system. Specifically, it causes a blockage in the nicotinergic neuronal pathway that is more abundant in insects than in warm-blooded animals, making the chemical much more toxic to insects than to warm-blooded animals. This

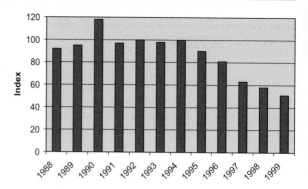

Fig. 44.1. Production of honey in west France. Note the decrease of after 1994, which could be explained partly by the use of the insecticide on sunflower. The vertical axis is graduated in % and 100% corresponds to the mean production between 1988 and 1994

binding on the nicotinic acetylcholine receptor (nAChR) leads to the accumulation of the acetylcholine neurotransmitter, resulting in the paralysis and death of the insect (Okazawa et al. 1998).

The seed-dressing for sunflowers was launched on the French market in 1994 with the name of Gaucho®. From 1995, beekeepers have observed the death of numerous bees and the decrease in honey production (Fig. 44.1). This problem has worsened with the increasing use of Gaucho® on sunflower and on maize. As a result, imidacloprid has been suspected as having harmful effects on honeybees in fields.

Imidacloprid exhibits a high oral toxicity to honeybees. The oral lethal dose 50% (LD_{50}) is observed between 49 and 102 ng per bee (Nauen et al. 2001), 3.7 and 40.9 ng per bee (Schmuck et al. 2001), 5 ng per bee (Suchail et al. 2001) or 40 and 60 ng per bee (Suchail et al. 2001). These values correspond to a lethal food concentration ranging between 0.1 and 1.6 mg kg^{-1} (Schmuck et al. 2001). The contact LD_{50} is about 24 ng/bee at 24 and 48 h (Suchail et al. 2001). The sub-lethal effect of imidacloprid on bees has not been investigated until recently. New studies have shown that the crucial functions of bees such as foraging are affected by sub-lethal doses of imidacloprid in the range from 1 to 20 µg kg^{-1} and from 0.1 to 2 ng/bee (Pham-Delegue and Cluzeau 1999; Colin and Bonmatin 2000). A dose of 0.1 ng per bee can also induce a decrease of habituation (Guez et al. 2001). Today, the action of imidacloprid on bees is not yet fully understood. There are at least two levels of toxicity, one centred at about 5 ng/bee and the other one at around 40 ng/bee. Moreover, imidacloprid and its metabolites would still be very toxic at much lower doses due to chronic intoxication. Briefly, studies converge and show a complex mechanism with an important toxicity at doses of imidacloprid in the µg kg^{-1} range, and even for lower amounts (Colin and Bonmatin 2000; Colin 2001; Guez et al. 2001; Belzunces 2001).

Nevertheless, data from Bayer AG indicated the absence of imidacloprid and its metabolite residues in the flowering of sunflowers raised from Gaucho® dressed seeds. This resulted from a sharp decrease of the imidacloprid content during the growth of treated plants. However, this data came from analysis using high performance liquid chromatography coupled with UV detection (HPLC/UV), which only allows quantification from 20–50 µg kg^{-1} (Placke and Weber 1993).

Based on the fact that harmful effects on bees appear in the µg kg^{-1} concentration range, a more sensitive methodology was clearly needed to precisely determine the insecticide content of flowers and pollens. Thus, new studies were launched in France by government agencies from 1998 (AFSSA, CNRS, INRA). One of the aims was to develop modern analytical methods allowing a limit of quantification (LOQ) near 1 µg kg^{-1}. For a comprehensive approach of the imidacloprid behaviour in fields, methods had to be efficient for soils, plants, flowers and pollens from field samplings.

Several methods have been reported for the analysis of imidacloprid. Although imidacloprid residues can be analysed by derivatization and gas chromatography methods (Macdonald and Meyer 1998; Uroz et al. 2001), HPLC appears to be a good alternative because of the thermolability and polarity of imidacloprid (Baskaran 1997; Yih-Fen and Powley 2000; Pous et al. 2001). HPLC gave sensitive results for imidacloprid in groundwater (Martinez-Galera et al. 1998), but the limit of detection (LOD) had to be lowered for the present study. Here, we developed analysis by high performance liquid chromatography coupled with mass spectrometry in tandem (HPLC/MS/MS) to detect and quantify imidacloprid in soils, plants, flowers and pollens with a LOD of 0.1 µg kg^{-1} and a LOQ of 1 µg kg^{-1} (Bonmatin et al. 2000a,c; Bonmatin et al. 2001).

44.2
Experimental

44.2.1
Samples and Extraction

Soils, plants and pollens, were sampled throughout France and were analysed to follow the fate of imidacloprid in the environment. The chemical composition of soils, plants and pollens is very heterogeneous. Each material contains organic compounds which can induce perturbations for the analysis. Thus, efficient extraction was needed to detect imidacloprid with a very low limit of detection. That is why three specific extraction methods were applied. Supplementary samples of soils, plants and pollens, which were totally free of imidacloprid, were spiked with known amounts of imidacloprid for calibration and assessment of the quality of our experiments. To identify the presence of imidacloprid, the treated samples were also compared with organically-farmed samples. The extraction schemes are shown for soils, plants and pollens in Fig. 44.2.

44.2.2
Analysis by Liquid Chromatography-Mass Spectrometry

Although the extraction steps are efficient and selective, it is necessary to perform a separation by high performance liquid chromatography. Imidacloprid is thermo-labile, thus HPLC allows direct injection of the extract, whereas gas chromatography (GC) requires derivatization. The detection by mass spectrometry coupled to LC allows clear identification of the parent ion at $m/z = 256$, a very high specificity using two or more ions products ($m/z = 209$ and $m/z = 175$), and a very low limit of detection as detailed in the next section.

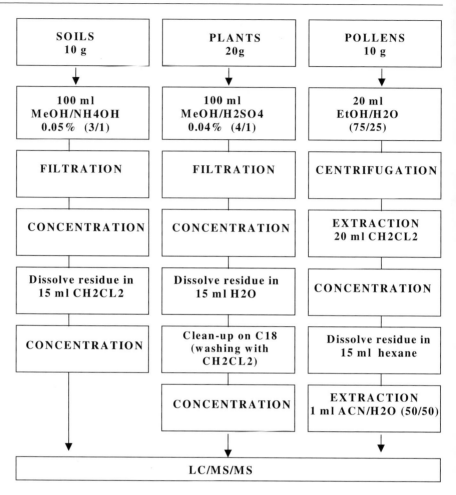

Fig. 44.2. Extraction schemes applied to soils, plants and pollens prior the separation and detection processes. Schemes were suited to obtain high recovery levels of imidacloprid at the level of the limit of quantification (1 µg kg^{-1})

Antipyrine was systematically used as an external standard between 10 injections of unknown samples. Unknown samples were also injected between series of standards and quality control samples. The retention time was constant with a deviation never exceeding 2.5%. The retention time was typically 2.4 min for the plants and 2.8 min for the pollen and soils samples depending on the chosen chromatographic flow. Calibration graphs were plotted for six standard solutions between 0.5 and 20 µg kg^{-1} of imidacloprid added in extracted materials (blanks). For each calibration point, three injections into the liquid chromatograph were performed. Linearity was found in the range between 0.5 and 20 µg kg^{-1}. The correlation coefficients were better than 0.99.

Fig. 44.3.
Scheme of the selected fragmentations and the mass spectra corresponding to the product ions of imidacloprid

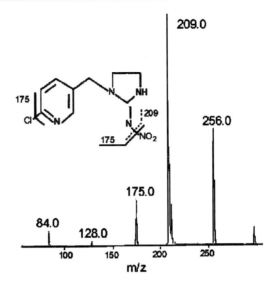

The specificity was performed by following two fragmentations of imidacloprid. The first fragment at $m/z = 209$ is due to the loss of NO_2. The second fragment at $m/z = 175$ is due to the losses of both NO_2 and Cl (Fig. 44.3). The chromatograms of product ions are clearly defined and are specific for imidacloprid. From multiple reaction monitoring (MRM) experiments, the ratio of the two product ion signals (175/209) gives an averaged value near 0.4. The latter value comes from the analysis of standards. Such an average value differs slightly depending on the starting material (soils, plants, pollens). However, when performing analysis of unknown samples, it did not vary by more than 20%, thus satisfying the corresponding quality criteria. Additionally, the ratio of $^{35}Cl/^{37}Cl$ imidacloprid signals was checked as a supplementary criteria of specificity.

The limit of detection was determined according to the signal of the less intense product ion. This limit was 0.1 µg kg^{-1} for soils and plants and 0.3 µg kg^{-1} for pollens. The limit of quantification was 1 µg kg^{-1} for each material type, with a S/N = 10, the latter being averaged from 10 injections. Reproducibility was tested six times for three concentrations of imidacloprid (1, 10, 18 µg kg^{-1}). The relative standard deviation values were lower than 18% for the plants and lower than 15% for pollens and soils. These values are in accordance with the quality criteria. The recovering rates (i.e. quantified quantity/theoretic quantity) are *(i)* in the 85–86% range for soils, *(ii)* in the 78–82% range for plants and *(iii)* in the 78–85% range for pollens. Note that relative concentrations of imidacloprid in the present paper were not recalculated to take account of the recovering rates, so these concentrations can be considered as being minimised by near 20%. As a matter of fact, the method respects the criteria (retention time, ratio between the products ions, signal-to-noise ratio…) of the directive 96/23 EC which is designed for banned substances. It is thus not surprising that manufacturers including Bayer AG, nowadays adopt such methods for a better characterisation of their products (Yih-Fen and Powley 2000; Schmuck et al. 2001).

44.3
Results and Discussion

44.3.1
Soil Analysis

French soils from numerous areas with varying climates, soil compositions and rain exposures were sampled. This approach gives our results a statistical dimension, allowing a general extrapolation in France (see Fig. 44.4). Among 74 samples of soils, 7 came from organic farming areas and did not reveal any signal of imidacloprid. These organically-farmed samples can be considered as control samples and demonstrate that the sampling procedure was performed without any external contamination. A total of 67 unknown samples were analysed. They are classified according to their imidacloprid content in Fig. 44.4 and we observed that 62 samples (91%) contain imidacloprid unambiguously (LOD of 0.1 µg kg^{-1}). In 65% of the cases, imidacloprid was quantified at more than 1 µg kg^{-1}. Interestingly, only 10 samples came from areas where seeds were treated the year of sampling. Such a difference (between 15% of treated area and 91% of positive soils) is not surprising and originates clearly from the long lifetime of imidacloprid in soils as outlined below.

To study the situation of areas with treated seed at the year of sampling, 10 soils were sampled after the cultivation of treated plants (seed treatment). Here, the plants were maize, wheat and barley. Among these 10 soils, 9 contained imidacloprid between 2 and 22 µg kg^{-1} with an average value of 12 µg kg^{-1}. Only 1 sample contained a concentration lower than 1 µg kg^{-1}. Thus, if the sampling is performed in the year of treatment, imidacloprid is always present in the soils and easily detectable after cultivation. For comparison, the amount of imidacloprid in soils can reach several hundred of µg kg^{-1} during the cultivation, depending on the passed time between sowing and sampling.

Soils were also analysed when they were exposed to imidacloprid one or two years before the sampling year. Although 11 soils were not exposed to imidacloprid during the two years preceding sampling, 7 soils contained imidacloprid between 0.1 (LOD) and 1 µg kg^{-1} (LOQ) and one sample contained 1.5 µg kg^{-1}. Subsequently, we considered soils for which no treated plants were cultivated the year of sampling (n) but which had received treated seeds one or two years before ($n - 1$ and/or $n - 2$). A set of 33 samples corresponds to such cases. Imidacloprid is present in 97% of these soils.

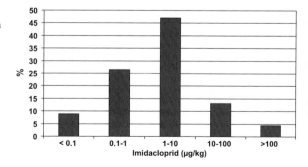

Fig. 44.4.
Sampled soils (%) as a function of the relative concentration of imidacloprid. Note that only 15% of these soils came from areas where seeds were treated

Moreover, its concentration ranges from 1.2 to 22 µg kg^{-1} in 78% of the samples and the mean value is 6 µg kg^{-1} in untreated soils (year n) which received treated seeds one year before sampling ($n - 1$). The mean value reaches 8 µg kg^{-1} in untreated soils which received treated seeds at years $n - 1$ and $n - 2$. This shows that a slight accumulation phenomenon in soils cannot be excluded.

Our data illustrates and confirms the strong retention of imidacloprid in soils. Note that the half life of imidacloprid in soils was already characterised from DT50 = 188 to 249 d by laboratory experiments (Belzunces anf Tasei 1997) while the DT50 of metabolites are still unknown.

44.3.2
Sunflower and Maize

A first set of 17 samples contained mature sunflowers coming from organically-farmed crops. These samples were analysed and no trace of imidacloprid was detected. Thus, these samples can be considered as control samples and demonstrate again that there was no external contamination during sampling.

Concerning untreated plants, sunflowers appear to be particularly capable of recovering the residual imidacloprid still present in soils from previous cultivation. Actually, with an average value of imidacloprid of about 6 µg kg^{-1} in soils at year $n + 1$, untreated sunflowers recover an average content of 1–2 µg kg^{-1} in flowering capitules. Imidacloprid is still detectable (LOD = 0.1 µg kg^{-1}) in untreated sunflowers even after two years of consecutive treated cultivation.

A second set of samples came from treated Gaucho® areas. The authorised dose for a Gaucho® treatment was 0.7 mg seed^{-1} for sunflowers. Despite the fact that the concentration of imidacloprid in growing plants sharply decreases with the time due to the increasing of biomass, a new phenomenon in sunflower capitulums was observed from all areas and for all studied sunflower varieties. Note that the part of sunflower named capitule is defined as a thick head of flowers on a very short axis, as a clover top, or a dandelion; a composite flower. Actually, from the appearance of the capitulums (about 40–50 d after sowing), the concentration of imidacloprid stops decreasing and, on the contrary, starts to increase. This increase of the imidacloprid concentration in capitulums is observed for 5 varieties of sunflowers. It is illustrated in Fig. 44.5 for one

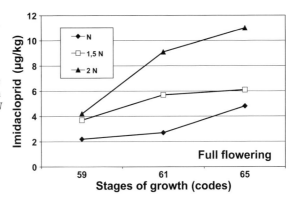

Fig. 44.5. Relative concentration (µg kg^{-1}) of imidacloprid in sunflowers capitutules as a function of the stages of growth. Codification of stages is defined according to Lancashire and Bleiholder 1991. Data are shown for 3 doses of seed-dressing: N, 1.5N and 2N (N: authorised dose on sunflowers). The sunflower variety is Rigasol. From the capitule formation (stage 59), the ascent of imidacloprid concentration is observed for the 3 doses of seed-dressing

variety, at 3 stages (59, 61, 65) of growth during the capitulum formation and for 3 doses of dressing: the authorised dose N, 1.5N and 2N. First, whatever the dose, the imidacloprid content significantly increases during flowering. Secondly, the concentration of imidacloprid in the capitulums, especially at full flowering, depends on the dose used for the seed-dressing, the contamination being greater when the dose is increased.

With regards to the behaviour of sunflowers depending on the variety, five varieties were treated with the authorised dose (N) and cultivated in the same area. The varieties were Pharaon, Rigasol, DK3790, Albena and Natil. Some varieties are rarely used in France (ex: Pharaon), while others are quite widely cultivated (ex: DK3790). Clearly, the decrease of imidacloprid quantities in sunflowers then followed by its ascent in the capitulum during flowering, were confirmed for all varieties of sunflowers, as seen in Fig. 44.6. The graph shows that values of imidacloprid in the capitulum during flowering, range from 2.5 µg kg^{-1} (Pharaon) to 9 µg kg^{-1} (Albena). Finally, the ascent of imidacloprid into the capitulums is a general phenomenon, leading to a final amount in the range of 1 to 10 µg kg^{-1} of the toxin in the flowering heads accessible to bees.

This new phenomenon is demonstrated in all sampled areas, varieties of sunflowers and doses of seed-dressing. Obviously, the ascent of imidacloprid during flowering is more pronounced when the doses of the seed dressing are high. With the authorised dose, the mean content of imidacloprid in sunflowers capitulums is 8 µg kg^{-1} during flowering. These results do not conflict with those from the manufacturer, which reported the absence of imidacloprid in flowers, since the limit of detection was unfortunately set at 20 µg kg^{-1} using HPLC/UV (Placke and Weber 1993). Our results are consistent with data from ^{14}C imidacloprid quantification in the head part of sunflowers. Here imidacloprid residues were found in the range from 5 to 30 µg kg^{-1} (Laurent

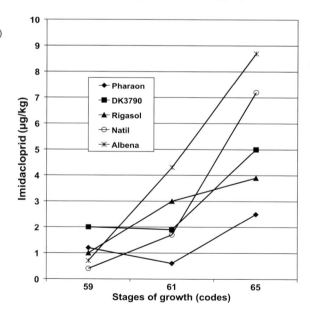

Fig. 44.6.
Relative concentration (µg kg^{-1}) of imidacloprid in sunflower capitules as a function of the stages of growth (Lancashire and Bleiholder 1991). Data are shown for 5 treated sunflowers varieties. Note that from stage 59, imidacloprid level increases until the full flowering (stage 65) for all varieties

and Scalla 2000). Note that ^{14}C methods are suited to laboratory or semi-laboratory investigations and cannot be used for investigations of a global and statistical description of the behaviour in fields.

44.3.3
Pollens

The study of pollens was performed on a set of sixty four sunflower pollen samples. A first set of eleven pollens came from organic farming crops and do not reveal the presence of imidacloprid. A second set of twenty nine pollens came from untreated Gaucho® crops in the year of sampling (n). Two of them showed a clearly positive result (LOD = 0.3 µg kg^{-1}) but concentrations were less than the LOQ = 1 µg kg^{-1}. As a matter of fact, the latter two pollens came from areas where Gaucho® crops were sowed the year before sampling ($n - 1$). This demonstrates that the persistence of imidacloprid in soils can sometimes lead to its recovery by untreated sunflowers sowed one year later. Thus, the toxin can reach the highest parts of the plant and can contaminate pollens. A third set of 24 samples of pollens from treated sunflowers were analysed. In only 17% of pollens, imidacloprid was not detected (LOD = 0.3 µg kg^{-1}). 25% of samples were positive with the amount of imidacloprid not exceeding 1 µg kg^{-1}. Other pollens (58%) contained imidacloprid at concentrations from 1 to 11 µg kg^{-1} and the mean value is centred at 3 µg kg^{-1}.

The mean value found in pollens from treated sunflowers corresponds to the quantity of imidacloprid inducing sub-lethal effects on foraging bees (Colin and Bonmatin 2000; Colin 2001). This means that the toxic is bioavailable during flowering and that foraging bees are first exposed to such doses by contact. Moreover, this means that pollens represent a way for the toxin to contaminate the beehive. Since pollens are stocked in the beehive and constitute a main source of nutriments (oral exposure). Our results on pollens are in agreement with the ^{14}C experiments from which the toxic content (imidacloprid and metabolites) was estimated at 13 µg kg^{-1}. Furthermore, our results are confirmed by Bayer AG (Schmuck et al. 2001). These authors recently used a similar LC/MS/MS technique to the one we developed, but with a 15-fold higher limit of detection at 1.5 µg kg^{-1}. They reported 3.3 µg kg^{-1} in pollens and 1.9 µg kg^{-1} in nectars.

44.4
Conclusion

We developed three extraction schemes followed by LC/MS/MS method to detect imidacloprid from field samples. These analytical methods are designed to reveal (limit of detection of 0.1 µg kg^{-1}) and quantify (limit of quantification of 1 µg kg^{-1}) very low concentrations of imidacloprid in soils, plants and pollens. To date, these methods are the most sensitive methods available to analyse such materials according to good laboratory practice and quality criteria from the directive 96/23/EC (Bonmatin 2002).

The long persistence, after one and two years, of imidacloprid in soils has been demonstrated in this study. Retention of imidacloprid in soils, coupled with the ability of sunflowers to recover the insecticide during the next cultivation, clearly explains the presence of imidacloprid in untreated plants. This situation is also observed for maize and several weeds or adventitious plants (plants which grow in fields but which have not been sown). For untreated wheat, rape and barley, imidacloprid is also recovered to a lesser extent from contaminated soils (Bonmatin et al. 2000b).

Seed treatment using imidacloprid protects plants against insects and is supposed to vanish before the arrival of pollinator insects. However, a new phenomenon has been demonstrated. We have shown that the relative amount of imidacloprid reaches a minimum, then increases in sunflowers from the time of the capitulum formation. As a consequence, relatively high levels are observed during flowering in the flowering heads. At this time, the capitulums of sunflowers contain a mean value of 8 μg kg^{-1} of imidacloprid. Another study on maize indicates a similar situation. The ascent of imidacloprid during flowering appears to be general behaviour, due to both enhanced metabolism and the strong mobilisation of resources for plants producing large amounts of grains such as sunflowers and maize.

Our data reveals the presence of imidacloprid in pollens with average values of 3 μg kg^{-1} (sunflowers and maize). Thus, imidacloprid appears to be bioavailable for bees in fields, in a range of concentrations corresponding to that of sub-lethal effects on bees and especially concerning the foraging activity (Colin and Bonmatin 2000; Colin 2001). This risk situation with respect to sunflowers and maize is worsened when considering *(i)* the additional toxic action of several imidacloprid metabolites (Nauen et al. 1998; Oliveira et al. 2000) as well as *(ii)* the very low concentrations inducing chronic mortality of bees which are in the 0.1–1 μg kg^{-1} range (Suchail et al. 2001; Belzunces 2001).

The commercialisation and the use of Gaucho® on sunflowers have been suspended in France since 1999 (J.O.R.F. 1999).

Acknowledgements

This expertise was carried out for the French Ministries of Agriculture and Environment with the financial support of the EC 1221/97 program. The authors would like to thank Dr Sophie Lecoublet and Valérie Charrier for their participation in this research.

References

Baskaran K, Kookana RS, Naidu RJ (1997) Determination of the pesticide imidacloprid in water and soil using high-performance liquid chromatography. J Chromatogr 787:271–275
Belzunces LP (2001) In: Rapport d'étude 2000-2001 au Ministère de l'Agriculture et de la Pêche, VIème Programme Communautaire pour l'Apiculture, Projet 2106
Belzunces LP, Tasei J-N (1997) Impacts sur les dépeuplements de colonies d'abeilles et sur les miellées. In: Rapport au Ministère de l'agriculture sur les effets des traitements de semences de tournesol au Gaucho® (imidacoprid)
Bonmatin J-M (2002) Insecticides et pollinisateurs: une dérive de la chimie? Science 2:42–46
Bonmatin J-M, Moineau I, Colin M-E, Bengsch ER, Lecoublet S and Fleché C (2000a) Effets des produits phytosanitaires sur les abeilles. Programmes 1999-2000 AFSSA-CNRS-INRA. In: Rapport de résultats N°3 au Ministère de l'Agriculture et de la Pêche

Bonmatin J-M, Moineau I, Colin M-E, Fleché C, Bengsch ER (2000b) Insecticide imidacloprid: availability in soils and plants, toxicity and risk for honeybees. EPRW 2000. Pesticides in Food and drink. p 134

Bonmatin J-M, Moineau I, Colin M-E, Fléché C, Bengsch ER (2000c) L'insecticide imidaclopride: Biodisponibilité dans les sols et les plantes, toxicité et risque pour les abeilles. Revue Française d'Apiculture 609:360-361

Bonmatin J-M, Moineau I, Lecoublet S, Colin M-E, Fléché C, Bengsch ER (2001) Neurotoxiques systémiques: biodisponibilité, toxicité et risque pour les insectes pollinisateurs – le cas de l'imidaclopride-, Produits Phytosanitaires. Eds. Presse Universitaires, Reims, France, pp 175-181

Colin M-E (2001) Influence des insecticides systémiques sur l'apprentissage spatio-temporel de l'abeille, Public conférence INRA/UAPV, 22 May 2001, Avignon, France

Colin M-E, Bonmatin J-M (2000) Effets de très faibles concentrations d'imidaclopride et dérivés sur le butinage des abeilles en conditions semi-contrôlées. In: Rapport au Ministère de l'agriculture et de la pêche

Directive 96/23/CE (1996) (Revision of commission decision 93/256/EC) Commission decision laying down analytical methods to be used for detecting certain substances and residues there of in live animals and animal products according to Council Directive 96/23/EC

Guez D, Suchail S, Gauthier M, Maleszka R, Belzunces L (2001) Contrasting effects of imidacloprid on habituation in 7 and 8 d old honeybees (Apis mellifera). Neurobiol Learning Memory 76:183-191

Journal Officiel de la République Française (1999) Avis aux détenteurs et aux utilisateurs de semences de tournesol. Ministère de l'agriculture et de la pêche, 14 février 1999, p 2413

Lancashire PD, Bleiholder H (1991) A uniform decimal code for growth stages of crops and weeds. Ann Appl Biol 119:561-601

Laurent F, Scalla R (2000) Transport et métabolisme de l'imidaclopride chez le tournesol. 3ème Programme Communautaire pour l'apiculture. Année 1999-2000. In: Rapport au Ministère de l'Agriculture et de la Pêche

MacDonald L, Meyer TJ (1998) Determination of imidacloprid and triadimefon in white pine by gas chromatography/mass spectrometry. Agric Food Chem 46:3133-3138

Martinez-Galera M, Garrido-Frenich A, Martinez-Vidal JL, Parrilla-Vasquez P (1998) Resolution of imidacloprid pesticide and its metabolite 6-chloronicotinic acid using cross-sections of spectrochromatograms obtained by high-performance liquid chromatography with diode-array detection. J Chromatogr A 799:149-154

Nauen R, Hungenberg H, Tollo B, Tietjen K, Elbert A (1998) Antifeedant effect, biological efficacy and high affinity binding of imidacloprid to acetylcholine receptors in Myzus persicae and Myzus nicotianae. Pest Managem Sci 53:133-140

Nauen R, Ebbinghaus-Kintscher U, Schmuck R (2001) Toxicity and nicotinic acetylcholine receptor interaction of imidacloprid and its metabolites in Apis mellifera (Hymenoptera: Apidae). Pest Managem Sci 57:577-586

Okazawa A, Akamatsu M, Ohoka A, Nishiwaki H, Cho WJ, Nakagawa N, Ueno T (1998) Prediction of the binding mode of imidacloprid and related compounds to house-fly head acetylcholine receptors using three-dimensional QSAR analysis. Pestic Sci 54:134-144

Oliveira RS, Koskinen WC, Werdin NR, Yen PY (2000) Sorption of imidacloprid and its metabolites on tropical soils. J Environ Sci Health B35:39-49

Pham-Delegue MH, Cluzeau S (1999) Effets des produits phytosanitaires sur l'abeille; incidence du traitement des semences de tournesol par Gaucho sur les disparitions de butineuses. Rapport final de synthèse au Ministère de l'Agriculture et de la Pêche

Placke FJ, Weber E (1993) Method for determination of imidacloprid residues in plant materials. Pflanzenschutz-Nachrichten Bayer 46/1993, p 2

Pous X, Ruiz M, Pico Y, Font G (2001) Determination of imidacloprid, metalaxyl, myclobutanil, propham, and thiabendazole in fruits and vegetables by liquid chromatography-atmospheric pressure chemical ionization-mass spectrometry. Fresenius J Anal Chem 371:182-189

Schmuck R, Schöning R, Stork A, Schramel O (2001) Risk posed to honeybees (Apis mellifera L., Hymenoptera) by an imidacloprid seed dressing of sunflowers. Pest Managem Sci 57:225-238

Suchail S (2001) Etude pharmacocinétique et pharmacodynamique de la létalité induite par l'imidaclopride et ses métabolites chez l'abeille domestique (*Apis mellifera* L.). Thèse, Université Claude Bernard-Lyon I, N°04-2001

Suchail S, Guez D, Belzunces LP (2001) Discrepancy between acute and chronic toxicity induced by imidacloprid and its metabolites in *Apis mellifera*. Environ Toxicol Chem 20:2482–2486

Uroz FJ, Arrebola FJ, Egea-Gonzales FJ, Martinez-Vidal JL (2001) Monitoring of 6-chloronicotinic acid in human urine by gas chromatography-tandem mass spectrometry as indicator of exposure to the pesticide imidacloprid. Analyst 126:1355–1358

Yih-Fen M, Powley CR (2000) LC/MS/MS: The method of choice in residue analysis. EPRW 2000. Pesticides in Food and Drink, p 102

Chapter 45

Impact of a Sulfonylureic Herbicide on Growth of Photosynthetic and Non-Photosynthetic Protozoa

T. Kawano · T. Kosaka · H. Hosoya

Abstract

We studied the impact of a sulfonylurea-based herbicide on the growth of aquatic microbes including photosynthetic and non-photosynthetic protozoa, and ex-symbiotic and non-symbiotic free living algae. This herbicide is designed to block the biosynthesis of branched amino acids in plants. A commercial sulfonylurea-based herbicide containing methyl 3-[[(4-methoxy-6-methyl-1,3,5-triazin-2-yl)carbamoyl]-sulfamoyl]-2-thiophenecarboxylate was added to the culture of *Paramecium trichium*, *P. caudatum*, *P. bursaria* (green paramecia), and *Euglena gracilis*, an ex-symbiotic algae isolated from green paramecia and free-living *Chlorella*, at various concentrations ranging from 0.01 to 1 000 mg l^{-1}. The viability of protozoa and algae was examined under microscopy, 1 week after addition of the herbicide. High concentrations of herbicide (100–300 mg l^{-1}) were shown to be inhibitory to the growth of symbiotic algae and free living *Chlorella*. The same range of herbicide concentrations was shown to be at lethal level for green paramecia and other non-photosynthetic paramecia. The herbicide showed no lethal effect in *Euglena gracilis*. Instead, the growth in *Euglena gracilis* was markedly enhanced. In addition, treatment with the herbicide resulted in marked changes in size and shape of the *Euglena* cells seriously affecting the euglenoid movement. Lastly we conclude that the sulfonylureic herbicides may not be harmless to aqueous environment.

Key words: algae, *Euglena*, herbicide, *Paramecium bursaria*, protozoa, sulfonylurea

45.1
Introduction

Protozoan cells are often used as bioindicators for chemical pollution, especially in the aqueous environment (Apostol 1973). Among protozoa, *Tetrahymena pyriformis* is the most commonly used ciliated model for laboratory research (Sauvant et al. 1999). Paramecia are also used to study the effect of water pollution by monitoring the cell motility and also by examining the frequency of the emergence of abnormal strains (Komala 1995). Cells of ciliates including *Paramecium* species and *T. pyriformis* are reportedly useful in the ciliate mobility test carried out in the presence and absence of various components determined as air pollutants (Komala 1995; Graf et al. 1999). In addition to the ciliate mobility test, Sauvant et al. (1999) described the methodological features of recently developed toxicological and eco-toxicological bioassays performed with ciliates, based on cell growth rate, biochemical markers and behavioural changes.

However, the data obtained for aqueous protozoa may not be applicable for other aqueous microorganisms such as algae. To survey the toxic chemicals targeting both or either of animals (protozoa) and plants (algae) in aqueous environment, we are now developing a sensitive biomonitoring system for the detection of toxic compounds using green paramecia and other photosynthetic protozoa.

In this study, we used the photosynthetic protozoa and ex-symbiotic and free living algae for assessing the impact of a sulfonylurea-based herbicide, which is designed to block the biosynthesis of branched amino acids in plants, on aqueous environment. It has been suggested that sulfonylureic herbicides mimic the endogenous amino acids and bind to acetolactate synthase (ALS), the key enzyme involved in biosynthesis of branch-chain amino acids (valine, leucine, and isoleucine) in plants (Stetter 1994; Whitcomb 1999). Since ALS is regulated through feed back control by valine and leucine, binding of the ALS inhibitors including sulfonylureas at the regulatory site on ALS results in inhibition of biosynthesis of branch-chain amino acids (Durner et al. 1991).

The sulfonylureic herbicides are known as growth inhibitors of many species of higher plants (Grossmann et al. 1992) as well as bacteria, fungi, yeast, and algae (Peterson et al. 1994; Wei et al. 1998; Whitcomb 1999). However, the impact of sulfonylureic herbicides on aqueous protozoa has not been reported to date. Here, a commercial sulfonylureic herbicide containing methyl 3-[[(4-methoxy-6-methyl-1,3,5-triazin-2-yl)carbamoyl]sulfamoyl]-2-thiophenecarboxylate (known as thifensulfuron methyl; 75% purity, w/w) was added to the culture of photosynthetic protozoa including *Euglena gracilis* and *Paramecium bursaria* (green paramecia), ex-symbiotic algae isolated from green paramecia and free-living *Chlorella*, at various concentrations ranging from 0.01 to 1000 mg l^{-1}. We present the data as a groundwork for providing useful preliminary knowledge for assessing the impact of a widely used herbicide to the aqueous microecosystem.

45.2
Experimental

45.2.1
Food Organism for Ciliates

Klebsiella pneumoniae was cultured on agar medium containing polypepton (10 g l^{-1}), meat extracts (2 g l^{-1}), NaCl (5 g l^{-1}) and glucose (1 g l^{-1}). Colonies of *E. aerogenes* growing on agar medium were picked with the tip of a platinum loop and dipped into 1000 ml of lettuce infusion and incubated statically for 3 h at ambient temperature (23 °C). This liquid culture of *E. aerogenes* was used as a food organism for paramecia.

45.2.2
Ciliates, Flagellate and Algae

Three *Paramecium* species, *Paramecium bursaria* (strains, MB-1, MBw-1 and EZ-22), *P. trichium* (strains, BD-2 and NJ-1) and *P. caudatum* (strain, KY-1) were used. *P. bursaria* is a photosynthetic species that harbours endosymbiotic green algae morphologically similar to *Chlorella* in its cytosolic space. *P. trichium* and *P. caudatum* are non-photosynthetic species harbouring no endosymbiotic algae. MBw-1 is an alga-free strain of *P. bursaria* screened from the paraquate-treated MB-1 strain (Nishihara et al. 1998; Hosoya et al. 1995). This alga-free strain of *P. bursaria* is morphologically similar to *P. trichium* strains. The paramecia were cultured in lettuce infusion containing 10% volumes of food organism culture (*E. aerogenes*) under a 12 h light/12 h dark regimen (ca. 2000 lux) at 23 °C. Paramecia were occasionally subcultured by inoculating the fresh lettuce infusion mixed with the food organism culture (100 ml), with 5 ml of confluent *Paramecium* cultures. *Euglena gracilis* was cultured in the fresh lettuce infusion without addition of food organism. The culture of *Euglena* was kept under the light/dark light regimen (ca. 2000 lux) at 23 °C.

The free-living *Chlorella* kessleri (C-531) was obtained in 1998 from the Institute of Applied Microbiology (IAM) culture collection at the University of Tokyo, and ex-symbiotic algal clone (SA-1) was isolated from *P. bursaria* (Nishihara et al. 1998). The algae were cultured in liquid CA medium as described (Nishihara et al. 1996, 1998). The algal cultures were kept under the light/dark light regimen (ca. 2000 lux) at 23 °C.

45.2.3
Herbicide

Figure 45.1 shows the structure of a sulfonylurea-based herbicide, methyl 3-[[(4-methoxy-6-methyl-1,3,5-triazin-2-yl)carbamoyl]sulfamoyl]-2-thiophenecarboxylate, known as thifensulfuron methyl. Harmony 75, a commercially available herbicide containing 75% (w/w) thifensulfulon methyl, was purchased from DuPont agricultural Caribe Industries Ltd. (Caribe, Puerto Rico). Harmony 75 was first dissolved in water and added to the culture of protozoa or algae at various concentrations ranging from 0.01 to 1000 mg l^{-1}, covering the recommended concentrations (0.01–0.1 mg l^{-1}) used for killing the sulfonylurea-sensitive plants (Gerwick et al. 1993). In addition, the experiments are designed to cover the higher doses used in the agricultural fields. According to the user's guideline supplemented with Harmony 75, plant surface and soil surface are likely exposed to high concentration of the herbicide. The herbicide manufacturer suggested the direct spraying of 1 m^3 of herbicide solution (30–50 mg l^{-1}) onto the leaves and seedlings of weeds growing in 1 ha of corn field (3–5 g ha^{-1}). Slightly higher doses (5–10 g ha^{-1}) are recommended for the weed control in wheat fields.

45.2.4
Herbicide and Growth of Algae and Protozoa

Paramecia were cultured on 24-well microplates. Each well was filled with 500 µl of fresh lettuce infusion without food organisms. The culture was initiated by addition of 200 cells of Paramecia to each well. Then cells were incubated in the presence and absence of various concentrations of the herbicide in microplates for seven days at 23 °C. After seven days of culture, the number of cells was counted under microscope using depletion slides.

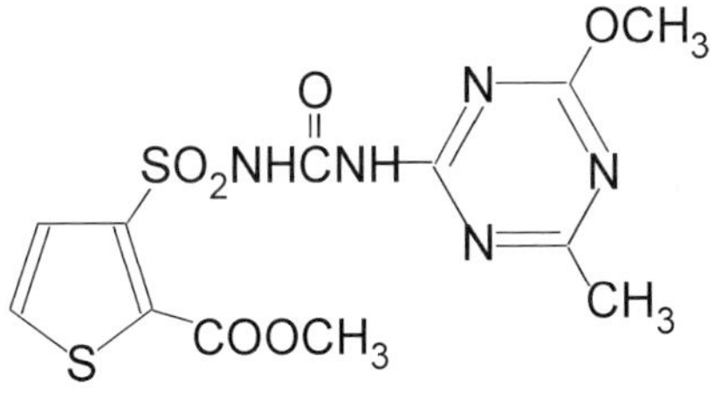

Fig. 45.1.
Structure of a sulfonylurea-based herbicide, methyl 3-[[(4-methoxy-6-methyl-1,3,5-triazin-2-yl)carbamoyl]-sulfamoyl]-2-thiophene-carboxylate, known as thifensulfuron methyl

Algal cells (ex-synbiotic and free-living *Chlorella*) and *Euglena* cells were also incubated in the fresh lettuce infusion without food organisms, containing various concentration of the herbicide. Initially, numbers of algal and *Euglena* cells were adjusted to be ca. 10^7 cells ml^{-1} and 3×10^3 cells ml^{-1}, respectively. Numbers of cells survived for seven days in the presence and absence of the herbicide were counted under microscopes using a hemocytometer. The number of the cells in the controls (without herbicide treatment) was expressed as 100% of cells, and the relative changes in cell numbers due to the action of herbicide was analysed.

45.2.5
Optical Measurements

Populational changes in morphology of *Euglena* cells were analysed on a FACSCalibur flow cytometer (Becton-Dickinson Immunocytometry Sytems, San Jose, CA) equipped with a 15 mW argon-ion laser (488 nm) as described previously (Gerashchenko et al. 2000, 2001). The fluorescence of endogenous chlorophyll was measured in the red fluorescence channel (FL-3) through a 650 nm long pass filter with logarithmic amplification. The forward (FSC) and 90° side scatter (SSC) signals were collected in linear mode. Approximately 1×10^4 events were measured. Analysis of the data was performed with CELLQuest software (Becton Dickinson). Cells were gated on the chlorophyll fluorescence signals to eliminate debris from the analysis. In addition, *Euglena* cells were morphologically characterised by bright field light microscopy (BH-2 microscope, Olympus, Japan). For detection of nuclei in the cells, the cells were stained with a fluorescent reagent, 4,6-diamino-2-phenylindole (DAPI) and localisation of the nuclei was determined (Gerashchenko et al. 2000, 2001).

45.3
Results and Discussion

45.3.1
Effect of the Herbicide on the Viability of Algae

As shown in Fig. 45.2, the effect of Harmony 75 treatment on the growth of free-living *Chlorella* (C-531) and ex-symbiotic algal clone (SA-1) isolated from *Paramecium*

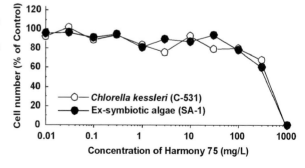

Fig. 45.2.
Effect of Harmony 75 treatment on the growth of free-living *Chlorella* and exsymbiotic algae isolated from *Paramecium bursaria*. Algal cells were counted under microscopes using a hemocytometer. Number of cells cultured in the absence of the herbicide is expressed as 100%. Number of cells in control culture is expressed as 100%

bursaria was examined. At lower concentration of Harmony 75, the growth of algae was not affected. In the presence of 300 mg l^{-1} of Harmony 75, the algal growth was slightly inhibited and in the presence of 1 000 mg l^{-1} of Harmony 75 growth of algae was completely inhibited. Since thifensulfulon methyl is reported as an inhibitory agent of the growth of plants and algae (Nystrom et al. 1999), we have expected that population of SA-1 and C-531 effectively decreases with increasing concentration of Harmony 75. However, unexpectedly, the toxic effect of Harmony 75 was observed only at very high concentrations between 300–1 000 mg l^{-1}.

45.3.2
Effect of the Herbicide on the Viability of Green Paramecia

Figure 45.3 shows the effect of Harmony 75 treatment on the growth of *P. bursaria*. Several strains of *P. bursaria* isolated in Japan were used and typical results from two green strains (MB-1 and EZ-22) are shown. An alga-free strain of *P. bursaria*, MBw-1, prepared from MB-1 by paraquat treatment was also used. In MB-1 and MBw-1, treatment with 30–100 mg l^{-1} of Harmony 75 enhanced the growth by 65–125%. By treatment with 300 mg l^{-1} of Harmony 75, growth of both MB-1 and MBw-1 were significantly inhibited and in the presence of 1 000 mg l^{-1} of Harmony 75, no living *Paramecium* was observed.

Growth in EZ-22 cells was not affected by treatment with 0.03–30 mg l^{-1} of Harmony 75. By 100 mg l^{-1} of Harmony 75, number of the live cells was lowered to 25% of controls. In the presence of 300 mg l^{-1} and higher concentrations of Harmony 75, no live cell was observed.

The behaviour of MB-1 and MBw-1, the two strains with and without endosymbiotic algae originally derived from the same clone, in the presence of the herbicide was almost identical. Therefore it has to be concluded that target of this sulfonylurea-based herbicide is not the endosymbiotic algae.

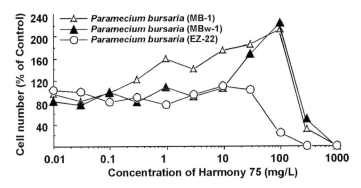

Fig. 45.3. Effect of Harmony 75 treatment on the growth of *Paramecium bursaria*. Several strains of *P. bursaria* isolated in Japan were used and typical results from two green strains (MB-1 and EZ-22) are shown. Alga-free sub-strain (MBw-1) prepared from MB-1 by paraquat treatment was also used. Paramecia sampled onto depletion slides were counted under microscopes. Number of cells cultured in the absence of the herbicide is expressed as 100%

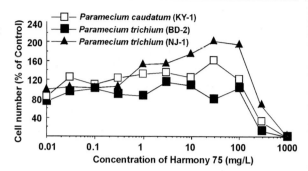

Fig. 45.4.
Effect of Harmony 75 treatment on the growth of *Paramecium caudatum* and *Paramecium trichium*. Paramecia sampled onto depletion slides were counted under microscopes. Number of cells in the control culture is expressed as 100%

45.3.3
Effect of the Herbicide on the Viability of Non-Photosynthetic Paramecia

Since the target of the Harmony 75 action in the green paramecia was shown to be the host *Paramecium*, the inhibitory effect of Harmony 75 on the growth of non-photosynthetic paramecia was also examined. Here, two species of paramecia, *P. caudatum* (KY-1) and *P. trichium* (BD-2, NJ-1) were used. Figure 45.4 shows the effect of Harmony 75 treatment on the growth of *P. caudatum* and *P. trichium*. The culture of *P. caudatum* KY-1 cells behaved similarly to *P. bursaria* MB-1 and MBw-1, in the presence of the herbicide. The growth of *P. caudatum* KY-1 was stimulated (by ca. 50–100%) by treatment with 1.0–100 mg l^{-1} of the herbicide. The growth of *P. trichium* BD-2 strain was also stimulated (to a minor extent) by the herbicide (1.0–100 mg l^{-1}). The impact of the herbicide treatment was much smaller in *P. trichium* NJ-1 strain. Up to 100 mg l^{-1} of Harmony 75 showed no significant effect on growth of the NJ-1 culture. The herbicide concentrations higher than 100 mg l^{-1} were shown to be toxic to all paramecia tested and 1 000 mg l^{-1} of Harmony 75 was shown to be absolutely lethal. Thus, live cells of *P. caudatum* and *P. trichium* were no longer observed in the presence of 1 000 mg l^{-1} of Harmony 75.

45.3.4
Effects of the Herbicide on Growth and Morphology of *Euglena gracilis*

A photosynthetic flagellate, *Euglena gracilis* was also cultured in the presence and absence of Harmony 75 for 7 d. Treatment of *Euglena gracilis* cells with the herbicide resulted in marked stimulation of growth (Fig. 45.5). In the presence of high concentrations of the herbicide (30–1 000 mg l^{-1}), the number of the cells were 2.5-fold greater than that in the control. At any concentration examined, no lethal effect of the herbicide was observed in *Euglena*.

Figure 45.6 shows the results from the flow cytometric analysis of the Harmony 75-induced cytological changes in *Euglena gracilis*. The cells of *Euglena gracilis* were incubated for 1 week in the presence and absence of the herbicide. Approximately 10^4 cells were analysed on a FACSCalibur flow cytometer (Becton-Dickinson Immunocytometry systems, San Jose, CA. USA). While the chlorophyll content in the cells (determined as changes in intensity of the chlorophyll-specific red fluorescence; Fig. 45.6, vertical scale) was not affected by any concentration of the herbicide, the

Chapter 45 · Impact of a Sulfonylureic Herbicide on Growth of Protozoa   501

Fig. 45.5.
Effect of Harmony 75 treatment on the growth of *Euglena gracilis*. Cells were counted under microscopes using a hemocytometer. Number of cells in the control culture is expressed as 100%

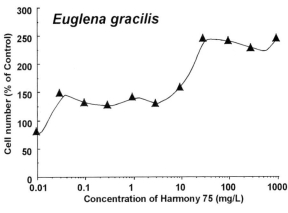

Fig. 45.6.
Flow cytometric analysis of the Harmony 75-induced cytological changes in *Euglena gracilis*. The cells of *Euglena gracilis* were incubated for 1 week in the presence and absence of the herbicide. Approximately 10 000 cells were analysed on a FACSCalibur flow cytometer (Becton-Dickinson Immunocytometry systems, San Jose, CA. USA). Two-parameter light scatter plots, forward scatters (indicating the cell size, *horizontal scale*) were plotted against chlorophyll fluorescence (FL-3, *vertical scale*). The herbicide-induced changes in cell size were clearly shown. Chlorophyll content in the cells was not affected by the herbicide

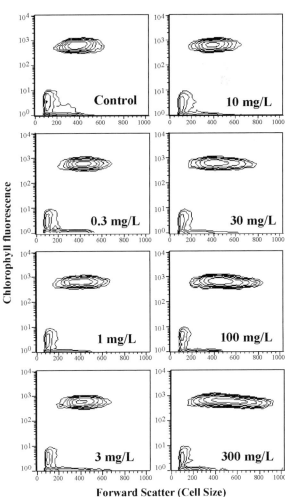

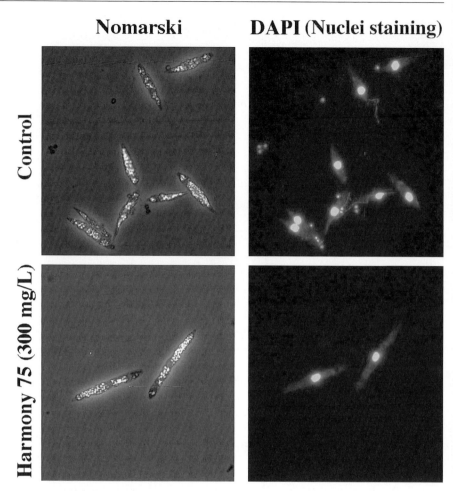

Fig. 45.7. Microscopic analysis of the cytological changes in *Euglena* cells. Harmony 75-treated resulted in irreversible elongation of cells. Nomarski differential interference contrast (*left*) and the fluorescence images (*right*) are shown. Localisation of the nuclei was determined by fluorescent staining with DAPI

size of cells (determined as the forward scatter; Fig. 45.6, horizontal scale) was markedly altered by high concentration of the herbicide.

As shown in Fig. 45.7, microscopic analysis revealed the Harmony 75 induced change in cell shape. Treatment with Harmony 75 (30–1 000 mg l^{-1}) of *Euglena* cells resulted in extreme elongation of cells: typical results in the presence of 300 mg l^{-1} Harmony 75 is shown in Fig. 45.7. Thus it has to be concluded that the Harmony 75-induced increase in the forward scatter value (cell size) obtained on flow cytometeric analysis must be due to the induced elongation of the cells. In addition, it has been clearly shown that cell elongation is not due to inhibition of cell division since the Harmony 75-treated elongating cell has single nucleus (Fig. 45.7) and the number of cells were increased by 2.5-fold (Fig. 45.5).

The euglenoid flagellates are able to change their shape rapidly in response to a variety of stimuli such as strong light. In this phenomenon called "euglenoid movement" or "metaboly", two extremes of shape can be identified: the "relaxed" form is cylindrical; the contracted form is round (Murray 1981). This type of movement can be observed in *Euglena gracilis* and *E. ehrenbergii*. It has been reported that contraction movement requires cytosolic increase in Ca^{2+} concentration and ATP (Murray 1981). The ionic requirement and involvement of ATP in the contraction movement has been confirmed by in-vitro experiments using detergent-extracted cells of euglenoid flagellates, *Astasia longa* (Suzaki and Williamson 1986) and *Euglena gracilis* (Murata and Suzaki 1998). It has been suggested that the mechanism for euglenoid movement is different from other known mechanisms of cellular movement such as flagellar movement (Suzaki and Williamson 1986). Surface of euglenozoan cells is covered by unique skeletal structure called pellicular strips that are tightly associated with plasma membrane and microtubules (Angeler et al. 1999). During the euglenoid movement (both contraction and relaxation), the pellicular strips dynamically interact with each other and generate the force required for rapid changes in cell shape (Suzaki and Williamson 1986). In this step, a plasma membrane integrated protein called crystallin may actively involved in cell shape change (Murata et al. 2000). The ultrastructure of the pellicle is currently considered as a valuable tool for systematic assessments within euglenoids and interspecific variations in pellicular strip construction may explain differences in form and degree of "metaboly" (Angeler et al. 1999).

45.4
Conclusion

The effect of a commercially available sulfonylureic herbicide, Harmony 75 which contains a sulfonylureic ALS inhibitor as the major component, on the growth of photosynthetic and non-photosynthetic protozoa, and ex-symbiotic and non-symbiotic free living algae was examined. The culture of *Paramecium trichium*, *P. caudatum*, *P. bursaria* (green *Paramecium*), *Euglena gracilis*, green *Paramecium* ex-symbiotic algae and free-living *Chlorella* were treated with various concentration of the herbicide. The growth of ex-symbiotic and free living algae was inhibited only when the concentration of added herbicide was higher than 300 mg l^{-1}. The same range of herbicide concentration was also lethal to the green paramecia and non-photosynthetic paramecia.

The growth of *Euglena gracilis* was stimulated by high concentrations of the herbicide (30–1000 mg l^{-1}). In addition, the herbicide induced marked changes in size and shape of the *Euglena* cells seriously affecting the euglenoid movement. Thus, sulfonylureic herbicides may not be harmless to *Euglena* cells.

Acknowledgement

The authors acknowledge Dr. B. I. Gerashchenko and Mr. K. Fumoto for their technical supports in optical measurements, and Mr. T. Kadono for his support in algal culture.

References

Angeler DG, Mullner AN, Schagerl M (1999) Comparative ultrastructure of the cytoskeleton and nucleus of *Distigma* (Euglenozoa). Eur J Protistol 35:309–318

Apostol S (1973) A bioassay of toxicity using protozoa in the study of aquatic environment pollution and its prevention. Environ Res 6:365–372

Durnner J, Gailus V, Boger P (1991) New aspects on inhibition of plant acetolactate synthase by chlorsulfuron and imazaquin. Plant Physiol 95:1144–1149

Gerashchenko BI, Nishihara N, Ohara T, Tosuji H, Kosaka T, Hosoya H (2000) Flow cytometry as a strategy to study the endosymbiosis of algae in *Paramecium bursaria*. Cytometry 41:209–215

Gerashchenko BI, Kosaka T, Hosoya H (2001) Growth kinetics of algal populations exsymbiotic from *Paramecium bursaria* by flow cytometry measurements. Cytometry 44:257–263

Gerwick BC, Mireles LC, Eilers RJ (1993) Rapid diagnosis of ALS AHAS-resistant weeds. Weed Technol 7:519–524

Graf W, Graf H, Wenz M (1999) *Tetrahymena pyriformis* in the ciliate mobility test. Validation and description of a testing procedure for the registration of harmful substances in the air as well as the effects of cigarette smoke on the human respiratory ciliated epithelium. Zentralbl Hyg Umweltmed 201:451–472

Grossmann K, Berghausr R, Retslaff G (1992) Heterotrophic plant-cell suspension cultures for monitoring biological-activity in agrochemical research – Comparison with screens using algae, germinating-seeds and whole plants. Pestic Sci 35:283–289

Hosoya H, Kimura K, Matsuda S, Kitamura M, Takahashi T, Kosaka T (1995) Symbiotic algae-free strains of the green *Paramecium Paramecium bursaria* produced by herbicide paraquat. Zool Sci 12:807–810

Komala Z (1995) Notes on the use of invertebrates, especially ciliates, in studies on pollution and toxicity. Folia Biol (Krakow) 43:25–27

Murata K, Suzaki T (1998) High-salt solutions prevent reactivation of euglenoid movement in detergent-treated cell models of *Euglena gracilis*. Protoplasma 203:125–129

Murata K, Okamoto M, Suzaki T (2000) Morphological change of cell-membrane-integrated crystalline structure induced by cell shape change in *Euglena gracilis*. Protoplasma 214:73–79

Murray JM (1981) Control of cell shape by calcium in the euglenophyceae. J Cell Sci 49:99–117

Nishihara N, Takahashi T, Kosaka T, Hosoya H (1996) Characterization of symbiotic algae-free strains of *Paramecium bursaria* by the herbicide paraquat. J Protozool Res 6:60–67

Nishihara N, Horiike S, Takahashi T, Kosaka T, Shigenaka Y, Hosoya H (1998) Cloning and characterization of endosymbiotic algae isolated from *Paramecium bursaria*. Protoplasma 203:91–99

Nystrom B, Bjornsater B, Blanck H (1999) Effects of sulfonylurea herbicides on non-target aquatic microorganisms – Growth inhibition of microalgae and short-term inhibition of adenine and thymidine incorporation in periphyton communities. Aquatic Toxicol 47:9–22

Peterson HG, Bouting C, Martin PA, Freemark KE, Ruecker NJ, Moody MJ (1994) Aquatic phyto-toxycity of 23 pesticides applied at expected environmental concentrations. Aquatic Toxicol 28:275–292

Sauvant MP, Pepin D, Piccinni E (1999) *Tetrahymena pyriformis*: a tool for toxicological studies. Chemosphere 38:1631–1669

Stetter J (1994) Chemistry of plant protection, vol. 10. Springer-Verlag, Berlin

Suzaki T, Williamson RE (1986) Reactivation of euglenoid movement and flagellar beating in detergent-extracted cells of *Astasia longa*: different mechanisms of force generation are involved. J Cell Sci 80:75–89

Wei LP, Yu HX, Sun Y, Fen JF, Wang LS (1998) The effects of three sulfonylurea herbicides and their degradation products on the green algae *Chlorella pyrenoidosa*. Chemosphere 37:747–751

Whitcomb CE (1999) An introduction to ALS-inhibiting herbicides. Toxicol Industr Health 15:231–239

Chapter 46

Abiotic Degradation of the Herbicide Rimsulfuron on Minerals and Soil

L. Scrano · S. A. Bufo · C. Emmelin · P. Meallier

Abstract

The photochemical behaviour of the sulfonylurea herbicide rimsulfuron, N-[[(4,6-dimethoxy-2-pyrimidinyl)amino]carbonyl]-3-(ethylsulfonyl)-2-pyridinesulfon amide, on silica and clay minerals, used as soil surrogates, was investigated and compared to a natural soil sample. The antagonistic behaviour of adsorption process and chemical degradation with respect to photodegradation was assessed and the formation of photoproducts was also determined. Results showed that all chemical and photochemical processes responsible for the disappearance of the herbicide follow a second order kinetic. The photochemical degradation of rimsulfuron was strongly affected by retention phenomena: with increasing of the adsorption capability of supports the photoreactivity of the herbicide decreased. The extraction rate of the herbicide covered the following values: soil 59.5%, illite 48.5%, aerosil 22.2%, montmorillonite 21.0%, showing that silica and clay minerals can retain and protect rimsulfuron from photodegradation much more than soil. Though, adsorption of the herbicide was always accomplished to a chemical reactivity of solid substrates. N-[(3-ethylsulfonyl)-2-pyridinyl]-4,6-dimethoxy-2-pyridineamine and N-(4,6-dimethoxy-2-pyrimidinyl)-N-[(3-(ethylsulfonyl)-2-pyridinyl)]urea were found both as photochemical and chemical metabolites.

Key words: rimsulfuron, soil adsorption, silica, clay minerals, chemical degradation, photodegradation

46.1
Introduction

Photochemical reactions contribute to abiotic degradation of pesticides in the environment (Scheunert 1992). The study of photodegradation processes is on different levels of development. The progress is highest for studies in the atmosphere, lower for surface water studies and lowest for investigations on soil compartment (Scheunert 1993). The main reason for the unsatisfactory status of development in the soil compartment is the inherent difficulty involved in working on non-homogeneous surfaces (Mingelgrin and Prost 1989). Therefore, it is recommended to simplify the investigation by using models which behaviour can be transferred to the natural environment (Klöpffer 1992). In the soil and colloidal fractions of soil, only surfaces exposed to solar irradiation can contribute to photodegradation (Albanis et al. 2002; Klöpffer 1992; Konstantinou et al. 2000). Considering the limitations and having in mind that adsorption is the most important reaction in soil, different surrogates can be used to simulate soil and soil component influences on pesticide fate (Jones 1991).

Generally, photolysis in soil will occur within a shallow surface zone, the depth of which depends on soil characteristics and photochemical properties of the target reactant. Direct absorption of light and photolysis of organic contaminants may be influenced by soil surface adsorption that is related to the content of colloidal materials. Indirect process could also be occurring depending on the presence of sensitising substances and singlet oxygen formation. Vertical depth for direct photolysis is generally restricted to 0.2–0.3 mm, indirect photolysis also below a layer of 0.7 mm (Herbert and Miller 1990). Humic substances in soil often act as sensitizers producing reactive intermediates such as singlet oxygen, hydroxyl radicals, superoxide anions, hydrogen peroxides and peroxy radicals. Such reactive species can potentially diffuse to a depth of 1 mm depending on soil moisture, porosity, and thermal gradient in sunlight exposed soil surface. Moreover, electronic structures, absorption spectra, and excited state lifetimes of soil adsorbed compounds are generally different from their solution properties, making it very difficult to predict what effects may result from soil-contaminant interactions (Albanis et al. 2002).

With respect to pesticide application on soil, a new class of agrochemicals named sulfonylureas has been developed in the last decade and offered on the market. The acceptance of sulfonylureas is mainly based on their low application rate (10–80 g ha^{-1}) and favourable environmental and toxicological properties (Beyer et al. 1988). Rimsulfuron is a selective sulfonylurea herbicide (compound 1 in Fig. 46.1) for the postemergence control of many crops. It was commercialised in Europe in 1992 by Du Pont de Nemours & Co. The pure active ingredient (a.i.) is a white odourless solid; its solubility is <10 mg l^{-1} in distilled water and 7 300 mg l^{-1} in buffer solution at pH 7 (25 °C); pK_a = 4.1; K_{OW} = 0.034 at pH 7; vapour pressure = 1.1 × 10^{-8} Torr at 25 °C. This product has little or no toxicological effects on mammals with oral LD$_{50}$ > 5 000 mg kg^{-1} in rats and with dermal LD$_{50}$ > 2 000 mg kg^{-1} in rabbits (Schneider et al. 1993). Using ^{14}C-labelled rimsulfuron Schneider et al. (1993) showed that degradation in aqueous solutions and soil environment does not depend on irradiation under natural sunlight, though it undergoes hydrolysis reactions. Besides, the same authors observed some effects of the natural light at pH 5. Neither rimsulfuron nor its metabolites were detected at soil depths lower than 8 cm in experimental fields under different crop management, in which manure treatments prolonged the herbicide half-life in the 0–8 cm surface soil

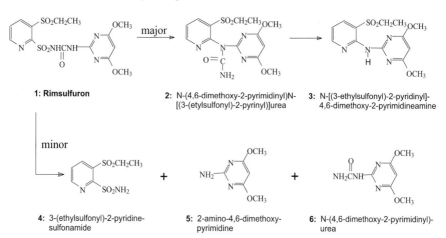

Fig. 46.1. Chemical structure and degradation pathways of rimsulfuron

layer from a minimum of 14 d (control) to a maximum of 46 d (Rouchaud et al. 1997). Scrano et al. (1999) showed that, under simulated sunlight in water, the half-life of photolysis reaction ranged from 1 to 9 d at pH 5 and 9, respectively. The hydrolysis rate was as high as the photolysis rate, and decreased on increasing pH values of the solution. The main metabolite identified in neutral and alkaline conditions as well as in acetonitrile was #3, *N*-[(3-ethylsulfonyl)-2-pyridinyl]-4,6-dimethoxy-2-pyridineamine, while #2, *N*-(4,6-dimethoxy-2-pyrimidinyl)-*N*-[(3-(ethylsulfonyl)-2-pyridinyl)]urea, and minor metabolites prevailed in acidic conditions (Fig. 46.1). Pantani et al. (1996) stated that rimsulfuron can be adsorbed on Al-hectorite (a smectite clay mineral), and decomposes on clay surfaces into two main metabolites.

The aim of the presented investigation was to study and compare the photochemical degradation of rimsulfuron adsorbed on a siliceous material and two clay minerals, used as soil surrogates, and a natural soil. The antagonistic behaviour of adsorption and other chemical processes with respect to photodegradation and the formation of photoproducts have been also considered.

46.2
Experimental

46.2.1
Materials

Solvent (pesticide grade), reagents (analytical grade) and filters (disposable sterilised packet) were purchased from Fluka and Sigma-Aldrich (Milan, Italy), aerosil 200 from Degussa (Dusseldorf, Germany), illite and montmorillonite from Ward's N.S.E. (Monterey, CA-USA). Ultrapure water was obtained with a Milli-Q system (Millipore, Bedford, MA-USA). Soil (Typic Rhodoxeralf) was sampled from *Sellata* area in Basilicata region, Southern Italy.

To avoid hydrolysis of rimsulfuron (Schneider et al. 1993), a stock solution (100 mg l^{-1}) of pure standard a.i. (98% – Dr. Ehrenstorfer GmbH, Augsburg, Germany) was prepared using anhydrous acetonitrile as solvent. This solution was maintained in the darkness at +4 °C and used to prepare working solutions (10 mg l^{-1}). Compounds #2 and #3 (Fig. 46.1) were prepared according to literature methods (Marucchini and Luigetti 1997; Rouchaud et al. 1997). All glass apparatus were heat sterilised by autoclaving for 60 min at 121 °C before use. Aseptic handling materials and laboratory facilities were used throughout the study to maintain sterility.

46.2.2
Adsorbed Phase Preparation

Physical and chemical properties of selected substrates are shown in Table 46.1. The soil was sieved (1 mm) and sterilised before use in order to avoid microbiological degradation (Cambon et al. 1998). Three replicates of each substrate were weighed and spiked drop by drop (with gentle stirring) with 0.2897 mmol kg^{-1} of rimsulfuron in acetonitrile (0.0232 mM). The paste thus prepared was spread on a glass plate in order to obtain 1 mm thick substrate layer. The plates were air dried in the darkness

Table 46.1. Composition of solid substrates. SS = specific surface; CEC = cationic exchange capacity; $pH(H_2O) = 6.50$; $pH(KCl\ 1N) = 5.63$

Adsorbent	Unit	Aerosil	Illite	Montmorillonite	Soil
SiO_2	%	99.55	54.71	54.30	45.82
TiO_2	%	0.00	0.91	0.71	0.81
Al_2O_3	%	0.36	18.90	18.66	19.69
Fe_2O_3	%	0.01	8.06	6.97	8.29
CaO	%	0.03	0.60	1.06	1.37
MgO	%	0.00	1.89	1.87	1.95
P_2O_5	%	0.00	0.21	0.02	0.08
K_2O	%	0.01	4.31	2.11	1.97
Na_2O	%	0.03	0.39	1.06	0.18
H_2O	%	0.45	10.01	13.22	19.44
SS	$m^2\ g^{-1}$	220	80	300	–
Size	µm	0.12	≤1	≤0.4	≤1 000
Org-C	%	–	–	–	4.78
Total-N	%	–	–	–	0.77
Sand	%	–	–	–	19.6
Silt	%	–	–	–	39.4
Clay	%	–	–	–	41.0
$pH_{(H_2O)}$	1:10 1:20	– 5.07	7.20 7.32	10.02 10.14	7.45 7.52
$pH_{(KCl\ 1N)}$	1:10 1:20	– 4.45	6.32 6.40	7.79 8.01	6.38 6.49
ΔpH	1:20	+0.62	+0.92	+2.13	+1.03
CEC	$cmol_+\ kg^{-1}$	51.5	18.8	56.9	12.4

at room temperature for one day. Adsorption supports were divided into two sub-sample groups. One group of sub-samples was used for irradiation experiments, and the other (control sub-samples) was kept in the dark at the same temperature.

46.2.3
Irradiation Experiments

Photochemical reactions were performed using a solar simulator (Suntest CPS+, Heraeus Industrietechnik GmbH, Hanau, Germany) equipped with a xenon lamp (1.1 kW), protected with a quartz plate (total passing wavelength: 280 nm < λ < 800 nm). The irradiation chamber was maintained at 20 °C by both circulating water from a thermostatic bath and through a conditioned airflow. Before the beginning of the experimental work the light emission effectiveness of the irradiation system was tested by using the uranyl oxalate method (Volman and Seed 1964; Murov et al. 1993). The disappearance of oxalate was 7.2×10^{-4} mol s^{-1}.

46.2.4
Extraction and Analysis

At the same prefixed times 1 × 1-cm width strips of the solid layers were scraped off from glass plates both irradiated and kept in the dark and rimsulfuron was extracted and detected. Extraction was carried out adding acetonitrile to solid materials at a ratio of 50/1 (v/w) and shaking for 30 min. After centrifugation at 5 000 rpm for 15 min the liquid phase was decanted. A second extraction was performed with another aliquot of acetonitrile (50/1, v/w), and the two liquid phases were combined and concentrated by fluxing nitrogen in a rotary evaporator; the final sample volume was adjusted to 5 ml with acetonitrile. The disappearance of rimsulfuron at various experimental times was determined by liquid chromatography after filtration over a 0.2 µm membrane. Analyses were performed on a HP 1090 (Hewlett Packard) liquid chromatograph equipped with a diode array detector (fixed at 230 nm), and a Dionex Omnipac PCX-500 5 µm packed column (18 cm long, 3.2 mm i.d.) + pre-column. The mobile phase used for all experiences was a acetonitrile-water mixture (1+1 by volume), containing a H_3PO_4 buffer (pH 3), at a flow rate of 1 ml min^{-1}. The retention time of rimsulfuron was 8.4 min. The calibration plot was performed in the concentration range 0.015–30 mg l^{-1} giving a linear correlation coefficient $r > 0.99$. At a signal-to-noise ratio of three, the limit of quantitation in the acetonitrile standard solutions was 0.011 mg l^{-1}. Metabolites #2 and #3 (Fig. 46.1) were also determined and confirmed by ion spray LC/MS/MS technique on a Perkin Elmer API 300 (coupled with a Waters 600 pump) using literature criteria (Li et al. 1996; Scrano et al. 1999).

46.3
Results and Discussion

46.3.1
Adsorption of Rimsulfuron on Soil and Soil Surrogates

To quantify and evaluate the influence of adsorption process, all supports were treated in triplicate with 0.2897 mmol kg^{-1} of rimsulfuron; but with the first extraction ($t = 0$) a large part of spiked herbicide was not recovered (C_0 values in Table 46.2). In order to improve the extraction efficiency, numerous solvents were furthermore applied: ethyl acetate; acetone; 1/1 (v/v) mixture of acetone/water; 2/1 (v/v) mixture of acetonitrile/water. But no significant increases of the herbicide recoveries were obtained. Such a disappearance of the herbicide can be slightly due to volatilisation and/or hydrolysis of rimsulfuron during the preparation of adsorption substrate layers on glass plates and 1-day air drying. In fact, vapour pressure of rimsulfuron is not very high (1.1 10^{-8} Torr at 25 °C), as compared to mostly used herbicides (Beyer et al. 1988), and hydrolysis is limited in our experimental conditions, being important in acidic aqueous solutions (Scrano et al. 1999). Though, the formation of bond residues reasonably plays a most important role in the limiting the extractability of rimsulfuron from our sorbent materials, as was previously ascertained for other sulfonylurea herbicides in soil (Albanis et al. 2002). The extraction rate covered the following values: soil 59.5%, illite 48.5%, aerosil 22.2%, montmorillonite 21.0%. The extraction efficiency for the herbicide was higher from soil sample as compared to the clay minerals and the siliceous material. The adsorption behaviour, and consequently the amount of the non-extractable herbicide are influenced by chemical and physico-chemical properties of the solid. With respect to aerosil and montmorillonite, they can be correlated to the high values of cationic exchange capacity (CEC) and specific surfaces (SS) shown in Table 46.1. Illite and the soil sample also retained the herbicide but at minor level. Illite is a clay mineral of the "smectites" family; it is a swelling material and can adsorb interlayer inorganic and organic molecules by means of cation or H^+ bridges as well as montmorillonite. The selected soil is a forestry soil rich in organic matter, which capability to retain organic chemicals is well known. Aerosil is a synthetic amorphous flame silica material used for its high retention properties.

Observing the behaviour of the herbicide kept in the dark, we note that it is characterised by a lasting reactivity (Figs. 46.2–46.5), since its extractability was going diminishing in the time. The interaction between rimsulfuron and sorbing materials does not occur very immediately, but shows varying durations depending on the chemical properties of the adsorbents. Kinetic parameters of this time dependent process were calculated using zero, first and second (Langmuir-Hinshelwood) order equations. The best fit was checked by statistical analysis using the determination coefficient (r^2) values. Apparently, all measured depletion rates of the extracted herbicide in the darkness were better described by a second order equation (Table 46.2). The rationale behind such a finding may be found considering that the amount of the xenobiotic disappeared at each time "t" is affected by its concentration in soil and also by the number of molecules which have reached the most effective steric arrangement on sorption sites (Mingelgrin and Prost 1989; Jones 1991); in turn this number depends again by the herbicide concentration in soil.

Table 46.2. Kinetic parameters of time dependent rimsulfuron depletion on adsorbed phase: n: reaction order; r^2: determination coefficient; C_0: extractable quantity at initial time ($t = 0$); Q_{max}: maximum amount of the adsorbed herbicide; τ: half-life; k: kinetic constant. Reported values are the mean of three replicate experiments

Support	Reaction or environment	n	r^2	C_0 (mmol kg^{-1})	Q_{max} (mmol kg^{-1})	τ (h)	k (mmol^{-1} kg^{-1} h^{-1})
Aerosil	Light	2	0.9987	0.0642	0.0439	3.95	5.765
	Darkness	2	0.9995	0.0642	0.0225	1.55	28.75
	Photolysis	2	0.9981	0.0642	0.0239	12.0	3.496
Illite	Light	2	0.9952	0.1404	0.0543	1.87	9.850
	Darkness	2	0.9917	0.1404	0.0097	4.30	23.90
	Photolysis	2	0.9999	0.1404	0.0447	1.55	14.40
Montmorillonite	Light	2	0.9983	0.0607	0.0433	1.66	13.94
	Darkness	2	0.9968	0.0607	0.0342	2.66	10.99
	Photolysis	2	0.9999	0.0607	0.0091	0.94	116.3
Soil	Light	2	0.9938	0.1725	0.1067	2.94	3.182
	Darkness	2	0.9986	0.1725	0.0590	5.35	3.174
	Photolysis	2	0.9998	0.1725	0.0492	1.52	13.40

46.3.2
Photochemical Degradation versus Adsorption

Kinetic parameters (Table 46.2) of rimsulfuron disappearance under irradiation conditions (Figs. 46.2–46.5) were calculated as mentioned above for samples kept in the dark. From r^2 values we stated that all measured reaction rates of adsorbed herbicide can be better described also in this case by a second order degradation equation:

$$-dC_t/dt = kC_t^2 \qquad (46.1)$$

where C_t is the amount (mmol) of the herbicide extracted at time t per kilogram of adsorbing phase, and k is the rate (or kinetic) constant.

The extent of Eq. 46.1 can be written as:

$$dQ_t/dt = k(Q_{max} - Q_t)^2 \qquad (46.2)$$

were Q_t is the quantity of disappeared (retained and/or degraded) herbicide per kilogram of adsorbent substrate, and Q_{max} is the maximum amount of the herbicide that disappears at the end of the process, i.e. if the reaction would be carried to completion. Integrating Eq. 46.2 and solving for Q_t it yields:

$$Q_t = Q_{max}t/(t+\tau) \qquad (46.3)$$

where τ = half-life = $1/Q_{max}k$.

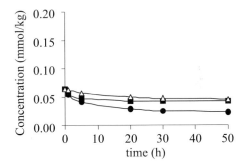

Fig. 46.2.
Rimsulfuron extracted from aerosil: (■) in the dark; (●) under light irradiation; (△) contribution to herbicide disappearance due to photolysis

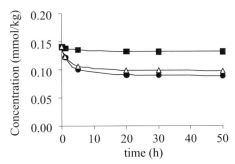

Fig. 46.3.
Rimsulfuron extracted from illite: (■) in the dark; (●) under light irradiation; (△) contribution to herbicide disappearance due to photolysis

Owing to the coexistence of adsorption (and other chemical reactions) and photolysis during the irradiation experiments, the contribution of the photolysis reaction to the disappearance of the herbicide was obtained adding the quantity $(C_0 - C_t)_{dark}$, which disappeared at each experimental time "t" in the darkness (adsorption), to the remaining $(C_t)_{light}$ concentration detected in the irradiated sub-sample at the same time (total reaction i.e. photolysis and adsorption); C_0 is the initial concentration (mmol kg^{-1}) of the herbicide extracted at beginning of the experiment ($t = 0$). This procedure is not fully rigorous since cannot take into account the synergistic effect between photolysis and adsorption during irradiation experiments. In fact, the herbicide adsorption continuously reduces the effective concentration of organic molecules, which can be photodegraded (see above); though, degraded rimsulfuron cannot be rapidly replaced by retained molecules. However, both processes occurring in the dark (adsorption) and under light irradiation (photolysis and adsorption) are of the same kinetic order and start from the same initial concentration (C_0). In these conditions calculations can meet theoretically a good approximation.

In all performed experiments the evolution of photodegradation processes, calculated as above, can be described by a second order kinetic. In this case the quantity Q_t in the Eq. 46.2 assumes the significance of "amount of photodegraded herbicide per kilogram of solid substrate at time t", and Q_{max} is the maximum degradable quantity. The photodegradation of rimsulfuron is strongly affected by retention phenomena. Generally, with increasing of the adsorption capability of supports the photo-

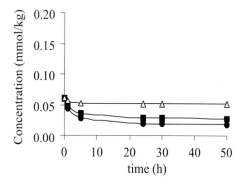

Fig. 46.4.
Rimsulfuron extracted from montmorillonite: (■) in the dark; (●) under light irradiation; (△) contribution to herbicide disappearance due to photolysis

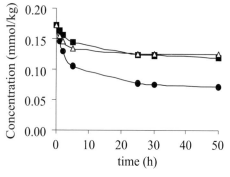

Fig. 46.5.
Rimsulfuron extracted from soil: (■) in the dark; (●) under light irradiation; (△) contribution to herbicide disappearance due to photolysis

reactivity of the herbicide decreases. Contribution of photodegradation to the disappearance of rimsulfuron adsorbed on aerosol seems to have the same importance as compared to the adsorption reaction (Fig. 46.2). The value of Q_{max}-*photolysis* is very close to Q_{max}-*adsorption* (darkness) in Table 46.2. But the photolytic reaction is slower than the adsorption (τ = 12.0 and 1.55 h, respectively). The photoreaction is very evident in the case of illite (Fig. 46.3), on which rimsulfuron shows a value of Q_{max}-*photolysis* much more higher than Q_{max}-*adsorption*. Also the photolysis half-life is more important with respect to adsorption half-life (τ = 1.55 and 4.30 h, respectively). The contribution of photolysis is almost negligible in the case of montmorillonite, on which adsorption is mostly responsible of the disappearance of rimsulfuron (Fig. 46.4). Soil shows a Q_{max}-*photolysis* value close to Q_{max}-*adsorption* as well as aerosol (Table 46.2). The contribution of photodegradation is almost equal to adsorption (Fig. 46.5), but the first reaction in soil is more important from a kinetic point of view (τ = 1.52 and 5.35 h, respectively). Moreover, soil represents the unique case in which the sum of the maximum amount of rimsulfuron extracted, C_0-*light*, and the maximum amount of herbicide disappeared, Q_{max}-*light*, is very close to the value of spiked rimsulfuron (0.2792 and 0.2897 mmol kg^{-1}, respectively). For the other substrates this sum ranged from 0.1040 to 0.1501 mmol kg^{-1}.

The differences between the spiked quantity and the values of the sum C_0-*light* + Q_{max}-*light* are due to the fact that Eq. 46.2 cannot consider the fraction of herbicide disappeared by volatilisation, hydrolysis and adsorption occurring between the preparation of spiked samples and the first extraction. In the case of soil, Eq. 46.2 can work better than other supports because of the presence of organic colloids. In fact, soil organic matter can adsorb the herbicide with a mechanism of solubilization/repartition, after that it becomes a sort of reservoir that supplies herbicide reactions (desorption, hydrolysis, photolysis) with "served" molecules.

46.3.3
Degradation Products

Metabolites #2 and #3 (Fig. 46.1) were extracted from adsorbing substrates and subsequently identified. They were also found as main hydrolysis products in soil by Shalaby et al. (1992) and Schneiders et al. (1993), and on Al-hectorite by Pantani et al. (1998). Table 46.3 shows the metabolite concentrations obtained at half-life time by LC/MS/MS technique. Compound #2, previously identified in acidic conditions (Scrano et al. 1999), was extracted from treated aerosol. This compound can derive from hydrolysis as well as photolysis reaction of rimsulfuron because it was also found in the sub-sample kept in the dark. The presence of such a product on aerosol is justified with the acidic value of measured pH for this substrate (Table 46.1). Compound #3, previously identified in neutral and alkaline conditions (Scrano et al. 1999), was found on the other treated substrates, which measured pH were neutral or sub-alkaline (Table 46.1). Also compound #3 can derive from both hydrolytic and photolytic reactions. Finally, from calculated mass balance we ascertained that a large molar fraction of metabolites formed during experiment time was not extracted because adsorption affected also the retention of these substances.

Table 46.3. Concentrations of rimsulfuron metabolites extracted at time of half-life from adsorbing supports and identified by LC/MS/MS technique (*hy:* hydrolysis; *ph:* photolysis)

Support	Environment	Reaction	Concentration (mmol kg^{-1}) / metabolite	
			#2	#3
Aerosil	Light	hy + ph	0.0114	–
	Darkness	hy	0.0056	–
Illite	Light	hy + ph	–	0.0629
	Darkness	hy	–	0.0012
Montmorillonite	Light	hy + ph	–	0.0165
	Darkness	hy	–	0.0140
Soil	Light	hy + ph	–	0.0489
	Darkness	hy	–	0.0293

46.4 Conclusion

On the basis of our findings and literature data, we realised that photolysis can be a way of degradation of rimsulfuron and similar pesticides as important as hydrolysis. This result is more important in the case of post-emergence herbicides because they are usually sprayed on soil surfaces and plant leaves. A variable fraction of herbicide could be retained by adsorbing surfaces, which can protect chemicals with respect to further degradation reactions because of their steric rearrangement into the adsorption sites. This process does not occur very immediately, but can have varying durations depending on the chemical properties of the xenobiotic substances and adsorbents. Moreover, adsorption process can be accomplished by other chemical reactions, which contribute in different extent to the degradation of the pesticide on solid surfaces. The antagonistic behaviour of adsorption, photolysis and other chemical degradation pathways is proven by the presence of metabolites both in dark conditions and under light irradiation. Obviously, the amount of each metabolite and the quantity and rate of degraded herbicide depend by the chemical and physico-chemical properties of adsorbing materials. The soil experimented in this investigation shows the most equilibrate situation with respect to soil surrogates.

References

Albanis TA, Bochicchio D, Bufo SA, Cospito I, D'Auria M, Lekka M, Scrano L (2002) Surface adsorption and photoreactivity of sulfonylurea herbicides. Int J Environ Anal Chem 82:561–569

Beyer EM, Duffy MJ, Hay JV, Schlueter DD (1988) Sulfonylurea herbicides. In: Kearney PC, Haufman DD (eds) Herbicides: chemistry, degradation, and mode of action, vol. 3. Dekker, New York, pp 117–189

Cambon JP, Bastide J, Vega D (1998) Mechanism of thifensulfuron-methyl transformation in soil. J Agric Food Chem 46:1210–1216

Herbert VR, Miller GC (1990) Dept dependence of direct and indirect photolysis on soil surfaces. J Agric Food Chem 38:913–918

Jones W (1991) Photochemistry and photophysics in clays and other layered solids. In: Ramamurthy V (ed) Photochemistry in organised and constrained media. VCH Publishers, New York, pp 303–358

Klöpffer W (1992) Photochemical degradation of pesticides and other chemicals in the environment: a critical assessment of the state of the art. Sci Total Environ 123/124:145–159

Konstantinou K, Zarkadis AK, Albanis TA (2000) Photodegradation of selected herbicides in various natural waters and soils under environmental conditions. J Environ Qual 30:121–130

Li LYT, Campbell DA, Bennett PK, Henion J (1996) Acceptance criteria for ultra trace HPLC – tandem mass spectrometry: quantitative and qualitative determination of sulfonylurea herbicide in soil. Anal Chem 68:3397–3404

Marucchini C, Luigetti R (1997) Determination of N-(3-ethylsulfuron-2-pyridinyl)-4,6-dimethoxy-2-pyridineamine in soil after treatment with rimsulfuron. Pestic Sci 51:102–107

Mingelgrin U, Prost R (1989) Surface interactions of toxic organic chemicals with minerals. In: Gerste Z, Chen Y, Mingelgrin U, Yeron B (eds) Toxic organic chemicals in porous media. Springer-Verlag, Berlin, pp 91–135

Murov SL, Carmichael I, Hug GL (1993) Handbook of photochemistry. Marcel Dekker, New York, pp 82–99

Pantani O, Pusino A, Calamai L, Gessa C, Fusi P (1996) Adsorption and degradation of rimsulfuron on Al-hectorite. J Agric Food Chem 44:617–621

Rouchaud J, Neus O, Callens D, Bulke R (1997) Soil metabolism of the herbicide rimsulfuron under laboratory and field conditions. J Agric Food Chem 45:3283–3291

Scheunert I (1992) Physical and physico-chemical processes governing the residue behaviour of pesticides in terrestrial ecosystems. In: Ebing W (ed) Chemistry of plant protection, vol. 8. Springer-Verlag, Berlin, pp 1–18

Scheunert I (1993) Transport, and transformation of pesticides in soil. In: Mansour M (ed) Fate and prediction of environmental chemicals in soils, plants and aquatic systems. Lewis Publishers, London, pp 1–22

Schneiders GE, Koeppe MK, Naidu MV, Horne P, Brown HM, Mucha CF (1993) Fate of rimsulfuron in the environment. J Agric Food Chem 41:2404–2410

Scrano L, Bufo SA, Meallier P, Mansour M (1999) Photolysis and hydrolysis of rimsulfuron. Pestic Sci 55:955–961

Shalaby Y, Bramble F, Lee P (1992) Application of thermospray LC/MS for residue analysis of sulfonylurea herbicides and their degradation products. J Agric Food Chem 40:513–517

Volman DH, Seed JR (1964) The photochemistry of uranyl oxalate. J Am Chem Soc 86:5095–5098

Part V

Chapter 47

Binding of Endocrine Disrupters and Herbicide Metabolites to Soil Humic Substances

A. Höllrigl-Rosta · R. Vinken · A. Schäffer

Abstract

The interactions of the two phenolic endocrine disrupters nonylphenol and bisphenol A and of two herbicide metabolites (hydroxydesethyl terbuthylazine and desethyl terbuthylazine) with dissolved humic and fulvic acids were investigated by means of dialysis, using ^{14}C-labeled compounds. Experiments were carried out at different pH values. The strength of xenobiotic-organic matter interactions was quantified by calculating organic carbon-normalised distribution coefficients K_{DOC}. The results show that pH changes in the range of 3 to 10 had little or no impact on the association of both phenols and desethyl terbuthylazine to humic acids. In contrast, the K_{DOC} values for hydroxydesethyl terbuthylazine association to humic acids at pH 3 and 4.5 were nearly one order of magnitude higher than those at neutral and basic pH. Association of the xenobiotics to fulvic acids was only observed at low pH and about one order of magnitude weaker than to humic acids under identical conditions. We conclude that the binding of the investigated xenobiotics occurs mainly via hydrophobic interactions. Results from additional soil sorption experiments indicated similar binding mechanisms to soil organic matter and dissolved humic acids for nonylphenol and bisphenol A. Due to strong interactions between dissolved humic acids and soil, the xenobiotic-organic matter interactions did not affect the sorption of the investigated compounds to soil.

Key words: dissolved humic substances, soil, phenols, triazines, association, dialysis, sorption

47.1 Introduction

47.1.1 Dissolved Humic and Fulvic Acids

The fate of xenobiotics in soil is basically determined by their interactions with matrix components. Degradability as well as leachability of organic compounds are strongly influenced by type and strength of their binding to natural organic matter (Kördel 1997). Besides the mineral-bound or solid soil organic matter, the dissolved organic matter (DOM) in the soil solution can interact with xenobiotics as well (Mingelgrin 2001). Furthermore, sorptive interactions between DOM and the solid soil matrix have to be considered as possible relevant parameters for the environmental assessment of chemicals (Rav-Acha and Rebhun 1992; Lee and Kuo 1999).

Humic substances are the major components of soil organic matter and by far the most abundant organic materials in the environment (Hayes and Clapp 2001). They originate from degradation and transformation reactions of organic material in soil and are generally accepted to be primarily responsible for the binding of organic xenobiotics (Bollag and Loll 1983; Bollag et al. 1992; Mingelgrin 2001). As polyelectrolytes containing carboxylic and phenolic functional groups, they can, on the one hand, interact with polar centres of xenobiotic molecules (Bollag et al. 1992; Senesi and Loffredo 2001). On the other hand, humic substances also contain lipophilic regions that enable sorption of nonpolar chemicals (Ragle et al. 1997; Senesi and Loffredo 2001). This means that humic substances are able to bind organic xenobiotics by a number of different mechanisms, ranging from strong chemical bonds (ionic, hydrogen, and covalent bonding), and charge transfer mechanisms to weaker interactions like van der Waals forces, ligand exchange, and hydrophobic bonding (Senesi and Miano 1995). Based on their solubility in different aqueous solutions, humic substances are operatively distinguished into insoluble humin, alkaline-soluble humic acids, and acid-soluble fulvic acids (Stevenson 1994; Piccolo et al. 2000). In general, humic acids are considered to contain higher amounts of aromatic structures as compared to the more acidic fulvic acids with more carboxylic and carbohydrate and fewer phenolic moieties (Liu and Ryan 1997; Christl et al. 2000; Ussiri and Johnson 2003).

47.1.2
Dialysis

Equilibrium dialysis is a convenient method for studying binding interactions between dissolved organic matter and organic xenobiotics. Separation of organic matter-containing and organic matter-free solutions by a semipermeable membrane allows easy quantification of free and bound analyte molecules. This avoids the need of additional phase separation steps that might disturb the equilibria between dissolved organic matter and organic chemicals. The relatively simple experimental design of dialysis experiments facilitates studies on the impact of various parameters like organic matter or analyte concentration, pH, and salinity on the association of various xenobiotics to dissolved organic matter (Lee and Farmer 1989; Clapp et al. 1997; De Paolis and Kukkonen 1997). Sample volumes can be significantly reduced and sampling improved by replacing dialysis bags with smaller dialysis chambers (Celis et al. 1998; Arnold et al. 1998; Williams et al. 1999). By combining dialysis and sorption experiments, complemental information on the complex interactions in the three-phase system of organic xenobiotics, soil, and dissolved organic matter can be obtained (Celis et al. 1998).

47.1.3
Aims and Scope

In our study, we investigated the association of two phenols used as industrial chemicals and of two triazine herbicide metabolites with peat- and soil-derived humic and fulvic acids. The strength of the respective interactions was quantified by calculating distribution coefficients (K_{DOC}) between aqueous phase and dissolved humic substances (normalised to their organic carbon content). By comparing K_{DOC} values for the different chemicals to humic or fulvic acids obtained at different pH values, conclusions can be drawn on the underlying binding mechanisms and possible implications for the fate of the investigated compounds in soil. Likewise, distribution coefficients to dissolved humic compounds can also be related to soil sorption coefficients.

The phenols used in the study, nonylphenol and bisphenol A, were chosen because of their endocrine disrupting properties. Nonylphenol ethoxylates that are used as surfactants and are degraded by microbial deethoxylation under aerobic conditions (Maguire 1999) are the main source of nonylphenol in the environment. Bisphenol A is used, among other applications, in coatings of thermoprint paper from which it can be released into the environment during paper recycling (Staples et al. 1998). Due to their lipophilicity, both compounds accumulate in the sludge fraction during sewage treatment and are introduced into soil by the agricultural use of sewage sludge.

Besides the two phenols, desethyl terbuthylazine and hydroxydesethyl terbuthylazine were tested. Terbuthylazine is a widely used herbicide in maize cultivation. Desethyl terbuthylazine and hydroxydesethyl terbuthylazine represent the important group of pesticide metabolites in this study. The concentrations of metabolites in soil often exceed those of the parent compounds. For an assessment of a pesticide, it is thus necessary to obtain data also on the environmental fate of its metabolites.

47.2
Experimental

47.2.1
Preparation of Humic Substances

To obtain fulvic acids from soil and humic acids from peat, standard extraction procedures from soil science (Swift 1996) were applied with minor modifications. Extraction was carried out with diluted sodium hydroxide (0.2 M). Humic acids were precipitated by acidification of the peat extract to pH 1 with hydrochloric acid (5 M), washed, and lyophilised. For the dialysis experiments, they were redissolved in diluted sodium hydroxide and the solution neutralised with sulphuric acid. After precipitating the humic acids from the soil extract, fulvic acids were obtained by concentration on DAX-8 adsorbent (Supelco, Bellefonte, USA). Both humic acid and fulvic acid solutions were dialysed against water through a membrane with a molecular weight cut-off of 1 kDa (Roth, Karlsruhe, Germany) before use in experiments, in order to remove salts and low-molecular fragments.

47.2.2
Xenobiotic Compounds

The structural formula of the ^{14}C-labelled chemicals used in the experiments are depicted in Fig. 47.1. Stock solutions of all compounds were prepared in methanol. Aqueous solutions were prepared from the stock solutions by removing the solvent in a stream of nitrogen and redissolving the chemicals in water.

47.2.3
Dialysis Experiments

The formation of organic matter adducts was studied in a specially constructed dialysis apparatus (Fig. 47.2). Dialysis takes place in polytetrafluorethene cells with an inner volume of approximately 18 ml. These consist of two cylindrical half cells which

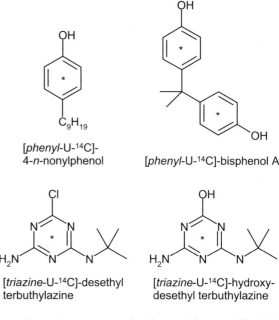

Fig. 47.1. Structural formula of the chemicals used in the experiments. The asterisk indicates the position of the ^{14}C-label (uniform labelling, i.e. statistical distribution of number and position of carbon atoms within the ring system)

[*phenyl*-U-^{14}C]-4-*n*-nonylphenol

[*phenyl*-U-^{14}C]-bisphenol A

[*triazine*-U-^{14}C]-desethyl terbuthylazine

[*triazine*-U-^{14}C]-hydroxy-desethyl terbuthylazine

are separated by a dialysis membrane (5 × 5 cm, molecular weight cut-off 1 kDa) clamped between tongue and groove. In each half cell, one hole covered with a polytetrafluorethene plug allows easy sampling by means of a pipette. Five to six complete cells are put into a holding device that is rotated around its cylindrical axis. Rotation ensures permanent wetting of the dialysis membrane for sample volumes down to 5 ml per half cell and also supports diffusion through the membrane.

Per single experiment, two corresponding half cells were filled with a solution of humic substances and water, respectively (7 ml each). Both solutions were stabilised with sodium azide (0.05%) and the water was spiked with an aqueous solution of the radio-

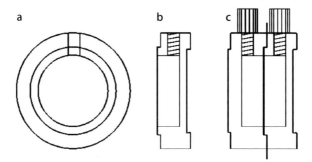

Fig. 47.2. Schematic representation of the dialysis cells used in the experiments. In the top view **a**, the half cell is depicted with the open side up, showing the positions of the outer wall, the groove (for a left half cell) or tongue (for a left half cell), and the inner void as well as of the plug hole. Cross section **b** shows a right half cell with tongue. Cross section **c** demonstrates the final assembly of one dialysis cell with plugs and the dialysis membrane clamped between groove and tongue of a left and right half cell, respectively

labelled analyte. Rotation speed was about 10 to 12 rpm. In defined time intervals, samples from both half cells were taken for determination of radioactivity via LSC. Equilibration times in organic matter-free systems had been determined previously and were below 24 h for all analytes. Experiments were typically carried out in duplicates.

47.2.4
Sorption to Soil

Soil sorption coefficients were determined according to a modified OECD standard method 106. Pre-soaked soil (25 g dry matter, moistened to maximum water holding capacity) was mixed with humic acid solution or 0.01 M calcium chloride (27 ml) and spiked with the radiolabelled compound (1 µg g^{-1} soil). Batches were shaken 24 h before phase separation by centrifugation. Radioactivity was determined in both phases for calculation of distribution coefficients.

47.3
Results and Discussion

47.3.1
Xenobiotic Interactions with Dissolved Humic and Fulvic Acids at Neutral pH

Aqueous solutions of nonylphenol, bisphenol A, desethyl terbuthylazine, and hydroxy-desethyl terbuthylazine were dialysed against solutions of humic or fulvic acids, in order to obtain information on their respective binding interactions with the dissolved humic substances. The concentrations of the xenobiotics in these dialysis experiments ranged from 3.4 to 21.7 µg l^{-1}, while the humic and fulvic acid solutions both had a dissolved organic carbon concentration of 190 mg l^{-1}.

The strength of interactions between xenobiotics and dissolved humic substances can be expressed as the compounds' distribution coefficients K_{DOC} between the aqueous phase and the organic carbon of the humic substances phase. The concentration c_{DOC} of bound molecules is calculated by subtracting the concentration of free analyte molecules in the water half cell (c_w) from the total concentration of free and humic substance-bound molecules in the humic substance half cell (c_{HA}). This value is divided by the concentration of free analyte (c_w). To normalise the result to the organic carbon content of the sorbent, it is divided by the dissolved organic carbon-concentration (DOC) of the humic substances solution.

$$K_{DOC} = \frac{c_{DOC}}{c_w} = \frac{c_{HA} - c_w}{c_w} \frac{1}{DOC}$$

Since the K_{DOC} values of the different compounds spread over several orders of magnitude, the logarithmic values $\log K_{DOC}$ are used in the further discussion.

In the experiments with fulvic acids at neutral pH, equilibria were reached that were characterised by identical analyte concentrations on both sides of the dialysis membrane. In contrast, dialysis of the compounds against a humic acid solution at neutral pH resulted in a surplus of radioactivity in the half cells containing the dis-

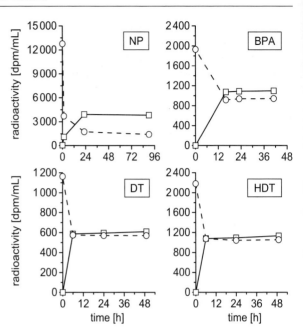

Fig. 47.3. Concentration curves for the dialysis of nonylphenol (NP), bisphenol A (BPA), desethyl terbuthylazine (DT), and hydroxydesethyl terbuthylazine (HDT) against dissolved humic acids at neutral pH. Experiments were carried out in duplicates. Error bars are smaller than symbols and therefore not shown. The dashed lines show the decreasing radioactivity concentrations over time in the half cells with water and spiked analyte, while the straight lines depict their increase in the half cells containing dissolved humic acids. Higher concentrations of radioactivity in the humic acid solutions at the end of the experiment indicate the formation of xenobiotic-humic acid associates

solved humic acids. Concentration curves for the dialysis of the four xenobiotics against humic acids at neutral pH are displayed in Fig. 47.3. They show an increase in radioactivity in the humic acid solution above the concentration in the water half cell for all compounds.

It can be concluded that no association of either compound with the fulvic acids had occurred. In the experiments with dissolved humic acids, associates between humic acid molecules and the chemicals had been formed, which, due to their size, could not pass the membrane pores. A loss in total dissolved radioactivity was observed in the case of nonylphenol. This could be attributed to strong sorption of the chemical to the cell walls and the dialysis membrane. It did, however, not affect the distribution between free and humic acid-bound nonylphenol molecules in the aqueous phase at equilibrium.

Mean $\log K_{DOC}$ values from duplicate experiments for the association to dissolved humic acids ranged from 2.56 and 2.60 for the triazine derivatives desethyl terbuthylazine and hydroxydesethyl terbuthylazine, respectively, over 2.94 for bisphenol A up to 3.96 for nonylphenol. The extent of association to dissolved humic acids is obviously related to the lipophilicity of the the xenobiotic compounds, indicating that binding is primarily due to hydrophobic interactions. This is in line with the observations of numerous other authors. Dialysis experiments conducted with pentachlorophenol revealed stronger binding to humic acids with a higher aromatic content than to fulvic acids (De Paolis and Kukkonen 1997). Similarly, Chen et al. (1992) observed higher equilibrium constants for the association of 1-naphthol to humic as compared to fulvic acids at neutral pH. In another study on the sorption of some chlorinated and/or alkylated phenols to various types of natural organic matter, binding was ascribed to the capability of the organic sorbent to form hydrophobic cavities, but also, depending on the chemical nature of the sorbent, on specific interactions with functional groups

(Ohlenbusch et al. 2000). However, such specific interactions with polar functional groups of analytes were not observed for the soil fulvic acids used in our experiments at pH 7. The key role of hydrophobic interactions in binding of organic xenobiotics to humic acids was confirmed also for various triazine herbicides and related compounds (Schmitt et al. 1997; Kulikova and Perminova 2002). Such kind of mechanism can thus explain the increasing strength of binding to dissolved humic acids from the more polar triazine derivatives to the lipophilic nonylphenol, as well as the observed absence of binding to dissolved fulvic acids for all investigated compounds at neutral pH.

47.3.2
Impact of pH on the Interactions of Xenobiotics with dissolved Humic and Fulvic Acids

To assess the impact of environmental conditions on the interactions between xenobiotics and dissolved humic acids, dialysis experiments were carried out at different pH values in the range of pH 3 to 12. Adjustment of pH was achieved by adding hydrochloric acid or sodium hydroxide to the respective test solutions.

The impact of the pH value on the association of the compounds with dissolved humic acids varied between the tested chemicals. An overview over the K_{DOC} values obtained for pH 3 to 10 with dissolved humic acids is given in Table 47.1. Virtually no pH dependency of binding interaction strength was observed for bisphenol A in this pH range, as shown by a mean $\log K_{DOC}$ of 2.94 ±0.11. In the case of nonylphenol, a mean $\log K_{DOC}$ of 3.85 ±0.12 was calculated for the pH values between 3 and 7, while the respective value at pH 10 was slightly decreased by about 0.3 units. A general tendency to weaker binding interactions at higher pH was confirmed by additional dialysis experiments with humic acids at pH 12 (data not included in Table 47.1), where no associates were found for any of the tested compounds. The two triazine herbicide metabolites desethyl terbuthylazine and hydroxydesethyl terbuthylazine showed association to humic acids only up to pH 7. While the $\log K_{DOC}$ values for desethyl terbuthylazine did not change significantly from pH 7 to pH 3, a remarkably stronger association to dissolved humic acids was observed for hydroxydesethyl terbuthylazine at acidic pH values (Fig. 47.4). From pH 7 to pH 4.5, mean $\log K_{DOC}$ values rose by about 0.8 units which is equivalent to a 6-fold increase in binding interactions.

Table 47.1. Binding coefficients $\log K_{DOC}$ for the binding of nonylphenol (NP), bisphenol A (BPA), desethyl terbuthylazine (DT), and hydroxydesethyl terbuthylazine (HDT) to dissolved humic acids vs. pH. Dialysis experiments lasted for 41 to 49 h, except for nonylphenol at pH 7 (90 h). All experiments were carried out in duplicates and both single values of $\log K_{DOC}$ are given. A significant increase of $\log K_{DOC}$ is visible for hydroxydesethyl terbuthylazine between pH 7 and pH 4.5

Compound	pH 3		pH 4.5		pH 7		pH 10	
NP	3.86	3.93	3.67	3.74	3.93	3.98	3.50	3.51
BPA	2.79	2.91	3.00	3.12	2.90	2.97	2.81	3.02
DT	2.28	2.64	not determined		2.53	2.58	no association	
HDT	3.27	3.46	3.35	3.38	2.57	2.62	no association	

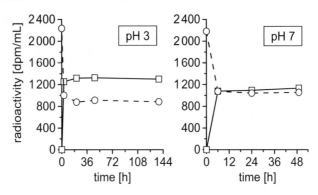

Fig. 47.4. Concentration curves for the dialysis of hydroxydesethyl terbuthylazine against dissolved humic acids at neutral (pH 7) and acidic pH (pH 3). Experiments were carried out in duplicates. Error bars are smaller than symbols and therefore not shown. The dashed lines show the decreasing radioactivity concentrations over time in the half cells with water and spiked analyte, while the straight lines depict their increase in the half cells containing dissolved humic acids. The greater difference between the two curves at equilibrium at pH 3 than at pH 7 corresponds to stronger binding interactions between hydroxydesethyl terbuthylazine and humic acids under acidic conditions

The effect of acidification was also investigated for the interactions of nonylphenol and hydroxydesethyl terbuthylazine with fulvic acids. At pH 3, the mean $\log K_{DOC}$ values at equilibrium amounted to 3.17 and 2.71 for nonylphenol and hydroxydesethyl terbuthylazine, respectively. In both cases, these values are about 0.7 units smaller than the corresponding values for the humic acids. Thus, the xenobiotic-DOM interactions at pH 3 are about five times weaker with fulvic than with humic acids for both tested compounds.

The increase or decrease of binding interactions depending on the solution pH can be attributed to molecular structure changes of either the humic acid sorbent or the xenobiotic sorbates. On the one hand, the negative charge of humic acids increases with pH as a consequence of deprotonation of first carboxylic and then phenolic moieties, and their conformation is changed due to disruption of intramolecular hydrogen bonds and electrostatic repulsion effects (Senesi and Loffredo 2001). We assume that such conformational changes of the dissolved humic acids are responsible for the absence of triazine-humic acid interactions at pH > 7. On the other hand, phenols as well as triazines contain functional groups that are protonated or deprotonated according to pH. It has been stated for pentachlorophenol that only the non-ionised form of the molecule is able to bind to dissolved humic acids (De Paolis and Kukkonen 1997). This effect can also explain the observed decrease in K_{DOC} for nonylphenol and bisphenol A at pH values of 10 and above. In the acidic pH range, hydroxydesethyl terbuthylazine displayed a markedly different behaviour than the other chemicals. The increase in binding interactions at pH 4.5 by nearly one order of magnitude is attributed to the high basicity of hydroxy triazines. With pK_a values around 5 (Schmitt et al. 1996), these compounds are protonated at lower pH, resulting in cationic species that can interact with electronegative functional groups of the dissolved humic or fulvic acids. The ability of hydroxy triazines to interact with humic substances on the basis of polar interactions like cation exchange or electron transfer has been emphasised by other authors, too (Lerch et al. 1997; Martin-Neto et al. 2001). In contrast, chlorinated triazines like desethyl terbuthylazine exhibit pK_a values around 2 and are thus not affected in their binding to humic substances by pH values down to 3. The results show that additional interaction mechanisms between xenobiotics and dissolved humic substances can become operative besides hydrophobic interactions at acidic or alkaline conditions and may substantially change the association behaviour of these compounds.

47.3.3
Soil Sorption Experiments

Complemental information to the picture on the interactions of xenobiotics with organic material in soil was added by carrying out sorption experiments with a loamy silt soil of low organic carbon content (1%) at the soil-inherent pH of 5.14 (KCl). Soil sorption coefficients normalised to the organic carbon content were determined with (log K_{OC}-HA) and without (log K_{OC}-Ref) addition of humic acids to the sorption solutions. The results are listed in Table 47.2. Most remarkably, nearly identical numerical values were observed for the log K_{DOC} and log K_{OC}-Ref values of the two phenols. Such close conformity of distribution coefficients possibly indicates that similar binding mechanisms were operative in both, soil sorption and dialysis experiments (Arnold et al. 1998). It can thus be concluded that nonylphenol and bisphenol A are complexed through hydrophobic interactions in a similar way by solid or mineral-bound as well as by dissolved humic acids. No such relationship was found for hydroxydesethyl terbuthylazine and desethyl terbuthylazine. Soil sorption coefficients (log K_{OC}-Ref) for these triazine derivatives were about 0.6 to 0.7 units lower than the respective distribution coefficients to dissolved humic acids. This unexpected behaviour might be explained by sorption sites that are only accessible in the dissolved state of the humic acids and are blocked in the solid organic matter.

Sorption experiments with humic acids added to the aqueous phase were conducted with nonylphenol and hydroxydesethyl terbuthylazine. For both chemicals, the additional dissolved organic sorbent had no impact on soil sorption coefficients. This agrees well with data from the literature, stating that natural dissolved organic matter had virtually no impact on the mobility of different phenols in soil (Lafrance et al. 1994; Busche and Hirner 1997). Complementary sorption experiments with pure humic acid

Table 47.2. Comparison of organic carbon-normalised binding coefficients for the association of nonylphenol (NP), bisphenol A (BPA), desethyl terbuthylazine (DT), and hydroxydesethyl terbuthylazine (HDT) to dissolved humic acids at pH 7 (log K_{DOC}) and for the sorption of the compounds to loamy silt soil with (log K_{OC}-HA) and without (log K_{OC}-Ref) addition of dissolved humic acids. Good agreement is found between log K_{DOC} and log K_{OC}-Ref values of both phenols and between log K_{OC}-Ref and log K_{OC}-HA values for nonylphenol and hydroxydesethyl terbuthylazine

Compound	log K_{DOC}	log K_{OC}-Ref	log K_{OC}-HA
NP	3.96 ±0.04 (n = 2)	3.95 ±0.01 (n = 3)	3.98 ±0.01 (n = 3)
BPA	2.93 ±0.04 (n = 2)	2.95 ±0.02 (n = 3)	not determined
DT	2.56 ±0.03 (n = 2)	2.12 ±0.00 (n = 2)	not determined
HDT	2.60 ±0.03 (n = 2)	1.94 ±0.05 (n = 7)	1.95 ±0.07 (n = 6)

solutions confirmed strong interactions between the dissolved humic acids and the loamy silt soil. Within few hours, the dark brown peat humic acids were removed from solution and replaced by light yellow water-soluble organic material from soil. However, the peat humic acids adsorbed by the soil did neither increase nor decrease its binding capacity for the xenobiotic chemicals.

47.4 Conclusions

Dialysis experiments with two phenols (nonylphenol and bisphenol A) and two herbicide metabolites (desethyl terbuthylazine and hydroxydesethyl terbuthylazine) against solutions of humic and fulvic acids at different pH values yielded an array of data on xenobiotic-DOM associates that allow to draw conclusions on type and mechanism of the binding interactions. At neutral pH, the association of the investigated chemicals to dissolved humic substances is primarily determined by hydrophobic interactions. As a consequence, no binding to fulvic acids occurs, whereas normalised association coefficients K_{DOC} to dissolved humic acids rise in the order desethyl terbuthylazine ≈ hydroxydesethyl terbuthylazine < bisphenol A < nonylphenol. In the acidic or alkaline pH range, the molecular structure of both xenobiotics and dissolved humic substances is changed due to protonation or deprotonation. Generally, binding interactions are decreased at higher pH, due to conformational changes of the humic substances, or to the formation of negatively charged, more hydrophilic and thus nonbinding phenolate ions from nonylphenol and bisphenol A. In contrast, K_{DOC} values for hydroxydesethyl terbuthylazin increase markedly at pH values near the compound's pK_a of approximately 5, indicating strong additional polar binding interactions. However, the possible formation of associates with dissolved humic acids had no impact on the fate of the compounds in soil sorption experiments with added peat humic acids, because these were virtually completely sorbed and replaced by water-soluble organic material from soil during equilibration. Since the sorption of the humic acids to the soil matrix does not interfere with the sorption of the xenobiotics and no binding of the chemicals occurs with the newly released organic matter in the aqueous phase, their soil sorption coefficients K_{OC} are not affected.

Acknowledgements

The authors wish to thank Syngenta Crop Protection (Basel, Switzerland) for supplying the terbuthylazine metabolites. We also thank M. Lenz and S. Antar for technical assistance.

References

Arnold CG, Ciani A, Müller SR, Amirbahman A, Schwarzenbach RP (1998) Association of triorganotin compounds with dissolved organic acids. Environ Sci Technol 32:2976–2983
Bollag JM, Loll MJ (1983) Incorporation of xenobiotics into soil humus. Experientia 39:1221–1231
Bollag JM, Myers CJ, Minard RD (1992) Biological and chemical interactions of pesticides with soil organic matter. Sci Total Environ 123/124:205–217

Busche U, Hirner AV (1997) Mobilization potential of hydrophobic organic compounds (HOCs) in contaminated soils and waste materials. Part 2: Mobilization potentials of PAHs, PCBs, and phenols by natural waters. Acta Hydrochim Hydrobiol 25:248–252

Celis R, Barriuso E, Houot S (1998) Sorption and desorption of atrazine by sludge-amended soil: dissolved organic matter effects. J Environ Qual 27:1348–1356

Chen S, Inskeep WP, Williams SA, Callis PR (1992) Complexation of 1-naphthol by humic and fulvic acids. Soil Sci Soc Am J 56:67–73

Christl I, Knicker H, Kögel-Knabner I, Kretzschmar R (2000) Chemical heterogeneity of humic substances: characterization of size fractions obtained by hollow-fibre ultrafiltration. Eur J Soil Sci 51:617–625

Clapp CE, Mingelgrin U, Liu R, Zhang H, Hayes MHB (1997) A quantitative estimation of the complexation of small organic molecules with soluble humic acids. J Environ Qual 26:1277–1281

De Paolis F, Kukkonen J (1997) Binding of organic pollutants to humic and fulvic acids: influence of pH and the structure of humic material. Chemosphere 34:1693–1704

Hayes MHB, Clapp CE (2001) Humic substances: considerations of compositions, aspects of structure, and environmental influences. Soil Sci 166:723–737

Kördel W (1997) Fate and effect of contaminants in soils as influenced by natural organic material – status of information. Chemosphere 35:405–411

Kulikova NA, Perminova IV (2002) Binding of atrazine to humic substances from soil, peat, and coal related to their structure. Environ Sci Technol 36:3720–3724

Lafrance P, Marineau L, Perreault L, Villeneuve JP (1994) Effect of natural dissolved organic matter found in groundwater on soil adsorption and transport of pentachlorophenol. Environ Sci Technol 28:2314–2320

Lee DY, Farmer WJ (1989) Dissolved organic matter interaction with napropamide and four other nonionic pesticides. J Environ Qual 19:567–573

Lee CL, Kuo LJ (1999) Quantification of the dissolved organic matter effect on the sorption of hydrophobic organic pollutant: application of an overall mechanistic sorption model. Chemosphere 38:807–821

Lerch RN, Thurman M, Kruger EL (1997) Mixed-mode sorption of hydroxylated atrazine degradation products in soil: a mechanism for bound residue. Environ Sci Technol 31:1539–1546

Liu X, Ryan KD (1997) Analysis of fulvic acids using HPLC/UV coupled to FT-IR spectroscopy. Environ Technol 18:417–424

Maguire RJ (1999) Review of the persistence of nonylphenol and nonylphenol ethoxylates in aquatic environments. Water Qual Res J Can 34:37–78

Martin-Neto L, Traghetta DG, Vaz CMP, Crestana S, Sposito G (2001) On the interaction mechanisms of atrazine and hydroxyatrazine with humic substances. J Environ Qual 30:520–525

Mingelgrin U (2001) Binding of small organic molecules by soluble humic substances. In: Clapp CE (ed) Humic substances and chemical contaminants. Soil Science Society of America, Madison, WI, pp 187–204

Ohlenbusch G, Kumke MU, Frimmel FH (2000) Sorption of phenols to dissolved organic matter investigated by solid phase microextraction. Sci Total Environ 253:63–74

Piccolo A, Celano G, Conte P (2000) Methods of isolation and characterization of humic substances to study their interactions with pesticides. In: Cornejo J, Jamet P (eds) Pesticide/soil interactions. INRA, Paris, pp 103–116

Ragle CS, Engebretson RR, von Wandruszka R (1997) The sequestration of hydrophobic micropollutants by dissolved humic acids. Soil Sci 162:106–114

Rav-Acha C, Rebhun M (1992) Binding of organic solutes to dissolved humic substances and its effect on adsorption and transport in the aquatic environment. Water Res 26:1645–1654

Schmitt P, Garrison AW, Freitag D, Kettrup A (1996) Separation of s-triazine herbicides and their metabolites by capillary zone electrophoresis as a function of pH. J Chromatogr A 723:169–177

Schmitt P, Freitag D, Trapp I, Garrison AW, Schiavon M, Kettrup A (1997) Binding of s-triazines to dissolved humic substances: Electrophoretic approaches using affinity capillary electrophoresis (ACE) and micellar electrokinetic chromatography (MEKC). Chemosphere 35:55–75

Senesi N, Loffredo E (2001) Soil humic substances. In: Hofrichter M, Steinbüchel A (eds) Biopolymers, vol. 1 (lignin, humic substances and coal). Wiley-VCH, Weinheim, pp 247–299

Senesi N, Miano TM (1995) The role of abiotic interactions with humic substances on the environmental impact of organic pollutants. In: Huang PM (ed) Environmental impact of soil component interactions, vol. 1. Lewis, Boca Raton, FL, pp 311–335

Staples CA, Dorn PB, Klecka GM, O'Block ST, Harris LR (1998) A review of the environmental fate, effects, and exposure of bisphenol A. Chemosphere 36:2149–2173

Stevenson FJ (1994) Humus chemistry, 2nd edn. Wiley, New York

Swift RE (1996) Organic matter characterization. In: Sparks DL, et al. (eds) Methods of soil analysis, part 3 – Chemical methods. Soil Science Society of America, Madison, WI, pp 1011–1069

Ussiri DAN, Johnson CE (2003) Characterization of organic matter in a northern hardwood forest soil by ^{13}C NMR spectroscopy and chemical methods. Geoderma 111:123–149

Williams CF, Farmer WJ, Letey J, Nelson SD (1999) Design and characterization of a new dialysis chamber for investigating dissolved organic matter complexes. J Environ Qual 28:1757–1760

Chapter 48

Potential Exposure to Pesticides during Amateur Applications of Home and Garden Products

P. Harrington · J. Mathers · R. Lewis · S. Perez Duran · R. Glass

Abstract

Volunteer "amateur" gardeners were observed, but not supervised, while they mixed, then applied carbendazim, and decontaminated their equipment. A whole body dosimetry method was used, including dosimeters on hands, feet and face. Personal air samplers collected airborne pesticide in the breathing zone. Analysis of the tank mixes prepared by volunteers indicated that the final concentration ranged from 55 to 177% of the intended concentration. Areas of body most heavily contaminated during mixing were the hands, with levels of up to 25 mg of active substance (a.s.) found due to spillages during measurement. Residues of pesticide within the measuring cap were up to 31 mg of a.s. During application the arms, hands, front torso and feet were most contaminated. Typical contamination rates during application were 20 ml h^{-1} of the diluted tank mix: up to 10 mg h^{-1} of active substance, with typical applications in these scenarios lasting 5 to 15 min.

48.1
Introduction

Studies to evaluate the exposure of operators during pesticide handling and application tend to concentrate on large scale commercial use (Glass et al. 2002; Wild et al. 2000). Recent studies have generated data that identify potential dermal and inhalation exposure during the application of amateur pesticides, specifically marketed for home and garden use in the UK. Volunteer "amateur" gardeners were observed, but not supervised, while they mixed, then applied a pesticide, and decontaminated their equipment. A whole body dosimetry method was used, including dosimeters on hands, feet and face. Personal air samplers collected airborne pesticide in the breathing zone while ground level collection media were used to measure ground contamination from run-off and spray drift.

Residues were extracted, using methanol, from a sectioned absorbent suit, boot covers and gloves worn by the operator over Personal Protective Equipment (PPE). The concentration of pesticide in the extracts was determined by HPLC with fluorescence detection. Spiked recovery samples of the clothing materials were analysed concurrently with the samples. The potential exposure to pesticides, for amateur users of home and garden products, is greatest during mixing and loading and is exacerbated by the use of the container cap as a measuring device. This method of concentrate measurement was found to be inaccurate when used by all but one of the volunteers, with the tank mix concentrations falling between 55 and 177% of that recommended.

Levels on the gloves of up to 25 mg of active substance were found, due to spillages and contamination during measurement, and residues of up to 31 mg were found within the measuring cap, representing a potential source of operator exposure. Potential dermal exposure during application is closely linked to the duration of application, with a maximum whole body contamination level of approximately 25 ml per hour being measured in three of the six test scenarios. In none of the six scenarios observed, were levels of carbendazim measured on the clothing worn by the operator considered to be sufficient to give rise to health concerns, even in situations where multiple applications might be used.

48.2
Experimental

Amateur gardener "volunteers" were asked to read the instructions set out by the manufacturer on the product label of a commercially available garden protection product, then prepare and apply the pesticide to suitable target plants. The pesticide used was a commercial formulation, marketed for amateur use, and bought from a local retail outlet.

The product used was: Doff "Plant Disease Control" containing 50 g l^{-1} carbendazim, recommended to be used at a dilution of 5 ml l^{-1}: equivalent to an active substance concentration of 250 µg ml^{-1}. The volunteers were supplied with a Tyvek® coverall, protective gloves and a face shield. Over the top of this Personal Protective Equipment they were dressed in a second, absorbent, coverall (Sontara®), cotton gloves and Tyvek® boot covers. Personal air samplers were attached to the outer suit in the breathing area, with a further air sampler placed approximately 5 m downwind of target plants, at a height of roughly 1.5 m to measure the potential air-borne exposure experienced by an observer. In all but the first two trials, Petri dishes, containing filter papers were placed under the target plants, in the areas that the operator, or other garden users, would walk through, to assess the ground deposition which would be an indication of pesticide deposits available to contaminate shoes.

The volunteers were asked to perform three tasks:

1. "Mix and Load" – Prepare an appropriate volume of pesticide formulation in the spraying equipment at a suitable concentration for the target plants.
2. "Application" – Apply the pesticide.
3. "Clean and decontaminate" – Complete the operation by cleaning the equipment used, ready for future applications.

At the end of the "mix and load" and "application" operations, samples were bagged and labelled and brought back to the laboratory for analysis together with field spikes. The suit was sectioned as indicated in the appendices and the pesticide extracted in methanol. Gloves, boots, filter papers from ground level, air sampler media and tissues from wiping the visor or spillages during mixing, were treated in a similar manner.

Carbendazim analysis used reverse-phase HPLC with isocratic elution with ammonium acetate/acetonitrile (pH 8) and fluorescence determination. The level of determination was 0.01 µg ml^{-1}, which would allow 5 µl of the formulation at the application concentration to be detected on the large sections of the suit and 1 µl elsewhere. Bracketed calibrations at a minimum of 4 levels were placed between a maximum of 10 samples. Results outside the calibration range were diluted and re-run.

48.3
Results

The contamination levels discussed refer to the total amount of active substance extracted from the personal protective equipment (PPE) and therefore can be regarded as potential dermal exposure of the operator to the pesticide.

48.3.1
Individual Results

Each scenario is represented diagrammatically (Fig. 48.1–48.6) in the Appendix.

Scenario 1

- One apple tree, in full leaf, approximately 5 m (h) by 6 m (w)
- Tank concentration: 138 µg ml^{-1} = 55% of recommended
- Application rate: 2.3 l in 17 min
- Comments:
  - *Mix and Load:* Significant spillage on cap, observed whilst measuring concentrate. One and a half capfuls were required to be measured, exacerbating inaccuracies in concentrate measurement. The drips around the cap gave 31 mg of active substance (a.s.) recovered from this area, none on gloves.
  - *Application:* Upward application gave the highest levels of exposure on head, upper torso and right (application) arm. Higher than anticipated levels of active substance were measured on feet. The highest residue of a.s. found from the application was 0.16 mg on the front torso. The contamination rate from this application was 22.3 ml h^{-1}, or 3.1 mg a.s. h^{-1}. The combined mix and load and application operations gave a total body exposure to the active substance of 0.87 mg per spraying operation.
  - Decontamination left less than 0.2 mg a.s. on clothing, but over 1 mg was left as a residue on the sprayer dip-tube.

Scenario2

- One apple tree, in full leaf, approximately 3 m (h) by 2 m (w)
- Tank concentration: 245 µg ml^{-1} = 97% of recommended
- Application rate: 0.25 l in 2.4 min
- Comments:
  - *Mix and Load:* Measurement of concentrate required exactly one full cap, therefore the final mix was more accurate. A total of 35 mg a.s. was recovered from gloves after mixing.

- *Application:* High levels on feet were observed, again. Separation of upper, from sole of the foot covers indicated that contamination was higher on the uppers. This suggests that drift/run-off from the target plants, makes a larger contribution than the pesticide on the ground. The highest residue of a.s. found from the application was 0.056 mg on the front torso. The contamination rate from this application was 15.4 ml h^{-1}, or 3.8 mg a.s. h^{-1}. The combined mix and load and application operations gave a total body exposure to the active substance of 35.5 mg per spraying operation, of which 35 mg was due to the mix and load stage.
- Decontamination/cleaning of equipment left residues of less than 0.2 mg in total.

Scenario 3

- Three apple trees, approximately 5 m (h) by 2 m (w), each, plus a currant bush
- Tank concentration: 170 µg ml^{-1} = 68% of recommended
- Application rate: 2.75 l in 15.4 min
- Comments:
  - *Mix and Load:* Contamination during mixing and loading was 13.5 mg of a.s. on the gloves, but with a further potential source of exposure of 25.5 mg a.s. recovered from the cap swab.
  - *Application:* Filter papers were placed out in the areas where the operator was walking, but not directly under the target plants. Access for the operator to get directly under the trees was difficult, so less exposure to head and visor. Although the feet show similar dosimetry to scenario 2, the carbendazim levels on the ground height filter papers were less than the detection limit. These observations suggest that the high levels found on the upper foot are directly from run-off, i.e. placing the lead foot under the target plant foliage, not from wind-borne drift. The highest residue of a.s. found from the application was 0.22 mg on the right glove. The contamination rate from this application was 25.0 ml h^{-1}, or 4.3 mg a.s. h^{-1}. The combined mix and load and application operations gave a total body exposure to the active substance of 14.6 mg per spraying operation, of which 1.1 mg was from spraying drift.
  - Decontamination/cleaning left combined residues on clothing and equipment of less than 0.15 mg a.s.

Scenario 4

- Five fruit trees, 2 m (h) by 2 m (w), each, plus rose bush 1.5 m (h) by 1 m (w)
- Tank concentration: 167 µg ml^{-1} = 67% of recommended
- Application rate: 4.1 l in 8.2 min
- Comments:
  - *Mix and Load:* Mixing and loading were performed with great care; less than 0.3 ml of concentrate was recovered from the left glove, but this still gives exposure levels during mixing and loading to 15 mg a.s.
  - *Application:* Spray was applied horizontally for these low level trees and shrubs, so levels on the subjects' head and shoulders were negligible. Nozzle adjustments were made with left hand, while spraying, hence high contamination levels to left arm. The operator walked close to bushes, leading with left foot, leading to high conta-

mination levels on the soles of the feet, particularly the left. The ground level filter papers, positioned near the bushes also received measurable dosages, in 4 out of 5 instances. The highest residue of a.s. found from the application was 0.050 mg on the sole of the left foot. The contamination rate from this application was 5.3 ml h^{-1}, or 0.9 mg a.s. h^{-1}. The combined mix and load and application operations gave a total body exposure to the active substance of 15.1 mg per spraying operation.
- Decontamination/cleaning left combined residues on clothing and equipment of 0.05 mg a.s.

Scenario 5

- Cucumber and tomato plants in conservatory/greenhouse. Plants 1.5 m high
- Tank concentration: 443 µg ml^{-1} = 177% of recommended
- Application rate: 0.57 l in 2.5 m
- Comments:
    - *Mix and Load:* Measurement of concentrate used a calibrated jug, but the operator inadvertently measured 1 pint instead of 1 l into the sprayer. A total of 34.5 mg of a.s. was recovered from mixing and loading, of which 18 mg was from the cap swab as potential exposure.
    - *Application:* Ten loose sheets of tissue, approximately 20 × 20 cm were used on the conservatory floor to assess the pesticide level that would be left on the ground in this family seating area. Five of these were in the immediate spraying area and five spaced out at about 1 m distance from the nearest target plant. The total residue from the 5 tissues nearest the target plants amounted to 0.3 mg of a.s. The tissues 1 m away had residue levels below the detection limit in 4 of 5 cases, the fifth having a residue equivalent to 0.005 mg. The operator used his left hand to hold each leaf to ensure coverage, this led to very high levels on left hand equating to 0.24 mg of a.s. over this relatively small surface area. The contamination rate from this application was 24.6 ml h^{-1}, or 10.9 mg a.s. h^{-1}. The combined mix and load and application operations gave a total body exposure to the active substance of 17 mg per spraying operation, of which 0.45 mg was from the application stage.
    - Decontamination/cleaning left combined residues on clothing and equipment of less than 0.05 mg a.s.

Scenario 6

- Seven apple and plum trees, in full leaf, 1.5 m to 2 m tall
- Tank concentration: 303 µg ml^{-1} = 121% of recommended
- Application rate: 4.5 l in 15.3 min
- Comments:
    - *Mix and Load:* Subject overfilled one capful, of the 2 measured, leading to high levels of a.s.: 21.9 mg, being found on the left glove. Only 5.5 mg a.s. was recovered from within the cap.
    - *Application:* Spray was applied horizontally, giving highest residues on arms torso and feet. The highest residue of a.s. found from the application was 0.26 mg on the front torso. The contamination rate from this application was 24.2 ml h^{-1}, or

7.3 mg a.s. h^{-1}. The combined mix and load and application operations gave a total body exposure to the active substance of 23.8 mg per spraying operation. The measured exposure to the active substance generated by the application stage of the spraying operation was 1.9 mg.
- Decontamination/cleaning left combined residues on clothing and equipment of slightly over 0.11 mg a.s.

48.3.2
General Observations

The potential exposure due to inhalation of the pesticide appears minimal, with only one air-borne sample (Scenario 1) registering a response on the analytical system, but still below the limit of determination of 0.01 µg ml^{-1}. This is likely to be due to the short sampling period and low concentration of the a.s. in the spray mixture.

The greatest risk of potential operator exposure occurs during the handling and measurement of the concentrate (mixing and loading). Accurate measurement of the concentrate is difficult to achieve using the measurement equipment supplied with the pesticide formulation. Only one test subject managed to prepare the pesticide for application with a less than 20% margin of error. In only two of the six cases the amount of concentrate measured exceeded the amount required, one of these was due to a confusion of measurement units. In all other cases the use of the cap as a measuring device resulted in less concentrate than intended being added to the mix.

Drips and spillages during measurement into the cap were monitored in 80% of cases. These have been excluded from the calculation of the potential operator exposure, but represent an additional contamination source. Gloves were contaminated during the mix and load operations in all but one case.

Potential exposure during application is closely linked to the duration of application, with a maximum whole body contamination level of approximately 25 ml per hour being measured in three of the six scenarios, above. This would equate to 6.25 mg of active substance per hour, at the recommended application strength. A typical application took between 5 and 15 min.

Extracts from collection media placed at ground level had high levels of active substance in those directly under the target plants. Negligible levels were measured on those placed more than 0.5 m from the immediate spray area. This indicates that run-off from the plant provides the major contribution to ground level contamination. The high levels measured on the uppers of the boot covers, confirm that drips directly from the spayed foliage offer a significant source of potential operator exposure. Spray drift contribution to ground level exposure levels appears minimal, or nil, in the operations observed.

Cleaning and decontamination of the equipment generally provides low exposure levels, between 0.01 and 0.1 ml found on gloves, but residues of up to 8 ml of tank mix were found on dip-tubes and plungers after use.

The estimated absorbed dose has been compared to the provisional Acceptable Operator Exposure Levels (AOEL) for carbendazim, supplied by the UK Pesticide Safety Directorate (PSD). For assessment of the worst case exposure levels in the scenarios, above, it has been assumed that no PPE is worn and that all of the pesticide measured from the whole body dosimetry is available to be absorbed. (A figure for dermal absorption for carbendazim of 1% has been used). The contribution to exposure to the active substance via inhalation is considered to be negligible in all of the above cases, as the levels measured were all below the level of determination.

In reality the clothing worn, could reduce the actual dermal exposure considerably. This would be particularly true if gloves were worn for the mix and load operations, which provide most of the contamination. However, the chest, hips, and feet are unlikely to have exposed skin during spray application, even in amateur pesticide application operations.

The provisional AOEL for carbendazim is 0.04 mg kg^{-1} per day. In the worst case scenario observed, Scenario 2, a 60 kg operator may receive a maximum potential dermal exposure of 35 mg during a single application cycle, of which 0.35 mg may be adsorbed = 0.006 mg kg^{-1} per application. More than six similar applications could be performed without the operator being exposed to carbendazim levels above the AOEL.

48.4 Conclusions

The potential exposure to pesticides, for amateur users of home and garden products, is greatest during mixing and loading and is exacerbated by the use of the container cap as a measuring device. This method of concentrate measurement was inaccurate in all but one instance, with the tank mix concentrations falling between 55 and 177% of that recommended. With this formulation the observed level of potential operator exposure during these application scenarios does not reach levels above acceptable operator exposure levels (AOEL). The total amount of active substance recovered from all clothing worn during a complete mix and load and application session was in the range from 1.0 to 35.0 mg. The wide range of values reflects the importance of avoiding contact with the concentrate during mixing and loading operations. There is no recommendation that gloves should be worn during this operation on the product label.

The maximum estimated exposure to carbendazim during the complete course of a single mix, load and application sequence is less than the AOEL by a factor of 6. Drips, spillages and residues found within the measuring cap may reduce this safety margin on subsequent operations. Therefore the use of carbendazim in the scenarios observed appears to give an adequate margin of safety.

Acknowledgement

This authors wish to acknowledge the funding provided for this work by the UK Pesticide Safety Directorate (DEFRA).

References

Glass CR, Gilbert AJ, Mathers JJ, Lewis RJ, Harrington PM, Perez-Duran S (2002) Potential for operator and environmental contamination during concentrate handling in UK agriculture. Aspects of Applied Biology 66:379–386

Wild SA, Mathers JJ, Glass CR (2000) The potential for operator hand contamination to pesticides during the mixing and loading procedure. Aspects of Applied Biology 57:179–183

Appendix

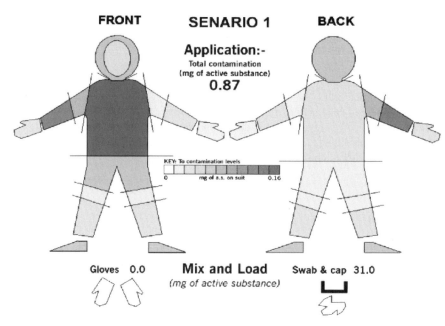

Fig. 48.1. Diagramme of scenario 1

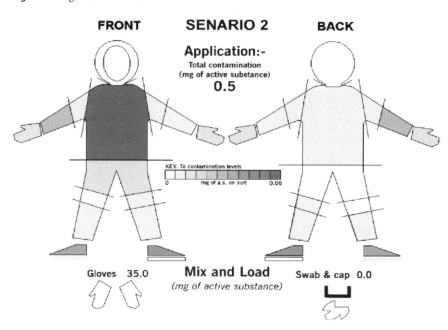

Fig. 48.2. Diagramme of scenario 2

Chapter 48 · Potential Exposure to Pesticides during Amateur Applications 537

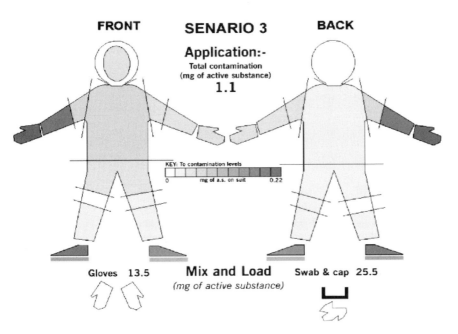

Fig. 48.3. Diagramme of scenario 3

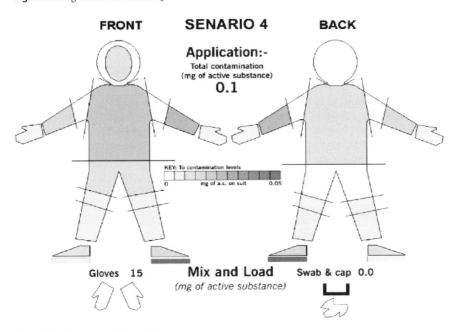

Fig. 48.4. Diagramme of scenario 4

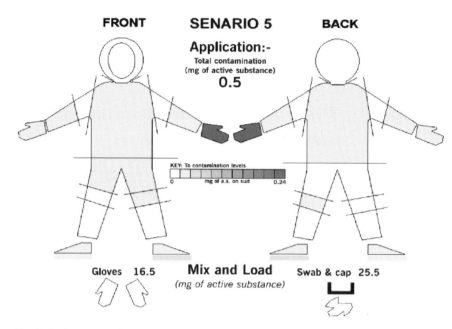

Fig. 48.5. Diagramme of scenario 5

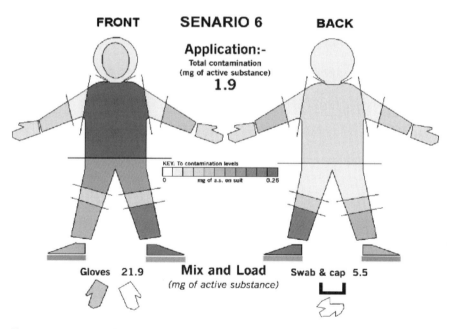

Fig. 48.6. Diagramme of scenario 6

Part VI
Green Chemistry

Part VI

Chapter 49

Carbon Dioxide, a Solvent and Synthon for Green Chemistry

D. Ballivet-Tkatchenko · S. Camy · J. S. Condoret

Abstract

Carbon dioxide is a renewable resource of carbon when we consider the reuse of existing CO_2 as a carbon source for producing chemicals. The development of new applications is of major interest from the point of view of carbon dioxide sequestration and within the scope of green chemistry. For example, using CO_2 instead of CO or $COCl_2$ for chemical synthesis constitutes an attractive alternative avoiding hazardous and toxic reactants. However, it has the lowest chemical reactivity, which is a serious drawback for its transformation. Supercritical CO_2 as a reaction medium offers the opportunity to replace conventional organic solvents. Its benign nature, easy handling and availability, non volatile emitting, and the relatively low critical point (P_c = 73.8 bar, T_c = 31 °C) are particularly interesting for catalytic applications in chemical synthesis, over a wide range of temperatures and pressures. The benefits of coupling catalysis and supercritical fluids are both environmental and commercial: less waste and less emission of volative organic compounds (VOCs), improved separation and recycling, and enhanced productivity and selectivity.

The case study described in this paper concerns the reaction of carbon dioxide with alcohols to afford dialkyl carbonates with special emphasis on dimethyl carbonate. It is of significant interest because the industrial production of this class of compounds, including polycarbonates, carbamates, and polyurethanes, involves phosgene with strong concerns on environmental impact, transport, safety and waste elimination. The future of carbon dioxide in green chemistry, including supercritical applications, is highly linked to the development of basic knowledge, know-how, and tools for the design of catalyst precursors and reactors.

Key words: carbon dioxide fixation, supercritical carbon dioxide, catalysis, dialkyl carbonate, dimethyl carbonate, phase equilibria, fractionation

49.1
Introduction

The development of environmentally-improved routes and the design of green chemicals are two facets of Green Chemistry devoted to reducing the impact of chemical processes and compounds on the environment (Anastas 2000). To reach the ultimate goals of cleaner production and sustainable development, new and improved syntheses of chemicals have to meet economics, ecological and social criteria (Vollenbroek 2002). The design of chemical reactions minimising waste, air emission, and hazardous reactants is on the way to environmentally benign technologies. For the academic chemist, minimisation of waste and emission relies on chemical yield, selectivity, and atom utilisation (Sheldon 1994). Therefore, catalytic processes likely to provide rapid and selective chemical transformations as well as effective recovery of both catalyst and product, without creating environmental problems, do have a significant impact in the chemical industry. In most cases catalyst recovery and recycling is of paramount importance for the economics of the process due to multistage, sophisticated, and costly recycling operations. The alternative of leaving-in the catalyst is acceptable in few cases. For example, the latest generations of Ziegler-Natta and metallocene-based catalysts are so productive that the catalyst present in concentration at the ppm level is left in the polyolefin product (Mülhaupt 1995).

49.1.1
Carbon Dioxide as a Solvent

In organic synthesis, most reactions are performed in the liquid state by the addition of a solvent with several beneficial purposes: *(i)* bringing the reactants to a single phase for rate enhancement, *(ii)* decreasing the viscosity of the reacting mixture for better diffusion of the reactive species, and *(iii)* providing easier mastering of heat effects. It should be pointed out that when one of the reactants is in the gaseous state (dihydrogen, carbon monoxide, dioxygen, ...), the gas-liquid mass transfer is often rate determining. The use of toxic solvents contributes to air emission of volatile organic compounds (VOCs) and of chloro hydrocarbons; they are now considered as environmentally unacceptable (Eckert and Chandler 1998). In the field of soluble metal-based catalysts, a considerable progress has been achieved through the concept of diphasic liquid-liquid catalysis: here the catalyst is recovered in one liquid phase and the product in the other one. This has led to drastic technological improvements but it has only reached single cases of industrialisation (Keim 1984; Cornils and Kuntz 1995; Driessen-Hölscher 1998). Further possibilities deserve attention with promising results in the use of fluorous solvents (Horvath 1998), non-aqueous ionic liquids (Wasserscheid and Keim 2000), and supercritical media (Jessop and Leitner 1999). These systems are potentially cost-effective by combining reaction and separation into a single reactor (Freemantle 2000).

The ideal supercritical fluid (SCF) should be non-toxic for man and the environment, non-carcinogen, non-flammable, and its intrinsic chemical reactivity should not be the source of runaway reactions. Concerning carbon dioxide, its impact on the global warming may be a handicap, but, when CO_2 applications are implemented, carbon dioxide is never released to the environment. While commercial developments of supercritical carbon dioxide ($scCO_2$) technology are successful for separation and extraction, its application as a reaction medium and as a reactant are still in its infancy (Jessop and Leitner 1999; Noyori 1999). The wide range of miscibility of gaseous reactants, e.g. dihydrogen and dioxygen, in supercritical fluids simplifies greatly the design of reactors in comparison with conventional ones which operate under di- or triphasic conditions. Furthermore the good transport properties of these fluids (viscosity, diffusivities) allow to design small and efficient reactors, where catalysts may have longer lifetime. The post-reacting separation is likely to be easy. Small temperature or pressure variations lead to solvent-free products. Fractionation of the effluent is also possible without involving any further separating agent. The greatest safety concern of supercritical fluids is the use of high pressure components in the process equipment.

49.1.2
Carbon Dioxide as a C1-Building Block for Chemicals

Carbon dioxide, one of the major man-made greenhouse gases, is a renewable resource of carbon when we consider the reuse of existing CO_2 as a carbon source for producing chemicals. The industrial syntheses of urea, cyclic carbonates, salicylic acid, and methanol already involve carbon dioxide as a reactant. Nowadays, the development of new applications is of major interest from the point of view of carbon dioxide sequestration and within the scope of green chemistry (Dinjus and Fornika 1996). Using CO_2 instead of CO or $COCl_2$ constitutes an attractive alternative avoiding hazardous and toxic components (Aresta and Quaranta 1997). However, it has the lowest chemical reactivity which is a serious drawback for its transformation and incorporation to organic molecules. Several ways of activation are currently under investigation; among them catalysis offers a number of options, but it still is a challenge to find activities of practical interest.

We focused our research on new syntheses for open chain organic carbonates because their production is historically dominated by phosgene chemistry (Fig. 49.1, route A). The unfavourable opinion on the phosgene industry concerns the environmental impact, transport, safety and waste elimination. The lethal concentration threshold for a 30 min exposure is 10 ppm, as compared to 870, 4 000, and 30 000 ppm for dichlorine, carbon monoxide, and ammonia, respectively. In addition, the wastes contain hydrogen chloride as well as chlorinated solvents, and they have to be treated prior to disposal. These environmental and safety handicaps motivate the development of new technologies for commodity chemicals (>10 000 t yr^{-1}). Its industrial use is likely to be confined to small groups of experts who are properly equipped (Senet 2000). For the production of dimethyl carbonate (DMC), catalytic alternatives via the oxidative carbonylation of methanol (Fig. 49.1, route B) have provided an answer (Rivetti 2000). However, the use of carbon monoxide imposes safety constraints and the use of dioxygen requires a strict control of the kinetics.

Dimethyl carbonate (DMC) is an alternative to phosgene in carbonylation reactions, and a substitute for methyl halides and dimethyl sulfate in methylation reactions (Tundo 2001). Its current production is estimated to be around 80 000 t yr^{-1} worldwide, with a growing demand for captive use by polycarbonate firms. DMC has a low toxicity, is not corrosive, and will not produce environmentally damaging by-products (no organics or salts). If the economics of the current technologies were more favourable, the market potential could be an order of magnitude higher because DMC is a good candidate for the replacement of methyl *tert*-butyl ether (MTBE) as a fuel additive (Pacheco and Marshall 1997). As a consequence, alternative reactions are under investigation. Among them, reacting methanol (or dimethyl acetals) and CO_2

Fig. 49.1.
Three reaction routes to open chain carbonates from alcohol and C1-building blocks

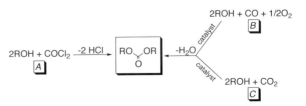

in the presence of a catalyst (Fig. 49.1, route C) leads to DMC under supercritical conditions (Isaacs et al. 1999; Ballivet-Tkatchenko et al. 2000); a drastic increase in activity is found if the experiments are conducted at ca. 2000 bar and 180 °C (Sakakura et al. 1999). There is a real challenge from these very first results to commercial application, because on one hand the activity is low and on the other hand the most active systems are tin(IV) compounds which are not environmentally friendly. An important task is the knowledge of the reaction mechanisms involved to get structure-activity relationships. This paper addresses some issues on the chemistry and emphasises the reactor design under supercritical conditions for route C (Fig. 49.1). It also gives an insight to the complete process, including the post-reacting fractionation using potentialities of supercritical state.

49.2
Experimental

49.2.1
General

All reactions were carried out under dry argon using Schlenk tube techniques. The solvents were purified by standard methods. The organotin compounds n-Bu$_3$SnCl, n-Bu$_2$SnCl$_2$, n-BuSnCl$_3$, (n-Bu$_3$SnO)$_2$, and (n-Bu$_2$SnO)$_n$ were used as received from Aldrich and Acros Chimica for the syntheses of the alkoxystannanes derivatives according to already published procedures (Ballivet-Tkatchenko et al. 2000). CO_2 N45 was purchased from Air Liquide. The ^{1}H, ^{13}C, and ^{119}Sn nuclear magnetic resonance (NMR) spectra were measured at 500.132, 125.770, and 186.501 MHz, respectively, on a Bruker DRX 500 spectrometer. For a detailed description and assignment of the experimental spectra see Ballivet-Tkatchenko et al. (2000). Infrared spectra were obtained with a FT-IR Bruker Vector 22 spectrometer, the sample being placed between NaCl windows either as neat or dispersed in Nujol. Elemental analysis was performed at the Laboratoire de Synthèse et Electrosynthèse Organométalliques, Université de Bourgogne, Dijon.

49.2.2
Volumetry

A Schlenk tube containing 1 mmol of the tin compound in 1 ml of toluene was connected to a pressure transducer and to a CO_2 reservoir of known pressure and volume, maintained at 19 °C. The amount of CO_2 gas absorbed by tin compound was calibrated by reference experiments. The calculated CO_2:Sn molar ratio was at ±0.05.

49.2.3
Reaction under CO_2 Pressure

In a 100-ml stainless steel batch reactor, a solution (10–30 ml) of the tin compound (4 mmol Sn) was introduced. The reactor was pressurised with CO_2, heated to 145 °C, then the CO_2 pressure was adjusted to the desired value, typically between 90 and 220 bar. After the run, the reactor was cooled to 0 °C, depressurised, and the liquid phase was analysed by gas chromatography (Fisons 8000, FID detector, J&W Scientific DB-WAX 15 m megabore column) and gas chromatography-mass spectrometry for identification (Fisons MD 800, EI 70 eV, J&W Scientific DB-1 60 m capillary column). Evaporation of the volatiles under vacuum at room temperature allowed to characterise the residue by NMR.

49.2.4
Experimental Set-up for the Fractionation Study

For the fractionation study, we have used an experimental process schematically shown on Fig. 49.2, comprising a 200 ml contacting vessel C (the reactor) followed by three cyclonic separators S1, S2, and S3 (ca. 20 ml each). As an efficient immobilised catalyst has not yet been found, the reaction could not be operated experimentally. We therefore worked with a mimicked reacting mixture in the contacting vessel where an intimate contact between a methanol-dimethyl carbonate-water mixture and CO_2 entering the process takes place. A constant feed flow rate of CO_2 of ca. 4 kg h^{-1} is kept. The outgoing fluid from the contactor C composed of a CO_2-methanol-dimethyl carbonate-water quaternary mixture undergoes a three successive depressurisation stages S1, S2, and S3, in order to recover the targeted components (DMC and water), and to recycle CO_2 and methanol.

Liquid phase analyses were done by gas chromatography with a Hewlett-Packard 5890 series II chromatograph equipped with a Supel-QTM Plot capillary column (30 m, 0.53 mm ID) from Supelco (USA) and a thermal conductivity detector (TCD). Synthetic mixtures were prepared from distilled water, dimethyl carbonate (Aldrich, 99%, D15,292-7), and methanol (Prolabo Chromanorm, min. 99.8%, 20834.291).

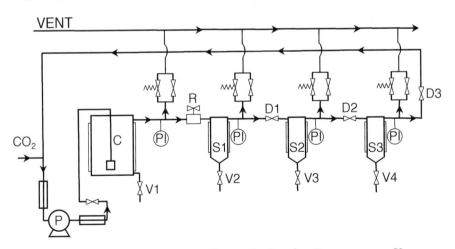

Fig. 49.2. Pilot SF200. *C*: contactor; *D1, D2, D3*: depressurizarion valves; *P*: pressure pump; *PI*: pressure indicators; *R*: back pressure regulator; *S1, S2, S3*: separators; *V1, V2, V3, V4*: sampling valves

49.2.5
Numerical Methods for Thermodynamic Calculations

In this study, phase equilibria are calculated using the cubic equation of state proposed by Redlich and Kwong (1949), modified by Soave (1972), and named SRK equation. To account for interaction taking place between components in the mixture, the "MHV-2" mixing rules, developed by Huron and Vidal (1979) and modified by Michelsen (Dahl and Michelsen 1990), are used. This type of modelling is now commonly recognised as a good choice for high pressure phase equilibrium calculations involving polar compounds, as it is the case in this study. Equations of state such as SRK equation give satisfying results at high pressure but have to be used only when dealing with apolar or poorly polar compounds. Conversely, activity coefficient models (differently named γ–ϕ approach) cannot be used at high pressure but are particularly appropriate for thermodynamic calculations involving polar compounds. The MHV-2 approach conciliates these two application domains, involving an activity coefficient model into mixing rules of an equation of state. The UNIQUAC activity coefficient model (Abrams and Prausnitz 1975) has been chosen to determine the value of the free excess Gibbs energy needed in the calculation of the mixture parameters.

Thermodynamic models have proved to be of great help in predicting thermodynamic behaviour of mixtures. However, these models need specific mixture parameters, reflecting interactions between the components in the mixture, that are obtained by fitting a limited set of experimental equilibrium data. The data used to determine mixture parameters can be equilibrium data from the binary or ternary sub-systems constituting the quaternary methanol-CO_2-water-DMC mixture. In our case, some binary or ternary data can be found in the existing literature. For DMC-water and DMC-CO_2 mixtures, no experimental work is available, and experiments have been necessary. Such new results are now available from our recent publication (Camy et al. 2002). Binary interaction parameters are calculated using the commercial software ProReg™ (PROSIM S.A., France). This model fed by adequate binary interaction coefficients allows us to describe the thermodynamic behaviour of the reacting mixture in the reactor itself and in the separation system.

49.3
Results and Discussion

49.3.1
Mechanistic Approach for Dimethyl Carbonate Formation

Our interest in the reaction of carbon dioxide with alcohols to afford dialkyl carbonates led us to focus on the mechanistic approach with tin compounds. These compounds were chosen because ^{119}Sn NMR spectrocopy provides useful information on the number of tin species and their coordination (Hani and Geanangel 1982) in addition to ^{1}H and ^{13}C spectra. IR spectroscopy was also used to monitor the presence of absorption bands of the carbonate moieties. It is not intended to report here in details the synthesis and characterisation of the compounds as preliminary results have already been published (Ballivet-Tkatchenko et al. 2000). It is more important to summarise and emphasise some elementary reaction steps for the purpose of reactor design.

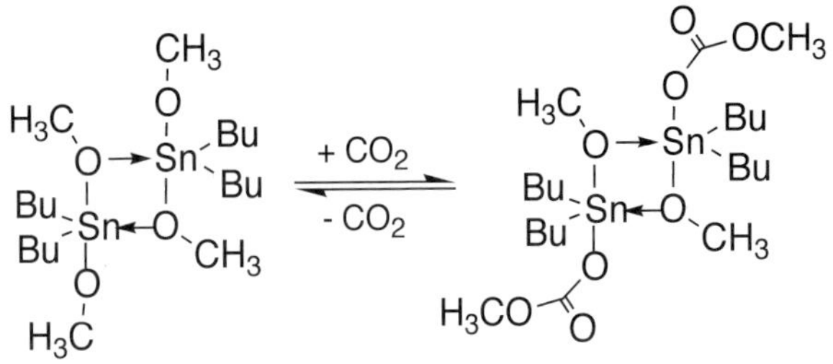

Fig. 49.3.
Dimeric arrangements for $n\text{-Bu}_2\text{Sn}(\text{OCH}_3)_2$ and its carbonated form

Several tin(IV) compounds were studied, $n\text{-Bu}_3\text{SnCl}$ (Bu = butyl), $(n\text{-Bu}_3\text{SnO})_2$, $n\text{-Bu}_3\text{SnOR}$ (R = phenyl, methyl), $n\text{-Bu}_2\text{SnCl}_2$, $(n\text{-Bu}_2\text{SnO})_n$, $n\text{-Bu}_2\text{Sn}(\text{OR})_2$ (R = phenyl, methyl, isopropyl, *tert*-butyl), $n\text{-BuSn}(\text{OCH}_3)_3$, $\text{Ph}_3\text{SnOCH}_3$ (Ph = phenyl), and $\text{Ph}_2\text{Sn}(\text{OCH}_3)_2$. The fixation of carbon dioxide was effective for compounds containing OR groups, preferably methoxy and isopropoxy. The new species formed is the result of carbon dioxide insertion in the Sn-OR bond, leading to the carbonate fragment Sn-OC(O)OR. At room temperature and atmospheric pressure, the reaction is reversible leading back to Sn-OR. Volumetric experiments showed that reaction stoichiometry is CO_2:Sn = 1, whatever the initial number of alkoxy fragments. Therefore, only one OR group is reacting. This result has been assigned to the dimeric structure of the polyalkoxystannes as shown in Fig. 49.3 for $n\text{-Bu}_2\text{Sn}(\text{OCH}_3)_2$. A recent X-ray single crystal structure determination of $(\text{CH}_3)_2\text{Sn}(\text{OCH}_3)_2$ and its carbonated form strengthens this assignment (Choi et al. 1999).

The formation of dialkyl carbonates was only found for the polyalkoxybutylstannanes. For example, starting from $n\text{-Bu}_2\text{Sn}(\text{OCH}_3)_2$ dimethyl carbonate is quantitatively produced (DMC:Sn = 1) upon heating a methanolic solution at 150 °C under 200 bar of carbon dioxide. Further experiments conducted in toluene instead of methanol point out that methanol is not directly involved in DMC formation (Ballivet-Tkatchenko et al. 2002). An intra-molecular rearrangement of the tin coordination sphere is taking place, but, in the overall reaction, methanol and carbon dioxide should be the reactants for the catalytic cycle (Fig. 49.1, route C). Under these experimental conditions, CO_2 reaches the supercritical state ($P_c = 73.8$ bar, $T_c = 31$ °C). Supercritical CO_2 as a solvent and a reactant has a great potential for optimising chemical reactions (Jessop and Leitner 1999); however, in our case the origin of the enhanced chemical reactivity is not clear as the phase behaviour under reaction conditions have a great influence. It can also be expected that CO_2 pressure has a beneficial effect on the carbonation reaction, which was found to be reversible at room temperature and atmospheric pressure.

49.3.2
Dimethyl Carbonate Synthesis Using Supercritical CO_2

Operation of chemical reactions in supercritical media is regularly emphasised by experts in the domain as a very promising field of research. Nevertheless, conception of reactors, as well as optimum choice of operating conditions, are still to be developed.

It is now well admitted in chemical engineering strategy, that conception of the post-reacting separation as early as the conception of the reactor, leads to more efficient processes. It is particularly true for supercritical media, because their adjustable solvation power is a very convenient way to operate the purification of products. In the case of the carbonation reaction of methanol (Fig. 49.1, route C), there is an additional reason, because one product of the reaction, water, has a detrimental effect on a possible reversal of carbonation reaction, as well as on the good operation of the catalyst. This implies that it would be preferable to eliminate reaction products as the reaction proceeds, in order to shift the equilibrium of the reaction. As a consequence the study of the entire process could be of a great help for optimising the reaction, and investigation of the fractionation process implemented at the output of the reactor has to be undertaken.

One first compulsory initial stage is to study the thermodynamic behaviour of the reacting mixture, i.e. the quaternary methanol-CO_2-water-DMC mixture. This knowledge is essential to characterize the possible coexisting phases in the reactor and define the optimum reaction conditions, as well as to design the post-reacting fractionation device.

49.3.2.1
Study of Operating Conditions in the Reactor

The main information needed is the description of phase equilibria existing in the reactor as a function of pressure and temperature. For this, a hypothetical perfectly mixed continuous reactor, fed with a methanol-CO_2 mixture, is considered. The DMC synthesis takes place with an arbitrarily chosen conversion, i.e. the concentrations inside the reactor are accordingly. Then the physical state of the mixture in this hypothetical reactor is calculated thanks to the thermodynamic model described above.

Figure 49.4 presents, for two different mixtures in the reactor, the coexistence zones of monophasic and diphasic states. In each diagram, qualitative information about the state of the mixture is marked for each domain. The letter L refers to a liquid phase, whose density is high (ca. 800–900 kg m^{-3}). Conversely, the vapour state, marked with the letter V, refers to a low-density phase (<100 kg m^{-3}). Finally, the letter F designates a fluid state whose density lies between that of a liquid and a gas. Because the critical line of this quaternary mixture was not calculated here, these denominations remain qualitative. Figure 49.4 shows that for a methanol rich mixture (diagram a), no fluid phase is present and there is a large diphasic liquid-liquid (L1-L2) zone, where a first liquid phase, mainly composed of water and methanol, is in equilibrium with a second liquid phase containing a large proportion of CO_2 and DMC. This equilibrium zone exists for low temperatures, at a pressure greater than about 80 bar. A second diphasic liquid-vapour zone (L-V) is present at low pressures, consisting in the equilibrium between a liquid phase and a CO_2 rich vapour phase. The main monophasic zone is a liquid (L); however, at high temperatures and low pressures, a restricted monophasic vapour phase zone (V) is evidenced. For a CO_2 rich feed (diagram b), the phase diagram undergoes perceptible modifications. Indeed, the diphasic zone is here further reduced and no liquid-liquid equilibrium is now observed. In this case, there is no longer liquid-liquid diphasic zone and a liquid-vapour zone (L-V) transforming into a liquid-fluid zone (L-F) is now present. At higher pressure and temperature we now observe a plain fluid phase.

Fig. 49.4.
Coexisting zones of monophasic and diphasic mixtures at 60% conversion for: **a** methanol rich feed, **b** CO_2 rich feed

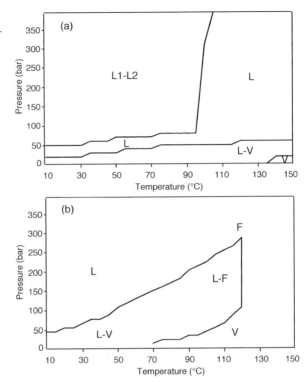

Study on dimethyl carbonate synthesis mechanism has shown that, for the moment, the most interesting results are obtained using polyalkoxybutylstannane catalysts, being active at about 150 °C. Moreover, to avoid mass transfer limitations which are likely to occur with the low-pressure diphasic system liquid mixture-gaseous CO_2 (Kizlink and Pastucha 1994, 1995), the reaction will advantageously be operated in a monophasic reacting mixture, and preferably in a "fluid" state, or vapour state. By comparison with the vapour state, the "fluid" state, that corresponds to higher pressures, is more interesting because *(i)* the volume of the reactor will be smaller *(ii)* the effect of pressure has been proved to be positive on the kinetics of the reaction, and *(iii)* the separation of the reaction products can be made easier by using the tunable solvent power of supercritical fluids. From Fig. 49.4 it can be concluded that, to be sure of running the reaction in a homogeneous fluid medium, a large excess of CO_2 is needed. Moreover, for such mixtures, a drop of pressure and temperature after the reactor would easily allow a return to a diphasic state, with a view to obtaining an efficient mixture fractionation. This last topic is the aim of the next part.

49.3.2.2
Study of the Mixture Fractionation

As previously mentioned, it is worthwhile to investigate the feasibility of the fractionation of the effluent methanol-DMC-water mixture, and to test a convenient and quite simple process, generally used to run post-extraction fractionation in supercritical technology (Fig. 49.2). Due to the great number of parameters, this study of the post-reacting fractionation cannot only lean on an experimental approach. Modelling of the process must be developed in order to represent the dynamic behaviour of the whole experimental process. Indeed, even if this kind of pilot is often used to operate extraction operations, the literature proposes very few results dealing with its dynamic modelling.

Our description of the process is based on the conventional chemical engineering concept of theoretical stage of equilibrium. A simplified numerical approach used to solve dynamic mass balances has been proposed and validated in comparison with a rigorous resolution of the algebro-differential system of equations (Camy and Condoret 2001). The results of the modelling were compared to experiments and gave good results for the contactor, while it was not very satisfactory for the description of the separation. Even if discrepancies between experimental and calculated results may probably originate from experimental procedure, and because we think that thermodynamic modelling is correct, hydrodynamic description of the separators is here likely to be oversimplified. The cyclonic separators cannot be regarded as simple theoretical stages as it is often done in the literature (Cesari et al. 1989). We proposed an other description (Camy and Condoret 2001), that, although more suitable, needs still to be improved by taking into account, for instance, the imperfect collection of droplets.

Our work has shown that is not realistic to envisage the separation of the resulting methanol-dimethyl carbonate-water mixture with such a simple device because components have too strong affinities. The use of simple separators in order to separate a liquid mixture when volatility of components, as well as their solubility into CO_2, are comparable, is not enough efficient. Under progress is now the use of modelling to investigate the feasibility of this separation, either using a greater number of separators, or by implementing a more complex strategy of separation. For instance a suitable way would be to consider the operation of one or several counter current columns of separation. Together with advances in the conception of an efficient catalyst, these approaches will be a useful contribution to the success of a viable process from CO_2.

49.4
Conclusion

Supercritical CO_2 as a reaction medium offers the opportunity to replace conventional organic solvents due to its adjustable solvation property as regards to both temperature and pressure, either pure or through the addition of solubilizers. The relatively mild critical point of CO_2, its benign nature, easy handling, availability, non volatile emitting are particularly attractive for catalytic applications in chemical synthesis, over a wide range of temperatures and pressures. The benefits of coupling catalysis and supercritical fluids are both environmental and commercial: less waste and VOCs emission, improved separation and recycling, and enhanced productivity and selectivity.

The case study described in this paper is of significant interest because the carbonation of alcohols to dialkyl carbonates replaces phosgene by carbon dioxide. Of course, carbon dioxide is much less reactive than phosgene, which means more energy input.

However, Life Cycle Analysis applied to the assessment of the environmental impact of the phosgene and carbon dioxide routes is greatly in favour of CO_2 (Aresta and Galatola 1999). The future of this molecule in green chemistry, including supercritical applications, is highly linked to the development of basic knowledge, know-how, and tools for the design of catalyst precursors as well as the technological conception of reactors and fractionation devices. The multi-component nature of a catalytic reaction (catalyst precursor, reactants, products), its changes upon time, and the pressure-temperature conditions make difficult the experimental approach. Synthetic chemistry, analytical chemistry, and chemical engineering are mandatory to cope with the above-mentioned requirements. Indeed, chemical engineering can bring better knowledge of the rather uncommon thermodynamics in the reactor, and has demonstrated the interest of studying the coupling of the reaction and the separation.

Acknowledgements

We are grateful for financial support of this work from the Centre National de la Recherche Scientifique under the programme "Catalyse et catalyseurs pour l'industrie et l'environnement".

References

Abrams DS, Prausnitz JM (1975) Statistical thermodynamics of liquid mixtures: a new expression for the excess Gibbs energy of partly or complete miscible systems. AIChE J 21:116–128

Anastas PT (2000) The role of catalysis in the design, development, and implementation of green chemistry. Catal Today 55:11–22

Aresta M, Galatola M (1999) Life cycle analysis applied to the assessment of the environmental impact of alternative synthetic processes. The dimethyl carbonate case. J Clean Prod 7:181–193

Aresta M, Quaranta E (1997) Carbon dioxide: a substitute for phosgene. Chemtech 27(3):32–40

Ballivet-Tkatchenko D, Douteau O, Stutzmann S (2000) Reactivity of carbon dioxide with butyl(phenoxy)-, (alkoxy)- and (oxo)stannanes: insight into dimethyl carbonate synthesis. Organometallics 19:4563–4567

Ballivet-Tkatchenko D, Chermette H, Jerphagnon T (2002) CO_2 as a C_1-building block for dialkyl carbonate synthesis. In: Maroto-Valer MM, Soong C, Song Y (eds) Environmental challenges and green gas control for fossil fuel utilization in the 21st century. Kluwer Acad/Plenum publishers, New York, pp 371–384

Camy S, Condoret JS (2001) Dynamic modelling of a fractionation process for a liquid mixture using supercritical carbon dioxide. Chem Eng Proc 40:499–509

Camy S, Pic JS, Badens E, Condoret JS (2003) Fluid phase equilibria of the reacting mixture in the dimethyl carbonate synthesis from supercritical CO_2. J of Supercritical Fluids 25:19–32

Cesari G, Fermeglia M, Kikic I, Policastro M (1989) A computer program for the dynamic simulation of a semi-batch supercritical fluid extraction process. Computers Chem Eng 13:1175–1181

Choi J-C, Sakakura T, Sako T (1999) Reaction of dialkyltin methoxide with carbon dioxide relevant to the mechanism of catalytic carbonate synthesis. J Am Chem Soc 121:3793–3794

Cornils B, Kuntz E (1995) Introducing TPPTS and related ligands for industrial biphasic processes. J Organomet Chem 502:177–186

Dahl S, Michelsen ML (1990) High-pressure vapor-liquid equilibrium with a UNIFAC-based equation of state. AIChE J 36:1829–1836

Dinjus E, Fornika R (1996) Carbon dioxide as a C1-building block. In: Cornils B, Herrmann WA (eds) Applied homogeneous catalysis with organometallic compounds. Wiley-VCH, Weinheim, pp 1048–1072

Driessen-Hölscher B (1998) Multiphase homogeneous catalysis. Adv in Catal 42:473–505
Eckert CA, Chandler K (1998) Tuning fluid solvents for chemical reactions. J of Supercritical Fluids 13:187–195
Freemantle M (2000) Eyes on ionic liquids. Chem Eng News 78(20):37–50
Hani R, Geanangel RA (1982) ^{119}Sn NMR in coordination chemistry. Coord Chem Rev 44:229–246
Horvath IT (1998) Fluorous biphase chemistry. Acc Chem Res 31:641–650
Huron MJ, Vidal J (1979) New mixing rules in simple equations of state for representing vapour-liquid equilibria of strongly non-ideal mixtures. Fluid Phase Equilibria 3:255–271
Isaacs NS, O'Sullivan B, Verhaelen C (1999) High pressure routes to dimethyl carbonate from supercritical carbon dioxide. Tetrahedron 55:11949–11956
Jessop PG, Leitner W (eds) (1999) Chemical synthesis using supercritical fluids. Wiley-VCH, Weinheim
Keim W (1984) Vor- und Nachteile der homogenen Übergangsmetallkatalyse, dargestellt am Shop-Prozess. Chem Ing Tech 56:850–853
Kizlink J, Pastucha I (1994) Preparation of dimethyl carbonate from methanol and carbon dioxide in the presence of organotin compounds. Collect Czech Chem Commun 59:2116–2118
Kizlink J, Pastucha I (1995) Preparation of dimethyl carbonate from methanol and carbon dioxide in the presence of Sn(IV) and Ti(IV) alkoxides and metal acetates. Collect Czech Chem Commun 60:687–692
Mülhaupt R (1995) Novel polyolefin materials and processes: overview and prospects. In: Fink G, Mülhaupt R, Brintzinger HH (eds) Ziegler catalysts. Springer-Verlag, Berlin, pp 35–56
Noyori R, guest editor (1999) Supercritical fluids. Chem Rev 99(2):353–634
Pacheco MA, Marshall CL (1997) Review of dimethyl carbonate manufacture and its characteristics as a fuel additive. Energy & Fuels 11:2–28
Redlich O, Kwong JNS (1949) On the thermodynamics of solution. V: An equation of state. Fugacities of gazeous solutions. Chem Rev 44:233–144
Rivetti F (2000) The role of dimethyl carbonate in the replacement of hazardous chemicals. CR Acad Sci Paris, Série IIC, Chemistry 3:497–503
Sakakura T, Choi JC, Saito Y, Masuda T, Sako T, Oriyama T (1999) Metal-catalyzed carbonate synthesis from carbon dioxide and acetals. J Org Chem 64:4506–4508
Senet JP (2000) Phosgene chemistry and environment, recent advances clear the way to clean processes: a review. CR Acad Sci Paris, Série IIC, Chemistry 3:505–516
Sheldon R (1994) Consider the environmental quotient. Chemtech 24(3):38–47
Soave G (1972) Equilibrium constants from a modified Redlich-Kwong equation of state. Chem Eng Sci 27:1197–1203
Tundo P (2001) New developments in dimethyl carbonate chemistry. Pure Appl Chem 73:1117–1124
Vollenbroek FA (2002) Sustainable development and the challenge of innovation. J Clean Prod 10:215–223
Wasserscheid P, Keim W (2000) Ionic liquids-New solutions for transition metal catalysis. Angew Chem Int Ed 39:3772–3789

Chapter 50

Mechanochemistry: An Old Technology with New Applications to Environmental Issues. Decontamination of Polychlorobiphenyl-Contaminated Soil by High-Energy Milling in the Solid State with Ternary Hydrides

M. Aresta · A. Dibenedetto · T. Pastore

Abstract

Mechanical energy was first used for running chemical reactions three centuries before Christ, for the preparation of mercury. The term mechanochemistry was introduced by Ostwald in 1893. The scientific principles of the technology were discussed by Heinicke in 1984. In this technology, energy transfer takes place through high energy milling of solids that undergo several transformations. Chemical reactions can also take place, that avoids the use of solvents with great benefit from the environmental point of view. As a matter of fact, mechanochemistry has been used so far for the preparation of new materials and running chemical reactions in absence of solvents. More recently, it has been used within the new perspective of application to solving environmental problems. In this paper, the utilization of mechanochemistry as a solid state technology for the dehalogenation of polychlorobiphenyls (PCBs) present in contaminated soil is described. The abatement of PCBs is quantitative. The high energy milling of soil with ternary hydrides represents a valid alternative to the technology based on the use of metal sodium and water, due to the higher safety, and more controlled reaction conditions.

Key words: mechanochemistry, high-energy milling, polychlorobiphenyls, solid state dechlorination, ternary hydrides

50.1 Introduction

The use of mechanical energy for driving a chemical reaction was first mentioned by Theophrastus (371–268 BC) in his treaty "De lapidibus", which means "about stones". The reaction described was the conversion of mercury sulphide into elemental mercury by reaction with copper metal (Eq. 50.1).

$$HgS + Cu \longrightarrow Hg + CuS \qquad (50.1)$$

The word mechanochemistry was introduced in 1893 by Ostwald, the 1909 Nobel Prize winner. The theoretical bases were discussed by Heinicke in 1984 (Heinicke 1984). The technology is based on the transfer of mechanical energy to reacting particles. During the collision, particles are subject to deformation, fracture, and welding. This allows reactions between solids to occur at a temperature close to room temperature. It can be questioned whether reacting particles are really in the solid state at the instant of collision, or if a liquid is formed. This could happen either because the transferred energy is converted into thermal energy that may melt the solids, or because under the action of pressure, and with the correct composition of the mix-

Table 50.1. Applications of mechanochemistry

Process	Product
Aromatic polycyclic hydrocarbons cleavage	Less toxic materials
Asbestos end-treatment	Inertization of microfibres
Ceramic-material manufacture	New materials from oxides
CuS ore work-up	CuS extracted from ores
Corundum grinding	Sized particles for industrial use
Reaction of solids	New products from no-solvent reactions
Cellulose fibers modification	Modified cellulose for specific purposes
Grinding of rocks or hard materials	Preparation of powders
Hard metal alloying	New alloys
Lignite coal grinding	Lignite particles for injection in boilers
Pigment production	New inorganic pigments
Plastics treatment	Waste plastics conversion into useful products

ture, an eutectic mixture is formed that may melt at a lower temperature than under ambient pressure. This point is not of secondary importance, because it raises the issue of assessing if the reaction takes place under heterogeneous conditions in the solid state, or if it occurs in a possibly homogeneous liquid system. These aspects, and many others, need further investigation in order to complete the understanding of the scientific basis of the technology and master its use. Mechanochemistry has been used for several applications since long time (Benjamin 1970) and has found an industrial utilization in the second half of the last century (Benjamin and Volin 1974). A summary of most common applications is given in Table 50.1.

Mechanochemical applications are known under several names: mechanical alloying (MA), high energy milling (HEM), and reactive milling (RM). Different bodies can be used for energy transfer, like spheres or rings. Recent applications of mechanochemistry occur in quite different areas (see Table 50.1). The most recent application is the field of contaminated soil treatment. We have applied mechanochemistry to the dechlorination of polychlorobiphenyls present in landfill soils (Aresta et al. 2001, 2003). Polychlorobiphenyls (PCBs) are among the most ubiquitous and persistent pollutants (Hutzinger et al. 1974; Lang 1992). Because of their inertness, such compounds are difficult to convert into less toxic species by using chemical processes or biological systems. The commonly adopted technologies for their disposal are incineration, that produces toxic compounds like dioxin, and land filling, that produces leacheates that can pollute water and soil. Therefore, "innovative technologies" for their inertization are needed. In the literature, only another application of mechanochemistry in the treatment of soil polluted with chlorinated compounds by reacting with metal sodium and hydrogen donor species (water) can be found (Birke 2001). In the present study we have used ternary hydrides and show that such species are able to convert quantitatively PCBs into biphenyl. $NaBH_4$ (Aresta et al. 2003) or $LiAlH_4$

(this study) contribute a metal ion that is used to fix the chloride eliminated from the organic substrate and provide the hydride ion that substitutes the chloride into the organic molecule. Solids like $NaBH_4$ and $LiAlH_4$ were separately mixed with a sample of soil and different weight ratios were used. Milling provides the energy necessary for the chemical reaction and induces an intimate contact between the reagents in the solid mixture, due to the resulting great specific surface. The products of the reductive process are inorganic chloride and biphenyl. The efficiency of the dechlorination/hydrogenation reaction was studied as a function of the milling time (up to 30 h). At fixed intervals of time, the total PCBs content and the inorganic chloride produced were measured. A different efficiency of abatement was observed per each hydride, starting from a total PCBs concentration of about 2 600 $mg\ kg^{-1}$. In this paper we report on the results of a laboratory scale experiment using a bench ring mill and compare $NaBH_4$ and $LiAlH_4$ as dehalogenation agents.

50.2
Experimental

50.2.1
Laboratory Technique and Equipment

The laboratory scale experiments were performed with a Fritsch (mod. Pulverisette 9) vibrating cup mill. The grinding set consists of two chrome steel rings. The mill has the following characteristics: variable rotational rings velocity (750/1 000 rpm), total mass of two rings 3 637 g, volumetric capacity 350 ml. The granulometry and composition of the soil sample, taken from a controlled landfill classified for toxic-harmful waste, show a typical sandy soil with more than 90% fine sand. Table 50.2 shows that the total PCBs concentration ranges around 2 600 $mg\ kg^{-1}$.

50.2.2
Analytical Techniques and Methods

The determination of chloride ions was carried out by using the method IV-2 (Soil Science 2000). The extraction of PCBs was carried out according to the EPA-3540 methodology (EPA 1995). The extracted solution was cleaned up following the EPA-3620-3630-3665 methodologies (EPA 1995). The determination of extracted PCBs was made according to EPA-8082 method by GC-MS (EPA 1995).

Table 50.2. Some analytical parameters of the used soil

Parameter	Unit	Value
Total PCBs (on the soil such it is)	$mg\ kg^{-1}$	1 520
Total PCBs (after 3 h of milling of neat soil)	$mg\ kg^{-1}$	2 600
BTX (benzene, toluene, xylene)	$mg\ kg^{-1}$	<0.1
Total hydrocarbons (C > 12)	$g\ kg^{-1}$	15.88
pH		6.84
Cr(VI)	$mg\ kg^{-1}$	<1

50.2.3
Pre-Conditioning of the Soil Sample

The contaminated soil sample was homogenized and dried at 323 K for 20 h in order to produce the optimal mechano-chemical conditions for the reductive dechlorination reaction to occur. It was then milled for about 3 h in order to convert it into small particles for both a more efficient performance of the solvent extraction and a more intimate contact with the reagent. We have found that the amount of extracted PCBs strongly depends on the soil particle size and three hours of preliminary milling are necessary for the correct quantification of PCBs in the sample (Table 50.2). Without the preliminary milling, the amount of extracted PCBs at $t = 0$ resulted to be lower than after a few hours of milling. The preliminary milling also reduces the induction time. Drying of the sample is necessary as humidity reduces the efficiency of milling due to the packing of the material on the mill surface and destroys the hydrides. The ternary hydrides are stored out of the contact with humidity and are handled with caution as they are flammable. During the reaction the added hydride was totally converted into inert species.

50.2.4
Test Parameters

In each run, samples were withdrawn at various times and PCBs and chloride were determined. The amount of soil used in each laboratory scale test (100 g) represents about 1/36 of the total weight of the mill rings and occupies about 2/3 of the total mill volume. Solid, pure hydrides were added to the soil at different weight ratios, namely 5%, 3%, 2.5% and 1%, w/w of hydride, respectively. The mixtures were homogenized by hand and introduced into the ring mill. Samples were monitored at different times: 3.5, 11, 18, 23, 30 h.

50.2.5
Extraction and Determination of PCBs

Ground samples were subjected to Soxhlet extraction (EPA Method 3540, 1995) for about 16 h with 200 ml of hexane/acetone mixture (1/1, v/v), then the extract was purified by absorption on dry Florisil and Silica gel columns followed by treatment with concentrate sulphuric acid. Analysis of samples was performed with a HP 6890 gas-chromatograph/electron capture detector GC/ECD with DB-5 capillary column (30 m × 0.2 mm id. and 0.33 µm film thickness). The gas-chromatograph was equipped with an auto-sampler and automatic integrator. The same samples were also analysed on GC/ECD Varian 3600 with IEC (International Electrotechnic Committee) method for the determination of every single PCB congener and verify their progressive degradation during the treatment. The PCBs content in the soil sample was quantitatively expressed by comparison with a standard mixture of Aroclor® 1254-1260 (1/1).

50.3 Results and Discussion

The general reaction of hydrogenation/dechlorination that occurs during the mechano-chemical treatment is represented in Eq. 50.2. It is a nucleo-philic substitution in which the chloride is replaced by the hydride with salt formation:

$$C_{12}H_xCl_y + MH \xrightarrow{E_{um}} C_{12}H_{10} + MCl \qquad (50.2)$$

where $C_{12}H_xCl_y$ represents the polychlorobiphenyls (with $0 < x < 8$ and $2 < y < 10$); MH is the hydride donor compound (NaBH$_4$, LiAlH$_4$); E_{um} is the energy given by milling; $C_{12}H_{10}$ is biphenyl and MCl is the chloride salt formed in the reaction.

An example of the abatement of total PCBs in the contaminated soil mixed with the hydride is shown in Fig. 50.1, where the chromatograms of the extract from soil after three hours of milling and at the end for the case of LiAlH$_4$ (5%, w/w) are reported. Monitoring the reaction during the milling time, shows that a greater abatement efficiency is observed for high chlorinated congeners during the first few hours of milling.

The trend of total PCBs concentration and that of inorganic-water soluble chloride with milling time is presented in Fig. 50.2. It is clear from Fig. 50.1a,b that LiAlH$_4$ is more efficient than NaBH$_4$ in the short time. In fact, with LiAlH$_4$ more than 90% of he total PCBs are dehalogenated within the first three hours. This is very important as a short milling time means a low energy input.

In general, the PCBs degradation trend is parallel to the free chloride ions production. For milling times around 18–23 h, the total PCBs concentration decreases from about 2 600 mg kg^{-1} to zero and the inorganic chloride concentration increases to the value expected for the total dechlorination of PCBs. With LiAlH$_4$ the dehalogenation is completed in 18 h, with respect to 23 h required by NaBH$_4$. We have tested LiAlH$_4$ at lower amounts and found hat it is very active also at a rate of 1:100 with respect to contaminated soil. Such amount is still above the stoichiometric mass and can be further reduced in order to minimize residual hydrides.

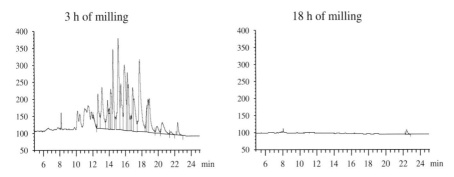

Fig. 50.1. Chromatograms of the initial and final sample. Note the total disappearance of the peaks due to the various PCBs congeners

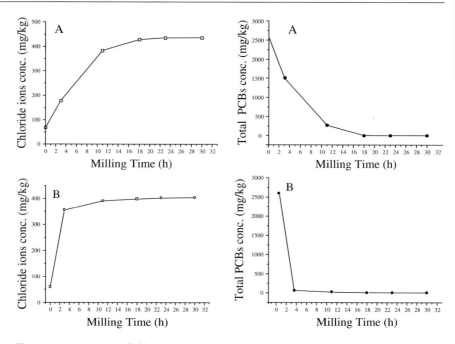

Fig. 50.2. Concentration of chloride ions (—●—) and total PCBs (—□—) vs time. The treatment with NaBH$_4$ (**a** 5%, w/w) or LiAlH$_4$ (**b** 5%, w/w) produces the formation of inorganic chloride with concomitant disappearance of PCBs

50.4
Conclusion

High Energy Milling (HEM) systems can be, thus, utilized as high efficiency chemical reactors for degrading contaminants as polychlorobiphenyls in complex matrices by adding the proper reducing agents. Both NaBH$_4$ and LiAlH$_4$ are effective for the total dehalogenation of PCBs. LiAlH$_4$ is more effective than NaBH$_4$, as it is able to reduce 90% of PCBs in three hours. The residual concentration of PCBs in the treated soil is in all cases below the limit stated by the Italian law (Gazzetta Ufficiale 1999) that fixes at 5 mg kg^{-1} the maximum PCBs concentration in industrial soil for reuse.

It should also be mentioned that non-toxic compounds are formed and residual salts can be washed out with water. For a number of factors such as confined reacting system and controlled conditions, low cost, clean process with no emissions, total abatement of PCBs, and non-toxic compounds formed this treatment appears a very attractive remedial technology.

Acknowledgements

Financial support by MURST, L488, Piano INCA, Project P6, coordinated by the METEA Research Centre of the University of Bari is acknowledged. Authors wish to thank ENEL Production Research Centre of Brindisi for the use of their premises.

References

Aresta M, Pastore T (2001) Solid state technology for the reductive elimination of polychlorobiphenyls (PCBs) from contaminated soils. ACE Meeting, Dijon, France, p 85
Aresta M, Caramuscio P, De Stefano L, Pastore T (2003) Solid state dehalogenation of PCBs in contaminated soil using $NaBH_4$. Waste Management 23(4):315–319
Benjamin JS (1970) Dispersion strengthened super-alloys by mechanical alloying. Met Trans 1(10):2943–2951
Benjamin JS, Volin TE (1974) Mechanism of mechanical alloying. Met Trans 5(8):1929–1934
Birke V (2001) Mechanochemical reductive dehalogenation: a novel approach for the destruction of PCBs. International Workshop on "Mechanochemical Processes for the Environment", Bari, Italy, p 6
EPA (1995) Tests methods for evaluating solid waste. USEPA SW-846, 3rd edn, Update III, U.S. GPO, Washington, DC
Gazzetta Ufficiale (1999) No. 293 del 15/12/1999, Ministero dell'ambiente DM del 25/10/1999, n. 471
Heinicke G (1984) Tribochemistry. Academy Verlag, Berlin
Hutzinger O, Safe S, Zitko V (1974) The chemistry of PCBs. CRC Press, Cleveland, Ohio
Lang N (1992) Polychlorinated biphenyls in the environment. Chromatographia 595:1–45

Part VI

Chapter 51

Development of a Bioreactor for Cometabolic Biodegradation of Gas-Phase Trichloroethylene

E. Y. Lee

Abstract

Novel biofilm reactor systems, a parallel trickling biofilter (TBF) system and a two-stage continuous stirred tank reactor (CSTR)/trickling biofilter system, were developed and operated for long-term continuous treatment of gas-phase trichloroethylene (TCE) by the bacterium *Burkholderia cepacia* G4. The effects of trichloroethylene concentrations on reactor performances were analyzed. The critical trichloroethylene elimination capacities were determined to be 8.6 and 25.3 mg trichloroethylene $l^{-1} d^{-1}$, respectively.

Key words: biofilter, *Burkholderia cepacia* G4, cometabolism, monooxygenase, trichloroethylene

51.1 Introduction

Trichloroethylene (TCE) has been widely used as industrial solvent and degreasing agent. Because of extensive use and improper disposal, it has become a widespread contaminant in soil and underground water (Fan 1988). One of the most promising treatment methods for trichloroethylene is microbial degradation in a trickling biofilter (TBF) (Chang and Alvarez-Cohen 1995; Sipkema et al. 1999; Sun and Wood 1997). Since trichloroethylene itself is not a growth substrate, it can be degraded via cometabolism, in which mono-oxygenase, the corresponding enzyme for initiating growth substrate oxidation, fortuitously transforms trichloroethylene (Ensley 1991; Kang et al. 2001). Trichloroethylene, however, is not easily treated by the simple trickling biofilter. This is mainly due to the toxicity of trichloroethylene and its degradation products to microbial cells, together with the competitive inhibition between primary substrate and trichloroethylene (Folsom et al. 1990; Oldenhuis et al. 1991). Therefore, thr removal efficiency of trichloroethylene in the simple trickling biofilter normally decreases with time. In this paper, we developed and operated a novel parallel trickling biofilter system consisting of two units of trickling biofilters in a parallel mode, one for trichloroethylene biodegradation and the other for biofilm reactivation for long-term stable treatment of trichloroethylene by the bacterium *Burkholderia cepacia* G4. A two-stage reactor system where a continuous stirred tank reactor (CSTR) with cell recycle from/to trickling biofilter was coupled to the trickling biofilter for the reactivation of the biofilms deactivated during trichloroethylene degradation was also developed. The effects of inlet trichloroethylene concentrations on trichloroethylene conversion and degradation rate were studied and reactor performances were evaluated for long-term stable continuous treatment of gas-phase trichloroethylene.

51.2 Experimental

51.2.1 Microorganism and Culture Conditions

B. cepacia G4 was employed in this study and M9 medium supplemented with phenol as a sole carbon source was used to culture the cells and develop biofilm in a trickling biofilter (Ye et al. 2000).

51.2.2 Analysis

Trichloroethylene concentrations were determined by analysis of 30 µl of inlet and outlet gas-phase samples on a gas chromatography (Hewlett Packard 5890 II plus, USA) equipped with an electron capture detector (ECD) and HP-5 capillary column (Alltech Inc., USA). Phenol concentrations were determined using the modified colorimetric assay, and the activity of toluene monooxygenase (TMO) was analyzed by modified naphthalene oxidation assay as described in the reference (Nelson et al. 1987).

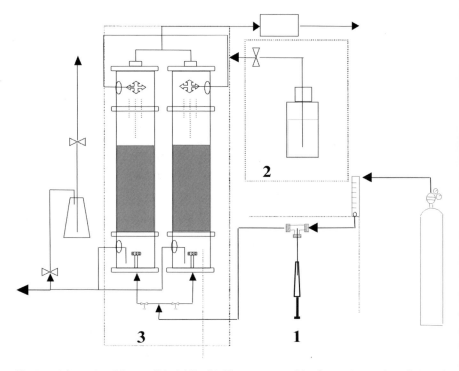

Fig. 51.1. Schematics of the parallel trickling biofilter system used for the continuous degradation of gas-phase trichloroethylene (*1.* Trichloroethylene supply unit; *2.* Parallel trickling biofilter; *3.* Medium supply unit)

51.2.3
Parallel Trickling Biofilter System

A schematic diagram of the parallel trickling biofilter system is shown in Fig. 51.1. The trickling biofilter unit consists of a 1.4-l glass cylinder (diameter: 5 cm, height: 39 cm) and ceramics as supporter matrix packed to a depth of 25 cm. Gas-phase trichloroethylene was introduced to the bottom of trickling biofilter by a syringe pump. All fittings and connectors were gas-tight and the temperature of trickling biofilter was controlled using a water circulator.

51.2.4
Two-Stage Continuous Stirred Tank Reactor/Trickling Biofilter System

A schematic diagram of the two-stage reactor system is shown in Fig. 51.2. *B. cepacia* G4 was cultured in a continuous stirred tank reactor at 300 rpm, 30 °C and a dilution rate of 0.02 h^{-1}. The operation condition of continuous stirred tank reactor was previously examined and optimized to achieve maximal levels of toluene monooxygenase activity (Lee and Park 2001). Culture broth containing *B. cepacia* G4 was pumped from the continuous stirred tank reactor to the trickling biofilter, and then the broth was circulated back to the continuous stirred tank reactor. Contaminated gas with given trichloroethylene concentrations was introduced to the bottom of trickling biofilter by a syringe pump.

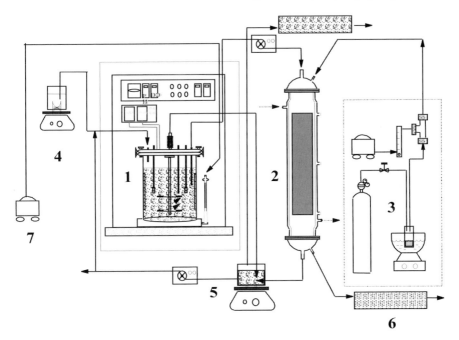

Fig. 51.2. Schematics of the two-stage continuous stirred tank reactor/trickling biofilter system for the continuous degradation of gas-phase trichloroethylene. *1.* Continuous stirred tank reactor; *2.* trickling biofilter; *3.* trichloroethylene supply unit; *4.* medium; *5.* stripping unit; *6.* trap; *7.* air pump

51.3
Results and Discussion

51.3.1
Trichloroethylene Biodegradation in a Simple Trickling Biofilter

Simple one-stage trickling biofilter has received much attention for treatment of gas-phase trichloroethylene, but treatment efficiency can be significantly reduced during long-term reactor operation mainly due to the deactivation of biofilm. The objective of this study is to develop a bioreactor system for the long-term stable treatment of gas-phase trichloroethylene.

In the operation of a simple one-stage trickling biofilter, the biofilm can be easily deactivated during the biodegradation of trichloroethylene due to the toxicity of trichloroethylene, and this deactivation results in the failure of long-term operation. As shown in Fig. 51.3, the toluene mono-oxygenase activity of *B. cepacia* G4 biofilm in a simple trickling biofilter clearly deactivated during the time course of trichloroethylene degradation. The relative toluene mono-oxygenase activity of the biofilm decreased down to 20% of initial value within 30 h. The toluene mono-oxygenase activity loss resulted in a marked decrease in the removal efficiency of trichloroethylene biodegradation from 70 to 20% with inlet trichloroethylene concentration of 12 ppm_v. These phenomena may limit the practical application of a simple trickling biofilter system for long-term treatment of industrial waste gas containing trichloroethylene.

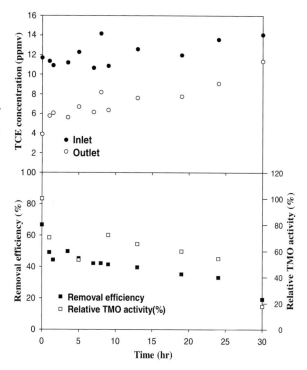

Fig. 51.3. Trichloroethylene biodegradation in the single-stage trickling biofilter system. The relative toluene monooxygenase activity of the biofilm and trichloroethylene removal efficiency continuously decreased, which clearly represented that simple trickling biofilter could not be applied for long-term treatment of trichloroethylene

51.3.2
Development and Operation of the Parallel Trickling Biofilter System

In order to overcome the decrease in the removal efficiency observed in the simple trickling-filter, a parallel trickling biofilter system consisting of two trickling-filters in a parallel mode was developed (Fig. 51.1). One trickling biofilter unit was used for trichloroethylene degradation and the other trickling biofilter unit for reactivation of the deactivated biofilm. When the biofilm activity in the trichloroethylene degradation trickling biofilter unit is deactivated down to a certain level, the reactor operation mode of the degradation unit is switched to a reactivation mode and the corresponding reactivation unit to reactor operation mode for trichloroethylene degradation, and vice versa. To determine the optimal switching time, the patterns of film deactivation in a trichloroethylene degradation operation mode and biofilm reactivation in a reactivation operation mode were analyzed. The toluene mono-oxygenase activity in the trichloro-ethylene degradation mode decreased down to about 30% of the initial level after 24 h operation. When the trickling biofilter operation mode was switched to a reactivation mode by supplying the growth medium, the toluene mono-oxygenase activity increased up to the initial level of activity within 12 h (data not shown). These results represented that long-term stable operation of trickling biofilter could be achieved by the parallel trickling biofilter system with switching time of 12 h.

The effects of inlet trichloroethylene concentrations on trichloroethylene conversion and degradation rate were investigated at concentrations ranging from 3 to

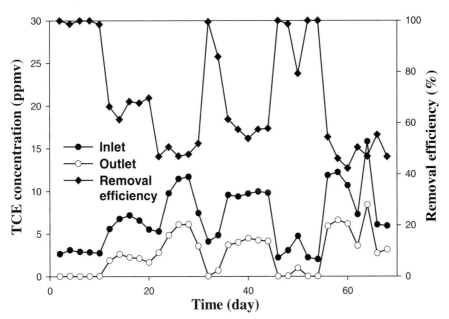

Fig. 51.4. Long-term continuous treatment of gas-phase trichloroethylene by the parallel trickling biofilter system. Trichloroethylene could be degraded with the removal efficiency above 50% by parallel operational mode for more than 3 months

15 ppmv. The flow rate of buffer containing 1 mg l^{-1} phenol, empty bed retention time, and the temperature of trickling biofilter were 0.86 ml min^{-1}, 4.9 min, and 30 °C, respectively. As shown in Fig. 51.4, more than 50% of trichloroethylene was degraded for feed concentrations ranging from 5 to 15 ppmv and almost 100% removal was achieved when trichloroethylene was introduced at concentration of less than 5 ppmv. Trichloroethylene could be degraded with the removal efficiency above 50% by the parallel operational mode with switching period of 12 h for more than 3 months.

51.3.3
Development and Operation of Two-Stage Continuous Stirred Tank Reactor/Trickling Biofilter System

For long-term stable operation of trickling biofilter, the microbial growth reactor of continuous stirred tank reactor with cell recycle from/to trickling biofilter was coupled for the reactivation of biofilm deactivated during trichloroethylene degradation in trickling biofilter (Fig. 51.2). It was prerequisite to achieve maximum expression of toluene monooxygenase in continuous stirred tank reactor, and operation conditions were previously optimized using batch culture results (Ye et al. 2000). Cell concentrations of optical density 0.8 (at 600 nm) were maintained in the continuous stirred tank reactor with a dilution of rate of 0.02 h^{-1}, temperature of 30 °C, phenol concentration of 5 mM, and oxygen flow rate of 0.33 vvm.

The effects of inlet trichloroethylene concentrations on trichloroethylene conversion and degradation rate were investigated at concentrations ranging from 7 to 15 ppm$_v$. As shown in Fig. 51.5, trichloroethylene conversion and degradation rate increased with increase in inlet trichloroethylene concentration up to 15 ppmv. Almost 100% removal was achieved when trichloroethylene was introduced at concentration of less than15 ppmv. In terms of trichloroethylene degradation rate, the critical elimination capacity of up to 25.3 mg trichloroethylene l^{-1} d^{-1} could be achieved and the reactor system was stably operated more than 2 months.

After long-term operation period of 2–3 months, the contaminations of the biofilm in the parallel and the two-stage systems were checked by counting the amounts of colonies on mineral salts medium agar plates to which phenol was supplemented as

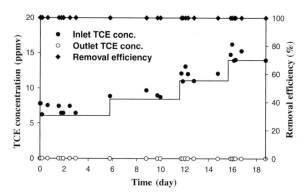

Fig. 51.5. Long-term continuous treatment of gas-phase trichloroethylene by the two-stage continuous stirred tank reactor/trickling biofilter system. Almost 100% removal was achieved and reactor system was stably operated more than 2 months

a sole carbon source. Although these systems were not operated aseptically, relative populations of contaminants were not exceeded more than 10% and contamination did not seem to affect significantly on trichloroethylene treatment. When these reactor systems were operated for the long-term treatment of trichloroethylene, semi-continuous or continuous supply of growth substrate led to the increase of large quantities of biomass, which clogged the system. Therefore, the method for the biomass reactivation under non- or limited-growth condition needs to be investigated.

51.4 Conclusion

In this study, novel bioreactor systems, parallel trickling biofilter and two-stage continuous stirred tank reactor/trickling biofilter, were developed and successfully operated for 2–3 months for trichloroethylene treatment. The critical elimination capacities of 8.6 and 25.3 mg trichloroethylene $l^{-1} d^{-1}$ could be obtained, respectively. These reactor systems could be applied for long-term continuous treatment of industrial waste gas containing trichloroethylene.

Acknowledgements

This study was supported by KOSEF through RRC-IETI (Project number: 99-10-03-02-A-3).

References

Chang H-L, Alvarez-Cohen L (1995) Model for the cometabolic biodegradation of chlorinated organics. Environ Sci Technol 29:2357–2367

Ensley BD (1991) Biochemical diversity of trichloroethylene metabolism. Annu Rev Microbiol 45:283–299

Fan AM (1988) Trichloroethylene: water contamination and health risk assessment. In: Ware GW (ed) Reviews of environmental contamination and toxicology, vol. 101. Springer-Verlag, New York

Folsom BR, Chapman PJ, Pritchard PH (1990) Phenol and trichloroethylene degradation by *Pseudomonas cepacia* G4; kinetics and interactions between substrates. Appl Environ Microbiol 56:1279–1285

Kang J, Lee EY, Park S (2001) Co-metabolic biodegradation of trichloroethylene by *Methylosinus trichosporium* is stimulated by low concentrations methane or methanol. Biotechnol Lett 23:1877–1882

Lee EY, Park S (2001) Development of two-stage CSTR/TBF system for the cometabolic degradation of gas-phase TCE by *Burkholderia cepacia* G4. Kor J Biotechnol Bioeng 16:511–515

Nelson MJ, Montgomery SO, Mahaffey WR, Pritchard PH (1987) Trichloroethylene metabolism by microorganisms that degrade aromatic compounds. Appl Environ Microbiol 53:604–606

Oldenhuis RJ, Oedzes Y, van der Waarde JJ, Janssen DB (1991) Kinetics of chlorinated hydrocarbon degradation by *Methylosinus trichosporium* OB3b and toxicity of trichloroethylene. Appl Environ Microbiol 57:7–14

Sipkema EM, de Koning W, van Hylckama Vlieg JE, Ganzeveld KJ, Janssen DB, Beenackers AA (1999) Trichloroethylene degradation in a two-step system by *Methylosinus trichosporium* OB3b. Optimization of system performance: use of formate and methane. Biotechnol Bioeng 63:56–68

Sun AK, Wood TK (1997) Trichloroethylene mineralization in a fixed film bioreactor using a pure culture expressing constitutively toluene *ortho*-monooxygenase. Biotechnol Bioeng 55:674–685

Ye BD, Yoon SJ, Park S, Lee EY (2000) Effect of growth substrates on cometabolic biodegradation of trichloroethylene by *Burkholderia cepaica* G4. Kor J Biotechnol Bioeng 15:474–481

Part VI

Chapter 52

Enhanced Solubilization of Organic Pollutants through Complexation by Cyclodextrins

S. Shirin · E. Buncel · G. W. vanLoon

Abstract

The cyclic glucose oligosaccharides known as cyclodextrins (CDs) are able to form host-guest inclusion complexes with a variety of organic compounds including a number of pollutants. Substituted cyclodextrins in which one or more hydroxyl groups have been modified with, e.g. $-CH_3$, $-CH_2CH(OH)CH_3$, $-CH_2COO^-$, $-COO^-$ and $-SO_3^-$, have many desirable properties compared with unmodified cyclodextrins, including greater aqueous solubility and binding ability. As part of studies on the abiotic degradation and soil-water interactions of pesticides and other hydrophobic organic compounds, in this work we have examined a number of substituted β-cyclodextrins for use as complexing agents for the enhancement of aqueous solubility of the organic pollutants trichloroethylene (TCE) and perchloroethylene (PCE). Solubility enhancement factors (P_t / P_0) up to 5.5 and 14 were determined for trichloroethyelene and perchloroethyelene respectively. Binding constants for trichloroehylene with the substituted cyclodextrins, evaluated using ^{1}H nuclear magnetic resonance (NMR), ranged from 3 M^{-1} to 120 M^{-1}. In experiments with soil-derived peat contaminated by perchloroethylene and trichloroethylene, it was shown that selected cyclodextrins are capable of effective removal of the contaminants, suggesting that a suitably substituted β-cyclodextrin may be a valuable additive in pump-and-treat protocols for site remediation of polychlorinated organics.

Key words: cyclodextrin, trichloroethylene, perchloroethylene, binding constant, solubility enhancement, soil/water remediation

52.1 Introduction

Contamination of soil and groundwater by organic pollutants such as trichloroethylene (TCE) and perchloroethylene (PCE) is widespread in industrial societies. Both trichloroethyelene and perchloroethyelene are understood to be potentially carcinogenic and mutagenic, and compounding the problem is the environmental resistance of these compounds to abiotic and biotic degradation. As a consequence of their serious hazard to health and environmental persistence, extending to decades or longer, removal of the two compounds from the subsurface has received urgency and numerous studies have focused on their environmental remediation. The overall goal of the present research has been to examine the possibility of using a variety of substituted β-cyclodextrins (βCDs) for enhancing the solubility of trichloroethylene and perchloroethylene with a view to developing methodologies that could maximize their removal from subsurface soils.

52.2
Review

52.2.1
Remediation Technologies

In order to remediate contaminated subsurfaces, several different technologies are currently in operation and others are in the developmental stage. The most important are pump-and-treat processes, soil vapour extraction, biodegradation, reactive barrier walls and in-situ oxidation methods. Pump-and-treat is the most widely used but as is true of all the methods, it has limitations which have led to ongoing research.

The pump-and-treat method involves pumping the groundwater from the subsurface soil, treating it to remove the chemicals and then returning the remediated groundwater to the soil (Fig. 52.1). The low aqueous solubility of trichloroethylene and perchloroethylene and other dense non-aqueous phase liquids, however, limits the effectiveness of the method because a large volume of water has to be pumped through the subsurface to remove the contaminant from a polluted site (Macky and Cherry 1989; National Academy of Science Report 1994). Current methods used to increase the aqueous solubility of hydrophobic pollutants involve employing surfactants or alcohols. Surfactants function to increase solubilities through formation of micelles in which the hydrophobic interior is compatible with many low-polarity organics. However, use of surfactants is limited by their tendency to sorb onto porous media (Pankow and Cherry 1996; Kan and Thomson 1990) and, moreover, surfactants usually lower the organic-water interfacial tension causing the contaminant material to

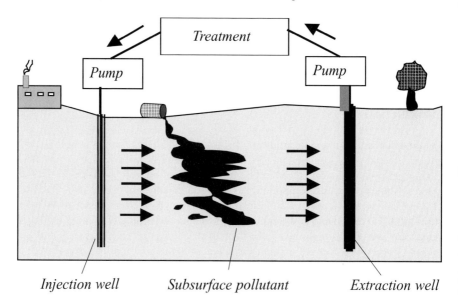

Fig. 52.1. The pump-and-treat method for subsurface soil/water remediation

disperse vertically in the soil column. Similar problems occur when using aqueous alcohol solutions.

52.2.2
Cyclodextrins as Solubilizing Agents

Current research aimed towards realizing alternative solubilizing agents has led to investigations of cyclodextrins as a possible viable alternative. Cyclodextrins are cyclic glucose oligosaccharides formed by action of *Bacillus macerans* on starch. The most common cyclodextrins are those containing 6, 7 or 8 glucose units and are termed α-, β- and γ-cyclodextrins. The illustration in Fig. 52.2 shows the toroidal shape and size of the three types of cyclodextrins. In these molecules, the secondary hydroxyls are directed outward from the narrow end while the primary hydroxyls also point outward from the wider opening. The exterior of the molecule is thus hydrophilic in nature due to its ability to form hydrogen bonds with the aqueous solvent while the interior of the cavity provides a hydrophobic environment. The internal diameter of β-cyclodextrin renders it highly suitable for inclusion of a large number of organic molecules of interest and, as well, β-cyclodextrin is the least expensive of the three. However, β-cyclodextrin has by far the lowest solubility, which will limit its ability to solubilize target contaminant organics.

Substitution of cyclodextrins through derivatization of the hydroxyl groups has provided a host of molecules of interest in different domains, including enhanced ability to interact with organic pollutants. At the same time, the derivatized species are generally significantly more soluble than β-cyclodextrin itself. Substituents that have been introduced include: $-CH_3$, $-CH_2CH(OH)CH_3$, $-COCH_3$, CH_2COO^-, $-SO_3^- Na^+$, and $-NH_2$. In principle, any or all of the three -OH groups in each glucose unit of the cyclodextrin molecule can be functionalized with these substituents. Hence, there can be 0 to 21 degrees of substitution in each β-cyclodextrin molecule (three -OH groups per glucose unit and seven glucose units in each molecule).

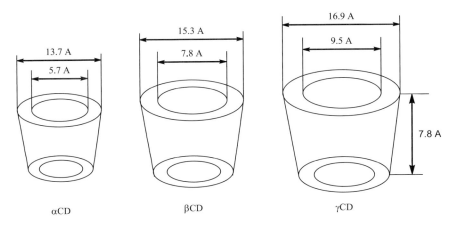

Fig. 52.2. Dimensions and shape of α-, β-, and γ-cyclodextrins

52.2.3
Substituted Cyclodextrins for Soil/Water Remediation of Organic Pollutants

A number of studies have described laboratory and field experiments that demonstrate the application of substituted cyclodextrins in environmental remediation strategies (Wang and Brusseau 1993; Brusseau et al. 1994; Wang and Brusseau 1995; Brusseau et al. 1997; McCray and Brusseau 1998; Boving et al. 1999). Using laboratory batch experiments, Wang and Brusseau (1993) showed that the aqueous solubility of several low-polarity compounds increased substantially upon addition of hydroxypropyl β-cyclodextrin (HPβ-CD) and that the increase is linear with respect to cyclodextrin concentration. Solubility enhancement is measured as the ratio of solubility in the presence of cyclodextrin: aqueous solubility in the absence of cyclodextrin. Using 5% HPβ-CD solubility enhancement ranges from ca. 3 (for TCE) to 147 (for anthracene) depending on the size and other properties of the guest molecule.

Brusseau et al. (1994) carried out a laboratory column study in which they measured the retardation factor (R_F) for passage of trichloroethylene through a column of Mt. Lemmon soil, whose organic matter content is 12.6%. Using water as the 'eluent', the R_F value (equal to the ratio of mobility of the mobile phase to mobility of the compound) was 8.0, but with a 1% (w/v) HPβ-CD solution, the value was reduced to 5.2, indicating enhanced mobility. The mobility of other organic compounds was also augmented when using an aqueous solution containing HPβ-CD. At the same time, quantitative recovery of the cyclodextrin from the column was achieved, indicating that the cyclodextrin itself was not sorbed by the soil media.

52.2.4
Field Studies Using Cyclodextrins

Based on information from the laboratory experiments, a field study was conducted at a waste disposal site located at Hill Air Force Base, Layton, Utah, where the soil had been contaminated with trichloroethylene and other light and dense non-aqueous phase liquid compounds (McCray and Brusseau 1998). Using four injection wells, a contaminated aquifer having 3×5 m surface area and 9 m depth was flushed with 10% aqueous HPβ-CD at a rate of 4.54 l min^{-1} over a 10-day period. The aqueous solubility of all the targeted compounds was significantly increased in the total 65 400 l of cyclodextrin solution, resulting in an average 41% concentration reduction in the contaminated zone.

In another laboratory study, Boving et al. (1999) employed HPβ-CD and Mβ-CD in a column system using subsoil from the Canadian Air Force Base, Borden, Ontario (organic carbon content = 0.29%) containing residual phase trichloroethylene and perchloroethylene to determine the effect on solubilization of chlorinated compounds. The resultant increases of their apparent aqueous solubility using 5% and 10% (w/v) of both HPβ-CD and Mβ-CD, were in good agreement with data obtained from separate batch experiments where the equilibrium solubility enhancement of trichloroethylene and perchloroethylene was determined using the same amounts of the cyclodextrins. None of the cyclodextrins showed any tendency to be sorbed by porous media. However, the trichloroethylene/perchloroethylene – water interfacial tension was found to decrease in the presence of 5% and 10% of both HPβ-CD and Mβ-CD, although the decrease was less than half of that observed when using surfactants.

As many contaminated sites contain mixed waste, it is of interest to ascertain whether cyclodextrin is able to extract metal ions while, at the same time, enhancing solubility of hydrophobic organic compounds. Using CMβ-CD, this has been shown to be a possibility in the case the simultaneous extraction of Cd^{2+} along with anthracene, trichlorobenzene, biphenyl and DDT (Wang and Brusseau 1995). There was no interference between these two processes, as metal ion complexation occurred via the substituted -CH_2COO^- groups, while that of the organic compound by host-guest complexation. Similar experiments were performed in column systems using three different soils and the results showed that CMβ-CD can effectively desorb both Cd^{2+} and a non-polar organic compound (phenanthrene) from soils in a flow-through system (Brusseau et al. 1997).

Ko et al. (1999) have compared HPβ-CD with two surfactants (Tween 80 and sodium dodecyl sulfate) in evaluating the ability to extract naphthalene and phenanthrene from kaolinite. Laboratory batch experiments as well as one-dimensional numerical model calculations in most cases showed higher extraction efficiency when using the surfactants compared with HPβ-CD; however, whereas both surfactants showed significant sorption, there was no tendency for the cyclodextrin to be sorbed by the clay mineral.

While the research described above has made use of different substituted β-cyclodextrins for enhancing removal of trichloroethylene and perchloroethylene from soil, there has been little attempt to relate performance to fundamental properties of the compounds. In the present work we examine, in a systematic manner, the ability of different β-cyclodextrins to solubilize these polychlorinated ethylenes and relate this ability to the binding constants of the host/guest complexes that are formed. This research is a part of ongoing studies in our laboratory concerned with the abiotic degradation and soil-water interactions of pesticides and other hydrophobic organic compounds (Annandale et al. 1998; Sha'ato et al. 2000; Omakor et al. 2001; Balakrishnan et al. 2001; Onyido et al. 2001).

52.3
Experimental

52.3.1
Materials

The following substituted β-cyclodextrins were used as received: hydroxypropyl-(HPβ-CD) (Aldrich), sulfated-(Sβ-CD1), sulfated-(Sβ-CD2), methyl-(Mβ-CD), and carboxymethyl-(CMβ-CD1) (Cerester). In addition, we used carboxymethyl-(CMβ-CD2) which had been synthesized in the laboratory of Professor G. R. Thatcher at Queen's University. The aqueous solubility of each of the substituted CDs was found to be >50% (w/v). Some properties of the selected cyclodextrins are given in Table 52.1.

Analytical grade trichloroethylene and perchloroethylene, with the reported purity of 99.5+% and 99+% respectively, were obtained from Aldrich. HPLC grade acetonitrile was purchased from Fisher. The D_2O (purity 99.9%) used in NMR experi-

Table 52.1. Substituted β-cyclodextrins used in this work: molar mass and degree of substitution

Compound	Substituent group	Degree of substitution (average)	Average molar mass (g mol^{-1})
Hydroxypropyl (HPβ-CD)	-CH$_2$CH(OH)CH$_3$	3 – 8 (5.5)	1 500
Methyl (Mβ-CD)	-CH$_3$	10 – 16 (12)	1 303
Carboxymethyl (CMβ-CD1)	-CH$_2$COO$^-$	3	1 375
Carboxymethyl[a] (CMβ-CD2)	-CH$_2$COO$^-$	1 – 11 (7)	1 671
Sulfated (Sβ-CD1)	-SO$_3^-$	3 – 16 (4)	1 543
Sulfated (Sβ-CD2)	-SO$_3^-$	3 – 16 (14)	2 563

[a] Molar mass was determined from the major peak obtained using FAB-MS. For all other cyclodextrins, average values provided by the supplier were used.

ments was obtained from Cambridge Isotopic Laboratories. International Humic Substance Society Peat had elemental composition as follows: C, 57.1%; H, 4.49%; O, 33.9%; N, 3.60%; S, 0.65%.

52.3.2
High Performance Liquid Chromatography

HPLC analysis was performed using a Varian 9002 series solvent delivery system fitted with a Varian 9050 variable wavelength UV-Visible detector and Varian Star Chromatographic Workstation Version 4.5 for peak integration. For separation, a C$_{18}$ APEX ODS 5 μ particle size column (Chromatographic Specialties, 4.6 × 250 mm) protected by a guard column was employed. The mobile phase was 70:30 acetonitrile:water that had been filtered through a Millipore system. For both PCE and TCE, the detector was set to a wavelength of 210 nm and the flow rate was 1.0 ml min^{-1}. Under these conditions, the TCE peak appeared at a retention time of ca. 6.5 min while PCE was detected at ca. 9.5 min. A large, irregular peak ascribed to CD also appeared between ca. 2 to 3 min; therefore, elution of the CDs did not interfere with the peak of interest.

52.3.3
Solubility Enhancement

To distilled de-ionized water (25 ml) containing 1% to 5% concentrations of cyclodextrin in a 100-ml volumetric flask was added a large excess (above the solubility limit) of trichloroethylene or perchloroethylene. A control flask that contained only water and the substrate was also prepared. Trichloroethylene- or perchloroethylene-aqueous mixtures containing cyclodextrin were then equilibrated by shaking on a wrist action shaker for 48 h at room temperature. Samples were transferred into capped glass centrifuge tubes and centrifuged at 3 500 rpm for 20 min. A 100 or 500 μl portion of the aqueous supernatant was withdrawn using a micropipette and diluted with water in a 10-ml volumetric flask. Samples of the diluted solution were then injected into the HPLC and both peak height and area of the substrate peak were recorded.

52.3.4
Binding Constant

The binding constant (K_{11}) was determined using a ^{1}H NMR titration method in which a 2×10^{-3} M trichloroethylene solution in D_2O and ca. 0.4 M cyclodextrin stock solutions in D_2O were initially prepared. A 2.00 ml portion of the trichloroethylene solution was placed in an NMR tube and the ^{1}H spectrum was recorded using a Bruker AM-400 spectrophotometer operating at 400.1 MHz. Different amounts of cyclodextrin stock solution were then added successively into the NMR tube containing trichloroethylene solution, using a graduated gas-tight Hamilton syringe. After each addition, the solution was thoroughly mixed before the spectrum was acquired. The chemical shift of the ^{1}H peak of trichloroethylene was monitored after each addition of cyclodextrin solution, using the residual solvent proton signal HDO peak (at 4.6 ppm relative to internal tetramethylsilane) as the reference. The final ratio of cyclodextrin to trichloroethylene was such that the cyclodextrin concentration was present in a 10:1 to 20:1 excess compared to the trichloroethylene concentration depending on the cyclodextrin being used.

52.3.5
Uptake and Release in Soil Organic Matter

A 50 to 100 mg portion of peat was placed in a 8-ml vial and the aqueous trichloroethylene (200 ppm) or perchloroethylene (165 ppm) solution was added without leaving any headspace. Each sample vial was prepared in duplicate along with one control omitting the peat, and these were allowed to equilibrate on the shaker for periods of either 1 or 7 d. After equilibration, the sample vials were centrifuged and the supernatant was withdrawn for HPLC analysis without any further dilution. A 5% cyclodextrin (w/v) solution was then added to the sample vials, which were again equilibrated for 18 h overnight before analysis. The sample vials were centrifuged and supernatant was withdrawn for analysis by HPLC.

52.4
Results and Discussion

52.4.1
Determination of Binding Constants from Solubility Measurements

By measuring the solubility of trichloroethylene and perchloroethylene in the presence of varying amounts of different cyclodextrins it is possible to measure the degree of solubility enhancement and to calculate the binding constant for the pollutant. Under equilibrium conditions and assuming the formation of a 1:1 complex between pollutant (P) and cyclodextrin (CD), one can write

$$P + CD \rightleftharpoons P:CD \qquad (52.1)$$

and define the binding constant for this reaction as

$$K_{11} = \frac{[P:CD]}{[P][CD]} \tag{52.2}$$

For an aqueous cyclodextrin solution, on addition of excess pollutant, both free pollutant, $[P]_o$, and complexed pollutant, $[P:CD]$, will exist in the solution. $[P]_o$ represents the aqueous solubility of the uncomplexed material, which remains constant upon addition of increasing amounts of cyclodextrin. The total aqueous solubility of pollutant at any cyclodextrin concentration, $[P]_t$, can then be expressed as

$$[P]_t = [P:CD] + [P]_o \tag{52.3}$$

Combining this equation with the expression for the binding constant, one obtains

$$[P]_t = [P]_o \{K_{11}[CD] + 1\} \tag{52.4}$$

Therefore

$$[P]_t / [P]_o = K_{11}[CD] + 1 \tag{52.5}$$

where $[P]_t$ is the aqueous concentration of the pollutant in presence of cyclodextrin and $[P]_o$ is the aqueous solubility, i.e. aqueous concentration of substrate in absence of cyclodextrin. The ratio of $[P]_t / [P]_o$ represents the 'aqueous solubility enhancement' of the compound for a given concentration of cyclodextrin. The 1:1 binding constant for the complex (K_{11}) can be obtained from the slope of plots of the equation.

In the present study, when increasing amounts of cyclodextrin were added to solutions containing excess trichloroethylene or perchloroethylene (i.e. amounts that exceeded the aqueous solubility), it was found that their solubility increased in a linear fashion (Fig. 52.3). In each experiment, the intercept of the linear regression line was found to be unity, within experimental error, in accord with Eq. 52.4. The solubil-

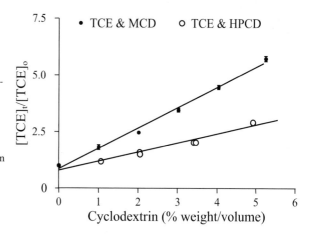

Fig. 52.3.
Enhanced solubilization of trichloroethylene (TCE) by methyl cyclodextrin (MCD) and hydroxypropyl cyclodextrin (HPCD). Note that solubility enhancement increases in linear fashion with increasing concentration of each cyclodextrin. Methyl cyclodextrin achieves greater enhancement than hydroxypropyl cyclodextrin

Table 52.2. Solubility enhancement factors and associated K_{11} values for trichloroethylene and perchloroethylene in the presence of 5% substituted cyclodextrins (*bracketed values* are standard errors). For calculation of K_{11}, see *Determination of binding constants*. HPβ-CD = hydroxypropyl cyclodextrin; Mβ-CD = methyl cyclodextrin; CMβ-CD1 = carboxymethyl cyclodextrin 1; CMβ-CD2 = carboxymethyl cyclodextrin 2; Sβ-CD1 = sulfated cyclodextrin 1; Sβ-CD2 = sulfated cyclodextrin 2

	HPβ-CD	Mβ-CD	CMβ-CD1	CMβ-CD2	Sβ-CD1	Sβ-CD2
Trichloroethylene						
Solubility enhancement	3.1 (0.2)	5.5 (0.4)	2.2 (0.2)	1.9 (0.1)	2.8 (0.3)	1.1 (0.2)
K_{11} (M^{-1})	62 (4)	119 (8)	38 (4)	29 (0.6)	60 (5)	5 (2)
Perchloroethylene						
Solubility enhancement	14 (0.4)	14 (0.4)	10.9 (0.5)	2.7 (0.6)	9.6 (0.9)	1.1 (0.1)
K_{11} (M^{-1})	404 (8)	341 (14)	301 (33)	66 (7)	284 (17)	2 (3)

ity enhancement factors, defined as $[P]_t / [P]_0$, calculated at 5% cyclodextrin concentration, and the binding constants, are listed in Table 52.2.

The K_{11} value depends on several factors, including size, structure, polarity and hydrophobicity of the guest molecule (Wang and Brusseau 1993) as well as size of the cavity. For complexation to occur, the entropy gain in the process by release of 'strained' (high energy, desolvated) water molecules from the cyclodextrin cavity upon inclusion of the hydrophobic molecule would have to be sufficient to overcome the hydrophobic interactions that operate between soil organic matter and low-polarity species like trichloroethylene and perchloroethylene. As perchloroethylene differs from trichloroethylene only by having one additional chlorine atom, size and structure may not be the major factor in this case. However, perchloroethylene is less polar and more hydrophobic than trichloroethylene, which would favour inclusion for the former. The binding constants and solubility enhancement factors for perchloroethylene were found to be larger than for trichloroethylene (Table 52.2). Similarly, Boving et al. (1999) showed that enhancement was greater for perchloroethylene than trichloroethylene with both 5% and 10% Hβ-CD and Mβ-CD. Factors determining binding constant values with different cyclodextrins will include the presence and nature of the substituent groups on the β-cyclodextrin molecule. Among the cyclodextrins used, Sβ-CD1, Sβ-CD2, CMβ-CD1 and CMβ-CD2 are modified by negatively charged substitutents groups.

The degree of substitution in the cyclodextrin molecule is another factor that may affect K_{11} values. A higher degree of substitution could provide steric inhibition for an incoming guest molecule. CMβ-CD2, which showed smaller aqueous solubility enhancement for both compounds, has higher degree of substitution than CMβ-CD1. Moreover, there was again only a small solubility increase in the presence of the highly substituted Sβ-CD2.

The presence of neutral substituent groups in cyclodextrin increases the hydrophobicity of the molecule as a whole, which leads to increased interaction with a hydrophobic guest molecule (Suzuki et al. 1996). With trichloroethylene, the enhancement factor was found to be largest for methylated-β-cyclodextrin. The hydroxypropyl group is also neutral but is larger in size compared to the methyl group, so steric interference may again be of importance here (Wilson and Verral 1998). In accord with this argument, trichloroethylene showed higher enhancement with Mβ-CD compared to HPβ-CD. Interestingly, perchloroethylene showed similar enhancement factors with both HPβ-CD and Mβ-CD.

52.4.2
Binding Constants from Proton NMR Measurements

As a consequence of the process of inclusion of the guest molecule inside the cyclodextrin cavity, both guest and host species experience an altered magnetic environment. This leads to changes in NMR chemical shifts of protons in the guest molecule and also in the H-3 and H-5 protons of the glucose sub-units that are projected inwards towards the cavity. Protons on the outer face of the cyclodextrin molecule are removed from binding and their chemical shift remains unaltered (Demarco and Thakker 1970). Inhomogeneity in substitution of the cyclodextrins used in the present work complicates the observed NMR spectra of the compound, causing broadening of some peaks and rendering it difficult to make assignments for individual protons. On the other hand, the guest trichloroethylene methine proton gave a sharp peak at 6.6 ppm and it was possible to follow the change in the chemical shift upon addition of cyclodextrin. The presence of a singlet can be attributed to a fast equilibrium between free and bound guest molecule as is applicable to loosely bound complexes (Connors 1987).

Therefore, in the NMR experiments, the concentration of cyclodextrin was increased stepwise and the position of the ^{1}H peak for trichloroethylene was measured relative to the HDO peak, itself referenced against an internal tetramethylsilane standard. With increasing cyclodextrin concentration, the ^{1}H peak for trichlrorethylene shifted downfield. The magnitude of chemical shift changes for both guest and host depends on various factors, especially the anisotropic effect, altered solvation, and the conformation of guest molecule inside the cavity; the total shift is usually found to be less than 0.3 ppm (Bergeron 1984). The maximum chemical shift changes observed in experiments carried out here are ca. 0.2 ppm.

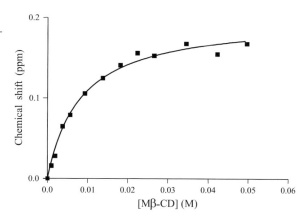

Fig. 52.4.
Plot showing change in chemical shift of trichloroethylene upon addition of increasing concentrations of Mβ-CD

The observed chemical shift (δ) is the average of the chemical shifts of the nuclei in TCE and the TCE-cyclodextrin complex, weighted by the fractional occupancy of these states, i.e.

$$\delta = f_{10}\delta_{TCE} + f_{11}\delta_{TCE\text{-}CD} \qquad (52.6)$$

where f_{10} = fraction of free trichloroethylene and f_{11} = fraction of complexed trichloroethylene. The chemical shift difference (Δ) between free and complexed trichloroethylene is given by $\Delta = \delta - \delta_{TCE}$ while at saturation, the maximum change in chemical shift (Δ_{11}) is described by $\Delta_{11} = \delta_{TCE\text{-}CD} - \delta_{TCE}$ (Connors 1987).

Combining these equations yields $\Delta = f_{11}\Delta_{11}$ which, when combined with the equation for the binding constant gives

$$\Delta = \frac{\Delta_{11} K_{11}[CD]}{1 + K_{11}[CD]} \qquad (52.7)$$

A representative plot of this equation for the trichloroethylene-Mβ-CD system is shown in Fig. 52.4. Values of individual analysis of each cyclodextrin are HPβ-CD (60), Mβ-CD (120), Sβ-CD1 (44), Sβ-CD2 (3.0), CMβ-CD (31), and CMβ-CD2 (21).

52.4.3
Stoichiometry of Complexes between Substituted Cyclodextrins and Trichloroethylene

The possibility of there being two guest molecules in each host cyclodextrin (i.e. a 2:1 complex) must be considered since trichloroethylene is a considerably smaller molecule (0.155 nm^3) compared to the size of the cyclodextrin cavity (0.346 nm^3). This was tested by attempting to fit the experimental data to equations for both 1:1 and 2:1 complexation (Connors 1987). Using the Graph Pad Prism program, converging fits could be obtained for both cases, but the 2:1 situation gave values with extremely large uncertainties, and in some cases K_{21} results were negative. On the other hand using the relation,

$$\Delta = \frac{\Delta_{11} K_{11}[CD]}{1 + K_{11}[CD]} \qquad (52.8)$$

which is based on a 1:1 model, the same data sets converged to consistent values for K_{11} with relatively small uncertainties. Based on duplicate measurements, K_{11} values (and average deviation) were for HPβ-CD, 60 (8); Mβ-CD, 120 (1); Sβ-CD1, 44 (3); Sβ-CD2, 3.0 (single measurement); CMβ-CD1, 31 (1); CMβ-CD2, 20 (3). The predicted 1:1 binding is supported in the hyperbolic curves generated, with correlation coefficients ranging from 0.98 to 0.99. We also used a different non-linear least squares program involving simplex optimization as a comparative method for obtaining binding constants (Lyon et al. 1998). When the 1:1 model was employed, results converged to give values similar to those obtained above using Graph Pad Prism. For the 2:1 case, convergence could not be achieved even after many iterations.

A similar situation was encountered in an investigation (Rekharsky et al. 1997) of the binding of both α- and β-cyclodextrins with a large range of small aliphatic and aromatic molecules. Using both calorimetric and ^{1}H NMR measurements, attempts to fit the data sets into a 2:1 model were unsuccessful as the generated parameters were associated with very large errors. Based on the poor curve fitting, the possibility of 2:1 binding was excluded. Because of the relatively small size of trichloroethylene compared to the host molecules, positioning of the trichloroethylene inside the cavity may not be specific and loose binding is expected. Notably, the magnitudes of the binding constant values found in this work are small compared to systems where a complementary fit between guest and host was postulated (Bender and Komiyama 1978; Bergeron 1984; Schneider et al. 1998).

52.4.4
Relation between Binding Constant and Solubility Enhancement

For any organic pollutant, it is expected that the degree of solubility enhancement would be correlated with the magnitude of the binding constant and this was found to be the case for both trichloroethylene and perchloroethylene, using the entire range of substituted cyclodextrins (Fig. 52.5).

As noted above for trichloroethylene, plots of P_t/P_0 against concentration of cyclodextrin were linear, with intercepts near unity, and slopes with values of K_{11}. An alternative method of treating the equation $[P]_t/[P]_0 = K_{11}[CD] + 1$ is to plot P_t against concentration of cyclodextrin. For such plots it has been shown (Connors 1987) that the slope ($= K_{11}P_0$) is always less than or equal to 1 if only 1:1 complexation occurs. In the present situation, slopes range from 0.2 to 0.8 depending on the cyclodextrin, again consistent with a 1:1 binding model. In most cases for trichloroethylene binding with various cyclodextrins, we found good agreement between the K_{11} values obtained from solubility and NMR experiments. For perchloroethylene, however, the NMR method could not be used because of the lack of a proton as a probe for binding. Theoretically, K_{11} could be measured using ^{13}C NMR, but the low aqueous solubility of perchloroethylene precluded such experiments. The good agreement between the K_{11} values for trichloroethylene using the two methods gives some confidence regarding the corre-

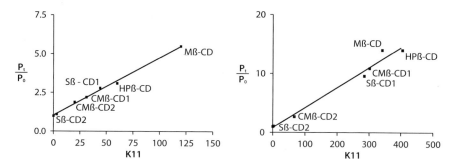

Fig. 52.5. Plot of aqueous solubility enhancement of trichloroethylene (*left plot*) and perchloroethylene (*right plot*) with various cyclodextrins vs. their respective K_{11} values obtained from the ^{1}H NMR experiment. The solubility enhancement, P_t/P_0 is the ratio of aqueous solubility in the presence of 5% cyclodextrin to the solubility in the absence of cyclodextrin

sponding values obtained for perchloroethylene by the single applicable procedure. The plot of P_t/P_0 for perchloroethylene against K_{11} according to $[P]_t/[P]_0 = K_{11}[CD] + 1$ was linear with a correlation coefficient of 0.981 (Fig. 52.5) and an intercept close to unity.

We conclude that aqueous solubility enhancement occurs due to formation of a host-guest complex between a cyclodextrin and trichloroethylene or perchloroethylene, and is driven primarily by hydrophobic forces. A high degree of substitution results in steric congestion for cavity opening and, consequently, leads to diminished inclusion of the polychloroethylenes examined. Non-polar substituents in the cyclodextrin host enhance binding by creating a more hydrophobic cavity. The interplay of these opposing influences gives rise to the magnitude of binding constants (K_{11}) and solubility enhancements $[P]_t/[P]_0$ found in the present work.

52.4.5
Soil Measurements

At the outset, we stated that the requirements for remediation reagents for subsurface soil and water include an ability to significantly enhance aqueous solubility of the target compound. In soil, however, increasing solubility also involves releasing the compound from its association with the soil materials and it is widely recognized that it is the organic fraction of the soil that plays a dominant role in retention of hydrophobic organic species such as the polychlorinated hydrocarbons. A major component of soil organic matter is termed humic material, which encompasses a class of polydisperse polymers composed of a three-dimensional aromatic and alkyl chain framework, modified by a variety of mostly oxygen-containing functional groups. The substantially hydrophobic nature of these materials contributes to their ability to sorb non-polar and low polarity organic compounds (Schwarzenbach et al. 1993). Peat, obtained from the International Humic Substances Society is a soil-derived substance having substantial humic and humic-like structure (vanLoon and Duffy 2000).

In the present study, removal of trichloroethylene or perchloroethylene from soil organic matter was determined by spiking slurries of IHSS peat in water with the two compounds and then measuring the amounts retained by the solid material after 24 h of equilibration (IHSS: International Humic Substances Society). Following this, cyclodextrins were added to the solution to give a concentration of 5%, and the aqueous concentrations of the polychloroethylene compounds were measured after a further 18 h equilibration. From this it was possible to determine the percentage of sorbed compound that had been extracted by the cyclodextrin. Experiments were done using the three cyclodextrins that exhibited the largest binding constants with trichloroethylene and perchloroethylene. The addition of 5% cyclodextrin (HPβ-CD or Sβ-CD1 or Mβ-CD) was able to effect efficient desorption of both trichloroethylene and perchloroethylene. Beginning with peat containing 2 200 µg g^{-1} of sorbed trichloroethylene, the percentage removal (and average deviation) using a single extraction was 73 (7) for HPβ-CD, 78 (7) for Mβ-CD and 93 (3) for Sβ-CD1. Corresponding percent removal efficiency for peat containing 6 600 µg g^{-1} of perchloroethyelene were 90 (2) using HPβ-CD, 98 (1) using Mβ-CD, and 93 (1) using Sβ-CD1. The more efficient extraction of perchloroethyelene is expected, based on the established K_{11} values. Thus a suitably substituted β-cyclodextrin may be a valuable additive for extracting and solubilizing polychlorinated organics during site remediation using pump-and-treat methods.

52.5
Conclusions

Enhancement of the solubility of trichloroethylene and perchloroethylene could be achieved through the formation of host – guest inclusion complexes between substituted cyclodextrins (host) and the polychloroethylene (guest), and is driven primarily by hydrophobic forces. Solubility enhancement factors up to 5.5 and 14 were determined for trichloroethylene and perchloroethylene, respectively. Binding constants (K_{11}) for complexes involving trichloroethylene were evaluated using a ^{1}H NMR technique; these range from 3 to 120 M^{-1}. Both solubility enhancement and the binding constant are dependent on the type and degree of substitution of the β-cycloextrin molecule. The cyclodextrins were found to be capable of effective removal of trichloroethylene and trichloroethylene retained by soil organic matter. Thus a suitably substituted β-cyclodextrin may be a valuable additive in pump-and-treat protocols for site remediation of polychlorinated organics.

Acknowledgements

Support of this work by the Natural Sciences and Engineering Research Council of Canada, Environmental Science and Technology Alliance Canada, as well as Queen's University is gratefully acknowledged.

References

Annandale MT, vanLoon GW, Buncel E (1998) Regioselectivity and steroelectronic effects in the reactions of the dinitroaniline herbicides trifluralin and benefin with nucleophiles. Can J Chem 76:873–883

Balakrishnan VK, Dust JM, vanLoon GW, Buncel E (2001) Catalytic pathways in the ethanolysis of fenitrothion, an organophosphorothioate pesticide. A dichotomy in the behaviour of crown/cryptand cation complexing agents. Can J Chem 79:157–173

Bender ML, Komiyama M (1978) Cyclodextrin chemistry. Springer-Verlag, New York

Bergeron RJ (1984) Inclusion complexes of aromatic compounds. In: Atwood JL, Davis JED, MacNicol DD (eds) Inclusion compounds, vol. 3. Academic Press, London, pp 391–443

Boving TH, Wang X, Brusseau ML (1999) Cyclodextrin-enhanced solubilization and removal of residual-phase chlorinated solvents from porous media. Environ Sci Technol 33:764–770

Brusseau ML, Wang X, Hu Q (1994) Enhanced transport of low-polarity organic compounds through soil by cyclodextrin. Environ Sci Technol 28:952–956

Brusseau ML, Wang X, Wang W (1997) Simultaneous elution of heavy metals and organic compounds from soil by cyclodextrin. Environ Sci Technol 31:1087–1092

Connors KA (1987) Binding constants: the measurement of molecular complex stability. John Wiley & Sons, New York

Demarco PV, Thakker AL (1970) Cyclohepta-amylose inclusion complexes: a proton magnetic resonance study. Chem Commun 1:2–4

Kan AT, Thomson MB (1990) Ground water transport of hydrophobic organic compounds in the presence of dissolved organic matter. Environ Toxicol Chem 9:253–263

Ko S, Schlautman MA, Carrraway ER (1999) Partitioning of hydrophobic organic compounds to hydroxypropyl-beta-cyclodextrin: experimental studies and model predictions for surfactant-enhanced remediation application. Environ Sci Technol 32:2765–2770

Lyon AP, Banton NJ, Macartney DH (1998) Kinetics of the self-assembly of alfa-cyclodextrin[2]pseudorotaxanes with polymethylene threads bearing quaternary ammonium and phosphonium end groups. Can J Chem 76:843–850

Macky DM, Cherry JA (1989) Groundwater contamination: pump-and-treat remediation. Environ Sci Technol 23:630–636

McCray JE, Brusseau ML (1998) Cyclodextrin-enhanced in-situ flushing of multiple-component immiscible organic liquid contamination at the field scale: mass removal effectiveness. Environ Sci Technol 32:1285–1293

National Academy of Science (1994) Performance of conventional pump-and-treat systems. Alternatives for ground water cleanup: report of the National Academy of Science Committee of Groundwater Cleanup Alternatives. National Academy Press, Washington, D.C

Omakor JE, Onyido I, vanLoon GW, Buncel E (2001) Nucleophilic displacement at the phosphorus centre of the pesticide fenitrothion [O,O-dimethyl O-(3-methyl-4-nitrophenyl) phosphorothioate] by oxygen nucleophiles in aqueous solution: α-effect and mechanism. J Chem Soc Perkin Trans 2:324–330

Onyido I, Omakor JE, vanLoon GW, Buncel E (2001) Competing pathways in the reaction of the pesticide fenitrothion [O,O-dimethyl O-(3-methyl-4-nitrophenyl)phosphorothioate] with some nitrogen nucleophiles in aqueous solution. Arkivoc, Part (xii) pp 134–142

Pankow JF, Cherry JA (1996) Dense chlorinated solvents in groundwater: background and history of the problem. Waterloo Press, Portland, OR

Rekharsky MV, Mayhew MP, Goldberg RN, Ross PD, Yamashoji Y, Inoue Y (1997) Thermodynamic and nuclear magnetic resonance study of the reactions of alfa- and beta-cyclodextrin with acids, aliphatic amines and cyclic alcohols. J Phys Chem (B)101:87–100

Schneider H, Hacket F, Rudiegr F (1998) NMR studies of cyclodextrins and cyclodextrin complexes. Chem Rev 98:1755–1785

Schwarzenbach RP, Gschwend PM, Imboden DM (1993) Environmental organic chemistry. John Wiley & Sons, New York

Sha'ato R, Buncel E, Gamble DG, vanLoon GW (2000) Kinetics and equilibria of metribuzin sorption on model soil components. Can J Soil Sci 80:301–307

Suzuki M, Takai H, Tanaka K, Narita K, Fujiwara F, Ohmori H (1996) Visible and ^{13}C nuclear magnetic resonance spectra of azo dyes and their complexes with cyclomalto-oligosaccharides. Carbohydrate Res 288:75–84

vanLoon GW, Duffy SJ (2000) Environmental chemistry – a global perspective. Oxford University Press, Oxford, UK

Wang X, Brusseau ML (1993) Solubilization of some low-polarity organic compounds by hydroxypropyl-beta-cyclodextrin. Environ Sci Technol 27:2821–2825

Wang X, Brusseau ML (1995) Simultaneous complexation of organic compounds and heavy metals by a modified cyclodextrin. Environ Sci Technol 29:2632–2635

Willson LD, Verral RE (1998) A ^1H NMR study of cyclodextrin-hydrocarbon surfactant inclusion complexes in aqueous solutions. Can J Chem 76:25–34

Part VI

Chemical Samples Recycling: The MDPI Samples Preservation and Exchange Project

S.-K. Lin

Abstract

Over the years, chemists have been known to make a significant contribution to science, and to generate new chemical substances as a result of their professional activities. Over 20 million compounds have been recorded in the literature, however, because of their use in chemical processes and their disposal in waste products, only a smaller number of these compounds are available for furture use. Therefore, there is a tremendous loss of molecular diversity and chemical heritage. Reproduction of samples, if successful, is expensive and time-consuming. It may also contribute to a substantial damage to the environment. Chemical sample archives are priceless resources that can tremendously facilitate and speed-up discovery of new drugs, and new compounds, as well as the development of many other chemical products, with industrial, pharmaceutical, agricultural, educational, and research applications. The Molecular Diversity Preservation International (MDPI) project has been successful in carrying out worldwide collections/deposits, exchanges, and "recycling" of a significant number of chemical samples.

53.1
Introduction

Since 1995, the collection and high throughput production of chemicals of high molecular diversity have been significantly increased, to cope with the crucial problem associated with chemical synthesis of new drugs and other chemical products needed in research and development. The related topic of molecular diversity has been so important that a journal *Molecular Diversity* (www.mdpi.org/modi) was launched (Lin 2003). In 1995, a nonprofit international organization, *Molecular Diversity Preservation International* (MDPI, www.mdpi.org) was founded in Switzerland, and started a worldwide repository and exchange of chemical samples as well as experimental data related to samples preparation and characterization. A chemistry journal, *Molecules* (www.mdpi.org/molecules) was launched by MDPI in 1996, with a unique editorial policy regarding the promotion these sample collection activities (Lin 1996a).

So far, we have already accepted for storage and registered more than 20 000 available chemical compound samples. *Molecules* is publishing its 8th volume in 2003. Many samples are supplied by the authors of *Molecules* and deposited as authentic reference samples.

53.2
Strategies and Activities of MDPI

53.2.1
Preserving Chemical Samples

The disposal of old chemical samples can be very expensive, and can also cause environmental damage. Recent studies have also demonstrated that environmental contamination by these chemicals can lead to a significant number of adverse health effects including various types of cancer in humans (Tchounwou 2003). For this and several other reasons, a new online periodical, *International Journal of Environmental Research and Public Health* (ISSN 1660-4601), has been launched (www.mdpi.net/ijerph).

Preservation of certain rare samples and their recycling is not only possible but also practically necessary. Similar to biospecies (WIPO 1994), samples of chemical species should be properly preserved due to many unique reasons. Chemical samples may decay. For this concern some scientists may not support the preservation of chemical samples. The preservation of biospecies also faced the sample stability problems. However, if some necessary measures are taken, the preservation of microorganisms (WIPO 1994) and chemical samples can be achieved. Regarding chemical samples, of course we do not collect the known unstable samples and known hazardous samples. Only adequately purified samples and carefully bottled samples are collected (e.g. if a compound is sensitive to light but otherwise very stable, this sample can be collected if it has been in a dark bottle). All sample vials should be properly sealed to prevent possible reaction with oxygen and moisture in the air. Stable samples, if carefully stored, can be preserved for many years. In pharmaceutical companies, almost all those properly archived organic samples prepared more than fifty years ago and stored even at room temperature so far have been found to be applicable and all can be used for such purposes as high throughput screening and useful for many

> **Box 53.1.** Some statistics (some of this data is the present author's estimation)
>
> - The average cost per sample of ordinary structural complexity prepared was estimated at around 7500 USD (Xu and Hagler 2002).
> - About 1/4 to 1/3 of a synthetic organic chemist's time is spent for preparing compounds according to literature. Samples are expensive, even prepared according to the literature, given that only 50 compounds are synthesized per chemist per year.
> - In principle, the samples of published molecules can be reproduced. However, it is certain that many samples will never be reproduced if doing so is too expensive (e.g. it takes more than 30 steps), or may never be reproduced because of the incomplete record of experimental data (see Sect. 53.3).
> - So far, more than 200 000 000 chemicals have been recorded according to Chemical Abstracts (see: www.cas.org/substance.html). There are 8 000 000 organic compounds recorded in Beilstein database (http://www.beilstein.com/products/xfire/). However, the number of chemicals commercially/readily available is much less than 100 000 before 1995, according to ACD (Available Chemicals Directory) database release. More than 99% of compounds recorded in literature exist only on paper; chemists discarded them.

other purposes. It is certain that most chemical samples, particularly organic compound samples can be stored for many years.

Many chemists, even many synthetic organic chemists, may insist that the publication of research papers is all they wanted. Traditionally a chemical synthesis is carried out for a target compound. Consequently no precursors prepared as intermediates are regarded as worthy to be preserved. Sometimes a compound is prepared for very specific academic and industrial purposes such as spectroscopic measurements or enzyme-ligand binding tests. Afterwards the samples may be regarded as useless and have to be discarded. In many cases such as the termination of a research project, graduation of a student, and retirement of a chemist or a professor, all the pertinent compounds accumulated may have to be discarded (Box 53.1 and 53.2). This is the loss of precious molecular diversity.

Molecular diversity has been a topic of increasing interest since 1995 (Waller 2002; Xu and Hagler 2002). Techniques of combinatorial chemical libraries have been developed to provide millions of compounds for pharmaceutical studies. The similarity and the gradual variation in the structures of chemical species in combinatorial chemical libraries permit adjustments designed to optimize structure-activity relationships. However, the high quality of a chemical library relies on the distinct differences of both the structures and properties of the collected samples (Lin 1996b,c). These high diversity compounds are isolated from natural sources, and/or traditionally or routinely prepared in the laboratories (Lin 1996a,b, 1997a,b, 1998).

Now, a rare sample as little as 1 mg can be used for dozens or even hundreds of bioactivity tests. Particularly in recent years, with the development of high throughput screening technology, the acquisition of chemical samples, even though the amount might be small, has become a bottleneck in the discovery process of new drugs and other chemical products. Absorption, distribution, metabolism, excretion, and toxicity (ADMET) studies of drugs in animal models can be performed before the active compound can be identified (Xu and Hagler 2002). Such studies may need large amounts of chemical samples. Many rare samples of smaller amounts should also be preserved for the purpose of pharmaceutical research.

Box 53.2. Several cases of samples loss

- *Case 1.* Summer 1994. The final work of a retired chemist in a pharmaceutical company (Ciba-Geigy) is to dissolve all of his 2000 samples into acetone, which took him and his assistant for one week.
- *Case 2.* October 1995. A recently retired professor called the present author from Denmark and told him that he just threw away more than 2000 samples because his successor needs the laboratory space and his wife did not permit him to move the sample home and place in the basement.
- *Case 3.* April 1997. One third of the samples had structural lists destroyed and ready to be trashed when we arrived in the chemistry department storage room. We were able to register 2/3 of all the 3000 old samples properly. MDPI still accepted the rest 1/3 samples and gave separate registration numbers (they will be distributed for random screenings. Active hits will be analyzed and the structures will be elucidated afterwards).

Fortunately most synthetic organic chemists and natural product chemists have a tradition to store their chemical samples. They would like to preserve them for other people to use in future. However, there is usually no room for them to store their samples, after retirement. MDPI has provided the service to store these samples, as well as to develop and maintain a database. Therefore the collection of isolated natural products and individual rare samples of synthetic source have been MDPI's main service. Box 53.3 lists the criteria for sample registration and deposition.

In order to encourage chemists to deposit their rare samples, the MDPI services are free of charge to its contributors. In addition, MDPI pays postage for delivery by express carrier (TNT, Federal Express, UPS, etc.) of the compounds to MDPI. Contributors are rewarded 50% of the net profit of sample distribution services. Samples are usually divided into at least 5 unit amounts. The last remaining part is further divided into several parts. A typical unit price per unit amount is 100 USD/100 mg, and may vary based on user's recommendation. MDPI distributes compound samples worldwide. Chemists may also store compounds themselves for the registered samples, also free of charge, provided that they send the unit amount of the ordered compound promptly. They may distribute directly their compounds by citing MDPI registration number.

In order to provide services fare to contributors and to customers, and insure that the samples are correctly characterized, we publish unit amount and the unit price and preferably contributor's name and address in our database release. If the contributors agree, other chemists may obtain samples free of charge for academic research. Now all the samples can be searchable on the www.molmall.org server. Most of the structures of MDPI samples not published in the literature are given CAS Reg. numbers and the MDPI collection is included in CHEMCATS – Online Chemical Catalogs (www.cas.org/CASFILES/chemcats.html).

Many samples have historical significance and related to chemical heritage. For example samples prepared by Nobel laureates can be preserved forever. A chemical museum will be built for this purpose.

Unlike handling of bulk chemical storage and transportation, the storage, package and transportation of chemical samples in small amount are easier. However, because of their large number and diversity, we will face other problems. The management of chemical sample archives itself has been an important topic of research. MDPI plans to carry out more studies on this topic.

Box 53.3. Samples criteria

- All compounds, whether they are new or old, synthetic or natural products, published or unpublished.
- Authentic supporting samples for papers published in chemistry journals.
- Samples should be reasonably stable at room temperature, as pure as specified, not hazardous and preferably in solid (crystal) form.
- Catalysts and enzymes, macromolecules and polymers as well as other samples are also acceptable.

53.2.2
Preserving Experimental Data

In a published paper, the "experimental design and methods" section is the part of great interest to synthetic chemists. Traditionally, our contribution of chemical information has relied on the full description of chemical preparation and compound characterization. However, more and more chemistry journals ask authors to shorten their papers, mainly the experimental part, or include the research data in the "so-called "supplementary material". This makes the record of the knowledge incomplete and the reproduction of chemical preparation and characterization inconvenient. For example, experimental details of R. B. Woodward's several most important synthetic works have never been properly published (Cornforth 1993). In such cases, if a synthesis is very difficult and if there is no one else as resourceful as the original author, it is almost impossible to prepare the reported samples again.

Generally an ordinary and average synthetic chemist working in industry will prepare and fully characterize about 1 000 samples in his whole carrier (Box 53.2). On average, only a small part of this precious data of chemical synthesis is ever recorded in chemistry journals and patents. Thus, the precious expertise of experienced chemists can seldom be fully recognized, appreciated and benefited by others (Lin 1996a, 1997a,b, 1998). Another urgent task of our organization MDPI is to record all the experimental data as a complementary measure of the normal publication of the existing journals. Any scattered unassembled experimental data for individual compounds, which are conventionally not publishable, are particularly welcomed; to be published in a special form of one molecule per short note and given special page numbers (M1, M2, etc.). A large volume of (to say 1 million) structures will be easily published in this way (see: www.molbank.org). If every synthetic chemist contributes 100 such posters, this number will be easily reached within several years. Afterwards, an online-searchable database constructed from this data will be open to the public free of charge and it will be a very useful treasure to all chemists and other related scientists (Lin and Patiny 2000). The detailed procedure of the chemical reactions can make the reproduction of chemicals easier.

The structures of MDPI samples not published in the literature are given CAS Reg. The publication of the samples database at www.molmall.org is also a pertinent way of recognizing sample contributor's synthetic works, if they were never published before. The information regarding the sample availability becomes an important part of Chemical Abstracts Service (www.cas.org).

53.3
Conclusion

Compared to scientists of other fields of study, chemists have created a vast volume of chemical samples and experimental data, which are definitely unique. We should take care of our complete intellectual properties and act properly to preserve worldwide high-quality molecular diversity of both chemical information and chemical samples generated by us. It is never too late to act. To support the MDPI activities, the recommendation is proposed: it should be an obligatory part of his/her final duty that a chemist finishing his/her works, and leaving the university or the company, deposits or registers all of his/her samples. He/she should also publish all his/her synthesis and structural characterization data. MDPI has been modifying both procedures for samples and data base development, to make it more and more convenient for chemists to make a significant contribution to chemical preservation, and scientific advancement.

Acknowledgement

The author wishes to thank Dr. Tchounwou for review and English corrections.

References

Cornforth JW (1993) The trouble with synthesis. Aust J Chem 46:157–170
Lin SK (1996a) A good yield and a high standard. Molecules 1:1–2
Lin SK (1996b) Molecular diversity assessment: logarithmic relations of information and species diversity and logarithmic relations of entropy and indistinguishability after rejection of Gibbs paradox of entropy of mixing. Molecules 1:57–67
Lin SK (1996c) Correlation of entropy with similarity and symmetry. J Chem Inf Comp Sci 36:367–376
Lin SK (1997a) Chemical information. Chem Eng News 75(21):4
Lin SK (1997b) Preserving and exploiting molecular diversity: deposit and exchange of chemical information and chemical samples. Molecules 2:1–2
Lin SK (1998) Chemical samples: chemists' and public concern, and preservation. Distributed at the MIT Forum on Chemicals and Society, Cambridge, MA, USA, 11–12 June 1998
Lin SK (2003) Editorial: what is molecular diversity. Mol Diversity 6:1
Lin SK, Patiny L (2000) MolBank: first fully web-based publication of chemical reaction data of individual molecules with structure search and submission. Internet J Chem 3:1
Tchounwou PB (2003) Editorial. Int J Environ Res Public Health 1:1
Waller, CL (2002) Recent advances in molecular diversity. Mol Diversity 5:173–174
World Intellectual Property Organization (WIPO) (1994) Guide to the deposit of microorganisms under the Budapest treaty. WIPO Publication No. 661 (E), Geneva
Xu J, Hagler A (2002) Chemoinformatics and drug discovery. Molecules 7:566–600

Chapter 54

Photodecomposition of Organic Compounds in Aqueous Solution in the Presence of Titania Catalysts

B. Malinowska · J. Walendziewski · D. Robert · J. V. Weber · M. Stolarski

Abstract

Materials prepared by sol-gel method combined with drying under supercritical conditions are called aerogels. Titania aerogels appear to be promising photocatalysts. In this study the aerogels were synthesized from tetraisopropyl orthotitanate (IZPT) diluted in anhydrous methanol or isopropanol. The properties of titania aerogels were changed by applying various preparation ways. The adsorption capability of the aerogels (kinetics and isotherms of adsorption) was checked. They revealed a multi-step adsorption, characteristic for mesoporous solids. The physicochemical properties of the catalysts were determined using X-ray Photoelectron Spectroscopy (XPS) and Scanning Electron Microscopy (SEM). The photocatalytic tests of the aerogels were conducted using aqueous solutions of 4-hydroxybenzoic acid and *p*-chlorophenol. The results prove that photocatalytic properties of the aerogels strictly depend on adsorption capacity.

Key words: aerogel, adsorption, 4-hydroxybenzoic acid, *p*-chlorophenol, photocatalysis

54.1 Introduction

The industrial era has brought about appalling conditions to the environment. Already existing contamination and prevention of new pollution are urgent problems that should be resolved and considered when new chemicals are being designed (Sedykh et al. 2001). Thus, extensive researches have been drawn up to find some effective methods for hazardous chemical compounds elimination from air, water (Kang et al. 2001) and soil. Among manifold possibilities the advanced oxidation processes (AOPs) appear to be highly efficient (Ilisz and Dombi 1999). In particular the heterogeneous photocatalysis has been extensively investigated for environment cleanup (Chun et al. 2001; Banhemann et al. 1999; Ollis and Al-Ekabi 1993; Pichat et al. 1997; Robert et al. 2000; Schiavello 1997). Photocatalytic oxidation provides a unique opportunity for the purification of environment. Commonly used photocatalysts are semiconductors. Among them titanium dioxide reveals the best photocatalytic properties (Axelsson and Dunne 2001; Fujishima et al. 2000). The first literature data concerning photocatalytic activity of TiO_2 date back to 1977, when Frank and Bard examined the possibilities of using TiO_2 to decompose cyanide in water (Frank and Bart 1977). Illumination of TiO_2 surfaces at energies above the band gap produces charge carries (electron and holes) which are the primary cause of the photochemical events (Mills and Hunte 1997; Hoffman et al. 1995). However, the attempts at preparing new photocatalysts have been realized. Sen and Vannice prepared titania aerogels and made use of them to degrade paraffins, olefins and alcohols by means of UV irradiation (Sen and Vannice 1988).

Sol-gel chemistry combining with supercritical drying is very powerful means of synthesizing catalytic materials (Pajonk 1997). Catalysts obtained this way are called aerogels. They are powders of very low densities, high porosity and large surface area (Walendziewski et al. 1999). One assumes that those kinds of catalysts are recent products of modern technology. In fact, the first aerogels were prepared in 1931 (Hunt and Ayers 2000). Steven S. Kristler set out to prove that "a 'gel' contained a continuous solid network of the same size and shape as the wet gel" (Kistler 1932). The preparation of aerogels can be divided into two steps: the formation of a wet gel and the drying of the wet gel to form an aerogel. An appropriate drying procedure is to evacuate the liquid phase under appropriate supercritical conditions (Pajonk 1999). Sol-gel chemistry either by itself or in combination with supercritical drying is a suitable means for the preparation of solids with novel structural and chemical properties (Schneider and Baiker 1997). The advantages of the sol-gel technology encompass: the usage of various wet-chemical tools for working out synthesis, flexibility of tailoring the physicochemical properties (density, surface area, pore volume) of aerogels. However, the technical application of the sol-gel catalyst is hampered by the costs, as to the aerogels in particular, by the manifold difficulties encountered in transferring them into an applicable form (Schneider and Baiker 1997).

In the present work preparation ways of titania aerogels and their application in photodecomposition processes are proposed. The model compound examined in this study was 4-hydroxy benzoic acid. Finally tests using p-chlorophenol were conducted since as a large amount of chlorinated compounds are currently produced and reach the environment (Ormand et al. 2001). Contamination by chlorinated compounds is one of the most serious environmental problems since they are toxic, non-biodegradable (Roques 1996) and found in natural water bodies (Bandara et al. 2001; Serpone et al. 2000). The aim of the research is to find optimal preparation way to obtain catalysts of better photocatalytic activities.

54.2
Experimental

54.2.1
Preparation of Titania Aerogels

In this study titania aerogels were prepared using a combination of sol-gel chemistry and supercritical drying of wet gel. First of all a precursor (IZPT – tetraisopropyl orthotitanate, Fluka) and an appropriate solvent (anhydrous methanol, isopropanol, Aldrich) were homogenized for 30 min at ambient temperature with stirring. The next step was hydrolysis for 2 h. The stoechiometric mole ratio H_2O: precursor with alcohol amounted to 4:1.

The obtained gels were aged 24 h at ambient temperature. The solvent was removed from the porous structure of the gel using supercritical drying. Figure 54.1 shows the equipment used for drying under supercritical conditions. Temperature and pressure cycles vs. time during supercritical drying are shown in the Fig. 54.2.

In order to prevent the gels from collapse at the heating-up stages, the initial step was to saturate the reactor with nitrogen to a pressure of about 80 atm (close to the critical parameters of the alcohol). The prepared aerogels were calcinated at 400 °C for 5 h.

Fig. 54.1.
Scheme of the equipment for aerogel drying under supercritical conditions

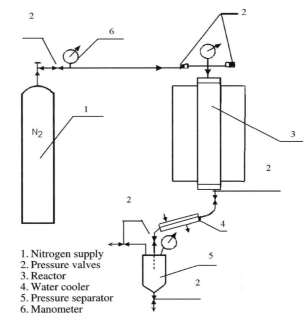

1. Nitrogen supply
2. Pressure valves
3. Reactor
4. Water cooler
5. Pressure separator
6. Manometer

Fig. 54.2.
Temperature and pressure cycles vs. time during supercritical drying

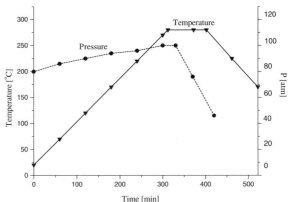

54.2.2
Adsorption Studies

The kinetics of adsorption of aerogels were determined using 100 ppm water solution of 4-hydroxybenzoic acid (4HBz acid, Aldrich). 100 ml of the solution and 0.1 g of aerogel were magnetically stirred during 60 min. Sampling was made in the equal time intervals. After filtration using Millipore 0.45 μm type HVLP filters the samples were analyzed by High Performance Liquid Chromatography (HPLC Waters 600 pump and 996 photodiode array detector).

Isotherms of adsorption were determined using different water solutions of 4HBz acid. 40 ml of each solution and 0.04 g of the aerogel (1 g l^{-1}) were magnetically stirred for 30 min as the equilibrium state was achieved. The samples were taken, filtrated (0.45 μm filters) and analyzed by HPLC. In the next step the aerogels were drained off, dried in a vacuum dryer at 110 °C and analyzed by Infrared Spectroscopy with Fourier Transformation (DRIFT technique) to confirm the adsorption phenomena. Diffuse reflectance spectra of aerogels were recorded on the BIO-RAD FTS-185 spectrometer equipped with a MCT detector and a Graseby-Specac "selector" optical accessory. The kinetic and isotherm studies were conducted in the dark to avoid random radiation.

54.2.3
Physicochemical Properties

The surfaces of aerogels were tested by X-ray photoelectron spectroscopy (XPS). XPS analyses were performed on RIBER MAC 2 spectrometer Al K α-radiation (12 kV and 25 mA). The spectra of C1s, O1s, Ti2s, Ti2p core levels were recorded. The surface of the aerogel was placed in a vacuum environment and then irradiated with photons in the X-ray energy range. The atoms comprising the surface emit electrons after direct transfer of energy from the photon to the core-level electron. These emitted electrons are subsequently separated according to energy and counted. The energy of the photoelectrons is related to the atomic and molecular environment from which they originated. The number of electrons emitted is related to the concentration of the emitting atom in the sample (Ratner and Castner 1997). XPS analyses give information about the quantitative and qualitative composition of elements on the surface of the catalyst. The size of particles and morphology of the aerogels were determined using Scanning Electron Microscopy (SEM). The apparatus used to determine these data was HITACHI S2500, 20 keV with secondary electron.

54.2.4
Photodegradation of 4HBz Acid and *p*-Chlorophenol Using Titania Catalyst

100 ppm water solution of 4HBz acid and 60 ppm water solution of *p*-chlorophenol were prepared. 500 ml each of the solution and 0.5 g of aerogel were homogenized using magnetic stirring for 30 min as equilibrium state was achieved. The photocatalytic processes were conducted in the Solar box ATLAS SUNTEST CPS+ simulating natural radiation (light source vapour Xenon lamp, 300 nm $< \alpha <$ 800 nm) and lasted 4 h. The mixtures of the contaminants and the catalysts were continuously magnetically stirred during photocatalytic tests. Sampling was in the equal intervals and filtrated (0.45 μm filters). In the case of 4HBz acid the samples were analyzed by HPLC and in the case of *p*-chlorophenol by UV-Vis spectroscopy (Shimadzu spectrometer, 200 nm $< \alpha <$ 500 nm).

54.3
Results and Discussion

The prepared aerogels are white powders of very low densities. Table 54.1. presents the prepared aerogels and their bulk densities. For comparison the bulk density of P25 TiO_2 Degussa was measured. It amounts to 4.38 g cm^{-3}.

Table 54.1. Prepared aerogels

Sample	Concentration of precursor in solvent (%)	Solvent	Bulk density (g cm³)
Z513	IZPT 30	Isopropanol	0.2067
Z514	IZPT 30	Methanol	0.1401
Z516	IZPT 60	Methanol	0.2099

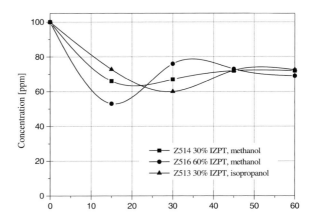

Fig. 54.3. Kinetics of adsorption

54.3.1
Adsorption Studies

In the case of adsorption kinetics the concentration of 4HBz acid in the aqueous solution after adsorption was measured (Fig. 54.3). The adsorption of 4HBz acid on the aerogel particles occurs very fast (considerable decrease of acid concentration in the solution after 15 min of stirring). The aerogel prepared from 60% of IZPT in methanol (Z516) indicates the best capability of adsorption. After 15 min of stirring an almost 50%-decrease of 4HBz acid concentration is observed. In turn the aerogel synthesized from 30% of IZPT in isopropanol (Z513) and the one synthesized from 30% of IZPT in methanol (Z516) indicate an almost equal capability of adsorption (30%-decrease of 4HBz acid) in the initial phase of the experiment. However, further results show that after 60 min of stirring the adsorption capabilities of Z516, Z514 and Z513 become equal (above 30%-decrease of 4HBz acid). These results constitute a proof of the adsorption-desorption phenomena.

Figure 54.4 depicts the isotherms of adsorption. The number of 4HBz acid moles (n_s) adsorbed per gram of aerogel are plotted vs. the equilibrium concentration of 4HBz acid solution (C_{eq}). The number of 4HBz acid moles (n_s) adsorbed per gram of aerogel was determined from the adsorption-induced decrease in the molarity (dC) and the volume (V) of 4HBz acid solution: $n_s = (VdC)/W$, where W is the weight of aerogel in grams (Robert and Weber 1998, 2000).

The shapes of isotherms show that multistep adsorption occurs. The isotherms observed with xerogels often fall into the type-I category (Langmuir type), which is generally found with microporous solids, whereas the isotherms observed with aerogels fre-

Fig. 54.4.
Isotherms of adsorption

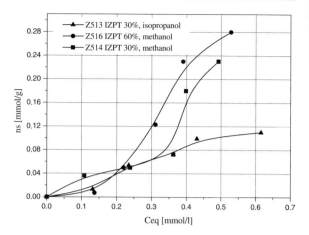

Fig. 54.5.
DRIFT spectra of 4HBz acid adsorbed on the titania catalysts

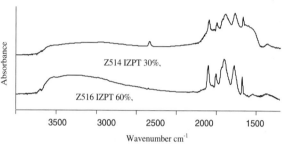

quently fall into the type-IV category (Fig. 54.4, the shapes of isotherms), which is very common among mesoporous solids. Thus, aerogels are usually meso- to macroporous (Schneider and Baiker 1997). Considering aerogels in this study the measurement of the size of the porous is in the progress now. The capability of adsorption follows the order: 60% of IZPT in methanol (Z516) > 30% of IZPT in methanol (Z514) > 30% of IZPT in isopropanol (Z513) whereas number of occupied adsorption sites per gram, on this level of examination, amounts to 2.8×10^{-4} mol g^{-1}; 2.3×10^{-4} mol g^{-1} and 1.5×10^{-4} mol g^{-1}, respectively.

The Diffuse Reflectance Infrared Fourier Transform Spectroscopy (DRIFTS) sampling technique has gained popularity for the study of powders, solids, and species adsorbed on solids (Culler 1993). In the present study DRIFT analyses were performed to confirm the adsorption of 4HBz acid on the aerogel surfaces. The samples were not diluted with KBr. Figure 54.5 shows obtained spectra for Z514 and Z516 catalysts.

Each spectrum in the Fig. 54.5 is obtained as a result of the difference between the catalyst spectrum and the spectrum after adsorption of 4HBz acid. In the case of aerogels there are absorption bands at the range of wavenumber from about 1700 cm^{-1} to 1100 cm^{-1}. They are associated with carboxylate and aromatic absorption bands (Robert and Weber 1998, 2000; Lee et al. 2000) and they prove the adsorption phenomenon of 4HBz acid on the aerogel surfaces. However, the aerogel prepared from 60% of IZPT in methanol (Z516) indicates a higher adsorption capacity than the one prepared for 30% of IZPT in methanol (Z514). Presumably, it is due to presence of a higher amount of basic sites on the surface of Z516.

54.3.2
XPS and SEM Analyses

The interesting peaks are in the range of binding energy 600–400 eV. There are peaks for Ti2p and O1s core levels. Obtained spectra helped to determine stoichiometric ratio of the elements (titanium, oxygen) on the surface of the aerogels (Table 54.2). It appeared that the surface of aerogel prepared from 30% of IZPT in methanol (Z514) includes the smallest amount of oxygen atoms whereas aerogel Z513 prepared from 30% of IZPT in isopropanol Z513 the biggest. The smallest amount of oxygen in the case of Z514 can be a proof that its surface structure has lots of defects. On the contrary bigger amount of oxygen in the case of Z516 and Z513 can be a proof of the orderly surface structure. Scanning electron microscopy images are shown in the Fig. 54.6.

There is an appreciable difference between two aerogels (Fig. 54.6). The aerogel prepared from methanol consists of smaller particles than the one prepared using of isopropanol. In the case of Z513 bigger clusters appeared. SEM analysis of Z516 (60% of IZPT in methanol) is in the course. Considering the obtained results for Z513 and Z514 aerogels a conclusion can be drawn that a structure of aerogel strongly depends on solvent used in the initial step of preparation.

Table 54.2.
Percentage composition of aerogel surface

Aerogel	O1s (%)	Ti2p (%)	TiO$_x$
Z513	67.7	25.7	TiO$_{2.63}$
Z516	67.3	26.3	TiO$_{2.56}$
Z514	58.8	31.5	TiO$_{1.86}$

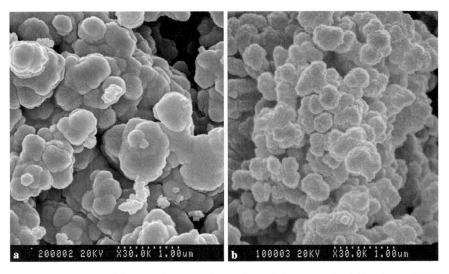

Fig. 54.6. SEM images of the aerogels. 6a: Z513 (30% of IZPT in isopropanol), 6b: Z514 (30% of IZPT in methanol)

54.3.3
Photocatalytic Tests

The activities of aerogels: Z513, Z516 in the photocatalytic processes have been checked. 4HBz acid has been chosen as a model contaminant of water solution. Figure 54.7 presents the obtained results. The first 30 min of each processes were conducted in the dark to obtain an adsorption equilibrium state. After that period of time the photocatalytic process started. It appeared that catalyst Z516 (60% of IZPT in methanol) indicates the best photocatalytic activity. That one reveals also the best adsorption capacity. It is associated with great amount of active sides on the Z516 surface. For comparison photodegradation of 4HBz acid in the presence of TiO_2 P25 Degussa was carried out. In the case of TiO_2 the final results are similar to Z513 (30% of IZPT in isopropanol).

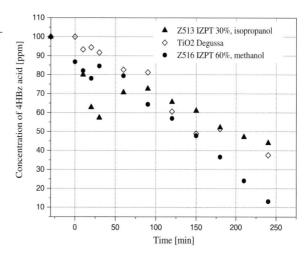

Fig. 54.7.
Kinetics of photocatalytic degradation of 4HBz acid in the presence of titania catalysts

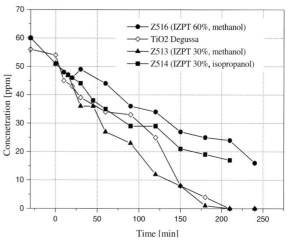

Fig. 54.8.
Kinetics of photocatalytic degradation of *p*-chlorophenol using titania catalysts

The next series of photocatalytic tests were conducted using p-chlorophenol. Three titania aerogels were tested: Z513 (30% of IZPT in isopropanol), Z514 (30% of IZPT in methanol), Z516 (60% of IZPT in methanol) and TiO_2 P25. In this case the best activity indicated aerogel Z513. Figure 54.8 shows the obtained results. Although the pathways of degradation kinetics of Z513 and TiO_2 P25 Degussa are not similar, the final results are the same. The kinetics of photocatalytic degradation for next two aerogels (Z514, Z516) indicated lower photocatalytic activity in the final part of the test. In particular a low activity was presented by Z516 aerogel, the most active catalyst in the photodegradation of 4HBz acid. One time more, the difference of the photocatalytic mechanisms is illustrated in ths case. It seems that the surface photodegradation of adsorbed molecules governs the overall degradation pathways (Robert and Weber 2000).

54.4 Conclusion

Prepared aerogels are powders of very low densities. Their structure depends on factors, percentage of precursor in solvent as well as kind of the solvent (differences in morphology and surface construction between aerogels). The aerogel prepared using of 60% of tetraisopropyl orthotitanate (IZPT) in methanol indicated the best 4-hydroxybenzoic acid (4HBz) adsorption capability. Thus it indicated the best photocatalytic activity in the case of 4HBz acid. However, while p-chlorophenol was used in the photocatalytic tests the best activity indicated aerogel prepared from 30% of IZPT in isopropanol. The final results for that aerogel are similar to the results obtained for TiO_2 P25 in both cases. The conclusion can be drawn that photocatalytic activities of aerogels to a large extend depend on the kind of pollutant. The aim of the research is to find optimal preparation conditions that finally will give aerogels of photocatalytic activity for broad list of contaminants and commercial applying.

Acknowledgement

We are very grateful to French Government for opportunity to carry out the research in the Laboratoire de Chimie et Applications in Metz University. We would like to thank: dr inż. Andrzej Krztoń, from Polish Academy of Sciences in Gliwice, for DRIFT analyses and helpful hits; Dr. Olivier Heintz, from the Université de Bourgogne (France), for XPS analyses and reports preparation; Dr. Nouari Chaoui from LCA Saint-Avold Metz University, for SEM analyses.

References

Axelsson A-K, Dunne LJ (2001) Mechanizm of photocatalytic oxidation of 3,4-dichlorophenol on TiO_2 semiconductor surfaces. J Photochem A Chem 144:205–213
Bandara J, Mielczarski JA, Kiwi J (2001) I. Adsorption mechanism of chlorphenos on iron oxides, titanium oxide and aluminum oxide as detected by infrared spectroscopy. Appl Cat B Environ 34:307–320
Bahnemann D (1999) In: Boule P (ed) Handbook of environmental photochemistry. Springer-Verlag, Heidelberg, p 285
Chun H, Yizhong W, Hongxiao T (2001) Influence of adsorption on the photodegradation of various dyes using surface bond-conjugated TiO_2/SiO_2 photocatalyst. Appl Cat B Environ 35:95–105

Culler SR (1993) Diffuse reflectance infrared spectroscopy: sampling techniques for qualitative, quantitative analysis of solids. In: Coleman PB (ed) Practical sampling techniques for infrared analysis. USA, pp 93–106

Frank SN, Bard AJ (1977) Heterogeneous photocatalytic oxidation of cyanide ion in aqueous solution at TiO_2 powder. J Am Chem Soc 99:303–304

Fujishima A, Rao TN, Tryk DA (2000) Titanium dioxide photocatalysis. J Photochem C Photochem Rev 1:1–21

Hoffman MR, Scot TM, Wonyong C, Bahneman DW (1995) Environmental application of semiconductor photocatalysis. Chem Rev 93(1):69–96

Hunt A, Ayers M (2000) A brief history of silica aerogels. Berkeley National Laboratory

Ilisz I, Dombi A (1999) Investigation of the photodecomposition of phenol in near-UV-irradiated aqueous TiO_2 suspensions. II. Effect of charge-trapping species on product distribution. Appl Cat A General 180:35–45

Kang M, Lee S-Y, Chung Ch-H, et al. (2001) Characterisation of a TiO_2 photocatalyst synthesised by the solvothermal method and its catalytic performance for $CHCl_3$ decomposition. J Photochem Photobiol A Chem 144:185–191

Kistler SS (1932) In: J Phys Chem 34:52

Lee SJ, Han SW, Yoon M, Kim K (2000) Adsorption characteristics of 4-dimethylaminobenzoic acid on silver and titania: diffuse reflectance infrared Fourier transform spectroscopy study. Vib Spectroscopy 24:265–275

Mills A, Hunte S (1997) An overview of semiconductor photocatalysis. J Photochem A Chem 108:1–35

Ollis DF, Al-Ekabi H (eds) (1993) Photocatalytic purification and treatment of water and air. Elsevier, Amsterdam

Ormad MP, Ovelleiro JL, Kiwi J (2001) Photocatalytic degradation of concentrated solutions of 2,4-chlorophenol using low energy light. Identification of intermediates. Appl Cat B Environ 32:157–166

Pajonk GM (1997) Catalytic aerogels. Catal Today 35:319–337

Pajonk GM (1999) Some catalytic applications of aerogels for environmental purposes. Catal Today 52:3–13

Pichat P (1997) In: Ert G, Knözinger H, Weitkamp J (eds) Handbook of heterogeneous photocatalysis. VCH-Wiley, Weiheim, p 2111

Ratner BR, Castner DG (1997) Electron spectroscopy for chemical analysis. In: Vickerman JC (ed) Surface analysis – the principal techniques, Wiley & Sons, Chichester, pp 43–93

Robert D, Weber JV (1998) Photocatalytic degradation of methylbutandioic acid (MBA) in aqueous TiO_2 suspension: influences of MBA adsorption on the solid semi-conductor. J Clean Prod 6:335–338

Robert D, Weber JV (2000) Study of adsorption of dicarboxylic acids on titanium dioxide in aqueous solution. Adsorption 6:175–178

Robert D, Parra S, Pulgarin C, Krzton A, Weber JV (2000a) Chemisorption of phenols and acids on TiO_2 surface. Applied Surf Scien 167(1–2):51–58

Robert D, Lede J, Weber JV (eds) (2000b) Chimie, soleil, energie environnement. Special issue in Entropie, No. 228

Roques H (1996) Chemical water treatment. VCH Verlag, Weinheim, Germany

Schiavello M (ed) (1997) Heterogeneous photocatalysis. Wiley & Sons, New York

Schneider M, Baiker A (1997) Titania-based aerogels. Catal Today 35:339–365

Sedykh A, Saiakhov R, Klopman G (2001) META V. A model of photodegradation for the prediction of photoproducts of chemicals under natural-like conditions. Chemosphere 45:971–981

Sen B, Vannice MA (1988) J Catal 113:52

Serpone N, Texier I, Emeline AV, Pichat P, Hidaka H, Zhao J (2000) Post-irradiation effect and reductive dechlorination of chlorophenols at oxygen-free TiO_2/water interfaces in the presence of prominent hole scavengers. J Photochem A Chem 136:145–155

Walendziewski J, Stolarski M, Steininger M, Pniak B (1999) Synthesis and properties of aluminia aeogels. React Kinet Catal Lett 66(1):71–77

Chapter 55

Depollution of Waters Contaminated by Phenols and Chlorophenols Using Catalytic Hydrogenation

D. Richard · L. D. Núñez · C. de Bellefon · D. Schweich

Abstract

A new process for the detoxification of phenol-containing waste waters based on catalytic hydrogenation suitable for a wide range of compounds is proposed. After an introduction pointing out the advantages of this approach, the principle of the multifunctional process is presented. Then two examples of application of this process for the treatment of polluted waste waters are reported; one containing chlorophenols and the other containing mixture of phenols of olive oil waste water. Using 4-chlorophenol as a model molecule of monoaromatic chlorophenols and tyrosol of phenols encountered in olive oil waste waters, the kinetics of catalytic hydrogenation was studied. In both cases, a scheme of the reaction was elucidated, different kinetics models were established and a dynamic simulation software was used for helping in their discrimination and for parameter estimation. In both cases, total conversion was achieved. The global rate of 4-chlorophenol conversion into cyclohexanol at 0.4 MPa and 353 K was 3.43 g h^{-1} g$^{-1}_{catalyst}$ while the rate of tyrosol conversion at 1 MPa and 353 K was 1.42 g h^{-1} g$^{-1}_{catalyst}$.

Key words: catalytic hydrogenation, phenol degradation, chlorophenol, olive oil waste water, tyrosol

55.1 Introduction

Phenols constitute a widespread important class of water pollutants. They are indeed discharged in the liquid effluents from various factories: chemical, petrochemical, paper, wood, metallurgy and cocking plants (Ramade 2000). They can also be originated from diffuse emissions, e.g. roads and pipes tar coatings and from the use of pesticides including their transformation products e.g. herbicides, fungicides like dinitroorthocresol (DNOC), and pentachlorophenol (PCP). Phenolic compounds are also found in the waste waters of agro-industrial processes like the olive oil mills, tomato processing and wine distilleries (Achilli et al. 1993; Hunt and Baker 1980).

Amongst numerous publications dedicated to the disposal of waste waters containing hazardous organic pollutants, catalytic processes have, so far, not been developed extensively. Furthermore most catalytic abatement research has been devoted to oxidation processes. On the contrary, there exists a wide perspective for the development of hydrogenation processes as they could avoid the drawbacks of oxidative treatments of water dispersed pollutants.

Matatov-Meytal and Sheintuch (1998) review catalytic oxidation methods including wet air oxidation (WAO) and super critical water oxidation (SCWO). They mention their high cost due to high-pressure equipment and high cost of running the reaction at high pressures and temperatures. The oxidation methods along with others including electrochemical and photochemical treatments have also been reviewed (Meunier and Sorokin 1997).

Anaerobic microbial decomposition of organic matter offers significant advantages over other biologic processes in lowering energy consumption and sludge production, but bacterial inhibition produced by polyphenols limits their application. Conventional aerobic treatment is not sufficient if the concentration of organic pollutants is too high (Pintar 1994) and most microbes are unable to degrade complex phenolic compounds (Capasso et al. 1992).

The usefulness of reductive processes in waste waters treatment was first reported by Kalnes and James (1988) who demonstrated for halogenated organic wastes, such as polychlorinated benzenes, that reductive treatment (especially hydrodechlorination) is economically more attractive than direct incineration. Catalytic hydrogenation using hydroprocessing catalysts, e.g. Ni-Mo/γ-Al$_2$O$_3$, was also proposed (Gioia 1991) as a possible alternative to thermal incineration for the disposal of hazardous organic waste liquids.

In contrast to the oxidative processes, catalytic hydrogenation cannot lead to ultimate harmless products like water and carbon dioxide. However, when a suitable catalyst is used, chlorophenols and phenols can be transformed into cyclohexanol. Depending on its concentration, cyclohexanol may then be either degraded by standard biological processes or recovered as valuable product. Two companies already propose processes based on catalytic hydrogenation, though they operate at very high conditions of temperature, up to 727 K, and pressure up to 7 MPa (Johnson 1988; Van Der Osterkamp 1987).

Catalytic hydrogenation of phenolic compounds has been studied in our laboratory in the last years. A specific ruthenium/charcoal catalyst and a three-step catalytic detoxification process were developed (Fouilloux et al. 1996). Ruthenium was chosen as a catalyst since it is a dechlorination catalyst and is also well known for its ring hydrogenation properties (Augustine 1995). This process is suitable for a wide range of phenolic compounds concentration, works under mild conditions: temperatures below 360 K and pressure less than 3 MPa, leads to a fairly concentrated solution of non toxic compound, can be eventually implemented on a polluted site, and can be adapted to different aromatic compounds.

This paper summarize some aspects about catalytic hydrogenation process and its application, performed by our research group, with the aim to show that this process could be an efficient and versatile method to deal with the depollution of different kind of phenolic compound-containing waste waters. Thus, first the principle of a proposed catalytic hydrogenation process is presented. Then, the kinetic studies of catalytic hydrogenation of water containing chlorinated phenolic compounds and olive oil mills waste waters (OMW) are reported.

55.2
Three Step Catalytic Depollution Process

The main drawback of catalytic depollution is the slow reaction rate due to the low pollutant concentration level of waste waters. This is easily overcome if the pollutant

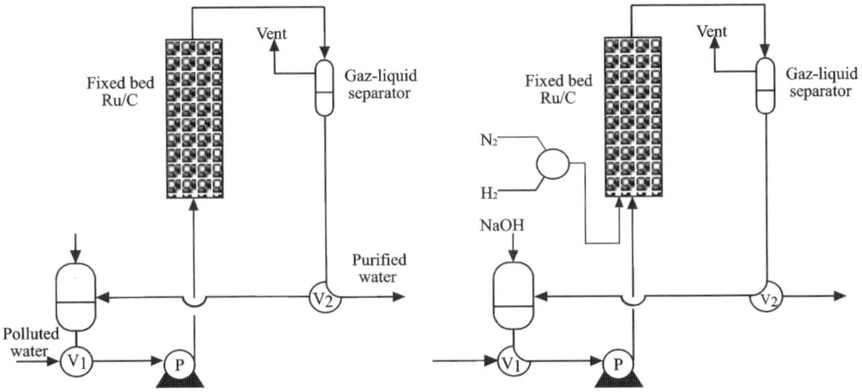

Fig. 55.1. Apparatus used for the decontamination of phenol-polluted waters by catalytic dehydrogenation. Schematic of the pilot-scale process, **a** flow during the adsorption-concentration step, **b** flow during the hydrogenation reaction

has been first concentrated by adsorption prior to hydrogenation. A convenient way is to trap the harmful compounds on the catalyst support itself (Félis et al. 1999b). The depollution process is performed in a single fixed bed using a periodic sequence of three steps: adsorption, catalytic hydrogenation and thermal regeneration. The solid in the fixed bed is 3% ruthenium loading activated carbon. This material is suitable for both liquid-solid adsorption (first step) and gas-liquid-solid catalytic hydrogenation (second step). Figure 55.1 illustrates the laboratory-scale process.

55.2.1
Adsorption-Concentration Step

Waste water is continuously fed (pump P) to a fixed bed of pollutant free activated carbon at ambient temperature and close to atmospheric pressure. As long as the outlet concentration is below the upper limit, a depolluted effluent is obtained. When the upper limit is reached, purification stops and two three-way valves (V1 and V2) allow recycling of the polluted solution through the fixed bed, condenser, gas-liquid separator, and liquid tank.

55.2.2
Hydrogenation Step

The activated carbon used for the adsorption step is also the catalyst support. When the reaction temperature is reached (353 K), hydrogen is fed to the fixed bed which is operated in a co-current upflow mode under mild pressure conditions (less than 3 MPa). In the case of chlorophenol hydrogenation, sodium hydroxide is added to the recycled solution since it promotes chlorophenol desorption as chlorophenolate ions. Sodium hydroxide is also used to neutralize the hydrochloric acid produced during the reaction.

55.2.3
Regeneration Step

Once the reaction is complete, the reaction products solution is discharged and the fixed bed is thermally regenerated to recover the initial state of the adsorbent and catalyst, and a new cycle is initiated.

Chaining the adsorption and reaction steps, this procedure has several decisive advantages which make the process original: first, the adsorption of aromatic compounds on activated carbon is efficient, even at ambient temperature. This ensures low energy costs. Second, prior to the reaction step, partial desorption by a base leads to a concentrated solution of pollutant. The reaction rates are thus much higher as compared to the treatment of the raw polluted effluent. Alternatively, the required amount of catalyst is much lower. Third, the reaction products are concentrated in a small volume of residual liquid. The lower the concentration of pollutant in the feed stream, the smaller the fraction of residual liquid, and the longer the adsorption step under ambient conditions. Conversely, the reaction step is independent of the concentration level of the polluted water. Fourth, using a single vessel and catalytic adsorbent for both steps ensures pollutant confinement. An appropriate fluid distribution to several fixed beds at various stages of the cycle would make the process quasi-continuous.

For the up-scale adsorption step, it is necessary to know the relation between the pollutant concentration in waste water, the flow rate of polluted water, and the critical time in which the critical concentration of pollutant is reached at the outlet of the adsorption fixed bed. Experimental breakthrough curves have been recorded under various conditions and a model has been built, based on the adsorption isotherm of the pollutant on the activated carbon, and of the kinetics of adsorption. This model relies upon the following assumptions: *(i)* one dimensional plug-flow prevails; *(ii)* the particles behave as a pseudo-homogeneous medium where the pollutant diffuses; *(iii)* external mass-transfer limitation is accounted for; *(iv)* adsorption equilibrium prevails at the fluid-solid external surface and gives global breakthrough curves in good agreement with experimental ones, however it tends to underestimate the critical time. The limitations of this model and the ways to overcome them in order to extrapolate the process are discussed in detail in a previous paper (Félis et al. 1999b).

In order to model reaction steps, it is necessary to know the kinetic of hydrogenation of specific penolic compound. In the sections below, 4-chlorophenol and tyrosol were taken as model molecules to the kinetic study of hydrogenation of monoaromatic chlorophenols and phenols of olive mill waste water, respectively.

55.3
Experimental

55.3.1
Chemicals

Chlorophenols, tyrosol, catechol and phenolic acids were purchased from Acros and used as received. A 30 g kg^{-1} Ru/C catalyst was prepared by incipient wetness impregnation of the activated charcoal by a RuCl$_3$ solution followed by drying and reduction by hydrogen at 573 K. Further detail can be found in previously published document (Félis et al. 1999a).

55.3.2
Catalytic Batch Test

The reaction was carried out in a stainless steel semi-batch, isothermal reactor of 0.15 l capacity. The temperature was controlled to ±1 K; baffles and magnetically coupled stirrer provided a good gas-liquid mass-transfer; a finely powdered catalyst (particle diameter ≈1 µm) was employed. The reactor was charged with 0.07 l of distilled water and a known quantity (0.1 to 1.0 g) of catalyst. When the temperature and pressure of the reactor were steady, the reaction was started by injection of the substrate into the reactor. The hydrogen consumption was evaluated by the decrease of pressure of an isothermal supply. The initial hydrogen consumption rate was calculated from the slope of the hydrogen consumption curve at the initial reaction time. Samples of the reaction medium were taken periodically for analysis by gas chromatography.

55.3.3
Analysis

Samples of reaction mixture were analyzed on a HP 5890 Series II gas chromatograph (Hewlett-Packard, Waldbronn, Germany) fitted with a FID detector and temperature programming (313 K to 523 K at 4 K min^{-1}) using a HP1-SE30 capillary column (10 m, 0.53 mm et 2.65 µm) and He (0.72 l h^{-1}) as carrier gas.

55.4
Results and Discussion

55.4.1
Chlorophenols Hydrogenation

Chlorophenols are chiefly used in agriculture as herbicides and pesticides, and for wood protection as bactericides and germicides. These compounds are very resistant to biodegradation and in addition they are subject to bioaccumulation in aquatic organisms.

The detoxification of water-containing monoaromatic chlorophenols bearing from 1 to 5 chlorine atoms in the aromatic ring has been studied. Out of the 19 possible chlorophenols, only 13 were commercially availables and were tested with Ru/C catalyst. Figure 55.2 shows the relative reactivity of the water soluble chlorophenols. 2-chlophenol and 4-chlorophenol are the most reactive; 2,3-dichloro-, 2,4-dichloro-, 2,6-dichloro- and 2,4,5-trichlorophenol require more time, while the others show intermediate reactivity. Studies carried out at lower temperature showed a change in the hierarchy of activities which is likely to be due to the difference between the activation energies of the dehydrochlorination and the hydrodearomatisation steps in the reaction mechanism.

Another set of experiments undertaken on the non-soluble chlorophenols (2,4,6-trichlorophenol, 2,3,4,6-tetrachlorophenol and pentaphenol) under the same conditions (298 K and 0.35 MPa H$_2$) demonstrated that the hydrogenation was also carried out with these compounds. Other pollutants, chloroaniline and dichlorodiphenyl were also tested and proved to give a mixture of cyclohexylamine and cyclohexanol for the first reaction and dicyclohexyl for the second.

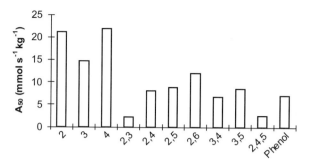

Fig. 55.2. Reactivity of chlorophenols for hydrogenation at 298 K under 0.35 MPa H_2 with 0.5 g of catalyst. 2-chlorophenol (2); 3-chlorophenol (3); 4-chlorophenol (4); 2,3-chlorophenol (2,3); 2,4-chlorophenol (2,4); 2,5-chlorophenol (2,5); 2,6-chlorophenol (2,6); 3,4-chlorophenol (3,4); 3,5-chlorophenol (3,5); 2,4,5-chlorophenol (2,4,5) and phenol

55.4.1.1
4-Chlorophenol Hydrogenation

The hydrogenation mechanism and kinetics were studied in detail for 4-chlorophenol (4CPOL). Analysis of reaction mixtures sampled during the reaction and the identification solely of phenol as intermediate species indicates a consecutive reaction as shown in Fig. 55.3.

The experiments were carried out under three different types conditions:

- Constant hydrogen partial pressure and initial 4-chlorophenol concentration (0.4 MPa and 100 mmol kg^{-1}) at three different temperatures: 313, 333 and 353 K which gives an apparent activation energy of 41 ±4 kJ mol^{-1} from the Arrhenius plot of the initial activity A_i.
- Constant hydrogen partial pressure and temperature (0.4 MPa and 353 K) but initial 4-chlorophenol concentration amounting to 101, 201 and 311 mmol kg^{-1} with initial hydrogen consumption fluxes of 296, 207, 137 mmol s^{-1} kg^{-1}. This suggests for Reaction (1) an order between –1 and 0 with respect to 4-chlorophenol.
- Constant initial 4-chlorophenol concentration and temperature (100 mmol kg^{-1} and 353 K) but hydrogen partial pressure of 0.40, 0.60, 0.89 and 1.08 MPa with initial hydrogen consumption fluxes of 262, 353, 569, 696 mmol s^{-1} kg^{-1}. Since the data (initial activity vs. hydrogen partial pressure) fall on a straight line passing through the origin, this suggests that Reaction (1) is in positive order, close to one, with respect to hydrogen.

It was assumed that Reactions (1) and (2) follow a Langmuir-Hinshelwood mechanism. This mechanism consists of three steps: adsorption of reactants on the catalytic surface, surface reaction between adsorbed species, desorption of products. Depending on the rate limiting step and the adsorption model (see Table 55.1), there are 12 rate equations for Reaction (1) and 24 for Reaction (2). Among the 288 possible couples of equations, the sets of experiments described above enabled us to eliminate the models which lead to a positive order with respect to 4-chlorophenol and cannot account for a positive order with respect to hydrogen. This leaves only 3 rate equations for

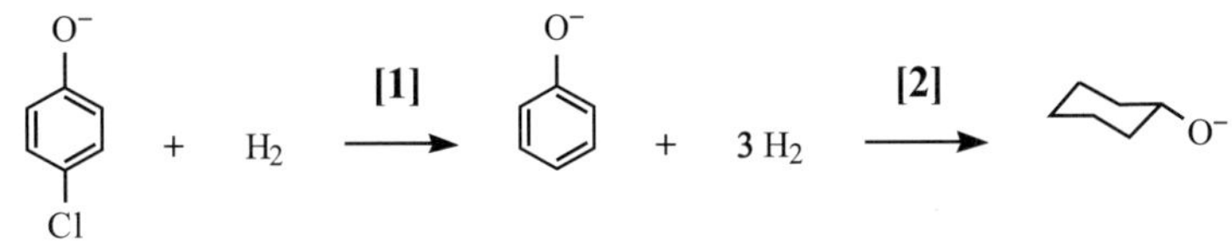

Fig. 55.3. The two successive reactions involved in 4-chlorophenol destruction by catalytic hydrogenation using Ru/C catalyst

Table 55.1. Plausible limiting steps and adsorption hypothesis for the kinetics of Reaction (1)

Limiting step	H$_2$ adsorption	
	Adsorption of 4-chlorophenol	
	Reaction between adsorbed species	
	Desorption of phenol	
H$_2$ adsorption	Dissociative:	H$_2$ + 2* $\longrightarrow$ 2H*
	Non-dissociative:	H$_2$ + * $\longrightarrow$ H$_2$*
H$_2$ and 4CPOL Adsorption	Competitive and non-dissociative:	H$_2$ + * $\longrightarrow$ H$_2$* and 4CPOL + * $\longrightarrow$ 4CPOL*
	Non-competitive and non-dissociative:	H$_2$ + *' $\longrightarrow$ H$_2$*' and 4CPOL + * $\longrightarrow$ 4CPOL*

Reaction (1) and 17 for Reaction (2). The Reactop computer program (Cheminform, St Petersbourg, Russia) described elsewhere (Bergault et al. 1998) was used to discriminate between the rival models. It allows to solve the mass balance equations and to adjust kinetic parameters to their optimum values.

The model which gave the smallest residual sum of squares was finally chosen. It assumes that adsorption is nondissociative and that 4-chlorophenol and hydrogen are competing for the same sites. The limiting step for Reaction (1) is the reaction of the two adsorbed species. For Reaction (2), the slow step is the reaction of an adsorbed hydrogen molecule with the intermediate complex result of the phenol and the H$_2$ molecule reaction. The optimal rate equations and parameters are:

$$r_1 = k_1 \cdot \frac{K_{4CPOL} \cdot [4COL] \cdot K_{H_2} \cdot [H_2]}{(1 + K_{H_2} \cdot [H_2] + K_{4CPOL} \cdot [4CPOL] + K_{PHEN} \cdot [PHEN] + K_{CyOH} \cdot [CyOH])^2}$$

$$r_2 = k_2 \cdot \frac{K_{PHEN} \cdot [PHEN] \cdot (K_{H_2} \cdot [H_2])^2}{(1 + K_{H_2} \cdot [H_2] + K_{4CPOL} \cdot [4CPOL] + K_{PHEN} \cdot [PHEN] + K_{CyOH} \cdot [CyOH])^3}$$

with

$\ln k_{10} = 17.5 \pm 0.5$ $\quad$ $\ln k_{20} = 12.7 \pm 0.3$
$E_1 = 45.0 \pm 2.7$ $\quad$ $E_2 = 33.2 \pm 2.4$
$K_{H_2} = 162 \pm 26$ $\quad$ $K_{PHEN} = 16.4 \pm 2.5$
$K_{4CPOL} = 4.0 \pm 1.8$ $\quad$ $K_{CyOH} = 7.4 \pm 1.9$

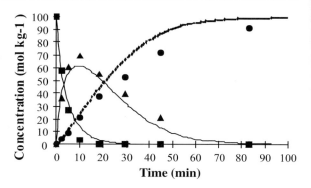

Fig. 55.4.
Agreement between experimental points (● 4-chlorophenol, ▲ phenol and ■ cyclohexanol) and the model estimate (*continued lines*)

where $k_i = k_{i0} \exp(-E_i/R \cdot T)$; k_{i0} being expressed in mol kg^{-1} min^{-1}, E_i in kJ mol^{-1} and K_i in kg mol^{-1}.

The satisfactory agreement between experimental and calculated curves (Fig. 55.4) and the plausible range of activation energies, give support to the model which is very well adapted for the design of a three-phase fixed bed reactor. Total conversion of 4-chlorophenol into cyclohexanol in 0.1 mol kg^{-1} aqueous solution was achieved after 1.5 h at 353 K and 0.4 MPa with a global rate of 3.43 g h^{-1} g$^{-1}_{catalyst}$.

55.4.2
Olive Oil Mill Waste Waters

In this study the catalytic hydrogenation technique was applied to another kind of phenols, those contained in the olive oil mill waste waters (OMW). This waste water also called alpechin in Spain, contains very high amounts of organic mater (up to 15% by weight), is resistant to degradation and constitutes an important environmental problem due to the content of phenolic compounds, which are antimicrobial and phytotoxic. These characteristics make them resistant to biological treatment (Hamdi 1992; Paixão et al. 1999; Ramos Comenzana et al. 1996). The multifunctional process, based in the catalytic hydrogenation of the phenolic compounds of alpechin, is proposed as a pretreatment of these waste waters, that according to Hamdi (1992) is indispensable to facilitate anaerobic digestion of OMW. The small size of the installations and simple operation required by this technology have suitable characteristics for the application in the olive oil industry, that is a seasonal activity and generally, many small factories are wide spread over the production area.

Amongst the phenolic compounds that have been identified in alpechin are: hydroxytyrosol, tyrosol, cathecol, resorcinol, o-vanillin, caffeic acid, elenolic acid, coumeric acid, gallic acid, vanillic acid, ferulic acid, syringic acid, 4-hydroxy-phenylacetic acid, 4-hydroxybenzoic acid, demethyl-oleuropeine, oleuropein, rutin, verbascoside and luteolin-7-glycosid acid (Catalano et al. 1999; Montedoro et al. 1992; Paredes et al. 1999; Perez et al. 1992; Servili et al. 1999). Some of these compounds are represented in Fig. 55.5. The phenolic compounds concentration depends on the kind of oil extraction technology and on the variety of olives used.

Since the olive oil mill waste waters are very complex mixtures, the kinetic study of its catalytic hydrogenation is a complex task that has to be based on model mol-

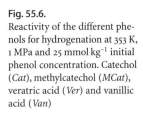

Fig. 55.5. Some of the phenolic compound encountered in olive oil mill waste water

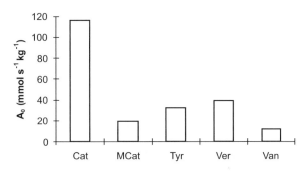

Fig. 55.6.
Reactivity of the different phenols for hydrogenation at 353 K, 1 MPa and 25 mmol kg^{-1} initial phenol concentration. Catechol (*Cat*), methylcatechol (*MCat*), veratric acid (*Ver*) and vanillic acid (*Van*)

ecules. Figure 55.6 shows the reactivity of different phenolic compounds encountered in olive mill waste water. Catechol (Cat), methylcatechol (MCat), veratric acid (Ver) and vanillic acid (Van) were tested in similar conditions (temperature: 353 K, Pressure: 1 MPa and initial phenolic concentration: 25 mmol kg^{-1}). This shows that tyrosol degradation by hydrogenation proceeds more slowly than these of a simple phenol like catechol and that the compounds which are more resistants to degradation are mainly the acidic-phenolic compounds. The hydrogenation of these acidic compounds is more difficult and will be addressed in another study.

55.4.2.1
Tyrosol Hydrogenation

Tyrosol (4-hydroxyphenethyl alcohol) was taken as representative of the phenolic compounds present in olive mill wastewater. Indeed it is one of the main phenol identified in vegetation waters (Capasso et al. 1992) and thus a valuable example of the simple phenolic compounds which according to Hamdi (1992) are responsible of the OMW

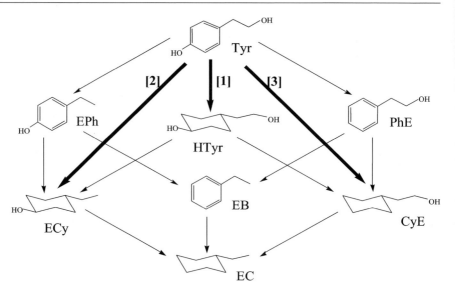

Fig. 55.7. General reaction scheme and actual reaction path for tyrosol catalytic hydrogenation

toxicity for methanogenic bacteria. Moreover, since it may undergo many reactions in the presence of hydrogen, namely aromatic ring hydrogenation, aromatic hydroxyl group hydrogenolysis and aliphatic hydroxyl group hydrogenolysis it is a good model molecule for the kinetic study.

The reaction products of tyrosol hydrogenation were identified by gas chromatography-mass spectroscopy technique. Since it was not commercially available product, the formation of hydrogenotyrosol was confirmed by ^{13}C and ^{1}H NMR (nuclear magnetic resonance). The reaction products identified were: ethylphenol (EPh), ethylcyclohexanol (ECy), phenylethanol (PhE), cyclohexylethanol (CyE), 4-(2-hydroxy-ethyl)-cyclohexanol or hexahydrogenotyrosol (HTyr), ethylbenzene (EB) and ethylcyclohexane (EC).

The general hydrogenation network for tyrosol is showed in Fig. 55.7. The hydrogenation experiences of EPh, PhE and HTyr carried out under the same conditions that the hydrogenation of tyrosol exhibited a reaction rate negligible, consequently the paths EPh ⟶ ECy, PhE ⟶ CyE, HTyr ⟶ CyE and HTyr ⟶ ECy were rejected. This fact in addition with the very small concentration levels (less than 2% of reaction products) of EPh, PhE, EB and EC, allowed to describe tyrosol hydrogenation by a simplified network consisting in the three reaction outlined in Fig. 55.7:

$$Tyr \longrightarrow HTyr \tag{1}$$

$$Tyr \longrightarrow ECy \tag{2}$$

$$Tyr \longrightarrow CyE \tag{3}$$

The kinetic experiments were carried out under the following conditions: Constant hydrogen pressure and temperature (1 MPa and 353 K) and tyrosol initial concentration amounting to 20, 40 and 65 mmol kg^{-1}. Constant initial tyrosol concentration and temperature (40 mmol kg^{-1} and 353 K) with hydrogen pressure of 0.4, 1, 2, 3 and 4 MPa. Constant hydrogen pressure and initial tyrosol concentration (1 MPa, 40 mmol kg^{-1}) at three different temperatures: 313, 333 and 353 K.

Using the same methodology described for 4-chlorophenol, a Langmuir-Hinshelwood kinetic formalism was assumed, the kinetic equations relatives to the different models (rate limiting step, mode of adsorption) were build and the composition of the reaction media computed vs. time for each experiment. The Reactop computer program was also used to discriminate between rival models.

The model which gave the smallest residual sum of squares was finally chosen. It assumes that adsorption of hydrogen is non-dissociative; tyrosol and hydrogen are not competing for the same sites but hydrogen competes with non-aromatic products (HTyr, ECy and CyE). For the three reactions, the limiting step is the surface reaction between adsorbed species. The optimal rate equations and parameters are:

$$r_1 = k_1 \cdot \frac{K_{Tyr} \cdot [Tyr]}{(1+K_{Tyr} \cdot [Tyr])} \cdot \frac{(K_{H_2} \cdot [H_2])^2}{(1+K_{H_2} \cdot [H_2]+K_{HTyr} \cdot [HTyr]+K_{CyE} \cdot [CyE]+K_{ECy} \cdot [ECy])^2}$$

$$r_2 = k_2 \cdot \frac{K_{Tyr} \cdot [Tyr]}{(1+K_{Tyr} \cdot [Tyr])} \cdot \frac{K_{H_2} \cdot [H_2]}{(1+K_{H_2} \cdot [H_2]+K_{HTyr} \cdot [HTyr]+K_{CyE} \cdot [CyE]+K_{ECy} \cdot [ECy])}$$

$$r_3 = k_3 \cdot \frac{K_{Tyr} \cdot [Tyr]}{(1+K_{Tyr} \cdot [Tyr])} \cdot \frac{(K_{H_2} \cdot [H_2])^2}{(1+K_{H_2} \cdot [H_2]+K_{HTyr} \cdot [HTyr]+K_{CyE} \cdot [CyE]+K_{ECy} \cdot [ECy])^2}$$

with

$\ln k_{1_0} = 17.5 \pm 1.6$ $\ln k_{2_0} = 26.9 \pm 1.8$ $\ln k_{3_0} = 22.6 \pm 2.1$
$E_1 = 43.3 \pm 4.3$ $E_2 = 77.9 \pm 5.0$ $E_3 = 67.8 \pm 6.0$
$K_{Tyr} = 246.0 \pm 3.5$ $K_{HTyr} = 50.0 \pm 1.0$ $K_{ECy} = 90.0 \pm 1.5$
$K_{CyE} = 1285 \pm 15$ $K_{H_2} = 52.7 \pm 16.7$

where $k_i = k_{i_0} \exp(-E_i/R \cdot T)$; k_{i_0} being expressed in mol kg^{-1} min^{-1}, E_i in kJ mol^{-1} and K_i in kg mol^{-1}.

Figure 55.8 shows the satisfactory agreement between experimental and calculated curves. Total conversion of tyrosol in 0.042 mol kg^{-1} aqueous solution was achieved after 3.3 h at 353 K and 1 MPa with a global rate of 1.42 g h^{-1} g$^{-1}_{catalyst}$. These results demonstrated that total conversion of tyrosol into nonphenolic compounds by catalytic hydrogenation is possible; the reaction mechanism and rate laws obtained allows for reactor design. This study is being completed by adsorption measurement in order to achieve the extrapolation of the adsorbent catalytic fixed bed. The catalytic hydrogenation of other phenolic compounds: veratric, caffeic and vanillic acids is being investigated and their reaction rates will be compared to the one of tyrosol hydrogenation. Hydrogenation of real OMW samples will have to be undertaken to valid the process.

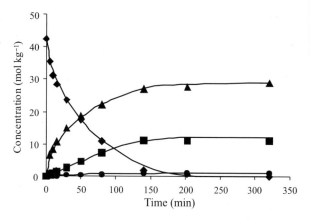

Fig. 55.8. Agreement between experimental points (♦ tyrosol, ▲ hydrogenotyrosol, ■ ethylcyclohexanol and ● cyclohexylethanol) and the model estimate (*continued lines*)

55.5 Conclusion

We have shown that combining adsorption and catalytic hydrogenation leads to an original and efficient detoxification process which work under mild conditions. It involves a "dual-purpose" catalyst which take parts in both adsorption and catalytic processes. Although some other points have to be investigated such as the poisoning of the catalyst and its deactivation, the regeneration of the adsorbent catalyst or the cost of the required infrastructure, the depollution of phenol-containing waste waters by catalytic hydrogenation is a process that should not be discarded and may compare favorably with the oxidative ones. Catalytic hydrogenation could be used in combination with biological depollution treatments to overcome their limitations i.e. when the pollutant concentration is too high or when the treated compounds are toxic for the microorganisms.

Acknowledgements

L. Delgado Núñez acknowledges financial support from CONACYT and UAM during the preparation of her Ph.D.

References

Achilli G, Cellerino GP, Gamache PH, Melzi D'eril GV (1993) Identification and determination of phenolic constituents in natural beverages and plants extracts by means of a coulometric electrode array system. J Chromatogr 632:111–117

Augustine RL (1995) Heterogeneous catalysis for the synthetic chemist. Marcel Dekker, New York pp 534

Bergault I, Fouilloux P, Joly-Vuillemin C, Delmas H (1998) Kinetics and intraparticle diffusion modelling of a complex multistep reaction: hydrogenation of acetophenone over a rhodium catalyst. J Catal 175:328–337

Capasso R, Cristinzio G, Evidente A, Scognamiglio F (1992) Isolation, spectroscopy and selective phytotoxic effects of polyphenols from vegetable waste waters. Phytochemistry 31(2):4125–4128

Catalano L, Franco I, Nobili MD, Leita L (1999) Polyphenols in olive oil mill waste waters and their depuration plant effluents: a comparison of the Folin-Ciocalteau and HPLC methods. Agrochimica 63(5–6):193–205

Félis V, Bellefon C, Fouilloux P, Schweich D (1999a) Hydrodechlorination and hydrodearomatization of monoaromatic chlorophenols into cyclohexanol on Ru/C catalyst applied to water depollution: influence of the basic solvent and kinetics of the reaction. Appl Cat B Environ 20:91–100

Félis V, Fouilloux P, de Bellefon C, Schweich D (1999b) Three-step catalytic detoxication of waste water containing chloroaromatics: experimental results and modelisaton issues. Ind Eng Chem Res 38(11):4213–4219

Fouilloux P, Félis V, de Bellefon C, to TREDI SA (1997) French patent 2 743 801, 25 july 1997, FR Appl 96/550 18 january 1996, CA 127:180 529b

Gioia F (1991) Detoxification of organic waste liquids by catalytic hydrogenation. J Hazard Mat 26:243–260

Hamdi M (1992) Toxicity and biodegradability of olive mill wastewaters in batch aerobic digestion. Appl Biochem Biotech 37:155–163

Hunt GM, Baker EA (1980) Phenolic constituents of tomato cuticle. Phytochemistry 19:1415–1419

Johnson RW (1988) Process for the removal of hydrocarbonaceous compounds from an aqueous stream and hydrogenating these compounds. US Patent 4,758,346

Kalnes TN, James RB (1988) Hydrogenation and recycle of organic waste streams. Environmental Progress 7(3):185–191

Matatov-Meytal Y, Sheintuch M (1998) Catalytic abatement of water pollutants. Ind Eng Chem Res 37(2):309–326

Meunier B, Sorokin A (1997) Oxidation of pollutants catalyzed by metallophtalocyanines. Acc Chem Res 30(11):470–476

Montedoro G, Servili M, Baldioli M, Miniati E (1992) Simple and hydrolyzable phenolic compounds in virgin olive oil. 2. Initial characterization of the hydrolyzable fraction. J Agric Food Chem 40:1577–1580

Paixão SM, Mendonça E, Picado A, Anselmo M (1999) Acute toxicity evaluation of olive oil mill wastewaters: a comparative study of three aquatic organisms. Toxicity of Wastewaters 263–269

Paredes C, Cegarra J, Roig A, Sanchez Monedero MA, Bernal MP (1999) Characterization of olive oil mill wastewater (alpechin) and its sludge for agricultural purposes. Biores Technol 67:111–115

Perez J, De La Rubia T, Moreno J, Martinez J (1992) Phenolic content and antibacterial activity of olive wastewaters. Environ Toxicol Chem 11:489–495

Pintar J (1994) Catalytic oxidation of aqueous p-chlorophenol and p-nitrophenol solutions. Chem Eng Sci 49:4391–4407

Ramade F (2000) Dictionnaire encyclopédique des pollutions. Ediscience international, Paris pp 690

Ramos Comenzana A, Juarez Jimenez B, Garcia Pareja MP (1996) Antimicrobial activity of olive mill wastewaters (alpechin) and biotransformed olive oil mill wastewater. Int Biodeter Biodegr 283–290

Servili M, Baldioli M, Selvaggini R, Miniati E, Macchioni A, Montedoro G (1999) High-performance liquid chromatography evaluation of phenols in olive fruit, virgin olive oil, vegetation waters, and pomace and 1D- and 2D-nuclear magnetic resonance characterization. J Am Oil Chem Soc 76(7):873–882

Van den Oosterkamp PF, Blomen LJMJ, Ten Doesschate HJ, Laghate AS, Schaaf R (1987) Dechlorination of PCB's dioxines and difuranes in organic liquid. In: Kolaczkowski ST, Crittenden BD (eds) Management of hazardous and toxic wastes in the process industries. Elsevier, pp 542–545

Part VI

Chapter 56

Treatment of Wastewater Containing Dimethyl Sulfoxide (DMSO)

P. Baldoni-Andrey · A. Commarieu · S. Plisson-Saune

Abstract

The increasing use of aprotic solvents in the industry leads to the set up of new wastewater treatment technologies adapted to the specificity of these compounds. The case of dimethyl sulfoxide (DMSO) is particularly interesting due to the specific properties of this component. Depending on concentration and the presence of other compounds in the effluent, several technologies can be chosen: concentration with reverse osmosis, biodegradation with specific process to avoid dimethyl sulfide (DMS) odours, oxidation in dimethyl sulfone ($DMSO_2$) by various oxidants. The choice has to be made according to economical and technical considerations. In this paper, different options are detailed with advantages and drawbacks.

Key words: DMSO, wastewater treatment, biodegradation, reverse osmosis, oxidation

56.1 Introduction

Dimethyl sulfoxide (Fig. 56.1) is a polar aprotic solvent which is very stable and has a low toxicity. This chemical has been produced by ATOFINA in Lacq since 1963. DMSO is mainly used as a reaction medium for organic synthesis, as a stripper for photoresists and as a mild oxidizing agent.

Figure 56.2 summarizes the reactivity of DMSO: Using chemical oxidation, such as AOPs (Advanced Oxidation Processes), dimethyl sulfoxide will be oxidised in dimethyl sulfone (Davies 1961; Reddy et al. 1992) before further oxidation or even mineralisation (formation of carbon dioxide).

The biological oxidation of DMSO leads to the formation of biomass, sulfuric acid and carbon dioxide. But, under reductive conditions, DMSO can be reduced in dimethyl sulfide (DMS) which induces odour nuisances (Sklorz and Binert 1994; Zinder and Brock 1978). That is particularly what happens if a classical biological treatment reaches anaerobic conditions. Therefore, DMS formation is the problem we want to avoid in biological wastewater treatment.

DMSO has a low toxicity with an IC_{50} on *Daphnia* of 16.3 g l^{-1}, when compared to toxic compounds which show an IC_{50} at the mg l^{-1} or even below (IC_{50}: inhibitory concentration 50% – the concentration required for 50% inhibition). DMSO is known to be readily biodegradable. Even if dimethyl sulfoxide induces no major impact on the natural environment, industrial wastewaters containing DMSO must be treated before its release in the natural medium. The objectives of the treatment are to remove the DMSO (like the other organic compounds) in the effluent or to recycle this compound because of its value.

Fig. 56.1.
Structure of dimethyl sulfoxide (DMSO)

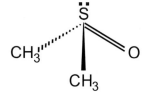

Fig. 56.2.
Main reactions of transformation of DMSO

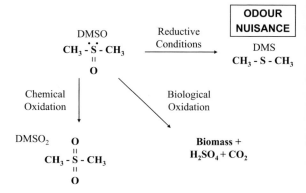

For solutions containing more than 15% of DMSO, distillation is considered as the most economical process to recycle this compound. For solutions containing less than 15% of DMSO we have studied biological treatment, reverse osmosis and chemical oxidation because these technologies are commonly used for industrial wastewater treatment.

56.2
Experimental

56.2.1
Biological Treatment

A pilot-scale activated sludge plant (2-m^3 activated sludge tank and 1-m^3 settlement tank) was employed to establish the operating conditions leading to satisfactory performance in terms of DMSO removal. Parameters such as pH, ORP (oxidation reduction potential) and temperature are measured continuously and pH values are maintained between 6.5 and 8.0. This pilot-scale reactor can be operated in nitrification/denitrification mode (sequenced reactor) and is equipped with a sludge recirculation device.

56.2.2
Reverse Osmosis

Reverse osmosis experiments have been carried out with a pilot-scale plant which is able to treat up to 250 l h^{-1}, equipped with a 4 inch spiral-wound reverse osmosis element (DESAL3 from SDI). For those trials, the influent is a 2.2% solution of DMSO in pure water.

56.2.3
Oxidation Processes

Several oxidation processes have been experimented on DMSO solutions in pure water. Ozonization (O_3) was run in a specific column on DMSO and DMS solutions. Solutions of DMSO and DMS were treated with sodium hypochlorite (bleach), using a molar ratio up to 1 (NaOCl/DMSO). Experiments were carried out using hydrogen peroxide (H_2O_2) activated by UV radiation (UV: ultraviolet). The photoreactor (volume 4.5 l) was equipped with a high pressure mercury vapour lamp (150 W) and a recirculation pump. The influent water solution contains 2 g l^{-1} DMSO and 1.2 g l^{-1} NaCl. For those experiments the molar ratio (H_2O_2/DMSO) ranged between 0 and 4.

56.3
Results and Discussion

56.3.1
Biological Treatment

DMSO represents 20% of the influent COD concentration (COD: chemical oxygen demand). The flow which is treated is equal to 40 l d^{-1} with a concentration of 1.5 g COD l^{-1}. The loading rate is maintained around 0.08 kg COD / kg TVSS / day (Total Volatil Suspended Solid).

In those operating conditions, DMSO is totally oxidised by the process (100% removal). However, these results bring two comments. First of all, we observed an acidification of the activated sludge tank; consequently, a pH control is necessary. Secondly, each time we had trouble with oxygen supply, we detected dimethyl sulfide (DMS) which induces odour nuisances. That means that DMSO was used as an electron acceptor (Zinder and Brock 1978).

The aim of the following experiments was to use nitrate as an electron acceptor in order to replace the DMSO in anoxic conditions. The DMS odour nuisance thresh-

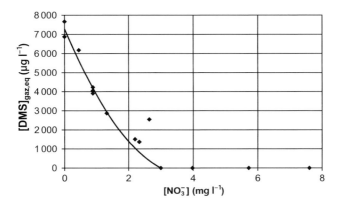

Fig. 56.3. DMS concentration (g) as a function of NO_3^- concentration (l)

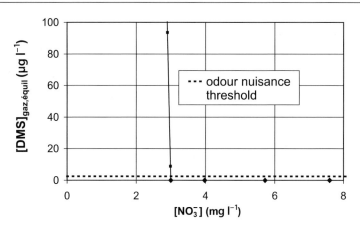

Fig. 56.4. DMS concentration (g) as a function of NO_3^- concentration (l)

old is very low (2.5 µg l^{-1}). Using the equation of HENRY, we determined the vapour/liquid equilibrium of DMS in water at 20 °C and 1 atm. Then, we expressed the DMS concentration in the effluent as gas phase concentrations. Figure 56.3, representing DMS concentration as a function of the nitrate concentration, shows a large decrease of DMS formation as the nitrate concentration increases.

Figure 56.4 is an enlargement of Fig. 56.3 to focus on the odour nuisance threshold (2.5 µg l^{-1}). Experimental plots demonstrate that for nitrate concentration higher than 3 mg l^{-1} the DMS concentration is under the odour nuisance threshold. It has to be noted that 3 mg l^{-1} of nitrate represents a low concentration of nitrogen, only 750 µg l^{-1}.

As a conclusion, a technical solution in the presence of DMSO, to avoid the formation of DMS in case of oxygenation troubles is to maintain a concentration of nitrate around 5 mg l^{-1}. This concentration is completely compatible with the European Environmental regulations.

Cost estimation: For such an effluent and a flow equal to 2 m^3/d, the estimated operational cost is equal to 2 € per day and the capital cost is around 3 000 €.

56.3.2
Reverse Osmosis

The objective of this treatment is the containment of DMSO for recycling or concentration before distillation. The reverse osmosis (Fig. 56.5) treatment of a DMSO solution with an entering flow of 6 m^3 d^{-1} enables a concentration of the DMSO from 2.2% to 13.3%. No higher retentate concentrations were observed in our conditions. From this retentate concentration, it will be necessary to use distillation to recycle DMSO. The rejection rate is equal to 80% with a Volumetric Concentration Factor of 7.3. The permeate flow still contains 0.4% of DMSO.

Cost estimation: The operated cost is around 30 € per day, due to the cartridge replacement once a year and especially due to the energy consumption (90%). The capital cost is around 30 000 €.

Fig. 56.5.
Reverse osmosis experiments

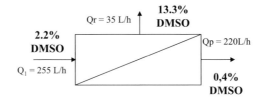

Table 56.1. Economical aspects of different technical processes for DMSO removal from waste water

Process	[DMSO] (%)	Flow (m³ d⁻¹)	Operational exp. (€ d⁻¹)	Capital exp. (€)
Biological	0.02	2	2	3 000
Reverse osmosis	2.2	6	30	30 000
H$_2$O$_2$/UV	0.1	2	70	30 000

56.3.3 Oxidation Processes

The objective of those treatments is the oxidation of dimethyl sulfoxide (DMSO) into dimethyl sulfone (DMSO$_2$), and maybe further into carbon dioxide and sulfate (mineralisation). Once DMSO$_2$ is generated, the formation of DMS becomes very difficult. Ozonization of DMSO solutions does not lead to the formation of dimethyl sulfone even if DMS is totally oxidised in DMSO by this process. The use of bleach leads to the formation of biocides (organonochlorinated species). The direct UV photolysis of hydrogen peroxide provides good results. With a DMSO solution of 0.1% and a flow of 2 m^3 d^{-1}, a conversion rate of 100% was obtained.

Cost estimation: the operational cost of this process is equal to 70 € per day including power, UV lamps and reagent (hydrogen peroxide). A rough estimation of the price of such a photoreactor is around 30 000 €.

56.4 Conclusion

Table 56.1 presents the treatment costs for each process, with the treated flow and the concentration of DMSO. Biological treatment is the most efficient and cheapest process for the treatment of solutions containing DMSO (small concentrations). However, the formation of DMS is possible in reducing conditions due to oxygenation troubles. Different technical solutions can be implemented to avoid the formation of DMS:

- to prevent anaerobic conditions
- to maintain concentration of nitrate above 3 mg l^{-1}

For solutions containing high concentrations of DMSO (1%–5%), reverse osmosis is efficient for DMSO containment. Then, the retentate may be distillated in order to recycle this chemical.

References

Davies AG (1961) Organic peroxides. Butterworth & Co, London
Reddy SR, Reddy JS, Kumar R, Kumar P (1992) Sulfoxidation of thioethers using titanium silicate molecular sieve catalysts. J Chem Soc Chem Commun 84-85
Sklorz M, Binert J (1994) Determination of microbial activity in activated sewage sludge by dimethyl sulfoxide reduction. Environ Sci & Pollut Res 1(3):140-145
Zinder SH, Brock TD (1978) Dimethyl sulfoxide as an electron acceptor for anaerobic growth. Arch Microbiol 116:35-40

Chapter 57

Productive Use of Agricultural Residues: Cements Obtained from Rice Hull Ash

L. B. de Paiva · F. A. Rodrigues

Abstract

This work discusses the use of rice hull ash as raw material to prepare cements, similar to Portland cement. Rice hull is an agricultural by-product containing about 20% of silica. Usually, this material is burned at the rice fields generating small silica particles, which may cause respiratory and environmental damage. We describe here the use of rice hull ash as a raw material to prepare β-Ca_2SiO_4, which is a component of commercial Portland cement. Heating rice hull at 600 °C gave silica with a surface area of 21 $m^2 g^{-1}$. Rice hull ash was mixed with CaO and $BaCl_2 \cdot 2H_2O$ in several proportions, added stoichiometricaly in order to keep a ratio (Ca + Ba) / Si = 2 or 1.95. The solids were mixed with water 1/20 (w/w) and sonicated for 60 min. The suspensions were dried, grounded and heated. Cements with structure similar to that of β-Ca_2SiO_4 were obtained at temperatures as low as 700 °C.

Key words: biomass, remediation, rice hull, cement

57.1 Introduction

There is a continuous interest in the use of biomass in productive processes with special emphasis on energy generation. Although there is still many different and sometimes controversial opinions about the issue, it seems clear that at some point in the future, the depletion of petroleum and natural gas will demand the use of alternative energy sources. As alternative energy we refer only to those renewable sources with potential for large-scale use, although at this moment, not fully exploited.

On the other hand, it is unquestionable that energy requirements will be higher in the forthcoming years due to continuous growing of world population and the need for achieving better living standards, especially in underdeveloped countries. Indeed it is well known that energy per capita consume is directly related to the quality of living standards. Rich and more industrialized countries consume much more energy than their counterparts, in the so-called "third world". In fact, although the United States account for only 5% of world population, they are responsible for about 25% of total energy consumption. Also, it is worthwhile to mention that in industrialized regions, the use of fossil fuels is much higher than all the other sources. The difference becomes much clearer if we compare the sources of energy among different regions, as presented in Table 57.1 (Klass 1995). Of course, in Africa the term biomass is predominantly related to the burning of wood while in North America and Europe much of fossil fuel is used in automotive vehicles.

Table 57.1. Contribution of main sources of energy in different areas around the world (1990)

Area	Fossil fuels (%)	Nonfossil electricity (%)	Biomass (%)
Africa	61.8	1.4	36.8
North America	90.7	5.2	4.1
South America	65.0	11.3	23.7
Asia	86.2	3.1	10.7
Europe	93.7	5.2	1.1

There are many arguments in favor of reducing the use of fossil fuels. From the environmental viewpoint, concerns about greenhouse effect and air quality are very often cited. Once again emissions of carbon dioxide are quite different among regions. While North America is responsible for about 30% of total CO_2 emissions, Africa contributes with only 3.5% (Klass 1995).

A recent paper (Cooney 1999) discusses the issues of clean and renewable energy taking in account economic considerations. It is clear that energy from biomass, solar, wind and geothermal sources still needs technological improvements in order to become economically viable.

This work presents specifically the utilization of rice hull, although it may be easily extended to some other biomass and/or agricultural residues. Rice hull is an abundant material, produced in many countries around the world and it contains approximately 20–30% of silica in its composition. In most underdeveloped countries, rice hull is usually discarded and burned at the fields. This common practice can lead to serious environmental damage, since silica particles remain suspended in the air, thus being a potential cause for respiratory diseases and soil impoverishment. In Brazil, for example, about 12×10^6 t of rice hulls are generated yearly. This amount is much higher in many southwest Asian countries. It seems a reasonable assumption that this agricultural waste presents a very high potential to be used in large scale. Probably the most successful use of rice hull is in thermoelectric plants. In fact there are some experiences already in progress. The main limitations are still the high costs and the presence of solid residues, the rice hull ash. Our work involves the recycling of silica derived from the burning of rice hull. The ideal scenario is the complete integration of rice hull ash into productive processes such as agricultural practices, e.g. production of rice; energetic process such as the use of rice hull to generate electric power and environmental process such as the use of rice hull ash for the production of Portland cement. Other possible applications of rice hull ash are described in literature (Proctor and Palaniappan 1990; Singh et al. 1993). Following, a short review of cement Chemistry is presented in order to clarify the goals of this work.

57.1.1
The Chemistry of Portland Cements

Portland cement is one of the most consumed materials; world per capita consume is estimated to be about 200 kg person^{-1} yr^{-1} or the equivalent to 1 t of concrete person^{-1} yr^{-1}

(Mehta 1985). The traditional method used for the production of cement is based on solid-state reactions, carried out at temperatures around 1 450 °C (Young and Mindess 1981). Portland cement is a complex material, composed basically by calcium silicates, calcium aluminates and calcium aluminoferrites, among others. The calcium silicates, Ca_3SiO_5 and β-Ca_2SiO_4, determine most of the adhesive properties of concrete as well as its strength and durability (Taylor 1990). These silicates account for nearly 75% of ordinary cement. Both silicates show about the same characteristics after complete hydration, such as physical and mechanical properties, although Ca_3SiO_5 hydrates much faster (Lea 1970).

Cements composed exclusively by β-Ca_2SiO_4 present great economic and environmental interest because they can be prepared at lower temperatures and consume less CaO. In conventional cement manufacture, $CaCO_3$ is the main source of CaO and during the burning of raw materials great amounts of CO_2 are released (Chatterjee 1996a,b). It is worth to mention that cement industry generates about 6% of total CO_2 emissions. Similarly, a decrease in the temperature needed to obtain the cement is very important since it will reduce the energy and fuels consumption. Many attempts have been made in order to synthesize β-Ca_2SiO_4 cement (Rodrigues and Monteiro 1999; Ishida et al. 1992; Jiang and Roy 1992). Those works have used hydrothermal treatment as synthetic method, since silica shows higher solubility under these conditions (Iler 1979).

Finally, Ca_2SiO_4 has five crystalline phases, and in general, the β-phase is the predominant one, although other phases are usually present. Also, between them, the β-phase shows the better hydraulic activity. Many studies have shown that addition of Ba in the crystalline structure acts as phase stabilizer for the β-Ca_2SiO_4. It was established that the partial replacement of Ca by Ba atoms was necessary in order to stabilize the formation of β-Ca_2SiO_4.

57.2
Experimental

Rice hull ash was obtained by heating the rice hull at 600 °C, in an open furnace to avoid the formation of CSi. The resulting material is a white-gray powder with surface area of 21 $m^2 g^{-1}$ (BET method) and identified as cristobalite.

CaO (Mallinckrodt, analytical grade reagent) was heated to 1 000 °C prior the use. In some preparations, CaO was obtained by thermal decomposition of $CaCO_3$, at 900 °C. $BaCl_2 \cdot 2H_2O$ (Synth), analytical grade reagent was used without further purification. Two types of cements are presented here: the first cement type is similar to the conventional cement and has a (Ca + Ba) / Si ratio of 2.0. The second cement type has a (Ca + Ba) / Si ratio of 1.95. Table 57.2 and 57.3 list the nominal chemical composition for each cement type. It is appropriate to observe that in the final composition, Ba concentration varies from 0 to 10% (in relation to total number of alkaline metals).

After the mixture, water was added to the solids to make a suspension that was sonicated for 60 min in an ultrasonic cleaner (Thornton GA, 25 kHz), at room temperature. The ratio water/solids in the suspension was kept at approximately 20:1. After

Table 57.2.
Nominal chemical composition of cements prepared with ratio Ca/Si or (Ca + Ba)/Si = 2.0

Nominal chemical composition	
$Ca_{2.0}SiO_4$	$(Ca_{1.88} + Ba_{0.12})SiO_4$
$(Ca_{1.96} + Ba_{0.04})SiO_4$	$(Ca_{1.84} + Ba_{0.16})SiO_4$
$(Ca_{1.92} + Ba_{0.08})SiO_4$	$(Ca_{1.80} + Ba_{0.20})SiO_4$

Table 57.3.
Nominal chemical composition of cements prepared with ratio Ca/Si or (Ca + Ba)/Si = 1.95

Nominal chemical composition	
$Ca_{1.95}SiO_4$	$(Ca_{1.83} + Ba_{0.12})SiO_4$
$(Ca_{1.91} + Ba_{0.04})SiO_4$	$(Ca_{1.79} + Ba_{0.16})SiO_4$
$(Ca_{1.87} + Ba_{0.08})SiO_4$	$(Ca_{1.75} + Ba_{0.20})SiO_4$

this treatment, an intermediate silicate was obtained, with a Ca/Si ratio of about 1.6, along with the corresponding hydroxides. The suspension was dried and grounded. The solid mixtures were heated from 400 to 1100 °C. The process was followed by Fourier transform infrared spectroscopy (FTIR, Perkin-Elmer 16 PC). Spectra were recorded using 32 accumulations. Samples were pressed in KBr discs and the spectra were obtained in the range of 400–4000 cm^{-1}. The crystalline structure of each silicate was studied by X-ray diffraction (Shimadzu). The analysis were performed using the powder method (CuKα-radiation, 40 kV, 40 mA), in the 2θ range from 5 to 50°.

57.3
Results and Discussion

57.3.1
Thermal Decomposition of Rice Hull

The first step in the synthesis is the separation between silica and the organic material present in the rice hull. It can be done in several ways. However since we are interested in the utilization of silica derived from thermoelectric plants, we have prepared this material by thermal degradation of rice hull. Samples were heated up to 800 °C then cooled at room temperature. The process was followed by FTIR spectroscopy (Fourier transform infrared). Results are shown in Fig. 57.1.

For convenience, only two spectra are presented here. After heating rice hull at 200 °C we still observe the presence of organic material in the region around 1640 cm^{-1} that may be assigned to C-O-C bending absorption band, present in crude fibers and cellulose. It may be noted that even after heating at 600 °C, an organic fraction still remains in the sample. However, further experimental development shows that it does not affect the overall synthesis, since this organic fraction is very small.

A relevant point to be added is that silica obtained from rice hull is a mixture between amorphous and cristobalite phase, although the most stable phase at this temperature range is quartz. Other authors corroborated these findings (Proctor 1990). Silica has three major peaks as can be observed in the spectrum after heating rice hull at 600 °C: at 1100 and 800 cm^{-1} relative to stretching modes of Si-O and the peak located at 480 cm^{-1} associated to Si-O bending mode.

Fig. 57.1.
FTIR spectra of rice hull after heating at 200 and 600 °C

Fig. 57.2.
FTIR spectra for the intermediate silicates with increasing amounts of barium replacement, for the cement type with ratio (Ca + Ba)/Si = 2.0

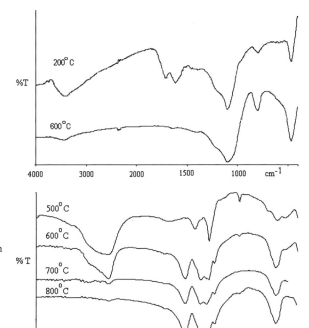

57.3.2
Preparation of the Intermediate Silicates

After mixing the solids in the proportions mentioned in Table 57.1 and 57.2, water was added and the suspensions were sonicated for one hour. After this procedure, in all cases, an intermediate silicate, with a (Ca + Ba)/Si ratio of about 1.6 was obtained, whose heating may lead to the cements. Some FTIR spectra are presented for the cement type with (Ca + Ba)/Si ratio of 2.0 in Fig. 57.2. Very similar results were obtained for cement type with (Ca + Ba)/Si ratio of 1.95. The % Ba displayed is equivalent to the chemical compositions presented in Table 57.1.

It is well known that silica solubility is strongly dependent on the pH. The mechanism of dissolution is very complex and many factors influence it, such as ionic strength and the nature of salts present, among others (Iler 1979). In this particular case, from the chemical viewpoint, no relevant difference is observed among the intermediate silicate, in respect to the barium chloride concentration. On the other hand, literature indicates that Ba^{+2} ions and silica (quartz) interact in a specific way (Rodrigues et al. 2001).

Apparently in the case presented here, this preferential pH-controlled adsorption does not interfere with the overall process. In fact, microscopic evaluation of intermediate silicate shows that particles are about one hundred times smaller than the starting silica particles. Probably, the utilization of ultrasonic cleaner improves the dissolution-precipitation process. Finally the strong band located at $1\,450\ cm^{-1}$ is due to Ca-O stretching mode caused by the presence of calcium oxide.

57.3.3
The Temperature Effect on the Synthesis of Cement

The intermediate silicates were heated from 500 to 1100 °C. Once again, FTIR spectroscopy has been used to monitor the progress of the synthesis as temperature is raised. The heating effect on the sample $(Ca_{1.92} + Ba_{0.08})SiO_4$ is presented in Fig. 57.3. The observed results are very similar for most of the remaining samples.

Two relevant points are observed in the spectra as temperature is raised: the disappearance of the band located at 1450 cm^{-1} and the better definition of three peaks at 520, 900 and 1000 cm^{-1}. As pointed out in the preceding section, calcium hydroxide is present in the mixture of solids, along with the intermediate silicate. As temperature is raised CaO enters into the structure of the calcium silicate. That is visualized by the disappearance of the peak at 1450 cm^{-1} (Ca-O stretching) after heating at 700 °C.

Simultaneously, we observe the formation of the peaks attributed to calcium silicates: at 900 and 1000 cm^{-1} (Si-O asymmetric stretching) and 520 cm^{-1} (Si-O out of plane bending vibration) which are in reasonable agreement with literature (Mollah 1998; Mollah 2000; Eitel 1964). Considering the intermediate silicate has an initial ratio Ca/Si ≈ 1.6, it may be assumed that CaO enters into the structure of the silicate, as a result of heating, rendering, in many cases, the desired calcium silicate with a (Ca + Ba)/Si ratio of 2 or 1.95. The synthesis is complete at 700 °C. Higher temperatures cause no relevant effect.

57.3.4
The Crystalline Structure

In order to confirm the formation of Ca_2SiO_4 it is important to know the crystalline structure of the synthesized silicate and the samples were analyzed by X-ray diffraction. Figure 57.4 shows selected results from silicates having a (Ca + Ba)/Si ratio of 1.95.

Ca_2SiO_4 has 5 different possible crystalline structures, some of which are undesirable in cement industry (Nettleship et al. 1993; Fukuda et al. 1994). The X-ray diffraction patterns obtained for the samples containing 2 and 4% of Ba correspond to that of β-Ca_2SiO_4 (according to JCPDS file 9-351, I-33-B3). The nominal chemical composition of these materials are respectively β-$(Ca_{1.91} + Ba_{0.04})SiO_4$ and β-$(Ca_{1.87} + Ba_{0.08})SiO_4$. Cements having a (Ca + Ba)/Si ratio of 2.0 were also synthesized in the same way as described for samples presented here.

On the other hand, samples without the addition of barium chloride give a mixture of products, probably β-Ca_2SiO_4 and a different calcium silicate. These results show the importance of adding phase stabilizers during the synthesis (Fukuda et al. 1992; Xin et al. 2000).

57.4
Conclusions

This work describes the synthesis of two kinds of cements obtained from rice hull ash. Their chemical properties and structure are very close to β-Ca_2SiO_4 that is

Fig. 57.3.
Effect of heating on the intermediate silicate. Spectra obtained for the sample $(Ca_{1.92} + Ba_{0.08})SiO_4$

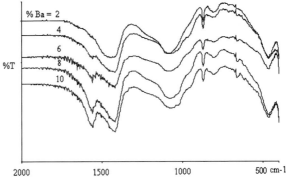

Fig. 57.4.
X-ray diffraction for samples with (Ca + Ba)/Si ratio of 1.95 after heating at 700 °C

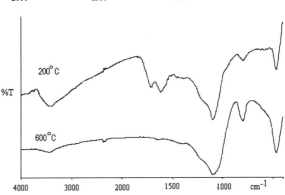

present in commercial portland cement. It was verified that small amounts of barium were introduced during the synthesis in order to obtain the desired crystalline structure. The method presented here allows the synthesis of β-Ca_2SiO_4 at temperatures as low as 700 °C, while in commercial production temperatures around 1 450 °C are needed.

The utilization of rice hull to produce cement could avoid the burning and wasting of this renewable material. It could also avoid the release of silica particles into the atmosphere whose consequences are the soil impoverishment, health problems and environmental damage. Finally the integration of productive processes is proposed: rice production, energy generation and cement manufacture. This is a complete cycle with no waste and directed towards the concept of sustainable development.

Acknowledgements

The authors wish to thank the Fundação de Amparo à Pesquisa do Estado de São Paulo (FAPESP) by financial support (grant: 98/09644-6). We are also in debt with Dr. Inés Joekes by helpful discussions and to Dr. F. M. Cassiola by transmission electronic microscopy studies. This work was also supported by Centro Interdisciplinar de Investigação Bioquímica (CIIB) from Universidade de Mogi das Cruzes (UMC).

References

Chatterjee AK (1996a) High belite cements – present status and future technological options: part I. Cem Concr Res 26:1213–1225

Chatterjee AK (1996b) Future technological options: part II. Cem Concr Res 26:1227–1237

Cooney CM (1999) Can renewable energy survive deregulation? Environ Sci Technol 33:494–497

Eitel WE (1964) Silicate science, vol. I: silicate structures. Academic Press, New York

Fukuda K, Maki I, Adachi K (1994) Structure change of Ca_2SiO_4 solid solutions with Ba concentration. J Am Ceram Soc 75:884–888

Fukuda K, Maki I, Ito S, Yoshida H, Aoki K (1992) Structure and microtexture changes in phosphorous-bearing Ca_2SiO_4 solid solutions. J Am Ceram Soc 77:2615–2619

Iler RK (1979) The chemistry of silica and silicates. John Willey & Sons, New York

Ishida H, Mabuchi K, Sasaki K, Mitsuda, T (1992) Low-temperature synthesis of β-Ca_2SiO_4 from hillebrandite. J Am Ceram Soc 75:2427–2432

Jiang W, Roy DM (1992) Hydrothermal processing of new fly ash cement. Cer Bull 71:642–647

Klass DL (1995) Biomass for renewable energy, fuels, and chemicals. Academic Press, San Diego

Lea FM (1970) The chemistry of cement and concrete. St. Martin's Press Inc, New York

Mehta KM (1986) Concrete: structure, properties and materials. Prentice Hall, New Jersey

Mollah MYA, Lu F, Cocke DL (1998) An X-ray diffraction (XRD) and fourier transform infrared spectroscopy (FT-IR) characterization of the speciation of arsenic (V) in portland cement type-V. Sci Total Environ 224:57–68

Mollah MYA, Yu W, Schennach R, Cocke DL (2000) A fourier transform infrared spectroscopic investigation of the early hydration of Portland cement and the influence of sodium lignosulfonate. Cem Concr Res 30:267–273

Nettleship I, Slavick KG, Kim YJ, Kriven WM (1993) Phase transformation in dicalcium silicate: III: effects of barium on the stability of fine-grained α_l' and phases. J Am Ceram Soc 76:2628–2634

Proctor A (1990) X-ray diffraction and scanning electron microscope studies of processed rice hull silica. J Am Oil Chem Soc 67:576–584

Proctor A, Palaniappan S (1990) Adsorption of soy oil free fatty acids by rice hull ash. J Am Oil Chem Soc 67:15–17

Rodrigues FA, Monteiro, PJM (1999) Hydrothermal synthesis of ccements from rice hull ash. J Mat Sci Letters 18:1551–1552

Rodrigues FA, Monteiro PJM and Sposito G (2001) The alkali-silica reaction: the effect of monovalent and bivalent cations on the surface charge of opal. Cem Concr Res 31:1549–1552

Singh SK, Stachowicz L, Girshick SL, Pfender E (1993) Thermal plasma synthesis of SiC from rice hull (husk). J Mat Sci Letters 12:659–660

Taylor HFW (1990) Cement chemistry. Academic Press, London

Xin C, Jun C, Lingchao L, Futian L, Bing T (2000) Study of Ba-bearing calcium sulphoaluminate minerals and cement. Cem Concr Res 30:77–81

Young JF, Mindess S (1981) Concrete. Prentice-Hall, New Jersey

Part VII
Ecotoxicology

Part VII

Chapter 58

Environmental Metal Cation Stress and Oxidative Burst in Plants. A Review

N. Kawano · T. Kawano · F. Lapeyrie

Abstract

Many reports suggest the involvement of reactive oxygen species (ROS) in salt stress in plants. To date, it has not been well documented how rapidly plant cells respond to the salt stress by producing reactive oxygen species. Under salt stress, plants are exposed to both the hyperosmotic shock and cation shock. Recently we have shown that treatment with metal cations such as Li^+, Na^+, K^+, Ca^{2+}, Mg^{2+}, La^{3+}, Gd^{3+} and Al^{3+} induces burst of $^{\cdot}O_2^-$ production in tobacco cells; whereas hyperosmotic treatments do not induce $^{\cdot}O_2^-$ production. Therefore, the salt-induced damages to plant cells mediated by reactive oxygen species may be due to cation shock. This chapter reviews the recent achievements in our understanding of plant response to salt stress. Lastly, we discuss the use of transition metal ions in mitigation of salt-induced reactive oxygen species.

Key words: aluminum, lanthanide, manganese, metal cation, *Nicotiana tabacum*, reactive oxygen species, salt stress, tobacco, zinc

58.1 Introduction

Recently, we have studied extracellular and intracellular oxidative burst associated with the production of reactive oxygen species (ROS) such as superoxide anion ($^{\cdot}O_2^-$), hydrogen peroxide (H_2O_2) and hydroxy radicals ($HO^{\cdot}$) in plants. Such oxidative bursts are known to be induced by various biotic chemical stimuli such as indole-3-acetic acid (Kawano et al. 2001a), salicylates (Kawano et al. 1998; Kawano and Muto 2000), aromatic monoamines (Kawano et al. 2000a,b) and fungal chito- and N-acetylchitooligosaccharides (Kawano et al. 2000c). We have proposed that the generation of reactive oxygen species is a rapid biochemical signal transduction event in plants. Currently, numerous studies on oxidative burst in plants has been focussed on plant-microbe interactions (Yoshioka et al. 2001). A dual function of reactive oxygen species, in which they behave as a signaling molecule at low level, and as a factor damaging the cells at high level, was first described in pathogenesis (Doke et al. 1996; Ryals et al. 1995; Chen et al. 1993).

In addition to plant responses to biotic stimuli, responses of plant cells to various abiotic environmental stresses including physical stresses such as exposure to strong light, and inorganic chemical stresses such as salt stress, also lead to oxidative bursts in plants. Although adaptation to such abiotic environmental stresses is crucial for plant growth and survival, the biochemical mechanisms of adaptation are still poorly understood and the signaling pathways involved remain elusive (Dat et al. 2000). Recently, reactive oxygen species have been proposed as a central component of plant adaptation to both biotic and abiotic stresses. There have been numerous reports indicating the production and involvement of reactive oxygen species in salt- and cation-induced injuries and growth inhibition in various plants.

According to Hernandez et al. (1993, 1995), the subcellular toxicity of salt stress added as NaCl in pea plants, at the level of the organelles such as mitochondria and chloroplasts, is linked to an oxidative stress mediated by production of $\cdot O_2^-$. In salt-tolerant plants, enzyme activities involved in detoxification of reactive oxygen species, such as superoxide dismutase, catalase, glutathione reductase, and monodehydroascorbate reductase, are potentially high and easily inducible by treatments with salts, compared to salt-sensitive plants (Hernandez et al. 1993; Shalata and Tal 1998). In many plant systems, treatment with low concentrations of NaCl results in increases in reactive oxygen species-scavenging enzyme activities, whereas lethal and sub-lethal concentrations of NaCl induces loss or a marked decrease in such enzyme activities (Gueta-Dahan et al. 1997, Lechno et al. 1997; Comba et al. 1998).

Tsugane et al. (1999) have analysed the pst1 *Arabidopsis* mutant which grows under intensive light in the presence of high concentrations of NaCl. It was shown that the salt tolerance was conferred in pst1 plants by the loss of a genetic factor that blocks the light-dependent induction of superoxide dismutase and ascorbate peroxidase. Thus salt tolerance in pst1 plants is due to the increased activities of superoxide dismutase and ascorbate peroxidase. According to Tanaka et al. (1999) overexpression of superoxide dismutase in rice confers salt tolerance to seedlings, further confirming that $\cdot O_2^-$ and derived reactive oxygen species are involved in salt-induced damages.

On the other hand, a positive role of the members of reactive oxygen species as key factors involved in plant adaptation to salt stress was also reported. Since salt damage to plants has been attributed not only to the accumulation of toxic ions but also to several other factors, mainly osmotic stress (Gueta-Dahan et al. 1997), adaptation to osmotic stress is also necessary for survival in the salt-stressed plants. Treatment of *Nicotiana plumbaginifolia* plants with $CuSO_4$, a compound used as a Fenton-type reagent which produces $HO\cdot$, prior to addition of NaCl leads to accumulation of an osmolyte proline, indicating that sublethal reactive oxygen species (in this case $HO\cdot$) mediates the induction of osmolyte production in plants, thus improving tolerance to osmotic stress (Savoure et al. 1999).

In support of many reports indicating the involvement of $\cdot O_2^-$ and derived reactive oxygen species in salt stress, we have shown the first direct evidence for cellular production of $\cdot O_2^-$ induced by treatments with various salts (Kawano et al. 2001b, 2002). In a series of experiments, we used an $\cdot O_2^-$-specific chemiluminescence method to obtain direct evidence that treatments with various salts of alkali metals, alkali earth metals and lanthanides stimulate the immediate production of $\cdot O_2^-$ in tobacco cell suspension culture. In the following sections, we would like to describe the background of experiments in details and propose a mechanism for induction of reactive oxygen species. Furthermore, novel approaches to prevent the salt-induced reactive oxygen species production are also reported.

58.2
Roles and Sources of Reactive Oxygen Species

Members of reactive oxygen species including $\cdot O_2^-$, H_2O_2 and $HO\cdot$ are known to be highly reactive causing oxidative damages to proteins, membrane lipids, and other cellular components (Asada 1994). On the other hand, plants are known to use oxygen species as defense-mediating agents. First, the oxygen species may directly participate in the mechanical strengthening of cell walls (Ogawa et al. 1997). Second, a protective mechanism that can limit the spread of pathogens depends on the production of oxygen species to initiate the localized cell death and a hypersensitive response to pathogen infection (Doke et al. 1996).

In plant cells, production of $\cdot O_2^-$ is an unavoidable process that is catalyzed by a one-electron reduction of O_2 (Asada 1994). Production of $\cdot O_2^-$ and its derivatives can be induced when plants are exposed to various biotic and abiotic stresses. In some occasions during plant defense induction against pathogens, plant peroxidases produce $\cdot O_2^-$ (Kiba et al. 1996; Kawano and Muto 2000; Kawano et al. 1998; 2000a,b; Vera-Estrella et al. 1992). However, in most other cases, induced production of $\cdot O_2^-$ is catalyzed by NAD(P)H-oxidizing enzyme systems localised in different cell compartments such as plasma membranes (Pinton et al. 1994; Doke et al. 1996; Murphy and Auh 1996) and peroxisomes (Del Rio et al. 1998; Lopez-Huertas et al. 1999).

In our recent studies, we have shown that the production of $\cdot O_2^-$ is induced in cultured tobacco cells, by cation-shocks (but not by the hyperosmotic shock) applied to cells by addition of various salts of alkali metals, alkali earth metals, lanthanides and aluminum (Kawano et al. 2001b). We also showed that the cation-induced production of $\cdot O_2^-$ could be inhibited in the presence of diphenyleneiodonium chloride, an NADPH oxidase inhibitor, indicating that NADPH oxidase is involved. Those results consistently support the view that certain metal ions with multiple valence effectively activate the $\cdot O_2^-$-generating activity of NADPH oxidase in tobacco cell suspension culture.

58.3
Plant Enzymes Involved in Production of Reactive Oxygen Species

Reactive oxygen species such as H_2O_2 are constitutively generated at low but significant levels in plant cells due to electron transport in chloroplasts and mitochondria, and enzymes involved in reduction/oxidation processes (Mehdy 1994; Wojtaszek 1997). The production of oxygen species during oxidative burst is dominantly associated with the extracellular compartment (extracellular matrix) and is regulated by developmental factors such as plant hormones, light, wounding and pathogen invasion (Wojtaszek 1997). Oxidative burst is generally defined as a rapid production of high levels of reactive oxygen species in response to external stimuli (Bolwell 1995; Mehdy 1994; Wojtaszek 1997). The first step in the pathogen-induced oxidative burst is believed to be the formation of $\cdot O_2^-$ by one-electron reduction of molecular oxygen (Mehdy 1994). The rapid production of reactive oxygen species, especially $\cdot O_2^-$ in plants was reported for the first time by Doke (1983a).

Plants react to microbial infection with a broad range of defense responses in an attempt to restrict or prevent pathogen growth (Bolwell and Wojtaszek 1997). The plant defense strategies include cell wall strengthening, production of microbicidal components and hydrolytic enzymes, and the induction of restricted cell death (Dixon and Lamb 1990; Dixon et al. 1994). These mechanisms are induced after specific recognition of the invading pathogens. Binding of elicitors to the plasma membrane-located receptor proteins occurs when the elicitor components are excised and released from pathogen bodies (Nürnberger et al. 1994). It has been clearly shown that once plant cells respond to the pathogen invasion, then rapid and transient production of reactive oxygen species occurs either in compatible and incompatible interactions (Baker and Orlandi 1995). A few hours after the initial transient response, sustained oxidative burst occurs in the incompatible interaction of plants and pathogens (Doke 1996). Reactive oxygen species generated during oxidative burst may act through microbial degradation, lignin formation and cell wall cross-linking (Bradley et al. 1992; Brisson et al. 1994; Ogawa et al. 1997; Wojtaszek et al. 1995). They may also function as intracellular signals, triggering two types of defense conditions known as hypersensitive reaction (Tenhaken et al. 1995) and systemic acquired resistance (Chen et al. 1993; Ryals et al. 1995).

To date, four groups of enzymes are considered to be involved in oxidative burst during the plant defense mechanism, namely NADPH oxidase, pH-dependent cell wall peroxidase, germin-like oxalate oxidase and amine oxidases. According to Bolwell (1995) and Bolwell and Wojtaszek (1997) NADPH oxidase and the pH-dependent cell wall bound peroxidase are considered to be the main sources of reactive oxygen species in plants.

58.4
NADPH Oxidase

The plant NADPH oxidase system is analogous to that of mammalian phagocytic cells catalyzing the following reaction (Eq. 58.1):

$$NADPH + 2O_2 \longrightarrow NADP^+ + 2·O_2^- + H^+ \tag{58.1}$$

The first observation that NADPH-dependent $·O_2^-$ generation occurs in plant system was reported by Doke (1983a,b). Using the nitroblue tetrazolium-reduction assay, NADPH-dependent $·O_2^-$ generation by potato tuber protoplasts treated with a fungal elicitor prepared from *Phytophthora infestans* hyphal wall component was detected. In TMV-infected tobacco leaves (Doke and Ohashi 1988) and in elicitor-treated tomato cotyledons (May et al. 1996), $·O_2^-$ generation around necrotic lesion was visualised using nitroblue tetrazolium. Using this nitroblue tetrazolium technique, it became clear that generation and accumulation of $·O_2^-$ are required before the onset of cell death initiating the lesion formation and subsequent $·O_2^-$ generation for limitless spread of cell death in *Arabidopsis* lsd1 mutant (Jabs et al. 1996). This process is similar to the involvement of $·O_2^-$ in programmed cell death in animal systems (Jacobson 1996).

The $·O_2^-$ generating activity measured with nitroblue tetrazolium staining was shown to be localized in the plasma membrane rich fraction prepared from potato tuber tissue treated with hyphal wall component (Doke and Miura 1995). Plasma membrane

NADPH-$\cdot O_2^-$ synthase of French bean (*Phaseolus vulgaris* L.) can be solubilized and separated from other NAD(P)H oxidoredactases through ion-exchange chromatography to a single peak (Gestelen et al. 1997).

Recent molecular biological studies revealed that *Arabidopsis thaliana* contains at least six homologs of the neutrophil NADPH oxidase gp91^{phox} subunit, which are mapped to different positions (Torres et al. 1998). A full-length homolog of the gp91^{phox} was also cloned from tomato (Amicucci et al. 1999). The N-terminal regions of the predicted proteins contain two EF hands Ca^{2+} binding motifs, indicating that direct regulation of the NADPH oxidase activity by Ca^{2+} may take place in plant cells (Keller et al. 1998; Torres et al. 1998; Amicucci et al. 1999). In *Arabidopsis* cell suspension culture, expression of the gp91^{phox} subunit is up-regulated by elicitation with harpin, a proteinaceous elicitor, or exposure to H_2O_2 (Desikan et al. 1998). Therefore the expression and activation of the plant NADPH oxidases are strikingly regulated by signal transduction pathway initiated by biotic chemicals.

58.5
Activation of NADPH Oxidase

Since the induction of $\cdot O_2^-$ production by treatments of tobacco cells with trivalent and divalent cations can be inhibited in the presence of an NADPH oxidase inhibitors (Kawano et al. 2001b, 2002), NADPH oxidase may be involved in the cation-induced $\cdot O_2^-$ production. It is likely that NADPH oxidase can be activated in the presence of metal cations. A similar phenomenon has been reported for NADPH oxidase from human neutrophils. Binding of divalent cations such as Ca^{2+} or Mg^{2+} to the membrane bound human neutrophil NADPH oxidase results in spontaneous activation of the $\cdot O_2^-$ producing activity (Suzuki et al. 1985; Cross et al. 1999). Furthermore, compared to divalent cations, monovalent cations have lesser effects on enhancement of $\cdot O_2^-$ producing activity of NADPH oxidase from human neutrophils (Suzuki et al. 1985). There is a possibility that cation treatment of tobacco cells directly activates the $\cdot O_2^-$ producing activity of NADPH oxidase by a similar manner to that proposed for human neutrophil enzyme. Thus signal transduction pathway may not participate in the mechanism for cation-dependent activation of $\cdot O_2^-$ production.

58.6
Targets and Action of Trivalent Cations

Recently Komiyama and his colleagues have shown that lanthanide ions possess various catalytic activities. For example, La^{3+}, Pr^{3+} and other lanthanides catalyze the formation of cAMP from ATP under physiological conditions (Yajima et al. 1994). Furthermore, in cooperation with other metals, lanthanide ions efficiently and nonenzymatically hydrolyze the phosphoester linkages in nucleic acids (Irisawa and Komiyama 1995), thus leading to degradation of DNA (Komiyama 1995) and RNA (Yashiro et al. 1996; Matsumura and Komiyama 1997).

However, the phenomenon observed in our system using tobacco cells may not be attributed to the above mechanisms since extracellularly added lanthanide ions hardly penetrate across the plasma membrane (Remedios 1977; van Steveninck et al. 1976);

thus the lanthanide ions may stay and act extracellularly. In addition, the induction of $^{\cdot}O_2^-$ by La^{3+} and Gd^{3+} that we observed was very rapid, and thus likely independent of gene expression events. Therefore, the involvement of lanthanide action on nucleic acids can be excluded in our experimental conditions.

It has been evidenced that lanthanides bind with high affinity to both plasma membrane and intracellular vesicle (van Steveninck et al. 1976), and it has been speculated that lanthanum regulates ABA-inducible gene expression by acting on ion channels (Lewis and Spalding 1998; Rock and Quatrano 1996). Actually, lanthanides inhibit Ca^{2+} channels in plants (Knight et al. 1997) and animals (Okamoto et al. 1998). Thus we have been employing La^{3+} and Gd^{3+} extensively as plasma membrane Ca^{2+} channel antagonists (Kawano et al. 1998, 2000a; Kawano and Muto 2000), though non specific inhibition of ion channels by treatment of *Arabidopsis* cells with lanthanum has been reported (Lewis and Spalding 1998). The action of lanthanide ions and other cations on $^{\cdot}O_2^-$ induction may not be due to blockage of Ca^{2+} channels since even addition of Ca^{2+} itself, and other cations with no inhibitory effect on Ca^{2+} channels, showed $^{\cdot}O_2^-$ inducing activity.

58.7
Action of Aluminum Ions

In addition to lanthanides, we have tested the effect of Al ions known to be toxic to plant roots. $AlCl_3$, optimally at 6.25 mM, induced a burst in production of $^{\cdot}O_2^-$ similarly to that of the lanthanides and the level of $^{\cdot}O_2^-$ production was greater than that induced by $LaCl_3$ (Kawano et al. 2003), further supporting our view that trivalent cations are very active in induction of $^{\cdot}O_2^-$ production.

58.8
Profiles of Zn^{2+} and Mn^{2+} Actions

Zinc is usually present in plants at high concentrations (Santa Maria and Cogliatti 1988). Zinc deficiency is one of the most widespread micronutrient deficiency in plants, causing severe reductions in crop production (Cakmak 2000). Increasing studies indicate that oxidative damage to cellular components caused in plants by reactive oxygen species is resulting from deficiency of zinc (Pinton et al. 1994; Cakmak 2000). We have tested the effect of extracellular supplementation of Zn^{2+} on cation-induced oxidative burst in tobacco cells. For comparison, we have also tested the activity of Mn^{2+}, another micronutrient reported to protect biological components and living cells from oxidative damages (Varani et al. 1991; Ledig et al. 1991).

Plants require trace metals including Mn and Zn for essential functions ranging from respiration to photosynthesis. Molecular biological studies on the mechanism for uptake of these metals by plants have recently been initiated (Delhaize 1996; Rogers et al. 2000). We have shown that extracellular supplementation of Zn^{2+} and Mn^{2+} inhibits the generation of $^{\cdot}O_2^-$ induced by addition of salts of non-transition metals. Although both Zn^{2+} and Mn^{2+} inhibited the salt-induced $^{\cdot}O_2^-$ generation, the modes of inhibition by those ions may be different since Mn^{2+} simply inhibited total $^{\cdot}O_2^-$ production while Zn^{2+} inhibited the immediate two phases of $^{\cdot}O_2^-$ production (phases 1 and 2) and allowed the delayed production of $^{\cdot}O_2^-$ to occur (phase 3) spending several minutes (Kawano et al. 2002).

58.9
Role of Mn^{2+}

The inhibitory role of Mn^{2+} might be related to the removal of $·O_2^-$ by oxidation of Mn^{2+} to Mn^{3+} by $·O_2^-$. There are some reports indicating the mechanism involved in removal of $·O_2^-$ by Mn^{2+}. Mn^{2+} is readily oxidized by $·O_2^-$ to Mn^{3+} (Coassin et al. 1992), and by similar mechanism some Mn-complexes detoxify $·O_2^-$ (Archibald and Fridovich 1982). Other inorganic biochemists also reported some evidence supporting the function of Mn^{2+} as an efficient scavenger of $·O_2^-$ (Fong et al. 1976; Jeffry 1983; Giuffrida et al. 1996). Thus Mn^{2+} is known to prevent free radical damage to living cells (Varani et al. 1991; Ligumsky et al. 1995; Ledig et al. 1991). In addition, Mn^{2+} is often employed as an $·O_2^-$ scavenger for preventing the biochemical reactions involving $·O_2^-$ (Momohara et al. 1990). Thus, we understood that inhibition of the cation-induced $·O_2^-$ generation is due to the action of added Mn^{2+} that simply removes the $·O_2^-$ produced in response to treatments with various metal cations. Although Mn^{2+} at sub-millimolar levels showed sharp inhibition of total $·O_2^-$ production, this high level of Mn^{2+} may be phytotoxic (Caldwell 1989), and it is not likely that plant cells can be exposed to such high concentrations of Mn^{2+} in normal physiological and environmental conditions.

58.10
Role of Zn^{2+}

In contrast to manganese, zinc is normally present in plants at high concentrations. For example, in roots of wheat seedlings, the cytoplasmic concentration of total Zn has been estimated at 0.4 mM (Santa Maria and Cogliatti 1988). According to Cakmak (2000), Zn-deficiency-related disturbances in cellular metabolism often resulting in inhibition of plant growth, are due to oxidative damages at membrane proteins, phospholipids, chlorophyll, DNA, SH-containing enzymes and indole-3-acetic acid. Increasing studies are confirming that Zn^{2+} acts against oxidative stress in cells, and the oxidative damages to cellular components resulting from attacks by reactive oxygen species are now considered to be the basis of such Zn-deficiency-related disturbances in plant (Pinton et al. 1994).

The effect of Zn^{2+} on protection of cells from damages by reactive oxygen species has also been reported in animal and human systems. According to Parat et al. (1997), inhibitory action of Zn^{2+} on apoptotic cell death is tightly related to its role in protecting cell membranes and DNA from damaging attack by reactive oxygen species. In addition, Zn^{2+} exerts a strong inhibitory effect on the generation of $·O_2^-$ by NADPH oxidase (Hammermüller et al. 1987; Bray and Bettger 1990). The $·O_2^-$-generating NADPH oxidase in membrane fraction from human phagocytic cells that is activated in the presence of divalent cations such as Ca^{2+}, is sensitive to Zn^{2+} and its activity can be strongly decreased by high Zn^{2+} concentrations (Suzuki et al. 1985).

Several reports have suggested that Zn^{2+} may inhibit the NADPH oxidase activity also in plants. Cakmak and his co-workers have surveyed the effect of Zn-deficiency on the level of NADPH dependent $·O_2^-$-generating activity (NADPH oxidase) in growing plants, and proposed that Zn^{2+} may have significant roles in plant protection by lowering the NADPH oxidase activity (Cakmak and Marschner 1993, 1988a,b; Pinton et al. 1994). McRae et al. (1982) also proposed that Zn^{2+} interferes with the $\cdot O_2^-$-generating activity of plant enzyme, based on the observation that Zn^{2+} inhibits an enzyme activity associated with the pea microsomal membrane, responsible for the NADPH-dependent increase in electron-paramagnetic signals for Tiron free radicals formed from Tiron, added to the system as a scavenger of $\cdot O_2^-$. However, it has not been clearly proved that the NADPH-dependent enzyme activity found in microsomal membrane fraction is representing the $\cdot O_2^-$-generating activity of NADPH oxidase and that the effect of Zn^{2+} is due to specific inhibition of the NADPH oxidase.

The impact of Zn^{2+} supplementation on NADPH oxidase-catalyzed generation of $\cdot O_2^-$ in living cells or plant tissue has not been reported, despite numerous studies examining the impact of zinc-deficiency. We have shown that Zn^{2+} supplementation inhibits and delays the metal cation-induced diphenyleneiodonium chloride-sensitive production of $\cdot O_2^-$ in living tobacco cells. Zn^{2+} may therefore protect the cells from the cation-induced NADPH oxidase-catalyzed production of $\cdot O_2^-$ and other reactive oxygen species derived from $\cdot O_2^-$.

In protection of plant cells from reactive oxygen species, zinc may act directly as inhibitor of reactive oxygen species production and also indirectly as a micronutrient that is utilized as a building block for Zn-containing proteins such as Cu,Zn-superoxide dismutase and zinc-finger proteins. Recently, a zinc-finger protein encoded by the *Arabidopsis* LSD1 gene was described as a negative regulator of $\cdot O_2^-$-dependent cell death in plants (Dietrich et al. 1997). However, we have brought evidence for the direct action of Zn^{2+} on reactive oxygen species production since the inhibitory effect of Zn^{2+} addition on $\cdot O_2^-$ generation was immediate.

58.11
Conclusion

We propose a model mechanism for the role of Zn^{2+} and Mn^{2+} in salt-induced and metal cation-induced cellular damages (Fig. 58.1). Plants are exposed to both the hyperosmotic shock (Savoure et al. 1999) and cation shock (due to Na^+) under salt stress (Kawano et al. 2001b). As illustrated in Fig. 58.1, there are two different flows of biochemical events leading to the salt-induced damages, namely cation shock-dependent and osmotic shock-dependent pathways. Since our recent works have clearly shown that only the cation treatments but not the hyperosmotic treatments resulted in burst of $\cdot O_2^-$ production in tobacco cells (Kawano et al. 2001b, 2002), damages to cells mediated by reactive oxygen species may be due to cation shock.

Similarly, metal toxicity including rhizotoxicity is brought about by the action of metal cations such as Al^{3+}. Low level of reactive oxygen species may contribute to adaptation of the cells to osmotic shock by stimulating the production of so-called compatible intracellular osmolytes such as proline and its derivatives (Savoure et al. 1999). However, excess of reactive oxygen species may lead to severe damages to cells. Low level of reactive oxygen species can be removed by endogenous reactive oxygen species-scavenging systems represented by Cu,Zn-superoxide dismutase and other enzymes, but high concentration of salts accompanied by high level of reactive oxygen species may inactivate the systems. Mn^{2+} may contribute to the removal of reactive oxygen species. Production of reactive oxygen species under salt stress may be

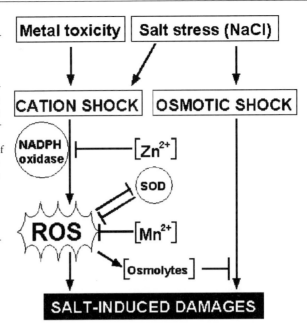

Fig. 58.1.
Model mechanism for the salt-induced and cation-induced cellular damages in plants. Toxic metals include Al, La, Gd, Mg, Ca, Na, K, and Li. Toxicity of transition metals such as Cu and Fe is brought about via different mechanisms. Protective role of micronutrients, Zn^{2+} and Mn^{2+} in prevention of reactive oxygen species (ROS) production is proposed. *Arrows* indicate the sequence of the events and the *T-shaped symbols* indicate the process inhibited in the presence of indicated chemicals. *SOD:* superoxide dismutase

due to activation of NADPH oxidase by excess of cations (of alkali metals, earth metals and lanthanides). This step may be the target of the action of Zn^{2+} to inhibit the salt-induced reactive oxygen species production, thus protecting the cells from salt-induced and metal cation-induced oxidative damages. Here we encourage the use of Mn^{2+} and Zn^{2+} in mitigation of reactive oxygen species-mediated damages to plants being exposed to salt stress.

References

Amicucci E, Gaschler K, Ward JM (1999) NADPH oxidase genes from tomato (*Lycopersicon esculentum*) and curly-leaf pondweed (*Potamogeton crispus*). Plant Biol 1:524–528
Archibald FS, Fridovich I (1982) The scavenging of superoxide radical by manganous complexes: in vitro. Arch Biochem Biophys 214:452–463
Asada K (1994) Production and action of active oxygen species in photosynthetic tissues. In: Foyer CH, Mullineaux PM (eds) Causes of photooxidative stress amelioration of defense system in plants. CRC press, London, pp 77–104
Baker CJ, Orlandi EW (1995) Active oxygen in plant pathogenesis. Annu Rev Phytopathol 33:299–321
Bolwell GP (1995) The origin of the oxidative burst in plants. Free Radic Res 23:517–532
Bolwell GP, Wojtaszek P (1997) Mechanism for the generation of reactive oxygen species in plant defense-a broad perspective. Physiol Mol Plant Pathol 51:347–366
Bradley DJ, Kjellbom P, Lamb CJ (1992) Elicitor- and wound-induced oxidative cross linking of a proline rich plant cell wall protein: a novel, rapid defense response. Cell 70:21–30
Bray TM, Bettger WJ (1990) The physiological role of zinc as an antioxidant. Free Radical Biol Med 8:281–291
Brisson LF, Tenhaken R, Lamb C (1994) Function of oxidative cross-linking of cell wall structural proteins in plant disease resistance. Plant Cell 6:1703–1712
Cakmak I (2000) Possible roles of zinc in protecting plant cells from damage by reactive oxygen species. New Phytol 146:185–205

Cakmak I, Marschner H (1988a) Enhanced superoxide radical production in roots of zinc deficient plants. J Exper Bot 39:1449–1460

Cakmak I, Marschner H (1988b) Zinc-dependent changes in ESR signals, NADPH oxidase and plasma membrane permeability in cotton roots. Physiol Plant 73:182–186

Cakmak I, Marschner H (1993) Effects of zinc nutritional status on activities of superoxide radical and hydrogen peroxide scavenging enzymes in bean leaves. Plant Soil 155/156:127–130

Caldwell CR (1989) Analysis of aluminum and divalent cation binding to wheat root plasma membrane proteins using terbium phosphorescence. Plant Physiol 91:233–241

Chen Z., Silva H, Klessig DF (1993) Active oxygen species in the induction of plant systemic acquired resistance induced by salicylic acid. Science 262:1883–1886

Coassin M, Ursini F, Bindoli A (1992) Antioxidant effect of manganese. Arch Biochem Biophys 299: 330–333

Comba ME, Benavides MP, Tomaro ML (1998) Effect of salt stress on antioxidant defense system in soybean root nodules. Austr J Plant Physiol 25:665–671

Cross AR, Erichson R, Eliss BA, Curnutte JT (1999) Spontaneous activation of NADPH oxidase in a cell-free system: unexpected multiple effects of magnesium ion concentrations. Biochem J 338: 229–233

Dat J, Vandenabeele S, Vranova E, Van Montagu M, Inze D, Van Breusegem F (2000) Dual action of the active oxygen species during plant stress responses. Cellul Mol Life Sci 57:779–795

Del Rio LA, Pastori GM, Palma JM, Sandalio LM, Sevilla F, Corpas FJ, Jiménez A, Lopez-Huertas E, Hernandez JA (1998) The activated oxygen role of peroxisomes in senescence. Plant Physiol 116:1195–1200

Delhaize E (1996) A metal-accumulator mutant of *Arabidopsis thaliana*. Plant Physiol 111:849–855

Desikan R, Burnett EC, Hancock JT, Neill SM (1998) Harpin and hydrogen peroxide induce the expression of a homologue of gp91-phox in *Arabidopsis thaliana* suspension cultures. J Exp Bot 49:1767–1771

Dietrich RA, Richberg MH, Schmidt R, Dean C, Dangl JL (1997) A novel zinc finger protein is encoded by the *Arabidopsis* LDS1 gene and functions as a negative regulator of plant cell death. Cell 88:685–694

Dixon RA, Lamb CJ (1990) Molecular communication in interactions between plants and microbial pathogens. Ann Rev Plant Physiol Plant Mol Biol 41:339–367

Dixon RA, Harrison MJ, Lamb CJ (1994) Early events in the activation of plant defense responses. Ann Rev Phytopathol 32:479–501

Doke N (1983a) Involvement of superoxide anion generation in the hypersensitive response of potato tuber tissues to infection with an incompatible race of *Phytophthora infestans* and to the hyphal wall components. Physiol Plant Pathol 23:345–357

Doke N (1983b) Generation of superoxide anion by potato tuber protoplasts upon the hypersensitive response to hyphal wall components of *Phytophthora infestans* and specific inhibition of the reaction by suppresser of hypersensitivity. Physiol Plant Pathol 23:359–367

Doke N, Miura Y (1995) In-vitro activation of NADPH-dependent O_2^- generating system in a plasma membrane-rich fraction of potato tuber tissues by treatment with an elicitor from *Phytophthora infestans* or with digitonin. Physiol Mol Plant Pathol 46:17–28

Doke N, Ohashi Y (1988) Involvement of O_2^- generating system in the induction of necrotic lesions on tobacco leaves infected with tobacco mosaic virus. Physiol Mol Plant Pathol 32:163–175

Doke N, Miura Y, Sanchez LM, Park HJ, Noritake T, Yoshioka H, Kawakita K (1996) The oxidative burst protects plants against pathogen attack: mechanism and role as an emergency signal for plant bio-defense – a review. Gene 179:45–51

Fong KL, McCay PB, Poyer JL (1976) Evidence for superoxide-dependent reduction of Fe^{3+} and its role in enzyme-generated hydroxyl radical formation. Chemico-Biol Interact 15:77–89

Gestelen PV, Asard H, Caubergs RJ (1997) Soublization and separation of a plant plasma membrane NADPH-O_2^- synthase from other NAD(P)H oxidoreductases. Plant Physiol 115:543–550

Giuffrida S, De Guidi G, Miano P, Sortino S, Condorelli G, Costanzo LL (1996) Molecular mechanism of drug photosensitization: VIII. Effect of inorganic ions on membrane damage photosensitized by naproxen. J Inorg Biochem 63:253–263

Gueta-Dahan Y, Yaniv Z, Zilinskas BA, Ben-Hayyim G (1997) Salt and oxidative stress: similar and specific responses and their relation to salt tolerance in citrus. Planta 203:460–469

Hammermüller JD, Bray TM, Bettger WJ (1987) Effect of zinc and copper deficiency on microsomal NADPH-dependent active oxygen generation in rat lung and liver. J Nutr 117:894–901

Hernandez JA, Corpas FJ, Gomez M, Del Rio LA, Sevilla F (1993) Salt-induced oxidative stress mediated by activated oxygen species in pea leaf mitochondria. Physiol Plant 89:103–110

Hernandez JA, Olmo E, Corpas FJ, Sevilla F, Del Rio LA (1995) Salt-induced oxidative stress in chloroplast of pea plants. Plant Sci 105:151–167

Irisawa M, Komiyama M (1995) Hydrolysis of DNA and RNA through cooperation of two metal ions: a novel mimic of phosphoesterases. J Biochem 117:465–466

Jabs T, Dietrich RA, Dangl JL (1996) Extracellular superoxide initiates runaway cell death in an *Arabidopsis* mutant. Science 273:1853–1856

Jacobson MD (1996) Reactive oxygen species and programmed cell death. Trends Biochem Sci 21:83–86

Jeffery EH (1983) The effect of zinc on NADPH oxidation and monooxygenase activity in rat hepatic microsomes. Mol Pharmacol 23:467–478

Kawano T, Muto S (2000) Mechanism of peroxidase actions for salicylic acid-induced generation of active oxygen species and an increase in cytosolic calcium in tobacco cell suspension culture. J Exper Bot 51:685–693

Kawano T, Sahashi N, Takahashi K, Uozumi N, Muto S (1998) Salicylic acid induces extracellular superoxide generation followed by an increase in cytosolic calcium ion in tobacco suspension culture: The earliest events in salicylic acid signal transduction. Plant Cell Physiol 39:721–730

Kawano T, Pinontoan R, Uozumi N, Miyake C, Asada K, Kolattukudy PE, Muto S (2000a) Aromatic monoamine-induced immediate oxidative burst leading to an increase in cytosolic Ca^{2+} concentration in tobacco suspension culture. Plant Cell Physiol 41:1251–1258

Kawano T, Pinontoan R, Uozumi N, Morimitsu Y, Miyake C, Asada K, Muto S (2000b) Phenylethylamine-induced generation of reactive oxygen species and ascorbate free radicals in tobacco suspension culture: Mechanism for oxidative burst mediating Ca^{2+} influx. Plant Cell Physiol 41:1259–1266

Kawano T, Sahashi N, Uozumi N, Muto S (2000c) Involvement of apoplastic peroxidase in the chitosaccharide-induced immediate oxidative burst and a cytosolic Ca^{2+} increase in tobacco suspension culture. Plant Peroxid Newslett 14:117–124

Kawano T, Kawano N, Hosoya H, Lapeyrie F (2001a) A fungal auxin antagonist, hypaphorine competitively inhibits indole-3-acetic acid-dependent superoxide generation by horseradish peroxidase. Biochem Biophys Res Commun 288:546–551

Kawano T, Kawano N, Muto S, Lapeyrie F (2001b) Cation-induced superoxide generation in tobacco cell suspension culture is dependent on ion valence. Plant Cell Environ 24:1235–1241

Kawano T, Kawano N, Muto S, Lapeyrie F (2001c) Retardation and inhibition of the cation-induced superoxide generation in BY-2 tobacco cell suspension culture by Zn^{2+} and Mn^{2+}. Physiol Plant 114:395–404

Kawano T, Kadono T, Furuichi T, Muto S, Lapeyrie F (2003) Aluminium-induced distortion in calcium signaling involving oxidative burst and channel regulations in tobacco BY-2 cells. Biochem Biophys Res Commun 308:35–42

Keller T, Damude HG, Werner D, Doerner P, Dixon R, Lamb C (1998) A plant homolog of the neutrophil NADPH oxidase gp91^{phox} subunit gene encodes a plasma membrane protein with Ca^{2+} binding motifs. Plant Cell 10:255–266

Kiba A, Toyota K, Ichinose Y, Yamada T, Shiraishi T (1996) Species-specific suppression of superoxide-anion generation of surfaces of pea leaves by the suppressor from *Mycosphaella pinodes*. Ann Phytopathol Soc Japan 62:508–512

Knight H, Trewavas AJ, Knight MR (1997) Calcium signalling in *Arabidopsis thaliana* responding to drought and salinity. Plant J 12:1067–1078

Komiyama M (1995) Sequence-specific and hydrolytic scission of DNA and RNA by lanthanide complex-oligoDNA hybrids. J Biochem 118:665–670

Lechno S, Zamski E, Tel-Or E (1997) Salt stress-induced responses in cucumber plants. J Plant Physiol 150:206–211

Ledig M, Tholey G, Megias-Megias L, Kopp P, Wedler F (1991) Combined effects of ethanol and manganese on cultured neurons and glia. Neurochem Res 16:591–596

Lewis BD, Spalding EP (1998) Nonselective block by La^{3+} of *Arabidopsis* ion channels involved in signal transduction. J Membrane Biol 162:81–90

Ligumsky M, Sestieri M, Okon E, Ginsburg I (1995) Antioxidants inhibit ethanol-induced gastric injury in the rat. Role of manganese, glycine, and carotene. Scand J Gastroenterol 30:854–860

Lopez-Huertas E, Corpas FJ, Sandalio LM, Del Rio LA (1999) Characterization of membrane polypeptides from pea leaf peroxisomes involved in superoxide radical generation. Biochem J 337:531–536

Matsumura K, Komiyama M (1997) Enormously fast RNA hydrolysis by lanthanide(III) ions under physiological conditions: eminent candidates for novel tools of biotechnology. J Biochem 122:387–394

May MJ, Hammond-Kosack KE, Jones JDG (1996) Involvement of reactive oxygen species, glutathione metabolism, and lipid peroxidation in *Cf*-gene-dependent defense of tomato cotyledons induced by race-specific elicitors of *Cladosporium fulvum*. Plant Physiol 110:1367–1379

McRae DG, Baker JE, Thompson JE (1982) Evidence for involvement of the superoxide radical in the conversion of 1-aminocyclopropane-1-carboxylic acid to ethylene by pea microsomal membranes. Plant Cell Physiol 23:375–383

Mehdy MC (1994) Active oxygen species in plant defense against pathogens. Plant Physiol 105:467–472

Momohara I, Matsumoto Y, Ishizu A (1990) Involvement of veratryl alcohol and active oxygen species in degradation of a quinone compound by lignin peroxidase. FEBS Lett 273:159–162

Murphy TM, Auh CK (1996) The superoxide synthase of plasma membrane preparation from cultured rose cells. Plant Physiol 110:621–629

Nürnberger T, Nennstiel D, Jabs T, Sacks WR, Hahkbrock K, Scheel D (1994) High affinity binding of fungal oligopeptide elicitor to parsley plasma membranes triggers multiple defense responses. Cell 78:449–460

Ogawa K, Kanematsu S, Asada K (1997) Generation of superoxide anion and localization of CuZn-superoxide dismutase in the vascular tissue of spinach hypocotyls: their association with lignification. Plant Cell Physiol 38:1118–1126

Okamoto T, Maeda O, Tsuzuike N, Hara K (1998) Effect of gadolinium chloride treatment on concanavalin A-induced cytokine mRNA expression in mouse liver. Jpn J Pharmacol 78:101–103

Parat M-O, Richard M-J, Poller S, Hadjur C, Favier A, Beani JC (1997) Zinc and DNA fragmentation in karatinocyte apoptosis: first inhibitory effect in UV-B irradiated cells. J Photochem Photobiol B Biol 37:101–106

Pinton R, Cakmak I, Marschner H (1994) Zinc deficiency enhanced NAD(P)H-dependent superoxide radical production in plasma membrane vesicles isolated from roots of bean plants. J Exper Bot 45:45–50

Remedios CD (1977) Lanthanide ions and skeletal muscle sarcoplasmic reticulum. I. Gadolinium localization by electron microscopy. J Biochem 81:703–708

Rock CD, Quatrano RS (1996) Lanthanide ions are agonists of transient gene expression in rice protoplasts and act synergy with ABA to increase *Em* gene expression. Plant Cell Rep 15:371–376

Rogers EE, Eide DJ, Guerinot ML (2000) Altered selectivity in an *Arabidopsis* metal transporter. Proc Natl Acad Sci USA 97:12356–12360

Ryals J, Lawton KA, Delaney TP, Friedrich L, Kessmann H, Neuenschwander U, Uknes S, Vernooij B, Weymann K (1995) Signal transduction in systemic acquired resistance. Proc Natl Acad Sci USA 92:4202–4205

Santa Maria GE, Cogliatti DH (1988) Bidirectional Zn-fluxes and compartmentation in wheat seedling roots. J Plant Physiol 132:312–315

Savoure A, Thorin D, Davey M, Hua X-J, Mauro S, Van Montagu M, Inze D, Verbruggen N (1999) NaCl and $CuSO_4$ treatments trigger distinct oxidative defense mechanisms in *Nicotiana plumbaginifolia* L. Plant Cell Environ 22:387–396

Shalata A, Tal M (1998) The effect of salt stress on lipid peroxidation and antioxidants in the leaf of the cultivated tomato and its wild salt-tolerant relative *Lycopersicon pennellii*. Physiologia Plantarum 104, 169–174

Suzuki H, Pabst MJ, Johnston Jr. RB (1985) Enhancement by Ca^{2+} or Mg^{2+} of catalytic activity of the superoxide-producing NADPH oxidase in membrane fractions in human neutrophils and monocytes. J Biol Chem 260:3635–3639

Tanaka Y, Hibino T, Hayashi Y, Tanaka A, Kishitani S, Yokota S, Takabe T (1999) Salt tolerance of transgenic rice overexpressing yeast mitochondrial Mn-SOD in chloroplasts. Plant Sci 148:131–138

Tenhaken R, Levine A, Brisson LF, Dixon RA, Lamb C (1995) Function of the oxidative burst in hypersensitive disease resistance. Proc Natl Acad Sci USA 92:4158–4163

Torres MA, Onouchi H, Hamada S, Machida C, Hammond-Kosack KE, Jones JDG (1998) Six *Arabidopsis thaliana* homologues of the human respiratory burst oxidase (gp91^{phox}). Plant J 14:365–370

Tsugane K, Kobayashi K, Niwa Y, Ohba Y, Wada K, Kobayashi H (1999) A recessive *Arabidopsis* mutant that grows photoautotrophically under salt stress shows enhanced active oxygen detoxification. Plant Cell 11:1195–1206

van Steveninck RFM, van Steveninck ME, Chescoe D (1976) Intracellular binding of lanthanum in root tips of barley (*Hordeum vulgare*). Protoplasma 90:89–97

Varani J, Ginsburg I, Gibbs DF, Mukhopadhyay PS, Sulavik C, Johnson KJ, Weinberg JM, Ryan US, Ward PA (1991) Hydrogen peroxide-induced cell and tissue injury: protective effects of Mn^{2+}. Inflamm 15:291–301

Vera-Estrella R, Blumwald E, Higgins VJ (1992) Effect of specific elicitors of *Cladosporium fulvum* on tomato suspension cells: evidence for involvement of active oxygen species. Plant Physiol 99:1208–1215

Wojtaszek P (1997) Oxiddative burst: an early plant response to pathogen infection. Biochem J 322:681–692

Wojtaszek P, Trethowan J, Bolwell GP (1995) Specificity in the immobilization of cell wall proteins in response to different elicitor molecules in suspension-cultured cells of french bean (*Phaseolus vulgaris* L.). Plant Mol Biol 28:1075–1087

Yajima H, Sumaoka J, Miyama S, Komiyama M (1994) Lanthanide ions for the first non-enzymatic formation of adenosine 3',5'-cyclic monophosphate from adenosine triphosphate under physiological conditions. J Biochem 115, 1038–1039

Yashiro M, Ishikubo A, Komiyama M (1996) Dinuclear lanthanum(III) complex for efficient hydrolysis of RNA. J Biochem 120:1067–109

Yoshioka H, Sugie K, Park HJ, Maeda H, Tsuda N, Kawakita K, Doke N (2001) Induction of plant gp91 phox homolog by fungal cell wall, arachidonic acid, and salicylic acid in potato. Mol Plant Microbe Interact 14:725–736

Part VII

Chapter 59

The LUX-FLUORO Test as a Rapid Bioassay for Environmental Pollutants

P. Rettberg · C. Baumstark-Khan · E. Rabbow · G. Horneck

Abstract

The bioassay LUX-FLUORO test was developed for the rapid detection and quantification of environmental pollutants with genotoxic and/or cytotoxic effects. This bacterial test system uses two different reporter genes whose gene products and their reactions, respectively, can be measured easily and simultaneously by optical methods. Genotoxicity is measured by the increase of bioluminescence in genetically modified bacteria which carry a plasmid with a complete *lux* operon for the enzyme luciferase from the marine photobacterium *P. leiognathi* under the control of a DNA-damage dependent so-called SOS promoter. If the desoxyribonucleic acid (DNA) in these bacteria is damaged by a genotoxic chemical, the SOS promoter is turned on and the *lux* operon is expressed. The newly synthesized luciferase reacts with its substrate thereby producing bioluminescence in a damage-proportional manner. In the second part of the system, genetically modified bacteria carry the gfp gene for the green fluorescent protein from the jellyfish *A. victoria* downstream from a constitutively expressed promoter. These bacteria are fluorescent under normal conditions. If their cellular metabolism is disturbed by the action of cytotoxic chemicals the fluorescence decreases in a dose-proportional manner. The combined LUX-FLUORO test can be used for the biological assessment of the geno- and cytotoxicity of a wide variety of organic and inorganic chemicals including complex mixtures in different matrices.

Key words: bacterial bioassay, genotoxicity, cytotoxicity, gfp, lux, LUX-FLUORO test, SOS-LUX test, whole-cell biosensor, DNA damage

59.1 Introduction

59.1.1 Genotoxicity and Cytotoxicity

Increasing levels of environmental pollution involve a risk not only for human health but also for the sensible balance of whole ecosystems. To identify sources of potential hazard, biological methods have been developed which are complementary to chemical and physical detection methods. Genotoxicity, the induction of deleterious changes in the genetic material of each organism, the desoxyribonucleic acid (DNA), as well as cytotoxicity, the break-down of the cellular metabolism, are both the result of complex cellular reaction cascades after the primary events of damage induction. The detection and quantification of both genotoxicity and cytotoxicity absolutely requires the application of biological methods, because physical and/or chemical methods are principally not suitable for the assessment of biological endpoints.

Genotoxicity tests can be defined as in-vitro and in-vivo tests designed to detect compounds which induce genetic damage directly or indirectly by various mechanisms. Manifestation and fixation of damage to DNA in the form of gene mutations, large scale chromosomal damage, recombination and numerical chromosomal changes are generally considered to be essential for heritable effects and in the multi-step process of malignancy, a complex process in which genetic changes may play only a part (ICH 1997). Compounds which are positive in tests that detect such kinds of damage have the potential to be human carcinogens and/or mutagens. In the last decades many different biological test systems for geno- and cytotoxicity measurements have been developed and used with a wide variety of test substances and newly synthesized chemicals. Examples of assay systems for the identification of mutagenic or carcinogenic agents are tumour induction in animals (Enzmann et al. 1998), the human somatic mutation test of the HPRT gene in peripheral blood cells (Albertini 2001) and induction of chromosome aberrations (Natarajan et al. 1996). Toxicity bioassays rely largely on lethality measurements. All such assays are generally time-consuming and expensive, and provide little information on the mechanisms of toxicity.

59.1.2
Bacterial Short-Term Tests

These constraints have led to the development of a number of short-term tests for detecting potential carcinogens. The most commonly used test is the Ames *Salmonella*/microsome mutagenicity assay, a bacterial reverse mutation assay. It employs several histidine dependent *Salmonella* strains each carrying different mutations in various genes in the histidine operon. These mutations act as hot spots for mutagens that cause DNA damage via different mechanisms and result in short frame shifts, transitions or transversions (Ames et al. 1973; Mortelmans and Zeiger 2000). Nearly 90% of the carcinogens that were positive in this short-tem test were mutagenic in higher animals.

In recent years genetic coupling of cellular metabolic functions to suitable reporter molecules has led to the development of toxicity tests measuring the alteration of the reporter by the toxic substance under examination. In the SOS chromotest a special strain of *Escherichia coli* has been used in which the structural gene for β-galactosidase, lacZ, is brought under control of sulA, a SOS-controlled gene involved in cell division inhibition (Quillardet and Hofnung 1993). In response to DNA-damaging agents, a set of functions known as SOS-response are induced which include synthesis of a number of proteins such as RecA and UmuC/D proteins related to mutagenesis (Fig. 59.1). The SOS chromotest provides a simple and direct colorimetric assay for β-galactosidase synthesised in response to a genotoxic agent. It also detects DNA damages induced by ionising and UV-radiation with high sensitivity. The SOS-induction potency is closely correlated to the mutagenic potency determined in the Ames test for most of the agents tested so far (White and Rasmussen 1996).

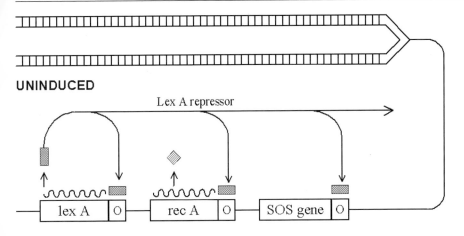

SOS induction

DNA damaged → signal induced → Rec A activated → Lex A cleaved

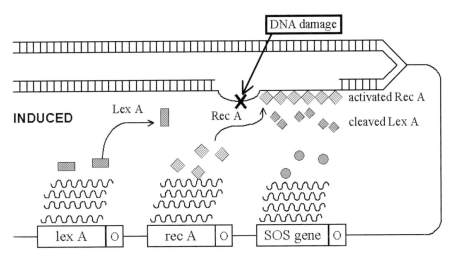

Fig. 59.1. Simplified scheme of the SOS system in *E. coli* (modified from Friedberg 1985)

59.1.3
Bioluminescence-Based Reporter Systems

In recent years, a new generation of methods, the use of living cells as luminescent protein biosensors, has evolved (D'Souza 2001). Bioluminescence as reporter signal offers the most advantages, because the reporter measurements are nearly instantaneous, they are exceptionally sensitive and quantitative, and typically there is no endogenous activity in the host cells to interfere with quantification. Luciferase genes have been cloned in between from many different species e.g. bacteria, beetles (including firefly), *Renilla* and *Aequorea*.

Bacterial luciferase was the first enzyme to be applied as genetic reporter. It is a dimeric protein of 80 kDa found predominantly in several marine bacteria (Baldwin et al. 1995). The luminescence is generated from an oxidation reaction involving the reduced flavin mononucleotide $FMNH_2$ and a long-chain aliphatic aldehyde to yield FMN, carboxylate and blue light of 490 nm. The genes encoding the bacterial luciferases are called *lux*. The *luxA* and *luxB* genes encode the α and β subunit of the enzyme. The primary advantage of working with bacterial *lux* genes is the ability to express the complete *lux* operon also in other bacteria, including the *luxC*, *luxD* and *luxE* genes, which encode the three proteins of the fatty acid reductase complex needed to recycle the reaction product back to the aldehyde substrate. Thus, autonomous expression of luminescence can be attained in bacterial cells in contrast to all other bioluminescent reporter systems which require the exogenous addition of the substrate luciferin.

59.1.4
The SOS-LUX- and LAC-FLUORO Test

Here we report on bioassays using genetically modified bacteria that have been developed by our group for rapid toxicity tests in aqueous environmental matrices (Ptitsyn et al. 1997; Horneck et al. 1998; Rettberg et al. 1999, 2001). The SOS-LUX test is based on the SOS induction in a similar way as compared to the SOS chromotest. A special *E. coli* plasmid, pPLS-1, has been constructed in which the promoterless *lux* operon (*luxCDABE*) of *Photobacterium leiognathi* is under control of the SOS-dependent *col* promoter and its synthesis is therefore regulated by the SOS system. DNA damage leads to an increased level of luciferase and therefore increased bioluminescence in presence of a variety of mutagens and also after exposure to ionising and UV radiation. The intensity of the emitted bioluminescence light is thereby proportional to the concentration of the DNA-damaging agent (Fig. 59.2a). With this system it is not only possible to quantify a genotoxic compound, but also to follow-up the kinetics of DNA-damage processing in the SOS system. However, if substances have to be tested which show simultaneously geno- and cytotoxicity in the same concentration range, the determination of the induction of the SOS system may be influenced by cell death. In a first approach the discrimination between the genotoxic and cytotoxic potency of the test substance was achieved by parallel measurements of the absorbance of the bacterial suspension as a measure for cell numbers. A test substance was considered to be neither genotoxic nor cytotoxic, if bioluminescence was not induced and the cell growth was comparable to that of the untreated control. If, however, bioluminescence and/or absorbance decrease during incubation then the test substance might be cytotoxic as well. In a further approach we therefore introduced a second method which reflects the metabolic activity of the bacteria. We now utilize the expression of green fluorescent protein (GFP) from the jellyfish *A. victoria* as reporter for protein synthesis in metabolically active cells. Green fluorescent protein can be quantitavely visualized by excitation with the appropriate radiation wavelengths and measurement of the resulting fluorescence in the absence of substrates and other cofactors without disturbing the cells. For our so-called LAC-FLUORO test we use a *gfp* gene which has been optimised for higher expression in bacteria and for maximal fluorescence yields using excitation wavelengths in the near UV region (360–400 nm) (Fig. 59.2b) (Crameri et al. 1996). This *GFPuv* gene was inserted in frame with the *lacZ* initiation codon of β-galactosidase from the *lac* operon of *E. coli*, so that a soluble β-galactosidase-GFPuv fusion protein is expressed which can easily be measured in a fluorometer. For the combined LUX-FLUORO test, recombinant *S. choleraesuis* subsp. *choleraesuis* strains carrying either the SOS-LUX plasmid pPLS-1 (TA1535-pPLS-1) or the lac-GFPuv plasmid (TA1535-pGFPuv) were used simultaneously to detect and quantify in parallel agents that exhibit either genotoxic, cytotoxic or geno- and cytotoxic effects (Baumstark-Khan et al. 2001; Rabbow et al. 2002) (Fig. 59.2).

59.2
Experimental

59.2.1
Plasmids, Bacteria and Growth Conditions

The construction of the plasmid pPLS-1 (DSMZ 10333) carrying the *luxCDABFE* genes downstream of a strong SOS-dependent promoter is already described (Ptitsyn et al. 1997; Horneck et al. 1998). This plasmid is the genotoxicity sensing reporter component of the combined test system. The plasmid pGFPuv (Clontech Laboratories Inc., CA, USA, 6079-1) carrying the optimised "cycle3" variant of green fluorescent protein (GFP) in frame with the lacZ initiation codon is the cytotoxicity sensing reporter component of the combined test system. The strain *S. choleraesuis* subsp. *choleraesuis* TA1535, one of the tester strains in the Ames test, was transformed with either the plasmid pPLS-1 or with the plasmid pGFPuv. Transformed bacteria were selected and cultivated at 37 °C in NB-medium supplemented with 50 µg ml^{-1} ampicillin.

59.2.2
Test Substances

The following compounds with differing genotoxic and cytotoxic effects were used to evaluate the LUX-FLUORO test (Merck and Sigma, Germany): mitomycin C (MMC), a DNA intrastrand cross-linking agent as a positive control for genotoxic compounds which do not require metabolic activation, chloramphenicol (CAP) and aureomycin (AM), both antibiotics as positive controls for cytotoxic compounds, and 4-nitro-quinoline-1-oxide (4-NQO) and N-methyl-N'-nitro-N-nitrosoguanidine (MNNG) as probable human carcinogens, as well as chromium(VI) ($K_2Cr_2O_7$), chromium(III) ($CrCl_3$), Zn(II) ($ZnCl_2$) as examples for genotoxic and/or cytotoxic heavy metal salts.

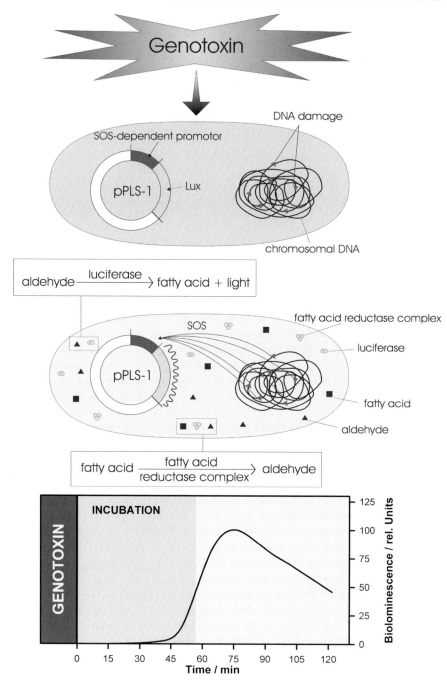

Fig. 59.2a. Schematic illustration of the principle of the combined LUX-FLUORO test. In the SOS-LUX test DNA-damages induced by genotoxic substances result in damage-proportional bioluminescence

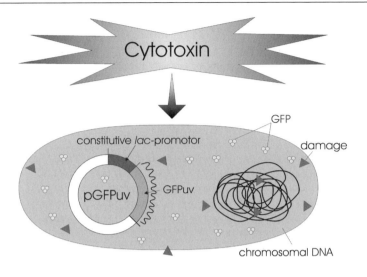
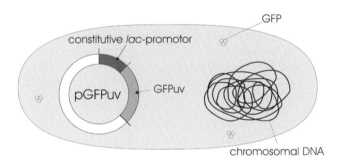
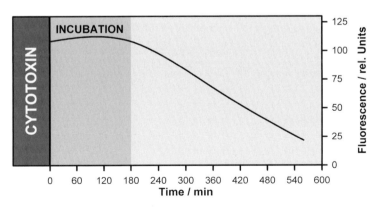

Fig. 59.2b. Schematic illustration of the principle of the combined LUX-FLUORO test. In the LAC-FLUORO test cytotoxic substances reduce the cellular fluorescence in a dose-proportional manner

59.2.3
Performance of the LUX-FLUORO Test

Log phase cultures of both bacterial strains were incubated in pre-warmed NB-medium containing 50 µg ml^{-1} ampicillin until the absorption at 600 nm (A_{600}) reached 0.2. 75 µl of TA1535-pPLS-1 or TA1535-pGFPuv cultures or a 1:1 mixture of both were added to each well of a white microplate with a transparent bottom (Isoplates™, Wallac, Turku, Finland) containing 75 µl of the diluted test samples or of the controls in 4 replicates each and placed into the microplate reader (Multilabel Counter 1420 Victor^2 form EG&G Wallac, Turku, Finland) with a controlled temperature of the plate of 30 °C. The reader was programmed to repeat the measurement cycle of 2 min orbital shaking, luminescence reading without filter for 0.2 s/well, followed by absorbance measurement for 0.1 s/well at 490 nm (20 nm band width) and by fluorescence reading for 0.1 s/well at 510 nm after excitation at 405 nm, every 10 min for 50 cycles adding up to 8 h kinetics.

59.2.4
Numerical Analysis

The raw data were transferred into an Excel macro sheet and luminescence response induction factors F_i were calculated for the genotoxic potential of the applied samples according to Eq. 59.1 with light emission data of the untreated culture (Lux_0), of the culture treated with the test samples (Lux_i) and cell growth (absorbance) of the untreated culture (A_0) and of the treated culture (A_i).

$$F_i = \frac{Lux_i \, A_0}{Lux_0 \, A_i} \tag{59.1}$$

The fluorescence response reduction factors F_r for cytotoxicity were calculated according to Eq. 59.2 with fluorescence data of the untreated culture (Flu_0) and of the cultures treated with the samples (Flu_r).

$$F_r = \frac{Flu_r}{Flu_0} \tag{59.2}$$

The threshold for a sample to be genotoxic was defined as a twofold increase of the luminescence response induction factor F_i, whereas a fluorescence reduction factor F_r of less than 0.5 is determined to be a signal for cytotoxicity in this test. The concentrations of the tested samples are final concentrations in the LUX-FLUORO test.

59.3
Results and Discussion

59.3.1
Kinetics of the LUX-FLUORO Test

Mitomycin C (MMC), a DNA intrastrand cross-linking agent (Iyer et al. 1964), is an example for a powerful chemical genotoxin. It is routinely used as a positive control for genotoxic directly acting substances in the SOS-LUX test. For the LAC-FLUORO

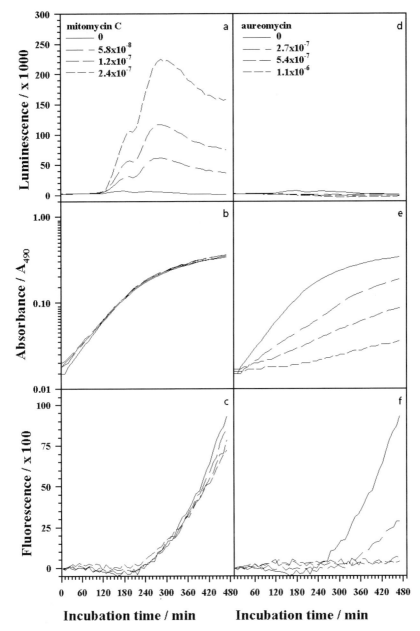

Fig. 59.3. Kinetics of the combined LUX-FLUORO test with control substances: **a** Kinetics of bioluminescence induction in the presence of mitomycin C, a genotoxin; **b** corresponding kinetics of the absorbance in the presence of mitomycin C; **c** corresponding kinetics of fluorescence induction in the presence of mitomycin C; **d** kinetics of bioluminescence induction in the presence of aureomycin, an antibiotic; **e** corresponding kinetics of the absorbance in the presence of aureomycin; **f** corresponding kinetics of fluorescence induction in the presence of aureomycin

test the antibiotics chloramphenicol (CAP) or aureomycin (AM) are used as positive controls for cytotoxic substances. In addition to these controls samples containing only the solvent, in this case H_2O, are always measured in parallel in each experiment. Figure 59.3 shows the kinetics of the LUX-FLUORO response during incubation of the cells with different concentrations of mitomycin C and aureomycin at 30 °C. In Fig. 59.3a the bioluminescence induction in the SOS-LUX test is increasing after a lag phase and reaches its maximum after about 5 h, whereas the bioluminescence in the presence of water only does not increase. The corresponding absorption curves are shown in Fig. 59.3b. The values for the samples with mitomycin C are nearly identical to those of the water samples because mitomycin C is not toxic under these conditions. In Fig. 59.3c the kinetics of the corresponding fluorescence in the presence of mitomycin C in the LAC-FLUORO test is shown, again compared to the solvent only. With water as well as with mitomycin C in these non-toxic concentrations the fluorescence is constantly increasing because the cells are dividing during the incubation and the number of fluorescent cells becomes higher. In the presence of aureomycin there is no bioluminescence induction (Fig. 59.3d) due to the lack of genotoxicity of this compound. The increase in adsorbance (Fig. 59.3e) as well as the increase in fluorescence (Fig. 59.3f) with incubation time becomes significantly smaller with higher concentrations of aureomycin. The cells grow slower or do not grow at all because of the toxicity of aureomycin. Similar results for the kinetics of cellular responses concerning cell number/biomass, bioluminescence and fluorescence were also obtained for other standard reagents used in geno- and cytotoxicity tests like chloramphenicol, N-methyl-N'-nitro-N-nitrosoguanidine, and 4-nitrochinoline-1-oxide (results not shown).

59.3.2
Dose-Response Curves of the LUX-FLUORO Test

Figure 59.4 gives the dose-response curves for the bioluminescence induction of the SOS *lux* reporter, the impairment of cell growth, and the expression of GFPuv after treatment with the indicated agents. The dose-response curves illustrating the genotoxic activities of these standard reagents are given in Fig. 59.4a. The cytotoxic potential of the compounds was assessed from the relative absorbance A_x/A_0 (Fig. 59.4b) and relative GFPuv fluorescence (Fig. 59.4c) with increasing concentrations. The simultaneous measurements of light emission, cell number/mass and GFPuv expression allows for a discrimination between genotoxic and cytotoxic potency of the test substances. A decrease in fluorescence and absorbance during incubation suggests the agent in question to be cytotoxic, as it is the case for aureomycin and N-methyl-N'-nitro-N-nitrosoguanidine in higher concentrations. For predominantly genotoxic agents, bioluminescence is induced in the concentration range, where cell mass production and GFPuv expression is barely influenced, as it is the case for mitomycin C concentrations up to 3 µM. Agents displaying a cytotoxic as well as a genotoxic capability identify themselves by an induced bioluminescence, and a reduction in cell mass as well as GFPuv expression. This could clearly be demonstrated for example for 4-nitrochinoline-1-oxide which is a potent genotoxin, having a high cytotoxic activity.

The same type of experiments was performed to investigate whether the LUX-FLUORO test is applicable for the quantification of the geno- and/or cytotoxic effects not only

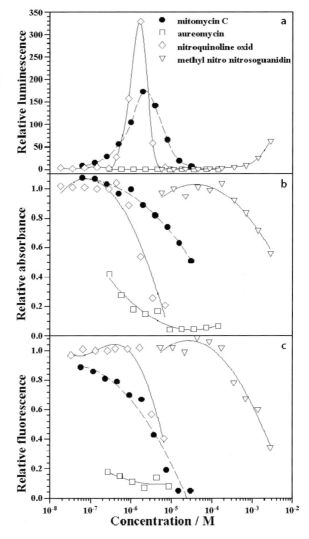

Fig. 59.4.
Dose-response curves of the combined LUX-FLUORO test for mitomycin C, aureomycin, 4-nitroquinoline-N-oxid and methyl nitro nitrosoguanidin: **a** relative bioluminescence; **b** relative absorbance; **c** relative fluorescence

of organic substances but also of heavy metal salts. In Fig. 59.5 the dose response curves for salts $K_2Cr_2O_7$, $CrCl_3$ and $ZnSO_4$ are given. From these three heavy metal salts only $K_2Cr_2O_7$ shows a clear genotoxic effect with a luminescence induction factor F_i of 8 (Fig. 59.5a) whereas $CrCl_3$ gives a weak signal with a F_i of nearly 3 at a concentration of 0.13 mM after 5 h of incubation, no F_i higher than 2 indicating a genotoxic effect could be found for $ZnSO_4$ at the examined concentrations. The lowest concentration of $K_2Cr_2O_7$ detected to be genotoxic was 0.68 mM, which was also the lowest concentration tested. All three heavy metal salts showed a decrease in fluorescence with an increasing concentration, indicating their cytotoxicity. The strong cytotoxic reaction is accompanied by a steep decrease of the luminescence below the value of the solvent control, H_2O (Fig. 59.5b), with $K_2Cr_2O_7$ beeing also the most cytotoxic of the three salts.

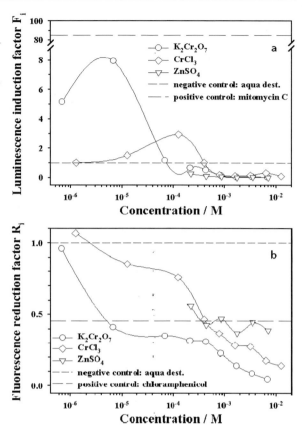

Fig. 59.5.
Dose-response curves of the combined LUX-FLUORO test for the heavy metal salts $K_2Cr_2O_7$, $CrCl_3$ and $ZnSO_4$: **a** luminescence induction factors; **b** fluorescence reduction factors

59.3.3
Advantages of the LUX-FLUORO Test

The SOS-LUX-TEST, as published Ptitsyn et al. 1997 and Horneck et al. 1998, is a receptor-reporter test, based on the plasmid pPLS-1, which combines a DNA-damage sensing promoter and the *lux* operon (*luxDCABE*) originating from *Photobacterium leiognathi*. This plasmid enables the transformed bacteria (here *Salmonella choleraesuis* TA1535) to express the luciferase and fatty acid reductase enzymes encoded by the SOS promoter controlled *lux* genes, leading to the production of bioluminescence in a dose dependent manner. In previous experiments, directly acting chemicals were tested with different *Escherichia coli* strains as host bacteria for the plasmid. Since then, the test has been improved by changing the host bacteria from *Escherichia coli* to *Salmonella choleraesuis* TA1535, one of the usual battery of tester strains in the Ames test, the most often used bacterial test for mutagenicity (Ames et al. 1973; Mortelmans and Zeiger 2000). TA1535 cells carry a *rfa* mutation leading to a defect in the outer cell membrane and a *uvrB* mutation which eliminates the excision repair system for DNA base damage. This host increased the sensitivity for the SOS-LUX test significantly, especially for hydrophobic genotoxic compounds (Rettberg et al. 2001).

Since some compounds exhibit a primary cytotoxic action, and since the induction of the SOS system and the expression of the *lux*-encoded genes may be influenced by cell death as a result of cytotoxicity, simultaneous measurements of cell mass production (A_{490}) had already been included as a parameter for testing of bacterial viability. This correction for cell mass is necessary because inhibition of growth influences the total light emission of the culture. As cell growth and cell mass production are not necessarily dependent upon each other, we then introduced a third parameter within the test, that is protein synthesis via plasmid controlled constitutive expression of GFPuv. This parameter determines metabolic activity on the base of functional protein synthesis. Green fluorescent protein (GFP) of the jellyfish *Aequorea victoria* (Chalfie et al. 1994) is a 27 kD monomer consisting of 238 amino acids. Unlike other bioluminescent reporters, the chromophore in green fluorescent protein is intrinsic to the primary structure of the protein. The fluorescence is independent of a special substrate or any cofactor, it only requires exposure to UV or blue light for emitting green light. Expression of green fluorescent protein is stable and species independent and can be monitored in living cells. The additional optimisation of wildtype green fluorescent protein for higher bacterial expression and for the maximal excitation wavelength of 395 nm renders the GFPuv variant (Crameri et al. 1996) highly favourable for the application described here. In the LAC-FLUORO test GFPuv fluorescence is mediated by the *lac*-promoter controlled plasmid pGFPuv, which is constitutively expressed in cells which lack the chromosomal repressor *lacI* as it is the case in *S. choleraesuis*. Reduction in GFPuv expression can thus well be used as an indicator for cytotoxicity.

In agreement with other tests like the MutatoxTM test (Arfsten et al. 1994), Ames test (Bianchi et al. 1983) and SOS-chromotest (Lantzsch and Gebel 1997), the results obtained with the LUX-FLUORO test displayed a clear genotoxic effect with the applied concentrations and incubation times of the choosen standard test substances with different genotoxic effects, while increasing cytotoxic effects could be shown in parallel for nearly all investigated chemicals.

59.4
Conclusion

In the combined LUX-FLUORO test the genotoxic potential of chemicals is quantitated by an increased bioluminescence and the cytotoxic potential by a decreased fluorescence in genetically modified *Salmonella typhimurium* TA1535 cells. This fast, sensitive, reproducible and inexpensive bioassay is suitable for the simultaneous determination of the genotoxic and cytotoxic potential of chemicals and mixtures of chemicals as they occur as pollutants in environmental samples, e.g. in water, air and soil. In addition to routine environmental monitoring, it can be applied for the first screening steps used for testing new substances during the development process in the pharmaceutical research.

Acknowledgements

The authors thank A. Rode, A. Backes and R. Nussbaum for assistence with the experiments.

References

Albertini RJ (2001) HPRT mutations in humans: biomarkers for mechanistic studies. Mutat Res 489:1–16

Ames BN, Lee FD, Durston WE (1973) An improved bacterial test system for the detection and classification of mutagens and carcinogens. Proc Natl Acad Sci USA 70:782–786

Arfsten DP, Davenport R, Schaeffer DJ (1994) Reversion of bioluminescent bacteria (Mutatox™) to their luminescent state upon exposure to organic compounds, munitions, and metal salts. Biomed Environ Sci 7:144–149

Baldwin TO, Christopher JA, Raushel FM, Sinclair JF, Ziegler MM, Fisher AJ, Rayment I (1995) Structure of bacterial luciferase. Current Opinion in Structural Biology 5:798–809

Baumstark-Khan C, Rode A, Rettberg P, Horneck G (2001) Application of the lux-fluoro-test as bioassay for combined genotoxicity and cytotoxicity measurements by means of recombinant *Salmonella typhimurium* TA1535 cells. Analyt Chim Acta 437:23–30

Bianchi V, Celotti L, Lanfranchi G, Majore F, Marin G, Montaldi A, Sponza G, Tamino G, Nenier P, Zantedechi A, Levis AG (1983) Genetic effects of chromium compounds. Mutat Res 117:279–300

Chalfie M, Tu Y, Euskirchen G, Ward WW, Prasher DC (1994) Green fluorescent protein as a marker for gene expression. Science 263:802–805

Crameri A, Whitehorn EA, Tate E, Stemmer WP (1996) Improved green fluorescent protein by molecular evolution using DNA shuffling. Nat Biotechnol 14:315–319

D'Souza SF (2001) Microbial biosensors. Biosens Bioelectron 16:337–353

Enzmann H, Bomhard E, Iatropoulos M, Ahr HJ, Schlueter G, William GM (1998) Short- and intermediate-term carcinogenicity testing – a review. Part 1: The prototypes mouse skin tumour assay and rat liver focus assay. Food Chem Toxicol 36:979–995

Friedberg EC (1985) DNA repair. Freeman and Company, New York

Horneck G, Ptitsyn LR, Rettberg P, Komova O, Kozubek S, Krasavin EA (1998) Recombinant *Escherichia coli* cells as biodetector system for genotoxins. In: Hock B, Barceló D, Cammann K, Hansen P-D, Turner APF (eds) Biosensors for environmental diagnostics. Taubner-Reihe UMWELT, Stuttgart, pp 215–232

ICH (1997) Harmonized Tripartite Guideline, 2SB, Genotoxicity: Standard Battery Tests, CPMP/ICH/174/95, http://www.ifpma.org/ich5s.html

Lantzsch H, Gebel T (1997) Genotoxicity of selected metal compounds in the SOS chromotest. Mutat Res 389:191–197

Mortelmans K, Zeiger E (2000) The Ames *Salmonella*/microsome mutagenicity assay. Mutat Res 455:29–60

Natarajan T, Balajee AS, Boei JJWA, Darroudi F, Dominguez I, Hande MP, Meijers M, Slijepcevic P, Vermeulen S, Xiao Y (1996) Mechanisms of induction of chromosomal aberrations and their detection by fluorescence in-situ hybridization. Mutat Res 372:247–258

Ptitsyn LR, Horneck G, Komova O, Kozubek S, Krasavin EA, Bonev M, Rettberg P (1997) A biosensor for environmental genotoxin screening based on an SOS lux assay in recombinant *Escherichia coli* cells. Appl Envir Microbiol 63:4377–4384

Quillardet P, Hofnung M (1993) The SOS chromotest: a review. Mutat Res 297:235–279

Rabbow E, Rettberg P, Baumstark-Khan C, Horneck G (2002) The SOS-LUX- and LAC-FLUORO-TEST for the quantification of genotoxic and/or cytotoxic effects of heavy metal salts. Analyt Chim Acta 456:31–39

Rettberg P, Baumstark-Khan C, Bandel K, Ptitsyn LR, Horneck G (1999) Microscale application of the SOS-LUX-TEST as biosensor for genotoxic agents. Anal Chim Acta 387:289–296

Rettberg P, Bandel K, Baumstark-Khan C, Horneck G (2001) Increased sensitivity of the SOS-LUX-test for the detection of hydrophobic genotoxic substances with *Salmonella typhimurium* TA1535 as host strain. Anal Chim Acta 426:167–173

White PA, Rasmussen JB (1996) SOS chromotest results in a broader context: empirical relationships between genotoxic potency, mutagenic potency, and carcinogenic potency. Environ Mol Mutagen 27:270–305

Chapter 60

Effects of Two Cyanotoxins, Microcystin-LR and Cylindrospermopsin, on *Euglena gracilis*

E. Duval · S. Coffinet · C. Bernard · J. Briand

Abstract

Freshwater eutrophisation causes blooms of cyanobacteria, including species that produce toxins harmful to animals. We present here a study of the effects of two hepatotoxins: microcystin-LR (heptapeptide) and cylindrospermopsin (alkaloid), on *Euglena*, a photosynthetic protist. Microcystin-LR (0.01–10 µg ml^{-1}) and cylindrospermopsin (0.13–12.5 µg ml^{-1}) showed no toxic effect on growth but significantly increased cell productivity. O_2 consumption was significantly stimulated half an hour after the toxin was added for microcystin and 48 h after for cylindrospermopsin for all the concentrations tested. In addition, a drastic inhibition of greening and photosynthesis as well as an 80% increase of reduced glutathione were observed at the high concentrations of cylindrospermopsin. Two-dimensional electrophoresis after ^{35}S amino acid labeling showed that with cylindrospermopsin a 23-kDa protein was induced in the first 2 h, whereas a 29-kDa protein was overexpressed with microcystin and cylindrospermopsin.

Key words: bioassay, cyanotoxins, *Euglena gracilis*

60.1
Introduction

Despite an increased awareness of the need for protection of freshwater habitats, human impacts continue to impair the services that these ecosystems provide. This constitutes a real problem for the supply of safe drinking water and wholesome food for human populations. Currently, in addition to chemical pollutants, drinking water or freshwater in recreation areas are also being subject to contamination by toxic cyanobacteria. Invasion of toxic cyanobacteria (*Nodularia spumigena*) was first documented by Francis in 1878 in Lake Alexandrina (Australia). Since then, they are increasingly held responsible for animals and human intoxications (Dietrich 2001).

Cyanobacteria are phototrophic prokaryotes that are among the most ancient living organisms on Earth. They exhibit a wide ecological tolerance that contributes to their competitive success in a broad spectrum of environments, allowing them to outcompete other phytoplankton organisms. Freshwater eutrophication causes the surface proliferation of cyanobacteria, leading to blooms and scums. Therefore, they are often considered as harmful organisms. Nearly half of the blooms are toxic, due to about 40 species or strains of the approximately 2 000 identified so far. These species produce toxins like neurotoxins (anatoxin-a, homoanatoxin-a, saxitoxin), skin irritants and hepatotoxins (mycrocystins, nodularins, cylindrospermopsin) (Carmichael 1992; Codd 1995). The toxins are often water-soluble and exhibit high chemical stabil-

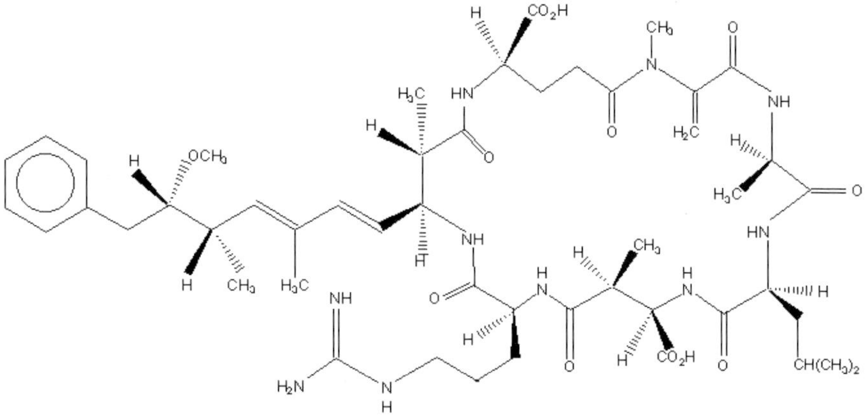

Fig. 60.1. Structure of microcystin-LR ($C_{49}H_{74}N_{10}O_{12}$, molecular weight = 995.2), an hepatotoxin produced by *Microcystis aeruginosa*

ity. Released in water in substantial amounts when the cyanobacterial cells die, their environmental persistence has important implications for public health.

Among these toxins, microcystins are a group of over 60 heptapeptide hepatotoxins, including microcystin-LR (Fig. 60.1). They have been identified among several cyanobacteria genera such as *Anabaena*, *Microcystis*, *Planktothrix* and *Nostoc*. Both lethal and sub-lethal effects have been demonstrated for a number of vertebrates and invertebrates (Falconer 1994). The liver is their main target in mammals. In 1996, 100 patients developed acute liver failure and 70 of them died at an hemodialysis center in Caruaru (Brazil). This was related to microcystins in the water used for hemodialysis (Carmichael 1996). Microcystins are inhibitors of serine/threonine protein phosphatases PP1 and 2A, leading to hyperphosphorylation of proteins (MacKintosh et al. 1990). Protein phosphorylation is the major postranslational modification performed by eukaryotic cells and is controlled by the balance between kinases and phosphatases. It is involved in the regulation of many cellular processes including metabolism, gene expression, extracellular signalling, cytosqueletal changes and cell cycle. This effect on cell division may explain the tumor-promoting activity of these toxins on the liver reported by Falconer (1991) and Nishiwaki-Matsushima et al. (1992). For a review of the toxic effects of microcystins see Dawson (1998).

In November 1979, an outbreak of hepatoenteritis and renal damages involved 148 people in Palm Island (Australia). A new toxin, cylindrospermopsin, was purified from the Palm Island strains of *Cylindrospermopsis raciborskii* (Ohtani et al. 1992). Cylindrospermopsin is a powerful hepatotoxin produced by cyanobacteria such as *Cylindrospermopsis raciborskii*, *Umezakia natans*, *Aphanizomenon ovalisporum*. It is an alkaloid with a cyclic guanidine system half bridged to a hydroxymethyluracil group (Fig. 60.2). The toxicity seems to be due to its inhibitory effect on protein synthesis (Terao et al. 1994) which may be the consequence of the binding of a cylindrospermopsin-activated metabolite to DNA (Shaw et al. 2000). Recently, indications of clastogenic and aneugenic action were noted (Humpage et al. 2000).

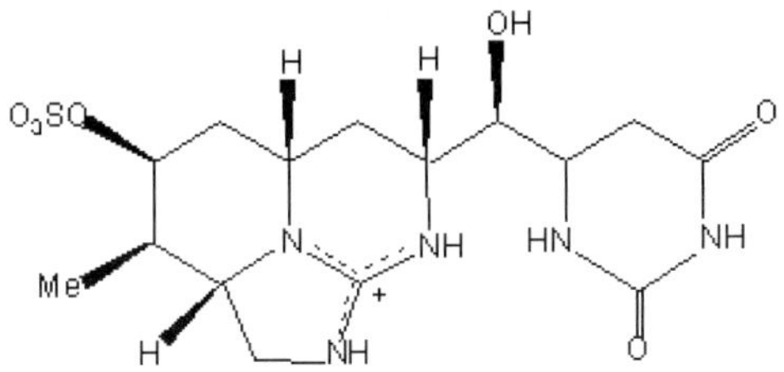

Fig. 60.2.
Structure of cylindrospermopsin ($C_{15}H_{21}N_5O_7S$, molecular weight = 415), an hepatotoxin produced by *Cylindrospermopsis raciborskii*

Here, we report the effects of two hepatotoxins (microcystin-LR and cylindrospermopsin) on *Euglena gracilis*, a flagellate protist. This eukaryotic cell can grow heterotrophically like animals cells, or phototrophically like vegetal cells. In the first case, because of its similarities with the hepatocyte, it is an interesting model cell of mammalian hepatic metabolism previously used in our laboratory for testing drugs (Briand et al. 1992). For example, the processes which transform lactate into polysaccharide reserves in *Euglena* and in hepatocytes are almost identical. Ethanol had similar effects in both cell types on lipid composition and cytochromes P450. Immuno-blotting analysis has demonstrated the presence of several isoenzymes of cytochromes P450 recognized by antibodies to human or rat cytochromes (Briand et al. 1993). As it plays an important role in fresh water as a component of the natural food chain therefore, phototrophically-grown *Euglena*, has also proved to be a good test organism in the study of the role of different cytotoxic substances including heavy metals (Fasulo et al. 1982). It grows easily on low cost medium and is very sensitive to environmental changes. For these reasons we use *Euglena*, in a multiparametric assay named "Euglenotox" to detect polluted waters (Briand 2002). It is therefore of considerable interest to establish whether *Euglena* reacts to cyanotoxins and specially to hepatotoxins and wether it will constitute a sensitive bioassay for the detection of these toxins in freshwater. Moreover, many studies have reported the effects of cyanotoxins on mammals and fish but few studies report the effects of cyanotoxins on protists.

60.2
Experimental

60.2.1
Cells Culture

Euglena gracilis cells (strain Z) were grown at 26 °C, heterotrophically in a mineral medium supplemented with lactate (33 mM) as carbon source, as well as vitamins B1 and B12 according to Calvayrac (1972). Toxins were added at different concentrations (microcystin-LR: 0.01–10 µg ml^{-1} and cylindrospermopsin: 0.013–12.5 µg ml^{-1}) after 48 h at the beginning of the exponential phase of growth. The effects on growth and cell viability, O_2 evolution, chlorophyll contents, proteins and reduced glutathione were studied for a period of seven days. Then the cultures were exposed to light for greening and photosynthesis studies. Microcystin-LR was obtained from Sigma. For cylindrospermopsin the toxin was extracted and partially purified from a 16-liter *Cylindrospermopsis raciborskii* strain AWQC CYP-026J culture. The cells were harvested by centrifugation

(20 min, 8 000 g) and lyophilized. Extraction was done in 15% methanol in Milli-Q water. The cells are disrupted by ultrasonication (ice bath, 10 min) and 3 cycles of freezing/thawing. The extracts were then centrifuged (10 min, 13 000 g) to remove cellular fragments and stored at –20 °C before use. Cylindrospermopsin was quantified with the HPLC system Varian 9010 constituted by an auto-sampler, a controlling unit and an UV detector Varian 9050 (HPLC: high performance liquid chromatography, UV: ultraviolet). The extract was eluted with a water/methanol gradient (98/2) on a C18 BDS Hypersil 5 µm cartridge from ThermoHypersil with a precolumn of the same composition. A single wavelength UV absorption (λ = 262 nm) was used for detection and quantification was done by comparison with a purified sample provided by Falconer.

60.2.2
Physiological Studies

Cell number was determined daily with a Malassez counting chamber after immobilization of the cells with a 5% solution of IK. Cell viability was estimated after Trypan blue coloration.

After seven days in the dark the cultures were exposed to continuous light (1 500 lux) for 48 h. Thereafter the chlorophylls were extracted with acetone/water (90/10, v/v) containing a small amount of $CaCO_3$ to avoid acidification. After centrifugation, the supernatant was collected. Total chlorophyll were determined by measuring fluorescence at 673 nm with a 428 nm excitation in a Hitachi F2000 spectrofluorometer. O_2 exchange was measured by polarography with an Hansatech oxygraph (DW2/2) at 25 °C, in the dark for O_2 uptake, and in the light (60 µE m^{-2} s^{-1}) for photosynthesis. Reduced glutathione was quantified with the bioxytech GSH-400 kit (Oxis).

All experiments were repeated three times independently. Results are expressed as mean ±SD (standard deviation), with statistical evaluation by analysis of variance followed by Fisher's least significant difference test. Differences were considered to be statistically significant at $P < 0.05$.

60.2.3
Protein Labeling and Electrophoresis

Cells were labelled with 0.37 MBq ml^{-1} of ^{35}S methionine and cysteine for two hours, added together with the toxins or 24 h later. Cells were harvested by centrifugation and washed twice with 10 mM Tris-HCl (pH 7.4). The pellets were incubated for 2 to 12 h in lysis buffer (5.10^4 cells µl^{-1}) (9.5 M urea, 2% NP40, 5% β-mercaptoethanol, 2% ampholines 5-7: 1.6% and 3-10: 0.4%). After 10 s of sonication, the lysates were centrifuged for 5 min at 5 000 g to release paramylon. Supernatants were analyzed by mono-dimensional electrophoresis in 12.5% polyacrylamide gels (Barques et al. 1983), and two-dimensional electrophoresis (Barques et al. 1990) with the DUAL mini-slab kit AE6450 (ATTO corp, Japan). For the first dimension a non-equilibrium pH gradient was used (NEPHGE) and for the second dimension the proteins were separated on a 12.5% polyacrylamide gel containing 0.1% SDS (sodium dodecyl sulfate). After staining (Coomassie blue R250) and drying, gels were exposed to Kodak X-OMAT AR-5 films at –20 °C. Proteins were quantified according to Bradford (1976) with BSA (bovine serum albumin) as standard.

60.3 Results and Discussion

The aims of this study were to investigate whether hepatotoxins could be detected by the bioassay "Euglenotox", and to explore some of the physiological effects of these toxins on *Euglena*.

60.3.1 Growth

The growth curve obtained with control cells presents *(i)* a lag phase, *(ii)* an exponential phase, *(iii)* a slackening phase and *(iv)* a stationary phase. The cell population density at phase iv gives the productivity of the culture. Drugs or pollutants often modify one or other of the characteristics of these phases. Therefore growth is one of the parameters of "Euglenotox".

We present in Fig. 60.3 the results obtained with microcystin. The duration of the lag phase was not modified. The generation time was decreased from 12 ±2 h for the control to 9 ±1 h for the highest concentration of toxin. However variance analysis indicated no significant difference from the control for all the concentrations studied. Cell productivity increased significantly from 2.6 ±0.1 to 3.2 ±0.08 × 10^6 cells ml^{-1}. The results were analogous for the two toxins. This increase of productivity may be due to the organic matter of the toxin or extract for two reasons: *(i)* it depends on the concentration of the toxin, *(ii)* the concentration of toxin at the end of the experiment (in the medium and the cells) is significantly diminished. Growth inhibition was found on other protists. On *Tetrahymena pyriformis*, Ward and Codd (1999) found an inhibition of growth rate with an IC$_{50}$ of 160 µg ml^{-1} and of maximum cell population density with an IC$_{50}$ of 85 µg ml^{-1}. Christoffersen (1996) found on heterotrophic nanoflagel-

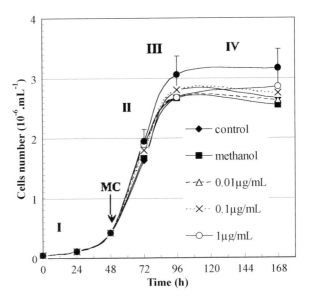

Fig. 60.3.
Growth of *Euglena* exposed to various concentrations of microcystin-LR during 7 d. *I*: lag phase; *II*: exponential phase; *III*: slackening phase; *IV*: stationary phase. Microcystin (*MC*) was added at the beginning of the exponential phase

lates an average reduction in growth rates of 49% after a toxic bloom which led to a maximum concentration in water of 141 µg l^{-1}. Here, in *Euglena*, no toxic effect on growth is shown either with Microcystin-LR or cylindrospermopsin. Microcystins are unable to cross the plasma membrane and to penetrate directly in cells. This limits the toxic effects to cells like hepatocytes that express a transporter. If *Euglena* expresses a transporter, toxic effects would be expected. On the contrary, if *Euglena* has no transporter, what mechanism could be responsible for the disappearance of the toxin from the medium? This led us to hypothesize that *Euglena* is able to detoxify the toxins.

60.3.2
Respiration

O_2 consumption was significantly stimulated in the presence of microcystin-LR (Fig. 60.4a). A 200% increase was observed at the highest concentrations. This stimulation operates very quickly, becoming measurable within half an hour after addition of the toxin.

For all cylindrospermopsin concentrations tested (Fig. 60.4b), a significant stimulation was observed after 48 h. Thus, both toxins increase O_2 consumption but not in the same manner.

Ward and Codd (1999) noticed a significant inhibition of respiration on *Tetrahymena* after 20 min exposure for all concentrations tested but they worked with much higher

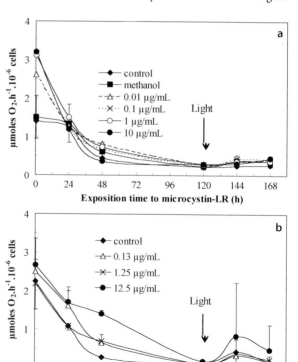

Fig. 60.4.
O_2 consumption of *Euglena gracilis* exposed to various concentrations of microcystin-LR (a) or cylindrospermopsin (b) during 7 d. O_2 consumption was significantly stimulated for all the concentrations tested half an hour after the toxin was added for microcystin and 48 h after for cylindrospermopsin

concentrations of microcystin-LR (25–500 µg ml^{-1}). Lawton (1992) found on *Paramecium primaurelia* that the respiration rate of cells exposed to 200 µg ml^{-1} of mycrocystin-LR was initially stimulated but after one hour the respiration was reduced. Pollutants often cause inhibition of sensitive steps of metabolic pathways that will result in reduced O_2 consumption and ATP production. In some cases an increase in respiration may be a consequence of an increased energy demand for ion uptake or export at the plasma membrane level (Meharg 1993) or for detoxification reactions mediated by cytochrome P450 monooxygenases. The uptake of microcystin-LR in hepatocytes occurs through ATP-dependent membrane transporters. As mycrocystin seems to be metabolized by *Euglena*, a possibility is that the increase in respiration could be linked to an increase of ATP consumption by the transporters. In the case of cylindrospermopsin, preliminary assays indicate that the increase in O_2 consumption seems to be correlated with an increase in KCN resistance. Further studies will be carried out to see if an alternative respiratory pathway is induced by the treatment, or if other reactions explain the increase in O_2 consumption.

60.3.3
Chlorophylls and Photosynthesis

Many pollutants such as heavy metals and herbicides strongly inhibit photosynthesis. These pollutants react with the photosynthetic apparatus at various levels of organization, causing alterations of the functions of chloroplast. They interact at the mo-

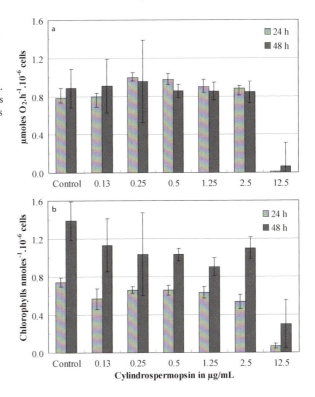

Fig. 60.5. Effects of cylindrospermopsin on greening and photosynthesis after 120 h of contact with the toxin in the dark then exposure to light for 24 h or 48 h. **a** Photosynthesis; **b** chlorophylls content. Significant differences ($P < 0.05$) were observed only at the highest concentration that induced a drastic inhibition of greening and photosynthesis

lecular level with protein complexes PSII and PSI and with photosynthetic enzymes of the carbon reducing cycle (Prasad and Strzalka 1999). They also act indirectly, for example by inhibiting the biosynthesis of chlorophylls or other pigments. For these reasons the study of chlorophylls and photosynthesis is an important part of our multiparametric test "Euglenotox".

Microcystin-LR has no significant effect on chlorophyll contents or photosynthesis. For cylindrospermopsin (Fig. 60.5), significant differences ($P < 0.05$) were observed only at the highest concentration that induced a drastic inhibition of greening (Fig. 60.5b) and photosynthesis (Fig. 60.5a). There is little information on the effects of microcystins or cylindrospermopsin on plants. In 1990 Siegl et al. noticed an inhibition of CO_2 incorporation in spinach leaf discs. In 1996, Toshihiko et al., on bean leaves treated with microcystin concentrations above 9 µg ml^{-1}, observed a 50% inhibition of photosynthetic rate within 8 h. At 9 µg ml^{-1} the inhibition was transient and the normal photosynthesis rate was recovered after 5 d. This could explain why we observed no effect at concentrations below 10 µg ml^{-1}, given that our photosynthesis measurements were performed on cells five days after addition of the toxin.

60.3.4
Reduced Glutathione

Glutathione has a key role in cellular defences against oxidative stress, and detoxifies numerous xenobiotics. With cylindrospermopsin at 2.5 and 12.5 µg ml^{-1}, it was possible to observe increases of reduced glutathione of 50% and 80% respectively (Fig. 60.6). This is not in agreement with the results of Runnegar et al. (1995) on cultured rat hepatocytes. They reported a decrease of reduced glutathione due to inhibition of its synthesis. An inhibitor of cytochromes P450 (α-naphtoquinone 10 µM) partially protected the cells from the fall of reduced glutathione and toxicity induced by cylindrospermopsin. An increase of reduced glutathione was observed on fish exposed to different pollutants. This could result from an increase of γ-glutamyl-cysteine synthetase, as in Cat-fish exposed to chlorothalonil, or from an inhibition of glutathione reductase (Cossu et al. 1997). In any case *Euglena* does not react to this toxin like hepatocytes but these results must be confirmed with pure toxin.

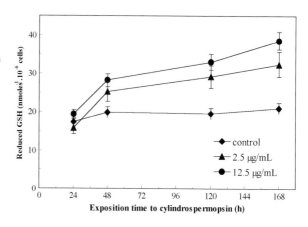

Fig. 60.6.
Effect of cylindrospermopsin on cellular content of reduced glutathione. The concentration of reduced glutathione is significantly increased after 24 h of contact with the toxin

60.3.5
Proteins

One-dimensional electrophoresis of proteins after two hours of contact with cylindrospermopsin showed a drastic inhibition of ^{35}S amino acid incorporation. Inhibition was not as strong after 24 h (Fig. 60.7a and d). This confirms the results of Terao et al. (1994) on mice. A new protein of 23 kDa (P23) was observed on the electrophoregrams after two hours of contact with cylindrospermopsin (Fig. 60.7a). This protein was not found on the auto-radiograms (Fig. 60.7b), which points to a low content in methionine and cysteine. This protein is transiently induced and it disappeared completely after 24 h of contact with the toxin (Fig. 60.7c).

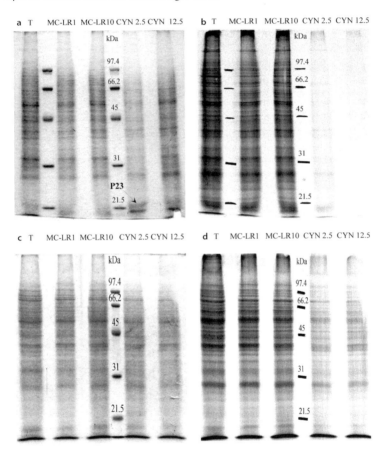

Fig. 60.7. One-dimensional gel electrophoresis of total proteins of *Euglena gracilis*. Cells were exposed to microcystin (*MC-LR*) or cylindrospermopsin (*CYN*) and labeled with ^{35}S amino acids for two hours at the same time or 24 h after. **a** and **c**: Electrophoregrams stained by Coomassie blue; **a** after two hours of contact with cylindrosperrmopsin a new protein (P23) is induced; **c** P23 has nearly disappeared after 24 h of contact with the toxin. **b** and **d** Autoradiograms of the same gels; **b** after two hours of contact with cylindrospermopsin no ^{35}S amino acids are incorporated in proteins; **d** 24 h after a weak incorporation takes place

Two-dimensional electrophoresis (Fig. 60.8), after ^{35}S amino acid labelling, showed that a 29-kDa protein (P29) is over-expressed with both microcystin-LR and cylindrospermopsin but with different kinetics in the two cases. For microcystin-LR the overexpression is observed at both time points. For cylindrospermopsin the over-expression is observed only after 24 h of exposure to the toxin. Compared to the control, cylindrospermopsin exposure for 2 h caused a general decrease in the intensity and number of spots. One exception is the 23-kDa protein previously found in the one-

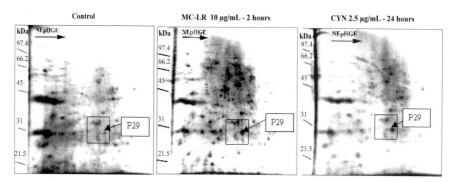

Fig. 60.8. Autoradiograms of two-dimensional gel electrophoresis. Cells were exposed to toxins – microcystin (*MC-LR*) or cylindrospermopsin (*CYN*) – during 2 h or 24 h and labeled for two hours with ^{35}S amino acids. A 29-kDa protein (P29) is overexpressed after 2 h of contact with *MC-LR* and after 24 h with *CYN*

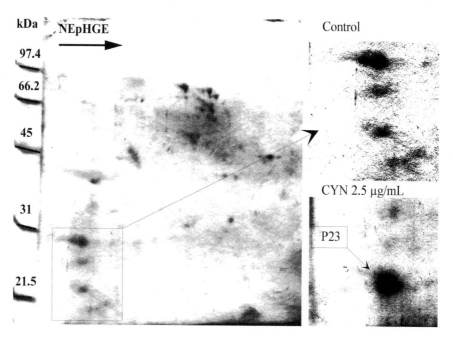

Fig. 60.9. Two-dimensional gels stained by Coomassie blue showing that P23 is induced by cylindrospermopsin (*CYN*)

dimensional gel stained with Coomassie blue and located in the acidic part of the gel (Fig. 60.9).

All prokaryotic and eukaryotic organisms that have been investigated respond to environmental stress such as exposure to heat, heavy metals, organic pollutants, and activated oxygen species (Banzet et al. 1998) by the immediate and transient synthesis of several specific proteins, historically called "heat-shock proteins" (hsp) and now more generally "stress proteins". Cessation of synthesis of most other proteins is often correlated with stress reactions. Hsp range in size from 16 kDa to 110 kDa. Some of them play an important role in the intracellular processing of proteins. They have been classified mainly into five families: hsp100, hsp90, hsp70, hsp60 and small hsp. Small hsp are more abundant in plants and some are specific to the chloroplast (hsp 22 and 25). Cylindrospermopsin is known to inhibit protein synthesis in liver cells (Terao et al. 1994) but no action on stress proteins has been reported. Two aspects of the response of *Euglena* to cylindrospermopsin exposure seem to point towards a stress reaction: *(i)* the transient inhibition of methionine and cysteine incorporation which indicates an inhibition of protein synthesis, and *(ii)* the fast induction of a specific protein (P23). The P29 protein induced by the two toxins appears to have an iso-electric point and a molecular weight that are very similar to those of a stress protein induced by heat shock in our strain of *Euglena*, although this remains to be confirmed. Stress proteins constitute good biomarkers for detection of polluted waters (Bonaly and Barque 1997). P23 and P29 could therefore constitute interesting biomarkers provided that their induction occurs at lower concentrations than those we have tested here.

60.4 Conclusions

Cylindrospermopsin and microcystin-LR show different types of effects on *Euglena*. O_2 consumption is increased. Greening and photosynthesis are not significantly modified except at the highest concentration of cylindrospermopsin. An increase in reduced glutathione and a strong transient inhibition of protein synthesis are observed with cylindrospermopsin (2.5 and 12.5 µg ml^{-1}). Two proteins are induced or overexpressed: P23 by cylindrospermopsin and P29 by both toxins. Thus, *Euglena* reacts differently to the toxins than hepatocytes or other protists such as *Tetrahymena*. Some of these responses, such as induction of proteins, may constitute potential biomarkers.

Acknowledgements

We are grateful to Dr. P. Baker – Australian Water Quality Centre, Salisbury, 5108 South Australia – who provided us with *Cylindrospermopsis raciborskii* strain AWQC cyp-026J; to V. Fessard and J.-M. Delmas – Laboratoire d'étude et de recherche sur les médicaments vétérinaires et désinfectants, AFSSA de Fougères, France – for the quantification by HPLC of cylindrospermopsin and to Dr. C. Butor and Dr. G. Noctor for reviewing the manuscript. This work was supported by Agence de l'Eau Loire Bretagne.

References

Banzet N, Richaud C, Deveaux Y, Kazmaier M, Gagnon J, Tryantaphyllidès C (1998) Accumulation of small heat shock proteins, including mitochondrial HSP22, induced by oxydative stress and adaptative response in tomato cells. Plant J 13:519–527

Barque J-P, Danon F, Peraudeau L, Yeni P, Larsen CJ (1983) Characterization by human antibody of a nuclear antigen related to the cell cycle. EMBO J 2:743–749

Barque J-P, Karniguian A, Brisson-Jeanneau C, Della-Valle V, Grelac F, Larsen CJ (1990) Human autoantibodies identify a nuclear chromatin-associated antigen (PSL or p55) in human platelets. Eur J Cell Biol 51:183–187

Bonaly J, Barque J-P (1997) Protéines de stress et microalgues. In: Lagadic L, Caquet T, Amiard J-C, Ramade F (eds) Biomarqueurs en écotoxicologie. Masson, Paris, pp 97–102

Bradford MM (1976) A rapid and sensitive method for the quantitation of microgam quantities of protein utilizing the principe of protein-dye binding. Anal Biochem 72:248–254

Briand J (2002) Euglenotox – utilisation of *Euglena* as a pollution biomarker. Cordis Focus 38:25

Briand J, Blehaut H, Calvayrac R, Laval-Martin D (1992) Use of a microbial model for the determination of drug effects on cell metabolism and energetics: study of citrulline-malate. Biopharm Drug Disp 13:1–22

Briand J, Julistiono H, Beaune P, Flinois JP, De Wazier I, Leroux JP (1993) Presence of different different ent forms of cytochrome P450 in *Euglena gracilis*. Biochim Biophys Acta 1203:199–204

Calvayrac R (1972) Le cycle des mitochondries chez *Euglena gracilis* en culture synchrone. Thèse de doctorat – Paris XI-University

Carmichael WW (1992) Cyanobacteria secondary metabolites – the cyanotoxins. J Appl Bacteriol 72: 445–459

Carmichael WW (1996) Liver failure and human deaths at a hemodialysis center in Brazil: microcystins as a major contributing factor. Harmful Algae News 15:11

Christoffersen K (1996) Effect of microcystin on growth of single species and of mixed natural populations of heterotrophic nanoflagellates. Nat Tox 4:215–220

Codd GA (1995) Cyanobacterial toxins: occurrence, properties and biological significance. Water Sci Technol 32:149–156

Cossu C, Doyotte A, Jacquin M-C, Vasseur P (1997) Biomarqueurs du stress oxydant chez les animaux aquatiques. In: Lagadic L, Caquet T, Amiard J-C, Ramade F (eds) Biomarqueurs en écotoxicologie. Masson, Paris, pp 149–163

Dawson RM (1998) The toxicology of microcystins. Toxicon 36:953–962

Dietrich RD (2001) Détecter les cyanotoxines des eaux. Biofutur 209:44–47

Falconer IR (1991) Tumour promotion and liver injury caused by oral consumption of cyanobacteria. Environ Toxicol Water Qual 6:177–184

Falconer IR (1994) Mechanism of toxicity of cyclic peptide toxins from blue-green algae. In: Falconer IR (ed) Algal toxins in seafood and drinking water. Academic Press, London, pp 177–187

Fasulo MP, Bassi M, Donini A (1982) Cytotoxic effects of hexavalent chromium in *Euglena gracilis*. I: First observations. Protoplasma 110:39–47

Humpage AR, Fenech M, Thomas P, Falconer IR (2000) Micronucleus induction and loss chromosome in transformed human white cells indicate clastogenic and aneugenic action of the cyanobacterial toxin, cylindrospermopsin. Mutat Res 472:155–161

Lawton LA (1992) Biological effets and significance of cyanobacterial peptides toxins. PhD Thesis, University of Dundee

MacKintosh C, Beattie C, Klumpp KA, Cohen P, Codd GA (1990) Cyanobacterial microcystin-LR is a potent and specific inhibitor of protein phosphatases 1 and 2A from both mammals and higher plants. FEBS Lett 264:187–192

Meharg AA (1993) The role of the plasmalemma in metal tolerance in angiosperms. Physiol Plant 88:191–198

Nishiwaki-Matsushima R, Ohta T, Nishiwaki S, Suganuman M, Koyama K, Ishiwaka T Carmichael WW, Fujiki H (1992) Liver tumor promotion by the cyanobacterial cyclic peptide toxin microcystin LR. J Cancer Res Clin Oncol 118:420–424

Ohtani I, Moore RE, Runnegar MTC (1992) Cylindrospermopsin: a potent hepatotoxin from the blue-green alga *Cylindrospermopsis raciborskii*. J Am Chem Soc 114:7941–7942

Prasad MNV, Strzalka K (1999) Impact of heavy metals on photosynthesis. In: Prasad MNV, Hagemeyer J (eds) Heavy metal stress in plants – from molecules to ecosystems. Springer-Verlag, Berlin Heidelberg New York, pp 117–138

Runnegar MTC, Kong S-M, Zhong Y-Z, Lu SL (1995) Inhibition of glutathione synthesis by cyanobacterial alkaloid cylindrospermopsin in cultured rat hepatocytes. Biochem Pharmacol 49:219–225

Shaw GH, Seawright AA, Moore MR, Lam PKS (2000) Cylindrospermopsin, a cyanobacterial alkaloid: evaluation of its toxicologic activity. Therapeutic Drug Monitoring 22:89–92

Siegl G, MacKintosh C, Stitt M (1990) Sucrose-phospate synthase is dephosphorylated by protein phosphatase 2A in spinach leaves. FEBS Lett 270:198–202

Terao K, Ohmori S, Igarashi K, Ohtani I, Watanabe MF, Harada KI, Ito E, Watanabe M (1994) Electron microscopic studies in experimental poisoning in mice induced by cylindrospermopsin isolated from blue-green algae *Umezakia natans*. Toxicon 32:833–843

Toshihiko A, Lawson T, Weyers JDB, Codd GA (1996) Microcystin-LR inhibits photosynthesis of *Phaseolus vulgaris* primary leaves: implications for current spray irrigation practice. New Phytol 133:651–658

Ward CJ, Codd GA (1999) Comparative toxicity of four microcystins of different hydrophobicities to the protozoan, *Tetrahymena pyriformis*. J Appl Microbiol 86:874–882

Part VII

A New Bioassay for Toxic Chemicals Using Green Paramecia, *Paramecium bursaria*

M. Tanaka · Y. Ishizaka · H. Tosuji · M. Kunimoto · N. Hosoya · N. Nishihara
T. Kadono · T. Kawano · T. Kosaka · H. Hosoya

Abstract

We designed a new toxic bioassay using the green paramecia *Paramecium bursaria* as testing organism. *P. bursaria* is a unicellular organism that occurs widely in rivers and ponds. Since *P. bursaria* uses metabolites of endosymbiotic green algae in the cytoplasm as a nutritive source, culturing *P. bursaria* is much easier than culturing mammalian cells. The use of *P. bursaria* will thus make quicker and more convenient evaluation of toxicity of various polluting chemicals. Here, we selected thirty-two pollutants such as pesticides, toxic metals and polycyclic aromatic hydrocarbons. Those substances were added at various concentrations to the culture medium of paramecia. Then the IC_{50} values, defined as the concentrations of chemicals inhibiting the growth of organisms by 50%, obtained for both paramecia and mammalian cell cultures were compared. We found that paramecia were much highly sensitive to some chemicals such as methylmercury chloride and mercuric chloride, compared to cultured mammalian cells. We conclude that *P. bursaria* is one of the best organism for assessing the effect of chemical pollutants in the aqueous environment.

Key words: bioassay, endosymbiotic algae, pesticides, heavy metals, PAHs, endocrine disruptors, green paramecia, IC_{50}, water pollution

61.1 Introduction

Environmental pollution has recently become a very complex issue due to the pollution of natural media by multiple chemicals. Many pollutants such as endocrine disruptors have been studied, as they are suspected of having an effect on the endocrine system. It has indeed been suggested that some pollutants act like a sex hormone, estrogenic hormone, and alter generative functions (Colborn et al. 1993; Guillette et al. 1994; Horiguchi et al. 1995; vom Saal et al. 1998). There are many chemical substances whose toxicity have not been detected or investigated in the environment. The analyses of toxicity of chemical substances on cells and organisms are urgently required. The bioassay system, a biological evaluation system, is useful to estimate the toxicity of chemical substances in living organisms. Several analytical methods using cultured mammalian cells have been developed (Soto et al. 1991; Shelby et al. 1996). However, the bioassay using cultured mammalian cells is inconvenient for rapid toxicological analyses and detection of chemical pollutants.

Here, we designed a new convenient bioassay using green paramecia, *Paramecium bursaria*, for rapidly assessing the toxicity of chemical pollutants. *P. bursaria* is a unicellular organism, which widely occurs in rivers and ponds. The name *P. bursaria* originates from the Latin word "*bursa*" meaning "a pocket" or "a pouch" since a single cell of *P. bursaria* packs several hundreds of green algae in its cytoplasm as endosymbionts. Noteworthy, *P. bursaria* can live on photosynthetic products supplied from the symbiotic algae (Weis 1979). Therefore culturing *P. bursaria* is much easier than culturing mammalian cells. Because *P. bursaria* is a unicellular organism, we can observe the impact of chemical substances not only on cells but also on individual organisms. With these advantages to the conventional methods, much quicker and more convenient evaluation of toxicity of various chemical pollutants can be achieved by the use of *P. bursaria*. In addition, it has been reported that endosymbiotic algae in *P. bursaria* can be removed from the host cells in the presence of a herbicide, paraquat, and alga-free "white" paramecia can be prepared (Hosoya et al. 1995; Nishihara et al. 1996). Thus, bleaching of green colour from *P. bursaria* could be a novel visual indication for monitoring of the contamination with certain herbicides.

In the present study, we selected thirty-two chemical substances as pollution indicators that are considered to be toxic to cultured mammalian cells (Soto et al. 1991; Shelby et al. 1996). The IC_{50} values indicating the concentration of chemicals lowering the growth rate of the culture by 50% were compared with the data obtained from the cultured mammalian cells. Here we found that *P. bursaria* is a sensitive pollution-indicating organism, thus suitable for evaluation of the toxic effects of various water pollutants.

61.2
Experimental

61.2.1
Chemicals

Thirty-two chemical substances including potential genotoxic carcinogens, herbicides, insecticides, anti-microbials and organic solvents were selected from the pollution-indicating substances listed by the National Institute for Environmental Studies, Japan. The names, formula (or abbreviations) and the range of concentrations (final concentrations in µM, unless indicated) of the substances tested are: mercuric chloride ($HgCl_2$, 0.02–200)*, cadmium chloride ($CdCl_2$, 0.01–100), nickel chloride ($NiCl_2$, 0.1–1 000), potassium dichromate ($K_2Cr_2O_7$, 0.001–10), triphenyltin chloride (Ph_3SnCl, 0.0001–1)*, tributyltin chloride (Bu_3SnCl, 0.0002–2)*, maneb (0.005–50)*, cupric sulphate ($CuSO_4$, 0.1–1 000), sodium arsenite ($NaAsO_2$, 0.01–100), potassium cyanide (KCN, 0.5–5 000), 2,4-dichlorophenoxyacetic acid (2,4-D, 0.2–2 000)*, benthiocarb (0.2–2 000)*, paraquat (0.1–1 000), malathion (0.2–2 000)*, thiuram (0.01–100)*, 2,4,5-trichlorophenol (2,4,5-TCP, 0.1–1 000)*, 2,5-dichlorophenol (2,5-DCP, 0.1–1 000)*, formaldehyde (HCHO, 0.05–500), di-(2-ethylhexyl)phthalate (DEHP, 0.25–2 500)*, 4-nitroquinoline-N-oxide (4NQO, 0.0005–5)*, 2-aminoanthracene (2-AA, 0.02–200)*, benzo[a]pyrene (B[a]P, 0.02–200)*, 1-methyl-5H-pyride(4,3-b)indol-3-amine (Trp-P-2, 0.01–100)* and dimethylsulfoxide (DMSO, 0.001–1%) were purchased from Wako Pure Chemical Industries, Osaka, Japan. Methylmercury chloride (MeHgCl, 0.002–20)*, pentachlorophenol (PCPhOH, 0.02–200)* and phenol (PhOH, 0.5–5 000)* were from Nakalai Tesque Inc., Kyoto, Japan. Gamma-hexachlorocyclohexane (lindane, 0.2–2 000)* and hexachlorophene (HCP, 0.02–200)* were from Sigma, St. Louis, Mo., USA. *p*-Nonyl-phenol (pNPhOH, 0.02–200)* and bis-phenol-A (BisPhA, 0.1–1 000)* were from Tokyo Chemical Industry Co., Tokyo, Japan. Triclosan (0.01–100)* was obtained from Ciba-Geigy Ltd., Basel, Switzerland. *: dissolved either in 1%, v/v DMSO or in the ultra-pure water (DMSO: dimethylsulfoxide).

61.2.2
Culture of Ciliates

One strain of *Paramecium bursaria* syngen 1 (KSK-103, mating type IV) collected in 1995 from Kawasa-kyo in Fuchu City, Hiroshima, Japan, was used here. The paramecia were propagated in lettuce infusion containing the food bacterium *Klebsiella pneumoniae*, with initial *Paramecium* density of 20 cells ml^{-1}. The culture was propagated under a light-dark cycle (LD; L:D = 12:12 h) at 23 °C. After 7 to 10 d of pre-culture, cells in the logarithmic growth phase were collected and used for further experiments.

61.2.3
Determination of Growth Rate and IC_{50} Value

For determination of growth rates, paramecia were cultured on 12-well microplates: IWAKI flat bottom, tissue culture-treated polystyrene, Asahi Techno Glass Corp., Tokyo, Japan. Each well on microplates was filled with 1 ml of fresh lettuce infusion containing various concentrations of environmental pollution-indicating substances and culture was started at the initial cell density of 20 cells ml^{-1}. After 1, 2 and 5 d of incubation at 23 °C, the number of cells in each well was counted under microscopes. The growth rates of paramecia determined for initial 1, 2 and 5 d of culture were expressed as relative increases, as compared to the initial number of cells at the beginning of the culture. Changes in growth rates were examined in the presence or absence of various substances and the IC_{50} values defined as the concentrations of the substances, at which proliferation of paramecia is inhibited by 50% was determined.

61.3
Results and Discussion

61.3.1
Preliminary Tests

In order to assess the effects of 23 chemicals out of 32 environmental pollution-indicating substances that require trace of organic solvent such as dimethylsulfoxide (DMSO), on the proliferation of *P. bursaria*, changes in growth rate of paramecia treated with and without DMSO were examined. The number of cells at 1, 2 and 5 d after initiation of the culture was counted and the growth rate was evaluated as the values relative to the number of cells right after the culture started. At the concentration of 1.0%, v/v or less, DMSO had no effect on the growth rate of *P. bursaria*. Therefore, we used DMSO as a solvent for some chemicals tested in the assay.

61.3.2
Toxicity of Agricultural Chemicals and Antimicrobials

We tested the inhibitory effect of environmental pollutants on the growth rate of *P. bursaria*. The results were then compared with the data obtained from the cultured mammalian cells such as human cervix carcinoma cell (HeLa), human neuroblast (NB-1) and rat kidney cell (NRK-52E). Firstly, the effect of various agrochemicals including organic herbicides, insecticides, fungicides and antimicrobials, were asssessed using *P. bursaria*. When *P. bursaria* were treated with 2,4-dichlorophenoxyacetic acid at 200 µM, benthiocarb at 200 µM, pentachlorophenol at 20 µM, lindane at 200 µM, malathion at 200 µM, *p*-nonylphenol at 20 µM, hexachlorophene at 2 µM, 2,4,5-trichlorophenol at 10 µM and 2,5-dichlorophenol at 100 µM, all the cells died out within 1 to 24 h after the culture started. Moderate concentration of 2,4-dichlorophenoxyacetic acid at 2–20 µM and malathion at 20 µM showed slight inhibitory effect on the proliferation rates in paramecia.

Table 61.1 shows the list of IC_{50} values in paramecia and cultured mammalian cells HeLa, NB-1 and NRK-52E. The IC_{50} values are defined as the pollutant concentrations, at which proliferation of paramecia or cultured mammalian cells is inhibited by 50%. As shown on Table 61.1, IC_{50} values of paramecia were much lower than IC_{50} values of cultured mammalian cells for most pollutants, except for thiuram. IC_{50} value for pentachlorophenol (PCPPhOH) in paramecia was notably lower than that of cultured

Table 61.1. Toxicity IC_{50} values in the green paramecia *P. bursaria* and mammalian cultured cells for agricultural chemicals and antimicrobials. The IC_{50} values, expressed in µM, of *P. bursaria* were determined as described in the experimental section, and the average values calculated from the 2–3 sets of replications are shown (a). The values obtained from a single measurement are marked as (b). IC_{50} values in µM in mammalian cell cultures such as (c) human cervix carcinoma cell (*HeLa*), (d) human neuroblast (*NB-1*) and (e) rat kidney cell (*NRK-52E*) were quoted from the report on the special research project "Development of comprehensive toxicity testing for the assessment of total risk from environmental chemicals (2001)" conducted by the National Institute for Environmental Studies, Japan. See experimental section for abreviations significance

Chemicals	IC_{50}			
	P. bursaria[a]	HeLa[c]	NB-1[d]	NRK-52E[e]
2,4-D	72.33	968.68	1 628.0	1 773.5
Benthiocarb	5.45	988.9	232.6	3 396.84
Paraquat	30.00[b]	1 587.17	134.6	373.17
PCPhOH	0.26	193.11	93.86	298.27
Lindane	5.85	1 019.44	835.5	1 728.67
Malathion	51.00	1 726.26	1 214.0	1 125.38
pNPhOH	6.10	180.0	62.77	166.68
HCP	0.34	18.8	27.89	144.3
Thiuram	2.69	85.29	0.14	0.32
2,4,5-TCP	2.55	151.84	72.54	244.19
2,5-DCP	43.75	572.52	469.3	710.18

mammalian cells. Pentachlorophenol is widely used in agricultural fields for multiple purposes such as herbicide, fungicide and insecticide. Thus novel bioassays for detection of this compound are needed. The use of *P. bursaria* is a possible solution to this problem since it is conclusive that the sensitivity of *P. bursaria*-based assay to pentachlorophenol is 10 000 times higher than that of conventional bioassays using mammalian cell cultures. Paraquat is also widely used as a photosynthesis-related herbicide. When the cells of *P. bursaria* were treated with paraquat in the 0.1–100 µM range, the herbicide turned the cells of "green paramecia" into "white" cells within 5 d. Indeed, the endosymbiotic green algae that normally multiply to several hundred cells within the cytoplasm of a single host *Paramecium*, are highly sensitive to the herbicide and were thus readily eliminated from the host cells (Fig. 61.1). In aqueous environments surrounded by agricultural fields, contamination with various agricultural chemicals including pentachlorophenol and paraquat may thus have a major impact on aqueous organisms. The bleaching of green paramecia could thus be a novel visible indicator of a small amount of herbicides present in water.

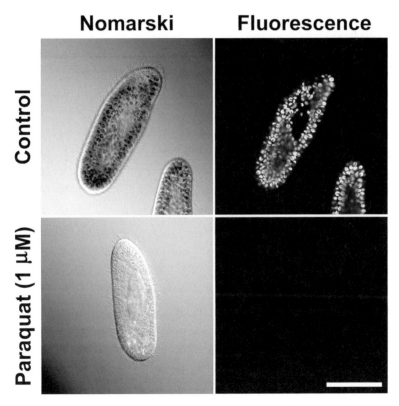

Fig. 61.1. Microscopic analysis of algal damage in the herbicide-treated *P. bursaria*. Nomarski differential interference contrast images (*left*) and the red fluorescence images (*right*). Images were obtained after 5 d of culture in the absence (*top*) or presence (*bottom*) of 1 µM paraquat. In the presence of paraquat, endosymbiotic algae were completely eliminated so that red fluorescence due to algal chlorophyll is no longer detectable in the treated paramecia (*bottom right*). *Scale bar:* 50 µm

61.3.3
Toxicity of Metallic and Metal-Containing Pollutants

Table 61.2 shows the list of IC_{50} values in *P. bursaria* and cultured mammalian cells, for metallic and metal-containing water pollutants. The data show that the sensitivities of paramecia to most chemicals, except Bu_3SnCl and $NaAsO_2$, were higher than that of cultured mammalian cells. Unexpectedly, $NaAsO_2$ showed a stimulatory effect on the growth of *P. bursaria* between 0.01 and 10 µM, and thus IC_{50} value could not be determined. When *P. bursaria* were treated with MeHgCl at 2 µM, $HgCl_2$ at 20 µM, $CdCl_2$ at 100 µM and $NiCl_2$ at 100 µM, all the cells acutely died out within 1 h after starting the culture. The IC_{50} values for MeHgCl, $HgCl_2$ and $CdCl_2$ in *P. bursaria* were much lower than that found in the cultured mammalian cells. These chemicals are typical, ubiquitous water pollutants that produce major pollution-related illness. Today, the use of mercury and mercury-containing chemicals in both industry and agriculture are restricted in most countries. However, it has been reported that loads of mercury used in the past have been accumulated by biological concentration (Zakova and Kockova 1999). Using *P. bursaria* for biomonitoring, it will thus be possible to detect the toxicity of such chemicals present in water at nano-molar levels.

For the cultured mammalian cells, KCN had no toxicity in the range of tested concentrations and, therefore, the IC_{50} values could not be determined. Although KCN showed no lethal effect for *P. bursaria* in the range of concentrations tested, KCN inhibited the proliferation of *P. bursaria* in a dose-dependent manner and IC_{50} value was determined to be about 340 µM, which is higher than the average KCN concentration in the environment.

61.3.4
Toxicity of Organic Solvents, Carcinogens, Mutagens and Metabolic Modulators

The IC_{50} values of organic solvents, carcinogens, mutagens and metabolic modulators, were determined for *P. bursaria*. Table 61.3 shows IC_{50} values determined in *P. bursaria* and cultured mammalian cells. When the cells were treated with formaldehyde at 50 µM, bis-phenol-A at 100 µM, 4-nitroquinoline-N-oxide at 0.5 µM and 2-aminoanthracene at 2 µM, all cells died out within 24 h after the culture started. Phenol at 5 000 µM and benzo[a]pyrene at 200 µM inhibited the proliferation of *P. bursaria* only at high concentrations. Di-(2-ethylhexyl)phthalate at 0.25–2 500 µM did not show any notable effect on the proliferation of *P. bursaria* within 5 d in the range of tested concentrations.

4-Nitroquinoline-N-oxide and 2-aminoanthracene were shown to be highly toxic to *P. bursaria* and therefore the IC_{50} values for those chemicals were found at very low levels, e.g. nano-molar levels, which were much lower than that found for cultured mammalian cells. 4-Nitroquinoline-N-oxide and 2-aminoanthracene are genotoxic carcinogens whose toxic actions are mediated by generation of reactive oxygen species. It is known that genotoxic carcinogens release oxygen radicals such as hydroxyl radicals in vivo (Botchway et al. 1998; Weisburger et al. 2001). It is possible that the toxic actions of 4-nitroquinoline-N-oxide and 2-aminoanthracene in *P. bursaria* are due to release of reactive oxygen species.

In addition to growth inhibition in the host *Paramecium* cells, 4-nitroquinoline-N-oxide and triclosan showed high toxicity to the endosymbiotic green algae in *P. bursaria*. When *P. bursaria* were treated with 4-nitroquinoline-N-oxide at 0.05 µM and triclosan at

Table 61.2.
IC_{50} values in *P. bursaria* and mammalian cultured cells for metallic and metal containing water pollutants. For details, see Table 61.1 caption

Chemicals	IC_{50}			
	P. bursaria[a]	HeLa[c]	NB-1[d]	NRK-52E[e]
MeHgCl	0.0082	14.47	1.78	2.33
$HgCl_2$	0.082	59.98	19.83	10.34
$CdCl_2$	0.28	>100.0	12.55	13.73
$NiCl_2$	5.10	319.35	529.6	785.79
$K_2Cr_2O_7$	0.96	4.07	3.37	9.85
Ph_3SnCl	0.036	0.36	0.21	0.5
Bu_3SnCl	0.64	0.67	0.48	0.71
Maneb	10.05	157.77	29.57	20.75
$CuSO_4$	3.55	4.91	27.02	34.39
$NaAsO_2$	>100.00[b]	>100.0	8.7	22.83
KCN	340.00	>5000.0	>5000.0	>5000.0

Table 61.3.
IC_{50} values in *P. bursaria* and mammalian cultured cells for organic solvents, carcinogens, mutagens and metabolic modulators. For details, see Table 61.1 caption

Chemicals	IC_{50}			
	P. bursaria[a]	HeLa[c]	NB-1[d]	NRK-52E[e]
HCHO	0	302.45	91.13	125.59
PhOH	1640.00	5606.04	8539.0	4716.83
BisPhA	23.50	198.8	135.0	275.32
DEHP	>2500.00[b]	>2500.0	4380.0	>2500.0
4NQO	0.06	1.28	0.39	0.47
Triclosan	4.30	54.43	27.3	139.77
2-AA	0.074	>1000.0	2849.0	>1000.0
B[a]P	90.00	>1000.0	>5000.0	>1000.0
Trp-P-2	2.93	12.12	14.05	18.67
DMSO (% v/v)	>1.0[b]	>1.0	>1.0	>1.0

1 µM, the colour of the organism turned into "white" in the same manner as the case of paraquat. Although 4-nitroquinoline-N-oxide and triclosan are not used as herbicides, it is possible that the water contaminated with these chemicals may cause major impacts on environment by damaging the photosynthetic microplankton acting as major primary producer in the aqueous ecosystem. Using green paramecia as an environmental bio-indicator, it will thus be possible to detect the toxic impact of small amounts of chemical pollutants contaminating the water. In some cases, responses of the endosymbiotic green algae, representing the responses of the photosynthetic microorganisms, to some chemicals are notably high, even at low level of pollution. Unfortunately, in this study, our tests covered only the effect of single-application of each chemical on the proliferation of paramecia. To establish the practical biomonitoring system using paramecia, it is necessary to investigate the impact of multiple applications of chemical pollutants to *P. bursaria*.

61.4 Conclusion

The toxical effect of thirty-two pollutants on the growth of *P. bursaria* was studied. The sensitivity of paramecia to chemicals was higher than that for cultured mammalian cells. In particular, the notably high sensitivity to MeHgCl, $HgCl_2$ and $CdCl_2$ was observed using paramecia. Furthermore, when *P. bursaria* were treated with paraquat in the 0.1–100 µM range, triclosan at 1 µM and 4-nitroquinoline-N-oxide at 0.05 µM, *P. bursaria* lost its green colour because the endosymbiotic green algae were damaged and removed. The bleaching of green paramecia should therefore be a useful indicator for detecting the effect of such chemicals present at low concentrations in aqueous environments. It is suggested that *P. bursaria* could be a very useful organism for developing a biomonitoring system for chemical water pollution.

Acknowledgements

We thank Ms. Toshiko Ohara for her helpful assistance. This work was supported in part by The Iwatani Naoji Foundation's Research Grant.

References

Botchway SW, Chakrabarti S, Makrigiorgos GM (1998) Novel visible and ultraviolet light photogeneration of hydroxyl radicals by 2-methyl-4-nitro-quinoline-N-oxide (MNO) and 4,4'-dinitro-(2,2')bipyridinyl-N,N'-dioxide (DBD). Photochem Photobiol 67:635–640

Colborn T, vom Saal FS, Sato AM (1993) Developmental effects of endocrine-disrupting chemicals in wildlife and humans. Environ Health Perspect 101:378–384

Guillette Jr. LJ, Gross TS, Masson GR, Matter JM, Percival HF, Woodward AR (1994) Developmental abnormalities of the gonad and abnormal sex hormone concentrations in juvenile alligators from contaminated and control lakes in Florida. Environ Health Perspect 102:680–688

Horiguchi T, Shiraishi H, Shimizu M, Yamazaki S, Morita M (1995) Imposex in Japanese gastropods (Neogastropoda and Mesogastropoda): effects of tributyltin and triphenyltin from antifouling pains. Mar Pol Bull 31:402–405

Hosoya H, Kimura K, Matsuda S, Kitaura M, Takahashi T, Kosaka T (1995) Symbiotic algae-free strains of green paramecia *Paramecium bursaria* produced by herbicide paraquat. Zool Sci 12:807–810

Nishihara N, Takahashi T, Takahashi T, Kosaka T, Hosoya H (1996) Characterization of symbiotic algae-free strains of *Paramecium bursaria* produced by the herbicide paraquat. J Protozool Res 6:60–67

Shelby MD, Newbold RR, Tully DB, Chae K, Davis VL (1996) Assessing environmental chemicals for estrogenicity using a combination of in-vitro and in-vivo assays. Environ Health Perspect 104:1296–1300

Soto AM, Justicia H, Wray JW, Sonnenshein C (1991) *p*-Nonylphenol: an estrogenic xenobitic released from modified polystyrene. Environ Health Perspect 92:167–173

vom Saal FS, Cooke P, Buchanan DL, Palanza P, Thayer KA, Nagel SC, Parmigiani S, Welshons WV (1998) A physiologically based approach to the study of bisphenol A and other estrogenic chemicals on the size of reproductive organs, daily sperm production, and behavior. Toxicol Industr Health 14:239–260

Weis DS (1979) Correlation of sugar release and Concanavalin A agglutinability with infectivity of symbiotic algae from *Paramecium bursaria* for aposymbiotic *P. bursaria*. J Protozool 26:117–119

Weisburger JH, Hosey JR, Larios E, Pittman B, Zang E, Hara Y, Kuts-Cheraux G (2001) Investigation of commercial mitolife as an antioxidant and antimutagen. Nutrition 17:322–325

Zakova Z, Kockova E (1999) Biomonitoring and assessment of heavy metal contamination of streams and reservoirs in the Dyje/Thaya River Basin, Czech Republic. Water Sci Technol 39:225–232

Chapter 62

Detection of Toxic Pollution in Waste Water by Short-Term Respirometry

S. Le Bonté · M. N. Pons · O. Potier · C. Plançon · A. Alinsafi · A. Benhammou

Abstract

A short-term batch respirometric test coupled with ultraviolet (UV) photometry was developed to detect the presence in waste water of toxic substances such as heavy metals, cleaning and sanitising agents, and textile dyes. Tests have been performed on a waste water plant containing various toxic substances in order to assess the usefulness of this global sensor for detecting toxic events. Short-term respirometry gives an estimation of the immediate biological activity, which is influenced by the consumption of rapidly biodegradable pollutants and by potential inhibition. To separate both effects, we combined short-term respirometry with a rapid estimation of pollution by UV-spectrophotometry.

Key words: waste water, respirometry, detection, toxics, sludge, UV

62.1 Introduction

Activated sludge systems for treatment of waste water are based on the removal of pollution from waste streams by bacterial degradation. The main drawback of this process is the potential inhibition of the sludge microflora by toxic chemicals. The presence of high levels of toxic chemicals in domestic waste water is usually accidental such as the leakage from storage tanks or pipes. Nonetheless, voluntary spillage by humans may also occur. Although most factories producing heavily polluted waters should now have their own treatment facilities, uncontrolled release could still pollute the sewer system. The following pollutants are usually found: heavy metals such as mercury, chromium, lead, cadmium, zinc, copper and aluminium; cleaning and sanitising agents such as acids and bases, solvents, bleaching compounds, and detergents; pesticides, herbicides, and motor fuels. Hydrocarbons from pavement, metals such as copper and zinc from roofs, aluminium from gates and window frames can also be transferred to the waste water treatment plant after rainfalls.

Once high levels of toxic substances have been introduced into the sewer system, a slowdown of the biological activity occurs, leading in turn to the decrease of the treatment efficiency. Then, partly treated waste water is discharged into natural waters such as rivers, lakes, and fjords, where it can be harmful to fauna. In most pollution cases, several weeks are needed to restore an efficient treatment after a toxic event. Furthermore, due to new regulations on water quality, bypassing the treatment is severely restricted, which means that all incoming waste water should be treated in a way or another in adequate facilities. Therefore, investigations are being carried out in order to develop sensors that are able to detect the presence of toxics.

Ideally sensors should be placed directly after the grit removal, or even further upstream in the sewer network, in order to give enough time to plant managers to take a decision adapted to the toxic event, e.g. diversion of waste water to storage tanks. Although toxic substances can be potentially harmful to human beings, the objective is to detect them before they reach and damage microorganisms at the biological treatment stage. Among the global parameters that are more or less routinely measured in waste-water treatment plants, only few are adequate for a rapid warning. Chemical oxygen demand (COD) requires at least a delay of two hours, whereas biological oxygen demand (BOD) requires at least five days. These measurements are thus still time-consuming and, therefore, inadequate for the continuous monitoring of the waste water quality. As there are many potential toxics, it is also difficult and expensive to install specific sensors at the plant inlet.

A global but rapid detection procedure is thus needed. Respirometry is based on the measurement and interpretation of the respiration of activated sludge. Respirometry measures the amount of oxygen that is consumed by microorganisms in response to a specially designed feeding protocol. The first investigation describing respirometry for controlling a biological process has been published by Petersack and Smith (1975). Thereafter, various studies were carried out to use respirometry in two ways: (1) the development of biosensors and (2) the calibration of activated sludge kinetic models such as those developed by the International Water Association (IWA) (Henze et al. 2000). Several respirometry-based systems have been proposed such as continuous vs. batch operation; fixed or suspended microorganisms; and spiking with rapidly biodegradable substrate such as acetate for calibration.

The development of respirometry-based biosensors for industrial effluent monitoring receives also a lot of interest. One of the first applications was the BASF toxicometer (Pagga and Günthner 1981). It has been developed to monitor the toxic properties of chemical waste water from BASF production plants involving a flow of approximately 500 000 $m^3 d^{-1}$ that is feeding a central waste water treatment plant. Since then, many other experiments have been conducted to test, for instance, acute toxicity on nitrifying sludge (de Bel et al. 1996; Gernaey et al. 1998; Kong et al. 1996; Grunditz and Dalhammar 2001), or to combine respirometry with other methods such as titrimetry (Gernaey et al. 2001). Here we present a short-term batch respirometric test aimed at rapidly detecting the presence of toxics in waste water. We also point out the necessity to combine short-term respirometry with a rapid estimation of pollution, such as the one provided by UV-spectrophotometry.

62.2
Experimental

Field tests were conducted on the 350 000 person-equivalent waste water treatment plant of Nancy-Maxéville, France, which receives mainly domestic effluent. At the time of the experiments, it was composed of a series of lamellar primary settlers, 9 000 m^3 biological tanks consisting of 100 m long, 4 m deep channels equipped with gas diffusers and 15 000 m^3 circular clarifiers.

Fig. 62.1.
Experimental set-up of respirometric tests. The 2 l-reactor is equipped with air supply, mechanical stirrer and oxygen probe connected to a computer

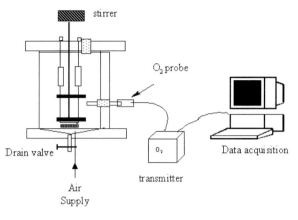

For respirometric tests, two in-house reactors have been built. The Plexiglas cylindrical vessels have working volumes of 1.7 and 2.3 l respectively. They are equipped with air supply and mechanical stirrers. Dissolved oxygen concentration is measured by oxygen probes (Orbisphere, Pack 02), connected to a computer for data acquisition (Fig. 62.1). The temperature is not controlled but is measured. This set-up has been chosen because respirometers can be continuously aerated, which avoids oxygen limitation. High sludge concentrations are used (Gernaey et al. 2001).

Sludge was sampled in the recycle line of the clarifiers to minimise residual concentrations of nitrogen-containing substances, in particular ammonium and nitrates, and of rapidly biodegradable carbon sources. For technical reasons waste water was sampled after the primary settler. A good compromise was found by using a sludge to waste water volume ratio equal to four. A waste water aliquot was 7 μm-filtrated and the UV-visible spectrum was obtained on an Anthelie Light spectrophotometer (Secomam, France) between 200 and 800 nm. When toxic substances were added, their concentration was always expressed in mg l^{-1} or ml l^{-1} of injected waste water.

62.3
Results and Discussion

62.3.1
Respirometry Principles

A short-term respirometric test proceeds as follows. The vessel is filled up with fresh sludge from the recycle line at a concentration of about 5 g suspended solids per liter. After a short delay of about 1–2 min, waste water is injected into the respirometer. Thereafter microorganisms increase their respiration due to biodegradation, which, in tun, decreases the dissolved oxygen concentration in the reactor (Fig. 62.2). When the rapidly biodegradable substrate is completely consumed, the dissolved oxygen concentration should increase up to about its initial value due to the constant aeration. The corresponding curve of oxygen utilisation rate (OUR) is presented in Fig. 62.3.

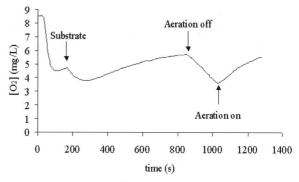

Fig. 62.2. Typical respirogram showing dissolved oxygen concentration vs. time in the respirometer

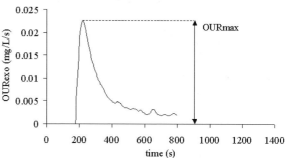

Fig. 62.3. Typical respirometric peak showing oxygen utilisation rate vs. time after injection of waste water in the respirometer. *OUR:* oxygen utilisation rate

The presence of a toxic component should thus be detected by a change of microbial behaviour, such as a decrease of oxygen utilisation rate of microorganisms. In case of complete poisoning leading to microbial death, no dissolved oxygen consumption should be observed. The respiration rate is calculated using a general mass balance for oxygen over the liquid phase (Eq. 62.1):

$$k_L a(c_s - c) = \frac{dc}{dt} + \text{OUR} \tag{62.1}$$

where c refers to the dissolved oxygen concentration in the vessel (mg l^{-1}), c_s denotes the dissolved oxygen concentration at saturation, and $k_L a$ is the oxygen mass transfer coefficient (s^{-1}).

The oxygen utilisation rate (OUR) (mg l^{-1} s^{-1}) is separated into two parts: the exogenous rate (OURexo) that results from the consumption of oxygen needed to degrade rapidly biodegradable substrates, and the endogenous rate (OURend) due both to the more slowly biodegradable substrates and to the maintenance of the microorganisms (Grady and Lim 1980; Spanjers et al. 1994; Gujer et al. 1996). By considering the two types of oxygen utilisation rate, Eq. 62.1 can be transformed into Eq. 62.2:

$$k_L a(c_s - c) = \frac{dc}{dt} + \text{OURend} + \text{OURexo} \tag{62.2}$$

Respirometric tests are performed with settled waste water that contains both rapidly and slowly biodegradable substrates. The rapid increase of respiration observed immediately after a waste water injection in the reactor corresponds to OURexo. To evaluate OURend, the aeration was shut down then turn on again. After aeration cut off (Fig. 62.2), the mass balance on dissolved oxygen becomes (Eq. 62.3):

$$0 = \frac{dc}{dt} + \text{OURend} \qquad (62.3)$$

OURend can thus be determined from the slope of c vs. time. It is assumed that microorganisms have then consumed all the rapidly biodegradable substrates. The re-aeration curve allows to determine c_s and $k_L a$ easily by least square regression (Eq. 62.4):

$$k_L a(c_s - c) = \frac{dc}{dt} + \text{OURend} \qquad (62.4)$$

Finally, Eq. 62.1 is integrated and OURexo is obtained (Fig. 62.3). OURmax is the maximal value taken by OURexo. The volume-specific total consumed oxygen (mg l^{-1}) used for the rapidly biodegradable substrate is calculated by integration of OURexo over time (Eq. 62.5):

$$V(O_2)^{t_2} = \int_{t_1}^{t_2} \text{OURexo} \, dt \qquad (62.5)$$

where t_1 refers to the injection time of the waste water sample and t_2 denotes the maximal time of monitoring. A t_1–t_2 delay of 5 min was chosen because it is sufficient to observe the consumption of all rapidly biodegradable substrates by microorganisms. Sludge is discarded after each test. As the aim is to obtain frequent measurements, the cycle time including filling, respirometric test, draining, and cleaning, has been set to 20 min.

62.3.2
Monitoring under Normal Conditions

Several experiments were carried out for long time periods, from 12 to 30 h, with a short-term respirometric test every 30 min. It was then possible to monitor the variations with time of the oxygen utilisation rates, as well as $V(O_2)^{5\text{min}}$ in absence of any toxic. Figure 62.4 presents the variations of OURend and of the maximum of OURexo (OURmax) from 7 June 2001, 7 A.M. to 8 June 2001, 12 P.M. OURend varies between 0.9 and 1.5 × 10^{-2} mg l^{-1} s^{-1} and OURmax between 1.7 and 3.2 × 10^{-2} mg l^{-1} s^{-1}. The latter is in the range of experimental values found in literature (Spanjers et al. 1994; Kong et al. 1996). These parameters are not constant over time as a minimal value is observed at 12 P.M. and a maximal value at 8 P.M. Thus, in absence of any toxic, since variations of respiration rates are still observed, they should not be mistaken as a response to toxic events. These variations are due to normal diurnal changes of pol-

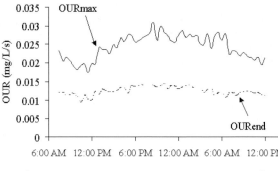

Fig. 62.4.
Respirometric test: oxygen utilisation rate (*OUR*) vs. time. *OURend* is the rate of slowly degradable substrates. *OURmax* is the maximum values of *OURexo*, the rate of rapidly degradable substrates (see text for details). Tests were performed using waste water samples every 30 min for 30 h

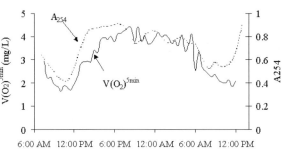

Fig. 62.5.
$V(O_2)^{5min}$ (- - -) and A_{254} (——) vs. time with one respirometric test every 30 min during 30 h with waste water. $V(O_2)$: volume-specific total consumed oxygen. Note that both parameters have similar variations. A_{254}: absorbance at 254 nm

lution. In order to separate this effect from a toxic effect, another information on the waste water composition should be provided.

As pointed out by many authors (Dobbs et al. 1972; Mrkva 1983), the absorbance at 254 nm, noted A_{254}, can be used as an indicator of the soluble organic pollution in waste water. Absorbance variation with time is related to changes in the human activity. Here the absorbance was obtained from the UV-visible spectrum. Time variations of $V(O_2)^{5min}$ and A_{254} are similar (Fig. 62.5). Depending on the pollution level and substrate quantity, microorganisms increase or decrease their respiration. This finding raises the need to combine respirometric measurements with another source of information on waste water. Toxics will indeed influence microbial respiration, but a decrease of $V(O_2)^{5min}$ would not be necessarily due to the presence of a toxic substance. For example, $V(O_2)^{5min}$ at 8 P.M. with a toxic substance could be equal to $V(O_2)^{5min}$ at 8 A.M. without any toxic.

62.3.3
Detection of Toxics

To demonstrate the effect of toxics on microorganism respiration, experiments were performed with two respirometers used in parallel. Both vessels were filled up with the same fresh sludge. Waste water was injected into the reference respirometer. A mixture of waste water with the toxic substance was injected in the second respirometer. Figure 62.6 represents the effect of sodium hydroxide (NaOH) on the oxygen utilisation rate (OUR). Sodium hydroxide, which is often used as a chemical cleaning agent, changes microbial activity and, as a result, we observe a decrease of both the OURmax and $V(O_2)^{5min}$. This finding confirms that a toxic substance can be detected by monitoring the oxygen utilisation rate of microorganisms.

Fig. 62.6.
Effect of NaOH on the oxygen utilisation rate (OUR). Two respirometric tests were run in parallel with same sludge. NaOH was injected in one of the two reactors. Note the sharp decrease of the rate following NaOH addition (lower curve)

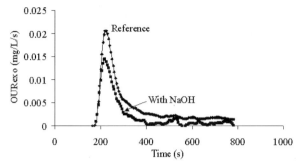

Fig. 62.7.
Inhibition levels of OURmax (diamonds) and $V(O_2)^{5min}$ (squares) vs. sodium hydroxide concentrations, determined by running two respirometric tests in parallel with the same operational conditions and with NaOH injection in one of the two respirometers. Note the strong inhibition above 100 mg l^{-1} NaOH

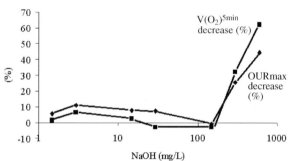

Experiments have been conducted with several potentially toxic components at different concentrations to establish their toxicity threshold. Respiration inhibition level is based on the difference between the OURmax or $V(O_2)^{5min}$ with or without toxic injection (Eq. 62.6 and 62.7).

$$\Delta OUR(\%) = \frac{OURmax_{reference} - OURmax_{toxic}}{OURmax_{reference}} 100 \qquad (62.6)$$

$$\Delta V(O_2)(\%) = \frac{V(O_2)^{5min}_{reference} - V(O_2)^{5min}_{toxic}}{V(O_2)^{5min}_{reference}} 100 \qquad (62.7)$$

Figure 62.7 represents the inhibition levels of the maximum oxygen utilisation rate (OURmax) and the volume-specific total consumed oxygen ($V(O_2)^{5min}$) at increasing concentrations of sodium hydroxide. Significant inhibition occurs for sodium hydroxide concentrations larger than 100 mg l^{-1}. This concentration corresponds to a pH value close to 11. At pH 12, the reduction of $V(O_2)^{5min}$ is larger than 70%. The OURmax is less affected as it reaches 50% at pH 12.

Similar inhibition experiments were performed to assess the effect of heavy metals on sludge response. Experiments were carried out with copper sulfate (CuSO$_4$). Above 7 mg CuSO$_4$ l^{-1} we observed a decrease of consumed oxygen $V(O_2)^{5min}$, reaching up to 60% for 150 mg l^{-1} of CuSO$_4$. Nonetheless, this is not the case with OURmax for which no threshold toxicity could be established, although the inhibition level varied between 5 and 20%. Experiments were also performed with bleach. Significant inhibition occurs

Table 62.1.
Toxic effect of textile dyes.
OUR: oxygen utilisation rate.
$V(O_2)$: volume-specific total oxygen consumed

Dye		Inhibition of OURmax (%)	Inhibition of $V(O_2)^{5min}$ (%)
Yellow 1	(50 mg l^{-1})	30	36
Yellow 2	(100 mg l^{-1})	36	31
Blue	(50 mg l^{-1})	47	45
Violet	(100 mg l^{-1})	40	40
Orange 1	(50 mg l^{-1})	43	40
Orange 2	(100 mg l^{-1})	39	33

for bleach concentrations above 2 ml l^{-1}. Bleach is a strong inhibitor since 6 ml l^{-1} are sufficient to induce a reduction of 50% of $V(O_2)^{5min}$. OURmax is less affected than $V(O_2)^{5min}$ as a 25% decrease is observed for a bleach concentration of 6 ml l^{-1}.

Table 62.1 presents tests performed in presence of several textile dyes. The purpose is here to test the effect of a small factory discharging directly its waste water into a municipal sewer system. As shown in Table 62.1, textile dyes induce inhibition since both OURmax and $V(O_2)^{5min}$ are reduced. An average reduction of 30% in is reached for OURmax and 37% for $V(O_2)^{5min}$.

Finally an experiment was carried out between 7 and 8 June 2001 with a respirometric test every 30 min by using the two respirometers in parallel and with the successive additions of various inhibitory substances in one of the two reactors (Fig. 62.8). Toxic components can be clearly detected by a sharp decrease of the consumed oxygen $V(O_2)^{5min}$. However the experiment with the paint solvent pointed out one of the difficulties of toxic detection. At very low concentration, as in this test, a solvent, i.e. a carbon source, increases the oxygen volume consumed as it can be used as a substrate. Any larger addition of this solvent induces in fact an immediate respiration decrease by microbial death.

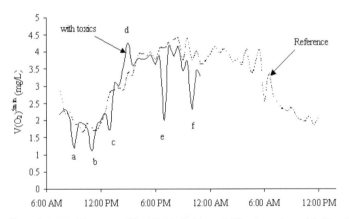

Fig. 62.8. Effect of CuSO$_4$ (*a*), NaOH (*b*), HCl (*c*), "white spirit" paint solvent (*d*), bleach (*e*) and gasoline (*f*) on respirometry at a waste water treatment plant. Successive respirometric test during 30 h with one test every 30 min. Two respirometers, "toxics" and "reference", were used in parallel with identical operational conditions. Toxics were added in one of the two reactors

62.4
Conclusions

Short-term respirometry represents a powerful means for the detection of toxic substances. However, respirometry alone is not sufficient to interpret the data with respect to inhibition of microbial activity. External information on the pollution level is necessary and can be provided rapidly by an UV absorbance sensor. It has to be noted that autotrophs are known to be more sensitive to toxic events than heterotrophs. It is unlikely that the short-term respirometric test used here is sensitive enough to protect autotrophs, which represent a few percents of the total biomass; although part of the ammonia contained in domestic waste water will be consumed during the test. But the protection of the largest part of the activated sludge is possible. No assumption of the inhibition mechanism was done. But it is certain that the tested inhibitors do not interfere with the microbial metabolism identically, as shown by the different variations of $V(O_2)^{5min}$ and OURmax. Based on this measurement procedure, a fault-detection method based on adaptive principal component analysis (APCA) has been proposed to take into account the normal variations of the respiration rate due to the normal variations of the biodegradable substrate present in waste water (Pons et al. 1999).

Acknowledgements

The authors wish to thank GEMCEA, Grand Nancy and the staff of Nancy-Maxéville waste water treatment plant for their support. Financial support for the French-Moroccan Committee is also acknowledged.

References

de Bel M, Stockes L, Upton J, Watts J (1996) Applications of respirometry based toxicity monitor. Water Sci Technol 33:289–296
Dobbs RA, Wise RH, Dean RB (1972) The use of ultra-violet absorbance for monitoring the total organic carbon content of water and waste water. Water Res 6:1173–1180
Gernaey K, Vanderhasselt A, Bogaert H, Vanrolleghem P, Verstraete W (1998) Sensors to monitor biological nitrogen removal and activated sludge settling. J Microbiol Methods 32:193–204
Gernaey K, Petersen B, Ottoy JP, Vanrolleghem P (2001) Activated sludge monitoring with combined respirometric-titrimetric measurements. Water Res 35:1280–1294
Grady Jr. CPL, Lim HC (1980) Biological waste water treatment, theory and applications. Marcel Dekker, New York
Grunditz C, Dalhammar G (2001) Development of nitrification inhibition assays using pure cultures of *Nitrosomonas* and *Nitrobacter*. Water Res 35:433–440
Gujer W, Henze M, Takashi M, van Loosdrecht M (1996) Activated sludge model No. 3. Wat Sci Technol 29:183–193
Henze M, Gujer W, Mino T, van Loosdrecht MCM (2000) Activated sludge models ASM1, ASM2, ASM2d, and ASM3. IWA Scientific and Technical Report No. 9, 2000, IWA Publishing, London
Kong Z, Vanrolleghem P, Willems P, Verstraete W (1996) Simultaneous determination of inhibition kinetics of carbon oxidation and nitrification with a respirometer. Water Res 30:825–836
Mrkva M (1983) Evaluation of correlations between absorbance at 254 nm and COD of river waters. Water Res 17:231–235

Pagga U, Günter W (1981) The BASF toxicometer. A helpful instrument to control and monitor biological waste water treatment plants. Water Sci Technol 13:233–238

Petersack JF, Smith RG (1975) Advanced automatic control strategies for the activated sludge treatment process. U.S. Environmental Protection Agency Report, EPA 67012-75-093

Pons MN, Roche N, Cécile JL, Potier O, Nieddu P, Prost C (1999) Fault detection for improved WWTPs control. Proc. 14th Triennal IFAC World Congress, Beijing, Vol O, pp 475–480

Spanjers H, Olsson G, Klapwijk A (1994) Determining short-term biochemical oxygen demand and respiration rate in an aeration tank by using respirometry and estimation. Water Res 28:1571–1583

Chapter 63
Environmental Biosensors Using Bioluminescent Bacteria

M. B. Gu

Abstract

Environmental biosensors to assess the toxicity of environmental media such as water, soil, and atmosphere have been developed using various recombinant bioluminescent bacteria. Those bacteria were constructed based on specific stress-responsive promoters in bacterial cells. They are thus activated by different groups of toxicity. For continuous monitoring of water toxicity, a multi-channel system having different stress-responsive strains in each channel, and composed of two-stage mini-bioreactors, was successfully developed. Soil toxicity was assessed using a soil biosensor based upon immobilization of recombinant bioluminescent bacteria that worked with the addition of rhamnolipids biosurfactant. An example of phenanthrene toxicity is shown. For the assessment of gas toxicity, an immobilization technique has been set up to allow the biosensor to come in direct contact with the toxic gas in the sensing chamber. An example of benzene toxicity is shown. This mini review will show how the recombinant bioluminescent bacteria can be utilized as environmental biosensors. With further findings and developments of new non-specific stress promoters, the potency and extensiveness of the information that can be obtained using these environmental biosensors is immense.

Key words: biosensor, luminescent bacteria, soil, water, air, PAHs

63.1 Introduction

Bioluminescence is being used as a prevailing reporter of gene expression in microorganisms and mammalian cells. Bacterial bioluminescence draws special attention from environmental biotechnologists since it has many advantageous characteristics such as no requirement of extra substrates, highly sensitive, and on-line measurability (Van Dyk et al. 1995; Vollmer et al. 1997; Gu and Choi 2001). Using bacterial bioluminescence as a reporter of toxicity has replaced the classical toxicity monitoring technology of using fish or *Daphnia* by cutting-edge technology. Fusion of bacterial stress promoters, which control the transcription of stress genes corresponding to heat-shock, DNA-, or oxidative-damaging stress (Van Dyk et al. 1995; Belkin et al. 1996; Vollmer et al. 1997) to the bacterial *lux* operon has resulted in the development of novel toxicity biosensors with a short measurement time, enhanced sensitivity, and ease and convenient usage. Therefore, these recombinant bioluminescent bacteria are expected to induce bacterial bioluminescence when the cells are exposed to stressful conditions, including toxic chemicals.

These recombinant bioluminescent bacteria have been used to develop toxicity biosensors in a continuous, portable, and in-situ measurement system for use in air, water, and soil environments (Gu et al. 1999, 2001a; Gu and Gil 2001; Gu and Chang 2001). All the data obtained from these toxicity biosensors within these environments were found to be repeatable and reproducible, and the minimal detectable level for the toxicity was found to be in the part per billion level for specific chemicals (Min et al. 1999; Choi and Gu 2001; Gu and Choi 2001; Gu et al. 2001c). Here, this short review will focus on how environmental biosensors and biomonitoring systems utilizing recombinant bioluminescent bacterial strains have been developed and implemented to detect the toxicity from air, water, and soil environments.

63.2
Experimental

63.2.1
Strains

Recombinant *Escherichia coli* DPD2794, containing a *recA::luxCDABE* fusion, as a model strain was used to monitor environmental damage to deoxyribonucleic acid (DNA), with mitomycin C as a model toxicant. And recombinant bacteria TV1061, containing a *grpE::luxCDABE* fusion, was used as another biosensor cell. This bacterial strain is responsive to toxicity due to protein-damaging agents. The DPD2540 strain, containing a *fabA::luxCDABE* fusion, was used as a biosensor cell responding to membrane-damaging agents. The GC2 strain, containing a *lac::luxCDABE* fusion, was used as a biosensor cell that is responsive to general toxicity causing cell death or luciferase inhibition.

63.2.2
System Development and Set-up Used in Those Biosensors

To minimize the operation cost and space for the set-up of this system, small bioreactors were fabricated with working volume of 10 or 20 ml. The mini-bioreactor has one side port, covered with glass, for holding a fiber optic probe. The highly sensitive lumino-meter (Model 20e, Turner Design, CA) was linked to the other side of fiber optic probe to measure the bioluminescence in the mini-bioreactor. The lumino-meter was connected to a personal computer through a RS232 serial connection in order to acquire the real time data. Oxygen was supplied through a head port with a sparge tube by using pressurized air with flow meter at 1 liquid volume/air volume/minute (v.v.m). Temperature was controlled by a water bath. After steady-state values of constant bioluminescence and cell density were obtained, the chemicals or test samples were injected into the second stage mini-bioreactor.

For the soil biosensor, the test reactor used is a 50 ml stainless steel cylinder having a water jacket to maintain a constant temperature using a heated circulation water bath (JEIO TECH, Korea). This reactor was filled with 25 ml of fresh Luria Bertani (LB) medium. Filtered air was supplied through a head port with a sparging tube while excess gas was vented through an outlet port. An injection port for the test samples was also included on the head port. The biosensor kit was connected to the end of a fiber optic probe by a black rubber tube and inserted through a hole in the reactor until submerged below the LB. The other side of fiber optic probe was connected to a highly sensitive luminometer (Model TD-20e, Turner Design, CA) to measure the bioluminescence level (BL) from the immobilized cells in the biosensor kit. The luminometer was linked to a personal computer through a RS232 serial connection in order to acquire real time data. Various concentrations of the test samples (rhamnolipid solutions, the extracted solutions from the flasks and contaminated soils) were prepared and stored in glass vials. After the bioluminescence level of the unit steadied, using a syringe, 5 ml of the test samples were injected into the reactor.

The gas biosensor kit in which the bioluminescent bacteria were immobilized was attached to the end side of a fiber optic light probe connected to a highly sensitive luminometer (Model 20c, Turner Design, CA) to measure the light output (bioluminescence) constitutively produced from the cells. The luminometer was linked to a computer through a RS232 serial connection for the purpose of real time data acquisition. The test chamber was a 100 ml stainless steel cylinder with a water jacket to maintain a constant temperature using a thermostatic water bath (VWR Scientific, USA). The biosensor kit with the fiber optic probe was connected to an upside port in the chamber by black rubber tubing so as to prevent the leakage of the bioluminescence or the entering of ambient light. After a steady-state bioluminescence level (BL) was reached, 3 ml of the benzene solution was injected into the reactor through the syringe port to provide various concentrations of vaporized benzene in the test chamber and the bioluminescence levels were measured every minute.

63.2.3
Chemicals

For the multi-channel continuous toxicity monitoring system, mitomycin C from Aldrich Chemical Co., USA and cerulenin from Aldrich Chemical Co. were dissolved in distilled water to make stock solutions. Stock concentrations: 0.4 mg ml^{-1} mitomycin C, 0.125 mg ml^{-1} cerulenin. These stock solutions were stored at 4 °C for mitomycin C and −20 °C for cerulenin until used. Phenol was purchased from Junsei Chemical Co., Japan. Contaminated water samples were artificially made by mixing various amount of mitomycin C, phenol, and cerulenin.

For the soil toxicity biosensor, the biosurfactant, rhamnolipid, was purchased from the Jeneil Biosurfactant Co., USA. This biosurfactant is a mixture of monorhamnolipid and dirhamnolipid at an approximate ratio of 1:1. The critical micelle concentration (CMC) of this biosurfactant was 0.037 mM (0.002 volume%). Phenanthrene was obtained from Wako Pure Chemical Industries, Ltd., Japan. Methylene chloride, HPLC-grade acetonitrile, and pure water were purchased from the Fisher Scientific Co. Ethanol was purchased from the Merck Chemical Co. For the gas toxicity biosensor, benzene, known as a representative volatile, toxic organic compound, was chosen and purchased from the Merck Company. Oleic acid was purchased from the Kanto Chemical, Japan. Benzene solutions were prepared with oleic acid as the solvent.

63.3
Results and Discussion

63.3.1
Water Toxicity Monitoring

Recently, several environmental biosensors and biomonitoring systems were successfully developed using recombinant bioluminescent bacteria for environmental toxicity monitoring. Among these monitoring systems, a continuous toxicity monitoring system has become a major tool in its ability to supply valuable information about pollutants in water. A two-stage mini-bioreactor system for continuous toxicity monitoring has been developed and improved continuously, and its results and application were found to be reproducible, repeatable, possible for long-term and real time operation for the detection of biological toxicity.

The two-stage mini-bioreactor system was successfully developed for continuous toxicity monitoring. This system consists of two mini-bioreactors in series (data not shown). The first stage allows for a continuous supplying of fresh cells to the second stage mini-bioreactor in which the biosensing cell and the sample water are mixed. If severely toxic chemicals are injected into the second bioreactor, which will lead to cell death, the system's ability to monitor toxicity will recover due to this fresh supply. This physical separation of the cell culture allows stable and reliable operation of the toxicity monitoring system, because the cell growth rate and cell concentration can be maintained in the first bioreactor. Therefore, this system can also be operated con-

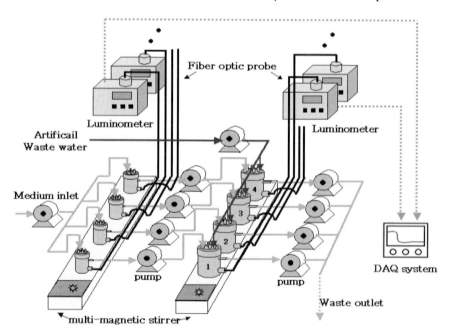

Fig. 63.1. Multi-channel continuous monitoring system

tinuously while maintaining its responsiveness to toxic chemicals. However, this system has a limitation in its qualitative classification of mixed toxicity samples. Therefore, a new multi-channel system has been developed for the continuous toxicity monitoring and classification of toxicity by combining two major components developed previously, the first being different recombinant bacteria and the second being the two-stage mini-bioreactor (Fig. 63.1).

The detection time and the detectable concentrations are lower than other systems. Pulse type exposures were used to evaluate the reproducibility and reliability. Step inputs of toxicants have been adopted to show the stability of the system. All data demonstrated that this two-stage mini-bioreactor system, using recombinant bacteria containing stress promoters fused with lux genes, is appropriate for continuous toxicity monitoring. Using this multi-channel continuous toxicity monitoring system, classification of toxicity in field samples was found to be possible. Thus, application to many different areas, including as an early warning system for wastewater biotreatment plant upsets and the monitoring and tracking of accidental spills, discharges or failures in plant operation are possible (Gu et al. 1999, 2000; Gu and Gil 2001; Gu et al. 2001b).

63.3.2
Soil Toxicity Monitoring

Polycyclic aromatic hydrocarbons (PAHs) are a group of solid phase organic chemicals containing two or more fused benzene rings. The occurrence of PAHs in soils, sediments, aerosols, animals and plants is of increasing environmental concern because some PAHs exhibit mutagenic and carcinogenic effects. Thus, it is necessary to monitor the hazard, and risk assessment of polluted soils using organisms. The detection of PAH toxicity in the environment is known to be restricted by their low solubility and high sorption to solids, thereby limiting their bioavailability. However, surfactants can enhance the rate of mass transfer from the solid and sorbed phases by increasing the rates of dissolution and desorption of PAHs. Therefore, in this study,

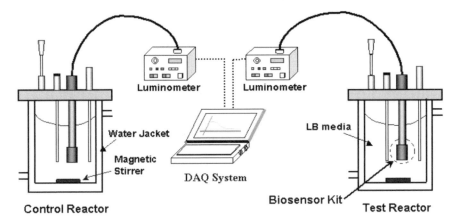

Fig. 63.2. Soil toxicity biosensor system. *LB:* Luria Bertani

Fig. 63.3.
Response of the biosensor to different phenanthrene concentrations, from 0 to 22.5 ppmw. *RBL*: ratio of bioluminescent level of induced cells vs. bioluminescent level of water control

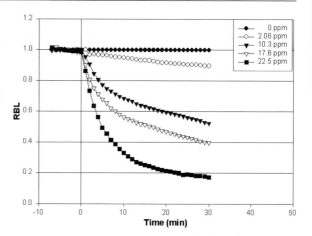

two different techniques, immobilization of the recombinant bioluminescent bacterial cells and use of a nontoxic biosurfactant, were combined to develop an in-situ toxicity biosensor system for soil, and then, using the developed soil biosensor, the possibility of detecting toxicity present in polluted soil was shown (Fig. 63.2).

A biosensor for assessing the toxicity of soils contaminated by polycylic aromatic hydrocarbon (PAHs) has been constructed using an immobilized recombinant bioluminescent bacterium GC2 (*lac::luxCDABE*), which constitutively produces bioluminescence. The biosurfactant, rhamnolipids, was used to extract a model PAH, phenanthrene, and was found to enhance the bioavailability of phenanthrene via an increase in its rate of mass transfer from the soil particles to the aqueous phase. The monitoring of phenanthrene toxicity was achieved through measuring the decrease in the bioluminescence when a sample extracted with the biosurfactant was injected into the mini-bioreactor (Fig. 63.3). The concentrations of phenanthrene in the aqueous phase were found to correlate well with the corresponding toxicity data obtained using this toxicity biosensor. In addition, it was also found that the addition of glass beads to the agar media enhanced the stability of the immobilized cells. This biosensor system, using a biosurfactant, may be applied as an in-situ biosensor to detect the toxicity of hydrophobic contaminants in soils and for performance evaluation of PAH degradation in soils (Gu and Chang 2001).

63.3.3
Gas Toxicity Monitoring

Serious discharges or leaks of volatile chemical compounds into the atmosphere have also been found to cause deleterious damage to lives and serious environmental problems. Some chemical compounds, such as volatile organic compounds, chlorinated derivative, hydrocarbons and so forth, can be very toxic to humans and cause abnormal health effects. Among them, toxic gaseous chemicals are present in such groups as carcinogenics, genotoxics, reproductives, systemic toxics, and skin/eye irritant. Therefore, studies on biosensors that use living organisms have been done to detect

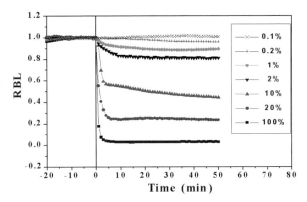

Fig. 63.4.
Response of the biosensor to different benzene concentrations, from 0.1 to 100%v. RBL: ratio of bioluminescent level of induced cells vs. bioluminescent level of water control

toxic gases. One such study used a gas biosensor utilizing immobilized bioluminescent cells and was used for the detection of gaseous toxicity. In this study, a whole-cell biosensor was developed using a recombinant bioluminescent *Escherichia coli* harboring a *lac::luxCDABE* fusion. Immobilization of the cells within Luria Bertani agar was chosen to maintain the activity of the microorganisms and to allow the bacteria to come into direct contact with the gas. Benzene, a volatile organic compound, was chosen as a toxic gas to evaluate the performance of this biosensor based on the bioluminescent response. This biosensor showed a dose-dependent response, and was found to be reproducible (Fig. 63.4) (Gil et al. 2000).

63.4 Conclusions

Various types of environmental biosensors have been developed and characterized using recombinant bioluminescent bacteria responsive to different specific stresses, such as DNA-, oxidative-, and protein-damaging agents. A multi-channel continuous toxicity monitoring system has been developed for monitoring and classification of water toxicity. A soil biosensor was also developed based upon immobilization of the bacterial cells and use of a biosurfactant as a non-toxic extracting agent. A gas toxicity biosensor was developed to measure toxicity of the gas samples via direct contact of the gas with the biosensor. Development of these various types of environmental biosensors using recombinant bioluminescent bacteria sensitive to specific group of toxicity should lead to further understanding of modes of toxic actions by many different environmental samples and make widen the application areas of the environmental samples.

Acknowledgement

This study was supported through the program of the National Research Laboratory (NRL) on Environmental Biosensors (KISTEP 10104000094-01J000004100). The author also thanks to Alexander von Humboldt Research Foundation for financial support to Dr. Man Bock Gu.

References

Belkin S, Smulski DR, Vollmer AC, Van Dyk TK, LaRossa RA (1996) Oxidative stress detection with *Escherichia coli* harboring a katG'::lux fusion. Appl Environ Microbiol 62:2252–2256

Choi SH, Gu MB (2001) Phenolic toxicity-detection and classification through use of a recombinant bioluminescent *Escherichia coli*. Environ Toxicol Chem 20:248–255

Gil GC, Mitchell RJ, Chang ST, Gu MB (2000) A biosensor for the detection of gas toxicity using a recombinant bioluminescent bacterium. Biosens Bioelectron 15:23–30

Gu MB, Chang ST (2001) Soil biosensor for the detection of PAH toxicity using an immobilized recombinant bacterium and a biosurfactant. Biosens Bioelectron 16:667–674

Gu MB, Choi SH (2001) Monitoring and classification of toxicity using recombinant bioluminescent bacteria. Water Sci Technol 43:147–154

Gu MB, Gil GC (2001) A multi-channel continuous toxicity monitoring system using recombinant bioluminescent bacteria for classification of toxicity. Biosens Bioelectron 16:661–666

Gu MB, Gil GC, Kim JH (1999) A two-stage minibioreactor system for continuous toxicity monitoring. Biosens Bioelectron 14:355–361

Gu MB, Mitchell RJ, Kim JH (2000) Continuous monitoring of protein damaging toxicity using a recombinant bioluminescent *Escherichia coli*. ACS Symposium Series: Recent Advances in Chemical sensors and Bio-sensors for Environmental monitoring 762:185–196

Gu MB, Choi SH, Kim SW (2001a) Some observations in freeze-drying of recombinant bioluminescent *Escherichia coli* for toxicity monitoring. J Biotechnol 88:95–105

Gu MB, Kim BC, Cho J, Hansen PD (2001b) The continuous monitoring of field water samples with a novel multi-channel two-stage mini-bioreactor system. Environ Monit Assess 70:71–81

Gu MB, Min J, Kim EJ (2001c) Toxicity monitoring and classification of endocrine-disrupting chemicals (EDCs) using recombinant bioluminescent bacteria. Chemosphere 46:289–294

Min J, Kim EJ, LaRossa RA, Gu MB (1999) Distinct responses of a recA::luxCDABE *Escherichia coli* strain to direct and indirect DNA damage agents. Mutat Res 442:61–68

Van Dyk TK, Smulski DR, Reed TR, Belkin S, Vollmer AC, LaRossa RA (1995) Responses to toxicants of an *Escherichia coli* strain carrying a uspA'::lux genetic fusion and an *E. coli* strain carrying a grpE'::lux fusion are similar. Appl Environ Microbiol 61:4124–4127

Vollmer AC, Belkin S, Smulski DR, Van Dyk TK, LaRossa RA (1997) Detection of DNA damage by use of *Escherichia coli* carrying recA'::lux, uvrA'::lux, or alkA'::lux reporter plasmids. Appl Environ Microbiol 63:2566–2571

Chapter 64

Evaluation of Water-Borne Toxicity Using Bioluminescent Bacteria

B. C. Kim · M. B. Gu · P. D. Hansen

Abstract

We investigated the toxicity of field waters using a multi-channel continuous monitoring system in Berlin, Germany. This system uses genetically engineered bioluminescent bacteria for the assessment of the toxicity of soluble chemicals. It showed easy and long-term monitoring without any system shut down due to pollution overloading. We used the bioluminescent bacterial strains DPD2794, DPD2540, TV1061 and GC2, which respond respectively to DNA-, cell membrane-, protein- and general cellular-damaging agents. The bioluminescent levels either increase for DPD2794, DPD2540 and TV1061 strains, or decrease for the GC2 strain after being mixed with toxic samples. We monitored the toxicity over a period of two to three weeks at three different sites: the Ruhleben wastewater treatment plant discharge flow, and river flows at the Teltowkanal and Fischereiamt in Berlin. At all sites the DPD2540 and TV1061 strains showed a significant increase of bioluminescence while bioluminescence decreased for the GC2 strain. This result demonstrates the occurrence of chemicals that affect the integrity of the cellular membrane; leading to either protein denaturation and inhibiting cellular metabolism or to cell death. Therefore, our findings suggest that the bioluminescent bacteria array may serve as a novel water toxicity monitoring system in outdoor fields.

Key words: potential toxicity, field water, multi-channel continuous monitoring, bioluminescent bacteria

64.1 Introduction

Water quality control is a major area of water management for maintaining a good public health and water ecology. Continuous monitoring of water is essential for good quality control because physico-chemical properties of water may change within very short periods of time, especially in river and sewage water. There are many real-time water-quality monitoring systems based on instruments and living organisms (Ahman and Reynolds 1999; Charef et al. 2000; van der Schalie et al. 2001). On one hand, instrumental analyses determine the physico-chemical factors, including conductivity, pH, temperature, and dissolved chemicals within the water. On the other hand, bioassays using living organisms such as fish or *Daphnia* provide information based either on the lethal dose or on behavioural patterns, thus giving a value for water toxicity (Radix et al. 1999; van der Schalie et al. 2001). Although instrumental and biological assays offer much information on the chemical and toxic nature of the samples, they do not provide nor predict the specific stress effects experienced by living organisms.

Recently, there has been eye opening progress in the development of microscale bioassays and biosensors to detect environmental toxicants, using recombinant or non-modified bioluminescent organisms (Van Dyk et al. 1995; Vollmer et al. 1997). Recombinant bioluminescent bacteria have plasmids in which the *luxCDABE* operon is fused with a stress promoter. As these promoters are induced in the presence of a specific group of chemicals or toxicants, it is possible to construct stress-specific strains. Some examples of these strains are DPD2540 (*fabA::luxCDABE*), DPD2794 (*recA::luxCDABE*), TV1061 (*grpE::luxCDABE*), and GC2 (*lac::luxCDABE*). These strains show a specific response for membrane damaging agents, DNA damaging agents, protein damaging agents and general cellular toxicity agents, respectively (Vollmer et al. 1997; Van Dyk et al. 1995).

To provide a rapid response as well as a long-term and real-time monitoring of potential toxicity in water samples, the two-stage mini-bioreactor system (single channel) has been developed (Gu et al. 1999) and expanded to a multi-channel system. This system employs a group of two-stage mini-bioreactor systems connected in parallel, and uses the genetically engineered bioluminescent bacteria mentioned above as indicators of specific stress responses (Gu and Gil 2001). The system has been continuously characterised using pure toxic chemicals, artificial wastewater and sampled field water, and proved the feasibility of classifying the toxicity of samples based upon their mode of toxic actions (Gu et al. 2001). The final purpose of this study was to develop an applicable on-line monitoring system for the detection of potential water-borne toxicity.

Here we report the first field application and operation of the multi-channel continuous monitoring system to investigate potential water toxicity. This system may serve as a promising water toxicity monitoring system using genetically engineered microbial strains and on-line based instruments.

64.2
Experimental

64.2.1
Recombinant Bioluminescent Strain

The recombinant bioluminescent *Escherichia coli* strains DPD2540, DPD2794, TV1061, and GC2 harbouring, respectively, a *fabA::luxCDABE* (Choi and Gu 2001), *recA::luxCDABE* (Min et al. 1999), *grpE::luxCDABE* (Gu et al. 2000), or *lac::luxCDABE* (Gil et al. 2000) fusion in host strain RFM443 were used as the biosensing cells within the system. All seed cultures were grown in 5 ml of Luria-Bertani medium (LB, initial pH of 7) containing 20 μg ml^{-1} ampicillin (Sigma Co., USA) in 15 ml Falcon tubes. The seed cultures were placed in a rotary incubator at 30 °C and at 250 rpm, except for the GC2 strain, which was grown at 37 °C. All strains were cultured overnight. From these cultures, 2 ml were then inoculated into the corresponding first stage mini-bioreactor using a sterilised syringe to begin system operation.

64.2.2
Multi-Channel Continuous Monitoring System

The set up of the multi-channel system and operation for field tests is described elsewhere (Gu et al. 2001; Gu and Gil 2001). This system was expanded from a one-

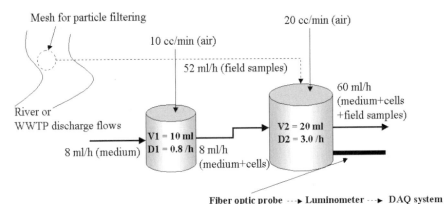

Fig. 64.1. The multi-channel continuous toxicity monitoring system and its operation parameters. At least three channels of one-set two-stage reactors were used for this research. After filtration of suspended particles using a mesh filter, waters from the wastewater treatment plant (*WWTP*) are continuously supplied to the second reactor, and fresh cultured cells are supplied to the second reactor from the first reactor (*V*: working volume, *D*: dilution rate)

channel system to multi-channel one. Each channel employs two mini-bioreactor series, the first reactor is used to grow cells while, in the second reactor, field water samples are pumped and mixed with the cells and bioluminescent changes are measured through a luminometer that is connected with the reactor through a fiber optic probe (Fig. 64.1). From the first reactor, fresh cells were continuously pumped into the second reactor, providing a fresh supply for monitoring (Gu et al. 2001; Daunert et al. 2000). Therefore, the system could provide a continuous long-term operation. The bioluminescent data was transferred from the luminometer to a computer-based automatic data acquisition program through a RS232 cable. All data is expressed as the ratio of bioluminescent level of induced cells vs. bioluminescent level of water control (RBL).

64.2.3
Test Sites

The biomonitoring system was set up at three different sites in Berlin, Germany, and was operated for a minimum of two days. The first place was Berlin Fischereiamt, which is responsible for supervising the water ecology and environment. The system was set up beside the river and had water pumped into the reactor directly. The Ruhleben wastewater treatment plant treats most of the wastewater from the residential areas and industrial plants of Berlin. This plant is located north of Fischereiamt on the Spree River. For water samples, the 'WaBoRu AQUATOX' system was used and the multi-channel continuous monitoring system was set beside it. This is a flow-through system that recognises and monitors the sub-lethal effects on individual species in aquatic communities that are caused by wastewater outlets, which transport critical and hazardous substances. Adjusting the valve of this system allows one to control the concentration of the treated outflow (Hansen 1986).

Our experiments were done at 100% purely undiluted treated wastewater, 40% treated water with 60% control water, 30%, and 10% outflow concentrations. As well, the AQUATOX system has four chambers, one to collect each water sample continuously. From each chamber, samples were pumped into the second reactors directly. The final system was set up in a monitoring station located beside the Teltow canal, located south of Berlin. This monitoring station is an unmanned system that simultaneously checks the physicochemical factors of the Teltow canal's water, including the conductivity, pH, temperature and chemical oxygen demand (COD). The water flows through a collection chamber within the station and, using bypass tubing, the water sample was pumped into the second reactors directly. To remove small, suspended particles, a mesh with a 0.1-inch whole was tied to the end of the collection tubes.

64.3
Results and Discussion

64.3.1
Toxicity of Field Waters – Berlin Fischereiamt

We used bioluminescent bacteria to test water toxicity. There are both 'lights-on' and 'lights-off' assays involving bioluminescent bacterium. The 'lights-on' assays are based upon cellular mechanisms that are activated when the bacteria are stressed by chemicals that cause damage to membrane damage, DNA, or protein. In contrast, the 'lights off' assay is based on the decrease of the bioluminescence resulting either from the inhibition of the luciferase activity or from cellular toxicity. Using the simultaneous operation of several channels containing different biosensing cells, the potential toxicity of a sample can thus be measured using two approaches: either an increase in the bioluminescence via stress responses or the inhibition of bioluminescence (Daunert et al. 2000). For each experiment done in this study, either three or four channels were set up, including the GC2 strain channel and at least two inducible strain channels.

In Fischereiamt's board, all four channels with DPD2540, DPD2794, TV1061, and GC2 bacterial strains were set up and used to monitor the toxicity of the river for 4 d from 10 July 2001 to 13 July 2001. We compared daily the bioluminescence levels of the GC2 channel with other one 'light on' assay channel: DPD2794 on the first day, DPD2540 on the second day, and TV1061 on the third day; in order to monitor general and specific toxicity of river water simultaneously. Although samples were injected into the channels at different days, we detected light increase after sample injection from the all 'light on' assay channels and bioluminescent level recover after injection of control water (Fig. 64.2).

The toxicity level, based on the bioluminescence induced, can be measured by comparing the ratio of bioluminescent level (RBL: see Sect. 64.2.2) with the control data for the same strain. Phenol was used as the model toxicant for DPD2540, TV1061 and GC2 strains, while mitomycin C was chosen for the DPD2794 strain. The increases in the bioluminescence from the DPD2794 and DPD2540 channels are below the detection limit (RBL = 2). However, the RBL from the TV1061 channel increased to a level 4.6-fold higher than before injection. Comparison of this result with the standard signature database using standard toxic chemicals gives a potential toxicity similar

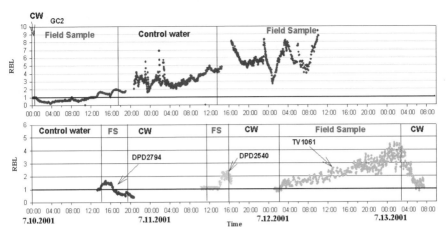

Fig. 64.2. Toxicity monitoring of Berlin Fischereiamt river water. *CW:* control water; *FS:* field sample; *WWTP:* wastewater treatment plant; *RBL:* ratio of bioluminescent level of induced cells vs. bioluminescent level of water control; *GC2:* strain of "light-off" bioluminescent bacteria. *DPD2540, DPD2794,* and *TV1061:* strains of "light-on" bioluminescent bacteria. Note the decrease of bioluminescence of the *GC2* strain when supplied with field waters (*top*). Note the increase of bioluminescence of *DPD2540, DPD2794* and *TV1061* strains when supplied with field waters (*bottom*)

with that of 300 ppmv phenol. Even though high bioluminescent levels from real field samples were seen, the biological toxicity can decrease or increase depending upon the other materials present, i.e. synergistic and antagonistic effects. Therefore, the exact classification and detection of a chemical's toxicity is essential and can be achieved using standard signature data. Using these signatures, the toxicity of the water can be measured.

The bioluminescence within the second reactor of the GC2 channel decreased to 0.5 RBL after addition of the field water samples (Fig. 64.2, *top left*). However, the bioluminescence of the control fluctuated. The GC2 strain has a plasmid-borne fusion of the *lac* promoter and *luxCDABE* operon. The dilution rate inside the second reactor is restricted to a value of about 3.0, a value that limit the nutrition available for microbial growth. As well, if water samples contain lactose-like materials, a common natural sugar, its presence may seriously affect the promoter's activity, leading to increases and fluctuations in the bacteria's bioluminescent levels. Although the GC2 strain has this limitation, this strain showed good responses when used in other biosensors, i.e. for gas-based or soil-based toxicity (Gil et al. 2000; Gu and Chang 2001). For this system, though, a strain that responds independently of the nutrients present, such as a phage promoter-lux fusion, should be used for the general toxicity assay channel.

64.3.2
Toxicity of Field Waters – Wastewater and Teltow Canal

In the Ruhleben wastewater treatment plant (3–5 August 2001) and the Teltow Canal monitoring station (14–15 August 2001), the DPD2540, TV1061 and GC2 channels were run simultaneously. The maximum increase or reduction of the bioluminescence (RBL)

Table 64.1. Summary of the continuous toxicity monitoring of field water in Berlin, Germany

Bacterial strain	Fisheriamt Berlin	Ruhleben WWTP Discharged flow conc. (Vol.-%)				Teltow canal	Potential toxic effects
		10	30	40	100		
DPD2540	++	N.C.	++	+	N.C.	++	Membrane damage
DPD2794	+	N.A.	N.A.	N.A.	N.A.	N.A.	DNA damage
TV1061	+++	N.C.	++	++	+++	++	Protein damage
GC2	F	N.C.	++	+++	N.C.	+++	Cellular toxicity

+: 1 < Max. RBL < 2.0 (for inducible channels), 0.8 < Max. Reduction of RBL < 1 (for GC2 channel).
++: 2.0 < Max. RBL < 4.0 (for inducible channels), 0.6 < Max. Reduction of RBL < 0.8 (for GC2 channel).
+++: 4.0 < Max. RBL (for inducible channels), Max. Reduction of RBL < 0.6 (for GC2 channel).
N.C.: No change in the bioluminescent level; N.A.: not available (not set up); F: the channel's bioluminescence was not stable; RBL: ratio of bioluminescent level of induced cells versus bioluminescent level of water control.

is summarised in Table 64.1. In the Ruhleben wastewater treatment plant, the concentration of the water samples was adjusted, using a valve from the plant's AQUATOX system, from 10% to 100%. No changes in the signal intensity occurred with the 10% outlet assay. Although it may appear that increases or decreases in the RBL level are not proportional to the outlet concentration, each concentration was done on different days due to system limitations, and represents separate experiments (Table 64.1). During the Teltow canal test, the GC2, DPD2540, and TV1061 channels were tested simultaneously. All channels showed significant bioluminescent changes, which suggest that this river is carrying chemicals involved in membrane damaging, protein damaging and bulk cellular toxicity (Table 64.1).

For all assays at all test sites, the bioluminescent levels of the TV1061 strain were the highest compared to controls. These bioluminescent increases suggest that protein damage or heat shock materials are the primary existing toxicants in the test samples. While any instrumental analysis cannot offer any toxic information for the samples being tested, this system does offer information on the potential toxicity while classifying the toxic effects.

64.4
Conclusion

This paper introduced a first field application of the multi-channel continuous toxicity monitoring system. This system is a novel solution for the detection of the potential toxicity of water and for water quality control. Through analysis of the bioluminescent data and toxicity of the water, an early warning can be declared owing to the fast bioluminescent signal change. Furthermore, in the case where a sample is toxic, a sample of the water can be taken for instrumental analysis. Although some unexpected results occurred, the feasibility of this system in real field applications was successfully demonstrated. Using predetermined operating protocols, a fully automatic data acquisition system, and an internet based data transfer system will allow to achieve a fully on-line water quality monitoring system. Also, extension of the channels using different recombinant bioluminescent strains can be a cost-effective solution for the simultaneous detection of potential water toxicity, regardless if it is from a wastewater treatment plant, river, lake, chemical plant or other source. Thus, application to many different areas, including as an early warning system of wastewater treatment plant failure or for the monitoring and tracking of accidental spills, is a viable possibility.

Acknowledgements

This research was supported by the Korea Science and Engineering Foundation (KOSEF) through the GIST/ADEMRC Europe Satellite Lab in the Technical University of Berlin. Authors are grateful for their support.

References

Ahmad SR, Reynolds DM (1999) Monitoring of water quality using fluorescence technique: prospect of on-line process control. Water Res 33:2069–2074
Charef A, Ghauch A, Baussand P, Martin-Bouyer M (2000) Water quality monitoring using a smart sensing system. Measurement 28:219–224
Choi SH, Gu MB (2001) Phenolic toxicity-detection and classification through use of a recombinant bioluminescent *Escherichia coli*. Environ Toxicol Chem 20:248–255
Daunert S, Barrett G, Feliciano JS, Shetty RS, Shrestha Smith-Spencer W (2000) Genetically engineered whole-cell sensing systems: coupling biological recognition with reporter genes. Chem Rev 100:2705–2738
Gil GC, Mitchell RJ, Chang ST, Gu MB (2000) A biosensor for the detection of gas toxicity using a recombinant bioluminescent bacterium. Biosens Bioelectron 15:23–30
Gu MB, Chang ST (2001) Soil biosensor for the detection of PAH toxicity using an immobilized recombinant bacterium and a biosurfactant. Biosens Bioelectron 16:667–674
Gu MB, Gil GC (2001) A multi-channel continuous toxicity monitoring system using recombinant bioluminescent bacteria for classification of toxicity. Biosens Bioelectron 16:661–666
Gu MB, Gil GC, Kim JH (1999) A two-stage minibioreactor system for continuous toxicity monitoring. Biosens Bioelectron 14:355–361
Gu MB, Mitchell RJ, Kim JH (2000) Continuous monitoring of protein damaging toxicity using a recombinant bioluminescent *Escherichia coli*. ACS Symposium Series: Recent Advances in Chemical sensors and Biosensors for Environmental Monitoring 762:185–196
Gu MB, Kim BC, Cho J, Hansen PD (2001) The continuous monitoring of field water samples with a novel multi-channel two-stage mini-bioreactor system. Environ Monit Assess 70:71–81
Hansen PD (1986) The 'WaBoLu-AQUATOX' for integral monitoring of water pollutants. Vom Wasser 67:221–235
Min J, Kim EJ, LaRossa RA, Gu MB (1999) Distinct responses of a *recA'::luxCDABE Escherichia coli* strain to direct and indirect DNA damaging agents. Mutat Res 442:61–68
Radix P, Lénard M, Papantoniou C, Roman G, Saouter E, Gallotti-Schmitt S, Thiébaud H, Vasseur P (1999) Comparison of *Brachionus calyciflorus* 2-D and MICROTOX chronic 22-h tests with *Daphnia magna* 21-D test for the chronic toxicity assessment of chemicals. Environ Toxicol Chem 18:2178–2185
van der Schalie WH, Shedd TR, Knechtges PL, Widder MW (2001) Using higher organisms in biological early warning systems for real-time toxicity detection. Biosens Bioelectron 7–8:457–465
Van Dyk TK, Smulski DR, Reed TR, Belkin S, Vollmer AC, LaRossa RA (1995) Responses to toxicants of an *Escherichia coli* strain carrying a *uspA'::lux* genetic fusion and an *E. coli* strain carrying a *grpE'::lux* fusion are similar. Appl Environ Microbiol 61:4124–4127
Vollmer AC, Belkin S, Smulski DR, Van Dyk TK, LaRossa RA (1997) Detection of DNA dmage by use of *Escherichia coli* carrying *recA'::lux, uvrA'::lux*, or *alkA'::lux* reporter plasmids. Appl Environ Microbiol 63:2566–2571

Part VII

Chapter 65

Bacteria-Degraders Based Microbial Sensors for the Detection of Surfactants and Organic Pollutants

A. N. Reshetilov · L. A. Taranova · I. N. Semenchuk · P. V. Iliasov · V. A. Borisov
N. L. Korzhuk · J. Emnéus

Abstract

We developped a microbial biosensor for detection of surfactants. The biosensor receptor is based on bacterial strains that are harboring plasmids for biodegradation of surfactants. We studied the sensitivity, stability and selectivity of the biosensor with respect to anionic, cationic and non-ionic surfactants, carbohydrates, alcohols, humic acids and toxic compounds such as aromatic xenobiotics. Under laboratory conditions the microbial biosensor was used to assess the purity level of water that was decontaminated from surfactants. On the basis of the obtained sensor characteristics, a portable microprocessor analyzer was set up. The biosensor device registers the transducer's signals and makes it possible to process calibration dependencies and determination of the concentration of target compound in a sample.

Key words: microbial biosensor, surfactants, environmental protection

65.1 Introduction

The development of biosensors and biosensor methods for detection of surfactants is currently emerging. Nevertheless, there are two main approaches that already have been outlined in the analysis of surfactants, namely immunoassays and microbial sensors. Immunoassays are using antibodies to different types of surfactants, whereas microbial sensors are using the bacterial cells as the recognition element of sensor. Immunoassays have high sensitivity and selectivity that are laid in the principle of immunoassay methods and undoubtedly relate to their positive properties. The immunodetection of some surfactants and their metabolites including linear alkyl benzene sulfonates, alkylphenol/alkylphenol ethoxylates has been described (Franek et al. 2001). The immunoassay format has, however, some drawbacks, requiring the selection of optimal conditions, sensor transducer type, immunogen synthesis and antibody production, which leads to a rather high cost.

The development of microbial sensors is another means for the detection of surfactants. Microbial sensors are based on simple analytical equipment; the cost of the cell receptor element is considerably lower compared to the cost of antibodies; and the assay can be performed in express-mode. The selectivity and sensitivity of microbial sensors are lower than those of immunosensors. However, in a number of cases, such as the initial assessment of surfactants concentration in effluents before their treatment, their characteristics meet practical requirements. Therefore, microbial biosensors should not be considered as an alternative but rather a useful complement to immunosensors and other existing methods aimed at detection of surfactants.

The described biosensors of microbial type employ bacteria-degraders of surfactants. One of the measurement principles is based on the registration of oxygen consumption during microbial transformation of an analyte. The detection of linear alkyl sulfonates using a microbial sensor based on sewage sludge bacteria was demonstrated by Nomura et al. (1994). The sensor was used to assess the concentration of linear alkyl sulfonates in Ayase River, one of the most polluted rivers in Japan. The bacteria of genus *Pseudomonas* isolated from natural sources polluted with surface active compounds, provided the basis for a microbial sensor model for detection of anionic surfactants (Reshetilov et al. 1997b). The lower limit of detection for sodium dodecyl sulfate reached about 1 µM, and under laboratory conditions the sensor allowed continuous measurements for 25 d.

Amperometric microbial sensors with screen-printed electrodes as the transducers are frequently applied for analytical purposes. The measurements of toxicity, determination of phenols and surfactants in tannery and textile wastewater by a commercial bacterial biosensor named Cellsense® was carried out. In the Cellsense® whole-cell bacterial biosensor, an electrical current is obtained from the respiratory chain of the bacteria using electron mediators. The current is proportional to the level of metabolic activity. The toxicity data obtained with an amperometric biosensor based on *E. coli* was correlated with chemical analysis of wastewaters. The samples were analyzed by the biosensor followed by liquid chromatography – mass spectrometry for the identification of organic pollutants. This system was effectively applied to real sample measurements of influent and effluent wastewater, industrial tannery wastes, and textile untreated wastewater (Farre et al. 2001).

Another efficient instrument for detection of toxic compounds as well as surfactants is an optical biotest based on registration of *Vibrio fischeri* bacteria bioluminescence, which is inhibited in the presence of toxic compounds. This test was used for the assessment of the toxicity of various samples of sewage sludge from different waste treatment plants located in North-East Spain, receiving domestic and industrial wastewater. Here, the authors combined both chemical analysis, such as liquid chromatography – mass spectrometry and the toxicity data from the bioluminescence inhibition of *Vibrio fischeri* using the ToxAlert 100® system to identify and quantify the polar organic toxicity caused by surfactants such as linear alkyl sulfonates, nonylphenol (NP), NP polyethoxylates, and NP carboxylates in sewage sludge. The use of the combined procedure allows the characterization of a sewage sludge organic toxicity and the quantification of their contribution to the total toxicity (Lacorte et al. 2000).

Thus, the efficient detection of surfactants can be performed using immuno- as well as microbial sensors. In some cases, the use of microbial sensors is preferred since they can be used directly to assess the concentration of surfactants in samples that contain high concentration of surfactants such as wastewaters, textile untreated waste and industrial tannery wastes water. The typical measurement time is within the range of 1–10 min which is considerably less than what is usually needed for immunosensors. Due to the simplicity of equipment and low cost of microbial bioreceptor, the analysis performed by microbial sensors is more economical than immunosensor analysis.

The aim of this study was to create a new model of a microbial biosensor for the detection of surface active compounds. This report presents the results of investigations that were directed to solving the following problems: *(1)* assessment of the substrate specificity of different bacterial strains that degrade surface active compounds for a wide range of compounds such as carbohydrates, alcohols, organic acids, xenobiotics; *(2)* investigation of the signal dependence on external conditions and optimization of biosensor performance; *(3)* assessment of influence of cultivation conditions as well as immobilization matrix on the signals of sensor; and *(4)* development of a portable microbial biosensor prototype for surface active compounds detection and its application for measurement of model samples in laboratory conditions.

65.2
Experimental

65.2.1
Microorganisms

Five bacterial strains, including degrader of volgonat *P. rathonis* T, degrader of sodium dodecyl sulfate (SDS) *Pseudomonas* sp. 2T/1, degrader of alkylsulfate and SDS *P. aeruginosa* 1C, degrader of metaupon *P. putida* K and degrader of sodium monoalkylsulfosuccinate *Achromobacter eurydice* TK were employed. All strains were cultivated under periodical conditions for 18 h at 35 °C on shaker (140 rpm) in 500-ml Ehrlenmeier flasks containing 0.2 liter of the medium having the following composition (g l^{-1}): Na$_2$HPO$_4$: 6, KH$_2$PO$_4$: 3, NaCl: 0.5, NH$_4$Cl: 1, MgCl$_2$: 0.1, CaCl$_2$: 0.01, SDS: 0.2. To perform the inoculation, a 24 h-aged bacterial cultures obtained on an agar medium of the same composition were used and ~ 10^8 cells were introduced into the medium. When studying the pH effect on bacterial growth rate, potassium-phosphate buffer was used. The bacteria were incubated under shaking conditions on a circular shaker (230 rpm). The concentration of the bacterial suspension was measured nephelometrically with a photocolorimeter FEK-56M (Russia) at $\lambda = 540$ nm. The biomass was collected by centrifugation (5 000 g, 20 min) and washed twice with 30 mM of potassium phosphate buffer (pH 7.6). The sodium dodecyl sulfate concentration in the medium was estimated on the basis of colorimetric method by reaction with methylene blue (Abbot 1962). The specific growth rate was estimated using optical methods.

65.2.2
Immobilization of Cells

The receptor element was fabricated by incorporating cells in different gels such as polyvinyl alcohol (PVA), by adhesion on chromatographic paper GF/A (Whatman, Great Britain) and cross-linking with bovine serum albumine (BSA) and glutaraldehyde by the methods described by Woodward (1985). The immobilized cells were stored at 4–5 °C and 100% humidity. The viability of *P. rathonis* T in different gels was estimated by the microculture method. The microculture was incubated on medium M9 including 0.5% glucose and containing 2% gel (agar, agarose, calcium alginate gel) in the chambers sealed with paraffin at 28 °C for 4 d (until the microcolonies were formed). The phase-contrasting study of culture growth dynamics (formation of microcolonies) was performed using light microscope ICM405 ("Opton", Germany).

The obtained membrane (receptor element) of 0.3–0.5 mm in thickness and having the size of 20 mm^2 was fixed on the measuring surface of a Clark electrode (Ingold 531-04, Switzerland) using a nylon net. The electrode was placed in an open cuvette having the working volume of 5 ml, which contained 20 mM of phosphate buffer (pH = 7.6, t = 20 °C). The measuring solution was saturated with ambient air. The rate of electrode output signal change (nA s^{-1}) with the addition of substrate, corresponding to the rate of oxygen concentration change in the layer of the immobilized cells, was used as a measured parameter. The receptor element activity (or stability) was estimated when measuring 200 mg l^{-1} (~0.73 mM) SDS concentrations. The operational stability of the biosensor(s), or operation time of one receptor element, was estimated by periodic measurement of SDS for 20–25 d. Between the measurements the biosensor was kept in the buffer solution at room temperature and constant stirring. The stability of cells during receptor element storage was estimated by measuring the responses of sensor to SDS for 20 d.

65.2.3
Chemicals

The compounds, used as test substrates during biosensor's specificity estimation, belonged to nine different groups. Most of them were obtained from Reakhim (Russia) while some surfactants (alkyl benzene sulfonates, volgonat, metaupon and some others) were common industrial products.

1. Surface active compounds:
   A Anionic: sodium dodecyl sulfate (SDS, 100% of total weight (TW)), volgonat (sodium alkane sulfonate ($C_nH_{2n+1}C_mH_{2m+1}$)CHSO$_3$Na, with $n + m$ = 11 to 17; 60% of TW), metaupon (sodium alkylmethyltaurine, 45% of TW), alkyl naphthalene sulfonate (50% of TW), disodium monoalkylsulfosuccinate (DSS-A; 35% of TW), decyl benzene sulfonate (100% of TW), two types of alkylbenzene sulfonates (ABS): ABS (90% of TW), ABS (40% of TW), chlorine sulfonol (40% of TW).
   B Cationic: alkamone (alcoxymethyl diethyl ammonium methylsulfate; 100% of TW), cetylpiridinium chloride (100% of TW), tetradecyl trimethylammonium chloride (100% of TW), tetradecyl trimethylammonium bromide (100% of TW).
   C ampholytic: sulfobetain SB14 (100% of TW).
   D Non-ionogenic: monoalkylphenyl ester of polyethylene glycol (OP-10; 99% of TW), Tween 60 (sorbitan monostearate 90% of TW), Tween 80 (sorbitan monooleate; 98% of TW), slovagen (70% of TW), dodecyl ether of polyethylene glycol (n = 10 and n = 14; 90% of TW), cetyl ether of polyethylene glycol (n = 6; 90% of TW), triton X-100 (n-(tret-octyl)phenol; 98% of TW).
2. *Aromatic and polyaromatic compounds and their sulfo- and halogen derivatives:* Phenol, bromophenol, chlorobenzene, naphthalene, naphthalene sulfonate, sulfosalicylate, sulfoadenylate, sulfobenzoate, toluene sulfonate, benzene sulfonate, aniline, aniline-N,N-diacetic acid, anthraquinone, naphthol.

3. *Amines, amides and their substituted derivatives:* urea, phenyl urea, *n*-phenylene diamine sulfate.
4. *Humic acids.*
5. *Fatty acids:* lauric acid, tridecanoic acid, linolenic acid, palmitic acid, myristic acid, margarine acid.
6. *Alkanes and chlorinated derivatives:* decane, heptane, chloroform.
7. *Phthalates:* dimethylphthalate, diethylphthalate, dibutylphthalate, di(2-ethylhexyl)-phthalate, dinonylphthalate, phthalic acid diamide.
8. *Carbohydrates:* glycerol, glucose, arabitol, arabinose, xylitol, xylose, galactose, sorbitol, sorbose, sucrose, fructose, maltose, raffinose.
9. *Alcohols:* methanol, ethanol, propanol, butanol.

65.2.4
Investigation of Sensor Characteristics

To investigate the pH dependence of sensor responses, 30 mM phosphate buffer was used. Varying the ratio of the buffer components, NaH_2PO_4 and Na_2HPO_4, we obtained solutions with the required pH value within a range of 4.5–8.0. The medium with pH 4.5 was obtained by use of 30 mM NaH_2PO_4 as the single component of the carrier solution. The temperature dependence of the biosensor responses was investigated within a range of 20 to 50 °C. The cuvette was thermostated by means of a Thermostat U1 device (Germany). The stabilizing accuracy of the device was ±0.5 °C. In ionic-strength studies, the measurements were carried out in sodium chloride solutions (pH 7.5) by varying the concentration of the salt from 1 to 500 mM.

65.3
Results and Discussion

65.3.1
Calibration Graphs

The development of biosensors implies the study of their characteristics, and, first of all, assessment of its sensitivity and selectivity that determine the number of substrates analyzed by the device, the range of their detection and the possibility of the biosensor application for an analysis of complex media. The results describing calibration dependencies of the models under study are presented in this subsection. The biosensors based on anionic surface active compounds-degrading bacterial strains were characterized by high sensitivity to these compounds with calibration curves for sodium dodecyl sulfate (SDS) shown on Fig. 65.1. All the curves had monotomous character, i.e. the increasing of the substrate concentration resulted in the signal increase. The lower limit of SDS detection was 0.25 mg l^{-1} (0.86 µM) for *P. rathonis* T, *P. putida* K and *A. eurydice* TK and 0.5 mg l^{-1} (1.73 µM) for *Pseudomonas* sp. 2T/1 and *P. aeruginosa* 1C. A linear dependence between current change (nA s^{-1}) and SDS concentration was observed within a range of 0.25–200 mg l^{-1} in semilogariphmic scale of the concentration axis. As shown on Fig. 65.1, all strains were characterized by similar signal values to SDS at the concentration of 734 µM that corresponds to mass concen-

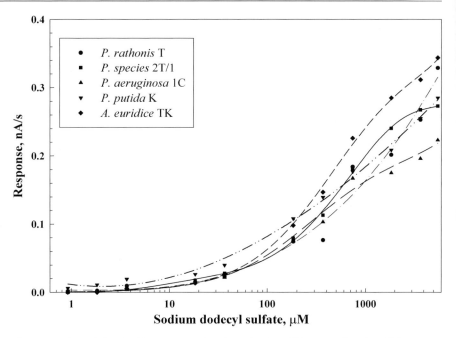

Fig. 65.1. Sodium dodecyl sulfate concentration dependencies of biosensors based on different surfactant-degrading strains

tration 200 mg l^{-1} used as the control probe for all surfactants in the present work. It should be noted that a wide range of measured SDS concentrations (0.25–1 500 mg l^{-1}) can be of great practical value for detection of anionic surface active compounds in both industrial and household waste waters and natural reservoirs.

65.3.2.
Substrate Specificity of Strains-Degraders of Surfactants

The results of evaluating the substrate specificity for the *P. rathonis* T-based sensor are presented in Fig. 65.2 and 65.3. The data for other sensors are presented in Table 65.1 and 65.2. The response values were normalized in relation to the sensitivity to SDS, i.e. the rate of the sensor response, which was taken as 100%. This normalization enables to estimate the ratio between the activities of strains during oxidation of different substrates.

At the assessment of substrate specificity towards surfactants (Fig. 65.2 and Table 65.1), the concentration of substrates was 200 mg l^{-1}, which corresponds to 734 µM for SDS. All strains were characterized by appreciably similar specificities and high sensitivity to most anionic, non-ionic and cationic surfactants. It should be noted that the stability of signals during measurement of cationic surface active compounds differed from the stability for other substrates. If the signals for anionic surface active compounds and non-ionogenic surface active compounds did not change for the whole operation time of receptor element, the responses to cationic surface active compounds were

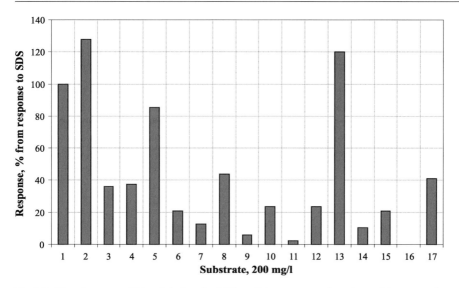

Fig. 65.2. The substrate specificity of the *P. rathonis* T-based biosensor towards surface active compounds. The x-axis designations: *1:* SDS; *2:* disodium monoalkylsulfosuccinate; *3:* alkyl naphthalene sulfonate; *4:* decyl benzene sulfonate; *5:* volgonat; *6:* metaupon; *7:* ABS (90% of TW); *8:* dodecyl ether of polyethylene glycol ($n = 10$); *9:* cetyl ether of polyethylene glycol ($n = 6$); *10:* OP-10; *11:* Triton X-100; *12:* Tween 80; *13:* sulfobetain SB14; *14:* alkamone; *15:* tetradecyl trimethylammonium bromide; *16:* cetylpiridinium chloride; *17:* humic acids

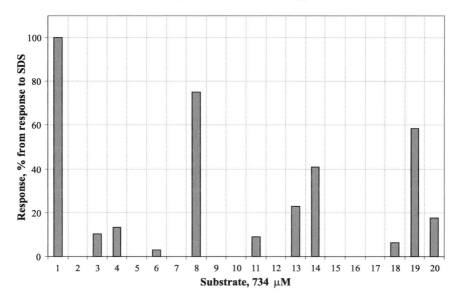

Fig. 65.3. The substrate specificity of the *P. rathonis* T-based biosensor towards aromatic compounds, organic acids, chloroform, some carbohydrates and alcohols. The x-axis designations: *1:* SDS, *2:* benzene sulfonate; *3:* toluene sulfonate; *4:* dimethylphthalate; *5:* diethylphthalate; *6:* dibutylphthalate; *7:* phthalic acid diamide; *8:* acetate; *9:* citrate; *10:* aniline; *11:* chloroform; *12:* phenol; *13:* p-bromophenol; *14:* n-phenylene diamine sulfate; *15:* salicylate; *16:* sodium benzoate; *17:* catechol; *18:* glucose; *19:* ethanol; *20:* butanol

Table 65.1. The substrate specificity of biosensors based on different strains towards surface active compounds. The first column values are identical to the x-axis designations in Fig. 65.2; the values in the other columns represent the responses for corresponding substrates normalized in relation to the response to sodium dodecyl sulfate which was taken as 100%

Substrate 200 mg l^{-1}	P. rathonis T	P. species	P. aeruginosa 1S	P. putida K	A. eurydice TK
1	100.0	100.0	100.0	100.0	100.0
2	128.0	184.0	100.0	159.0	130.9
3	36.0	4.3	0	14.8	27.0
4	37.5	7.1	10.1	14.5	11.8
5	85.4	82.4	123.0	82.0	83.8
6	20.8	2.0	7.0	28.9	6.9
7	12.5	3.9	0	29.5	5.0
8	44.0	16.1	33.0	24.3	28.0
9	6.0	0	0	9.8	18.8
10	23.6	8.0	51.0	6.8	13.5
11	2.3	0	20.0	0	0
12	23.6	42.0	43.0	17.3	34.0
13	120.0	30.0	58.0	46.0	96.0
14	10.6	0	21.0	0	0
15	20.7	10.4	64.0	0	0
16	0	0	0	0	0
17	40.9	6.5	27.2	13.6	9.0

obtained only during the first measurement, whereupon the receptor element of the biosensor lost its activity. Possibly, this is due to that cationic surface active compounds possess a pronounced antimicrobial effect that damages the cells. Aside from surfactants, the strains demonstrated significant responses only to naphthalene, phenylene diamine sulfate and some organic acids, whereas among carbohydrates and alcohols the responses were obtained only for glucose, ethanol and propanol (Fig. 65.3).

Characterizing the selectivity of the degrading bacteria, one should note that *P. rathonis* T and *Pseudomonas* sp. 2T/1 strains possess high selectivity and sensitivity to SDS and some surfactants of other classes (Fig. 65.2). The biosensor based on *P. rathonis* T and *Pseudomonas* sp. 2T/1 provided a specific response to disodium monoalkylsulfosuccinate (approximately 100% of the response to SDS for both strains), alkyl sulfonate (approximately 50% of the response to SDS for both strains), dodecyl ether of polyethylene glycol ($n = 10$, 41% for *P. rathonis* T), Tween 80 (approximately 20% for *Pseudomonas* sp. 2T/1), alcoxymethyl methyldiethylammonium methylsulfate (100% for *P. rathonis* T and 15% for *Pseudomonas* sp. 2T/1), humic acids (20% for *P. rathonis* T), naphthalene (70% for *P. rathonis* T), some fatty acids and carbohydrates. The responses of these strains to alcohols did not exceed 60% of the response to SDS

Table 65.2. The substrate specificity of the biosensors based on surface active compounds degrading strains towards aromatic compounds, organic acids, chloroform, some carbohydrates and alcohols. The first column values are identical to the x-axis designations on the Fig. 65.3; the values in the other columns represent the responses for corresponding substrates normalized in relation to the response to sodium dodecyl sulfate which was taken as 100%

Substrate 734 µM	P. rathonis T	P. species	P. aeruginosa 1S	P. putida K	A. eurydice TK
1	100.0	100.0	100.0	100.0	100.0
2	0	0	0	0	0
3	10.4	0	0	14.5	0
4	13.2	0	0	12.6	22.1
5	0	0	0	0	0
6	3.0	0	0	0	0
7	0	0	0	11.8	0
8	75.2	94.9	47.9	59.1	52.0
9	0	0	0	0	0
10	0	0	0	0	0
11	8.9	15.2	4.7	0	0
12	0	0	0	0	0
13	23.0	0	0	0	0
14	41.0	44.0	32.5	12.8	20.0
15	0	2.4	0	12.0	3.8
16	0	0	0	0	0
17	0	0	40.3	0	0
18	6.3	23.1	13.4	20.3	6.9
19	58.3	9.8	23.1	63.8	165.6
20	17.7	14.5	22.3	25.0	53.7

for ethanol, 5% for propanol and 15% for butanol when using *Pseudomonas* sp. 2T/1. Thus, considering the possibility for practical application of these microbial biosensors for detection of surfactants, basing it on the obtained data of their selectivity and calibration characteristics, it can be concluded that satisfactory detection characteristics are obtained and that theses microbial strains can be used in biosensors to detect surfactants.

The main disadvantage of microbial biosensors is their low selectivity. At the same time, in the course of the environmental monitoring, the detection of a whole pool of xenobiotics present in water ecosystems rather then a specific pollutant is often very important. The application of differential registration scheme and pattern recognition approaches (Weimar et al. 1990; Reshetilov et al. 1998), e.g. by means of measuring an analyte with several electrodes in one cell, will enable not only to enhance the selectivity of detection, but also identify surfactants of different classes. Thus, using a biosensor system consisting of three electrodes based on *P. rathonis* T, *Pseudomonas* sp. 2T/1 and *P. aeruginosa* 1C, we assume the possibility to make statements about the presence or absence of e.g. dodecyl ether of polyethylene glycol ($n = 14$) (from the response of sensors based on *P. rathonis* T, *P. aeruginosa* and the absence of response from the sensor based on *Pseudomonas* sp. 2T/1) or tetradecyl trimethylammonium bromide (from the response of sensors based on *P. aeruginosa* 1C and the absence of response from the sensor based on *P. rathonis* T and *Pseudomonas* sp. 2T/1).

The responses to decane and heptane were measured and compared to the responses to SDS concentration of 200 mg l^{-1} that corresponds to 734 µM. The sensitivity to decane for the *P. rathonis* T strain is approximately 20% of the sensitivity to SDS, approximately 15% for *Pseudomonas* sp. 2T/1 and 20% for *P. aeruginosa* 1C. Among all studied strains only *P. aeruginosa* 1C generated a signal for heptane (approximately 20% of the response to SDS). At the concentration of 5 mg l^{-1} (corresponding to 18.3 µM SDS) decane and heptane, no responses were obtained for all studied strains, while signals to SDS and a number of other surfactants were registered even in the concentration range of 1–10 µM.

Thus, the comparison of signals from biosensors based on *Pseudomonas* and *Achromobacter* bacteria demonstrated a considerable similarity of parameters during the detection of different organic substrates as well as high sensitivity to SDS, volgonat and some non-ionogenic surface active compounds. The studied strains can be used as a base of receptor element in the development of microbial biosensor for detection of surfactants. From this viewpoint the strain that seems to be the most prospect is *P. rathonis* T. A biosensor model based on this strain enables the detection of surfactants and some aromatic hydrocarbons, providing rather high selectivity, sensitivity and reproducibility of signals during long periods of time. Further experiments were thus aimed at optimizing the operation of the biosensors.

65.3.3
Influence of External Conditions and Optimization of Performance

The biosensor signal dependence on external conditions (pH, temperature, salt concentration) was studied for clarification of their effect on the output signals and to find the optimal conditions for measurements. The pH-dependence curve was of non-monotonous character (Fig. 65.4). The maximum signals were observed at pH 7.5. All sensor characteristics given further in this work were obtained at this pH value. The temperature dependence of the biosensor response was studied within a range of 20–50 °C. The plot was of the parabola type with the maximum reached between 35–40 °C. At 60 °C an irreversible loss of the biosensor activity occurred. Nevertheless, the range of room temperatures (20–25 °C) was optimal for the measurements, since the appreciable growth of the biosensor signal at a temperature of 35–40 °C was followed by a 3- to 4-fold increase of the recovery time.

The dependence of responses on the salt concentration (actually, ionic strength) in the carrier electrolyte was characterized by a sharp growth of the signal while increasing the sodium chloride concentration from 1 to 30–50 mM followed by a monotonous signal decrease. At high salt concentrations of 500 mM, the sensor responses

Fig. 65.4. Dependency of the *P. rathonis* T-based sensor on pH of carrier solution

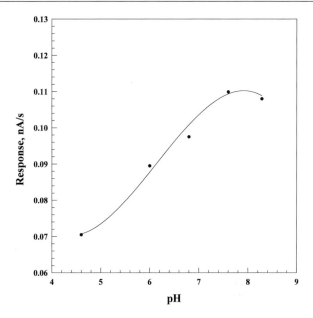

almost declined to zero. The obtained salt dependence of the signals could be explained by the properties of *P. rathonis* T used in the study and should be taken into account at the measurement of real samples.

The preliminary assessment of the operational stability of the sensor, i.e. operation time for one receptor element, was performed by multiple measurement of SDS (200 mg l^{-1}) for 3–5 d. During this period the biosensor was kept in buffer solution at room temperature and constant stirring. The decline in signal rate during this period did not exceed 5–10% for *P. rathonis* T, *Pseudomonas* sp. 2T/1 and *P. aeruginosa* 1C and 20% for *P. putida* K and *A. eurydice* TK for 24 h. The stability of strains during storage of the receptor elements was estimated by measuring the signals of the different receptor elements to a SDS concentration of 200 mg l^{-1} for 5–6 d. The maximum activity during storage in agar was maintained in *P. rathonis* T, *Pseudomonas* sp. 2T/1 and *P. aeruginosa* 1C strains. All further experiments were performed using the *P. rathonis* T strain.

65.3.4
Influence of Cultivation Conditions

The efficiency of the bacterial process of surfactants degradation is influenced by factors such as the nature of microorganism, its physiological condition, substance concentration, pH, osmotic pressure, incubation temperature and mode of cell cultivation. In this connection, studies of the growth regularity of *P. rathonis* T strain were performed on a medium containing SDS as a single carbon source. The cells growth curve seen in Fig. 65.5 represents a classical S-shaped curve. While constructing the biosensor receptor element, the cells at different growth stages were used. It was established that the maximum biosensor signal for 200 mg l^{-1} SDS concentration was obtained when using cells harvested after 5–6 h, corresponding to the exponential

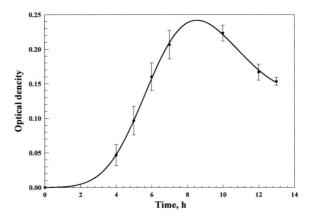

Fig. 65.5.
The biomass growth curve for *P. rathonis* T cells

growth stage of the cells. At the same time, reliable signal differences for microorganisms obtained in the middle and end of exponential growth stage were not revealed. The effect of cultivation temperature on the degrading activity of *P. rathonis* T strain with the bacteria cultivated at 20 °C, 28 °C, 35 °C, 40 °C was studied. The optimal temperature for strain growth and SDS utilization was established to be 35 °C at which the degradation of substrate occurred at a high rate, reaching the maximum value during 5–6 h of incubation. The maximum specific growth rate at this temperature was 0.51 h^{-1}. Total SDS degradation occurred within 13 h. The change in cultivation temperature was accompanied by the decline in SDS degradation rate from 0.45 h^{-1} to 0.23 h^{-1} and, correspondingly, the period of total substance destruction was prolonged from 5 to 10 h.

As the abrupt decrease in pH of the medium is known to be a factor limiting the growth of degrading bacteria on surfactant-containing synthetic medium at conditions of periodical cultivation, it was of interest to study the pH effect on the growth of this strain. The results of this experiment showed that the bacterial growth in the medium having pH 8.0 occurred faster than at pH 7 and that the SDS concentration of 200 mg l^{-1} is completely utilized by the culture within 16 h. In a buffer medium of pH 6.0, the growth of *P. rathonis* T was absent. The data obtained matches fairly well the results of the biosensor pH-dependence study.

The present study describes a case when the culture growth substrate and biosensor test substrate is represented by the same compound that implies identical enzyme systems ensuring its utilization and formation of biosensor response. From our point of view, this accounts for practically absolute compliance of temperature and pH optima both for culture growth and biosensor functioning. At the same time, microbial biosensors for detection of compounds that cannot be used as growth substrates are known. As an example a biosensor based on bacterial cells of *Gluconobacter oxydans* is characterized by high sensitivity to glucose, but this strain cannot use glucose as carbon source due to its metabolic properties (Reshetilov et al. 1997a). It is obvious that in these cases the sensor characteristics may differ from the respective parameters of culture growth. This fact should be considered during the development of new biosensors and optimization of their operation conditions. Thus, the results of the study indicate that the respiratory activity level of the receptor element of the biosensor based on *P. rathonis* T was maximal when the cells were collected during the exponential growth stage (the age of culture is 5–6 h). The compliance of temperature (35 °C) and pH optima (pH 7.8) of culture growth and biosensor response was registered. The data obtained are important for the optimization of surfactants detection sensitivity.

65.3.5
Influence of Immobilization Matrix

The analysis of literature data published in recent years revealed that the immobilization by including of cells into the membranes of polyvinyl alcohol and calcium alginate gel is most frequently used in microbial biosensors, especially in environmental monitoring-oriented ones (Racek 1995). Therefore, at the first stage of our work, the operational stability and storage stability in agar, agarose and calcium alginate gels, was studied. The concentration of cells under all immobilization methods was 1.8 g of dried cells l^{-1}. The experimental data demonstrated that when the cells were included in agar gels the receptor element maintained its operation stability for more than 20 d and, however, with a decline in signal of 20% for the first 3 d of measurements, whereupon the signal became stabilized. The decline in signal during the same period was not more than 4% for agarose gel.

The study of storage stability demonstrated that the activities of the cells immobilized into agar and agarose gels were practically the same. The decline in signal activity during storage in agarose gel occurred a little faster, which probably was due to lower oxygen permeability of agarose as compared to agar gel. The inclusion of cells in calcium alginate gel proved to be less efficient, which may due to both higher molar concentration of the buffer used and the chemical and structural properties of the calcium alginate gel. The receptor element constructed in this way operated for 9 d and the decline in operation activity was approximately 11%. The receptor element recovery time after measurement was 1.5 h. The cells lost their activity completely during storage, on day 7 after immobilization. Thus, the immobilization into agar gels provided the possibility to maintain higher activity and stability of cells.

When immobilizing the cells by their inclusion into a membrane of poly(vinyl alcohol) (PVA), it was found that *P. rathonis* T maintained its operational stability under these conditions for more than 11 d and a storage stability for about 6 d. The decline in operational stability was approximately 8% and storage stability – 16% after 1 d. The receptor element recovery time was approximately 1 h.

The biosensor receptor element stability with *P. rathonis* T being immobilized by the adhesion on chromatographic paper GF/A was also performed. However, a reproducible signal was not obtained in this case. The decline in operational stability and storage stability of the biosensor was more than 80% for 1 d. The receptor element recovery time was 0.5–0.6 h after measuring 0.7 mM SDS. This low stability level may be due to washing-out of the cells from the carrier surface due to the structural properties of cells that do not provide a robust attachment during their interaction with the carrier. The immobilization of *P. rathonis* T by cross-linking with glutaraldehyde and BSA proved to be unacceptable. Respiratory activity of cells in this biosensor receptor element configuration was basically absent.

The comparison of literature data with the obtained results showed that the inclusion in gels provides a number of advantages in contrast to other methods of immobilization. Under certain conditions the carrier can protect the cells from the effect of unfavorable factors and contribute to long-term maintenance of their biochemical activity (Racek 1995; D'Souza 2001). In most works, dealing with biosensor detection of xenobiotics, the authors consider the immobilization into calcium alginate gel and PVA to be more advantageous (Nomura et al. 1994; Rainina et al. 1996). In our experiments, the stability of the biosensor receptor element with *P. rathonis* T in agar gel was 2.5 times higher as compared to being immobilized in calcium alginate gel and PVA membrane, and the reproducibility of signals was highest when immobilized in agarose gel. Practically all methods used in this work provided a receptor element recovery time within a shorter time, as compared to data given in work where the surfactant assay was carried in a reactor-type receptor element (Nomura et al. 1994). Rapid receptor element recovery was observed with cells immobilized on chromatographic paper GF/A, in a membrane of PVA, and agar gel.

The influence of cell loading on the biosensor responses showed that the sensor output signals increased with the biomass content in the receptor element. At cell concentrations higher than 1200 mg l^{-1} (dry weight), saturation occurred and the signal amplitude increased insignificantly. This concentration was used in all further studies.

65.3.6
Application of Biosensors for Detection of Model Samples at Laboratory Conditions

The microbial biosensor based on *P. rathonis* T was tested on samples of model and real wastes containing anionic surfactants both before and after their treatment in bioreactors with bacteria-destructors of surfactants. The standard chemical measurement of anionic surface active compounds by methylene blue was used as a control method. The results of the testing as well as the data obtained using the control method are presented on Fig. 65.6. Three bioreactors (R1–3) were used for treatment of the samples. R1 included a mixture of strains obtained from Anox AB (Lund, Sweden), R2: a mixture of the surfactant degrading strains *P. rathonis* T, *Pseudomonas* sp. 2T/1, *P. putida* K, *A. eurydice* TK,

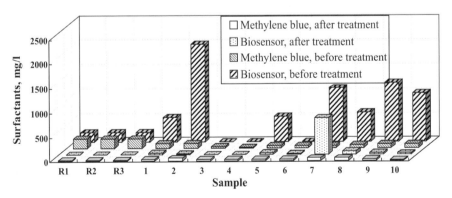

Fig. 65.6. Results of biosensor measurements on model- and real samples. *R1–3:* The concentration of surfactants in model medium; *1–10:* the concentration of surfactants in real samples of wastes

P. alcaligenes, P. fluorescence, and R3: a mixture of both the strains from Anox AB and the cultures used in R2. All strains were immobilized on suspended plastic carrier. The total anionic surfactants concentration entering the reactors was 100 mg l^{-1} (50 mg l^{-1} of alkane sulfonate, 20 mg l^{-1} of alkane sulfate and 30 mg l^{-1} of alkylbenzene sulfonate). The reactor volume was 2 l and the flow rate made up 300–400 ml/24 h. The experiment demonstrated a high level of model media degradation in all of the bioreactors (R1–R3). The results of biosensor detection were coinciding very well with the results of control method. The Student's *t*-test did not revealed significant differences between the results of the two methods with the probability of 0.99.

The next step of biosensor examination included the measurement of surfactants concentration in 10 samples of wastes obtained from various sources, before and after their treatment in bioreactor R3. The high level of biodegradation was demonstrated for samples No. 1, 2, 5, 8–10. The biosensor did not detect the presence of alkane sulfate and alkane sulfonate in samples No. 3, 4 and 6; which indicates that these samples contain other surfactants that cannot be detected by the sensor. The Student's *t*-test carried out for this step showed the absence of significant differences between the results of the two methods, with the probability equal to 0.8. The high values of biosensor response for non-treated wastes are possibly due to the presence of significant concentration of organics. Thus, it was shown that the biosensor based on *P. rathonis* T cells is characterized by high sensitivity and selectivity and seems to be promising for surfactants assaying in model and real wastewater samples.

65.3.7
Portable Microprocessor-Based Biosensors

The results presented in previous sections were obtained using a laboratory biosensor model based on a Clark-type electrode with a preamplifier (Ingold, Switzerland). The biosensor signal was transferred to the computer through analog-to-digital/digital-to-analog converter (ADC/DAC) adapter and processed by the software "Sensor for Windows". As the optimized characteristics of the sensor (sensitivity, selectivity, stability etc.) were shown to be satisfactory from the standpoint of its practical application, it was used as a base for the development of a portable electronic block that should provide registration and total processing of the biosensor signal and could be considered as a prototype of an industrial class device.

The microprocessor device AB-1 (Analyser Biosensor-1st version) was created. The microprocessor device AB-1 is the electronic part of the microbial biosensor for surfactants detection. Its purpose is the registration and processing of signals of the Clark amperometric electrode, containing immobilized surfactant-degrading bacteria. To extend the device functions it was constructed so that it may be used as a measuring and processing unit in biosensors based on pH-sensitive field-effect transistors also. This makes it more universal and provides a base for microbial cell, enzyme types and immunosensors, where the signals generated by a receptor element is accompanied with pH changes. The device includes a power supply unit, analog type amplifier, ADC (analog-digital converter) unit and programmable microprocessor based controller, liquid-crystal display for indication of results of analysis. The device represents a compact apparatus and can be regarded as a prototype for industrial production, see Fig. 65.7a.

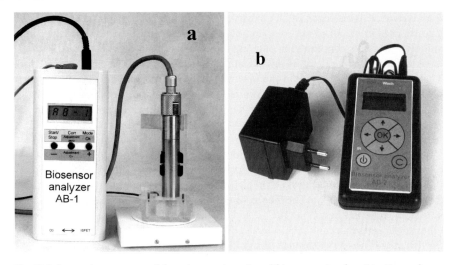

Fig. 65.7. External appearance of the microprocessor based biosensor signal registration and processing unit. **a** AB-1 model. *On the left:* Processing unit possessing calibration, calculations and displaying the concentration of an analyte in the sample; *on the right:* measuring unit (electrode with the immobilized cells) in the cuvette. **b** AB-2 model

The developed laboratory prototype possesses biosensor functions enabling: *(1)* to perform electronic calibration ("zero" output adjustment in the absence of transducer; 100% output adjustment when biorecognition element is conditioned for measurement), *(2)* to make biosensor calibration ("one-point" mode of calibration is used), *(3)* to process biosensor signals by their amplitude and initial rate; and *(4)* to possess a digital output for switching ON and OFF peristaltic pump to refresh the carrier buffer. Further developments resulted in creation of new versions of the device, AB-2 and AB-3 (Fig. 65.7b and Fig. 65.8, respectively). The new features realized in these models and absent in the first version of the device are the possibility for multi-point calibration, reduced level of noise and high accuracy of measurements, and the ability to use the device in combination with a personal computer. The biosensor analyzer AB-3 represents a unit for amplification, registration and processing of signals from amperometric biosensor transducers (in this case, Clark electrode; the analyzer also can be used for registration and processing of signals from screen-printed electrodes without any modification). It's minimal re-equipment with the connecting amplification unit enables registration of signals of any biosensor transducer type – field effect transistors, optical sensor, thermistor etc. The dimensions of the device are $14 \times 8 \times 3.8$ cm with mass of 190 g.

The results of laboratory trials showed that the constructed prototype could be effectively used for the measurement of signals of the sensors based on bacterial cells and Clark type electrode. The biosensor prototype demonstrated high stability of performance of the electronic part; high accuracy of signal measurement; and simple mode of operation. The constructed device was used for surfactants measurements at laboratory conditions to study the stability of receptor element prepared from *P. rathonis* T at different modes of immobilisation and storage conditions. It can be considered as a prototype of an industrial class.

Fig. 65.8. External appearance of the microprocessor based biosensor signal registration and processing unit AB-3 attached to the computer, peristaltic pump and measuring unit

65.4
Conclusion

The efficiency of strains oxidizing surfactants was assessed for further selection of bacteria to construct receptor elements of microbial biosensors. The substrate specificity of 5 different strains was studied and these strains were shown to have individual substrate specificity patterns with approximately the same sensitivity to SDS, a model surfactant of anionic type. The most prospect strain for construction of a receptor element was *P. rathonis* T that had the highest sensitivity to anionic surfactants. The technology for construction of sensor receptor element based on *P. rathonis* T cells was developed. Using this biosensor, the concentration of surfactants was assessed under laboratory conditions before and after treatment of water in column-type bioreactor. The biosensor could be used in monitoring of wastewaters during their treatment in model conditions. At the same time, to obtain an unequivocal answer for the question related to the concentration of surfactants in real samples on the initial monitoring stage, the control of obtained values using standard liquid chromatography-mass spectrometry methods will be required. A portable microprocessor device was developed on the basis of parameters obtained during the study of the laboratory model. The device represents the electronic part of the biosensor and is designed for registration of bioreceptor signals, calculation of calibration dependence and measurement of surfactants concentration in a sample. Being combined with the receptor element, the device represents a laboratory biosensor prototype for detection of surfactants.

Acknowledgments

This study was performed within the framework of INCO-Copernicus projects "Biological tools for a sustainable water management" (BIOTOOLS, IC15-CT98-0138) and "Biosensor feed-back control of wastewater purification: photooxidation followed by biological degradation using highly active strains of surfactant degrading bacteria" (BIOFEED, ICA2-CT-2000-10033) as well as INTAS project "Biosensor feed-back control of wastewater purification: photooxidation followed by biological degradation using highly active strains of surfactant degrading bacteria" (BIOFEED, N 99-0995). Authors express gratitude to the Thomas Welander, Head of SME "Anox AB" (Lund, Sweden) for target setting, results discussion and providing of equipment and samples for biosensor examination of wastes.

References

Abbot D (1962) The colorimetric determination of anionic surface-activity materials in water. Analyst 87:287–293

D'Souza S (2001) Microbial biosensors. Biosens Bioelectron 16:337–353

Farre M, Pasini O, Alonso M, Castillo M, Barceló D (2001) Toxicity assessment of organic pollution in wastewaters using a bacterial biosensor. Anal Chim Acta 426:155–165

Franek J, Zeravík S, Eremin A, Yakovleva J, Badea M, Danet A, Nistor C, Ocio N, Emnéus J (2001) Antibody-based methods for surfactant screening. Fresenius J Anal Chem 371:456–466

Lacorte S, Guiffard I, Fraisse D, Barceló D (2000) Broad spectrum analysis of 109 priority compounds listed in the 76/464/CEE council directive using solid-phase extraction and GC/EI/MS. Anal Chem 72:1430–1440

Nomura Y, Ikebukuro K, Yokoyama K, Takeuchi T, Arikawa Y, Ohno S, Karube I (1994) A novel microbial sensor for anionic surfactant determination. Anal Lett 27:3095–3108

Racek J (1995) Cell-based biosensors. Technomic Publishing Company, Lancaster, Basel

Rainina E, Efremenco E, Varfolomeyev S, Simonian A, Wild J (1996) The development of a new biosensor based on recombinant *E. coli* for the direct detection of organophosphorus neurotoxins. Biosens Bioelectron 11:991–1000

Reshetilov A, Donova M, Dovbnya D, Boronin A, Iliasov P, Leathers T, Greene R (1997a) Evaluation of *Gluconobacter oxydans* whole cell biosensors for amperometric detection of xylose. Biosens Bioelectron 12:241–247

Reshetilov A, Semenchuk I, Iliasov P, Taranova L (1997b) The amperometric biosensor for detection of sodium dodecyl sulfate. Anal Chim Acta 347:19–26

Reshetilov A, Lobanov A, Morozova N, Gordon S, Greene R, Leathers T (1998) Detection of ethanol in a two-component glucose/ethanol mixture using a nonselective microbial sensor and a glucose enzyme electrode. Biosens Bioelectron 13:787–793

Weimar U, Schierbaum KD, Gopel W, Kowalkowski R (1990) Pattern recognition methods for gas mixture analysis: application to sensor arrays based upon SnO_2. Sens Actuators B 1:93–96

Woodward J (ed) (1985) Immobilized cells and enzymes. Practical approach, IRL Press Ltd., Oxford – Washington DC

Study of Cr(VI) and Cd(II) Ions Toxicity Using the Microtox Bacterial Bioassay

E. Fulladosa · I. Villaescusa · M. Martínez · J.-C. Murat

Abstract

The Microtox® bioassay, based upon the fading of light emitted by the luminescent bacteria *Vibrio fischeri* when exposed to noxious substances, was used for studying the changes in speciation and the related changes in toxicity of two metals known as environmental pollutants. It was verified that modifications of pH and of ionic composition of the incubation medium did not affect the standard toxicity of phenol. By contrast, Cr(VI) toxicity was found to decrease as pH increased, underscoring that hydrogenchromate anion, the dominant species at low pH, is the most harmful. Cr(VI) toxicity was not modified when changing the medium composition, as this metal does not form chloro-complexes in the presence of sodium chloride. Conversely, Cd(II) toxicity was almost unaffected by pH within the 5.0–7.0 range. Replacing sodium chloride either by sodium nitrate or by sodium perchlorate resulted in changes of the measured cadmium toxicity, due to changes in speciation. Free Cd^{2+} ion was found to be the most harmful toward the *Vibrio fischeri* bacteria. In conclusion, both pH and ionic composition are factors that strongly influence the measured toxicity of environmental samples containing hexavalent chromium and/or cadmium when using the Microtox® bioassay.

Key words: Microtox®; metal speciation; water pollution; pH effect, ionic strength effect; toxicity

66.1 Introduction

Increasing need for monitoring environmental pollution in urban or industrialized areas leads to the development of very sensitive detectors of harmful substances. It is an accepted assumption that measurement of some chemical concentrations in reference to established regulations does not give an accurate account of the environmental noxiousness. Therefore, much attention was paid to biological sensors, markers or detectors. Biological models for this purpose must be suitable for routine tests, easy to keep at the laboratory, posing few ethical problems and as much as possible standardized for reproducibility. Measurement of biological indices must be reliable, easy to carry out in laboratory with medium-range equipment and as unequivocal as possible. Biological models such as microorganisms and cultured cells were frequently studied during the past years as they meet the required criteria. In order to test the toxicity of a given pollutant, classical indices used in toxicology, such as EC_{50} (being the concentration which induces 50% of a maximal effect), LD_{50} (being the concentration which kills 50% of the exposed organisms) and inhibition of cell growth, were widely used (Fergusson 1991; Crosbi 1998). In the past years, a variety of biological models adapted to toxicology were proposed and several indices of biological suffering during or after exposure to a variety of contaminants were investigated. Organisms from different trophic levels have been used, viz, bacteria *Bacillus cereus*, *Vibrio fischeri* (Khangarot and Ray 1987; Ribó 1983; Ho et al. 1999; Castillo et al. 2000); nematode *Panagrellus redivivus* (Castillo et al. 2000); cladocerans *Daphnia magna*, *Daphnia similis*, *Ceriodaphnia dubia* (Khangarot and Ray 1987; Dierickx and Bredael-Rozen 1996; Castillo et al. 2000; Fochtman et al. 2000; Choi and Meier 2001); fish *Oncorhynchus mykiss*, *Cyprinus carpio*, *Pimephales promelas*, *Salmo gairdneri*, *Fathead minnow* (Khangarot and Ray 1987; Dierickx and Bredael-Rozen 1996; Castillo et al. 2000; Fochtman et al. 2000; Choi and Meier 2001); amphipod *Ampelisca abdita* (Ho et al. 1999); algae *Lemna minor*, *Scenedesmus quadricauda* (Ince et al. 1999; Fochtman et al. 2000); plant *Lactuca sativa* (Castillo et al. 2000); HT29, A549 and HepG2 cultured cell lines (Delmas et al. 1996, 1998; Gaubin et al. 2000).

The model described in the present article consists of a suspension of marine bacteria *Vibrio fischeri*, that emit light under standard conditions. The light producing mechanism is tied to the metabolic processes within the cell. If a toxic substance disturbs the metabolic processes, a reduction of the light output will result. This method is commercialized under the brand name of Microtox® (Bulich 1986). The Microtox system is a tool used for screening or monitoring the presence of pollutants in environmental samples. As well, it was successfully used for studying the toxicity of a large number of single chemicals (Ribó and Kaiser 1983; Kaiser and Ribó 1988) and some metallic salts (Hindwood and McCormick 1987; Ribó et al. 1989). In order to understand the impact of toxic elements on aquatic organisms, it is important to know both the chemical speciation of the metal in solution and the toxicity of each chemical species. Total concentration of the element, ionic composition of the medium and pH are the main variables influencing the speciation.

For several elements like chromium or arsenic the chemical speciation will change depending on the pH (Puigdomènech 1983). Generally, divalent metals exist as free ions and as different species or complexes, depending on the physical properties of the medium (pH, ionic strength, etc). In the past, it was presumed that concentration of the free ionic forms determined the toxicity of metals toward aquatic organisms (Sunda et al. 1978; Gadd and Griffiths 1978). However, recent studies have demonstrated large difference in effects of the metal according to their speciation (Villaescusa et al. 1996; Sorvari and Sillanpää 1996; Villaescusa et al. 1997, 2000).

In the standard Microtox test, it is recommended to use a 2% (w/v) NaCl solution (0.34 mol l^{-1}) for incubation and to keep the pH value within the 5.5–6.5 range (Bulich 1979). However, it was demonstrated by Krebs (1983) that the sensitivity of luminescent bacteria used in the Microtox test was not affected by pH within the 4.5–9.5 range and that these bacteria could withstand large variations in ionic composition. Thus, it seemed to be interesting to use the Microtox assay for studying the change in toxicity of some metallic salts as a function of pH and medium composition. Taking into account that many metals form chloro-complexes in the presence of chloride anions and that few metals are complexed in the presence of nitrate or perchlorate (Sillen and Martell 1982), a set of experiments was carried out by replacing sodium chloride by either sodium nitrate or sodium perchlorate. In the present article, results concerning Cr(VI) and Cd(II) toxicity upon the *Vibrio fischeri* bacteria, under different conditions of pH and ionic composition, are reported.

66.2 Experimental

66.2.1 Test Reagents and Chemicals

The Microtox test reagent is a preparation of a specially developed strain of the luminescent marine bacteria *Vibrio fischeri* supplied by Azur Environmental (Carlsbad, CA, USA). Potassium dichromate, cadmium chloride, cadmium nitrate, cadmium perchlorate and phenol were reagent grade and purchased from Merck (Darmstadt, Germany). Dilution solutions of different ionic strengths were prepared by dissolving either NaCl, $NaNO_3$ or $NaClO_4$ (Merck reagent grade) in ultra-pure water (Milli-Q system, Millipore, Bedford, MA,USA).

66.2.2 Apparatus

The test was performed using the Microtox Model-500 Toxicity Analyzer system from Microbics Corporation (Carlsbad, CA, USA). The total metal concentration was determined with an Atomic-Absorption Spectrophotometer (Varian Techtron, AA-1275/1475 model, Springvale, Australia). The pH values were recorded using a Digilab-517 pH-meter (Crison, Barcelona, Spain).

66.2.3 Metal Species Distribution Diagrams

Species distribution diagrams for Cr(VI) and Cd(II) as a function of pH, at constant NaCl concentration, were established by using a special computerized program (Puigdomènech 1983) based upon the equilibrium constants given in the literature (Baes and Mesmer 1976; Sillen and Martell 1982). The same computerized program was used for establishing the distribution diagrams of Cd(II) and Cr(VI) species, at pH 6.0, as a function of NaCl, $NaNO_3$ or $NaClO_4$ concentrations.

66.2.4 Determination of the Effective Concentration (EC_{50})

In order to establish the most suitable range of Cr(VI) and Cd(II) concentrations for the toxicity determination, preliminary tests were conducted in each ionic media as described elsewhere (Villaescusa et al. 1996, 1997). The effective concentration (EC_{50}), at which a 50% loss of light emission is observed, was determined using the gamma (Γ) function, which is defined as the ratio of lost to remaining light. The EC_{50} value is the concentration at which $\Gamma = 1$ (Ribó and Rogers 1990).

66.2.5
Phenol Toxicity

Phenol is considered as a standard for toxicity in Microtox tests. Therefore, in order to verify that the principle of the assay would not be affected by pH or ionic composition, a set of preliminary experiments was performed with phenol solutions prepared in different media (namely, 2.0–5.0% (w/v) NaCl, 3.0–6.0% (w/v) $NaNO_3$, 4.0–8.0% (w/v) $NaClO_4$) or at different pH values within the 5.0–9.0 range.

66.2.6
Chromium(VI) Species Toxicity

In order to study Cr(VI) toxicity as a function of pH (ranging from 4.6 to 9.3), potassium dichromate solutions were prepared in 2% (w/v) NaCl. On the contrary, when studying the influence of ionic strength (NaCl concentrations ranging from 1.0 to 3.0%, w/v), the pH value was kept constant at 6.0.

66.2.7
Cadmium(II) Species Toxicity

In order to investigate Cd(II) toxicity as a function of pH (ranging from 5.0 to 7.0), solutions of cadmium chloride were prepared in 2% (w/v) NaCl. Conversely, the effect of medium composition and ionic strength (namely, 1.5–7.0% (w/v) NaCl, 2.5–7.0% (w/v) $NaNO_3$ and 4.0–6.0% (w/v) $NaClO_4$ solutions, containing different concentrations of $CdCl_2$, $Cd(NO_3)_2$ and $CdClO_4$, respectively) on toxicity was evaluated at constant pH value (6.0).

66.3
Results and Discussion

66.3.1
Phenol Toxicity in Different Media

As phenol is a standard in the Microtox assay, phenol toxicity in different media was determined in order to check that changes in pH and medium composition would not affect the observed toxicity. Phenol EC_{50} values (defined as the concentrations of phenol that produce a 50% decrease in light emission by the luminescent bacteria) were determined in 2% (w/v) NaCl solutions adjusted at different pHs. Within the studied 6.0–9.0 pH range, the average EC_{50} for phenol after 5 min exposure was 29.94 ±2.4 mg l^{-1} indicating the absence of significant change in toxicity due to pH. Our value is in accordance with the EC_{50} values reported in literature (Hindwood and McCormick 1987; Ribó and Rogers 1990). When evaluating the effect of ionic strength, phenol EC_{50} values were determined in different media. Table 66.1 shows the phenol EC_{50} values at different concentrations of NaCl, $NaNO_3$ and $NaClO_4$ adjusted at pH 6.0 after 5 min exposure.

Table 66.1.
Effect of medium composition on phenol toxicity measured using the Microtox assay. Phenol EC_{50}-5 min values at pH 6.0

Medium			EC_{50}-5 mim (mg l^{-1})
	(% w/v)	(mol l^{-1})	
NaCl	2.0	0.34	30.39
	3.0	0.51	25.52
	5.0	0.85	27.18
NaNO$_3$	3.0	0.34	29.10
	3.5	0.41	25.12
	4.0	0.47	24.40
	6.0	0.71	27.83
NaClO$_4$	4.0	0.34	39.43
	5.0	0.41	29.36
	6.0	0.49	29.72
	8.0	0.78	33.57

It appears that similar values for phenol toxicity were obtained in all tested media. These results confirm *(i)* that there is no effect of pH on phenol toxicity, *(ii)* that NaNO$_3$ and NaClO$_4$ are acceptable as alternative media and *(iii)* that the ionic strength does not affect the measured EC_{50} values. Other authors have already reported similar findings when studying the toxicity of metals and other substances (Hindwood and McCormick 1987). Hence, the stability of the Microtox bioassay within a rather wide range of pH and medium composition was stated and further experiments were designed to study the toxicity of metals on *Vibrio fischeri* as a function of the speciation and/or complexation.

66.3.2
Effect of pH on Cr(VI) Toxicity

Cr(VI) toxicity as a function of pH, using the bacterial Microtox assay, was evaluated in 2% (w/v) NaCl solutions adjusted at different pHs within the 4.6–9.3 range. Cr(VI) EC_{50} values determined after 15 min exposure to the bacteria vs. pH values are plotted in Fig. 66.1. As can be seen, EC_{50} values were found to increase, meaning that toxicity decreases, when pH increases. A maximum was observed at pH 7.0–7.5. The change in toxicity can be attributed to changes in the Cr(VI) species distribution as a function of pH, as seen in Fig. 66.2. At pH values under 6.0 the main species is the hydrogenchromate anion whereas at neutral and basic pH values the predominant species is the chromate anion.

When comparing Fig. 66.1 and 66.2, it appears that toxicity decreases (EC_{50} increases) when the percentage of hydrogenchromate anion also decreases. It suggests that this anion is the most responsible species for Cr(VI) toxicity in a 2% (w/v) NaCl medium. These results show that the toxicity of Cr(VI) depends on the speciation, which changes with the pH. Not paying attention to the pH, when analyzing a sample containing hexavalent chromium, could lead to an incorrect determination of toxicity.

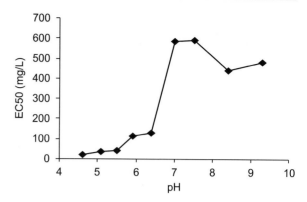

Fig. 66.1.
Effect of pH on Cr(VI) toxicity. [NaCl] = 0.34 mol l^{-1}

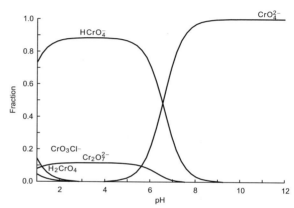

Fig. 66.2.
Cr(VI) species distribution diagram. [Cr(VI)] = 1.92 mmol l^{-1}. [NaCl] = 0.34 mol l^{-1}

66.3.3
Effect of Ionic Strength on Cr(VI) Toxicity

As mentioned above, the Microtox assay is currently performed in 2% (w/v) NaCl for the osmotic protection of the bacteria. In order to check whether ionic strength could influence the Cr(VI) toxicity to the luminescent bacteria, a set of experiments was carried out at pH 6.0 at NaCl concentrations varying from 0.17 to 0.51 mol l^{-1}. EC_{50} values after 15 min exposure showed no significant variation (25.1 ±2.3 mg l^{-1}), indicating that variations in NaCl concentration does not affect Cr(VI) toxicity. These results can be explained by the fact that Cr(VI) does not form chloro-complexes at pH 6.0 and that there is no alteration of the speciation.

66.3.4
Effect of pH on Cd(II) Toxicity

The species distribution diagram corresponding to cadmium in 2% (w/v) NaCl as a function of pH (Fig. 66.3) reveals that there is no change of cadmium speciation within the studied pH range (5.0–7.0). Free Cd^{2+} and $CdCl^+$ species are present in the same percentage at both pHs.

Fig. 66.3.
Cd(II) species distribution diagram. [Cd(II)] = 10 mmol l^{-1}. [NaCl] = 0.34 mol l^{-1}

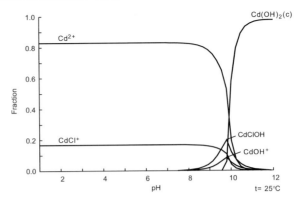

As a consequence, Cd(II) toxicity was not expected to vary with the pH. In order to check that Cd(II) toxicity was not influenced by pH, EC$_{50}$ values for Cd(II) were measured in 2% (w/v) NaCl adjusted at pH 5.0, 6.0 or 7.0. The obtained EC$_{50}$-15 min values were 10.1, 10.9 and 11.0 mg l^{-1}, respectively, confirming that cadmium toxicity measured with Microtox is independent of pH within the studied range. The absence of pH effect on toxicity was also found by Iglesias (1996) when Pb(II) and Ni(II) toxicity was evaluated at pHs ranging from pH 5.5 to pH 7.0 using the Microtox assay. Nevertheless, these results cannot be generalized to all divalent metals. For instance, in the case of mercury, pH changes result in changes in the species distribution and, consequently, in the final toxicity (Ribó et al. 1989).

66.3.5
Effect of Ionic Strength on Cd(II) Toxicity

Cadmium does form complexes in NaCl and NaNO$_3$ solutions as can be seen in the species distribution diagrams shown in Fig. 66.4a,b. Nevertheless, no complexes are formed between cadmium and the perchlorate anion (Sillen and Martell 1982). Consequently, only free Cd^{2+} is present in NaClO$_4$ solutions within the tested concentrations range.

Figure 66.4 shows that Cd(II) is totally complexed by the chloride ions, forming the CdCl$^+$, CdCl$_2$ and CdCl$_3^-$ species. When in nitrate solution, free Cd^{2+} species decreases from 60% to 40% as nitrate ion concentration and CdNO$_3^+$ species increases. Cd(II) EC$_{50}$ values after 15 min exposure, measured in NaCl, NaNO$_3$ and NaClO$_4$ at different ionic strengths are presented in Table 66.2.

It is found that EC$_{50}$ values increase sharply at high ionic concentrations, indicating a decrease of toxicity. Table 66.2 indicates that in NaCl concentrations ranging from 0.34 to 1.10 mol l^{-1}, cadmium toxicity decreased as the CdCl$^+$ species decreased and the CdCl$_3^-$ species increased, whereas the CdCl$_2$ form remained constant, as shown in Fig. 66.4a. Thus, the CdCl$^+$ species seems to be the most toxic as compared with the other chloro-complexes.

When sodium nitrate solution was used as medium, only a slight increase in EC$_{50}$ values was found as a function of the increase of salt concentration up to 0.80 mol l^{-1}. From Fig. 66.4b, it must be pointed out that free Cd^{2+} was always present

Fig. 66.4.
Cadmium species distribution diagrams at pH 6.0 as a function of NaCl (a) and NaNO$_3$ (b) concentrations

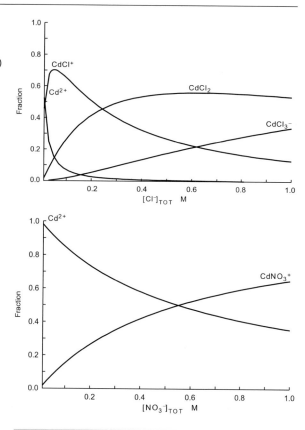

Table 66.2.
Effect of media and ionic strength on Cd(II) toxicity measured using Microtox assay. Cd(II) EC$_{50}$-15 min values at pH 6.0

Medium (mol l^{-1})	EC50-15 min (mg l^{-1})		
	NaCl	NaNO$_3$	NaClO$_4$
0.34	12.1	3.1	3.5
0.41	18.6	4.7	3.3
0.49	26.9	4.7	3.9
0.60	42.3	4.8	a
0.80	131.0	8.6	a
1.10	334.0	a	a

[a] Values not determined due to the low salt solubility.

whatever the nitrate concentration. When comparing Table 66.2 and Fig. 66.4b it can be observed that the EC$_{50}$ values were almost constant although the percentage of either Cd^{2+} or CdNO$_3^+$ changed, suggesting that not only free metal but also CdNO$_3^+$ species could be responsible for the observed toxicity. Table 66.2 also indicates that cadmium toxicity was independent of the NaClO$_4$ concentration, what is consistent with the fact that no complexes are formed between cadmium and perchlorate anion.

Changes in speciation and toxicity of some other metals as a function of the medium are reported in the literature. In the case of mercury, the speciation is influenced by the high concentration of chloride, leading to formation of $HgCl_2$ which modifies the toxicity of the sample (Ribó et al. 1989). Zinc or lead toxicity was also evaluated as a function of the sodium chloride concentration and was found to be correlated to change in speciation and to the formation of chloro-complexes (Villaescusa et al. 2000).

66.4
Conclusions

Our study shows that the toxicity of metal-containing samples, measured by the Microtox bacterial bioassay, is influenced by pH when metal speciation is pH-sensitive. It is the case of Cr(VI), the toxicity of which is mainly due to the hydrogenchromate anion. By contrast, Cd(II) toxicity was unaffected by pH within the 5.0–7.0 range. The reported data also demonstrate that, when using the Microtox bacterial assay, the ionic strength may influence the observed toxicity when metal ions can form complexes with the salts counteranions. Unlike Cr(VI), it is the case of Cd(II), the toxicity of which is influenced by the chloride concentration. In this respect, $CdCl^+$ appears to be the most toxic species. $NaNO_3$ and $NaClO_4$ solutions have been tested as alternative media to study the effect of medium composition on Cd(II) toxicity. In the presence of nitrate, the EC_{50} values were almost constant within the studied concentration range. In the presence of perchlorate, cadmium toxicity appears to be independent of the salt concentration. Finally, when determining the toxicity of a sample containing metallic compounds, $NaClO_4$ medium and constant pH are the recommended conditions. Attention should be paid to the fact that modification of pH before analysis would lead to an incorrect evaluation of toxicity.

References

Baes CF, Mesmer RE (1976) The hydrolysis of cations. Wiley Ltd., New York
Bulich AA (1979) Use of luminescent bacteria for determining toxicity in aquatic environment. In: Marking LL, Kimberley RA (eds) Aquatic toxicology ASTM 667. American Society for Testing and Materials, pp 98–106
Bulich A (1986) Bioluminescence assays. In: Bitton G, Dutka B (eds) Toxicity testing using microorganisms, vol. 1. CRC Press, Inc., Boca Raton, Florida
Castillo G, Vila E, Neild E (2000) Ecotoxicity assessment of metals and wastewater using multitrophic assays. Environ Toxicol 15:370–375
Choi K, Meier PG (2001) Toxicity evaluation of metal plating wastewater employing the Microtox assay: a comparison with cladocerans and fish. Environ Toxicol 16:136–141
Crosbi D (1998) Environmental toxicology and chemistry. Oxford University Press, UK
Delmas F, Trocheris V, Villaescusa I, Murat JC (1996) Expression of stress proteins in cultured human cells as a sensitive indicator of metal toxicity. Fresenius J Anal Chem 354:615–619
Delmas F, Schaak S, Gaubin Y, Croute F, Arrabit D, Murat JC (1998) Hsp72 mRNA production in cultured cells submitted to nonlethal aggression by heat, alcohols or chromium(VI). Cell Biol Toxicol 14:39–46
Dierickx PJ, Bredael-Rozen E (1996) Correlation between the in-vitro cytotoxicity of inorganic metal compounds to cultured fathead minnow fish cells and the toxicity to Daphnia magna. Bull Environ Contam Toxicol 57:107–110

Fergusson JE (1991) The heavy metals elements: chemistry, environmental impact and health effects. Pergamon Press, UK

Fochtman P, Rasca A, Nierzedska E (2000) The use of conventional bioassays, microbiotests, and some rapid methods in the selection of an optimal test battery for the assessment of pesticides toxicity. Environ Toxicol 15:376–384

Gadd GM, Griffiths AJ (1978) Microorganisms and heavy metal toxicity. Microbiol Ecol 4:103–317

Gaubin Y, Vaissade F, Croute F, Beau B, Soleilhavoup JP, Murat JC (2000) Implication of free radicals and glutathione in the mechanism of cadmium-induced expression of stress proteins in the A549 cell-line. Biochim Biophys Acta 1495:4–13

Hinwood AL, McCormick MJ (1987) The effect of ionic solutes on EC_{50} values measured using the Microtox test. Toxicity Assessment 2:449–461

Ho K, Kuhn A, Pelletier M, Hendricks T, Helmstetter A (1999) pH dependent toxicity of five metals to three marine organisms. Environ Toxicol 14:235–240

Iglesias M (1996) Avaluació de la toxicitat de plom(II) i de níquel(II) en NaCl i $NaClO_4$, mitjançant la utilització del bioassaig Microtox. Final project. Escola Politécnica Superior, Universitat de Girona, Spain

Ince N, Dirilgen N, Apikyan I, Tezcanli G, Ustun U (1999) Assessment of toxic interactions of heavy metals in binary mixtures: a statistical approach. Arch Environ Contam Toxicol 36:365–372

Kaiser KL, Ribó JM (1988) *Photobacterium phosphoreum* toxicity bioassay I. Test procedures and applications. Toxicity Assessment 2:305–323

Khangarot BS, Ray PK (1987) Correlation between heavy metal acute toxicity values in *Daphnia magna* and fish. Bull Environ Contam Toxicol 38:722–726

Krebs F (1983) Toxicity test using freeze-dried luminescent bacteria. Gewässerschutz Wasser Abwasser 63:173–230

Puigdomènech I (1983) TRIT A-00K-3010. Royal Institute of Technology, Stockholm

Ribó JM, Kaiser KL (1983) Effects of selected chemicals to photoluminescent bacteria and their correlations with acute and sublethal effects on other organisms. Chemosphere 12:1424–1442

Ribó JM, Rogers (1990) Toxicity of mixtures of aquatic contaminants using the luminescent bacteria bioassay. Toxicity Assessment 5:135–152

Ribó JM, Yang JE, Huang PM (1989) Luminescent bacteria toxicity assay in the study of mercury speciation. Hydrobiologia 188/189:155–162

Sillen LG, Martell AE (1982) Stability constants of metal-ion complexes. Pergamon Press, Oxford, UK

Sorväri J, Sillanpää M (1996) Influence of metal complex formation on heavy metal and free EDTA and DTPA acute toxicity determined by *Daphnia magna*. Chemosphere 33(6):1119–1127

Sunda WG, Engel DW, Thuotte RM (1978) Effect of chemical speciation on toxicity of cadmium to grass shrimp palaemonetes pugio: importance of freecadmium ion. Environ Sci Technol 12(4):409–413

Villaescusa I, Martinez M, Pilar M, Murat JC, Hosta C (1996) Toxicity of cadmium species in luminescent bacteria. Fresenius J Anal Chem 354:566–570

Villaescusa I, Marti S, Matas C, Martinez M, Ribó JM (1997) Chromium(VI) toxicity to luminescent bacteria. Environ Toxicol Chem 16(5):871–874

Villaescusa I, Casas M, Martinez M, Murat JC (2000) Effect of zinc chloro-complexes to photoluminescent bacteria: dependence of toxicity on metal speciation. Bull Environ Contam Toxicol 64:729–734

Chapter 67

Cultured Human Cells as Biological Detectors for Assessing Environmental Toxicity

E. Fulladosa · Y. Gaubin · D. Skandrani · I. Villaescusa · J.-C. Murat

Abstract

The presented investigations were carried out in order to detect environmental pollutants using the HT29, HepG2 and A549 cultured human cells as biological sensors of toxicity. We measured the growth rate inhibition and the expression level of stress proteins after exposure to cadmium, nickel, chromium, ethanol, 1-propanol, benzene, toluene, xylene, dichlorobenzene and trichlorobenzenes. Threshold concentrations were determined for selected pollutants and significance as well as reliability of the results were discussed. New perspectives for developing improved biodetectors are reported.

Key words: environmental toxicity; biodetectors; cultured cells; stress proteins; heavy metals, organic pollutants

67.1 Introduction

Dramatic expansion of industrial and urban areas in recent decades has caused increasing environmental and health burden. As a consequence, there is an increasing demand for toxicity assays in order to provide information concerning the potential impact of isolated chemical or complex mixtures of contaminants, especially on aquatic ecosystems and human health. Indices of water contamination merely relying on chemical analysis are generally considered as insufficient since they concern a limited number of suspected contaminants, i.e. the ones which are looked for, and refer to official normative regulations which are subjected to frequent changes.

Contrary to a popular belief, scientific knowledge indicates that the natural environment, i.e. without any impact of human industry, is not harmless. In fact, life started and has evolved in a highly "polluted" environment and the present forms of life do exist only because they have developed defense and adaptive mechanisms at both organism and molecular levels to counteract the deleterious effects of many natural factors. For that reason, aggression of living beings by industrial contaminants can be considered as just additive to aggression due to natural factors, either physical such as cosmic rays, sun rays, radioactivity, drought, dust, heat, freezing temperatures, etc, or chemical such as oxygen, ozone, sulfides, heavy metals, arsenic, selenium, etc, or biological such as viruses, parasitic organisms, toxic substances from plants, fungi and bacteria, etc. Considering human health, it frequently happens that the aggressive effect of a pollutant combines with the deleterious effects of natural aggressors, which results in large differences in susceptibility of individuals to a same level of pollution. Besides, individual genome in a given species, including *Homo sapiens*, does not express a similar level of efficiency in defense mechanisms. It means that precise evaluation of environmental hazard cannot be stated for a single person but it will rather apply to a global population on the basis of epidemiological studies and official safety regulations.

Nevertheless, investigations for new biological models and for very sensitive biological indices of environmental contamination are very much needed, especially for testing very low concentrations of pollutants. A good understanding of the physiological and molecular mechanisms underlying the biological response is also welcome in order to properly interpret the measured shifts of the biological parameters.

For ethical reasons, tests on blood parameters or biopsies from humans or higher vertebrates, as performed in post traumatic check-up or in epidemiological studies, cannot be used when investigating the effects of experimental exposures to potentially harmful or deadly substances. Therefore, biological models such as microorganisms or cultured cells were very much studied during the past years as they meet the required criteria. In order to test the toxicity of a given pollutant, classical indices used in toxicology, such as EC_{50} (concentration which provokes a 50% decrease in a biological process), LD_{50} (concentration which provokes a 50% mortality) or growth arrest, were formerly used (Tamborini et al. 1990; Bull et al. 1993; Apostoli 2002). More recently and as a consequence of new discoveries, attention was paid to several functional indices of cell suffering such as growth rate, membrane integrity, as accounted for by lactate-dehydrogenase (LDH) leakage or dye exclusion, decreased oxidative phosphorylation rate, reduced glutathione level, DNA fragmentation, changes in enzyme activities, overexpression of stress proteins and/or metallothioneins, etc.

In the past years, our team has developed an in-vitro model consisting in human cells kept in permanent culture, namely, the HT29, HepG2 and A549 cell-lines (Delmas et al. 1995, 1996). Indices of cell suffering after exposure to an aggressor were the cell growth-rate (GR) and the expression level of several stress proteins (SP). It was especially interesting to investigate the stress proteins since these proteins, mostly of the molecular chaperones family, are over-produced as a response to aggressions. It appears to be a large family of proteins, to which is attributed the general role of repairing molecular lesions, of protecting cell essential functions and of attenuating the effects of toxic substances (Welch 1993). The stress proteins were first identified as proteins specifically induced by sub-lethal heat shocks (Ritossa 1962). Later, the stress proteins family was much enlarged and comprised the so-called heat shock proteins (HSP), the glucose regulated proteins (GRP), the metallothioneins (MTs), the enzymes counteracting the aggressive effect of free radicals and reactive oxygen species and some other proteins such as ubiquitin (Feder and Hofmann 1999). Although the stress proteins were discovered to be overexpressed after a specific aggression, and named accordingly, it has been demonstrated that a variety of different aggressors can trigger their synthesis (Feder and Hofmann 1999).

67.2
Experimental

67.2.1
Cell Culture

Human cell-lines, namely, HT29 from colon mucosa (Fogh et al. 1977), HepG2 from liver (Davit-Spraul et al. 1994) and A549 from lung alveolar epithelium (Lieber et al. 1976), were routinely cultured in Petri dishes at 37 °C in Dulbecco's modified Eagle's medium (DMEM) supplemented with 5% fetal calf serum under air:CO_2 (19:1) atmosphere. Under the specified conditions, confluence was reached within 7 d.

67.2.2
Exposure Protocols

Either during the exponential phase of growth (measurement of the growth rate) or when reaching confluence, cells were submitted to different concentrations of pollutants for a given time. In order to avoid any binding of the metal to any component of the culture medium, exposure to metals was carried out in protein-free Hanks'saline. The proper effect of replacing the DMEM by the protein-free Hank's medium for the duration of the exposure was verified to be without consequence on either growth-rate or stress proteins over-expression (Delmas et al. 1996). Pollutants were given at sublethal concentrations as indicated in the tables (see results section, below). They were dissolved in pure water except aromatic compounds which were diluted in a dimethylcetone-water mixture. At the used concentration, dimethylcetone was verified to exert no effect on the cells. Negative controls were untreated cells whereas positive controls were cells submitted to a standard heat shock (45 °C for 30 min) known to trigger a strong expression of stress proteins in these cells (Delmas et al. 1996, 1998).

67.2.3
Evaluation of Growth Rate

Growth rate was estimated both by cell counting: cell suspension obtained by trypsin-EDTA treatment was introduced into a Coulter counter (Coultronics, France); and by measuring at intervals the total proteins content according to the Bradford (1976) method. No discrepancy between the methods was found.

67.2.4
Quantification of Stress Proteins Expression

Electrophoretically separated stress proteins were specifically identified on Western blots by using the corresponding monoclonal antibodies. The densitometric analysis of autoradiographs obtained from the chemi-luminescence due to the antibody-coupled peroxidase activity, was carried out as previously described (Gaubin et al. 2000).

67.2.5
Expression of Results

Concerning the expression of the results, threshold concentrations instead of the classical EC_{50} were used. EC_{50}, being the concentration that produces 50% of a maximal effect, is more accurate and easier to calculate. However, over-expression of stress proteins is a complex response involving, at least, DNA transcription, mRNA maturation and polysomal traduction. Hence, a maximal response cannot be stated and a half effect cannot be calculated. Threshold concentration can be defined as the concentration of a pollutant, a chemical or a quantified physical agent which induces the first significant shift of a parameter as compared to controls.

67.3
Results and Discussion

67.3.1
Metal Ions Threshold Concentrations

The effect of some metals on either growth rate or stress proteins over-expression in the three human cell-lines was investigated. Threshold concentrations for Cd(II), Ni(II) and Cr(VI), defined as the concentration which provokes a slight adverse effect on cell growth, and their effect on either growth rate or stress proteins over-expression in the three human cell-lines are presented in Table 67.1.

Cells growth-rate was found to decrease after the exposure to either cadmium or nickel. As seen in Table 67.1, when comparing threshold values, cadmium was found to be much more harmful than nickel (1 vs. 500 µmol l^{-1}, HepG2, 6 h exposure; 50 vs. 500 µmol l^{-1}, HT29, 6 h exposure). This confirms that, among heavy metals, cadmium is very aggressive.

Concerning stress proteins over-expression, cadmium was found to trigger a significant increase of stress proteins in HepG2 cells after a 6 h exposure at a concentration as low as 0.5 µmol l^{-1}. Nevertheless, for much shorter exposure (30 min), the concentration of this metal must be 10 µmol l^{-1} to be detected in the same model. In HT29

Table 67.1. Threshold concentrations (µmol l^{-1}) of some metals inducing a significant effect on cell growth rate (*GR*) and stress protein over-expression (*SP*) after exposure for a given duration, to HepG2, HT29 or A549 cells

Treatment		Threshold concentration (µmol l^{-1})					
Metal	Exposure time	HepG2 cell line		HT29 cell line		A549 cell line	
		GR	SP	GR	SP	GR	SP
Cd(II)	6 h	1	0.5	50	10	50	10
Cd(II)	30 min	50	10	50	50	–	–
Ni(II)	6 h	500	500	500	500	–	–
Cr(VI)	6 h	–	500	–	1 000	–	–

These results are from Delmas et al. (1996).

cells, Cd detection threshold appears at 10 µmol l^{-1} for 6 h. Again, nickel appears much less aggressive since the detection threshold was found at 500 µmol l^{-1} for a 6 h exposure. Chromium was tested as dichromate, which is considered as an aggressive anion. Threshold concentrations for chromium were found to be 500 and 1000 µmol l^{-1} for HepG2 and HT29 cells, respectively. In general, HepG2 cells were found to be more sensitive than HT29 or A549 cells. This is probably due to the fact that, unlike hepatocytes-derived HepG2 cells, HT29 and A549 cells, derived from gut mucosa and respiratory tract respectively, produce some mucus that may trap some amount of metal.

Under our experimental conditions, the threshold concentration producing an effect on either cell growth rate or SP over-expression are, in general, of the same order of magnitude. However, cadmium induces stress proteins at concentrations significantly lower than the concentrations affecting cell growth. This fact might indicate that stress proteins exert a protective effect towards the very high noxiousness of cadmium.

67.3.2
Alcohols Threshold Concentrations

Alcohols are widely used as industrial solvents. Low concentrations of ethanol are usually considered harmless since this chemical can be metabolized into acetaldehyde and acetic acid which may form acetyl-CoA. From this point of view, other alcohols are much more toxic. Besides, alcohols can perturb the cell membrane function as they modify the fluidity of the phospholipids bi-layer. Some results related to ethanol and 1-propanol are shown in Table 67.2.

Table 67.2 shows that cultured cells are rather tolerant towards alcohols. Ethanol, given for 15 min, must be at concentration as high as 5% or 8% (v/v) to produce an adverse effect on cell growth or to induce stress proteins over-expression in HepG2 or HT29 cells, respectively. 1-propanol appears to be somewhat more harmful, displaying an adverse effect on HepG2 cells at 1% (v/v) concentration. This is consistent with the known high metabolic toxicity of this alcohol. Again, as it was the case with metals, it appears that the HepG2 cell-line is more sensitive than the other tested cell lines.

67.3.3
Benzene and Derivatives Threshold Concentrations

Noxiousness of benzene and its derivatives, chemicals that are commonly present in household products and in unleaded gasoline, was also investigated. These chemicals were tested on the A549 cell line and the results are shown in Table 67.3.

Table 67.2.
Threshold concentrations (% v/v) of alcohols inducing a significant effect on cell growth rate (*GR*) and stress protein over-expression (*SP*) in HepG2 or HT29

Treatment		Threshold concentrations (% v/v)			
Chemical	Exposure time	HepG2 cell line		HT29 cell line	
		GR	SP	GR	SP
Ethanol	15 min	5	5	8	8
1-Propanol	15 min	2	1	4	2

These results are from Delmas et al. (1998).

Table 67.3.
Threshold concentrations (μmol l^{-1}) for a significant effect on cell growth rate (*GR*) and stress protein overexpression (*SP*), after exposure of A549 cells to benzene and derivatives for 96 h

Chemical treatment	Threshold concentration (μmol l^{-1}) A549 cell line	
	GR	SP
Benzene	500	2000
Toluene	100	500
Xylene	50	500
1,2,4-Trichlorobenzene	10	100
1,2–Dichlorobenzene	5	50
1,3,5–Trichlorobenzene	1	30

Benzene and benzene derivatives were found to be poor inducers of stress proteins. As shown in Table 67.3, stress proteins, specifically GRP78, was significantly over-expressed as a consequence of exposure of A549 to benzene, chlorinated benzenes, toluene and xylene. As GRP78 is a molecular chaperone acting within the endoplasmic reticulum (Dorner et al. 1992), it is postulated that benzene and its derivatives are not toxic at the metabolic level but, rather, exert their adverse effect at the structural level.

Table 67.3 shows that chlorinated benzenes are the most toxic benzene derivatives and indicates that in all the cases an effect on growth rate is obtained at concentrations 4- to 30-fold lower than those triggering the GRP78 over-expression. This result suggests that benzene and its derivatives exert manifold effects upon biological processes within the cell. It is important to underscore that our results cannot be directly extrapolated to effects on animals of actual pollution by aromatic chemicals. Our model consists of naked cells immediately submitted to the chemicals. In animal organisms, such pollutants are subjected to several defense mechanisms resulting in active detoxification and excretion.

67.4
Conclusions

From the presented data it can be concluded that *(i)* the most sensitive index of biological suffering for pollutants detection depends on the pollutant itself: SP over expression for metals, GR inhibition for benzene or derivatives and similar sensitivity of both indexes in the case of alcohols, *(ii)* cadmium and chlorinated benzenes are the most toxic compounds analyzed in this study and *(iii)* HepG2 is usually more sensitive than A549 or HT29.

Nevertheless, these indexes do not allow the identification of the pollutant or the mixture of pollutants responsible for the shift of the measured index. In this respect, some improvement in detection of pollutants may be expected from the development of the cDNA microarrays method. As the functional role of each of the genes-encoded proteins is or will be known, it should be possible to comprehend the mechanisms underlying the cell reaction and get some idea about the identity of the responsible aggressor(s). This remark underscores the need for further researches in order to develop satisfactory biodetectors of environmental pollution.

References

Apostoli P (2002) Elements in environmental and occupational medicine. J Chromatogr B Analyt Technol Biomed Sci 778:63–97

Bradford MM (1976) A rapid sensitive method for the quantitation of protein utilizing the principle of protein-dye binding. Analyt Biochem 72:248–254

Bull RJ, Conolly RB, Ohanian EV, Swenberg JA (1993) Incorporating biologically based models into assessments of risk from chemical contaminants. J Am Water Work Assoc 83:89–52

Davit-Spraul A, Pourci ML, Soni T, Lemonnier A (1994) Regulatory effect of galactose on gal-1-P uridyl-transferase activity in human hepatoblastoma HepG2 cells. FEBS Lett 354:232–236

Delmas F, Trocheris V, Murat JC (1995) Stress proteins in cultured HT29 human cell-line: a model for studying environmental aggression. Int J Biochem 27:395–391

Delmas F, Trocheris V, Villaescusa I, Murat JC (1996) Expression of stress proteins in cultured human cells as a sensitive indicator of metal toxicity. Fresenius J Anal Chem 354:615–619

Delmas F, Schaak S, Gaubin Y, Croute F, Arrabit C, Murat JC (1998) Hsp72 mRNA production in cultured cells submitted to nonlethal aggression by heat, alcohols or chromium(VI). Cell Biol Toxicol 14:39–46

Dorner AJ, Wasley LC, Kaufman RJ (1992) Overexpression of GRP78 mitigates stress induction of glucose regulated proteins and blocks secretion of selective proteins in Chinese hamster ovary cells. EMBO J 11:1563–1571

Feder ME, Hofmann GE (1999) Heat shock proteins, molecular chaperones and stress response. Annu Rev Physiol 61:243–282

Fogh J, Fogh JM, Orfeo T (1977) 127 cultured human tumor cell-lines producing tumors in nude mice. J Natl Cancer Inst 59:221–226

Gaubin Y, Vaissade F, Croute F, Beau B, Soleilhavoup JP, Murat JC (2000) Implication of free radicals and glutathione in the mechanism of cadmium-induced expression of stress proteins in the A549 cell-line. Biochim Biophys Acta 1495:4–13

Lieber M, Smith B, Skazal A, Nelson-Rees W, Todoro G (1976) A continuous tumor cell line from a human lung carcinoma with properties of a type II alveolar epithelial cells. Int J Cancer 17:62–70

Ritossa F (1962) A new puffing pattern induced by a temperature shock and DNP in *Drosophila*. Experientia 18:571–573

Tamborini P, Sigg H, Zbinden G (1990) Acute toxicity testing in the nonlethal range. Regul Toxicol Pharmacol 12:69–87

Welch WJ (1993) How cells respond to stress. Scientific American May:34–41

Part VII

Chapter 68

Genotoxic Impact of 'Erika' Petroleum Fuel on Liver of the Fish *Solea Solea*

A. Amat · T. Burgeot · M. Castegnaro · A. Pfohl-Leszkowicz

Abstract

On 12 December 1999, one third of the load of the tanker 'Erika', amounting to about 10 000 t crude oil flowed into sea waters close to the French Atlantic Coast (Finistère region). The spilled oil was fuel No. 6, a heavy or residual fuel that is a complex mixture of polycyclic aromatic compounds (PAC) of high molecular weight. Some polycyclic aromatic compounds are genotoxic substances that induce carcinogenic lesions in laboratory animals. DNA adducts, reflecting genotoxic effects, are used as biomarker of early pollution. In this study, the genotoxic impact of the 'Erika' oil was assessed by studying the presence of DNA adduct in the liver of immature fishes (*Solea solea*) from four locations of the French Brittany coasts, two, six and nine months after the disaster. Two months after the spill, a high amount of DNA adducts was found in samples from all locations (92 to 290 DNA adduct/10^9 nucleotides). DNA adducts were more persistent in the North than in the South of the affected French Brittany coasts. In September, no significant difference was observed between the locations. When incubated in presence of an 'Erika' fuel extract, DNA adduct patterns similar to those obtained from the liver of *Solea solea* liver are observed both, in cell culture (HePG2) and, in presence of fish liver microsomes.

Key words: DNA adduct; ^{32}P-postlabelling; biomarker; genotoxicity; fuel No. 2; Polycyclic Aromatic Hydrocarbons (PAH); Polyclyclic Heterocyclic Hydrocarbons (PHH); fish: *Solea solea*

68.1
Introduction

On 12 December 1999, the tanker 'Erika' broke into two parts approximately 30 miles off the French coast (at "Point of Penmarc'h", South Finistère, France). This accident occurred in rough weather conditions with wind forces from 8 to 9, and waves high up to 6 m. About 10 000 t of fuel was released in the sea. The oil pollution was first observed on the shore of South Finistère on 23 December, 11 d after the accident. Four hundred kilometers of Atlantic coast (from Morbihan to Vendée) were impacted between 23 December 1999 and February 2000 (see Fig. 68.1).

Various analyzes performed by independent institutes identified the spilled oil as fuel No. 6 (CAS No. 68553-00-4 or "Bunker C" or fuel No. 2 in French nomenclature), which is the highest boiling fraction of the heavy distillates from petroleum; its boiling point is >200 °C (and generally < 400 °C). This fuel is known as "residual oil", because they are produced from distillation residues from refinery processing. This product has a high viscosity and is a complex mixture of relatively high molecular weight component that is difficult to characterize in detail (IARC 1989). In addition to paraffinic, cycloparaffinic, olefinic compounds, the 'Erika' fuel oil contained a mixture of polycyclic aromatic hydrocarbons (PAH), polyclyclic heterocyclic hydro-

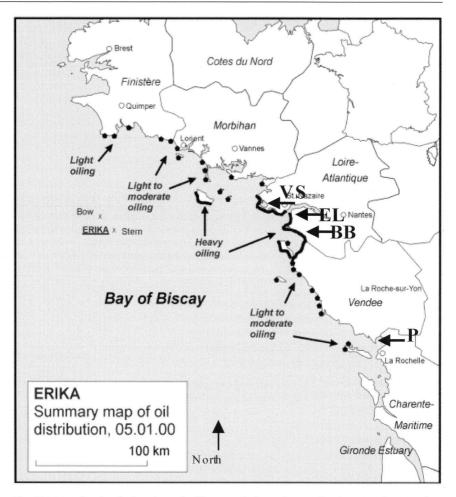

Fig. 68.1. Map showing the locations of spill. *Arrows* indicate the sampling locations from North to South of the impacted region. *VS:* Vilaine south; *EL:* Estuary of the Loire; *BB:* Bay of Bourgneuf, *P:* Pertuis. *Black points* represent the distribution of oil loaded on the coast on 5 January 2000 (ITOPF, International Tanker Owners Pollution Federation Limited)

carbons (HPH) including carbazoles, thiophenes and heavy metals (IARC 1989; Mazeas and Budzinski 2002; Boudet et al. 2000).

The "*Institut National de l'Environnement Industriel et des Risques*" (INERIS) and the French Ministry of Environment reported high mortality of birds (7 380 alive, 53 118 dead) due to the direct physical toxic effect of the fuel (Lacroix 2000). 'Erika' oil has been detected on the feather of bird collected on Atlantic shore after the 'Erika' accident (Mazeas and Budzinski 2002) and also in sediment (Mazeas and Budzinski 2001). They also pinpointed the potential risks of the 'Erika' fuel on the aquatic ecosystem exposed to the water soluble fraction of the fuel (Cicollela 2000). This oil spill can induce ecological and economical disturbances for fishery resources because this contamination arose near hatcheries. Fuel No. 6 has a low evaporation or dissolution potential (Irwin et al. 1997). Emulsion formed by Fuel No. 6 is very stable so it may be highly persistent in the sediments and coastal rocks impacting benthic and sessile organisms. Its composition makes it resistant to a microbial degradation with a biodegradation rate around 11% in laboratory cultures over 80 d (Oudot 2000). According to the literature, the $\log K_{OW}$ of the fuel oil No. 6 varies between 2.7 and 6.6 (CONCAVE 1998 in Baars 2002) and/or 3.3 and 7.1 (Irwin et al. 1997), indicating a high bioaccumulation potential of at least a part of its components. As observed previously for the accident of the *Exxon Valdez* oil (Carls et al. 2001), Mazeas and Budzinski (2001) confirmed that the sediments were contaminated by the petroleum product spilled by the 'Erika' tanker.

The 'Erika' petroleum is a complex mixture, which belongs to petroleum fuels classified by IARC in Group 2B, possibly carcinogenic to humans, which thus can react with genetic material such deoxyribonucleic acid (DNA). The presence of a DNA adduct in a critical gene provides the potential for occurrence of a mutagenic event, resulting in subsequent alterations in gene expression and a loss of growth control (for a review see Poirier et al. 2000). When assessing the dose/response relationship of toxic or carcinogenic compounds, it is necessary to consider not only the external dose, estimated by determination of the concentrations of chemicals in the environment, but also the internal dose, measured by the amount of toxicant and its metabolite(s) present in cells, tissues or body fluids. All these measurements represent biomarkers of exposure but are not relevant to the ability of the substance to induce damage in living systems. The biological effective dose reflects the amount of toxicant that has interacted with cellular macromolecules at a target site. Mechanistically, the biological dose is more relevant to disease than the internal dose, because it takes into account the differences of metabolism as well as the extent repair mechanisms. It is called, biomarker of effect.

For numerous chemicals, DNA is the prime target. DNA adducts are thus probably one of the most relevant indicators of genetic damage due to exposure to toxins. The detection of DNA adducts is widely used as biomarker of aquatic contamination (Dunn et al. 1987; Kurelec et al. 1989a; Varanasi et al. 1989; Ray et al. 1991; Stein et al. 1994; French et al. 1996; El Adlouni et al. 1995; Harvey et al. 1997; Pfau et al. 1997; Burgeot et al. 1996; Boillot et al. 1997; Ericson et al. 1998; Ericson and Lasson 2000; Lyons et al. 1999, 2000; Stephensen et al. 2000). Recently, Rice et al. (2000) used English Sole (*Pleuronectes vetulus*) to test the genotoxic impact of polluted sediment. Petrapiana et al. (2002) demonstrated bottom dwelling flatfish, such as *Lepidhorombus boscii*, are sensitive species to evaluate histopathologic damages. Nine year after the *haven* oil spill (Ligurian Sea, Italy, 1991) they found positive response in liver of this flatfishe living in sites contaminated by hydrocarbons residues. Moreover, DNA adducts and the prevalence of degenerative hepatic lesions in aquatic species living in contaminated areas have been established (Myers et al. 1998; Reichert et al. 1998; Baumann 1998; Van Schooten et al. 1995; Vincent et al. 1998; Ericson et al. 1998).

To investigate whether fishes living in the vicinity of the 'Erika' tanker oil spill suffered genotoxic damage, we analyzed the DNA adducts in liver of fish *Solea solea* (a bottom-dwelling species), at three different time periods (February, July, September 2000). To confirm that the DNA adducts observed are due to the impact of 'Erika' fuel, we compared them to DNA adducts formed in vitro by incubating an 'Erika' fuel extract with fish liver microsomes and DNA. We also analyzed the DNA adduct formed in HePG2 cells treated by an 'Erika' fuel extract.

68.2 Experimental

68.2.1 Field Sampling

Groups of 10 juvenile sole fish (*Solea solea*) were collected at four locations along the French Brittany coasts (Fig. 68.1). This species has been chosen because it was present at this time in the locations. These fishes live solitary in burrows into sandy and muddy bottoms and eat worms, mollusc and small crustaceans during the night. Livers were immediately excised, frozen and stored at –80 °C prior to analysis. Two pools of 5 livers were analyzed at each locations. Site 1 (Pertuis), was selected as potential control area, being about 110 km away from the other 3 sampling and 250 km far from the 'Erika' accident.

68.2.2 Extraction of 'Erika' Fuel

Two types of extraction have been performed.

68.2.2.1 Extraction with DMSO

Polycyclic aromatic hydrocarbons (PAH) contained in 'Erika' fuel (100 mg) were extracted with DMSO (1 ml) under agitation (1 h) at room temperature. This extract was used for microsome incubation (DMSO: dimethylsulfoxide).

68.2.2.2 Asphaltene Precipitation of Petroleum

For HepG2 cells culture, we used maltene extract as described by Mazeas and Budzinski (2001). Briefly, 500 µl of pentane were added to a 20 mg amount of fuel. The sample is slowly shaken for 10 min and the supernatant cetrifugated for 3 min at 700 rpm. The extraction is repeated 20 times. The pooled collected fractions were evaporated to 500 µl under a stream of nitrogen.

68.2.3 In-vitro Incubation

68.2.3.1 Chemicals

Nicotinamide adenine dinucleotide phosphate reduced ($NADPH_2$), bovine serum albumin (BSA), aprotinine, phenylmethylsulfonylfluorid (PMSF) and arachidonic acid (AA) were obtained from Sigma (St. Quentin Fallavier, France).

68.2.3.2
Purification of Microsomes

Livers from fishes were homogenised in a buffer solution [potassium chloride (KCl) 1.15%, 50 mM Na_2KPO_4, pH 7.4], with phenylmethylsulfonyl fluoride (10 µg ml^{-1}) and aprotinine (5 µg ml^{-1}). The buffer volume was three times the organ weight. After an initial centrifugation at 9 000 g for 20 min, the supernatant was taken and ultracentrifuged at 105 000 g for 1 h. The pellets were homogenised in 1 to 2 ml of pH 7.4 buffer containing Na_2KPO_4 (50 mM), KCl (0.15 M), EDTA (1 mM), dithiothreitol (DTT) (1 mM), glycerol (20%). The mixture was centrifuged again at 105 000 g for 1 h. Finally, microsomes were suspended in the same buffer to obtain a final concentration of about 10 mg of protein and stored at –80 °C prior to analysis. All steps of the isolation were carried out at 4 °C.

68.2.3.3
Incubation of Microsomes

68.2.3.3.1
Measurement of Protein Levels

Bradford's method (Bradford 1976) was used to measure the protein levels in microsomes. An aliquot of each microsomal preparation was diluted with a sodium chloride (NaCl, 0.15N) for determination of protein concentration as follows. To 100 µl of this dilution, 1 ml of the reagent was added. The blue coloration of the mixture was measured with a spectrophotometer (595 nm). The absorbency was compared with a standard curve calculated with 25 µg ml^{-1} to 100 µg ml^{-1} solutions of bovine serum albumin (BSA).

68.2.3.3.2
In-vitro Incubation to Determine DNA Adduct Formation

0.5 mg of microsomal proteins were incubated in vitro in the presence of 70 µg DNA and 'Erika' fuel extract (10 µl) in 500 µl (final volume) of Tris-HCl 50 mM, EDTA 1 mM, pH 7.4. The mixture was incubated at 37 °C for 3 min before addition of co-substrate. For measurement of cytochrome P 450 (CYP)-dependent DNA adduct formation, $NADPH_2$ (10 µl, 10 mg ml^{-1}) was added. For cyclooxygenase (COX) and lipoxygenases (LIPOX) dependent activity, arachidonic acid (AA) (10 µl, 1 mg ml^{-1}) was added. The mixture was then incubated at 37 °C for 45 min.

Two controls were added: *(1)* incubation of DNA without microsomes and *(2)* incubation of microsomes alone. All incubations were performed in triplicate.

68.2.4
HePG2 Cell Culture

68.2.4.1
Chemicals

The growth culture media Eagle's minimum essential medium (D-MEM with glutamax, 4 500 mg l^{-1} de D glucose, sodium pyruvate), phosphate-buffer saline (PBS) and

Trypsin were obtained from Gibco (Cergy pontoise, France). Hepatic cell line (HePG2) were obtained from ATCC (American Type Culture Collection, Mannass, USA). Benzo[a]pyrene was purchased from Sigma.

68.2.4.2
Cell Culture Conditions

Hepatic cell line (HePG2) were cultured in 75 cm² flasks with 10 ml of culture medium supplemented with 10% fetal bovine serum and 1% penicillin/streptomycin, at 37 °C under 5% CO_2 in sterile conditions. After trypsin digestion, the cells were resuspended in this medium to obtain 1×10^6 cells per ml. At 4th to 5th day after seeding, the medium was replaced with fresh medium (5% foetal bovine, 1% penicillin/streptomycin) before treating cells with 'Erika' fuel extract and BaP dissolved in DMSO. Pentane and DMSO never exceed 0.1% of total incubation volume. The cells were incubated for 24 h in presence of 10 µg ml^{-1} of 'Erika' "fuel" extract and 25 µg ml^{-1} of B(a)P. For each treatment, three flasks were required. At the end of the treatment, cells of each flask are harvested in a total of 8 ml of PBS. The cells harvested are pooled and centrifuged at 700 rpm at 4 °C and suspended in 700 µl of SET (NaCl: 0.1 M; EDTA: 20 mM; Tris-Hcl: 50 mM; pH 8). After treatment with sodium dodecyl sulfate (SDS, 10%) and Potassium acetate (5 M), DNA is purified by phenol/chloroform extraction (Pfohl-Leszkowicz et al. 1991).

68.2.5
^{32}P-Postlabelling Method of Analysis of DNA Adducts

68.2.5.1
Chemicals

Proteinase K, Ribonuclease A and T1 (RNase A and RNase T1), spleen phosphodiesterase and microccocal nuclease were purchased from Sigma (L'Isle d'Abeau, France); T4 polynucleotide kinase and [γ^{32}P-ATP], 370 Tbq mmol^{-1} (5 000 Ci mmol^{-1}) were from Amersham (Les Ullis, France); nuclease P1 from Boehringer (Manheim, Germany); rotiphenol from Rothsichel (Lauterbourg, France); cellulose MN 301 was from Macherey Nagel (Düren, Germany); the polyethyleneimine (PEI) was from Corcat (Virginia Chemicals, Portsmouth, VA, USA). The PEI-cellulose plates for thin layer chromatography (TLC) were made in the laboratory.

68.2.5.2
The ^{32}P-Postlabelling Procedure

For the ^{32}P-postlabelling method, samples were pooled in two duplicates. DNA were extracted and purified as described previously (Pfohl-Leszkowicz et al. 1991). The ^{32}P-postlabelling method is that originally described by Reddy and Randerath (1986) with minor modifications. In brief, DNA (7 µg) was digested at 37 °C for 4 h with micrococcal nuclease (183 mU) and spleen phosphodiesterase (12 mU) in a reaction mixture (total volume 10 µl) containing 20 mM sodium succinate and 10 mM $CaCl_2$, pH 6. Subsequently, adducted nucleotides are enriched. Digested DNA was treated with nuclease P1 (6 µg) at 37 °C for 45 min before ^{32}P-postlabelling. Normal nucleotides, pyrophosphate and excess ATP were removed by chromatography on polyethyleneimine cellulose plates in 2.3 M NaH$_2$PO$_4$, pH 5.7 (D1) overnight. Origin areas containing labelled adducted nucleotides were cut out and transferred onto another polyethyleiminecellulose plate, which was run in urea 8.5 M, lithium formate 5.3 M, pH 3.5 (D2). Two further migrations (D3 and D4) were performed perpendicularly to D2. The solvent for D3 was lithium chloride 1 M, urea 8 M, tris 0.5 M, pH 8, and the buffer D4 was 1.7 M NaH$_2$PO$_4$, pH 6. Autoradiography was carried out at –80 °C for 24 or 48 h in the presence of an intensifying screen. The radioactivity of the spots is analyzed by a phosphoimager equipped with an Ambis software treatment system. In the analysis of each batch of liver DNA, a BaP modified standard obtained during the European Union collaborative study on ^{32}P-postlabelling validation method (Phillips et al. 2000) was used as positive control. A negative control was also analyzed each time. The ^{32}P-postlabelling assay is a highly sensitive method (limit of detection 1 adduct per 10^{10} nucleotides), in which the adducted nucleotides obtained by DNA digestion, were radioactively labelled (Phillips et al. 1999, 2000). The number of adduct is expressed in Relative Adduct Level (RAL)/10^9 nucleotides.

68.3
Results and Discussion

68.3.1
Field Study

The aim of this study was to investigate and follow the genotoxic impact of 'Erika' fuel on marine environment. Part of 'Erika' oil has been trapped by sediment and consequently could contaminate the animals living in sediment for a long time. For this reason, we analyzed liver DNA-adducts of sole living in the areas impacted by the spillage. These fishes are exposed to sediments because they made burrows into sandy and muddy bottoms, and eat worms, molluscs and crustaceans.

Figure 68.2 presents typical liver DNA adduct patterns from soles sampled at four locations along the French Brittany coast – Perthuis (P), Bay of Bourgneuf (BB), estuary of Loire (EL) and Vilaine South (VS) – and at three time periods February, July and September 2000. Two different radioactive zones can be delimited on the autoradiogram (Fig. 68.2b): a diagonal radioactive zone (DRZ) and zone 1 (Z1). In February, DRZ and Z1 are observed in all samples, even in Perthuis location (P), which was expected to be a non-contaminated area. Already in July, only few adducts, mainly Z1, persisted in EL and VS locations, and this continued in September. On the contrary, in the two other locations, DRZ and Z1 persisted in July and significantly decreased only in September.

The quantification of the DNA-adducts are presented in Fig. 68.3. The highest total DNA adduct levels (Fig. 68.3a) are detected in February, two months after the accident in all areas in February, the total DNA adducts levels reached 286 and 210 adducts/10^9 nucleotides, respectively in liver of fishes caught in Bay of Bourgneuf (BB) and Vilaine South (VS). In July, an important decrease of the total DNA adduct levels was observed in VS (76.4%) and EL (85.2%). Only a moderate decrease (27%) of the total DNA adducts levels was

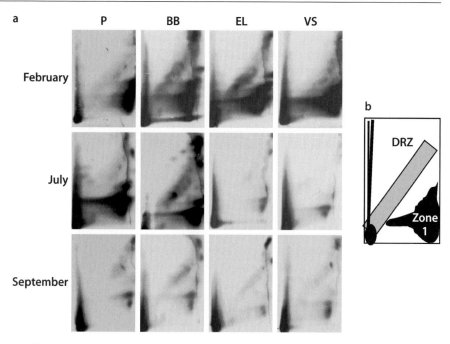

Fig. 68.2. a Typical autoradiogram of adducts detected in the liver of *Solea solea* after the 'Erika' oil spill. Sampling time period: February, July and September 2000. *P:* Perthuis; *BB:* Bay of Bourgneuf; *EL:* Estuary of Loire; *VS:* Vilaine South. **b** Scheme and numbering of the bulky DNA adducts zones detected by the ^{32}P-postlabelling method

observed in July in Bay of Bourgneuf. In Perthuis (P), the average of total DNA adduct remained similar to that observed in February with a level of 160 adducts/10^9 nucleotides. In September, the average of the total DNA adduct levels at all locations were very low (about 50 adducts/10^9 nucleotides) compared to those observed in February.

Each radioactive zones (DRZ and Z1) represented a lot of different individual DNA adducts, generated by the cross reactivity of genotoxic compounds with DNA. The DRZ is typical of a contamination by polycyclic aromatic hydrocarbons (PAH) as described by Randerath et al. (1988), Lyons et al. (1999), Schilderman et al. (1999) and Ericson and Larson (2000). Both radioactive zones have been separately quantified (Fig. 68.3b). The largest amount of DNA adducts in DRZ was observed in February in Bay of Bourgneuf. For this period, it is 3 to 5 times higher than those from the other three locations. In fish livers from Bay of Bourgneuf, the DRZ is only significantly reduced, in samples caught in September. In all areas, the levels of Z1 are very high in February, ranging from 111 to 165 adducts/10^9 nucleotides. In July, Z1 is significantly reduced, in Vilaine South and Estuary of Loire, whereas these adducts persisted in Bay of Bourgneuf and even increased in Perthuis (P). On September, the residual DNA adduct is very low at all locations.

Our results demonstrated that the 'Erika' fuel have a genotoxic impact on fish liver. Presence of DNA adduct in Perthuis location is not surprising. Indeed, ITOPF have published on 5 January a map showing the presence of 'Erika' oil also in the region (Fig. 68.1). The faster DNA adduct decrease in Estuary of Loire and Vilaine South, in

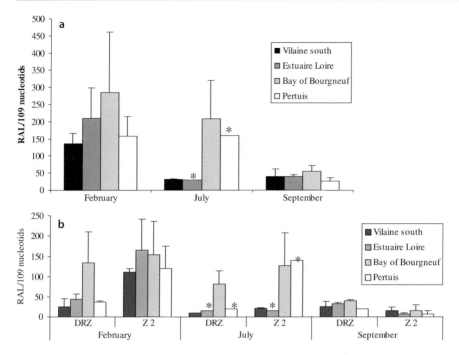

Fig. 68.3. Quantification of DNA adducts in liver of fish exposed to 'Erika' fuel in marine environment. **a** Total DNA adduct level expressed as number of adduct per 10^9 nucleotides. **b** DNA adduct level of DRZ and Z1. *Note a:* Only one pool of fish liver were analyzed

contrast to the long persistence of DNA adduct in Bay of Bourgneuf and Perthuis could be explained as follow. First, flow from the river enhanced the removal of the contaminant from the sediment or diluted the contamination with input of new river sediments. Second, the fishes are not at the same step of development in July and September than in February. In their early development stage, the *Solea solea* lived in the bay, but as soon as they convert into juvenile (stage 5 of the maturation), they migrate in the estuary of the river (Amara et al. 2000). This occur between April and June. Thus, the sampling in July and September included fishes coming not necessary from areas highly contaminated. Third, in Bay of Bourgneuf and Perthuis, in addition to the contamination of the sediments, a high amount of oil has been trapped on the rocks delimiting the bay. Observations made by C.E.D.R.E. indicate that the bay of Bourgneuf have been strongly and regularly impacted by oil slick at least until April 2000 (www.le-cedre.fr).

Altogether, our data confirm the interest of the analysis of DNA adduct in fish liver to follow the impact of a pollution by PAC. They are in accordance with those from Lyons et al. (1997) and Harvey et al. (1999) who analyzed the genotoxic impact of the *Sea Empress* crude oil on fish tissues (*Lipophorys pholis*) and on *Limanda limanda*), as well as invertebrate species (*Halichondira panicea* and *Mytilus edulis*) respectively. Similar to the results of this study, the DNA adduct patterns of fish liver exhibited a typical diagonal zone (DRZ). DNA adducts persisted in vertebrates but not in invertebrates species 12–17 months after the spill.

Authors have demonstrated that DNA adduct in fish liver represent an exposure to genotoxic substance over a long period as compared to their measurement in blood which represents a very early exposure (Telli-Karakok 2001; Ericson et al. 1999). It has also been found that DNA adducts in fish liver are very persistent (Ericson et al. 1999; Stein et al. 1993; Varanasi et al. 1989 in Ericson and Larson 2000). Indeed, French et al. (1996) observed a steady increase in DNA adduct levels during a chronic exposure of English sole (*Pleuronectes vetulus*) to PAH-contaminated sediment for 5 weeks and that they are very persistent even after a depuration period. In the same way, Aas et al. (2000), have also observed that hepatic DNA adduct formed in Atlantic cod (*Gadus morhua*) appeared 3 d after exposure to low concentration of crude oil and increased steadily during the 30 d following exposure. Over 60% of the DNA adducts remained after 60 d. Our first sampling was about 2 months after the accident, falling between these two time point. The data from February sampling presents thus the situation between the maximum impact on DNA formation and its reduction by 40%.

68.3.2
In-vitro Biotransformation of 'Erika' Petroleum

The 'Erika' oil contains high amount of naphtalene, phenanthrene, chrysene and dibenzothiophene (Mazeas and Budzinski 2002). We have previously demonstrated that depending of the PACs, genotoxic compounds are generated during biotransformation by CYP, COX and LIPOX (Genevois et al. 1998). To confirm that fish liver have the ability to biotransform the 'Erika' fuel, we incubated fish liver microsomes, in presence of an extract of an extract of the 'Erika' fuel and DNA. Two different co-substrate have been used: $NADPH_2$, needed for cytochrome P450 activity (CYP) and arachidonic acid (AA), needed for cyclooxygenase (COX) and lipoxygenase (LOX) activities. No DNA adducts are observed when the extract of the 'Erika' fuel is incubated in presence of microsomes alone (Fig. 68.4a). In the same way, no DNA adducts are observed when microsomes are incubated alone (Fig. 68.4b), nor when microsomes are incubated with DNA alone (Fig. 68.4c). A diagonal radioactive zone (DRZ) is observed when the 'Erika' fuel was incubated in presence of microsomes, DNA and $NADPH_2$ (Fig. 68.4d). Several individual DNA-adducts are observed when extract of the 'Erika' fuel was incubated in presence of microsomes, DNA and AA (Fig. 68.4e). Comparison of these patterns with those obtained in liver of *Solea solea* after the 'Erika' spill, confirmed that the DRZ is due to biotransformation of components from the extract of the 'Erika' fuel by cytochrome P 450 into genotoxic metabolites reactive with DNA. This ability of the microsomal fraction to

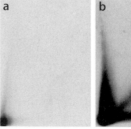
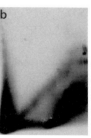
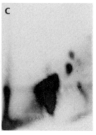

Fig. 68.4.
In-vitro incubation of 'Erika' fuel in presence of fish liver microsomes and DNA.
a Microsome + 'Erika' fuel; **b** microsome alone; **c** microsome + DNA; **d** microsome + 'Erika' fuel + DNA + NaDPH2; **e** microsome + 'Erika' fuel + DNA + arachidonic acid

transform PAH into species which can form DNA adducts has been reported by Peters et al. (2002) in fish liver microsome incubated with BaP. Biotransformation of the components of the extract of the 'Erika' fuel in presence of arachidonic acid by cycloxygenase and/or lipoxygenase induces the formation of several DNA adduct having the same chromatographic properties than DNA adducts, observed in *Solea solea*, notably Z1.

These results demonstrated that 'Erika' fuel is genotoxic after metabolic activation by cytochrome P 450, cylooxygenase and lipoxygenase.

68.3.3
Genotoxic Impact of 'Erika' Fuel on Hepatocytes

Unfortunately, no sampling had been made before the 'Erika' oil spill, therefore we have no basis for comparison with the situation before the accident. To confirm that the genotoxic effect observed in fish liver is related to 'Erika' petroleum, we have incubated HePG2 (human hepatocyte) for 24 h in presence of an extract of the 'Erika' fuel or benzo(a)pyrene (Fig. 68.5).

No DNA adduct are observed when HepG2 are incubated only with the vehicle (Fig. 68.5a). Incubation of these cells with the 'Erika' fuel extract lead to the formation of 2 radioactive zones (Fig. 68.5b). The pattern is similar to with those obtained in Sole exposed to 'Erika' oil in marine environment. The levels of DNA adduct reached 106 adducts/10^9 nucleotides and 152 adducts/10^9 nucleotides in diagonal zone (DRZ) and Zone 1, respectively. Incubation of HePG2 in presence of BaP induce the formation of one major adduct reaching a level of 331 adducts/10^9 nucleotides (Fig. 68.5c). These data indicates that human hepatocyte biotransformed 'Erika' fuel into genotoxic metabolites similarly to hepatic cell of fishes and confirmed that the adducts observed in field study are related to the contamination of sediment by the 'Erika' fuel.

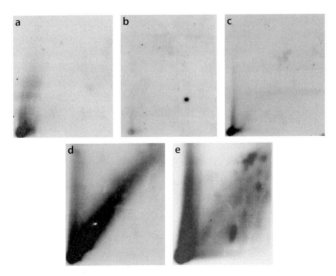

Fig. 68.5. DNA adduct pattern after incubation of HePG2 cells for 24 h. **a** Control cells; **b** cells incubated with 'Erika' fuel; **c** cells incubated with benzo[a]pyrene

68.4 Conclusion

Despite the fact that we have no basis for comparison with the situation prior to the accident because no sampling had been made before the 'Erika' oil spill, these results demonstrate a genotoxic event to the liver of the fishes living in the coast impacted by the 'Erika' spill. The sequence of the events depended of the location and the period of sampling. For the earlier period, two months after the spill, a high amount of DNA adducts is detected at all locations. Then, in July, high level of DNA adduct persisted only in the south locations, Bay of Bourgneuf (BB) and Pertuis (P), whereas for the north sites, Estuary of the Loire River (EL) and Vilaine South (VS), a strong decrease of DNA adduct was observed. Finally in September a low residual DNA adduct level was observed for all locations. We have confirmed by in-vitro experiments, that fish livers and human hepatocytes are able to biotransform components of the 'Erika' fuel into genotoxic compounds leading to similar DNA adducts than those observed in livers of fish living in polluted area.

Acknowledgments

This study was supported by IFREMER. We also thank the team of the oceanographic boat Gwen-Drez.

References

Aas E, Baussant T, Balk L, Liewenborg B, Andersen OK (2000) PAH metabolites in bile, cytochrome P4501A and DNA adducts as environmental risk parameters for chronic oil exposure: a laboratory experiment with Atlantic cod. Aquatic Toxicol 51:241–258

Amara R, Lagardere F, Desaunay Y, Marchand J (2000) Metamorphosis and estuarine colonisation in the common sole, *Solea solea* (L.): implications for recruitment regulation. Oceanologica Acta 23(4):469–484

Baars B-J (2002) The wreckage of the tanker 'Erika' – human health risk assessment of beach cleaning, sunbathing and swimming. Toxicol Lett 128:55–68

Baumann PC (1998) Epizootics of cancer in fish associated with genotoxins in sediment and water. Mutat Res 411:227–233

Boillot K, Burgeot T, Arnould JP, Pfohl-Leszkowicz A (1997) DNA adduct detection in the liver of the dragonnet *Callionymus lyra* and the flounder *Platchthys flesus*: a potential bioindicator of pollutant. Mutat Res 379:116, (Abstract)

Boudet C, Chemin F, Bois F (2000) Evaluation du risque sanitaire de la marée noire consécutive au naufrage de l'ERKA, rapport 6. Institut national de l'environnement industriel et des risques, Ministère de l'Aménagement du Territoire et de l'Environnement, Unité d'Evaluation des Risques Sanitaires, Direction des Risques Chroniques

Bradford MM (1976) A rapid and sensitive method for the quantification of microgram quantities of protein utilizing the principe of protein dye binding. Anal Biochem 72:248–254

Burgeot T, Bocquéné G, Porte C, Dimeet J, Santella RM, Garcia de la Parra LM, Pfohl-Leszkowicz A, Raoux C, Galgani F (1996) Bioindicators of pollutant exposure in the northwestern Mediterranean Sea. Mar Ecol Prog Ser 131:125–141

Carls MG, Babcock MM, Harris PM, Irvine GV, Cusick JA, Rice SD (2001) Persistence of oiling in mussel beds after the *Exxon Valdez* oil spill. Mar Environ Res 1:167–190

Cicollela A (2000) Evaluation des risques sanitaires et environnementaux résultant du naufrage de l'Erika et des opérations de nettoyage des cotes, rapport de synthèse, Institut national de l'environnement industriel et des risques. Ministère de l'Aménagement du Territoire et de l'Environnement, Unité d'Evaluation des Risques Sanitaires, Direction des Risques Chroniques

Dunn BP, Black J, Maccubin A (1987) ^{32}P-postlabelling analysis of aromatic DNA adducts in fish from polluted areas. Cancer Res 47:6543–6548

El Adlouni C, Tremblay J, Walsh P, Lageux J, Bureau J, Laliberte D, Keith G, Nadeau D, Poirier GG (1995) Comparative study of DNA adduct levels in white sucker fish (*Catostomus commersoni*) from the basin of the St. Lawrence River (Canada). Molecular Cell Biochem 1148:133–138

Ericson G, Larsson A (2000) DNA adduct in perch (*Perca fluviatilis*) living in coastal water polluted with bleached pulp mill effluents. Ecotox Environ Safety 46:167–173

Ericson G, Lindesjöö E, Balk L (1998) DNA adducts and histopathological lesions in perch (*Perca fluviatilis*) and northern pike (*Esox lucius*) along a polycyclic aromatic hydrocarbon gradient on the Swedish coastline of the Baltic Sea. Can J Fish Aquat Sci 55:815–824

Ericson G, Noaksson E, Balk L (1999) DNA adduct formation and persistence in liver and extrahepatic tissues of northern pike (*Exos lucius*) following oral exposure to benzo[a]pyrene, genzo(k)fluoranthrene and 7H-dobenzo(c, g) carbazole. Mutat Res 427:135–145

French B, Reichert WL, Hom HR, Nishimoto HR, Stein JE (1996) Accumulation and dose response of hepatic DNA adducts in English sole (*Pleuronectes vetulus*) exposed to a gradient of contaminated sediments. Aquatic Toxicol 36:1–16

Genevois C, Pfohl-Leszkowicz A, Boillot K, Brandt H, Castegnaro M (1998) Implication of Cytochrome P450 1A isoforms and the Ah receptor in the genotoxicity of coal-tar fume condensate and bitume fumes condensates. Environ Toxicol Pharmacol 5:283–294

Harvey JS, Lyons BP, Waldock M, Parry JM (1997) The application of the ^{32}P-postlabelling assay to aquatic biomonitoring. Mutat Res 378:77–88

Harvey JS, Lyons BP, Page TS, Stewart C, Parry JM (1999) An assessment of the genotoxic impact of the *Sea Empress* oil spill by the measurement of DNA adduct levels in selected invertebrate and vertebrate species. Mutat Res 103–114

IARC (1989) Occupational exposures in petroleum refining crude oil and major petroleum fuels. International Agency for Research on Cancer, monographs on the evaluation of carcinogenic risks to humans. Vol. 45, Lyon, France

Irwin RJ, VanMourik M, Stevens L, Seese MD, Basham W (1997) Environmental contaminants encyclopaedia-entry on fuel oil n°6. National Park Service, Water Resources Division, Fort Collins, CO, USA

Kurelec B, Garg A, Krea S, Gupta RC (1989a) DNA adducts as biomarkers in genotoxic risk assessment in the aquatic environment. Mar Environ Res 28:317–321

Kurelec B, Garg A, Krca S, Chacko M, Gupta RC (1989b) Natural environment surpasses polluted environment in inducing DNA damage in fish. Carcinogenesis 7:1337–1339

Lacroix G (2000) Evaluation sanitaire des risques lors des soins apportées aux oiseaux mazoutés, rapport 2, Institut national de l'environnement industriel et des risques. Ministère de l'Aménagement du Territoire et de l'Environnement, Unité d'Evaluation des Risques Sanitaires, Direction des Risques Chroniques

Lyons BP, Harvey JS, Parry JM (1997) An initial assessment of the genotoxic impact of the Sea Empress oil spill by the measurement of DNA adduct levels in the interdidal teleost *Lipophorys pholis*. Mutat Res 263–268

Lyons RA, Temple JMF, Evans D, Palmer SR (1999a) Acute health effects of the *Sea Empress* oil spill. J Epidemiol Community Health 53:306–310

Lyons BP, Stewart C, Kirby MF (1999b) The detection of biomarkers of genotoxin exposure in the European flounder (*Platichthys flesus*) collected from the River Tyne Estuary. Mutat Res 446:111–119

Lyons BP, Stewart C, Kirby MF (2000) ^{32}P-postlabelling analysis of DNA adducts and EROD induction as biomarkers of genotoxin exposure in dab (*Limanda limanda*) from British coastal waters. Mar Environ Res 50:575–579

Mazeas L, Budzinski H (2001) Polycyclic aromatic hydrocarbon ^{13}C/^{12}C ratio measurement in petroleum and marine sediments; application to standard reference materials and a sediment suspected of contamination from the *Erika* oil spill. J Chromatogr A 165–176

Mazeas L, Budzinski H (2002) Molecular and stable carbon isotopic source identification of oil residues and oiled bird feathers sampled along the Atlantic coast of France after the *Erika* oil spill. Environ Sci Technol 36(2):130–137

Myers MS, Johnson LL, Hom T, Collier TK, Stein JE, Varanasi U (1998) Toxicopathic hepatic lesions in subadult english sole (*Pluronestes vetulus*) from Puget Sound, Washington, USA: relationship with other biomarkers of contaminant exposure. Mar Environ Res 45(1):47-67

Oudot J (2000) Biodegradability of the *Erika* fuel oil. C.R. Acad. Sci. Paris, Sciences de la vie/Life Sciences 323:945-950

Peters LD, Telli-Karrakok F, Hewer A, Philips DH (2002) In-vitro mechanistic differences in bezo(a)pyrene DNA adduct formation using fish liver and digestive gland microsomal activating system. Mar Environ Res 54:499-503

Petrapiana D, Modena M, Guidetti, Faluga C, Vacchi M (2002) Evaluating the genotoxic damage and hepatic tissue alterations in demersal fish species: a case study in Ligurian Sea (NW-Mediterranean). Mar Pol Bull 44:238-243

Pfau W (1997) DNA adducts in marine and freshwater fish as biomarkers of environmental contamination. Biomarkers 2:145-151

Pfohl-Leszkowicz A, Chakor K, Creppy EE, Dirheimer G (1991) DNA-adducts formation and variation of DNA-methylation after treatment of mice with ochratoxine A. In: Castegnaro M, Plestina R, Dirheimer G, Chernozemsky IN, Bartsch H (eds) Mycotoxins, endemic nephropathy and urinary tracts tumours. IARC Scientific Public. Lyon 115:245-253

Phillips DH, Castegnaro M, et al. (1999) Standardization and validation of DNA adduct postlabelling methods: report of interlaboratory trials and production of recommended protocols. Mutagenesis 14:301-315

Phillips DH, Farmer PB, Beland FA, Nath RG, Poirier MC, Reddy MV, Turteltaub KW (2000) Methods of DNA adduct determination and their application to testing compounds for genotoxicity. Environ Mol Mutagen 35:222-233

Poirier MC, Beland FA (1992) DNA adduct measurements and tumour incidence during chronic carcinogen exposure in animals models:implication for DNA adduct based human cancer risk assessment. Chem Res Toxicol 5:749-755

Randerath E, Mittal D, Randerath K (1988) Tissue distribution of covalent DNA damage in mice treated dermally with cigarette 'tar'. Preference for Lung and Carcinogenesis 9:75-80

Ray S, Dunn BP, Payne JF, Fancey L, Helbig R, Beland P (1991) Aromatic DNA-carcinogen adducts in bluga whales from the Canadian Artic and Golf of St. Lawrence. Mar Pol Bull 22:392-395

Reichert WL, Myers MS, Peck-Miller K, French B, Anulacion BF, Collier TK, Stein JE, Varanasi U (1998) Molecular epizootiology of genotoxic events in marine fish: linking contaminant exposure, DNA damage, and tissues-level alterations. Mutat Res 411:215-225

Rice CA, Myers MS, Willis ML, French BL, Casillas E (2000) From sediment bioassay to fish biomarker-connecting the dots using simple trophic relationships. Mar Environ Res 50:527-533

Schilderman PAEL, Moonen AJC, Maas LM, Welle I, Kleinjans JCS (1999) Use of crayfish in bimonitoring studies of environmental pollution of the river Meuse. Ecotox and Env Saf 44:241-252

Stein EJ, Reichert WL, Varanasi U (1994) Molecular epizootiology: assessment of exposure to genotoxic compounds in teleost. Environ Health Perspect 102:20-23

Stephensen E, Svavarsson J, Sturve J, Ericson G, Adolfsson-Erici M, Förlin L (2000) Biochemical indicators of pollution exposure in shorthorn sculpin (*Myoxocephalus scorpius*), caught in four harbours on the southwest coast of Iceland. Aquatic Toxicol 431-442

Telli-Karakok F, Gaines A, Hewer A, Philips D (2001) Differences between blood and liver aromatic DNA adduct formation. Environ Int 26:143-148

Van Schooten FJ, Maas LM, Moonen EJC, Kleijans JCS, Vanderoots R (1995) DNA dosimetry in biological indicator species living on PAH-contaminated soils and sediments. Ecotox and Env Saf 30(2):171-179

Varanasi U, Reichert WL, Stein JE (1989) ^{32}P-postlabelling analysis of DNA adducts in liver of wild English sole (*Parophrys vetulus*) and winter flounder (*Pseudopleuronectus americanue*). Cancer Res 49:1171-1177

Vincent F, Boer J, Pfohl-Leszkowicz A, Cherel Y, Galgani F (1998) Two cases of ras mutation are associated with hyperplasia in callionymus lyra exposed to polychlorinated biphenyls and polycyclic aromatic hydrocarbons. Molecular Carcinogenesis 21(2):121-127

Chapter 69

Heavy-Metal Resistant Actinomycetes

M. Siñeriz Louis · J. M. Benito · V. H. Albarracín · Thierry Lebeau · M. J. Amoroso
C. M. Abate

69.1
Introduction

Water and soil pollution has become a major concern in the world, as much of the population relies on groundwater as its major source of drinking water as well as on soil as cultivable land. Heavy-metal contamination brings a potential health hazard that can cause metal toxicoses in animals and humans (Volesky and Holan 1995).

Microorganisms play an important role in the environmental fate of toxic metals and radionuclides because various biological mechanisms transform soluble and insoluble forms of xenobiotics. These mechanisms are part of natural biogeochemical cycles. They are potentially useful for both in-situ and ex-situ bioremedial treatment processes for solid and liquid wastes (Gadd 2000).

The use of microorganisms for recovering of metals from waste streams, as well as the employment of plants for landfill application (Watanabe 1997), has raised growing attention. A wide variety of fungi, algae, and bacteria are now under study or are already in use as biosorbents for heavy-metal remediation (Gadd 1992; Diels et al. 1993; Kotrba et al. 1999). Bacteria have evolved strategies to cope with toxic metals in the environment, and bacterial operons that confer resistance to cadmium, mercury, copper and arsenic have been described (Silver and Phung 1996; Nies and Brown 1998; Xu et al. 1998). Recently some chromosomal genes conferring metal tolerance have been detected in several bacterial species as well (Cai et al. 1998; Xiong and Jayaswal 1998; Hassan et al. 1999).

Actinomycetes are Gram-positive bacteria with high content of guanine and cytosine (G + C) (55–75 mol%) that generally exhibit a filament growth with ramifications. They are metabolically versatile being a significant component of soil microbial population and they were proved to habit in marine and aquatic sediments (Peczynska-Czoch and Mordarski 1984). Another characteristic of actinomycetes is the great variation of its secondary metabolism with active compounds as final products. Actinomycetes stand out within the prokaryotes microorganisms by their wide morphologic diversity, presenting a continuous transition from round cells, rod-shaped cells, to hyphae that are fragmented with ramifications or stable with ramifications (Goodfellow et al. 1988; Larpent and Sanglier 1989).

Copper is a heavy metal that has been used for years as an ingredient of bactericides and fungicides and as a growth enhancer of pigs (Trajanovska et al. 1997). Copper is found in high concentrations (0.1 to 20 mM) in contaminated soils thus being a threat to soil bacteria population. There is a copper filter plant near an agricultural area of Tucumán, Argentina that flows its wastewater to a natural channel. It has been determined that during the process copper contamination is produced in the environment (Amoroso et al. 1996).

Copper is needed at low concentration (<0.1 µM) to cell normal functions but it is very toxic at high concentrations because its production of free radicals causes a serious damage to the cell (Gutteridge and Wilkins 1983; Simpson et al. 1988). Nevertheless, bacteria exposure to toxic levels of copper has led to the development and evolution of various genetics mechanisms that regulate the uptake and resistance to copper (Trevors 1987). Plasmid pRJ1004, that confers copper resistance to *E. coli*, is one example of these mechanisms (Tetaz and Luke 1983). pRJ1004 is a conjugate 78-megadalton plasmid found in an *E. coli* strain isolated from the effluent of a piggery where pigs were fed a supplemented diet with copper sulphate. It was shown to carry a copper-resistance determinant, *pco* (plasmid-borne copper resistance) (Tetaz and Luke 1983). Brown et al. (1995) characterised and described the nucleotide sequence of the *pco* resistance determinant, together with copper transport studies.

There is a lot of information available on copper resistance genetic mechanisms in Gram-negatives bacteria (Williams et al. 1993; Harwood and Gordon 1994; Munson et al. 2000) but little has been done in Gram-positives. In spite of this, Trajanovska et al. (1997) reported the detection of copper and other heavy metals resistance genes in Gram-positives and Gram-negatives isolated from a lead contaminated area. There is still less information about the genes that codifies heavy metal resistance in actinomycetes. Recent experiences indicate that mercury resistance in *Streptomyces* is related to the presence of giant linear plasmids (Ravel et al. 1998, 2000a,b).

Cadmium is a very toxic metal and has been found in the environment at increased concentrations producing different pathologies in humans and animals (Friberg et al. 1979).The accelerated growing of industrial activities producing this contamination has increased the cadmium liberation at a higher rate than the one of the natural geochemical processes (Nriagu and Pacyma 1988). The monitoring of the viability of cadmium resistant actinomycetes in culture medium and in soil samples is of considerable importance because of the potential capacity of these strains in the bioremediation of cadmium (Amoroso et al. 1998).

One of the easier ways is to follow the viability by fluorescence. The green fluorescent protein (GFP) has proved to be a particularly useful and sensitive gene reporter. Recently a red-shifted variant of GFP was developed, which gives brighter fluorescence and higher levels of expression in mammalian cells. Enhanced Green Fluorescent protein (EGFP) shows a 35-fold enhancement of fluorescence over wild-type GFP when excited at 488 nm, and possesses excitation and maximal emission of 488 nm and 507 nm, respectively (Sun et al. 1999).

Using actinomycetes strains isolated from heavy metals contaminated soils we studied their resistance to copper at different concentrations and then assayed an hybridisation experience with probes constructed using two copper resistance genes (*pco*R and *pco*A) from pRJ1004 of *E. coli*. With the aim to study the survival of the cadmium resistance strain *Streptomyces* R25 (Amoroso et al. 1998) in cadmium contaminated soil we carried out transformations experiments using pIJ8660 that contain the EGFP gene to transform this strain and analyze their survivals.

69.2 Experimental

69.2.1 Samples

Sediment samples were collected from the water reservoir El Cadillal (not contaminated area), and from a drainage channel that receives effluents from a copper mine. Both places are located in Tucumán, Argentina. A sediment sample was also collected from a uranium mine, Wismut, in eastern Thuringia, Germany. Each sample was aseptically collected using sterile test tubes, and kept at 5 °C until they were dried at 30 °C to constant weight. Samples were diluted with sterile water prior inoculation onto isolation plates in duplicate.

69.2.2 Microorganisms

Isolation of microorganisms was carried out in Starch-Casein Agar (SC) medium containing per litre: starch: 10.0; casein: 1.0; K_2HPO_4: 0.5; agar: 15.0. The pH was adjusted prior sterilisation to 7.0. The medium was supplemented with 10.0 µg ml^{-1} nalidixic acid (NA) and cycloheximide to inhibit growth of Gram-negative bacteria and fungi, as previously reported for actinomycetes isolations (Amoroso et al. 1998). Plates were incubated at 25 °C and colonies were inoculated by streaking on agar medium without the antibiotics.

Streptomyces R25, previously isolated by Amoroso et al. (1998) was selected because of its cadmium resistance. Dr. Jill Williams gently provided *Escherichia coli* ED8739 pRJ1004 from the Department of Genetics of the University of Melbourne, Australia. *Streptomyces* MC2 and Cad1 were used as negative controls in the heavy metal assays because they showed to be sensitive to all the concentrations tested.

69.2.3 Qualitative Assays of Copper Resistance

Primary qualitative screening assays were carried out in square plates containing MM agar medium (composition in g l^{-1}: L-asparagine: 0.5; K_2HPO_4: 0.5; $MgSO_4 \cdot 7H_2O$: 0.2; $FeSO_4 \cdot 7H_2O$: 0.01; glucose: 10.0; agar: 15.0) amended with three different $CuSO_4$ concentrations: 0.1, 0.25 and 0.5 mM respectively. The isolated strains were spread onto the plates and microbial growth was used as the qualitative parameter of metal resistance.

69.2.4 Copper Analysis

Copper concentrations were determined by atomic absorption spectrometry. Solid-phase samples were first digested by dissolving sediments in concentrate nitric acid (American Public Health Association 1992).

69.2.5
Plasmid and DNA Extraction

Selected strains: C14, C31, C43 and C61 isolated from El Cadillal; M3, M9 and M26 isolated from the copper and gold mine; QL1-3, QL1-4 and QL1-8 isolated from the uranium mine were grown at 28 °C during 4–5 d in MM amended with 80 mg ml^{-1} copper. The pellets were collected by centrifugation and washed twice with sterile distilled water. Total genomic DNA extraction was carried out according with the technique described by Fisher (1985) modifying the lisozyme concentration used in 8 mg ml^{-1} instead of 2 mg ml^{-1} (DNA: desoxyribonucleic acid). Plasmid DNA was isolated by alkaline lysis method (Birnboim and Doly 1979).

69.2.6
Dot Blot

Templates used in the hybridisations were the DNA isolated from the selected strains; DNA from actinomycete MC2, Cad1 strains and *Bacillus* sp. O9 were used as negative controls. The 600 and 1800 bp PCR amplified fragments obtained from the plasmid pRJ1004 using the primers pairs pcoA1-pcoA2 and pcoR1-pcoR2 were used as positive controls (PCR: polymerase chain reaction). These fragments were visualized after the PCR reaction in a 0.8% agarose gel and purified from the gel using the kit Wizard PCR Preps DNA Purification System from Promega (Madison, USA). Hybridization signals were detected using a Phototope Detection Kit (Biolabs).

69.2.7
Probe Labelling

The probes were constructed using *pco*A (600 bp), *pco*R (1800 bp) gene fragments and the plasmid pIJ8660. They were labelled using the kit NEBlot-Phototope™ (BioLabs).

69.2.8
Oligonucleotide Primers

The oligonucleotides were synthesized by Bio-Synthesis, Inc. (Lewisville, USA). Universal primers fD1 (5'-AGAGTTTGATCCTGGCTCAG-3') and rD1 (5'-AAGGAGGTGATCCAGCCGCA-3') were used to check that all the isolated DNAs were amplifiable by PCR and they are designed to amplify a fragment of approximately 1.5 kb corresponding to the 16S rDNA (Weisburg et al. 1991). Chromosomal DNA of *Bacillus* sp. O9 strain was used as positive control of this reaction. Primers pairs pcoA1-pcoA2 (pco A1 5'-CGTCTCGACGAACTTTCCTG-3' and pcoA2 5'-GGACTTCACGAAACATTCCC-3') and pcoR1-pcoR2 (pcoR1 5'-CAGGTCGTTACCTGCAGCAG-3' and pcoR2 5'-CTCTGATCTCCAGGACATATC-3') were design to specifically amplified *pco*A and *pco*R genes respectively, which are part of the *pco* operon carried in the plasmid pRJ1004 that was therefore used as a positive control (Brown et al. 1995; Trajanovska et al. 1997).

69.2.9
PCR Amplification

Templates for PCR reaction included the following: total genomic DNA from the strains C43, M26, QL1-3, QL1-4, QL1-8, *Bacillus* sp. O9 and the plasmid pRJ1004. Amplifications were performed in 25 µl reaction volumes. The following PCR program was used: initial denaturation at 94 °C for 5 min followed by 35 cycles of 94 °C for 45 s, 55 °C for 45 s, 72 °C for 1 min 30 s, with a final extension step at 72 °C for 7 min. For the PCR reaction with the primers pairs pcoA1-pcoA2 and pcoR1-pcoR2, a program developed by Trajanovska et al. (1997) was used. Sterile bidistilled water was used as negative control. Amplifications reactions were carried out in a GeneAmp PCR 9700 system (Applied Biosystems).

69.2.10
Purification, Cloning and Sequencing of a PCR Product

The 600 bp PCR amplification product of strain M26 was subjected to electrophoresis in 1% agarose gels, and recovered from there using the Wizard PCR Preps DNA Purification System (Promega) and cloned into p-Gem Teasy plasmid (Promega) following the manufacture protocol. DNA sequencing on both strands was performed by the dideoxy chain termination method with an ABI prism 3700 DNA Analyzer, using the ABI Prism BigDye Terminator Cycle Sequencing Ready Reactions kit (PE Biosystems).

69.2.11
Transformation Procedures for *Streptomyces* R25 Cadmium Resistant

The plasmid pIJ8660 used in the transformation experience was a gift of Dr. Jongho Sun, Department of Genetics, John Innes Centre, Norwich NR4 7UH, UK. This plasmid contains the pUC18 replication origin and apramycin-resistant gene, *aac* (3) IV, for maintenance and selection in *E. coli*, respectively. It possesses the ϕC31 *int* gene and *att* P site, allowing insertion of the plasmid at the chromosomal ϕC31 attachment site, using apramycin resistance for selection in streptomycetes.

For the mycelium transformation a modification of Kieser et al. (2000) method was used. *Streptomyces* R25 was grown at 30 °C during 72 h. Twenty µl of the competent cell were mixed with 2 µl of plasmid DNA (0.1 µg). The electroporation was carried out in a Cell Porator (GIBCO BRL) with the following conditions: 1.5 kV, 330 µF, 4 KΩ. The cells were grown on TSB medium and incubated 3 d at 30 °C. Transformant colonies were detected with fluorescence microscope. Transformation results were confirmed by Dot Blot assays using the plasmid pIJ8660 as a probe. Plasmid pIJ8660 and *Streptomyces* R25 not transformed were used as positive and negative control respectively.

69.2.12
Visualization of Enhanced Green Fluorescent Protein (EGFP)

Fluorescence microscopy was carried out on a Zeiss Axiolab equipped with a filter set, using a ×100 objective for examination of culture on cover slips. The wavelength of light used for fluorescence detection was 450–490 nm.

69.3
Results and Discussion

69.3.1
Qualitative Assays of Copper Resistance

In order to estimate the metal stress that the isolated strains support in their natural environment, copper concentration determination was performed in the sediments of the uranium mine, drainage channel and El Cadillal samples. 0.5 and 2.0 mg l^{-1} of copper concentrations were found for the two first samples respectively but no copper was detected in El Cadillal sample.

Forty-four actinomycete strains were isolated, thirty-one from El Cadillal, ten from the drainage channel and three from the uranium mine. The colonies morphology were typical of actinomycetes group characterized by the formation of vegetative mycelium and a secondary or aerial mycelium in permanent contact with the air.

These strains were tested for qualitative copper resistance by plate analysis (Table 69.1). One hundred percent isolated actinomycetes strains from El Cadillal were capable to grow at 0.1 mM of Cu^{2+}, 77% at 0.25 mM and only 39% at 0.5 mM, suggesting an inhibition growth more than 50% when the copper concentration increased five-folds. On the other hand, all the isolated strains from both mines were able to grow at all Cu^{2+} tested concentrations (Table 69.1).

These results indicate that copper resistance may be widespread amongst actinomycetes growing in contaminated environments, validating the hypothesis that bacteria expose to toxic levels of a metal, can evolve biochemical mechanisms to regulate copper uptake and resistant toxicity (Yang et al. 1993).

69.3.2
Dot Blot

The presence of DNA sequences that codifies copper resistance was evaluated in ten actinomycete strains (C14, C31, C43, C61 from El Cadillal; M3, M12, M26 from the drainage channel;.QL1-3, QL1-4 QL1-8 from the uranium mine) that presented the highest biomass concentration (data not show) in culture medium with 0.5 mM copper concentration (Table 69.1). Dot Blot assays were carried out using *pco*A and *pco*R genes of the operon *pco* described in *Escherichia coli* and also detected in Gram-positive bacteria (Trajanovska et al. 1997). DNA sequence of this genes were used as positive controls, DNA samples isolated from the actinomycete strains sensitive to all metal concentrations tested and

Table 69.1. Qualitative screening of copper resistance

	Total strains	0.1 mM Cu^{2+a}	0.25 mM Cu^{2+a}	0.5 mM Cu^{2+a}
El Cadillal	31	100	77	39
Drainage channel	10	100	100	100
Uranium mine	3	100	100	100

^{a} The results are expressed in % of the total isolated strains from each sample.

Bacillus sp. O9 strain were used as negative controls. Similar positive hybridisation signals were obtained when the two probes were used for the DNA samples of C43, M26, QL1-3, QL1-4 strains, whereas no signal was observed for DNA of QL1-8 strain (Fig. 69.1).

The positive hybridisation signals obtained (Fig. 69.1) indicate the presence of DNA homologous sequences among the actinomycete and *E. coli* assayed strains. These homology sequences could be part of consensus sequence of *pco* operon, and indicate that these actinomycete strains may have a similar copper resistance mechanism to *E. coli*. Moreover bacteria resistance to copper conferred by plasmids has been described in *Pseudomonas* (Cooksey 1994), *Xhantomonas* (Lee et al. 1994) and *E. coli*

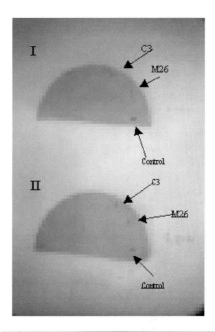

Fig. 69.1a.
Autoradiography of the hybridisation assay. Strains C14, C31, C43, C61, M3, M12 and M26.
I: pcoA probe. *II:* pcoR probe

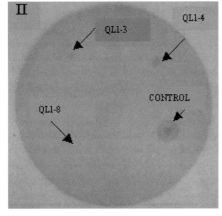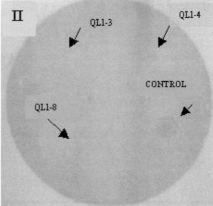

Fig. 69.1b. Autoradiography of the hybridisation assay. Strains QL1-3, QL1-4 and QL1-8. *II:* pcoR probe

(Brown et al. 1995). These systems are highly homologous (Brown et al. 1994; Cooksey 1994) and contain the same genes. Contrary to the obtained results the QL1-8 copper resistant strain showed no signal in Dot Blot assay; indicating that this strain may have different copper resistant DNA codify sequences than the other actinomycete strains assayed (Fig. 69.1). Further experiences should be done with QL1-8 strain using primers and probes constructed with genes involved in copper resistance in another bacteria. At this moment, in our laboratory we are identifying the genes involved in the genetic regulation of actinomycetes copper resistance.

69.3.3
PCR Amplifications

In order to determinate if the homology found above was a partial or complete homology of copper resistant sequences, PCR reactions were carried out using total genomic DNA as templates of the strains that gave positive hybridisation signals in the Dot Blot assay (Fig. 69.1). A 1.5 kb fragment was amplified from all DNA samples when the primers fD1 and rD1 were used. When the primers pair pcoA1-pcoA2 were tested, none of the templates showed amplification, with the exception of the positive control (plasmid pRJ1004) that gave a fragment of the expected size (1.8 kb). The hybridisation signals obtained in Dot blot assay with C43, M26, QL1-3 and QL1-4 strains could indicate the presence of homology sequences between *E. coli* and actinomycetes present in the inner of the gene, which do not hybridise with the primer pairs pcoA1-pcoA2 used in the PCR reaction.

When the primer pairs pcoR1-pcoR2 were used in the PCR assay, it was only obtained a 600 bp amplification product for the strain M26 like the positive control (Fig. 69.2). The M26 amplification product was purified, cloned and sequenced. The sequence analysis showed 100% of similarity with the *pco*R gene of the *pco* operon previously described in *E. coli* (Brown et al. 1995). The 600 bp consensus fragment will be used in future for identification copper resistant microorganisms as target sequence in PCR and hybridisation assays.

69.3.4
Transformation Procedures for *Streptomyces* R25 Cadmium Resistant

For evaluating the survival and bioremedial cadmium soil activity of R25 strain, transformation assay with the plasmid pIJ8660 were made. This plasmid containing the green fluorescent gene (EGFP) was introduced by electroporation into *Streptomyces* R25 strain. The fluorescence of the transformant colonies was observed in substrate hyphae by microscopy (data no shown).

Dot Blot hybridization was performed in order to confirm that the fluorescence of the colonies was due to the insertion of plasmid pIJ8660 carrying the EGFP gene into *Streptomyces* R25. In order to do this, we used this plasmid as a probe. Results are shown in Fig. 69.3 where it is possible to observe hybridizations signals in the five colonies tested.

The fluorescence of R25 strain was observed after many generations in culture medium without apramycin, indicating the presence of EGFP gene. Thus, the gene sequence is probably integrated in the actinomycetes chromosome because the pIJ8660 plasmid has a *E. coli* replication origin only.

Fig. 69.2.
Agarose gel electrophoresis showing PCR product using pcoR1-pcoR2 as primers.
a Negative control (sterile water); **b** DNA of strain M26; **c** positive control (pRJ1004 plasmid); **d** MWM 1 kb DNA Ladder (PROMEGA)

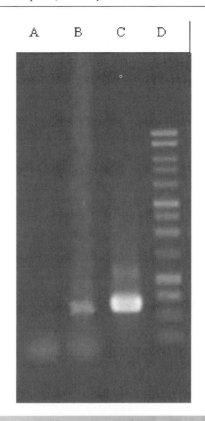

Fig. 69.3.
Dot blot hybridisation of transformed *Streptomyces* colonies using pIJ8660 as probe. Transformed colonies indicated as *1, 2, 3, 4, 5*. Positive and negative control were DNA of pIJ8660 and *Streptomyces* R25 respectively

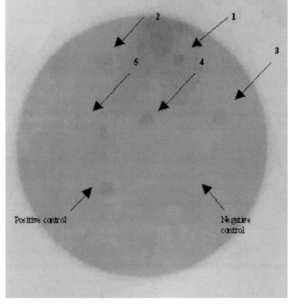

69.4 Conclusions

To our knowledge, these are the first results showing evidences of genetic systems involved in copper resistance mechanisms in actinomycete strains, as well as the characterization of 600 bp sequence that may be related with the copper actinomycete resistance mechanism. The resistant copper and cadmium actinomycetes strains presented in this work, as well as the use of plasmid pIJ8660 for evaluating the survival and bioremedial cadmium soil activity of R25 strain constitute potentially useful tools to design and develop bioremediation experiences at copper and cadmium polluted areas.

References

American Public Health Association (1992) Standard methods for the examination of water and wastewater, 18th edn. American Public Health Association, Washington, DC

Amoroso MJ, Romero N, Carlino F, Hill R, Oliver G (1996) Estudios de la contaminación de aguas en un embalse de la región por desechos industriales. V Conferencia Internacional sobre Ciencia y Tecnología de los Alimentos. (PSMmr), Archivo Med004. wp5, (disco 1)

Amoroso MJ, Castro GR, Carlino FJ, Romero NC, Hill RT, Oliver G (1998) Screening of heavy metal-tolerant actinomycetes isolated from the Salí River. J Gen Appl Microbiol 44:129–132

Birnboim HC, Doly J (1979) A rapid alkaline extraction procedure for screening recombinant plasmid DNA. Nucleic Acids Res 7(6):1513–1523

Brown NL, Lee BTO, Silver S (1994) Bacterial transport of and resistance to copper. In: Sigel H, Sigel A (eds) Metal ions in biological systems, vol. 30. Dekker, New York, pp 405–434

Brown NL, Barrett SR, Camakaris J, Lee BTO, Rouch DA (1995) Molecular genetics and transport analysis of the copper-resistant determinant (pco) from *Escherichia coli* plasmid pRJ1004. Mol Microbiol 17:1153–1166

Cai J, Salmon K, DuBow MS (1998) A chromosomal *ars* operon homologue of *Pseudomonas aeruginosa* confers increased resistance to arsenic and antimony in *Escherichia coli*. Microbiology 144:2705–2713

Cooksey DA (1994) Copper uptake and resistance in bacteria. Mol Microbiol 7:1–5

Diels L, Van Roy S, Mergeay M, Doyen W, Taghavi S, Leysen R (1993) Immobilization of bacteria in composite membranes and development of tubular membranes reactor for heavy metal recuperation. In: Peterson R (ed) Effective membrane processes: new perspectives. Kluwer Academic Publisher, Dordrecht, pp 275–293

Fisher SH (1985) Rapid small scale isolation of *Streptomyces* total DNA: procedure 4. In: Hopwood DA, Bibb MJ, Chater KF, Kieser T, Bruton CJ, Kieser HM, Lydiate DJ, Smith CP, Ward JM (eds) Genetic manipulation of *Streptomyces* – A laboratory manual. The John Innes Foundation

Friberg L, Kjellstrom T, Nordberg G, Piscator M (1979) Cadmium. In: Friberg R, Nordberg GF, Vouk VB (eds) Handbook on the toxicology of metals. Elsevier/North-Holland Biomedical Press, Amsterdam New York Oxford, pp 355–381

Gadd GM (1992) Microbial control of heavy metal pollution. In: Fry JC, Gadd GM, Herbert RA, Jones CW, Watson-Craik IA (eds) Microbial control of pollution. Cambridge University Press, Cambridge, UK, pp 59–87

Gadd GM (2000) Bioremedial potential of microbial mechanisms of metal mobilization and immobilization. Environ Biotech 11:271–279

Goodfellow M, Williams ST, Mordarski M (1988) Actinomycetes in biotechnology. Academic Press, San Diego

Gutteridge JMC, Wilkins S (1983) Copper salt dependent hydroxyl radical formation damage to proteins acting as antioxidants. Biochim Biophys Acta 759:38–41

Hassan MT, van der Lelie D, Springael D, Roemling U, Ahmed N, Mergeay M (1999) Identification of a gene cluster, *czr*, involved in cadmium and zinc resistance in *Pseudomonas aeruginosa*. Gene 238:417–425

Kieser T, Bibb MJ, Buttner MJ, Chater KF, Hopwood DA (2000) Practical *Streptomyces* genetics. The John Innes Foundation, Norwich

Kotrba P, Doleckova L, De Lorenzo V, Ruml T (1999) Enhanced bioaccumulation of heavy metal ions by bacterial cells due to surface display of short metal binding peptides. Appl Environ Microbiol 65:1092–1098

Larpent JP, Sanglier JJ (1989) Biotechnologie des antibiotiques. Masson, Paris

Lee Y-A, Hendson M, Panopoulus NJ, Schroth MN (1994) Molecular cloning, chromosomal mapping, and sequence analysis of copper resistance genes from *Xhantomonas campestris* p.v. juglandis: homology with blue copper proteins and multicopper oxidase. J Bacteriol 176:173–188

Munson GP, Lam DL, Outlen FW, O'Halloran TV (2000) Identification of a copper-responsive two-component system on teh chromosome of *Escherichia coli* K-12. J Bacteriol 182:5864–5871

Nies DH, Brown NL (1998) Two-component systems in the regulation of heavy metal resistance. In: Silver S, Walden W (eds) Metal ions in gene regulation. Chapman & Hall, New York

Nriagu JO, Pacyma JM (1988) Quantitative assessment of world-wide contamination of air, water and soilsby trace elements. Nature 333:134–139

Peczynska-Czoch W, Mordarski M (1984) Transformation of xenobiotics. In: Goodfellow M, Mordarski M, Williams ST (eds) The biology of the actinomycetes. Academic Press, London, UK, pp 287–336

Ravel J, Schrempf H, Hill RT (1998) Mercury resistance is encoded by transferable giant linear plasmids in two Chesapeake Bay *Streptomyces* strains. Appl Envir Microbiol 64:3383–3388

Ravel J, Wellington EMH, Hill RT (2000a) Interspecific transfer of *Streptomyces* giant linear plasmids in sterile amended soil microcosms. Appl Envir Microbiol 66:529–534

Ravel J, DiRuggiero J, Robb FT, Hill RT (2000b) Cloning and sequence analysis of the mercury resistance operon of *Streptomyces* sp. strain CHR28 reveals a novel putative second regulatory gene. J Bacteriol 182:2345–2349

Silver S, Phung LT (1996) Bacterial heavy metal resistance: new surprises. Ann Rev Microbiol 50:753–789

Simpson JA, Cheeseman DH, Smith SE, Dean RT (1988) Free-radical generation by copper ions and hydrogen peroxide. Biochem J 254:519–523

Sun J, Kelemen GH, Fernández-Abalos JM, Bibb MJ (1999) Green fluorescent protein as a reporter for spatial and temporal gene expression in *Streptomyces coelicolor* A3(2). Microbiology 145:2221–2227

Tetaz TJ, Luke RKJ (1983) Plasmid-contolled resistance to copper in *Escherichia coli*. J Bacteriol 154:1263–1268

Trajanovska S, Britz ML, Bhave M (1997) Detection of heavy metal ion resistance genes in Gram-positive and Gram-negative bacteria isolated from a lead-contaminated site. Biodeg 8:113–124

Trevors JT (1987) Copper resistance in bacteria. Microbiol Sci 4:29–31

Volesky B, Holan ZR (1995) Biosorption of heavy metals. Biotechnol Prog 11:235–250

Watanabe ME (1997) Phytoremediation on the brink of commercialisation. Environ Sci Technol 31:182–186

Weisburg WG, Barns SM, Pelletier DA, Lane DJ (1991) 16S ribosomal DNA amplification for phylogenetic analysis. J Bacteriol 173:697–703

Williams JR, Morgan AG, Rouch DA, Brown NL, Lee BTO (1993) Copper-resistant enteric bacteria from United Kingdom and Australian piggeries. Appl Environ Microbiol 59:7027–7033

Xiong A, Jayaswal RK (1998) Molecular characterization of a chromosomal determinant conferring resistance to zinc and cobalt in *Staphylococcus aureus*. J Bacteriol 180:4024–4029

Xu C, Zhou T, Kuroda M, Rosen BP (1998) Metalloid resistance mechanism in prokaryotes. J Biochem 123:16–23

Yang CH, Menge JA, Cookey DA (1993) Role of copper resistance in competitive survival of *Pseudomonas fluorescens* in soil. Appl Environ Microbiol 59:580–584

Index

A

AAS (see atomic absorption spectroscopy)
accelerated solvent extraction 433
acetate 288
acetic acid 630
Achromobacter 708, 715
 –, *eurydice* 708, 710, 719
acid
 –, acetic 630
 –, alkylsulfonic 344
 –, aniline-N,N-diacetic 709
 –, chlorobenzoic acid 344
 –, 4-chloro-2-methylphenoxyacetic acid (MCPA) 462
 –, dichlorophenoxyacetic 673
 –, ethylenediaminetetraacetic acid (EDTA) 107
 –, fatty 710
 –, fulvic 516, 518, 522
 –, humic 204, 516, 518, 522, 706, 710
 –, hydroxybenzoic 29, 590
 –, lauric 710
 –, linolenic 710
 –, margarine 710
 –, monocarboxylic acids 375
 –, myristic 710
 –, nalidixic 758
 –, palmitic 710
 –, tridecanoic 710
Acinetobacter 32–34, 38–40
actinomycetes 214, 756
 –, heavy-metal resistant 756
adipocytes 352
adsorbent, low-cost 250
adsorption 259, 462, 510, 590, 592
advanced oxidation process (AOP) 590, 614
Aequorea 647, 656
 –, *victoria* 656
aerogel 590
 –, adsorption 592
Aeromonas 431

aerosol 372, 408
 –, organic 370
Agrobacterium radiobacter 477
air 690
 –, pollutants 382
 –, pollution 370, 408
aluminium 630
alcohol, threshold concentration 738
algae 92, 125, 494, 495, 725
 –, endosymbiotic 672
Algiers 370
alkamone 709
alkanes 370, 710
alkylbenzenes, linear 344
alkylphenols 304
alkylsulfonic acid 344
aluminum 630
 –, -27 463
 –, ions 635
Ames 645
amides 710
amines 630, 710
 –, aromatic 268, 278
aminobenzothiazole
 –, biodegradation 297
 –, metabolite 299
amitrol 329
ammonia 99, 327
Ampelisca abdita 725
amphipod 725
Anabaena 659
analysis
 –, cadmium standard 78
 –, double spike 76
 –, in-field 13
 –, microscopic 91
 –, of PAHs 433
 –, trace 99
anatoxin 658
aniline 709
 –, -N,N-diacetic acid 709

Index

anthracene 673
anthrachinoid 275
  –, dyes 275
anthraquinone 709
antimicrobials 675
AOP (*see* advanced oxidation process)
Aphanizomenon ovalisporum 659
Aporrectodea giardi 473
Arabidopsis 137, 631, 633–635, 637
  –, *thaliana* 634
arabinose 710
arabitol 710
arsenic 146, 152, 229, 238, 673
  –, particulate 157
  –, remediation 228, 229
Aspergillus
  –, *foetidus* 280
  –, *niger* 280
asphaltene, precipitation 745
atomic absorption spectrometry (AAS) 9, 146, 198, 251, 258, 473, 758
atrazine 329, 472
  –, biodegradation 472, 477
  –, metabolism 474
  –, photodegradation 472
azo dyes 268, 273

B

Bacillus 214–220, 278, 570, 725, 759, 760, 762
  –, *cereus* 278, 725
  –, *macerans* 570
  –, *simplex* 215
  –, *subtilis* 278, 279
bacteria 31, 125, 214, 268, 495, 650, 648, 725
  –, bioluminescent 690, 698
  –, luminescent 690
  –, pathogenic 31
barium 136, 152
  –, anthropogenic input 160
  –, particulate 157
bark 250
barley (*see Hordeum vulgare*)
bee 482
beetle 647
benthiocarb 673
benzene 734
  –, threshold concentrations 738
  –, sulfonate 709
benzo[a]pyrene 673
benzothiazole 294
  –, biodegradation 294, 298
  –, metabolites 300
binding

–, constant 568, 574
  –, determination 574
  –, heterogeneity 120
bioassay 644, 658, 662, 672, 698, 724
  –, bacterial 644
bioavailability 187, 418, 423, 424, 462, 694
  –, PAHs 430
biodegradation 268, 294, 316, 472, 477, 614, 706
  –, experiment 474
  –, test 307
biodetector 734
biofilm 560
biofilter 560
biological oxygen demand (BOD) 681
bioluminescence 646, 653, 690, 699
biomarker 31, 742
biomass 620
bioreactor 560
  –, continuous 430
bioremediation 132, 214, 268, 271, 316
  –, by microorganisms 214
biosensor 132, 681, 690, 695, 698, 714, 719
  –, microbial 706
  –, environmental 690
biosorption 280
biotransformation 304
bisphenol 516
Bjerkandera 275, 431
  –, *adusta* 275
blood 645
BOD (*see* biological oxygen demand)
boron tribromide 343
bound residues 329
Brassica oleracea acephala 186
Brassicaceae 186, 192
bromate 360
  –, control 368
bromophenol 709
Burkholderia cepacia 281, 560–563
butanol 710
by-products, halogenated 360

C

cadmium 4, 136, 146, 186, 197, 204, 214, 258, 673, 724
  –, -110 75, 85
  –, -111 78
  –, -112 75, 85
  –, -113 79
  –, -114 75, 85
  –, -116 75, 78
  –, -116/-110 ratio 83, 84
  –, analysis 216
  –, concentration 80

–, isotopic
  –, composition 75
  –, data 80
–, removal with coffee beans 258
–, resistance 760, 763
–, standard analyses 78
–, toxicity 727, 729
calcium 136, 139, 146, 242, 630
–, carbonate 440
calmodulin (CaM) 132, 133
–, peptides 136
cancer 305
candytuft 188
capillary electrophoresis 107
carbendazim 528
carbohydrates 710
carbon
–, -13 43, 65, 318, 331, 463, 465, 543, 545
  –, delta 46, 47, 53, 65, 66, 68, 69
–, -13/-12 ratio 318
–, -14 304, 307–309, 331, 352–354, 418, 421, 516
  –, labelling 421–423, 426, 427
  –, dating 67
–, activated 259, 360
–, dioxide 540
  –, as solvent 541
  –, fixation 540
  –, pressure 544
  –, supercritical 540, 546
–, isotopes 65
–, monoxide (CO) 20
carcinogens 677
catalysis 540
cationic exchange capacity (CEC) 509
CCA (*see* chromated copper arsenate)
CD (*see* circular dichroism)
CEC (*see* cationic exchange capacity)
cell
  –, cultured 734
  –, human 734
  –, immobilization 708
  –, lines, human 736
  –, mammalian 672
cement 75, 620, 621
  –, works 419
Ceriodaphnia dubia 725
cerite 259
cesium 136
  –, -137 197
char 89
chelation 132
chemical oxygen demand (COD) 681, 701
chemiluminescence 13, 15
chitin 259

chitosan 259
Chlonis barbata 200
Chlorella 494, 495, 497
–, kessleri 496
chloride 316
–, concentration technique 316
chlorinated phenols 329
chlorine 360
–, -35 486
–, dioxide 360
chlorobenzene 709
chlorobenzoic acid 344
4-chloro-2-methylphenoxyacetic acid (MCPA) 462
chlorophenol 590, 600
–, hydrogenation 604, 605
chlorophyll 664
chromated copper arsenate (CCA) 234
chromium 146, 152, 181, 229, 238, 724, 727
–, particulate 157
–, remediation 229
–, solid phase 157
–, toxicity 727–729
ciliates 495, 674
Ciperus rotundus 200
circular dichroism (CD) 132
cladocerans 725
clay 462, 473
–, anionic 462
–, minerals 504
–, pesticide adsorption 464
climate 57, 65
–, past 65
–, semi-arid 57
clustering, functional 176
coal 75, 89, 440
–, ash 144
  –, leaching 144
–, combustion 145
–, environmental issues 146
–, minerals 145
–, treatment, environmental impact 145
coating process 7
cobalt 13, 16, 18, 124, 152, 197, 372
–, anthropogenic input 160
–, particulate 157
CO (*see* carbon monoxide)
COD (*see* chemical oxygen demand)
coffee beans 258
coke 89
color 268
cometabolism 560
complexation 126, 568
composition, isotopic 75
compounds

–, aromatic 709
  –, polycyclic 390
–, mutagenic 418
–, organic, photodecomposition 590
–, polyaromatic 709
–, xenobiotic 518
conditions
  –, aerobic 430
  –, methanogenic 430
contaminant
  –, airborne 89
  –, organic 328
contamination 196
copper 16, 75, 146, 152, 181, 197, 204, 229, 238, 250, 258, 673, 756
  –, (II) 124
  –, analysis 758
  –, anthropogenic input 160
  –, particulate 157
  –, remediation 228, 229
  –, removal with coffee beans 258
  –, resistance 758, 761
Coriolus versicolor 279
cork 250
Corynebacterium 276, 279
Coulter counter 736
coupling, oxidative 304
cow 419
crandallite 242
Cunninghamella 310
cyanobacteria 658
cyanotoxin 658
Cyathus 279
cyclodextrin 568, 570
cycloheximide 758
cylindrospermopsin 658
Cylindrospermopsis raciborskii 659, 661, 668
Cynodon dactylon 200
Cyprinus carpio 725
cytotoxicity 644

D

dairy ruminant 418, 419
Daphnia 725
  –, *magna* 725
  –, *similis* 725
DDT 336, 338
decolorization 268
decomposition 590
defense 633
degradation 452
  –, abiotic 504
  –, chemical 504
  –, microbial 269
  –, photo- 504
  –, photochemical 510
  –, products 512
dehalogenation agent 554
depollution 600
  –, catalytic 601
detection 680
  –, window, analytical 120, 128
detector, biological 734
deuterium 43, 59, 331
  –, delta- 47, 51, 66
dialkyl carbonate 540
dialysis 516, 517
  –, experiment 518
dichloroaniline 329
dichlorobenzene 734
dichlorophenoxyacetic acid 673
diffraction, X-ray 4
diffuse infrared Fourier transform spectroscopy (DRIFTS) 19, 21
dihydroxybenzothiazole metabolites 300
dilution 316
dimethyl carbonate (DMC) 540, 542
  –, synthesis 546
dimethyl sulfide (DMS) 614
dimethyl sulfone ($DMSO_2$) 618
dimethyl sulfoxide (DMSO) 614, 673
  –, oxidation 618
dioxins 99, 352, 418
  –, bioavailability 423
  –, transfer to human 418
disrupter, endocrine 304, 516
dissolved organic matter (DOM) 516
DMC (*see* dimethyl carbonate)
DMSO (*see* dimethyl sulfoxide)
$DMSO_2$ (*see* dimethyl sulfone)
DNA
  –, adduct 742
  –, damage 644
  –, extraction 759
DOM (*see* dissolved organic matter)
double spike analysis 76
DRIFTS (*see* diffuse infrared Fourier transform spectroscopy)
dye 268, 288
  –, anthrachinoid 275
  –, azo 268
  –, decolorization 268
  –, enzymatic cleavage 273
  –, Indigoid 274
  –, phtalocyanine 275
  –, textile 268
  –, triphenyl methane 274

Index   773

E

EA (*see* elemental analyser)
earthworm 473
–, cast, composition 473
earthworm casts 472
EC_{50} 735
ecosystem, adapted methanogenic 430
ecotoxicology 628
EDTA (*see* ethylenediaminetetraacetic acid)
effect
  –, cytotoxic 644
  –, genotoxic 644
EGFP (*see* enhanced green fluorescent protein)
electrophoresis 661
  –, capillary 107
electrospray mass spectrometric (ES-MS) 132, 134
elemental analyser (EA) 46
endocrine disrupter 516, 672
enhanced green fluorescent protein (EGFP) 760
enrichment factor 157
Enterobacter aerogenes 495
Enterococcus 278
enzyme 271, 269, 495, 632
  –, remediation 269
equilibrium, dynamic 382, 384
Erika (vessel) 742, 751
Escherichia coli 32–34, 36, 38–40, 277, 645, 647, 655, 691, 696, 699, 707, 757, 758, 760–763
ES-MS (*see* electrospray mass spectrometry)
ethanol 710, 734
ethylenediaminetetraacetic acid (EDTA) 107
Euglena 494, 497, 663
  –, *ehrenbergii* 502
  –, *gracilis* 494–496, 499, 502, 658, 660
europium 135–137
eutrophisation 658
EXAFS (*see* X-ray absorption fine structure)
exposure
  –, dermal 528
  –, inhalation 528
extraction
  –, accelerated solvent 433
  –, phyto- 186
  –, sequential 4, 186
exudates 207

F

fat factory 370
fathead minnow 725
fatty acid 710
Fenton's reagent 448
fertiliser 43, 47
field water, toxicity 701
fish 725, 742, 745
flagellate 495, 660
Flavobacterium 279
fluorescence 132
  –, microscopy 760
  –, spectroscopy 134
fluorine-19 463
Fomitopsis pinicola 215–220
formaldehyde 673
fractionation 540
fructose 710
fuel
  –, burning 418
  –, N°2 742
fulvic acids 516, 518, 522
fungi 268, 304, 495
  –, isolation 307
fungicide 294, 343
fungus 214
furfural 336
Fusarium 304
fuzzy set 176

G

Gadus morhua 751
galactose 710
gallium 630
gas toxicity 695
gas chromatography-mass spectrometry (GC/MS) 335
Gaucho® 482
GC/MS (*see* gas chromatography-mass spectrometry)
genotoxicity 644, 742
geopolymers 328
Geotrichum candidum 279
gfp 644
GHMBC (*see* gradient heteronuclear mutiple bond correlation)
Gluconobacter oxydans 717
glucose 710
glutathione 658, 665
glycerol 710
gradient heteronuclear mutiple bond correlation (GHMBC) 296
grape stalks 250
groundwater 43, 57, 319
  –, mixing 316
  –, perchloroethylene contamination 568
  –, trichloroethylene contamination 568
growth rate 674

H

Halichondira panicea 750
Hawai'i 152
heavy metal (*see also individual elements*) 75, 132, 176, 197, 222, 672, 734
-, biosorption 214
-, concentration, model 176
-, pollution 192, 222
-, remediation 222
herbicide 294, 494, 496
  -, degradation 504
  -, effect on algae viability 497
  -, effect on green paramecia viability 498
  -, effect on non-photosynthetic paramecia viability 499
  -, effects on *Euglena gracilis* 499
  -, metabolites 516
  -, sulfonylureic 494
high performance liquid chromatography (HPLC) 288, 294, 296, 393, 410, 484
high resolution magic angle spinning nuclear magnetic resonance (HR-MAS) 462
Homo sapiens 735
Hordeum vulgare 198, 200, 202
-, mercury loading 201
HPLC (*see* high performance liquid chromatography)
HR-MAS (*see* high resolution magic angle spinning nuclear magnetic resonance)
humic
  -, acids 204, 516, 522, 706, 710
  -, substances 123, 516
    -, preparation 518
hydrocarbons
  -, aromatic
  -, polycyclic 392
hydrodechlorination 601
hydrogen
  -, -1 294, 296, 543
  -, -2 49, 57, 463
    -, delta 321
  -, -3 352
  -, peroxide 630
hydrogenation 600
  -, catalytic 600
Hydrogenophaga palleronii 277
hydrolysis 332
hydroxides, layered double 462
hydroxy radical 630
hydroxybenzoic acid 590
hydroxyl radical 505
hypochlorous acid 360

I

Iberis intermedia 186
IC_{50} value 672, 674, 677
ICP-AES (*see* inductively coupled plasma-atomic emission spectroscopy)
ICP-MS (*see* inductively coupled plasma mass spectrometry)
imidacloprid 482
immunoassay 706
Indigo 274
  -, carmine 288
  -, degradation 290
  -, dye 274, 288
indium-113 78
inductively coupled plasma mass spectrometry (ICP-MS) 156
inductively coupled plasma-atomic emission spectroscopy (ICP-AES) 207
in-field analysis 13
inhalation, pesticides 528
insecticide 482
iodine-3 299
ionisation, chemical 392
IRMS (*see* isotope ratio mass spectrometer)
iron 16, 75, 146, 204, 449
Irpex lacteus 279
isotope ratio mass spectrometer (IRMS) 47
isotopes (*see also individual elements*) 43, 57
  -, environmental 57
  -, geochemistry 75

J

jellyfish 656

K

kale 188
kerogen 440
Klebsiella pneumoniae 674

L

laccases 308
LAC-FLUORO test 647
Lactuca sativa 725
Laetiporus sulphureus 279
landfill 370
lanthanide 139, 630
  -, binding 137
  -, specificity 139
lanthanium 630, 634
lauric acid 710

LC/APCI-ITMS (*see* liquid chromatography/
   atmospheric pressure chemical ionisation-
   ion trap mass spectrometry)
LD_{50} 735
leaching 144
lead 4, 5, 146, 152, 181, 186, 197, 230
   –, -(II) 123
   –, anthropogenic input 160
   –, particulate 157
   –, remediation 222
Lemna minor 725
Lepidhorombus boscii 744
light microscopy 497
Limanda limanda 750
lindane 344, 673
linolenic acid 710
lipid 632
lipogenesis 352, 356
lipolysis 352, 353, 356
Lipophorys pholis 750
liquid chromatography/atmospheric pressure
   chemical ionisation-ion trap mass
   spectrometry (LC/APCI-ITMS) 392
lithium 630
   –, -AlH_4 554
liver 742
luciferase 647
   –, bacterial 647
luminol (5-aminophthalhydrazide) 16
   –, reaction 16
Lupinus luteus (yellow lupin) 198, 200, 202
   –, mercury loading 201
lux (bacterial luciferases) 644, 647, 655, 690, 699
LUX-FLUORO test 644, 651
   –, dose-response curves 653
   –, performance 651

M

magnesium 136, 146, 630
Maillard reactions 336
maize 488
major elements 144
malathion 673
maltose 710
mammalian cell 672
maneb 673
manganese 146, 181, 630, 635, 636
margarine acid 710
mass spectrometry 31, 392, 482
material
   –, bituminous 370
   –, carbonaceous 89
matrix-assisted laser desorption 31

matter
   –, organic 89
   –, macromolecular 328
   –, particulate 328, 408
   –, suspended 152
MCPA (*see* 4-chloro-2-methylphenoxyacetic acid)
mechanochemistry 552
mercury 196
   –, availability 196
   –, contamination 196
   –, phytoremediation 196
metal
   –, accumulation in plants 204
   –, binding 136, 137
   –, cation 13, 630
   –, chelation 111, 132
   –, complexation 126
   –, heavy 3, 132
   –, ion ligands 204
   –, particulate 152
   –, phases 3
   –, transformation 3
   –, removal 250
   –, speciation 724
   –, threshold concentrations 737
   –, toxic 132, 756
   –, analysis 13
   –, trace 120
   –, speciation 120
   –, uptake 250
metaupon 709
methanol 542, 710
method, in-situ 3
methoxychlor 338
methylfurfural 336
methylmercury 243
mercury 196, 197, 243, 673
   –, speciation 199
micro total analytical system (μTAS) 13
Micrococcus 431
microcystin 658
Microcystis 659
microelements 144
microorganism 214, 561
   –, adaptation 276
   –, isolation 276
   –, pathogenic 31
   –, sources 276
microscopy
   –, fluorescence 760
   –, light 497
microsome
   –, incubation 746
   –, purification 746

microtox 724, 726
milk 418, 419
    –, PAH contamination 419
milling, high energy 552
mining 75
mixture fractionation 548
moanatoxin 658
model 176
    –, transport 382
modulator, metabolic 677
molybdenum 75
monitoring 698
monocarboxylic acids 375
monooxygenase 560
montmorillonite 440
Morocco 57, 58
motorway 419
Mucor 304, 310
mutagenicity 645
mutagen 677
Mycobacterium 279, 295
mycrocystin 658
myristic acid 710
Mytilus edulis 750

N

NaBH$_4$ 554
NADPH 633
    –, oxidase 634
nalidixic acid 758
n-alkanes 374
naphthalene 329, 709
    –, sulfonate 709
naphthol 709
natural organic matter (NOM) 120, 362
Neurospora crassa 279
neurotoxin 658
nickel 146, 152, 180, 181, 186, 197, 204, 250, 673
    –, -(II) 124
    –, particulate 157
    –, solid phase 157
Nicotiana
    –, *plumbaginifolia* 631
    –, *tabacum* 630
nitrate 43
    –, contamination 43
nitrated polycyclic aromatic hydrocarbons (NPAH) 370
nitrofluoranthene 379
nitrogen
    –, oxides (NO$_x$) 20, 372, 408
    –, -15 43, 47, 52, 294, 296, 300, 463
        –, delta 43, 47, 51–53, 55

nitro-polycyclic aromatic hydrocarbons (NPAH) 379
nitropyrene 379
NMR (*see* nuclear magnetic resonance)
Nocardioides 477
Nodularia spumigena 658
nodularin 658
NOM (*see* natural organic matter)
nonylphenol 304, 516, 674
    –, biotransformation 308, 312
    –, metabolites 308
    –, surfactants 304
Nostoc 659
NPAH (*see* nitrated polycyclic aromatic hydrocarbons)
nuclear magnetic resonance (NMR) 294, 296, 462, 465, 468, 568

O

oil 75
oligosaccharide 568, 630
olive oil 600
    –, mill waste waters 607
    –, pits 250
OM (*see* organic matter)
Oncorhynchus mykiss 725
optimization 31
organic matter (OM) 89
    –, macromolecular 328
    –, natural 360, 362
osmosis, reverse 614, 615, 617
oxidation 448, 614
    –, advanced 590, 614
    –, catalytic 600
    –, processes 618
oxygen
    –, -18 43, 44, 47, 49, 57, 58
    –, delta 43, 47, 51–53, 59, 321
    –, reactive species 630
    –, singlet 505
ozone 360, 408

P

PAHs (*see* polycyclic aromatic hydrocarbons)
palladium 78
palmitic acid 354, 710
Panagrellus redivivus 725
paramecia 672
    –, green 498, 672
    –, non-photosynthetic 499
Paramecium
    –, *bursaria* 494–499, 502, 672–677, 679

–, *caudatum* 494, 495, 499, 502
–, *primaurelia* 664
–, *tetraurelia* 132, 136
–, *trichium* 494, 495, 499, 502
paraquat 673
particulate matter 408
pathogen 633
pathogenic microorganisms 31
PCA (*see* principal components analysis)
PCBs (*see* polychlorobiphenyls)
p-chlorophenol 28
PCR
  –, amplification 760, 763
  –, product 760
  –, reaction 760
peat 65
  –, cores 65
pendimethaline 329
Penicillium 279
pentachlorophenol 673
peptide 132
  –, metal-binding 136
  –, synthesis 134
perchloroethylene 568
pesticides 329, 462, 672
  –, exposure 528
  –, mobility 462
petroleum 440, 742, 745
  –, biotransformation 751
petrology, organic 89
Phanerochaete 274, 279, 280, 431
  –, *chrysosporium* 274, 279, 280, 307, 310
phase equilibria 540
Phaseolus vulgaris 634
phenanthrene 690
phenol 28, 329, 516, 673, 709, 727
  –, degradation 600
  –, toxicity 727
PHH (*see* polyclyclic heterocyclic hydrocarbons)
Phlebia tremellosa 279
phosgene 542
phosphorus-32 742
  –, post-labelling 747
Photobacterium leiognathi 647, 655
photocatalysis 590
photocatalyst 590
photodecomposition 590
photodegradation 440, 472, 504
photometry 680
photooxidation 440
photoproduct 444
photosynthesis 664
phtalocyanine 275

–, dyes 275
phthalate 673, 710
p-hydroxybenzoic acid 29
phytoextraction 186, 196
Phytophthora infestans 633
phytoplankton 125
phytoremediation 186, 196, 204, 269
Picea abies 237
pig 352, 423
Pimephales promelas 725
Piptoporus betulinus 279
Planktothrix 659
plant 186, 725
  –, metal accumulation 204
  –, salt stress 630
  –, stress 630
plasmid 648, 706
  –, extraction 759
plasticizer 343
Pleuronectes vetulus 744, 751
Pleurotus ostreatus 277, 279
Poland 65
pollen 92, 490
pollutant
  –, environmental 644
  –, metallic 677
  –, organic 89, 99, 370, 706, 734
  –, solubilization 568
  –, transport model 382
pollution 408
  –, toxic 680
polychlorobiphenyls (PCBs) 552
  –, decontamination 552
polyclyclic heterocyclic hydrocarbons (PHH) 742
polycyclic aromatic hydrocarbons (PAHs) 370, 378, 392, 418, 672, 690, 694, 742, 745
  –, analysis 433
  –, atmospheric 408
  –, bioavailability 423, 430
  –, biodegradation 431
  –, contamination 419
  –, degradation 448, 452
  –, extraction 433
  –, persistence 431
  –, photolysis 440
  –, removal 430
  –, sampling 408
  –, transfer to human 418
polyethylene glycol 709
polymer 3
pore water 99
Portland cement 621
potabilisation 360

potassium 630
praseodymium 634
principal components analysis (PCA) 152, 166
proline 637
propanol 710, 734
protein 133, 666, 746
  –, damages 632
  –, labeling 661
Proteus 278
protozoa 494
  –, non-photosynthetic 494
  –, photosynthetic 494
Pseudomonas 215–217, 219, 270, 277–279, 431, 474, 475, 477, 479, 707, 708, 710, 713, 715, 716, 719, 762
  –, *aeruginosa* 216–220, 708, 710, 715, 716
  –, *alcaligenes* 720
  –, *fluorescence* 720
  –, *luteola* 277, 278
  –, *pseudomallei* 279
  –, *putida* 708, 710, 716, 719
  –, *rathonis* 708, 710, 711, 713, 715–722
Pycnoporus
  –, *cinnabarinus* 277, 279
  –, *sanguineus* 279
pyrene 440
  –, photodegradation 440
  –, photolysis 445
pyrolysis 335
pyrrol-2-carboxaldehyde 336

R

raffinose 710
rainwater 16
reactor, biological 432
recharge 57
recycling 584
remediation 133, 196, 448, 568, 620
  –, electrochemical 222
  –, electrodialytic 222, 224, 234
  –, enzyme 269, 281
  –, phyto- 269
  –, technologies 569
Renilla 647
residues, agricultural 620
respiration 663
respirometry 680, 682
Rhizobium 305
Rhodococcus 279, 294, 295, 298, 302
  –, *erythropolis* 294–296, 298, 300
  –, *pyrinidovorans* 302
  –, *rhodochrous* 294–297, 299, 301
rice hull 620
  –, ash 620

–, thermal decomposition 623
rimsulfuron 504
  –, degradation 504
risk evaluation 418
root exudates 204, 207
runoff 152
ruthenium odixe 334

S

salicylates 630
Salmo gairdneri 725
Salmonella 32–34, 39, 40, 645, 655, 656
  –, *choleraesuis* 648, 655, 656
  –, *enteritidis* 39
  –, *typhimurium* 656
salt 31, 637
  –, stress 630, 631
sample recycling 584
sampling 107
saxitoxin 658
scanning electron microscopy (SEM) 6, 590
Scenedesmus quadricauda 725
SCWO (*see* super critical water oxidation)
sediment 89
  –, contaminants 89
  –, dioxin contaminated 99
  –, riverine 328
Sellata 506
SEM (*see* scanning electron microscopy)
semipermeable membrane devices (SPMDs) 392, 394
sensor
  –, in-situ 13
  –, microbial 706
sewage sludge 304, 430, 448, 456
  –, management 430
  –, PAHs 448
  –, treatment 430
silica 504
silicon
  –, -29 463
  –, dioxide 440
slovagen 709
sludge 304, 305, 680, 707
  –, activated 31
  –, contaminated 304
  –, PAH removal 430
  –, treatment 430
sodium 630
  –, hydroxide 685
soil 89, 1756, 196, 304, 305, 472, 516, 690
  –, adsorption 504
  –, contaminants 89

Index 779

–, contamination 196
  –, with cadmium 214
  –, with mercury 196
  –, with perchloroethylene 568
  –, with polychlorobiphenyl 552
  –, with thallium 186
  –, with trichloroethylene 568
–, organic matter 574
–, particulate metal phases 3
–, pollution 3, 222
  –, with heavy metals 222, 228–230
–, remediation 568
–, rimsulfuron adsorption 508
–, sorption 520
  –, experiments 524
Solea solea 742, 744, 745, 750
solid phase extraction 433
solid state dechlorination 552
solubility enhancement 568, 573
solvent 31
  –, organic, toxicity 677
soot 372
sorbose 710
sorption 250, 516, 520
  –, capacity 89
SOS-LUX test 644, 647
Souss-Massa basin 58
speciation 199
spectrometry
  –, mass 31
  –, transmittance 22
Sphagnum 65, 69
SPMDs (*see* semipermeable membrane devices)
spores 92
steelworks 419
storm water 154
strength, ionic 724, 729, 730
Streptococcus faecalis 278
Streptomyces 215–220, 279, 757, 758, 760, 763
  –, cadmium resistance 763
stress proteins 734, 736
strontium 136, 242
  –, -90 197
subsampling 99
substance
  –, humic 120, 123, 124, 206, 473, 505, 516
    –, dissolved 516
    –, preparation 518
sucrose 710
sulfoadenylate 709
sulfobenzoate 709
sulfobetain 709
sulfonate 706, 709

sulfonylurea 494
sulfosalicylate 709
sulfur
  –, -34 43
    –, delta 43, 46, 47, 55
  –, -35 658, 661, 666, 667
  –, dioxide 372
sulphate 43
sunflower 488
super critical water oxidation (SCWO) 600
superoxide 630
surface water 360
surfactant 706, 709
  –, detection 706
suspended particulate matter 152
system
  –, aquatic, trace metal speciation 120
  –, isotopic 75

T

TCDD 352
temperature 65
terbium 135–137
terbuthylazine 329, 516
ternary hydrides 552
test, respirometric 680
tetrachlorobiphenyl 329
Tetrahymena 494, 662, 663, 668
  –, *pyriformis* 494, 662
textile
  –, dye 268, 288
  –, effluent 268
    –, treatment 288
thallium 75, 186
  –, binding forms 190
  –, phytoextraction 192
thin layer chromatography (TLC) 296
thiuram 673
time resolved laser induced fluorescence analysis (TRLIF) 135
tin 543, 546
  –, -112 78
  –, -114 78
  –, -116 78, 83
  –, -118 85
  –, -119 543, 545
titania
  –, aerogel 591
  –, catalyst 590
titanium dioxide 21
TLC (*see* thin layer chromatography)
tobacco 630, 632
toluene 382, 385, 734

–, sulfonate 709
toxicity 631, 677, 701, 724
  –, assessment 734
  –, environmental 734
  –, potential 698
  –, water-borne 698
trace
  –, analysis 99
  –, metal 120, 152
  –, speciation 120
Trametes 277, 279, 280, 304
  –, *hirsuta* 279
  –, *versicolor* 279, 280, 304, 307, 310–312, 314
transmittance spectrometry 22
transport model 382
treatment, biological 616
triazines 516
trichlorobenzene 734
trichloroethene 316
  –, biodegradation 316
  –, dilution 316
trichloroethylene 560, 568
  –, biodegradation 560, 563
trichlorophenol 673
tridecanoic acid 710
trihalomethane 360
  –, control 365
triphenyl methane 274
  –, dye 274
Triticum aestivum 198, 200, 202
triton 709
TRLIF (*see* time resolved laser induced fluorescence analysis)
tunnel 408
Tween 709
tyrosol 600
  –, hydrogenation 608

U

Umezakia natans 659
uranium 132
  –, -234 197
  –, -238 197
  –, binding 139
  –, dioxide 139
urea 710

V

vanadium 152
  –, solid phase 157
Vibrio fischeri 707, 724–726, 728
volgonat 709

W

WAO (*see* wet air oxidation)
waste
  –, incineration 419
  –, streams 756
  –, vegetable 250
  –, water 680
    –, mercury contaminated 242
    –, treatment 31, 242, 614
water 690, 698
  –, chlorophenol contaminated 600
  –, depollution 600
  –, drinking 360
  –, natural 120
  –, phenol contaminated 600
  –, pollution 672, 724
  –, remediation 568
  –, potabilisation 360
  –, surface w. 360
watershed 152
wet air oxidation (WAO) 600
wheat (*see Triticum aestivum*)
whole-cell biosensor 644
wood 75, 234
  –, burning 418

X

Xhantomonas 762
XPS (*see* X-ray photoelectron spectroscopy)
X-ray
  –, absorption fine structure (EXAFS) 4
  –, diffraction 4
  –, fluorescence (XRF) 5
  –, fluorescence spectrometry 207
  –, photoelectron spectroscopy (XPS) 590
XRF (*see* X-ray fluorescence)
xylene 734
xylitol 710
xylose 710

Y

yeast 495
Yugoslavia 146

Z

zeolite 259
zinc 4, 75, 78, 146, 152, 181, 186, 197, 204, 630, 635
  –, anthropogenic input 160
  –, particulate 157
  –, remediation 222

Periodic Table

1 IA	2 IIA	3 IIIB	4 IVB	5 VB	6 VIB	7 VIIB	8
1 1.0079 **H** HYDROGEN							
3 6.941 **Li** LITHIUM	**4** 9.0122 **Be** BERYLLIUM						
11 22.990 **Na** SODIUM	**12** 24.305 **Mg** MAGNESIUM						
19 39.098 **K** POTASSIUM	**20** 40.078 **Ca** CALCIUM	**21** 44.956 **Sc** SCANDIUM	**22** 47.867 **Ti** TITANIUM	**23** 50.942 **V** VANADIUM	**24** 51.996 **Cr** CHROMIUM	**25** 54.938 **Mn** MANGANESE	**26** 55.845 **Fe** IRON
37 85.468 **Rb** RUBIDIUM	**38** 87.62 **Sr** STRONTIUM	**39** 88.906 **Y** YTTRIUM	**40** 91.224 **Zr** ZIRCONIUM	**41** 92.906 **Nb** NIOBIUM	**42** 95.94 **Mo** MOLYBDENUM	**43** (98) **Tc** TECHNETIUM	**44** 101.07 **Ru** RUTHENIUM
55 132.91 **Cs** CAESIUM	**56** 137.33 **Ba** BARIUM	57-71 **La-Lu** Lanthanide	**72** 178.49 **Hf** HAFNIUM	**73** 180.95 **Ta** TANTALUM	**74** 183.84 **W** TUNGSTEN	**75** 186.21 **Re** RHENIUM	**76** 190.23 **Os** OSMIUM
87 (223) **Fr** FRANCIUM	**88** (226) **Ra** RADIUM	89-103 **Ac-Lr** Actinide	**104** (261) **Rf** RUTHERFORDIUM	**105** (262) **Db** DUBNIUM	**106** (266) **Sg** SEABORGIUM	**107** (264) **Bh** BOHRIUM	**108** (277) **Hs** HASSIUM

— atomic mass (mean relative)

LANTHANIDE

57 138.91 **La** LANTHANUM	**58** 140.12 **Ce** CERIUM	**59** 140.91 **Pr** PRASEODYMIUM	**60** 144.24 **Nd** NEODYMIUM	**61** (145) **Pm** PROMETHIUM

ACTINIDE

89 (227) **Ac** ACTINIUM	**90** 232.04 **Th** THORIUM	**91** 231.04 **Pa** PROTACTINIUM	**92** 238.03 **U** URANIUM	**93** (237) **Np** NEPTUNIUM